Combinatorics, Automata and Number Theory

This collaborative volume presents recent trends arising from the fruitful interaction between the themes of combinatorics on words, automata and formal language theory, and number theory. Presenting several important tools and concepts, the authors also reveal some of the exciting and important relationships that exist between these different fields. Topics include numeration systems, word complexity function, morphic words, Rauzy tilings and substitutive dynamical systems, Bratelli diagrams, frequencies and ergodicity, Diophantine approximation and transcendence, asymptotic properties of digital functions, decidability issues for D0L systems, matrix products and joint spectral radius. Topics are presented in a way that links them to the three main themes, but also extends them to dynamical systems and ergodic theory, fractals, tilings and spectral properties of matrices.

Graduate students, research mathematicians and computer scientists working in combinatorics, theory of computation, number theory, symbolic dynamics, fractals, tilings and stringology will find much of interest in this book.

ENCYCLOPEDIA OF MATHEMATICS AND ITS APPLICATIONS

All the titles listed below can be obtained from good booksellers or from Cambridge University Press. For a complete series listing visit

http://www.cambridge.org/uk/series/sSeries.asp?code=EOM

80 O. Stormark *Lie's Structural Approach to PDE Systems*
81 C. F. Dunkl and Y. Xu *Orthogonal Polynomials of Several Variables*
82 J. P. Mayberry *The Foundations of Mathematics in the Theory of Sets*
83 C. Foias et al. *Navier–Stokes Equations and Turbulence*
84 B. Polster and G. F. Steinke *Geometries on Surfaces*
85 R. B. Paris and D. Kaminski *Asymptotics and Mellin–Barnes Integrals*
86 R. J. McEliece *The Theory of Information and Coding, 2nd edn*
87 B. A. Magurn *An Algebraic Introduction to K-Theory*
88 T. Mora *Solving Polynomial Equation Systems I*
89 K. Bichteler *Stochastic Integration with Jumps*
90 M. Lothaire *Algebraic Combinatorics on Words*
91 A. A. Ivanov and S. V. Shpectorov *Geometry of Sporadic Groups II*
92 P. McMullen and E. Schulte *Abstract Regular Polytopes*
93 G. Gierz et al. *Continuous Lattices and Domains*
94 S. R. Finch *Mathematical Constants*
95 Y. Jabri *The Mountain Pass Theorem*
96 G. Gasper and M. Rahman *Basic Hypergeometric Series, 2nd edn*
97 M. C. Pedicchio and W. Tholen (eds.) *Categorical Foundations*
98 M. E. H. Ismail *Classical and Quantum Orthogonal Polynomials in One Variable*
99 T. Mora *Solving Polynomial Equation Systems II*
100 E. Olivieri and M. Eulália Vares *Large Deviations and Metastability*
101 A. Kushner, V. Lychagin and V. Rubtsov *Contact Geometry and Nonlinear Differential Equations*
102 L. W. Beineke and R. J. Wilson (eds.) with P. J. Cameron *Topics in Algebraic Graph Theory*
103 O. J. Staffans *Well-Posed Linear Systems*
104 J. M. Lewis, S. Lakshmivarahan and S. K. Dhall *Dynamic Data Assimilation*
105 M. Lothaire *Applied Combinatorics on Words*
106 A. Markoe *Analytic Tomography*
107 P. A. Martin *Multiple Scattering*
108 R. A. Brualdi *Combinatorial Matrix Classes*
109 J. M. Borwein and J. D. Vanderwerff *Convex Functions*
110 M.-J. Lai and L. L. Schumaker *Spline Functions on Triangulations*
111 R. T. Curtis *Symmetric Generation of Groups*
112 H. Salzmann et al. *The Classical Fields*
113 S. Peszat and J. Zabczyk *Stochastic Partial Differential Equations with Lévy Noise*
114 J. Beck *Combinatorial Games*
115 L. Barreira and Y. Pesin *Nonuniform Hyperbolicity*
116 D. Z. Arov and H. Dym *J-Contractive Matrix Valued Functions and Related Topics*
117 R. Glowinski, J.-L. Lions and J. He *Exact and Approximate Controllability for Distributed Parameter Systems*
118 A. A. Borovkov and K. A. Borovkov *Asymptotic Analysis of Random Walks*
119 M. Deza and M. Dutour Sikirić *Geometry of Chemical Graphs*
120 T. Nishiura *Absolute Measurable Spaces*
121 M. Prest *Purity, Spectra and Localisation*
122 S. Khrushchev *Orthogonal Polynomials and Continued Fractions*
123 H. Nagamochi and T. Ibaraki *Algorithmic Aspects of Graph Connectivity*
124 F. W. King *Hilbert Transforms I*
125 F. W. King *Hilbert Transforms II*
126 O. Calin and D.-C. Chang *Sub-Riemannian Geometry*
127 M. Grabisch et al. *Aggregation Functions*
128 L. W. Beineke and R. J. Wilson (eds.) with J. L. Gross and T. W. Tucker *Topics in Topological Graph Theory*
129 J. Berstel, D. Perrin and C. Reutenauer *Codes and Automata*
130 T. G. Faticoni *Modules over Endomorphism Rings*
131 H. Morimoto *Stochastic Control and Mathematical Modeling*
132 G. Schmidt *Relational Mathematics*
133 P. Kornerup and D. W. Matula *Finite Precision Number Systems and Arithmetic*
134 Y. Crama and P. L. Hammer (eds.) *Boolean Functions*
135 V. Berthé and M. Rigo (eds.) *Combinatorics, Automata and Number Theory*
136 A. Kristály, V. D. Rădulescu and C. Varga *Variational Principles in Mathematical Physics, Geometry, and Economics*

Combinatorics, Automata and Number Theory

Edited by

VALÉRIE BERTHÉ
LIAFA, Université Paris 7 - CNRS, France

MICHEL RIGO
Université de Liège, Belgium

CAMBRIDGE UNIVERSITY PRESS
Cambridge, New York, Melbourne, Madrid, Cape Town, Singapore,
São Paulo, Delhi, Dubai, Tokyo, Mexico City

Cambridge University Press
The Edinburgh Building, Cambridge CB2 8RU, UK

Published in the United States of America by Cambridge University Press, New York

www.cambridge.org
Information on this title: www.cambridge.org/9780521515979

© Cambridge University Press 2010

This publication is in copyright. Subject to statutory exception
and to the provisions of relevant collective licensing agreements,
no reproduction of any part may take place without the written
permission of Cambridge University Press.

First published 2010

Printed in the United Kingdom at the University Press, Cambridge

A catalogue record for this publication is available from the British Library

ISBN 978-0-521-51597-9 Hardback

Cambridge University Press has no responsibility for the persistence or
accuracy of URLs for external or third-party internet websites referred to
in this publication, and does not guarantee that any content on such
websites is, or will remain, accurate or appropriate.

Contents

List of contributors		*page* ix
Preface		xi
Acknowledgements		xix

1	**Preliminaries**	**1**
	V. BERTHÉ, M. RIGO	
	1.1 Conventions	1
	1.2 Words	3
	1.3 Languages and machines	13
	1.4 Associated matrices	22
	1.5 A glimpse at numeration systems	26
	1.6 Symbolic dynamics	27
	1.7 Exercises	32
	1.8 Notes	33

2	**Number representation and finite automata**	**34**
	CH. FROUGNY, J. SAKAROVITCH	
	2.1 Introduction	34
	2.2 Representation in integer base	37
	2.3 Representation in real base	53
	2.4 Canonical numeration systems	77
	2.5 Representation in rational base	84
	2.6 A primer on finite automata and transducers	95
	2.7 Notes	103

3	**Abstract numeration systems**	**108**
	P. LECOMTE, M. RIGO	
	3.1 Motivations	108

	3.2 Computing numerical values and S-representations	117
	3.3 S-recognisable sets	122
	3.4 Automatic sequences	138
	3.5 Representing real numbers	152
	3.6 Exercises and open problems	158
	3.7 Notes	160
4	**Factor complexity**	**163**
	J. CASSAIGNE, F. NICOLAS	
	4.1 Introduction	163
	4.2 Definitions, basic properties, and first examples	163
	4.3 The theorem of Morse and Hedlund	166
	4.4 High complexity	168
	4.5 Tools for low complexity	171
	4.6 Morphisms and complexity	181
	4.7 The theorem of Pansiot	185
	4.8 Complexity of automatic words	214
	4.9 Control of bispecial factors	215
	4.10 Examples of complexity computations for morphic words	221
	4.11 Complexity computation for an s-adic family of words	226
	4.12 Exercises and open problems	239
5	**Substitutions, Rauzy fractals and tilings**	**248**
	V. BERTHÉ, A. SIEGEL, J. THUSWALDNER	
	5.1 Introduction	248
	5.2 Basic definitions	250
	5.3 Tilings	259
	5.4 Ancestor graphs and tiling conditions	273
	5.5 Boundary and contact graphs	284
	5.6 Geometric coincidences	290
	5.7 Overlap coincidences	296
	5.8 Balanced pair algorithm	309
	5.9 Conclusion	315
	5.10 Exercises	316
	5.11 Notes	317
6	**Combinatorics on Bratteli diagrams and dynamical systems**	**324**
	F. DURAND	
	6.1 Introduction	324
	6.2 Cantor dynamical systems	325

6.3	Bratteli diagrams	325
6.4	The Bratteli-Vershik model theorem	330
6.5	Examples of BV-models	336
6.6	Characterisation of Strong Orbit Equivalence	351
6.7	Entropy	357
6.8	Invariant measures and Bratteli diagrams	360
6.9	Eigenvalues of stationary BV-models	364
6.10	Exercises	370

7 Infinite words with uniform frequencies, and invariant measures — 373
S. FERENCZI, T. MONTEIL

7.1	Basic notions	374
7.2	Invariant measures and unique ergodicity	376
7.3	Combinatorial criteria	383
7.4	Examples	388
7.5	Counter-examples	395
7.6	Further afield	405
7.7	Exercises	406
7.8	Note: Dictionary between word combinatorics and symbolic dynamics	409

8 Transcendence and Diophantine approximation — 410
B. ADAMCZEWSKI, Y. BUGEAUD

8.1	The expansion of algebraic numbers in an integer base	412
8.2	Basics from continued fractions	427
8.3	Transcendental continued fractions	430
8.4	Simultaneous rational approximations	436
8.5	Explicit examples for the Littlewood conjecture	443
8.6	Exercises and open problems	448
8.7	Notes	449

9 Analysis of digital functions and applications — 452
M. DRMOTA, P. J. GRABNER

9.1	Introduction: digital functions	452
9.2	Asymptotic analysis of digital functions	456
9.3	Statistics on digital functions	480
9.4	Further results	499

10	**The equality problem for purely substitutive words**	**505**
	J. HONKALA	
	10.1 Purely substitutive words and D0L systems	505
	10.2 Substitutive words and HD0L sequences	507
	10.3 Elementary morphisms	509
	10.4 Nearly primitive D0L systems	512
	10.5 Periodic and nearly periodic words	515
	10.6 A balance property for ω-equivalent 1-systems	519
	10.7 The equality problem for purely substitutive words	523
	10.8 Automatic words	525
	10.9 Complexity questions	526
	10.10 Exercises	527
11	**Long products of matrices**	**530**
	V. D. BLONDEL, R. M. JUNGERS	
	11.1 The joint spectral characteristics	531
	11.2 Applications	541
	11.3 The finiteness property	550
	11.4 Approximation algorithms	552
	11.5 Conclusions	558
	11.6 Exercises	559
	11.7 Notes	561

References	563
Notation index	594
General index	599

List of contributors

Boris Adamczewski
CNRS, Université de Lyon, Université Lyon 1, Institut Camille Jordan,
43 boulevard du 11 novembre 1918, F-69622 Villeurbanne cedex, France.

Valérie Berthé
LIAFA, Université Paris 7 – CNRS UMR 7089
Case 7014, F-75205 Paris cedex 13, France.

Vincent D. Blondel
Department of Mathematical Engineering,
Université catholique de Louvain,
Avenue Georges Lemaître 4, B-1348 Louvain-la-Neuve, Belgium.

Yann Bugeaud
IRMA – Université de Strasbourg-Mathématiques – CNRS UMR 7501
7 rue René Descartes, F-67084 Strasbourg cedex, France.

Julien Cassaigne
Institut de mathématiques de Luminy
CNRS UMR 6206 – Université Aix-Marseille II
Case 907, 163, avenue de Luminy, F-13288 Marseille cedex 09, France.

Michael Drmota
Institut für Diskrete Mathematik und Geometrie, TU Wien,
Wiedner Hauptstrasse 8–10, A-1040 Wien, Austria.

Fabien Durand
LAMFA – Université de Picardie Jules Verne – CNRS UMR 6140
33 rue Saint Leu, F-80039 Amiens cedex 1, France.

Sébastien Ferenczi
Institut de mathématiques de Luminy
CNRS UMR 6206 – Université Aix-Marseille II
Case 907, 163, avenue de Luminy, F-13288 Marseille cedex 09, France.

Christiane Frougny
Université Paris 8 and
LIAFA, Université Paris 7 – CNRS UMR 7089
Case 7014, F-75205 Paris cedex 13, France.

Peter J. Grabner
Institut für Analysis und Computational Number Theory, TU Graz,
Steyrergasse 30, A-8010 Graz, Austria.

Juha Honkala
Department of Mathematics, University of Turku, 20014 Turku, Finland.

Raphaël M. Jungers
Department of Mathematical Engineering,
Université catholique de Louvain,
Avenue Georges Lemaître 4, B-1348 Louvain-la-Neuve, Belgium.

Pierre Lecomte
Université de Liège, Institut de Mathématiques,
Grande Traverse 12 (B 37), B-4000 Liège, Belgium.

Thierry Monteil
LIRMM – Université Montpellier II – CNRS UMR 5506
161 rue Ada, F-34392 Montpellier cedex 5, France.

François Nicolas
Department of Computer Science, University of Helsinki
P.O. box 68, FI-00014 University of Helsinki, Finland.

Michel Rigo
Université de Liège, Institut de Mathématiques,
Grande Traverse 12 (B 37), B-4000 Liège, Belgium.

Jacques Sakarovitch
LTCI, CNRS/ENST – UMR 5141
46 rue Barrault, F-75634 Paris cedex 13, France.

Anne Siegel
CNRS – Université Rennes 1 – INRIA, IRISA UMR 6074
Campus de Beaulieu, F-35042 Rennes cedex, France.

Jörg Thuswaldner
Chair of Mathematics and Statistics
University of Leoben, A-8700 Leoben, Austria.

Preface

As the title may suggest, this book is about *combinatorics on words, automata and formal language theory*, as well as *number theory*. This collaborative work gives a glimpse of the active community working in these interconnected and even intertwined areas. It presents several important tools and concepts usually encountered in the literature and it reveals some of the exciting and non-trivial relationships existing between the considered fields of research. This book is mainly intended for graduate students or research mathematicians and computer scientists interested in combinatorics on words, theory of computation, number theory, dynamical systems, ergodic theory, fractals, tilings and stringology. We hope that some of the chapters can serve as useful material for lecturing at master level.

The outline of this project has germinated after a very successful international eponymous school organised at the University of Liège (Belgium) in 2006 and supported by the European Union with the help of the European Mathematical Society (EMS). Parts of a preliminary version of this book were used as lecture notes for the second edition of the school organised in June 2009 and mainly supported by the European Science Foundation (ESF) through the AutoMathA programme. For both events, we acknowledge also financial support from the University of Liège and the Belgian funds for scientific research (FNRS).

We have selected ten topics which are directed towards the fundamental three themes of this project (namely, combinatorics, automata and number theory) and they naturally extend to dynamical systems and ergodic theory (see Chapters 6 and 7), but also to fractals and tilings (see Chapter 5) and spectral properties of matrices (see Chapter 11). Indeed, as it will be shown in particular in Chapter 7 there exist tight and fruitful links between properties sought for in dynamical systems and combinatorial properties of the corresponding words and languages. On the other hand, linear algebra and extremal matrix products are important tools in the framework of this book: some matrices are canonically associated with morphisms and graphs and a notion like joint spectral radius introduced in (Rota and Strang 1960) has therefore applications in automata theory or combinatorics on words.

Each chapter is intended to be self-contained and relies mostly on the introductory Chapter 1 presenting some preliminaries and general notions. Some of the major links existing between the chapters are given in the figure below.

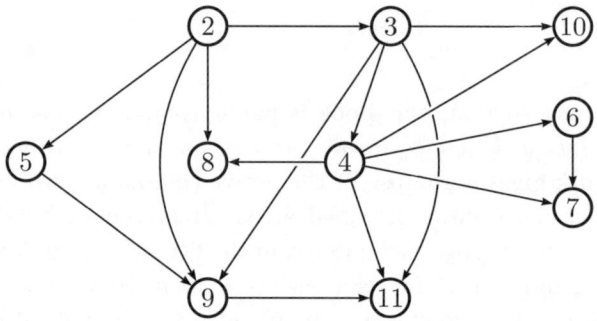

Fifteen authors were collaborating on this volume. Most of them kindly served as lecturers for the CANT schools.

Let us succinctly sketch the general landscape without any attempt at being exhaustive. Short abstracts of each chapter are given below.

Combinatorics on words is a quite recent topic in (discrete) mathematics, and in the category of 'Concrete Mathematics' according to the terminology introduced by (Graham, Knuth, and Patashnik 1989). It deals with problems that can be stated in a non-commutative monoid such as estimates on the factor complexity function for infinite words, construction and properties of infinite words, the study of unavoidable regularities or patterns, substitutive words, *etc*. In the spirit of Lothaire's seminal book series, see (Lothaire 1983), (Lothaire 2002) and (Lothaire 2005), but with a different focus put on interactions between fields of research, we will deal in this book with the complexity function counting factors occurring in an infinite word, properties and generalisations of automatic sequences in the sense of (Allouche and Shallit 2003) and also the equality problem for substitutive (or also called morphic) words, see Chapters 3, 4 and 10. Motivations to study words and their properties are coming, for instance, from the coding of orbits and trajectories by words. This constitutes the basis of symbolic dynamical systems (Lind and Marcus 1995). This explains why dynamical systems enter the picture, mainly in Chapters 6 and 7, and are at the origin of the introduction of the fractals studied in Chapter 5. A historical example is the study by M. Morse of recurrent geodesics on a surface with negative curvature (Morse 1921). As another example, similar ideas are found in connection with the word problem in group theory (Epstein, Cannon, Holt, et al. 1992). Moreover the use of combinatorics is sought in the analysis of algorithms, initiated by D. E. Knuth, and which greatly relies on number theory, asymptotic methods and computer algebra (Lothaire 2005), (Greene and Knuth 1990), (Knuth 2000). Readers interested in asymptotics methods and limiting properties of digital functions should in particular read Chapter 9.

Keep in mind that both combinatorics on words and theory of formal languages have important applications and interactions in computer science (Perrin and Pin 2003) and physics. To cite just a few: study and models of quasi-crystals, aperiodic order and quasiperiodic tilings, bio-informatics and DNA analysis, theory of parsing, algorithmic verification of large systems, coding theory, discrete geometry and more precisely discretisation for computer graphics on a raster display, *etc*. This shows that algorithmic issues have also an important role to play.

Two chapters of this book, Chapters 2 and 3, deal with **numeration systems**. Such systems provide a main bridge between number theory on the one hand, and words combinatorics and formal language theory on the other hand. Indeed any integer can be represented in a given numeration system, like the classical integer base q numeration system, as a finite word over a finite alphabet of digits $\{0,\ldots,q-1\}$. This simple observation leads to the study of the relationships that can exist between the arithmetical properties of the integers and the syntactical properties of the corresponding representations. One of the deepest and most beautiful results in this direction is given by the celebrated theorem of Cobham (Cobham 1969) showing that the recognisability of a set of integers depends on the considered numeration system. This result can therefore be considered as one of the starting points of many investigations, for the last thirty years, about recognisable sets of integers and about non-standard or exotic numeration systems. Surprisingly, a recent extension of Cobham's theorem to the complex numbers leads to the famous Four Exponentials Conjecture (Hansel and Safer 2003). This is just one example of the fruitful relationship between formal language theory (including the theory of automata) and number theory. Many such examples will be presented here.

Numeration systems are not restricted to the representation of integers. They can also be used to represent real numbers with infinite words. One can think of continued fractions, integer or rational base representations, beta-expansions, *etc.* Again it is remarkable that some syntactical properties of the representations of reals may reflect number-theoretical properties, like transcendence, of the represented numbers. These questions are also treated in this book, see in particular Chapters 2 and 8. About Diophantine analysis or approximations of real numbers by algebraic numbers, striking developments through a fruitful interplay between Diophantine approximation and combinatorics on words can be observed, see again Chapter 8. Analogously, a rich source of challenging problems in analytic number theory comes from the study of digital functions, *i.e.*, functions defined in a way that depends on the digits in some numeration system. They are the object of Chapter 9.

The study of simple algorithmic constructions and transformations of infinite words plays here an important role. We focus in particular on the notion of **morphic words**, also called **substitutive words**. They are obtained iteratively by replacing letters with finite words. These words, as well as their associated symbolic dynamical systems, present a very rich behaviour. They occur in most of the chapters, see in particular Chapters 3, 4, 5, 6, 8 and 10. In the case where we replace letters with words of the same length, we obtain so-called automatic sequences. Several variations around the notion of morphic words are presented, as D0L systems (see Chapter 10), or else as adic words and transformations, and linearly recurrent subshifts. They occur in particular in Chapters 6 and 7. Note that most of the symbolic dynamical systems considered are of zero entropy, such as substitutive dynamical systems, odometers (see Chapter 6 and 9) or linearly recurrent systems (see Chapter 6).

Graphs and automata appear to be a very natural and powerful tool in this context. This is illustrated *e.g.* in Chapter 2 with special focus on operations performed on expansions of numbers realised by automata or transducers, or in Chapter 5 which is devoted to tilings by fractals whose boundary is described in terms of graphs. Graphs associated with substitutions appear ubiquitously, for instance, under the form of prefix-suffix graphs, of Rauzy graphs of words, or of the automata generating automatic sequences. Incidence matrices of graphs also play here an important role, hence the recurrence of notions like the spectral radius and its generalisations (see Chapter 11), or the importance of Perron–Frobenius' Theorem.

We are very pleased that Cambridge University Press proposed to consider this book, as Lothaire's books, as part of the *Encyclopedia of Mathematics and its Applications* series.

Let us present the different contributions for this book.

Chapter 2 by Ch. Frougny and J. Sakarovitch
Number representation and finite automata

In this chapter, numbers are represented by their expansion in a base, or more generally, with respect to a basis, hence by words (finite or infinite) over an alphabet of digits.

Is the set of expansions for all integers or all reals (within an interval) recognised by a finite automaton? Which operations on numbers translate into functions on number expansions that are realised by finite transducers? These are some of the questions that are treated in this chapter. The classical representation in an integer base is first considered, then the representation in a real base and in some associated basis. Finally, representations in canonical number systems and in rational bases are briefly studied.

Chapter 3 by P. Lecomte and M. Rigo
Abstract numeration systems

The motivation for the introduction of abstract numeration systems stems from the celebrated theorem of Cobham dating back to 1969 about the so-called recognisable sets of integers in any integer base numeration system. An abstract numeration system is simply an infinite genealogically ordered (regular) language. In particular, this notion extends the usual integer base numeration systems as well as more elaborated numeration systems such as those based on a Pisot number. In this general setting, we study in details recognisable sets of integers, *i.e.*, the corresponding representations are accepted by a finite automaton. The main theme is the link existing between the arithmetic properties of integers and the syntactical properties of the corresponding representations in a given numeration system. Relationship with automatic sequences and substitutive words is also investigated, providing an analogue to another famous result of Cobham from 1972 about k-automatic sequences. Finally, the chapter ends with the representation of real numbers in an abstract numeration system.

Chapter 4 by J. Cassaigne and F. Nicolas
Factor complexity

The factor complexity function $p(n)$ of an infinite word is studied thoroughly. Tools such as special factors and Rauzy graphs are introduced, then applied to several problems, including practical computation of the factor complexity of various kinds of words, or the construction of words having a complexity asymptotically equivalent to a specified function.

This chapter includes a complete proof of Pansiot's characterisation of the complexity function of purely morphic words, and a proof of a conjecture of Heinis on the limit $p(n)/n$.

The authors would like to thank Jean-Paul Allouche for his bibliographic help, Juhani Karhumäki for his kind hospitality during the redaction of this chapter, and Christian Mauduit for his participation in the proof of Theorem 4.7.15. F. Nicolas was supported by the Academy of Finland under the grant 7523004 (Algorithmic Data Analysis).

Chapter 5 by V. Berthé, A. Siegel and J. Thuswaldner
Substitutions, Rauzy fractals and tilings

This chapter focuses on a multiple tiling associated with a primitive substitution σ. We restrict to the case where the inflation factor of the substitution σ is a unit Pisot number. This multiple tiling is composed of tiles which are given by the unique solution of a set equation expressed in terms of a graph associated with the substitution σ: these tiles are attractors of a graph-directed iterated function system (GIFS). Each of these tiles is compact, it is the closure of its interior, it has a non-zero measure and it has a fractal boundary that is also a solution of a graph-directed iterated function system defined by the substitution σ. These tiles are called *central tiles* or *Rauzy fractals*, according to G. Rauzy who introduced them. The aim of this chapter is to list several tiling conditions, relying on the use of various graphs associated with σ.

The authors would like to thank W. Steiner for his efficient help for drawing pictures of fractals, as well as J.-Y. Lee and B. Solomyak for their precious comments on Section 5.7.

Chapter 6 by F. Durand
Combinatorics on Bratteli diagrams and dynamical systems

The aim of this chapter is to show how Bratteli diagrams are used to study topological dynamical systems. Bratteli diagrams are infinite graphs that provide a very efficient encoding of the dynamics that transform some dynamical properties into combinatorial properties on these graphs. We illustrate their wide range of applications through classical notions: invariant measures, entropy, expansivity, representation theorems, strong orbit equivalence, eigenvalues of the Koopman operator.

Chapter 7 by S. Ferenczi and T. Monteil
Infinite words with uniform frequencies, and invariant measures

For infinite words, we study the properties of uniform recurrence, which translates the dynamical property of minimality, and of uniform frequencies, which corresponds to unique ergodicity; more generally, we look at the set of invariant measures of the associated dynamical system. We present some achievements of word combinatorics, initiated by M. Boshernitzan, which allow us to deduce information on these invariant measures from simple combinatorial properties of the words. Then we review some known examples of words with uniform frequencies, and give important examples which do not have uniform frequencies. We finish by hinting how these basic notions have given birth to very deep problems and high achievements in dynamical systems.

The first author wishes to thank the MSRI for its hospitality during the redaction of this chapter. The second author wishes to thank the Poncelet Laboratory and the Asmus Family for their hospitality during the redaction of this chapter.

Chapter 8 by B. Adamczewski and Y. Bugeaud
Transcendence and Diophantine approximation

Finite and infinite words occur naturally in Diophantine approximation when we consider the expansion of a real number in an integer base or its continued fraction expansion. The aim of this chapter is to present several number-theoretical problems that reveal a fruitful interplay between combinatorics on words and Diophantine approximation. For example, if the decimal expansion of a real number viewed as an infinite word on the alphabet $\{0, 1, \ldots, 9\}$ begins with arbitrarily large squares, then this number must be either rational or transcendental.

Chapter 9 by M. Drmota and P. J. Grabner
Analysis of digital fuctions and applications

The aim of this chapter is to study asymptotic properties of digital functions (like the sum-of-digits function) from different points of view and to survey several techniques that can be applied to problems of this kind. We first focus on properties of average values where we explain periodicity phenomena in the 'constant term' or the main term of the corresponding asymptotic expansions. We compare the classical approach by Delange, a Dirichlet series method, and a measure-theoretic method. Secondly, we discuss distributional properties like Erdős–Wintner-type theorems and central limit theorems that work for very general q-additive functions and even if these functions are only considered for polynomial subsequences. These general results are complemented by very precise distributional results for completely q-additive functions which are based on a generating function approach. A final section discusses some further problems like the recent solution of the Gelfond problems on the sum-of-digits function and dynamical aspects of odometers.

The authors are supported by the Austrian Science Foundation FWF, projects S9604 and S9605, parts of the Austrian National Research Network 'Analytic Combinatorics and Probabilistic Number Theory'.

Chapter 10 by J. Honkala
The equality problem for purely substitutive words

We prove that the equality problem for purely substitutive words is decidable. This problem is also known as the D0L ω-equivalence problem. It was first solved by Culik and Harju. Our presentation follows a simpler approach in which elementary morphisms play an important role. We will also consider the equality problem for sets of integers recognised by finite automata in various ways.

Chapter 11 by V. D. Blondel and R. M. Jungers
Extremal matrix products and the finiteness property

We introduce and study questions related to long products of matrices. In particular, we define the joint spectral radius and the joint spectral subradius which characterise, respectively, the largest and smallest asymptotic rate of growth that can be obtained by forming long products of matrices. Such long products of matrices occur naturally in automata theory due to the possible representation of automata by sets of adjacency matrices.

Joint spectral quantities were initially used in the context of control theory and numerical analysis but have since then found applications in many other areas, including combinatorics and number theory. In the chapter we describe some of their fundamental properties, results on their computational complexity, various approximation algorithms, and three particular applications related to words and languages.

Acknowledgements

The editors would like to express their gratitude to Shabnam Akhtari, Nicolas Bédaride, Emilie Charlier, Timo Jolivet, Tomi Kärki, Sébastien Labbé, Marion Le Gonidec, Geneviève Paquin and N. Pytheas Fogg who were kind enough to read drafts of this book. They pointed out mathematical and stylistic errors and suggested many improvements. The editors also would like to thank their former editor Peter Thompson and their present editor Clare Dennison whose constant support has been a precious help through all this project, and Pierre Lecomte for his invaluable contribution in the organisation of the CANT schools.

1
Preliminaries

Valérie Berthé,

Michel Rigo

The aim of this chapter is to introduce basic objects that are encountered in the different parts of this book. In the first section, we start with a few conventions. Section 1.2 presents finite and infinite words and fundamental operations that can be applied to them. In particular important concepts like eventually periodic words, substitutive words or factor complexity function are introduced (more material is given in Chapter 4). Sets of words are languages. They are presented in Section 1.3 together with regular languages, finite automata and transducers (more material is presented in Section 2.6). Section 1.4 introduces some matrices naturally associated with automata or morphisms. Section 1.5 presents basic results on numeration systems that will be developed in Chapter 2. Finally, Section 1.6 introduces concepts from symbolic dynamics.

1.1 Conventions

Let us start with some basic notation used throughout this book. We assume the reader to be familiar with usual basic set operations like union, intersection or set difference: $\cup$, $\cap$ or $\setminus$. Sets of numbers are of particular interest. The set of non-negative integers (respectively integers, rational numbers, real numbers, complex numbers) is $\mathbb{N}$ (respectively $\mathbb{Z}$, $\mathbb{Q}$, $\mathbb{R}$, $\mathbb{C}$). Let a be a real number and $\mathbb{K} = \mathbb{N}$, $\mathbb{Z}$, $\mathbb{Q}$ or $\mathbb{R}$. We set

$$\mathbb{K}_{\geq a} := \mathbb{K} \cap [a, +\infty), \ \mathbb{K}_{>a} := \mathbb{K} \cap (a, +\infty),$$

$$\mathbb{K}_{\leq a} := \mathbb{K} \cap (-\infty, a], \ \mathbb{K}_{<a} := \mathbb{K} \cap (-\infty, a).$$

Combinatorics, Automata and Number Theory, ed. Valérie Berthé and Michel Rigo. Published by Cambridge University Press. ©Cambridge University Press 2010.

For instance, $\mathbb{N}_{>0}$ can indifferently be written $\mathbb{N} \setminus \{0\}$ or $\mathbb{N}_{\geq 1}$. Let $i, j \in \mathbb{Z}$ with $i \leq j$. We use the notation $[\![i, j]\!]$ for the set of integers $\{i, i+1, \ldots, j\}$.

Let X, Y be two sets. The notation $X \subseteq Y$ stands for the fact that every element of X is an element of Y, whereas $X \subset Y$ stands for the strict inclusion, i.e., $X \subseteq Y$ and $X \neq Y$. Let X^Y denote the set of all mappings from Y to X. Therefore the set of sequences indexed by $\mathbb{N}$ (respectively by $\mathbb{Z}$) of elements in X is denoted by $X^{\mathbb{N}}$ (respectively by $X^{\mathbb{Z}}$). As a particular case, 2^X is the power set of X, i.e., the set of all subsets of X. Indeed, 2 can be identified with $\{0, 1\}$ and maps from X to $\{0, 1\}$ are in one-to-one correspondence with subsets of X. In particular, if X is finite of cardinality Card $X = n$, then 2^X contains 2^n sets. The Cartesian product of X and Y is denoted by $X \times Y$. It is the set of ordered pairs (x, y) for all $x \in X$ and $y \in Y$. For a subset X of a topological space, $\text{int}(X)$ stands for the *interior* of X, $\overline{X}$ for the closure of X, and ∂X for its *boundary*, that is, $\partial X = \overline{X} \setminus \text{int}(X)$.

The floor of a real number x is $\lfloor x \rfloor = \sup\{z \in \mathbb{Z} \mid z \leq x\}$, whereas $\{x\} = x - \lfloor x \rfloor$ stands for the fractional part of x. For $\mathbf{x} = (x_1, \ldots, x_n) \in \mathbb{R}^n$, we will use the following set of notation for the most usual norms

$$||\mathbf{x}||_1 := \sum_{i=1}^n |x_i|, \quad ||\mathbf{x}||_\infty = \max_i |x_i|, \quad ||\mathbf{x}||_2 = \sqrt{\sum_{i=1}^n x_i^2},$$

and will denote the corresponding open ball with radius R and centre $\mathbf{x}$ as $B_1(\mathbf{x}, R)$, $B_\infty(\mathbf{x}, R)$, $B_2(\mathbf{x}, R)$, respectively. For more on vector norms, see Section 4.7.2.2.

It is a good opportunity to recall here notation about asymptotics. Let $f, g : \mathbb{R} \to \mathbb{R}$ be two functions. The definitions given below can also be applied to functions defined on another domain like $\mathbb{R}_{>a}$, $\mathbb{N}$ or $\mathbb{Z}$. We assume implicitly that the following notions are defined for $x \to +\infty$. We write $f \in \mathcal{O}(g)$, if there exist two constants x_0 and $C > 0$ such that, for all $x \geq x_0$, $|f(x)| \leq C|g(x)|$. We also write $f \ll g$ or $g \gg f$, or else $g \in \Omega(f)$. Note that we can write either $f \in \mathcal{O}(g)$ or $f = \mathcal{O}(g)$. Be aware that in the literature, authors sometimes give different meanings to the notation $\Omega(f)$. Here we consider a bound, for all large enough x, but there exist variants where the bound holds only for an increasing sequence $(x_n)_{n \geq 0}$ of reals, i.e., $\limsup_{x \to +\infty} |g(x)|/|f(x)| > 0$.

If g belongs to $\mathcal{O}(f) \cap \Omega(f)$, i.e., there exist constants x_0, C_1, C_2 with $C_1, C_2 > 0$ such that, for all $x \geq x_0$, $C_1|f(x)| \leq |g(x)| \leq C_2|f(x)|$, then we write $g \in \Theta(f)$. As an example, the function $x^2 + \sin 6x$ is in $\Theta(x^2)$ and $x^2|\sin(4x)|$ is in $\mathcal{O}(x^2)$ but not in $\Theta(x^2)$. In Figure 1.1, we have represented the functions $x^2 + \sin 6x$, $x^2|\sin(4x)|$, $4x^2/5$ and $6x^2/5$.

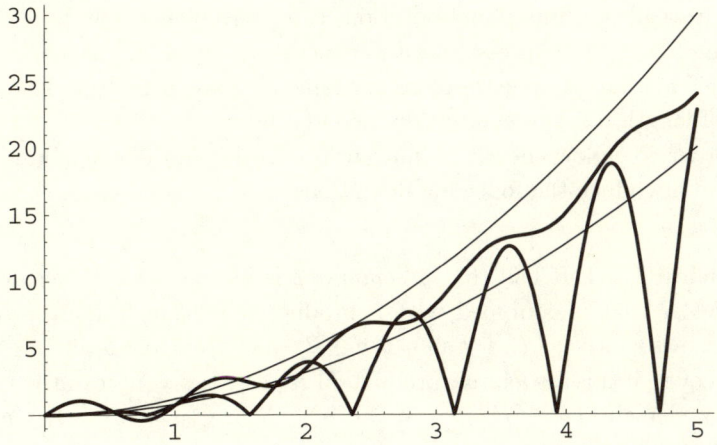

Fig. 1.1 The functions $x^2 + \sin 6x$, $x^2|\sin(4x)|$, $4x^2/5$ and $6x^2/5$.

If $\lim_{x \to +\infty} \frac{f(x)}{g(x)} = 0$, we write $f = o(g)$. Finally, if $\lim_{x \to +\infty} \frac{f(x)}{g(x)} = 1$, we write $f \sim g$. For more on asymptotics, see for instance (de Bruijn 1981) or the first chapter of (Hardy and Wright 1985).

Lastly, we will use the notation $\log = \log_e$ for the natural logarithm, whereas $\log_2$ will denote the binary logarithm.

1.2 Words

This section is only intended to give basic definitions of concepts developed later on. For material not covered in this book, classical textbooks on finite or infinite words and their properties are (Lothaire 1983), (Lothaire 2002), (Lothaire 2005). See also the chapter (Choffrut and Karhumäki 1997) or the tutorial (Berstel and Karhumäki 2003). The first chapters of the books (Allouche and Shallit 2003) and (Pytheas Fogg 2002) also contain many references for further developments in combinatorics on words.

1.2.1 Finite words

An *alphabet* is a finite set of *symbols* (or *letters*). Usually, alphabets will be denoted using Roman upper case letters, like A or B. The most basic and fundamental objects that we shall deal with are *words*.

Let A be an alphabet. A *finite* word over A (to distinguish with the infinite case that will be considered later on) is a finite sequence of letters in A. In a formal way, a word of length $n \in \mathbb{N}$ is a map u from $[\![0, n-1]\!]$

to A. Instead of a functional notation, it is convenient to write a word as $u = u_0 \cdots u_{n-1}$ to express u as the concatenation of the letters u_i. The *length* of u, that is, the size of its domain, is denoted by $|u|$. The unique word of length 0 is the *empty word* denoted by ε.

In order to endow the set of finite words with a suitable algebraic structure, we introduce the following definitions.

Definition 1.2.1 Recall that a *semigroup* is an algebraic structure given by a set R that is equipped with a product operation from $R \times R$ to R which is associative, *i.e.*, for all $a, b, c \in R$, $(a \cdot b) \cdot c = a \cdot (b \cdot c)$.

Moreover, if this associative product on R possesses a (necessarily unique) identity element $1_R \in R$, *i.e.*, for all $a \in R$, $a \cdot 1_R = a = 1_R \cdot a$, then this algebraic structure is said to be a *monoid*. For instance the set $\mathbb{N}^d$, with $d \geq 1$, of d-tuples of non-negative integers with the usual addition component-wise is a monoid with $(0, \ldots, 0)$ as identity element.

Definition 1.2.2 Let $(R, \cdot)$ and $(T, \diamond)$ be monoids with respectively 1_R and 1_T as identity element. A map $f : R \to T$ is a *monoid morphism* (or *homomorphism of monoids*) if $f(1_R) = 1_T$ and for all $a, b \in R$, $f(a \cdot b) = f(a) \diamond f(b)$.

Let $u = u_0 \cdots u_{m-1}$ and $v = v_0 \cdots v_{n-1}$ be two words over A. The *concatenation* of u and v is the word $w = w_0 \cdots w_{m+n-1}$ defined by $w_i = u_i$ if $0 \leq i < m$, and $w_i = v_{i-m}$ otherwise. We write $u \cdot v$ or simply uv to express the concatenation of u and v. Notice that this operation is associative. Let u be a word and $n \in \mathbb{N}$. Naturally, let u^n denote the concatenation of n copies of u and we set $u^0 = \varepsilon$. A *square* is a word of the form uu, where $u \in A^*$.

The set of all (finite) words over A is denoted by A^*. Endowed with the concatenation of words as product operation, A^* is a monoid with ε as identity element. It is the *free* monoid generated by A (freeness means that any element in A^* has a unique factorisation as product of elements in A). Notice that the length map $|\cdot| : (A^*, \cdot) \to (\mathbb{N}, +)$, $w \mapsto |w|$ is a morphism of monoids. Let $A^+ = A^* \setminus \{\varepsilon\}$ denote the free semigroup generated by A. Finally, for $n \in \mathbb{N}$, A^n is the set of words of length n over A and $A^{\leq n}$ is the set of words over A of length less or equal to n.

The *mirror* (sometimes called *reversal*) of a word $u = u_0 \cdots u_{m-1}$ is the word $\tilde{u} = u_{m-1} \cdots u_0$. It can be defined inductively on the length of the word by $\tilde{\varepsilon} = \varepsilon$ and $\widetilde{au} = \tilde{u}a$ for $a \in A$ and $u \in A^*$. Notice that for $u, v \in A^*$, $\widetilde{uv} = \tilde{v}\tilde{u}$. A *palindrome* is a word u such that $\tilde{u} = u$. For instance, the palindromes of length at most 3 in $\{0,1\}^*$ are

$$\varepsilon, 0, 1, 00, 11, 000, 010, 101, 111.$$

We end this section about finite words with the notion of code.

Definition 1.2.3 A subset $Y \subset A^+$ is a *code* if, for all $u_1, \ldots, u_m, v_1 \ldots, v_n \in Y$, the equality $u_1 \cdots u_m = v_1 \cdots v_n$ implies $n = m$ and $u_i = v_i$ for $i = 1, \cdots, m$. A code is said to be a prefix code if none of its elements is a prefix of another one.

1.2.2 Infinite words

To define infinite words, we consider maps taking values in an alphabet but defined on an infinite domain. A *(one-sided) infinite word* over an alphabet A is a map from the set $\mathbb{N}$ of non-negative integers to A. Using the same convention as for finite words, we write $x = x_0 x_1 x_2 \cdots$ to represent an infinite word. It is sometimes convenient to use a notation like $x = (x_n)_{n \geq 0}$. If the domain is the set $\mathbb{Z}$ of integers, then we speak of *bi-infinite word* (in the literature, we also find the terminology of *two-sided infinite words*). In this latter situation, a convenient notation is to use a decimal point to determine the position of the image of 0 like $\cdots x_{-2} x_{-1}.x_0 x_1 x_2 \cdots$.

In what follows if no explicit mention is made then we shall be dealing with one-sided infinite words and we will omit reference to it.

The set of infinite words over A is denoted by $A^{\mathbb{N}}$. We can define a concatenation operation from $A^* \times A^{\mathbb{N}}$ to $A^{\mathbb{N}}$ as follows. The concatenation of the finite word $u = u_0 \cdots u_{n-1}$ and the infinite word $x = x_0 x_1 \cdots$ is the infinite word $y = y_0 y_1 \cdots$ denoted by ux and defined by $y_i = u_i$ if $0 \leq i \leq n-1$, and $y_i = x_{i-n}$ if $i \geq n$.

Example 1.2.4 Consider the infinite word $x = x_0 x_1 x_2 \cdots$ where the letters $x_i \in \{0, \ldots, 9\}$ are given by the digits appearing in the usual decimal expansion of $\pi - 3$,

$$\pi - 3 = \sum_{i=0}^{+\infty} x_i \, 10^{-i-1},$$

i.e., $x = 14159265358979323846264338327950288419\cdots$ is an infinite word.

Definition 1.2.5 Any subset X of $\mathbb{N}$ (respectively $\mathbb{Z}$) gives rise to an infinite (respectively bi-infinite) word over $\{0,1\}$, namely its *characteristic word*. Let x be this word. It is defined as follows

$$x_n = \begin{cases} 1, & \text{if } n \in X, \\ 0, & \text{otherwise.} \end{cases}$$

It also refers to the *indicator function* of the set X, denoted by $\mathbb{1}_X(n)$.

Example 1.2.6 Consider the characteristic sequence of the set of prime numbers $x = x_0 x_1 \cdots = 0011010100010100010100010000\cdots$.

1.2.3 Factors, topology and orderings

The following notions can be defined for both finite and infinite words. Let us start with the finite case. Let $u = u_0 \cdots u_{n-1}$ be a finite word over A. If u can be factorised as $u = vfw$ with $v, f, w \in A^*$, we say that f is a *factor* of u. If $f = u_i \cdots u_{i+|f|-1}$, then f is said to *occur* at position i in u. For convenience, $u[i, i + \ell - 1]$ denotes the factor of u of length $\ell \geq 1$ occurring at position i. The number of occurrences of f in u is denoted by $|u|_f$. In particular, if $a \in A$, then $|u|_a$ denotes the number of letters a occurring in u. If u is a finite or infinite word over A, then alph(u) is the set of letters which occur in u. If u is the empty word, then alph(u) is the empty set. One has alph$(u) \subseteq A$.

Assume that $A = \{a_1 < \cdots < a_n\}$ is totally ordered. The map $\mathbf{P} : A^* \to \mathbb{N}^n$, $w \mapsto {}^t(|w|_{a_1}, \ldots, |w|_{a_n})$ is called the *abelianisation map*. It is trivially a morphism of monoids. Notice that in the literature, this map is also referred to as the *Parikh mapping*. Note that for a matrix $\mathbf{M}$, ${}^t\mathbf{M}$ is the transpose of $\mathbf{M}$.

If $u = fw$ (respectively $u = vf$) then f is a *prefix* (respectively a *suffix*) of u. A word $u = u_0 \cdots u_{n-1}$ of length n has exactly $n+1$ prefixes: ε, u_0, $u_0 u_1$, ..., $u_0 \cdots u_{n-2}$, u. The same holds for suffixes. A *proper* prefix (respectively *proper* suffix) of u is a prefix (respectively suffix) different from the full word u. Let us observe that a factor of u is obtained as the concatenation of consecutive letters occurring in u. By opposition a *scattered subword* of $u = u_0 \cdots u_{n-1}$ is of the form $u_{i_0} u_{i_1} \cdots u_{i_k}$ with $k < n$ and $0 \leq i_1 < i_2 < \cdots < i_k < n$.

Example 1.2.7 Let $A = \{0, 1\}$ be the binary alphabet consisting of letters 0 and 1. The set A^* contains all the finite words obtained by concatenating 0's and 1's. The concatenation of the words $u = 1001$ and $v = 010$ is the word $w = uv = 1001010 = w_0 \cdots w_6$. The word v occurs twice in w at positions 2 and 4. We have $w[1, 3] = 001$ and the suffix 1010 is a square, i.e., $(10)^2$. To conclude with the example, $|w|_0 = |u|_0 + |v|_0 = 2 + 2 = 4$.

The notions of *factor*, *prefix* or *suffix* as well as the relevant notation introduced for finite words can be extended to infinite words. Factors and prefixes are finite words, but a suffix of an infinite word is also infinite.

Let $x = x_0 x_1 x_2 \cdots$ be an infinite word over A. For instance, for $\ell \geq 0$, $x[0, \ell - 1] = x_0 \cdots x_{\ell-1}$ is the prefix of length ℓ of x. We denote by $x[i, i + \ell - 1] = x_i \cdots x_{i+\ell-1}$ the factor of length $\ell \geq 1$ occurring in x at position $i \geq 0$. For $n \geq 0$, the infinite word $x_n x_{n+1} \cdots$ is a suffix of x. See the relationship with the notion of shift introduced in Section 1.6.

Definition 1.2.8 The *language* of the infinite word x is the set of all its factors. It is denoted by $L(x)$. The set of factors of length n occurring in x is denoted by $L_n(x)$.

Definition 1.2.9 An infinite word x is *recurrent* if all its factors occur infinitely often in x. It is *uniformly recurrent* (also called *minimal*), if it is recurrent and for every factor u of x, if $T_x(u) = \{i_1^{(u)} < i_2^{(u)} < i_3^{(u)} < \cdots\}$ is the infinite set of positions where u occurs in x, then there exists a constant C_u such that, for all $j \geq 1$,

$$i_{j+1}^{(u)} - i_j^{(u)} \leq C_u.$$

An infinite set $X \subseteq \mathbb{N}$ of integers having such a property, *i.e.*, where the difference of any two consecutive elements in X is bounded by a constant, is said to be *syndetic* or with *bounded gap*. Otherwise stated, an infinite word x is uniformly recurrent if, and only if, for all factors $u \in L(x)$, the set $T_x(u)$ is infinite and syndetic.

Definition 1.2.10 One can endow $A^\mathbb{N}$ with a *distance* d defined as follows. Let x, y be two infinite words over A. Let $x \wedge y$ denote the longest common prefix of x and y. Then the distance d is given by

$$d(x, y) := \begin{cases} 0, & \text{if } x = y, \\ 2^{-|x \wedge y|}, & \text{otherwise.} \end{cases}$$

It is obvious to see that, for all $x, y, z \in A^\mathbb{N}$, $d(x, y) = d(y, x)$, $d(x, z) \leq d(x, y) + d(y, z)$ and $d(x, y) \leq \max(d(x, z), d(y, z))$. This last property is not required to have a distance, but when it holds, the distance is said to be *ultrametric*.

This notion of distance extends to $A^\mathbb{Z}$. Notice that the topology on $A^\mathbb{N}$ is the product topology (of the discrete topology on A). The space $A^\mathbb{N}$ is a compact *Cantor set*, that is, a totally disconnected compact space without isolated points. Since $A^\mathbb{N}$ is a (complete) metric space, it is therefore relevant to speak of convergent sequences of infinite words. The sequence $(z_n)_{n \geq 0}$ of infinite words over A *converges* to $x \in A^\mathbb{N}$, if for all $\varepsilon > 0$, there exists $N \in \mathbb{N}$ such that, for all $n \geq N$, $d(z_n, x) < \varepsilon$. To express the fact that a sequence of finite words $(w_n)_{n \geq 0}$ over A converges to an infinite word

y, it is assumed that A is extended with an extra letter $c \notin A$. Any finite word w_n is replaced with the infinite word $w_n ccc\cdots$ and if the sequence of infinite words $(w_n ccc\cdots)_{n \geq 0}$ converges to y, then the sequence $(w_n)_{n \geq 0}$ is said to converge to y.

Let $(u_n)_{n \geq 0}$ be a sequence of non-empty finite words. If we define, for all $\ell \geq 0$, the finite word v_ℓ as the concatenation $u_0 u_1 \cdots u_\ell$, then the sequence $(v_\ell)_{\ell \geq 0}$ of finite words converges to an infinite word. This latter word is said to be the concatenation of the elements in the infinite sequence of finite words $(u_n)_{n \geq 0}$. In particular, for a constant sequence $u_n = u$ for all $n \geq 0$, $v_\ell = u^{\ell+1}$ and the concatenation of an infinite number of copies of the finite word u is denoted by u^ω.

Definition 1.2.11 An infinite word $x = x_0 x_1 \cdots$ is *(purely) periodic* if there exists a finite word $u = u_0 \cdots u_{k-1} \neq \varepsilon$ such that $x = u^\omega$, *i.e.*, for all $n \geq 0$, we have $x_n = u_r$ where $n = dk + r$ with $r \in [\![0, k-1]\!]$. An infinite word x is *eventually periodic* if there exist two finite words $u, v \in A^*$, with $v \neq \varepsilon$ such that $x = uvvv\cdots = uv^\omega$. Notice that purely periodic words are special cases of eventually periodic words. For any eventually periodic word x, there exist words u, v of shortest length such that $x = uv^\omega$, then the integer $|u|$ (respectively $|v|$) is referred to as the *preperiod* (respectively *period*) of x. An infinite word is said to be *non-periodic* if it is not ultimately periodic. A set $X \subseteq \mathbb{N}$ of integers is *eventually periodic* if its characteristic word is eventually periodic. Otherwise stated, X is eventually periodic if, and only if, it is a finite union of arithmetic progressions. Recall that an arithmetic progression is a set of integers of the kind $p\mathbb{N} + q = \{pn + q \mid n \in \mathbb{N}\}$.

Definition 1.2.12 The *complexity function* of an infinite word x maps $n \in \mathbb{N}$ onto the number $p_x(n) = \operatorname{Card} L_n(x)$ of distinct factors of length n occurring in x.

This function will be studied in detail in Chapter 4.

Definition 1.2.13 An infinite word x is *Sturmian* if $p_x(n) = n + 1$ for all $n \geq 0$. In particular, Sturmian words are over a binary alphabet.

From the developments in Chapter 4 and in particular thanks to the celebrated theorem of Morse and Hedlund, Sturmian words are non-periodic words of smallest complexity.

A survey on Sturmian words by J. Berstel and P. Séébold can be found in (Lothaire 2002); the chapter by P. Arnoux in (Pytheas Fogg 2002) is also of interest.

The complexity function counts the number of different factors of a given length in an infinite word x. Each distinct factor u of length n increments $p_x(n)$ by one whether it occurs only once in x or conversely occurs many times. So to speak, $p_x(n)$ does not reveal the frequency of occurrences of the different factors. We might need more precise information concerning the frequency of a factor.

Definition 1.2.14 Let x be an infinite word. The *frequency* $f_x(u)$ of a factor u of x is defined as the limit (when n tends towards infinity), if it exists, of the number of occurrences of the factor u in $x_0 x_1 \cdots x_{n-1}$ divided by n, i.e., provided the limit exists,

$$f_x(u) = \lim_{n \to +\infty} \frac{|x[0, n-1]|_u}{n}.$$

Let us now introduce orders on words. The sets A^* and $A^\mathbb{N}$ can be ordered as follows.

Definition 1.2.15 Assume that $(A, <)$ is a totally (or linearly) ordered alphabet. Then the set A^* is totally ordered by the *radix order* (or sometimes called *genealogical order*) defined as follows. Let u, v be two words in A^*. We write $u \prec v$ if either $|u| < |v|$, or if $|u| = |v|$ and there exist $p, q, r \in A^*$, $a, b \in A$ with $u = paq$, $v = pbr$ and $a < b$. By $u \preceq v$, we mean that either $u \prec v$ or $u = v$. The set A^* can also be totally ordered by the *lexicographic order* defined as follows. Let u, v be two words in A^*, we write $u < v$ if u is a proper prefix of v or if there exist $p, q, r \in A^*$, $a, b \in A$ with $u = paq$, $v = pbr$ and $a < b$. By $u \leq v$, we mean that either $u < v$ or $u = v$. .

Observe that on a unary (*i.e.*, single letter) alphabet, the two orderings over $\{a\}^*$ coincide but if the cardinality of the alphabet A is at least 2, then the radix order is a well order (*i.e.*, every non-empty subset of A^* has a least element for this order) but the lexicographic order is not. For instance, the set of words $\{a^n b \mid n \geq 0\}$ does not have a least element for the lexicographic order.

Definition 1.2.16 Notice that the lexicographic order introduced on A^* can naturally be extended to $A^\mathbb{N}$. Let $x, y \in A^\mathbb{N}$. We have $x < y$ if there exist $p \in A^*$, $a, b \in A$ and $w, z \in A^\mathbb{N}$ such that $x = paw$, $y = pbz$ and $a < b$.

1.2.4 Morphisms

Particular infinite words of interest can be obtained by iterating morphisms (or homomorphisms of free monoids). A survey on morphisms is given in (Harju and Karhumäki 1997). Again the textbooks like

(Queffélec 1987), (Pytheas Fogg 2002), (Lothaire 1983), (Lothaire 2002) or (Berstel, Aaron, Reutenauer, et al. 2008) are worth reading for topics not considered here.

Let A and B be two alphabets. A *morphism* (also called *substitution*) is a map $\sigma : A^* \to B^*$ such that $\sigma(uv) = \sigma(u)\sigma(v)$ for all $u, v \in A^*$ (see also Definition 1.2.2). Note that the terminology substitution often refers in the literature to non-erasing endomorphisms. We similarly define the notion of *endomorphism* if $A = B$. Notice that in particular, $\sigma(\varepsilon) = \varepsilon$. Usually morphisms will be denoted by Greek letters. To define completely a morphism, it is enough to know the images of the letters in A, the image of a word $u = u_0 \cdots u_{n-1}$ being the concatenation of the images of its letters, $\sigma(u) = \sigma(u_0) \cdots \sigma(u_{n-1})$. Otherwise stated, any map from A to B^* can be uniquely extended to a morphism from A^* to B^*.

Definition 1.2.17 Let $k \in \mathbb{N}$. A morphism $\sigma : A^* \to B^*$ is *uniform* (or *k-uniform*) if for all $a \in A$, $|\sigma(a)| = k$. A 1-uniform morphism is often called *coding* or *letter-to-letter* morphism. If for some $a \in A$, $\sigma(a) = \varepsilon$, then σ is said to be *erasing*, otherwise it is said to be *non-erasing*.

If $\sigma : A^* \to B^*$ is a non-erasing morphism, it can be extended to a map from $A^{\mathbb{N}}$ to $B^{\mathbb{N}}$ as follows. If $x = x_0 x_1 \cdots$ is an infinite word over A, then the sequence of words $(\sigma(x_0 \cdots x_{n-1}))_{n \geq 0}$ is easily seen to be convergent towards an infinite word over B. Its limit is denoted by $\sigma(x) = \sigma(x_0)\sigma(x_1)\sigma(x_2)\cdots$. We similarly extend σ to a map from $A^{\mathbb{Z}}$ to $B^{\mathbb{Z}}$ as follows. If $x = \cdots x_{-2} x_{-1}.x_0 x_1 x_2 \cdots$ is a bi-infinite word over A, then the sequence of words $(\sigma(x_{-n} \cdots x_{-1}.x_0 \cdots x_{n-1}))_{n \geq 0}$ is easily seen to be convergent towards a bi-infinite word over B. Its limit is here again denoted by $\sigma(x)$. Consequently, the definition of morphisms extends from A^* to $A^* \cup A^{\mathbb{N}} \cup A^{\mathbb{Z}}$. For the sake of simplicity, we define morphisms on A^*, but we consider implicitly their action on infinite and bi-infinite words. Notice that if σ is erasing, then the image of an infinite or bi-infinite word could be finite.

Let $\sigma : A^* \to A^*$ be a morphism. A finite, infinite or bi-infinite word x such that $\sigma(x) = x$ is said to be a *fixed point* of σ.

Definition 1.2.18 If there exist a letter $a \in A$ and a word $u \in A^+$ such that $\sigma(a) = au$ and moreover, if $\lim_{n \to +\infty} |\sigma^n(a)| = +\infty$, then σ is said to be (right) *prolongable* on a. Let $\sigma : A^* \to A^*$ be a morphism prolongable on a. We have

$$\sigma(a) = a\,u, \ \sigma^2(a) = a\,u\,\sigma(u), \ \sigma^3(a) = a\,u\,\sigma(u)\,\sigma^2(u), \ \ldots.$$

Since, for all $n \in \mathbb{N}$, $\sigma^n(a)$ is a prefix of $\sigma^{n+1}(a)$ and because $|\sigma^n(a)|$ tends to infinity when $n \to +\infty$, the sequence $(\sigma^n(a))_{n \geq 0}$ converges to an infinite word denoted by $\sigma^\omega(a)$ and given by

$$\sigma^\omega(a) := \lim_{n \to +\infty} \sigma^n(a) = a\, u\, \sigma(u)\, \sigma^2(u)\, \sigma^3(u) \cdots.$$

This infinite word is a fixed point of σ. An infinite word obtained in this way by iterating a prolongable morphism is said to be *generated by* σ, and more generally, *purely substitutive* or *purely morphic*. In the literature, one also finds the term *pure morphic*. If $x \in A^\mathbb{N}$ is purely morphic and if $\tau : A \to B$ is a coding, then the word $y = \tau(x)$ is said to be *morphic* or *substitutive*.

Let $A = \{a, b, c\}$ and $\sigma \colon A^* \to A^*$ be the endomorphism defined by $\sigma(a) = a$, $\sigma(b) = bb$, $\sigma(c) = aab$. The morphism σ is not prolongable on the letter c but the sequence of words $(\sigma^n(c))_{n \geq 0}$ converges to the infinite word aab^ω, which is morphic but not purely morphic. For other examples of morphic words that are not purely morphic, see Example 4.6.5, Exercise 4.12, Exercise 4.13, and Proposition 4.7.2. See also Exercise 10.1.2 in the same vein. For an example of a purely morphic word that is fixed by no non-erasing endomorphism other than the identity, see Exercise 4.11.

We also consider bi-infinite morphic words.

Definition 1.2.19 If there exist a letter $b \in A$ and a word $u \in A^+$ such that $\sigma(b) = ub$ and moreover, if $\lim_{n \to +\infty} |\sigma^n(b)| = +\infty$, then σ is said to be left *prolongable* on b. We have

$$\sigma(b) = u\,b, \quad \sigma^2(a) = \sigma(u)\,u\,b, \quad \sigma^3(a) = \sigma^2(u)\,\sigma(u)\,u\,b, \ \ldots.$$

Let $\sigma : A^* \to A^*$ be a morphism that is both right prolongable on a and left prolongable on b. The sequence $(\sigma^n(b).\sigma^n(a))_{n \geq 0}$ converges to a bi-infinite word denoted by $\sigma^\omega(b).\sigma^\omega(a)$ which is a fixed point of σ. If furthermore, there exist a letter $c \in A$ and $\ell \in \mathbb{N}$ such that ba is a factor of $\sigma^\ell(c)$, then the bi-infinite word $\sigma^\omega(b).\sigma^\omega(a)$ is said to be generated by σ, and more generally *purely substitutive* or *purely morphic*. We similarly define as in Definition 1.2.18 a *substitutive* or *morphic* bi-infinite word.

Definition 1.2.20 For each morphism $\sigma : A^* \to B^*$, we define the *width* of σ, denoted by $\|\sigma\|$, as $\|\sigma\| := \max_{a \in A} |\sigma(a)|$.

It is clear that $|\sigma(w)| \leq \|\sigma\|\,|w|$ for every $w \in A^*$.

Let us note that a morphism $\sigma : A^* \to B^*$ is injective if, and only if, letters of A are mapped to distinct words and the language $\sigma(A)$ is a code. See (Lothaire 2002, Proposition 6.13).

Example 1.2.21 (Thue–Morse word) Consider the 2-uniform morphism defined over the alphabet $\{a,b\}$ by $\sigma : a \mapsto ab, b \mapsto ba$. The infinite (purely morphic) word

$$\sigma^\omega(a) = abbabaabbaababbabaababbaabbabaab \cdots$$

is the celebrated *Thue–Morse word*. This word can also be obtained as follows. Consider the morphism $\gamma : a \mapsto b, b \mapsto a$ and define the sequence of finite words $u_0 = a$ and, for all $n \geq 1$, $u_n = u_{n-1}\gamma(u_{n-1})$. It is an exercise to show that the sequence $(u_n)_{n \geq 0}$ converges to the Thue–Morse word. For more details on the Thue–Morse word, see Section 4.10.4.

Many properties of the Thue–Morse word can be found in the paper (Allouche and Shallit 1999). In several chapters of (Pytheas Fogg 2002), the Thue–Morse word or the Fibonacci word introduced below are also discussed in detail.

Example 1.2.22 (Fibonacci word) Another significant example of a purely morphic word is the *Fibonacci word*. It is obtained from the non-uniform morphism defined over the alphabet $\{a,b\}$ by $\sigma : a \mapsto ab, b \mapsto a$,

$$\sigma^\omega(a) = (x_n)_{n \geq 0} = abaababaabaababaababaabaababaabababaa \cdots.$$

It is a Sturmian word and can be obtained as follows. Let $\varphi = (1+\sqrt{5})/2$ be the Golden Ratio. For all $n \geq 1$, if $\lfloor(n+1)\varphi\rfloor - \lfloor n\varphi \rfloor = 2$, then $x_{n-1} = a$, otherwise $x_{n-1} = b$. For more details, see Section 4.10.3.

Example 1.2.23 (Squares) Consider the alphabet $A = \{a,b,c\}$ and the morphism $\sigma : A^* \to A^*$ defined by $\sigma : a \mapsto abcc, b \mapsto bcc, c \mapsto c$. We get

$$\sigma^\omega(a) = abccbcccbcccccbcccccccbcccccccccbcc \cdots.$$

Using the special form of the images of b and c, it is not difficult to see that the difference between the position of the nth b and the $(n+1)$st b in $\sigma^\omega(a)$ is $2n+1$. Since the difference between two corresponding consecutive squares $(n+1)^2 - n^2$ is also $2n+1$, if we define the coding $\tau : a, b \mapsto 1, c \mapsto 0$, we get exactly

$$\tau(\sigma^\omega(a)) = 1100100001000000100000000100000000000100 \cdots$$

which proves that the characteristic sequence of the set of squares is morphic. One can show that this morphic sequence cannot be generated using a uniform morphism, for instance see (Eilenberg 1974) where it is shown that the set of squares is not k-recognisable. Also see Example 1.3.16.

Example 1.2.24 (Powers of 2**)** Consider the 2-uniform morphism defined over the alphabet $\{a,b,c\}$ by $\sigma : a \mapsto ab, b \mapsto bc, c \mapsto cc$ and the coding $\tau : a, c \mapsto 0, b \mapsto 1$. We have

$$\sigma^\omega(a) = abbcbcccbccccccbccccccccccccccccbcc\cdots$$

and

$$\tau(\sigma^\omega(a)) = 01101000100000001000000000000000100\cdots.$$

Developing the same kind of arguments as in the previous example, this latter morphic word is easily seen to be the characteristic word of the set of powers of two. For more details on this morphic word, see Section 4.10.2.

For complements on morphisms, see Section 4.6.1. These notions will also be extended to the framework of D0L and HD0L systems in Chapter 10, see also Section 3.4.2, where a *D0L system* is a triple of the form (A, σ, w) where A is an alphabet, σ is a morphism of A^* and w is a word over A. A language D, that is a set of words, is called a *D0L language* if there exists a D0L-system (A, σ, w) such that $D = \{\sigma^k(w) \mid k \in \mathbb{N}\}$. A language H is called a *HD0L language* if there exist two alphabets A and B, a D0L language D over A and a morphism $\tau : A^* \to B^*$ such that $H = \tau(D)$.

1.3 Languages and machines

Formal languages theory is mostly concerned with the study of the mathematical properties of sets of words. For an exhaustive exposition on regular languages and automata theory, see (Sakarovitch 2003), or in the same spirit (Eilenberg 1974). See also the chapter (Yu 1997), or (Sudkamp 2005), (Hopcroft and Ullman 1979) and the updated revision (Hopcroft, Motwani, and Ullman 2006) for general introductory books on formal languages theory. In (Perrin 1990), the relationship of automata with recognisable sets of integers is presented. In this section, we do not present languages of infinite words and the corresponding automata crafted to recognise these languages, a reference is the book (Perrin and Pin 2003), see also (Thomas 1990). Notions presented here have been kept minimal, more definitions and results on finite automata and transducers can be found in Section 2.6.

1.3.1 Languages of finite words

Let A be an alphabet. A subset L of A^* is said to be a *language*. Note for instance that this terminology is consistent with the one of Definition 1.2.8.

Since a language is a *set* of words, we can apply all the usual set operations like union, intersection or set difference: ∪, ∩ or \. The concatenation of words can be extended to define an operation on languages. If L, M are languages, LM is the language of the words obtained by concatenation of a word in L and a word in M, i.e.,

$$LM = \{uv \mid u \in L, v \in M\}.$$

We can of course define the concatenation of a language with itself, so it permits us to introduce the power of a language. Let $n \in \mathbb{N}$, A be an alphabet and $L \subseteq A^*$ be a language. The language L^n is the set of words obtained by concatenating n words in L. We set $L^0 := \{\varepsilon\}$. In particular, we recall that A^n denotes the set of words of length n over A, i.e., concatenations of n letters in A. The *(Kleene) star* of the language L is defined as

$$L^* = \bigcup_{i \geq 0} L^i.$$

Otherwise stated, L^* contains the words that are obtained as the concatenation of an arbitrary number of words in L. Notice that the definition of Kleene star is compatible with the notation A^* introduced to denote the set of finite words over A. We also write $L^{\leq n}$ as a shorthand for

$$L^{\leq n} = \bigcup_{i=0}^{n} L^i.$$

Note that if the empty word belongs to L, then $L^{\leq n} = L^n$. We recall that $A^{\leq n}$ is the set of words over A of length at most n. If L is a language, then alph(L) is the set of all letters which occur in the words of L, i.e., alph(L) = $\cup_{u \in L}$ alph(u).

Example 1.3.1 Let $L = \{a, ab, aab\}$ and $M = \{a, ab, ba\}$ be two finite languages. We have $L^2 = \{aa, aab, aaab, aba, abab, abaab, aaba, aabab, aabaab\}$ and $M^2 = \{aa, aab, aba, abab, abba, baa, baab, baba\}$. One can notice that Card(L^2) = (Card L)^2 but Card(M^2) < (Card M)^2. This is due to the fact that all words in L^2 have a unique factorisation as concatenation of two elements in L but this is not the case for M, where $(ab)a = a(ba)$. We can notice that

$$L^* = \{a\}^* \cup \{a^{i_1} b a^{i_2} b \cdots a^{i_n} b a^{i_{n+1}} \mid \forall n \geq 1, i_1, \ldots, i_n \geq 1, i_{n+1} \geq 0\}.$$

Since languages are sets of (finite) words, a language can be either *finite* or *infinite*. For instance, a language L differs from $\emptyset$ or $\{\varepsilon\}$ if, and only if, the language L^* is infinite. Let L be a language, we set $L^+ = LL^*$. The mirror operation can also be extended from words to languages: $\tilde{L} = \{\tilde{u} \mid u \in L\}$.

Definition 1.3.2 A language is *prefix-closed* (respectively *suffix-closed*) if it contains all prefixes (respectively suffixes) of any of its elements. A language is *factorial* if it contains all factors of any of its elements.

Obviously, any factorial language is prefix-closed and suffix-closed. The converse does not hold. For instance, the language $\{a^n b \mid n > 0\}$ is suffix-closed but not factorial.

Example 1.3.3 The set of words over $\{0,1\}$ containing an even number of 1's is the language

$$\begin{aligned} E &= \{w \in \{0,1\}^* \mid |w|_1 \equiv 0 \pmod{2}\} \\ &= \{\varepsilon, 0, 00, 11, 000, 011, 101, 110, 0000, 0011, \ldots\}. \end{aligned}$$

This language is closed under mirror, *i.e.*, $\tilde{L} = L$. Notice that the concatenation $E\{1\}E$ is the language of words containing an odd number of 1's and $E \cup E\{1\}E = E(\{\varepsilon\} \cup \{1\}E) = \{0,1\}^*$. Notice that E is neither prefix-closed, since $1001 \in E$ but $100 \notin E$, nor suffix-closed.

Similarly as for infinite words (see Definition 1.2.12), we can count the number of words of a language of a given length. A language L of A^* is said to have to have *bounded growth* if for every n there are less than k words of length n in L, for a fixed integer k. Such languages are also called *slender*. For more on slender languages, see, *e.g.*, Proposition 2.6.3 and Section 3.3.2.

If a language L over A can be obtained by applying to some finite languages a finite number of operations of union, concatenation and Kleene star, then this language is said to be a *regular language*. This generation process leads to *regular expressions* which are well-formed expressions used to describe how a regular language is built in terms of these operations. From the definition of a regular language, the following result is immediate.

Theorem 1.3.4 *The class of regular languages over A is the smallest subset of 2^{A^*} (for inclusion) containing the languages $\emptyset$, $\{a\}$ for all $a \in A$ and closed under union, concatenation and Kleene star.*

Example 1.3.5 For instance, the language L over $\{0,1\}$ whose words do not contain the factor 11 is regular. This language can be described by the regular expression $L = \{0\}^*\{1\}\{0, 01\}^* \cup \{0\}^*$. Otherwise stated, it is generated from the finite languages $\{0\}$, $\{0, 01\}$ and $\{1\}$ by applying union, concatenation and star operations. Its complement in A^* is also regular and is described by the regular expression $A^*\{11\}A^*$. The language E

from Example 1.3.3 is also regular, we have the following regular expression $\{0\}^*(\{1\}\{0\}^*\{1\}\{0\}^*)^*$ describing E.

1.3.2 Automata

As we shall briefly explain in this section, the regular languages are exactly the languages recognised by finite automata.

Definition 1.3.6 A *finite automaton* is a labelled graph given by a 5-tuple $\mathcal{A} = (Q, A, E, I, T)$ where Q is the (finite) *set of states*, $E \subseteq Q \times A^* \times Q$ is the finite set of *edges* defining the *transition relation*, $I \subseteq Q$ is the set of *initial states* and T is the *set of terminal (or final) states*. A *path* in the automaton is a sequence

$$(q_0, u_0, q_1, u_1, \ldots, q_{k-1}, u_{k-1}, q_k)$$

such that, for all $i \in [\![0, k-1]\!]$, $(q_i, u_i, q_{i+1}) \in E$, $u_0 \cdots u_{k-1}$ is the *label* of the path. Such a path is *successful* if $q_0 \in I$ and $q_k \in T$. The language $L(\mathcal{A})$ *recognised* (or *accepted*) by $\mathcal{A}$ is the set of labels of all successful paths in $\mathcal{A}$.

Any finite automaton $\mathcal{A}$ gives a partition of A^* into $L(\mathcal{A})$ and $A^* \setminus L(\mathcal{A})$. When depicting an automaton, initial states are marked with an incoming arrow and terminal states are marked with an outgoing arrow. A transition like (q, u, r) is represented by a directed edge from q to r with label u, $q \xrightarrow{u} r$.

Example 1.3.7 In Figure 1.2 the automaton has two initial states p and r, three terminal states q, r and s. For instance, the word ba is recognised by the automaton. There are two successful paths corresponding to the label ba: (p, b, q, a, s) and (p, b, p, a, s). For this latter path, we can write $p \xrightarrow{b} p \xrightarrow{a} s$. On the other hand, the word $baab$ is not recognised by the automaton.

Example 1.3.8 The automaton in Figure 1.3 recognises exactly the language E of the words having an even number of 1 from Example 1.3.3.

Definition 1.3.9 Let $\mathcal{A} = (Q, A, E, I, T)$ be a finite automaton. A state $q \in Q$ is *accessible* (respectively *co-accessible*) if there exists a path from an initial state to q (respectively from q to some terminal state). If all states of $\mathcal{A}$ are both accessible and co-accessible, then $\mathcal{A}$ is said to be *trim*.

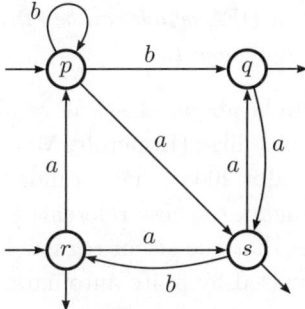

Fig. 1.2 A finite automaton.

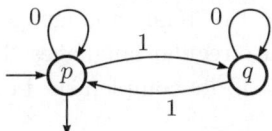

Fig. 1.3 An automaton recognising words with an even number of 1.

Definition 1.3.10 A finite automaton $\mathcal{A} = (Q, A, E, I, T)$ is said to be *deterministic* (*DFA*) if it has only one initial state q_0, if E is a subset of $Q \times A \times Q$ and for each $(q, a) \in Q \times A$ there is at most one state $r \in Q$ such that $(q, a, r) \in E$. In that case, E defines a partial function $\delta_{\mathcal{A}} : Q \times A \to Q$ that is called the *transition function* of $\mathcal{A}$. The adjective *partial* means that the domain of $\delta_{\mathcal{A}}$ can be a strict subset of $Q \times A$. To express that the partial transition function is total, the DFA can be said to be *complete*. To get a total function, one can add to Q a new 'sink state' s and, for all $(q, a) \in Q \times A$ such that $\delta_{\mathcal{A}}$ is not defined, set $\delta_{\mathcal{A}}(q, a) := s$. This operation does not alter the language recognised by $\mathcal{A}$. We can extend $\delta_{\mathcal{A}}$ to be defined on $Q \times A^*$ by $\delta_{\mathcal{A}}(q, \varepsilon) = q$ and, for all $q \in Q$, $a \in A$ and $u \in A^*$, $\delta_{\mathcal{A}}(q, au) = \delta_{\mathcal{A}}(\delta_{\mathcal{A}}(q, a), u)$. Otherwise stated, the language recognised by $\mathcal{A}$ is $L(\mathcal{A}) = \{u \in A^* \mid \delta_{\mathcal{A}}(q_0, u) \in F\}$ where q_0 is the initial state of $\mathcal{A}$. If the automaton is deterministic, it is sometimes convenient to refer to the 5-tuple $\mathcal{A} = (Q, A, \delta_{\mathcal{A}}, I, T)$.

As explained by the following result, for languages of finite words, finite automata and deterministic finite automata recognise exactly the same languages.

Theorem 1.3.11 (Rabin and Scott 1959) *If L is recognised by a finite* automaton $\mathcal{A}$, there exists a DFA which can be effectively computed from $\mathcal{A}$ and recognising the same language L.

A proof and more details about classical results in automata theory can be found in textbooks like (Hopcroft, Motwani, and Ullman 2006), (Sakarovitch 2003) or (Shallit 2008). For standard material in automata theory we shall not refer again to these references below.

One important result is that the set of regular languages coincides with the set of languages recognised by finite automata.

Theorem 1.3.12 (Kleene 1956) *A language is regular if, and only if, it is recognised by a (deterministic) finite automaton.*

Observe that if L, M are two regular languages over A, then $L \cap M$, $L \cup M$, LM and $L \setminus M$ are also regular languages. In particular, a language over A is regular if, and only if, its complement in A^* is regular.

Example 1.3.13 The regular language $L = \{0\}^*\{1\}\{0, 01\}^* \cup \{0\}^*$ from Example 1.3.5 is recognised by the DFA depicted in Figure 1.4. Notice that the state s is a *sink*: non-terminal state and all transitions remain in s.

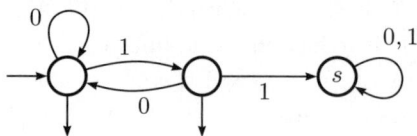

Fig. 1.4 A DFA accepting words without factor 11.

The following result is often useful to prove that a given language is not regular. It appeared first in (Bar-Hillel, Perles, and Shamir 1961).

Lemma 1.3.14 (Pumping lemma) *Let $L \subseteq A^*$ be a regular language accepted by a DFA with ℓ states. If $t \in L$ is a word of length $|t| \geq \ell$, then there exist $u, v, w \in A^*$ such that $t = uvw$, $|uv| \leq \ell$, $v \neq \varepsilon$ and $uv^*w \subseteq L$.*

The idea of the proof (pigeonhole principle) is that any path of length at least ℓ must contain a cycle. Let us first consider a simple example showing an application of the pumping lemma.

Example 1.3.15 Let us show that the language P of all the palindromes over an alphabet A of cardinality at least 2 is not regular. Assume that P is regular and accepted by a DFA with ℓ states. Consider the word $a^\ell b a^\ell \in P$, with a, b letters in A. With notation of Lemma 1.3.14, there exist i, j with $i \geq 0$, $j > 0$ and $i + j \leq \ell$, such that $u = a^i$, $v = a^j$, $w = a^{\ell-i-j}ba^\ell$ and for all $n \in \mathbb{N}$, $uv^n w = a^i a^{nj} a^{\ell-i-j} ba^\ell \in P$ which is a contradiction.

Example 1.3.16 The language of the decimal representations of squares $R = \{1, 4, 9, 16, 25, 36, \ldots\} \subset \{0, \ldots, 9\}^*$ is not regular. Assume to the contrary that R is regular. Notice that the square of the integer with $10^n 1$ as decimal representation has $10^n 20^n 1$ as decimal representation. Since regular languages are closed under intersection, $R' = R \cap \{1\}\{0\}^*\{2\}\{0\}^*\{1\}$ must be regular. A careful inspection shows that

$$R' = \{10^n 20^n 1 \mid n \geq 0\}\,.$$

Indeed, it is left as an exercise to show that $10^i 20^j 1$ is the decimal representation of a square if, and only if, $i = j$. Assume that R' is accepted by a DFA with ℓ states. Let us apply the pumping lemma to R'. The word $10^\ell 20^\ell 1$ belonging to R' can be factored as uvw with $|uv| \leq \ell$. But $uv^k w$ does not belong to R' for $k > 1$ which leads to a contradiction.

The special case of morphic words obtained by q-uniform morphism were introduced by A. Cobham in his seminal paper (Cobham 1972). These infinite words are usually referred to as q-*automatic sequences*. We conclude this section on automata by explaining where does the term 'automatic' come from. See also (Hopcroft and Ullman 1979) and the surveys (Allouche 1987), (Allouche and Mendès France 1995).

Definition 1.3.17 A *deterministic finite automaton with output (DFAO)* over an alphabet A is a 6-tuple $\mathcal{A} = (Q, A, \delta, \{q_0\}, B, \tau)$ where Q, δ and $\{q_0\}$ are defined as for DFA, $\delta : Q \times A \to Q$ being a total function, B is a finite alphabet and $\tau : Q \to B$ is the *output function*.

A DFAO acts like a map from A^* to B. With any word $w \in A^*$ is associated the output $\tau(\delta(q_0, w))$. If $B = \{b_1, \ldots, b_t\}$ then the DFAO $\mathcal{A}$ corresponds to a partition of A^* into t (regular) languages

$$L_i = \{w \in A^* \mid \tau(\delta(q_0, w)) = b_i\}, \quad i = 1, \ldots, t.$$

An infinite word $x = (x_n)_{n \geq 0} \in B^\mathbb{N}$ is said to be k-*automatic* if there exists a DFAO $(Q, \{0, \ldots, k-1\}, \delta, \{q_0\}, B, \tau)$ such that, for all n,

$$x_n = \tau(\delta(q_0, \mathrm{rep}_k(n)))$$

where $\mathrm{rep}_k(n)$ denotes the k-ary representation of n (see Section 1.6). Roughly speaking, the nth term of the sequence is obtained by feeding a DFAO with the k-ary representation of n. For a complete and comprehensive exposition on k-automatic sequences and their applications see the book (Allouche and Shallit 2003).

Theorem 1.3.18 (Cobham 1972) *Let $k \geq 2$. An infinite word $x \in A^\mathbb{N}$ is k-automatic if, and only if, there exist a k-uniform morphism $\sigma : B^* \to B^*$ prolongable on a letter $b \in B$ and a coding $\tau : B \to A$ such that $x = \tau(\sigma^\omega(b))$.*

Example 1.3.19 We have seen in Example 1.2.21 that the Thue–Morse word is generated using a 2-uniform morphism. This word is also 2-automatic. Indeed, we can consider the automaton in Figure 1.3 as a DFAO where the output of the states p and q are respectively 0 and 1.

1.3.3 Transducers

Let A, B be two alphabets. A *transducer* is an automaton given by a 6-tuple $\mathcal{T} = (Q, A, B, E, I, T)$ whose transitions are labelled by elements in $A^* \times B^*$ instead of considering a unique alphabet A. We can therefore use the terminology introduced for automata. Notice that to obtain the label of a path, if $(u, v), (w, x) \in A^* \times B^*$, then the product in $A^* \times B^*$ is the concatenation component-wise, i.e., $(u, v)(w, x) = (uw, vx)$. The *language accepted* by $\mathcal{T}$ is a subset of $A^* \times B^*$, i.e., a relation from A^* into B^*.

It is common to encounter special cases of transducers. First, if labels of transitions belong to $A \times B$, then the transducer is said to be *letter-to-letter*. From a given transducer $\mathcal{T} = (Q, A, B, E, I, T)$, we get an automaton $\mathcal{T}' = (Q, A, E', I, T)$, the *underlying input automaton* of $\mathcal{T}$, where $(q, u, q') \in E'$ if, and only if, there exists $v \in B^*$ such that $(q, (u, v), q') \in E$. If $\mathcal{T}'$ is deterministic, the transducer $\mathcal{T}$ is said to be *sequential*.

When depicting an automaton recall that terminal states are marked by an outgoing arrow. We consider here transducers where the outgoing arrows, i.e., the edges designating terminal states, can be labelled with pairs of the form (ε, w). The convention to define the relation realised by $\mathcal{T}$ is that if a path from $i \in I$ to $t \in T$ is labelled by (u, v) and if t has an outgoing arrow labelled by (ε, w), then (u, vw) belongs to the relation realised by $\mathcal{T}$. This can be seen as a shortcut to describe the following construction. Note that the labels of outgoing arrows are not involved in the possible sequentiality of the transducer as defined above. First, add a new terminal state t' that does not belong to the set of states of $\mathcal{T}$ and having only incident edges. For all terminal states $t \in T$ having a labelled outgoing arrow, add an edge with that label from t to t'. The new set of terminal states is the subset of T made of the terminal states with unlabelled outgoing arrow and t'. An example is given below, compare Figures 1.5 and 1.6. Another way of performing this is to to consider terminal states as functions, as in Section 2.6.

Finally, it is implicitly understood that concatenation of labels of transitions is read from left to right. But if words are read from right to left, we speak of *right transducers*.

There is a primer on finite automata and transducers in Section 2.6.

Example 1.3.20 Consider the sequential right transducer depicted in Figure 1.5. Recall that the adjective 'right' means that entries are read from right to left. For instance $(101, 1000)$ and $(1001, 1010)$ belong to the relation

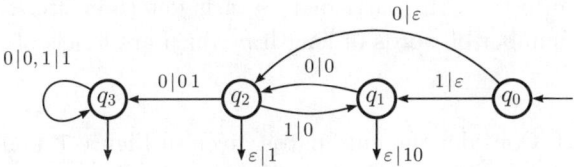

Fig. 1.5 A transducer.

realised by the right transducer. Indeed, we have the successful paths

$$q_0 \xrightarrow{1|\varepsilon} q_1 \xrightarrow{0|0} q_2 \xrightarrow{1|0} q_1 \xrightarrow{\varepsilon|10}$$

and

$$q_0 \xrightarrow{1|\varepsilon} q_1 \xrightarrow{0|0} q_2 \xrightarrow{0|01} q_3 \xrightarrow{1|1} q_4$$

where consecutive labels $(u_1, v_1), \ldots, (u_k, v_k)$, $u_i, v_i \in \{0,1\}^*$ are concatenated from the right: $(u_k \cdots u_1, v_k \cdots v_1)$. An equivalent right transducer realising the same relation is given in Figure 1.6.

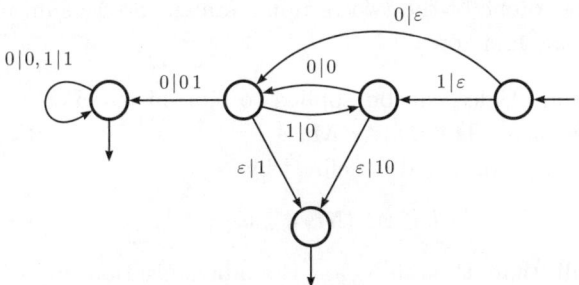

Fig. 1.6 Another transducer realising the same relation.

1.4 Associated matrices

Let $\mathbb{K}$ be a field. The set of matrices with r rows and c columns having entries in $\mathbb{K}$ is denoted by $\mathbb{K}^{r \times c}$. If entries are indexed by elements belonging to two finite sets S and T, we write $\mathbb{K}^{S \times T}$. Let $\mathcal{A} = (Q, A, E, I, T)$ be an automaton such that $E \subseteq Q \times A \times Q$, i.e., labels of edges are letters. An automaton being a directed graph, we can define its *adjacency matrix* $\mathbf{M} \in \mathbb{N}^{Q \times Q}$ indexed by $Q \times Q$ by

$$\mathbf{M}_{q,r} = \mathrm{Card}\{a \in A \mid (q, a, r) \in E\}.$$

If we are dealing with more than one automaton, we use notation like $\mathbf{M}(\mathcal{A})$ to specify the considered automaton. Using classical arguments from graph theory, one can show that, for all $n \geq 0$, $(\mathbf{M}^n)_{q,r}$ counts the number of paths of length n from q to r. In particular, if $\mathcal{A}$ is deterministic, the element $(\mathbf{M}^n)_{q,r}$ is the number of words of length n which are labels of paths from q to r.

Example 1.4.1 Consider the automaton given in Figure 1.4. If states are ordered from left to right, we get the adjacency matrix

$$\mathbf{M} = \begin{pmatrix} 1 & 1 & 0 \\ 1 & 0 & 1 \\ 0 & 0 & 2 \end{pmatrix}.$$

It is easy to see that, for all $n \geq 1$,

$$\mathbf{M}^n = \begin{pmatrix} F_n & F_{n-1} & * \\ F_{n-1} & F_{n-2} & * \\ 0 & 0 & 2^n \end{pmatrix}.$$

where $F_{-1} = 0$, $F_0 = 1$ and $F_j = F_{j-1} + F_{j-2}$ for all $j \geq 1$. In particular, the number of words of length $n \geq 0$ over $\{0, 1\}$ not containing the factor '11' is $F_n + F_{n-1} = F_{n+1}$. Indeed, one has to count paths of length n from the initial state to one of the two terminal states. So we sum up the first two entries on the first row.

The same kind of idea can be applied to morphisms. Let $\sigma : A^* \to A^*$ be an endomorphism. The matrix $\mathbf{M}_\sigma \in \mathbb{N}^{A \times A}$ associated with σ is called the *incidence matrix* of σ and is defined by

$$\forall a, b \in A, \ (\mathbf{M}_\sigma)_{a,b} = |\sigma(b)|_a.$$

Let us recall that $\mathbf{P}$ stands for the abelianisation map. If $A = \{a_1, \ldots, a_d\}$, then the matrix $\mathbf{M}_\sigma$ can be defined by its columns:

$$\mathbf{M}_\sigma = \begin{pmatrix} \mathbf{P}(\sigma(a_1)) & \cdots & \mathbf{P}(\sigma(a_d)) \end{pmatrix}$$

and it satisfies:
$$\forall w \in A^*, \ \mathbf{P}(\sigma(w)) = \mathbf{M}_\sigma \mathbf{P}(w).$$

A square matrix $\mathbf{M} \in \mathbb{R}^{n \times n}$ with entries in $\mathbb{R}_{\geq 0}$ is *irreducible* if, for all i, j, there exists k such that $(\mathbf{M}^k)_{i,j} > 0$. A square matrix $\mathbf{M} \in \mathbb{R}^{n \times n}$ with entries in $\mathbb{R}_{\geq 0}$ is *primitive* if there exists k such that, for all i, j, we have $(\mathbf{M}^k)_{i,j} > 0$. Similarly, a morphism over the alphabet A is *irreducible* if its incidence matrix is irreducible. A substitution is *primitive* if its incidence matrix is primitive.

The terminology irreducible comes from the fact that a matrix $\mathbf{M}$ is irreducible if, and only if, if it has no non-trivial invariant space of coordinates. Primitive matrices are also called *irreducible and aperiodic matrices*. See (Gantmacher 1960) or (Seneta 1981) for details on matrices with non-negative entries.

One checks that a primitive morphism always admits a power that is (left and right) prolongable, and which thus generates both an infinite word and a bi-infinite word which are fixed points of σ. See (Queffélec 1987, Proposition V.1).

Theorem 1.4.2 (Perron–Frobenius' theorem) *Let $\mathbf{M}$ be an irreducible matrix with non-negative entries. Then $\mathbf{M}$ admits a positive eigenvalue α which is larger than or equal in modulus to the other eigenvalues λ: $\alpha \geq |\lambda|$. The eigenvalue α and its algebraic conjugates (that is, the roots of the minimal polynomial of α) are simple roots of the characteristic polynomial of $\mathbf{M}$ and thus are simple eigenvalues. Furthermore, there exists an eigenvector with positive entries associated with α. The eigenvalue α is called the* Perron–Frobenius eigenvalue *of $\mathbf{M}$.*

Furthermore, if $\mathbf{M}$ is primitive, *then the eigenvalue α dominates (strictly) in modulus the other eigenvalues λ: $\alpha > |\lambda|$.*

Definition 1.4.3 Let σ be a morphism. If its incidence matrix $\mathbf{M}_\sigma$ is irreducible, then the Perron–Frobenius eigenvalue of $\mathbf{M}_\sigma$ is called the *inflation factor* of the morphism σ.

Lemma 1.4.4 *(Gantmacher 1960) Let $\mathbf{M}$ be an irreducible matrix with non-negative entries, and let α be its Perron–Frobenius eigenvalue. The inequality $\alpha \mathbf{v} \leq \mathbf{M}\mathbf{v}$ (considered as a component-wise inequality) either implies that $\mathbf{v}$ is an eigenvector associated with the Perron–Frobenius eigenvalue, or that $\mathbf{v} = \mathbf{0}$. In either case, we have $\mathbf{M}\mathbf{v} = \alpha \mathbf{v}$.*

As an application of Perron–Frobenius' theorem, we deduce the existence of frequencies for every factor of an infinite word generated by a primitive morphism. For a proof, see (Queffélec 1987) or Chapter 5 in (Pytheas Fogg 2002). For general results on frequencies of factors, see Chapter 7.

Theorem 1.4.5 *Let σ be a primitive prolongable morphism. Let u be an infinite word generated by σ. Then every factor of u has a frequency. Furthermore, all the frequencies of factors are positive. The frequencies of the letters are given by the coordinates of the positive eigenvector associated with the Perron–Frobenius eigenvalue, renormalised in such a way that the sum of its coordinates equals 1.*

The positive eigenvector associated with the Perron–Frobenius eigenvalue, renormalised in such a way that the sum of its coordinates equals 1 is usually called the *normalised Perron–Frobenius eigenvector* or *Perron–Frobenius eigenvector*. This is the normalisation choice that will be used in Chapter 10.

Furthermore, words generated by primitive morphisms are uniformly recurrent:

Proposition 1.4.6 *Let σ be a primitive morphism. For every $k \in \mathbb{N}$, any fixed point of σ^k is uniformly recurrent. Let σ be a morphism prolongable on the letter $a \in A$. We assume furthermore that all the letters in A actually occur in $\sigma^\omega(a)$ and that $\lim_{n \to +\infty} |\sigma^n(b)| = +\infty$ for all $b \in A$. The morphism σ is primitive if, and only if, the fixed point of σ beginning by a is uniformly recurrent.*

For an example of a morphism which is not primitive with a uniformly recurrent fixed point, consider $\sigma : 0 \mapsto 0010, 1 \mapsto 1$. The infinite word $\sigma^\omega(0)$ is called the *Chacon word* (see Exercise 1.8). One has $\lim_{n \to +\infty} |\sigma^n(1)| = 1$. Also see connections with Section 6.5.

The case where the Perron–Frobenius eigenvalue of the incidence matrix of a primitive morphism is a Pisot number is of particular interest.

Definition 1.4.7 *An algebraic integer $\alpha > 1$, i.e., a root of a monic polynomial with integer coefficients, is a Pisot-Vijayaraghavan number or a Pisot number if all its algebraic conjugates λ other than α itself satisfy $|\lambda| < 1$. An algebraic integer is a unit if its norm equals 1, i.e., if the constant term of its minimal polynomial equals 1 in absolute value.*

We recall that the algebraic conjugates of an algebraic integer are the roots of its minimal polynomial.

A primitive morphism σ is said to be *Pisot* if its Perron–Frobenius eigenvalue is a Pisot number.

Preliminaries

A Pisot morphism σ is said to be *unit* if its Perron–Frobenius eigenvalue is a unit Pisot number.

Example 1.4.8 Consider the Fibonacci morphism σ introduced in Example 1.2.22. The incidence matrix of σ is

$$\mathbf{M}_\sigma = \begin{pmatrix} 1 & 1 \\ 1 & 0 \end{pmatrix}.$$

Since $\mathbf{M}_\sigma^2$ contains only positive entries, the morphism is primitive. The Perron–Frobenius eigenvalue of $\mathbf{M}_\sigma$ is the Golden Ratio $\varphi = (1+\sqrt{5})/2$ satisfying $\varphi^2 - \varphi - 1 = 0$. This algebraic integer has $(1-\sqrt{5})/2$ as Galois conjugate which is of modulus less than 1. Consequently, we have a unit Pisot morphism.

A Pisot morphism σ is said to be a *Pisot irreducible substitution* if the algebraic degree of the Perron–Frobenius eigenvalue of its incidence matrix is equal to the size of the alphabet. This is equivalent to the fact that the characteristic polynomial of its incidence matrix is irreducible. A Pisot morphism which is not a Pisot irreducible morphism is called a *Pisot reducible morphism*. Examples of Pisot reducible morphisms are $1 \to 12$, $2 \to 3$, $3 \to 4$, $4 \to 5$, $5 \to 1$ and the Thue–Morse morphism $1 \to 12$, $2 \to 21$. Indeed, the characteristic polynomial of the incidence matrix is respectively equal to $X^5 - X^4 - 1 = (X^2 - X + 1)(X^3 - X - 1)$ and to $X^2 - 2X = X(X - 2)$.

Theorem 1.4.9 *Let σ be a morphism such that its incidence matrix $\mathbf{M}_\sigma$ is irreducible. If its Perron–Frobenius eigenvalue α is such that for every other eigenvalue λ of $\mathbf{M}_\sigma$ one has $\alpha > 1 > |\lambda| > 0$, then σ is primitive and Pisot irreducible.*

For a proof, see (Canterini and Siegel 2001b) and (Pytheas Fogg 2002, Chapter 1).

We end this section with the notion of spectral radius that will be developed in Chapter 11, also see Section 4.7.2.2. Let $\|\cdot\|$ be a submultiplicative matrix norm. That is a vector norm that satisfies for all square matrices $\mathbf{A}$, $\mathbf{B}$, $\|\mathbf{AB}\| \leq \|\mathbf{A}\| \cdot \|\mathbf{B}\|$. Note that some authors use the terminology matrix norm only for those norms which are submultiplicative. The *spectral radius* of the complex square matrix $\mathbf{A}$ is defined as the largest modulus of its eigenvalues. It is proved to represent the asymptotic growth rate of the norm of the successive powers of A:

$$\rho(\mathbf{A}) = \lim_{t \to \infty} \|\mathbf{A}^t\|^{1/t}.$$

This quantity does provably not depend on the used norm.

1.5 A glimpse at numeration systems

Various numeration systems will be considered in detail in this book, see mainly Chapters 2 and 3: integer base, real base, rational base, canonical number systems, abstract numeration systems. As a short appetiser, we merely recall in this section how to write down non-negative integers in the usual p-ary numeration system, $p \geq 2$ being an integer. More details are given in Section 2.2.1.

For any positive integer n, there exist $\ell \geq 0$ such that $p^\ell \leq n < p^{\ell+1}$ and unique coefficients $c_0, \ldots, c_\ell \in \{0, \ldots, p-1\}$ such that

$$n = \sum_{i=0}^{\ell} c_i \, p^i \text{ and } c_\ell \neq 0.$$

The coefficients $c_\ell, \ldots, c_0$ can be computed by successive Euclidean divisions. Set $n_0 := n$. We have $n_0 = c_\ell \, p^\ell + n_1$ with $n_1 < p^\ell$ and for $i = 1, \ldots, \ell$, $n_i = c_{\ell-i} \, p^{\ell-i} + n_{i+1}$ with $n_{i+1} < p^{\ell-i}$. The word $c_\ell \cdots c_0$ is said to be the *p-ary representation* or *p-expansion* of n (sometimes called *greedy* representation) and we write

$$\mathrm{rep}_p(n) = c_\ell \cdots c_0.$$

Longer developments are given in Section 2.2.1 of Chapter 2 where $\mathrm{rep}_p(n)$ is denoted by $\langle n \rangle_p$. We set $\mathrm{rep}_p(0) = \varepsilon$. So rep_p is a one-to-one correspondence between $\mathbb{N}$ and $\{\varepsilon\} \cup \{1, \ldots, p-1\}\{0, \ldots, p-1\}^*$.

Let $A \subset \mathbb{Z}$ be a finite alphabet and $u = a_0 \cdots a_\ell$ be a word over A. We set

$$\mathrm{val}_p(a_0 \cdots a_\ell) = \sum_{i=0}^{\ell} a_{\ell-i} \, p^i.$$

We say that $\mathrm{val}_p(u)$ is the *numerical value* or *evaluation* of u, also denoted by $\pi_p(u)$ (see, *e.g.*, Chapter 2).

The restriction of $\mathrm{rep}_p \circ \mathrm{val}_p$ to the set of words over A having a non-negative numerical value is the *normalisation*:

$$\nu_{A,p} : u \in A^* \mapsto \mathrm{rep}_p(\mathrm{val}_p(u)).$$

Again, reference to the alphabet A can be omitted if the context is clear.

Example 1.5.1 (Signed digits) Let $A = \{\bar{1}, 0, 1\}$ where $\bar{1}$ stands for -1. We have $\mathrm{val}_2(100\bar{1}) = 7$ and $\mathrm{val}_2(\bar{1}01) = -3$. In particular, $\mathrm{rep}_2(\mathrm{val}_2(100\bar{1})) = 111$, *i.e.*, $\nu_2(100\bar{1}) = 111$.

Definition 1.5.2 A set $X \subseteq \mathbb{N}$ of integers is *p-recognisable* if the language
$$\mathrm{rep}_p(X) = \{\mathrm{rep}_p(n) \mid n \in X\}$$
is regular. Observe that a set X is p-recognisable if, and only if, its characteristic word is p-automatic.

Proposition 1.5.3 *Let $p \geq 2$. Any eventually periodic set of integers is p-recognisable.*

It is an easy exercise. See for instance (Sakarovitch 2003, Prologue) for a proof. The realisation of division by finite automata (together with arithmetic operations modulo q) is discussed in Chapter 2. Also see Proposition 3.1.9 for similar considerations.

Definition 1.5.4 Two integers $p, q \geq 2$ are *multiplicatively independent* if the only integers m, n satisfying $p^m = q^n$ are $m = n = 0$. Otherwise, p and q are said to be *multiplicatively dependent*. In other words, p and q are multiplicatively dependent if, and only if, $\log p / \log q$ is rational.

Theorem 1.5.5 (Cobham 1969) *Let $p, q \geq 2$ be two multiplicatively independent integers. If $X \subseteq \mathbb{N}$ is both p-recognisable and q-recognisable, then X is eventually periodic.*

Many efforts have been made to get a simpler presentation of Cobham's theorem, as G. Hansel did in (Hansel 1982). Also see (Perrin 1990), (Allouche and Shallit 2003) and (Rigo and Waxweiler 2006).

Several aspects of numeration systems are treated in Chapter 2. Various numeration systems for the representation of integers are discussed in (Fraenkel 1985). The chapter by Ch. Frougny in (Lothaire 2002) presents non-standard numeration systems for the representations of integers as well as β-numeration systems. The surveys (Barat, Berthé, Liardet, et al. 2006) and (Bruyère, Hansel, Michaux, et al. 1994) are also of interest and contain many pointers to the existing literature. The latter one develops also a logical characterisation of p-recognisable sets in terms of an extension of the Presburger arithmetic $\langle \mathbb{N}, + \rangle$ and extension of Cobham's theorem on the base dependence to the multidimensional case.

1.6 Symbolic dynamics

Let us introduce some basic notions in symbolic dynamics. For expository books on the subject, see (Cornfeld, Fomin, and Sinaĭ 1982), (Kitchens 1998), (Lind and Marcus 1995), (Perrin 1995b), (Queffélec 1987) and (Kůrka 2003).

1.6.1 Subshifts

Let S denote the following map defined on $A^{\mathbb{N}}$, called the *one-sided shift*:

$$S((x_n)_{n\geq 0}) = (x_{n+1})_{n\geq 0}.$$

In particular, if $x = x_0 x_1 x_2 \cdots$ is an infinite word over A, then, for all $n \geq 0$, its suffix $x_n x_{n+1} \cdots$ is simply $S^n(x)$. Note that for convenience, the shift is sometimes denoted by σ, when no misunderstanding with morphisms on words can be made. This latter convention is used in Chapter 2. The map S is uniformly continuous, onto but not one-to-one on $A^{\mathbb{N}}$. This notion extends in a natural way to $A^{\mathbb{Z}}$. In this latter case, the shift S is one-to-one. The definitions given below correspond to the *one-sided shift*, but they extend to the *two-sided shift*.

Definition 1.6.1 Let x be an infinite word over the alphabet A. The *orbit* of x under the action of the shift S is defined as the set

$$\mathcal{O}(x) = \{S^n x \mid n \in \mathbb{N}\}.$$

The *symbolic dynamical system* associated with x is then defined as $(\overline{\mathcal{O}(x)}, S)$, where $\overline{\mathcal{O}(x)} \subseteq A^{\mathbb{N}}$ is the closure of the orbit of x.

In the case of bi-infinite words we similarly define $\mathcal{O}(x) = \{S^n x \mid n \in \mathbb{Z}\}$ where the (two-sided) shift map is defined on $A^{\mathbb{Z}}$. The set $X_x := \overline{\mathcal{O}(x)}$ is a closed subset of the compact set $A^{\mathbb{N}}$, hence it is a compact space and S is a continuous map acting on it. One checks that, for every infinite word $y \in A^{\mathbb{N}}$, the word y belongs to X_x if, and only if, $L(y) \subseteq L(x)$. For a proof, see (Queffélec 1987) or Chapter 1 of (Pytheas Fogg 2002). Note that $\overline{\mathcal{O}(x)}$ is finite if, and only if, x is eventually periodic.

More generally, let Y be a closed subset of $A^{\mathbb{N}}$ that is stable under the action of the shift S. The system (Y, S) is called a *subshift*. The *full shift* is defined as $(A^{\mathbb{N}}, S)$.

A subshift (X, S) is said to be *periodic* if there exist $x \in X$ and an integer k such that $X = \{x, Sx, \ldots, S^k x = x\}$. Otherwise it is said to be *aperiodic*.

If (Y, S) is a subshift, then there exists a set $X \subseteq A^*$ such that for every $u \in A^{\mathbb{N}}$, the word u belongs to Y if, and only if, $L(u) \cap X = \emptyset$. A subshift Y is said to be of *finite type* if the set $X \subseteq A^*$ is finite. A subshift is said to be *sofic* if the set X is a regular language.

Example 1.6.2 The set of infinite words over $\{0,1\}$ which do not contain the factor 11 is a subshift of finite type, whereas the set of infinite words over $\{0,1\}$ having an even number of 1's between two occurrences of the letter 0 is a sofic subshift which is not of finite type.

Definition 1.6.3 Let $x \in A^{\mathbb{N}}$. For a word $w = w_0 \cdots w_r$, the *cylinder set* $[w]_x$ is the set $\{y \in X_x \mid y_0 = w_0, \cdots, y_r = w_r\}$. If the context is clear, the subscript x will be omitted.

The cylinder sets are *clopen* (open and closed) sets and form a basis of open sets for the topology of X_x. Furthermore, one checks that a clopen set is a finite union of cylinders. In the bi-infinite case the cylinders are the sets $[u.v]_x = \{y \in X_x \mid y_i = u_i, y_j = v_j, -|u| \leq i \leq -1, 0 \leq j \leq |v|-1\}$ and the same remarks hold.

1.6.2 Dynamical systems

We have introduced the notions of a symbolic dynamical system and of a subshift. Such discrete systems belong to the larger class of topological dynamical systems, which have been intensively studied in topological dynamics. See for instance (Cornfeld, Fomin, and Sinaĭ 1982). For references on ergodic theory, see *e.g.* (Walters 1982) or (Silva 2008).

Definition 1.6.4 A *topological dynamical system* (X, T) is defined as a compact metric space X together with a continuous map T defined onto the set X.

A topological dynamical system (X, T) is *minimal* if, for all x in X, the orbit of x, *i.e.*, the set $\{T^n x \mid n \in \mathbb{N}\}$, is dense in X.

Let us note that if (X, S) is a subshift, and if X is furthermore assumed to be minimal, then X is periodic if, and only if, X is finite.

The symbolic dynamical system (X_x, S) associated with the infinite word x is minimal if, and only if, for every $y \in X_x$, $L(y) = L(x)$. More generally, properties of symbolic dynamical systems associated with an infinite word are strongly related to its combinatorial properties. For a proof of Theorem 1.6.5 below, see for instance Chapter 5 of (Pytheas Fogg 2002).

Theorem 1.6.5 *Let x be an infinite word. If x is recurrent, then the shift $S \colon X_x \to X_x$ is onto. Furthermore, (X_x, S) is minimal if, and only if, x is uniformly recurrent.*

In other words, x is uniformly recurrent if, and only if, $L(y) = L(x)$ for every y such that $L(y) \subseteq L(x)$. The idea of the proof of the equivalence in Theorem 1.6.5 can be sketched as follows: if w is a factor of x, we write

$$\overline{\mathcal{O}(x)} = \bigcup_{n \in \mathbb{N}} S^{-n}[w],$$

and we conclude by a compactness argument.

Two dynamical systems (X_1, T_1) and (X_2, T_2) are said to be *topologically conjugate* (or *topologically isomorphic*) if there exists an homeomorphism f from X_1 onto X_2 which conjugates T_1 and T_2, that is:

$$f \circ T_1 = T_2 \circ f.$$

Let (X, T) be a topological dynamical system. Let $\mathcal{M}(X)$ stand for the set of Borel probability measures on X. A Borel measure μ defined over X is said T-*invariant* if $\mu(T^{-1}(B)) = \mu(B)$, for every Borel set B. The map T is said to preserve the measure μ. This is equivalent to the fact that for any continuous function $f \in \mathcal{C}(X)$, then $\int f(Tx) \, d\mu(x) = \int f(x) \, d\mu(x)$. A topological system (X, T) always has an invariant probability measure. For more details, see Proposition 7.2.4.

The case where there exists only one T-invariant measure is of particular interest. A topological dynamical system (X, T) is said to be *uniquely ergodic* if there exists one and only one T-invariant Borel probability measure over X.

We have considered here the notion of dynamical system, that is, a map acting on a given set, in a topological context. This notion can be extended to measurable spaces: we thus get measure-theoretic dynamical systems. For more details about all of the notions defined in this section, one can refer to (Walters 1982).

Definition 1.6.6 A *measure-theoretic dynamical system* is defined as a system $(X, T, \mu, \mathcal{B})$, where $\mathcal{B}$ is a σ-algebra, μ a probability measure defined on $\mathcal{B}$, and $T : X \to X$ is a measurable map which preserves the measure μ, i.e., for all $B \in \mathcal{B}$, $\mu(T^{-1}(B)) = \mu(B)$.

A measure-theoretic dynamical system $(X, T, \mu, \mathcal{B})$ is *ergodic* if for every $B \in \mathcal{B}$ such that $T^{-1}(B) = B$, then B has either zero measure or full measure.

In particular, a uniquely ergodic topological dynamical system yields an ergodic measure-theoretic dynamical system.

A measure-theoretic ergodic dynamical system satisfies the *Birkhoff ergodic theorem*, also called *individual ergodic theorem*. Let us recall that the abbreviation a.e. stands for 'almost everywhere': a property holds almost everywhere if the set of elements for which the property does not hold is contained in a set of zero measure.

Theorem 1.6.7 (Birkhoff Ergodic Theorem) *Let* $(X, T, \mu, \mathcal{B})$ *be a measure-theoretic dynamical system. Let* $f \in L^1(X, \mathbb{R})$. *Then the sequence* $(\frac{1}{n} \sum_{k=0}^{n-1} f \circ T^k)_{n \geq 0}$ *converges a.e. to a function* $f^* \in L^1(X, \mathbb{R})$. *One has*

$f^* \circ T = f^*$ a.e. and $\int_X f^* d\mu = \int_X f d\mu$. Furthermore, if T is ergodic, since f^* is a.e. constant, one has:

$$\forall f \in L^1(X, \mathbb{R}), \quad \frac{1}{n}\sum_{k=0}^{n-1} f \circ T^k \xrightarrow[n \to \infty]{\mu-a.e.} \int_X f d\mu.$$

The notion of conjugacy between two topological dynamical systems extends in a natural way to this context. Two measure-theoretic dynamical systems $(X_1, T_1, \mu_1, \mathcal{B}_1)$ and $(X_2, T_2, \mu_2, \mathcal{B}_2)$ are said to be *measure-theoretically isomorphic* if there exist two sets of full measure $B_1 \in \mathcal{B}_1$, $B_2 \in \mathcal{B}_2$, and a measurable map $f : B_1 \to B_2$ called *conjugacy map* such that

- the map f is one-to-one and onto,
- the reciprocal map of f is measurable,
- $f \circ T_1(x) = T_2 \circ f(x)$ for every $x \in B_1 \cap T_1^{-1}(B_1)$,
- μ_2 is the image of the measure μ_1 with respect to f, that is,

$$\forall B \in \mathcal{B}_2, \; \mu_1(f^{-1}(B \cap B_2)) = \mu_2(B \cap B_2).$$

If the map f is only onto, then $(X_2, T_2, \mu_2, \mathcal{B}_2)$ is said to be a *measure-theoretic factor* of $(X_1, T_1, \mu_1, \mathcal{B}_1)$.

1.6.3 Substitutive dynamical systems

As a class of examples, let us consider symbolic dynamical systems associated with purely substitutive words. Note that for such symbolic dynamical systems, one uses the terminology 'substitution' rather than the terminology 'morphism'.

First we recall that if σ is a primitive substitution, then there exists a power of σ that is prolongable, and thus, which generates an infinite word (see Definition 1.2.18). Similarly, there exists a power of σ which generates a bi-infinite word which is purely morphic in the sense of Definition 1.2.19. For more details, see (Queffélec 1987, Proposition V.1). We then deduce from Proposition 1.4.6 and Theorem 1.6.5 that if σ is a primitive prolongable substitution, then all the (infinite or bi-infinite) words generated by σ are uniformly recurrent, and thus have the same language. In other words, all the symbolic dynamical systems associated with any of the words generated by one of the powers of σ do coincide. Hence, we can associate in a natural way with a primitive substitution a symbolic dynamical system.

Definition 1.6.8 Let σ be a primitive substitution. The *symbolic dynamical system associated with* σ is the system associated with any of the (infinite or bi-infinite) words generated by one of the powers of σ, according respectively to Definitions 1.2.18 and 1.2.19. We denote it by (X_σ, S).

Let us quote an interesting property of substitutive dynamical systems associated with a primitive substitution. For more details, see (Queffélec 1987) and (Pytheas Fogg 2002).

Theorem 1.6.9 *Let σ be a primitive substitution. The system (X_σ, S) is uniquely ergodic.*

The corresponding invariant measure is uniquely defined by its values on the cylinders: the measure of the cylinder $[w]$ is defined as the frequency of the finite word w in any element of X_σ, which does exist and does not depend on the choice of the fixed point, according to Theorem 1.4.5.

1.7 Exercises

Exercise 1.1 Show that over a binary alphabet, any word of length ≥ 4 contains a square as factor.

Exercise 1.2 Show that over a ternary alphabet A, it is possible to build an infinite word avoiding squares. See for instance (Lothaire 1983). As a by-product, show that the set of (finite) words over A having a non-empty factor which is a square, is not regular.

Exercise 1.3 Show that a word u of even (respectively odd) length is a palindrome if, and only if, there exists a word v such that $u = v\tilde{v}$ (respectively there exist a word v and a letter a such that $u = va\tilde{v}$).

Exercise 1.4 Show that if L and M are regular languages then $L \cap M$ is also regular.

Exercise 1.5 Let L be a language over the unary alphabet $\{a\}$. Show that L is regular if, and only if, there exists an eventually periodic set $X \subseteq \mathbb{N}$ such that $L = \{a^i \mid i \in X\}$.

Exercise 1.6 Let L be a regular language over A. Show that $|L| = \{|u| : u \in L\} \subseteq \mathbb{N}$ is eventually periodic. Give a counter-example illustrating that the converse does not hold.

Exercise 1.7 Prove that a Sturmian word is recurrent. Give an example of a bi-infinite word x with complexity function satisfying $p_x(n) = n+1$ for all n that is not recurrent.

Exercise 1.8 (Chacon word) We recall that the Chacon morphism σ is defined over the alphabet $\{0,1\}$ by $\sigma : 0 \mapsto 0010,\ 1 \mapsto 1$. Prove that the Chacon word $\sigma^\omega(0)$ begins with the following sequence of words $(b_n)_{n\geq 0}$:

$$b_0 = 0, \text{ and } \forall n \in \mathbb{N},\ b_{n+1} = b_n b_n 1 b_n.$$

Deduce that the Chacon word is uniformly recurrent.

Exercise 1.9 Give an example of a morphism that is irreducible but not primitive.

1.8 Notes

The study of combinatorics on words can be traced back to the work of A. Thue (Thue 1906) in 1906 and later on (Thue 1912) where he investigated repetitions in words, also see (Berstel 1995), then rooted in the papers (Morse and Hedlund 1938), (Morse and Hedlund 1940). Later on, the subject was taken forward on the one hand by M.-P. Schützenberger in France and on the other hand by P. S. Novikov and S. I. Adjan in Russia. Now *combinatorics on words* is considered as a research topic by itself and has received classification subject **68R15** by the American Mathematical Society. For a comprehensive survey on the origins of combinatorics on words, see (Berstel and Perrin 2007).

For a nice account of the history of automata theory, see (Perrin 1995a). A first reference to automata can be traced back to (McCulloch and Pitts 1943). The notion of regular expressions goes back to (Kleene 1956) and non-deterministic automata were introduced in (Rabin and Scott 1959).

2
Number representation and finite automata

Christiane Frougny,

Jacques Sakarovitch

2.1 Introduction

Numbers do exist – independently of the way we represent them, of the way we write them. And there are many ways to write them: integers as a finite sequence of digits once a base is fixed, rational numbers as a pair of integers or as an eventually periodic infinite sequence of digits, or reals as an infinite sequence of digits but also as a continued fraction, just to quote a few. *Operations on numbers* are defined – independently of the way they are computed. But when they are computed, they amount to be algorithms that work on the representations of numbers.

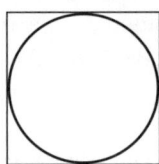

Fig. 2.1 Numbers do exist, a Greek view: $\frac{\pi}{4} = \frac{C}{P} = \frac{D}{S}$. Numbers are then *ratio* between measures (C = length of the circle, P = perimeter of the square, D = surface of the disk, S = surface of the square).

In this chapter, numbers will be represented by their expansion in a base, or more generally, with respect to a basis, hence by *words* over an alphabet of digits. The algorithms we shall consider are those that can be performed by finite state machines, that is, by the simplest machines one can think of. Natural questions then arise immediately. First, whether or not the whole set of expansions of all the positive integers, or the integers, or the real numbers (within an interval) is itself a set of finite, or infinite, words that is recognised by finite automata. Second, which operations on numbers can

Combinatorics, Automata and Number Theory, ed. Valérie Berthé and Michel Rigo. Published by Cambridge University Press. ©Cambridge University Press 2010.

thus be defined by means of finite automata? how is this related to the chosen base? how, in a given base, may the choice of digits influence the way the operations can be computed? These are some of the questions that will be asked and, hopefully and to a certain extent, answered in this chapter.

It is not only these questions, repeated in every section, that will give this chapter its unity but also the methods with which we shall try to answer them. In every numeration system, defined by a base or a basis, we first consider a trivial infinite automaton – the *evaluator*, whose states are the values of the words it reads – from which we define immediately the *zero automaton* which recognises the words written on a signed digit alphabet and having value 0. From the zero automaton we then derive transducers, called *digit-conversion transducers*, that relate words with same values but written differently on the same or distinct alphabets of digits and, from these, transducers for the normalisation, the addition, *etc*. Whether all these latter transducers are finite or not depends on whether the zero automaton is finite or not and this question is analysed and solved by combinatorial and algebraic methods which depend on the base.

We begin with the classical – one could even say basic – case where the base is a positive integer. It will give us the opportunity to state a number of elementary properties which we nevertheless prove in detail for they will appear again in the other forthcoming parts. The zero automaton is easily seen to be finite, as are the adder and the various normalisers. The same zero automaton is the socle on which we build the *local* adder for the Avizienis system, and the normaliser for the *non-adjacent forms* which yield representations of minimal weight.

The first and main non-classical case that will retain our attention is the one of numeration systems often called *non-standard*: a non-integer real β is chosen as a base and the (real) numbers are written in this base; a rather common example is when β is equal to the *Golden Ratio* φ. Such systems are also often called *beta-numeration* in the literature. In contrast with the integer-base case, numbers may have several distinct representations, even on the canonical alphabet, and the *expansion* of every number is computed by a *greedy* algorithm which produces the digits from left to right, that is, *most significant digit first*. The *arithmetic* properties of β, that is, which kind of algebraic integer it is, are put into correspondance with the properties of the system such as for instance the rationality† of the set of expansions. The main result in that direction is that the zero automaton is finite if, and only if, β is a *Pisot number* (Theorem 2.3.31).

† We use 'rational set' as a synonym of 'regular set', see Section 1.3 and Section 2.6.1.

Another property that is studied is the possibility of defining from β a sequence of integers that will be taken as a *basis* and that will thus yield a numeration system (for the positive integers), in the very same way as the *Fibonacci numeration system* is associated with the Golden Ratio. Although restricted to the integers, these systems happen to be more difficult to study than those defined by a real base, and the characterisation of those for which the set of expansions is rational is more intricate (Theorem 2.3.57).

Section 2.4 is devoted to *canonical numeration systems* in algebraic number fields. In these systems, every integer has a unique finite expansion, which is not computed by a greedy algorithm but by a right-to-left algorithm, that is, by an algorithm which computes the least significant digit first. The main open problem in this area is indeed to characterise such canonical numeration systems. A beautiful result is the characterisation of Gaussian integers as a base of canonical numeration systems (Theorem 2.4.12).

The third and last kind of numeration systems which we consider is the one of systems with a base that is not an algebraic integer but a *rational number*. First the non-negative integers are given an expansion which is computed *from right to left*, as in the case of canonical numeration sytems. The set of all expansions is not a rational language anymore; it is a very intriguing set of words indeed, a situation which does not prevent the zero automaton to be still finite, and so is the digit-converter from any alphabet to the canonical one. The expansions of real numbers are not really 'computed' but defined *a priori* from the expansions of the integers. The matter of the statement is thus reversed and what is to be proved is not that we can compute the expansion of the real numbers but that every real number is given a representation (at least one) by this set brought from 'outside' (Theorem 2.5.23). This topic has been explored by the authors in a recent paper (Akiyama, Frougny, and Sakarovitch 2008) and is wide open to further research.

In Section 2.6 (before the Notes section) we have gathered definitions† and properties of finite automata and transducers that are not specific to the results on numeration systems but relevant to more or less classical parts of automata theory, and currently used in this chapter.

From this presentation, it appears that we are interested in the way numbers are written rather than in the definition of set of numbers *via* finite automata. And yet the latter has been the first encounter between finite automata theory and number representation, namely, Cobham's Theorem (we mention it only as it stands in the background of the proof that the

_____
† Notions defined in that Section 2.6 are shown *slanted* in the text.

map betwen the representations of numbers in different bases cannot be realised by finite automata). Speaking of this theorem, it is interesting to quote this seminal paper (Cobham 1969):

This adds further evidence [...] that, insofar as the recognition of set of numbers goes, finite automata are weak, and somewhat unnatural.

We think, and we hope the reader will be convinced, that the matter developed in this chapter supports the view that finite automata are on the contrary a natural and powerful concept for studying numeration systems.

2.2 Representation in integer base

We first recall how numbers, integers or real numbers, may be represented in an integer base, like 2 or 10, that is, in the way that everyone does in everyday life. The statements and proofs in these first two subsections are thus simple and well-known, when not even trivial. We nevertheless write them explicitly for they allow us to see how the several generalisations to come in the sections below differ from, and are similar to, the basic case of integer base numeration systems.

Let p be a fixed integer greater than 1, which we call the *base* (in our running examples, we choose $p = 2$, or $p = 3$ when 2 differs from the general case). The *canonical alphabet* of digits A_p associated with p is $A_p = \{0, 1, \ldots, p-1\}$. The integer p together with A_p defines the *base p numeration system*.

Note that A_p is naturally (and totally) ordered and thus A_p^* is naturally (and totally) ordered by the *lexicographic* and by the *radix* orders.

2.2.1 Representation of integers

The choice of the base p implicitly gives every word of A_p^* an integer value, via the *evaluation map* π_p: for every word w of A_p^*, we have

$$w = a_k\, a_{k-1} \cdots a_1\, a_0 \quad \longmapsto \quad \pi_p(w) = \sum_{i=0}^{k} a_i p^i.$$

This definition of π_p implies that numbers are written with the *most significant digit* on the left.[†]

Lemma 2.2.1 *The map π_p is injective on A_p^k, for every integer k.*

[†] A convention which certainly is the most common one, even in languages written from right to left, but not universal, in particular among computer scientists (see (Cohen 1981) on the endianness problem).

Proof Let $u = a_{k-1}a_{k-2}\cdots a_0$ and $v = b_{k-1}b_{k-2}\cdots b_0$ be two distinct words of A_p^* of length k such that $\pi_p(u) = \pi_p(v)$. Hence

$$\sum_{i=0}^{k-1} a_i p^i - \sum_{i=0}^{k-1} b_i p^i = 0 \quad \text{and therefore} \quad P(X) = \sum_{i=0}^{k-1}(a_i - b_i)X^i$$

is a polynomial in $\mathbb{Z}[X]$ vanishing at $X = p$. By Gauss Lemma on primitive polynomials, $P(X)$ is divisible by the minimal polynomial $X - p$. Contradiction, since $|a_0 - b_0|$ is strictly smaller than p. □

The map π_p is not injective on the whole A_p^* since $\pi_p(0^h u) = \pi_p(u)$ holds for any u in A_p^* and any integer h. On the other hand, Lemma 2.2.1 implies that this is the only possibility and we have:

$$\pi_p(u) = \pi_p(v) \text{ and } |u| > |v| \implies u = 0^h v \text{ with } h = |u| - |v|.$$

Conversely, every integer N in $\mathbb{N}$ can be given a representation as a word in A_p^* which, thanks to the foregoing, is unique under the condition it does not begin with a zero. This representation can be computed in two different ways, which we call, for further references, the *greedy algorithm* – which computes the digits *from left to right*, that is, *most significant digit first* – and the *division algorithm* – which computes the (same) digits *from right to left*, that is, *least significant digit first*.

The greedy algorithm. Let N be any positive integer. There exists a unique k such that $p^k \leq N < p^{k+1}$. We write $N_k = N$ and, for every i, from $i = k$ to $i = 0$,

$$a_i = \left\lfloor \frac{N_i}{p^i} \right\rfloor \quad \text{and} \quad N_{i-1} = N_i - a_i p^i.$$

Then, a_i is in A_p, a_k is different from 0 and $N_i < p^i$. It holds:

$$N = \sum_{i=0}^{k} a_i p^i = \pi_p(a_k \cdots a_0).$$

The division algorithm. Let N be any positive integer. Write $N_0 = N$ and, for $i \geqslant 0$, write

$$N_i = p N_{i+1} + b_i \tag{2.1}$$

where b_i is the remainder of the division of N_i by p, and thus belongs to A_p. Since N_{i+1} is strictly smaller than N_i, the division (2.1) can be repeated only a finite number of times, until eventually $N_\ell \neq 0$ and $N_{\ell+1} = 0$ for some ℓ (and thus $b_\ell \neq 0$). The sequence of successive divisions (2.1) for $i = 0$ to $i = \ell$ produces the digits $b_0, b_1, \ldots, b_\ell$, and it holds:

$$N = \sum_{i=0}^{\ell} b_i \, p^i = \pi_p \, (b_\ell \cdots b_0) \, .$$

The integer N can also be written as

$$N = ((\cdots (b_\ell \, p + b_{\ell-1}) \cdots) p + b_1) p + b_0,$$

that is, as the evaluation of a polynomial by a *Horner scheme*. By Lemma 2.2.1, $k = \ell$ and $a_k \cdots a_0 = b_\ell \cdots b_0$. We have thus proved the following.

Theorem 2.2.2 *Every non-negative integer N has a unique representation in base p which does not begin with a zero. It is called the p-expansion of N and denoted by $\langle N \rangle_p$.*

Note that the representation of 0 is the empty word ε. It also follows that the *set of p-expansions* is the rational language

$$L_p = \{\langle N \rangle_p \mid N \in \mathbb{N}\} = A_p^* \setminus 0\, A_p^* = \{A_p \setminus \{0\}\} A_p^* \cup \{\varepsilon\}.$$

The map π_p is not only a bijection between L_p and $\mathbb{N}$ but also a morphism of ordered sets (when L_p is ordered by the trace of the *radix order* $\prec$ on A_p^*).

Proposition 2.2.3 *For all n and m in $\mathbb{N}$, $\langle n \rangle_p \prec \langle m \rangle_p$ holds if, and only if, $n < m$.*

Remark 2.2.4 It also follows from Proposition 2.2.3 that for any two words v and w of A_p^* and of the *same length*, $v \prec w$ if, and only if, $\pi_p(v) < \pi_p(w)$.

A first finite transducer: the divider by q. Let q be a fixed positive integer and let $[q] = \{0, \ldots, q-1\}$ be the set of remainders modulo q. For every integers s and a, the Euclidean division by q yields unique integers b and r such that

$$ps + a = qb + r. \tag{2.2}$$

If s is in $[q]$ and a is in A_p, then b is in A_p – and by definition r is in $[q]$. Equation (2.2) thus defines a transducer:

$$\mathcal{Q}_{p,q} = (\,[q], A_p, A_p, E, \{0\}, [q]\,) \text{ with } E = \{(s,(a,b),r) \mid ps + a = qb + r\}.$$

The transducer $\mathcal{Q}_{p,q}$ is sequential. (Indeed, $\mathcal{Q}_{p,q}$ is co-sequential as well if p and q are co-prime.) Figure 2.2 shows $\mathcal{Q}_{2,5}$.

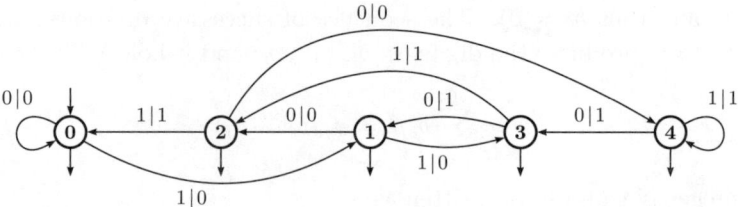

Fig. 2.2 The divider $\mathcal{Q}_{2,5}$.

The realisation of division by finite automata (together with arithmetic operations modulo q) is exactly what is behind computation rules such as the *casting out nines* or divisibility criteria such as the divisibility by 11 (a number is divisible by 11 if, and only if, the sum of digits of odd rank is equal to the sum of digits of even rank). That such criterium exists in every base for any fixed divisor was already observed by Pascal (*cf.* (Pascal 1654, pp. 84–89), see also (Sakarovitch 2003, Prologue)).

2.2.2 The evaluator and the converters

Finite automata really come into play with number representation when we allow ourselves to use sets of digits *larger than the canonical alphabet*. Let p be the base fixed as before but the digits be a priori *any integer*, positive or negative. Consider then the (doubly infinite) automaton $\mathcal{Z}_p$ whose states are the integers, that is, $\mathbb{Z}$, which reads (from left to right) the numbers (thus written on the 'alphabet' $\mathbb{Z}$), and which runs in such a way that, at every step of the reading, the reached state indicates the value of the portion of the number read so far. The initial state of $\mathcal{Z}_p$ is thus 0 and its transitions are of the form:

$$\forall s,t,a \in \mathbb{Z} \qquad s \xrightarrow[\mathcal{Z}_p]{a} t \quad \text{if, and only if,} \quad t = ps + a, \qquad (2.3)$$

from which we get the expected behaviour:

$$\forall w \in \mathbb{Z}^* \qquad 0 \xrightarrow[\mathcal{Z}_p]{w} \pi_p(w).$$

It follows from (2.3) that $\mathcal{Z}_p$ is both *deterministic* and *co-deterministic*. It is logical to call $\mathcal{Z}_p$ the *evaluator*.

In fact, we shall consider only finite parts of $\mathcal{Z}_p$. First, we restrict our alphabet to be a *finite* symmetrical part B_d of $\mathbb{Z}$: $B_d = \{-d, \ldots, d\}$ where d is a positive integer, $d \geq p-1$ and thus $A_p \subset B_d$. Second, we choose 0 as a unique final state and we get an automaton $\mathcal{Z}_{p,d} = (\mathbb{Z}, B_d, E, \{0\}, \{0\})$ where the transitions in E are those defined by (2.3). This automaton accepts thus the writings of 0 (in base p and on the alphabet B_d) and we call it a *zero automaton*. It is still infinite but we have the following.

Proposition 2.2.5 *The trim part of $\mathcal{Z}_{p,d}$ is finite and its set of states is $H = \{-h, \ldots, h\}$ where h is the largest integer (strictly) smaller than $d/(p-1)$.*

Proof As B_d contains A_p and is symmetrical, every z in $\mathbb{Z}$ is *accessible* in $\mathcal{Z}_{p,d}$.

If m is a positive integer larger than, or equal to, $d/(p-1)$, the 'smallest' reachable state from m is $mp - d$, which is also larger than, or equal to, $d/(p-1)$: m is not co-accessible in $\mathcal{Z}_{p,d}$ and the same is true if m is smaller than, or equal to, $-d/(p-1)$.

If m is a positive integer smaller than $d/(p-1)$, then the integer $k = m(p-1)+1$ is smaller than, or equal to, d, and $m \xrightarrow{\overline{k}} (m-1)$ is a transition in $\mathcal{Z}_{p,d}$. (A signed digit $-k$ is denoted by $\overline{k}$.) Hence, by induction, a path from m to 0 in $\mathcal{Z}_{p,d}$. The same is true if m is a negative integer strictly larger than $-d/(p-1)$. □

Figure 2.3 shows $\mathcal{Z}_{2,2}$. By definition, the trim part of the automaton $\mathcal{Z}_{p,d}$ is the *strongly connected component* of 0. From now on, and unless otherwise stated, we let $\mathcal{Z}_{p,d}$ denote the automaton reduced to its trim part only. The automaton $\mathcal{Z}_{p,d}$ is not so much interesting in itself but as the core of the construction of a series of transducers that transform representations of a number into others and that we call by the generic name of *digit-conversion transducers*, or *converters* for short.

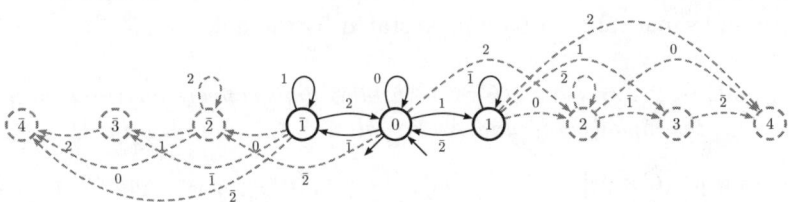

Fig. 2.3 A finite view on $\mathcal{Z}_{2,2}$. The part which is not co-accessible is shown in grey and dashed.

2.2.2.1 The converters and the normalisers

We need some more elementary notation and definitions. From any two alphabets of integers C and A, we build the alphabet $C + A$ (resp. $C - A$)

of all sums (resp. differences):

$$C + A = \{z \mid \exists c \in C, \exists a \in A \quad z = c + a\},$$
$$C - A = \{z \mid \exists c \in C, \exists a \in A \quad z = c - a\}.$$

Let $u = c_{k-1}c_{k-2}\cdots c_0$ and $v = a_{k-1}a_{k-2}\cdots a_0$ be two words of length k of C^* and A^* respectively. The *digitwise addition* of u and v is the word $u \oplus v = s_{k-1}s_{k-2}\cdots s_0$ of $(C+A)^*$ such that $s_i = c_i + a_i$ (resp. the *digitwise subtraction* is the word $u \ominus v = d_{k-1}d_{k-2}\cdots d_0$ of $(C-A)^*$ such that $d_i = c_i - a_i$), for every i, $0 \leq i < k$. If u and v have not the same length, the shortest is silently padded on the left by 0's and both $u \oplus v$ and $u \ominus v$ are thus defined for all pairs of words. In any case, the following obviously holds:

$$\pi_p(u \oplus v) = \pi_p(u) + \pi_p(v) \quad \text{and} \quad \pi_p(u \ominus v) = \pi_p(u) - \pi_p(v).$$

Let $B_d = \{-d, \ldots, d\}$ be the (smallest) symmetrical part of $\mathbb{Z}$ that contains $C - A$: $d = \max\{|c-a| \mid c \in C, \ a \in A\}$. From $\mathcal{Z}_{p,d} = (H, B_d, E, \{0\}, \{0\})$, we then define a letter-to-letter (left) transducer $\mathcal{C}_p(C \times A) = (H, C, A, F, \{0\}, \{0\})$, whose transitions are defined by

$$s \xrightarrow[\mathcal{C}_p(C \times A)]{c \mid a} t \quad \text{if, and only if,} \quad s \xrightarrow[\mathcal{Z}_{p,d}]{c-a} t, \qquad (2.4)$$

for every s and t in H, c in C, and a in A, that is, if, and only if,

$$ps + c = t + a. \qquad (2.5)$$

Both (2.4) and (2.5) show that a given transition in $\mathcal{Z}_{p,d}$ may give rise to no or several transitions (or to a transition with several labels) in $\mathcal{C}_p(C \times A)$. This transducer relates every u in C^* with all words in A^* with the same length and same value in base p, as stated by the following.

Proposition 2.2.6 *Let $\mathcal{C}_p(C \times A)$ be the digit-conversion transducer in base p for the alphabets C and A. For all u in C^* and all v in A^*,*

$$(u,v) \in |\mathcal{C}_p(C \times A)| \quad \text{if, and only if,} \quad \pi_p(u) = \pi_p(v) \text{ and } |u| = |v|.$$

Proof If (u,v) is in $|\mathcal{C}_p(C \times A)|$, then $|u| = |v|$ as $\mathcal{C}_p(C \times A)$ is letter-to-letter and, on the other hand, the successful computation labelled by (u,v) in $\mathcal{C}_p(C \times A)$ maps onto a successful computation labelled by $u \ominus v$ in $\mathcal{Z}_{p,d}$ and thus $\pi_p(u \ominus v) = 0$, that is, $\pi_p(u) = \pi_p(v)$.

Conversely, if $u = c_{k-1}c_{k-2}\cdots c_0$ and $v = a_{k-1}a_{k-2}\cdots a_0$ are in C^* and A^* respectively, $u \ominus v$ is in B_d^* and if $\pi_p(u) = \pi_p(v)$, then $u \ominus v$ is the label of a successful computation of $\mathcal{Z}_{p,d}$, every transition (s, d_i, t) of which is the image of a transition (s, c_i, a_i, t) in $\mathcal{C}_p(C \times A)$. These transitions form a successful computation whose label is (u, v). □

If $A = A_p$, $\mathcal{C}_p(C \times A_p)$ is *input co-deterministic*, or *co-sequential*, since $ps + c = t + a$ and $ps' + c = t + a'$ would imply $p(s - s') = a - a'$, and then $s = s'$ as both a and a' are in A_p. Every word u in C^*, padded on the left by the number of 0's necessary to give it the length of $\langle \pi_p(u) \rangle_p$ is thus the input of a unique successful computation in $\mathcal{C}_p(C \times A_p)$ whose output is the unique p-expansion of $\pi_p(u)$.

This is the reason why $\mathcal{C}_p(C \times A_p)$ is rather called *normaliser* (in base p and for the alphabet C), denoted by $\mathcal{N}_p(C)$, and more often described by its *transpose*, a letter-to-letter *right* transducer, which is thus input deterministic, or *sequential*. In order to keep (2.5) valid, we also change the sign of the states in the transpose. Finally, every state is given a *final function* which outputs the p-expansion of the value of the state: it is equivalent to reaching the state 0 by reading enough leading 0's on the input. In conclusion, we have shown the following.

Theorem 2.2.7 *Normalisation in base p for any input alphabet of digits is realised by a finite letter-to-letter sequential right transducer.*

Figure 2.4 shows $\mathcal{N}_2(C_2)$ and its transpose, where $C_2 = A_2 + A_2 = \{0, 1, 2\}$ is the alphabet on which are written words obtained by *digitwise* addition of two binary expansions of integers: $\mathcal{N}_2(C_2)$ realises the addition in base 2.

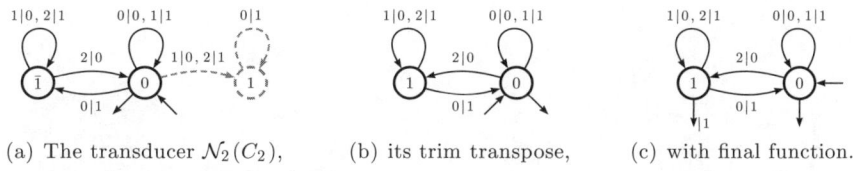

(a) The transducer $\mathcal{N}_2(C_2)$,   (b) its trim transpose,   (c) with final function.

Fig. 2.4 A normaliser in base 2.

2.2.2.2 The signed-digit representation

The zero automaton uses negative digits as well as positive ones; we can make use of these digits not only as *computational means* but for the *representation* of numbers as well.

Let us first remark that if an alphabet of integers A contains a complete set of representatives of $\mathbb{Z}/p\mathbb{Z}$, all of which are smaller than p in modulus, then the division algorithm (2.1) may be run with digits taken in A instead of A_p in such a way that it terminates, which proves that every positive integer has a p-representation as a word in A^*, that is, $\pi_p \colon A^* \to \mathbb{N}$ is surjective. On the other hand, π_p is injective if, and only if, there is at most one digit in A for every representative of $\mathbb{Z}/p\mathbb{Z}$. Both conditions are met if $p = 2q+1$ is odd and A is the symmetric alphabet $B_q = \{-q, \ldots, q\}$. The first case, $p = 3$ and $A = \{-1, 0, 1\}$, yields a beautiful numeration system, celebrated in (Knuth 1998). But now we are more interested in systems where numbers may have indeed *several representations*. In what follows, we choose A to be a symmetric alphabet B_h:

$$B_h = \{-h, \ldots, h\} \qquad \text{with} \qquad h \geq \left\lfloor \frac{p+1}{2} \right\rfloor.$$

As B_h is symmetrical, every integer, positive *and* negative, has a p-representation as a word in B_h^*. Equation (2.3) and the construction of $\mathcal{Z}_p$ immediately yield the following.

Proposition 2.2.8 *In base p, and with the symmetric digit alphabet B_h, the sign of a number is always given by (the sign of) its left-most digit if, and only if, h is less than p.*

More important, $\pi_p \colon B_h^* \to \mathbb{Z}$ is not only surjective, but also *not injective*. The converter $\mathcal{C}_p(B_h \times B_h)$ maps every word in B_h^* to all words of B_h^* that have the same value (modulo some possible padding on the left by 0's): we call it the *redundancy transducer* (in base p on the alphabet B_h) and denote it by $\mathcal{R}_p(B_h)$. If $h = \lfloor \frac{p+1}{2} \rfloor$, it follows from Proposition 2.2.5 that $\mathcal{R}_p(B_h)$ has three states. Figure 2.5 shows $\mathcal{R}_2(B_1)$ and $\mathcal{R}_3(B_2)$.

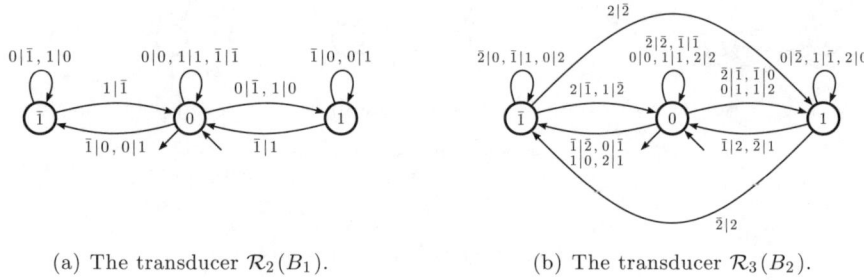

(a) The transducer $\mathcal{R}_2(B_1)$.   (b) The transducer $\mathcal{R}_3(B_2)$.

Fig. 2.5 Two redundancy transducers.

This symmetric representation of numbers is an old folklore.† It has been given a renewed interest in computer arithmetic for the redundancy in the

---

† It was known (at least) as early as Cauchy who advocated such system for $p = 10$ and $h = 5$ with the argument that it makes the learning of addition and multiplication easier: the size of the tables is roughly divided by 4 (see (Cauchy 1840)).

representations allows us to improve the way operations are performed, as we shall see now. The following is to be found in (Avizienis 1961) for bases larger than 2, in (Chow and Robertson 1978) for the binary case – although the original statements and proofs are not formulated in terms of automata.

Theorem 2.2.9 *In base $p \geq 3$ with the symmetric digit alphabet B_h, where $h = \lfloor \frac{p}{2} \rfloor + 1$, the addition may be realised by a 1-local letter-to-letter transducer, and by a 2-local one if $p = 2$ and $h = 1$.*

Note that a '1-local letter-to-letter transducer' is by definition a 'sequential letter-to-letter transducer', that a '2-local letter-to-letter transducer' is equivalent to a 'sequential transducer' but not necessarily to a 'sequential letter-to-letter transducer' (*cf.* Section 2.6).

Proof We assume first that $p \geq 3$; the cases of odd $p = 2q + 1$ and of even $p = 2q$ will induce slight variations in the definitions but the core will be the same. In both cases, $h = q + 1$. The alphabet for digitwise addition is B_{2h} with $2h = p + 1$ if p is odd, $2h = p + 2$ if p is even. Let

$$V_p = \begin{cases} \{-(h-1), \ldots, h-1\} & \text{if } p \text{ is odd,} \\ \{-(h-2), \ldots, h-1\} & \text{if } p \text{ is even.} \end{cases}$$

In both cases, Card $V_p = p$ and is a set of representatives of $\mathbb{Z}/p\mathbb{Z}$. Not only do we have $V_p \subset B_h$, but for any s in V_p, $s+1$ and $s-1$ belong to B_h as well (this is the condition which is not verified when $p = 2$).

Let $\mathcal{V}_p$ be the subautomaton of $\mathcal{Z}_p$, with V_p as set of states, 0 as initial state, and every state is final. We turn $\mathcal{V}_p$ into a transducer $\mathcal{W}_p$ with input alphabet B_{2h}. Every transition

$$s \xrightarrow[\mathcal{V}_p]{d} t = ps + d \quad \text{gives}$$

$$s \xrightarrow[\mathcal{W}_p]{t|ps} t \quad \text{and also} \quad s \xrightarrow[\mathcal{W}_p]{t+p|p(s+1)} t \quad \text{or} \quad s \xrightarrow[\mathcal{W}_p]{t-p|p(s-1)} t,$$

or both, according to whether $t+p$, $t-p$, or both, are in B_{2h}. By construction, the input automaton of $\mathcal{W}_p$ is
(i) deterministic,
(ii) complete (over the alphabet B_{2h}) and
(iii) 1-local (that is, the end of a transition is determined by the label).
Since $t = ps + d$, $\mathcal{W}_p$ is a converter and if (u, v) is the label of a computation of $\mathcal{W}_p$ which (begins in 0 and) ends in t, then

$$\pi_p(u) = \pi_p(v) + t.$$

Let now $\mathcal{W}_p'$ be the transducer obtained from $\mathcal{W}_p$ by replacing every transition

$$s \xrightarrow[\mathcal{W}_p]{m|pn} t \quad \text{by} \quad s \xrightarrow[\mathcal{W}_p']{m|n} t,$$

and by setting the final function T as $T(t) = t$ for every t in V_p. By construction, and the above remark, the output alphabet of $\mathcal{W}_p'$ is B_h. If (u, v') is the label of a computation of $\mathcal{W}_p'$ which (begins in 0 and) ends in t, then

$$\pi_p(u) = p\, \pi_p(v') + t = \pi_p(v't).$$

As $v't$ is the output of $\mathcal{W}_p'$ for the input u, $\mathcal{W}_p'$ answers the question.

For $p = 2$, the foregoing construction, starting from $V_2 = \{0, 1\}$, works perfectly well, but for the fact that $\mathcal{W}_2'$ contains one, and only one, transition whose output is not in B_1:

$$1 \xrightarrow[\mathcal{W}_2']{2|2} 0.$$

The same construction is then carried out again, but starting from $\overline{V_2} = \{\bar{1}, 0\}$, which yields a transducer $\overline{\mathcal{W}_2'}$ which contains one, and only one, transition whose output is not in B_1:

$$\bar{1} \xrightarrow[\overline{\mathcal{W}_2'}]{\bar{2}|\bar{2}} 0.$$

The composition $\mathcal{W}_2'' = \mathcal{W}_2' \circ \overline{\mathcal{W}_2'}$ is a 2-local letter-to-letter sequential transducer in which no transition has an output outside B_1 since no transition in $\mathcal{W}_2'$ has a transition with output $\bar{2}$ and no transition in $\overline{\mathcal{W}_2'}$ with input 2 has an output outside B_1: $\mathcal{W}_2''$ answers the question. $\square$

Figure 2.6 shows $\mathcal{W}_3'$ and $\mathcal{W}_2''$.

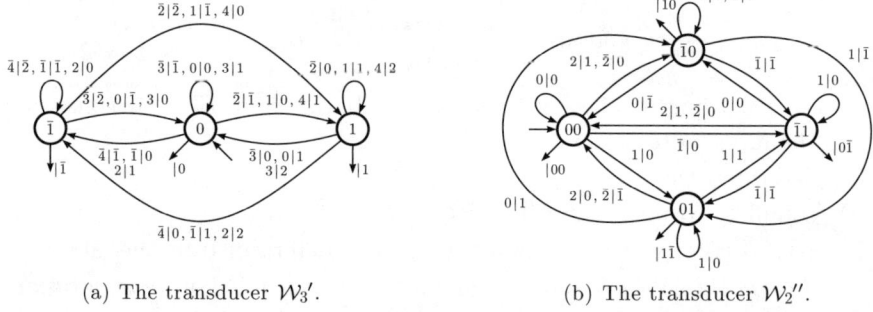

(a) The transducer $\mathcal{W}_3'$.   (b) The transducer $\mathcal{W}_2''$.

Fig. 2.6 Two local adders.

Remark 2.2.10 The same construction as the one for $p = 2$ can be carried out for any even $p = 2q$ and would yield a 2-local automaton for the addition if the numbers are written on the smaller alphabet B_h with $h = q$.

2.2.2.3 The minimal weight representation

Multiplication by a fixed integer obviously falls in the case of normalisation but, in contrast with addition, multiplication (between two numbers) cannot be realised by a finite automaton. However, redundant alphabets and redundancy transducers are not irrelevant to the subject for they allow useful preprocessing to efficient multiplication algorithms.

Let A be a digit alphabet and $u = u_k u_{k-1} \cdots u_1 u_0$ be in A^* with the u_i in A. The *weight* of u is the *absolute sum of digits* $\|u\| = \sum_{i=0}^{k} |u_i|$. The *Hamming weight* of u is the *number* of non-zero digits in u. Of course, when $A \subseteq \{-1, 0, 1\}$, the two definitions coincide; as, for sake of simplicity, we consider here this case only, we do not introduce another notation and speak simply of 'weight'.

The multiplication of two numbers represented by u and v respectively amounts to a series of addition of u with shifted copies of u itself, as many times as there are non-zero digits in v: the smaller the weight of v, the more efficient the multiplication by $\pi_p(v)$. Hence the interest for representation of minimal weight. The following statement and proof is an 'automata translation' of a classical description of (binary) representations of minimal weight as 'non-adjacent form' due to Booth (Booth 1951) and Reitwiesner (Reitwiesner 1960).

Theorem 2.2.11 *The computation of a 2-representation of minimal weight over the alphabet $B_1 = \{-1, 0, 1\}$ from the 2-expansion of an integer x is realised by a finite sequential right transducer. The result is a representation with no adjacent non-zero digits.*

Remark 2.2.12 The study of minimal weight representations goes on with the computation of the mean weight (that gives an evaluation of the benefits of the construction). These minimal weight representations have also applications to cryptography. See also Section 9.2.4.4.

2.2.3 Representation of reals

Real numbers from the interval $[0, 1)$ are traditionally represented as infinite sequences of digits (infinite on the right), that is, by elements of $A_p^{\mathbb{N}}$. *By convention*, and although $\mathbb{N}$ contains 0, we consider, in this context and for sake of simplicity of the writing, that an element u of $A_p^{\mathbb{N}}$ is a sequence of digits whose *indices begin with* 1: $u = (u_i)_{i\geq 1}$ where every u_i is in A_p.

The set $A_p^{\mathbb{N}}$ is naturally a *topological space* equipped with the (total) *lexicographic order*: for u and v in $A_p^{\mathbb{N}}$, $u < v$ if, and only if, if $w = u \wedge v$ is the longest common prefix to u and v, then $u = w a u'$ and $v = w b v'$ with a and b in A_p and $a < b$. With our convention, the evaluation map, still denoted by π_p, gives every word u of $A_p^{\mathbb{N}}$ a real value:

$$u = u_1 u_2 \cdots \longmapsto \pi_p(u) = \sum_{i=1}^{\infty} u_i\, p^{-i}.$$

When finite and infinite words are mixed in the same context, the latter are prefixed with the *radix point* inside the function π_p. For instance, it holds:

$$\forall u = (u_i)_{i\geq 1} \in A_p^{\mathbb{N}} \qquad \pi_p(.u) = \lim_{n\to +\infty} \frac{1}{p^n}\, \pi_p(u_1 u_2 \cdots u_n), \qquad (2.6)$$

$$\forall u \in A_p^{\mathbb{N}},\ \forall w \in A_p^{*} \qquad \pi_p(.w u) = \frac{1}{p^{|w|}}\,(\pi_p(w) + \pi_p(.u)). \qquad (2.7)$$

Proposition 2.2.13 *The map $\pi_p \colon A_p^{\mathbb{N}} \to [0,1]$ is a continuous and order-preserving function. Moreover, for u and v in $A_p^{\mathbb{N}}$, $u < v$, and $w = u \wedge v$, $\pi_p(u) = \pi_p(v)$ if, and only if, $u = w\,a\,(p{-}1)^{\omega}$ and $v = w\,(a{+}1)\,0^{\omega}$.*

Proof Let us first make the obvious remark – which will be used silently in the sequel – that if u and v are such that for every i, $u_i \leq v_i$ and if there exists at least one j such that $u_j \neq v_j$, then $\pi_p(u) < \pi_p(v)$.

Next, the not less obvious identity

$$\sum_{i=1}^{+\infty}(p-1)p^{-i} = (p-1)\left(\frac{\frac{1}{p}}{1-\frac{1}{p}}\right) = 1 \qquad (2.8)$$

implies in particular that $\pi_p\left(A_p^{\mathbb{N}}\right) \subseteq [0,1]$.

The set $A_p^{\mathbb{N}}$ is a metric space with $d(u,v) = 2^{-|u\wedge v|}$ if $u \neq v$ (and $d(u,u) = 0$ of course). Then, again by (2.8), $|\pi_p(u) - \pi_p(v)| \leq 2p^{-(|u\wedge v|-1)}$ and π_p is Lipschitz, hence continuous.

Let then u and v be in $A_p^{\mathbb{N}}$, $u < v$, and let k be the smallest index such that $u_k \neq v_k$, that is, $u_k \leq v_k - 1$. Let

$$u' = u_1 u_2 \cdots u_{k-1} u_k\, p{-}1\, p{-}1\cdots \quad \text{and} \quad v' = v_1 v_2 \cdots v_{k-1} (u_k + 1)\, 0\,0 \cdots$$

By the foregoing, $\pi_p(u') = \pi_p(v')$ and, if $u \neq u'$, then $\pi_p(u) < \pi_p(u')$, and if $v \neq v'$, then $\pi_p(v') < \pi_p(v)$, which shows that π_p is order-preserving. $\square$

Let x be a non-negative real number. If $x \geq 1$, a first way for representing x is to treat its *integral part* $\lfloor x \rfloor$ and its *fractional part* $\{x\}$ separately, to compute $\langle \lfloor x \rfloor \rangle_p$ as we have done in the previous section, to compute $\langle \{x\} \rangle_p$ as we shall see below, and to combine them with the radix point:

$$\langle x \rangle_p = \langle \lfloor x \rfloor \rangle_p \cdot \langle \{x\} \rangle_p.$$

Another way is to determine the (unique) integer k such that $p^{k-1} \leq x < p^k$ first, to consider the real $y = \dfrac{x}{p^k}$ which belongs to $[0,1)$, to compute $\langle y \rangle_p = u_1 u_2 \cdots$ and to recover the representation of x by setting the radix point at the right place: $\langle x \rangle_p = u_1 u_2 \cdots u_k \cdot u_{k+1} u_{k+2} \cdots$. We shall obviously take the second option, and from now on consider real numbers from $[0,1)$ only. Given such an x in $[0,1)$ which is then likely to have a p-representation which is an infinite sequence on the right, there is no hope to have *an algorithm* which computes the digits *from right to left*, and we are left with the right algorithm which computes the digits *from left to right*.

The greedy algorithm. Let x be in $[0,1)$. Write $z_0 = x$ and, for every $i \geq 1$, let

$$u_i = \lfloor p z_{i-1} \rfloor \quad \text{and} \quad z_i = \{p z_{i-1}\}. \tag{2.9}$$

Every u_i is in A_p, and it holds

$$z_0 = u_1 p^{-1} + z_1 p^{-1} = u_1 p^{-1} + u_2 p^{-2} + z_2 p^{-2} = \cdots = \sum_{i=1}^{+\infty} u_i p^{-i}, \tag{2.10}$$

that is, the infinite word $u = (u_i)_{i \geq 1}$ in $A_p^{\mathbb{N}}$ is a p-representation of x. It is *the p-expansion* of x, denoted by $\langle x \rangle_p$ or $\mathsf{d}_p(x)$ (when a more functional notation is needed). The computation described by (2.9) is referred to as the *greedy algorithm*.

By convention (and by abuse), we say that a p-representation u is *finite* if it ends with the infinite word 0^ω: $u = w 0^\omega$ with w in A_p^* (and indeed the finite word w is sufficient to compute $\pi_p(u)$). An x in $[0,1)$ is said to be *p-decimal* if x has a finite p-representation, that is, if, and only if, x is an integer divided by a (sufficiently large) power of p.

Corollary 2.2.14 *The map $\pi_p \colon A_p^{\mathbb{N}} \to [0,1]$ is a surjective function. An x in $[0,1)$ has more than one p-representation in $A_p^{\mathbb{N}}$ if, and only if, it is p-decimal, in which case it has only two of them, and its p-expansion is the finite one, which is larger in the lexicographic order than the other infinite one.*

It also follows that the *set of p-expansions* is the rational language (of infinite words):

$$D_p = \{\langle x \rangle_p \mid x \in [0,1)\} = A_p^{\mathbb{N}} \setminus A_p^* (p-1)^{\omega}.$$

Figure 2.7 shows a finite Büchi automaton which recognises D_2.

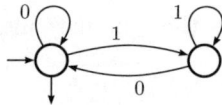

Fig. 2.7 A finite Büchi automaton for the language of 2-expansions D_2.

A first finite transducer over infinite words: the divider by q.

Let us consider the transducer $\mathcal{Q}_{p,q}$ of Section 2.2.1 again (q is a fixed integer and $[q]$ the set of remainders modulo q):

$$\mathcal{Q}_{p,q} = (\,[q], A_p, A_p, E, \{0\}, [q]\,) \text{ with } E = \{\bigl(s, (a,b), r\bigr) \mid ps + a = qb + r\},$$

where a and b are in A_p and s and r in $[q]$ (see Figure 2.2). We are now interested in the *infinite* computation of $\mathcal{Q}_{p,q}$.

Let u be the p-expansion of an x in $[0,1)$, let c be *the* computation of $\mathcal{Q}_{p,q}$ with input u (it exists as $\mathcal{Q}_{p,q}$ is input-complete and is unique as $\mathcal{Q}_{p,q}$ is input-deterministic), and let v be the output of c, in $A_p^{\mathbb{N}}$. Let $r_0 = 0$ and for every $i \geq 1$ it holds:

$$p\, r_{i-1} + u_i = q v_i + r_i.$$

Equation (2.10) then becomes

$$x = u_1 p^{-1} + z_1 p^{-1} = q v_1 p^{-1} + p^{-1}(r_1 + z_1) = \cdots = q \sum_{i=1}^{+\infty} v_i p^{-i},$$

that is, $\pi_p(u) = q\, \pi_p(v)$. And $\mathcal{Q}_{p,q}$ realises the division by the integer q over the p-representations of the reals of $[0,1)$.

As a rational number is the quotient of an integer by another integer, and since $\mathcal{Q}_{p,q}$ is input-deterministic, a computation whose input is ultimately a sequence of 0's ends in a circuit, therefore the description of the division as a finite sequential transducer is a proof of the following classical statement.

Proposition 2.2.15 *The p-expansion of a rational number r/q, in any integer base p, is eventually periodic (of period less than q).*

The zero (Büchi) automaton and (Büchi) converters

The 'zero-automaton' for real number representations is basically the same as the one we have built for the representations of the integers, that is, it is based upon the automaton $\mathcal{Z}_p$ (cf. Section 2.2.2). As above, let $B_d = \{-d, \ldots, d\}$ be a finite symmetrical part of $\mathbb{Z}$ with $d \geq p - 1$.

Proposition 2.2.16 *An infinite word u in $B_d^{\mathbb{N}}$ has value 0 in base p if, and only if, it is accepted by the Büchi automaton $\mathcal{Z}'_{p,d} = (H', B_d, E, \{0\}, H')$ with $H' = \{-h', \ldots, h'\}$ where h' is the largest integer smaller than, or equal to, $d/(p-1)$.*

Proof By the definition of $\mathcal{Z}'_{p,d}$, every infinite word u that labels an infinite computation in $\mathcal{Z}'_{p,d}$ is accepted by $\mathcal{Z}'_{p,d}$. For every (finite) prefix w of u, $|\pi_p(w)| \leq h'$ and then, by (2.6), $\pi_p(.u) = 0$.

Conversely, let u in $B_d^{\mathbb{N}}$ which does not label a computation in $\mathcal{Z}'_{p,d}$, that is, there exists a prefix w of u such that

$$0 \xrightarrow[\mathcal{Z}_p]{w} t = \pi_p(w) \qquad \text{with} \qquad t > d/(p-1).$$

We have, on the one hand, $\pi_p(.u) \geq \pi_p(.w\overrightarrow{d}^\omega)$ and, on the other hand,

$$\pi_p(.w\overrightarrow{d}^\omega) = \frac{1}{p^{|w|}} \left(t - \frac{d}{p-1} \right) > 0.$$

(The case $t < -d/(p-1)$ is identical.) $\square$

The proof also yields the characterisation of $\mathcal{Z}'_{p,d}$ as the (full) $\mathcal{Z}'_{p,d}$ restricted to its (non-trivial) *strongly connected components*. Figure 2.8 shows $\mathcal{Z}'_{2,1}$ and $\mathcal{Z}'_{2,2}$.

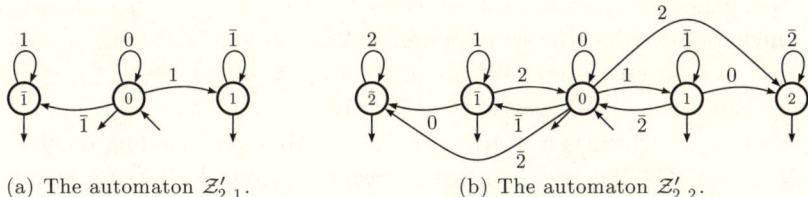

(a) The automaton $\mathcal{Z}'_{2,1}$. (b) The automaton $\mathcal{Z}'_{2,2}$.

Fig. 2.8 Two 'zero automata' for binary representations of reals.

From the zero automaton for real representations, one derives *converters* and *normalisers*, as in the case of the representations of integers, but for the point that not every word in $A_p^{\mathbb{N}}$ is a p-expansion and that there exists thus a distinction between a converter to the canonical alphabet and a normaliser to the same alphabet. For instance, Figure 2.9 shows the converter and normaliser from and to the canonical alphabet, in the binary case.

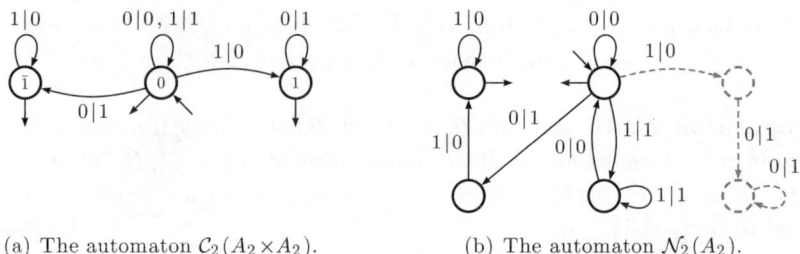

(a) The automaton $\mathcal{C}_2(A_2 \times A_2)$.   (b) The automaton $\mathcal{N}_2(A_2)$.

Fig. 2.9 The converter and normaliser over the canonical alphabet for binary representations of reals.

2.2.4 Base changing

As soon as we want *to compare* the representation of integers in different bases, finite automata show a kind of weakness, that is, no finite transducers exist in general which transform the p-expansion of an integer N into its q-expansion. This follows in fact from the fundamental theorem, due to Alan Cobham, which we referred to in the introduction and which has been presented in Chapter 1 (see Theorem 1.5.5).

This deep result obviously implies, and stays behind, the fact that no finite transducer $\mathcal{T}$ may relate the expansions of integers in base p and in base q, for multiplicatively independent p and q. For, if there was one such $\mathcal{T}$, the image of the p-expansions of a p-recognisable set X by $\mathcal{T}$ would be a rational set of A_q^* and X would thus be q-recognisable as well. It is not necessary however to establish Cobham's Theorem in order to prove the non-existence of such a transducer $\mathcal{T}$. For the latter, it is sufficient for instance to prove that the set of powers of 2 is not 3-recognisable – a simple, and classical, exercise (see (Eilenberg 1974)).

On the other hand, every integer in $\{0, 1, \ldots, p^k - 1\}$ has a p-representation which is a word of A_p^* of length k (by padding on the left with enough 0's) and this defines a *morphism* τ from $A_{p^k}^*$ to A_p^* such that $\pi_p\left(\tau\left(\langle N \rangle_{p^k}\right)\right) = N$ for every N in $\mathbb{N}$. Using inversion and composition of finite transducers, we then get the following.

Proposition 2.2.17 *If p and q are two multiplicatively dependent positive integers, then there exists a finite transducer from A_p^* to A_q^* which maps the p-expansion of every positive integer onto its q-expansion.*

Corollary 2.2.18 *If p and q are two multiplicatively dependent positive integers, then the p-recognisable sets and q-recognisable sets of positive integers coincide.*

2.3 Representation in real base

This section is about the so-called *beta-expansions* where the base is a real number $\beta > 1$. By a greedy algorithm producing the most significant digit first, every positive real number is given a β-expansion, which is an infinite word on a canonical alphabet of integer digits. The main difference with the case where β is an integer is that a number may have several representations on the canonical alphabet, the greedy expansion being the greatest in the lexicographic order.

The set of greedy β-expansions forms a symbolic dynamical system, the β-shift, and we start this chapter by establishing some properties of symbolic dynamical systems defined by means of the lexicographic order, and not related to numeration systems. From this, we derive some properties of the β-shift. We then describe several properties of β-expansions in the important case where β is a Pisot number.

Instead of taking a base, which is a number, it is also possible to take a *basis*, that is, a sequence of integers, like the sequence of Fibonacci numbers. This allows any non-negative integer to be represented. We study these systems more particularly when the basis is a linear recurrent sequence and investigate the conditions under which the set of greedy expansions is recognisable by a finite automaton.

We also consider the problem of changing the basis and describe cases where the conversion between the expansions in the two numeration systems is realisable by a finite transducer.

2.3.1 Symbolic dynamical systems

Definitions for symbolic dynamical systems have been given in Chapter 1; we briefly recall some of them as we adopt slightly different notation (see also (Lothaire 2002, Chapter 1)). Let A be a finite alphabet. A word s in $A^\mathbb{N}$ *avoids* a set $X \subset A^+$ if no factor of s is in X. Denote $S(X)$ the set of words of $A^\mathbb{N}$ which avoid X.

A (one-sided) *symbolic dynamical system*, or *subshift*, is a subset of $A^\mathbb{N}$ of the form $S(X)$ for some $X \subset A^+$. Equivalently, it is a closed shift-invariant subset of $A^\mathbb{N}$. In this chapter, the *shift* on $A^\mathbb{N}$ is denoted σ, and is implicit in all our notations.

A subshift S of $A^\mathbb{N}$ is of finite type if $S = S(X)$ for a finite set $X \subset A^+$.

A subshift S of $A^{\mathbb{N}}$ is sofic if $S = S(X)$ for a rational set $X \subset A^+$, or, equivalently, if $L(S)$ is rational.

A subshift S of $A^{\mathbb{N}}$ is *coded* if there exists a prefix code $Y \subset A^*$ such that $S = \overline{Y^\omega}$, or, equivalently, if the language of S is equal to the set of factors of Y^*, that is, $L(S) = F(Y^*)$, (Blanchard and Hansel 1986).

In the remaining of this section, A is a totally ordered alphabet.

Definition 2.3.1 A word v in $A^{\mathbb{N}}$ is said to be a *lexicographically shift maximal* word (lsm-word for short) if it is larger than, or equal to, any of its shifted images: for every $k \geq 0$, $\sigma^k(v) \leq v$.

Definition 2.3.2 Let $v = (v_i)_{i \geq 1}$ in $A^{\mathbb{N}}$. We denote by

(i) $v_{[n]}$ the prefix of length n of v: $v_{[n]} = v_1 v_2 \cdots v_n$. By convention, $v_{[0]} = \varepsilon$.

(ii) $S_v = \{u \in A^{\mathbb{N}} \mid \forall k \geq 0,\ \sigma^k(u) \leq v\}$, the set of words in $A^{\mathbb{N}}$, all the shifted images of which are smaller than, or equal to, v.

(iii) $D_v = \{u \in A^{\mathbb{N}} \mid \forall k \geq 0,\ \sigma^k(u) < v\}$, the set of words in $A^{\mathbb{N}}$, all the shifted images of which are smaller than v.

(iv) $Y_v = \{v_{[n]}a \in A^* \mid \forall n \geq 0, \forall a \in A,\ a < v_{n+1}\}$.

Proposition 2.3.3 *If v in $A^{\mathbb{N}}$ is an lsm-word, then S_v is a subshift coded by Y_v.*

Proof From their definition follows that S_v is shift-invariant and closed and that Y_v is a prefix code. Let w be in $L(S_v)$; then $w \leq v_{[n]}$ with $n = |w|$. Either $w = v_{[n]}$ and thus a prefix of a word in Y_v or $w < v_{[n]}$ and thus of the form $w = v_1 \cdots v_{n_1-1} a_1 w_1$, with $a_1 < v_{n_1}$ and $w_1 \leq v_1 \cdots v_{|w_1|}$, that is $w = y_1 w_1$ with y_1 in Y_v and w_1 in $L(S_v)$. Iterating this process, we see that w belongs to $F(Y_v^*)$. Conversely, let $w = (w_n)_{n \geq 1} = y_1 y_2 \cdots$ be in Y_v^ω, with y_i in Y_v. Then $w < v$. For each k, $w_k w_{k+1} \cdots$ begins with a word of the form $v_{j_k} v_{j_k+1} \cdots v_{j_k+r-1} a_{j_k+r}$ with $a_{j_k+r} < v_{j_k+r}$, thus $w_k w_{k+1} \cdots < v_{j_k} v_{j_k+1} \cdots \leq v$, and thus w is in S_v. □

Proposition 2.3.4 *Let v be an lsm-word in $A^{\mathbb{N}}$. Then, the following conditions are equivalent:*

(i) *the subshift S_v is recognised by a finite Büchi automaton and, thus, is sofic;*

(ii) *the set D_v is recognised by a finite Büchi automaton;*

(iii) *the word v is eventually periodic.*

Proof [Sketch] Let $\mathcal{S}_v$ be the (infinite) automaton whose states are the $v_{[n]}$ for all n in $\mathbb{N}$, and whose transitions are $v_{[n]} \xrightarrow{v_{n+1}} v_{[n+1]}$ and $v_{[n]} \xrightarrow{a} v_{[0]}$ for every $a < v_{n+1}$. All states are final and $v_{[0]}$ is initial. This automaton $\mathcal{S}_v$ recognises $\mathrm{Pref}(Y_v^*)$, which is equal to $F(Y_v^*)$. As a Büchi automaton, $\mathcal{S}_v$ recognises S_v.

Let $\mathcal{D}_v$ be the automaton obtained from $\mathcal{S}_v$ by taking $v_{[0]}$ as unique final state. As a Büchi automaton, $\mathcal{D}_v$ recognises D_v (*cf.* Figure 2.10).

Now, the automata $\mathcal{S}_v$ and $\mathcal{D}_v$ have both finite minimal quotients, $\mathcal{S}'_v$ and $\mathcal{D}'_v$ respectively, if, and only if, v is eventually periodic. These automata $\mathcal{S}'_v$ and $\mathcal{D}'_v$ recognise the same sets of finite words and the same sets of infinite words as $\mathcal{S}_v$ and $\mathcal{D}_v$ respectively. $\square$

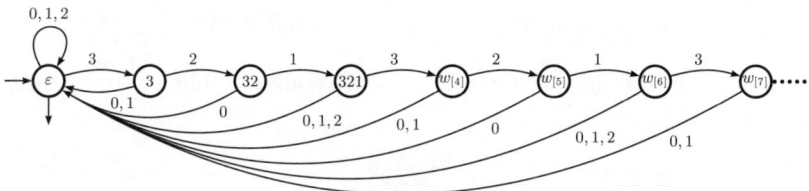

Fig. 2.10 The infinite automaton $\mathcal{D}_w$, for $w = (3\,2\,1)^\omega$.

Remark 2.3.5 In the case where v is eventually periodic but not purely periodic, the minimal quotients $\mathcal{S}'_v$ and $\mathcal{D}'_v$ have the same underlying graph, and $\mathcal{D}'_v$ can also be obtained from $\mathcal{S}'_v$ by taking the image of $v_{[0]}$ as unique final state, see Figure 2.11.

In the case where v is purely periodic, of the form $(v_1 v_2 \cdots v_p)^\omega$, the situation is slightly different and $\mathcal{S}'_v$ and $\mathcal{D}'_v$ have not the same underlying graph. However, $\mathcal{D}'_v$ can also be obtained from $\mathcal{S}'_v$ by performing an in-splitting of the image of $v_{[0]}$ and by keeping as a unique final state the one that does not belong to the loop labelled by $v_1 \cdots v_p$, see Figure 2.12.

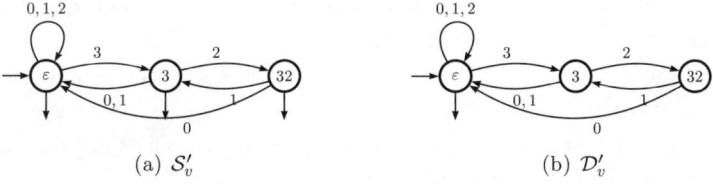

Fig. 2.11 Finite automata for S_v and D_v, $v = 3\,(2\,1)^\omega$.

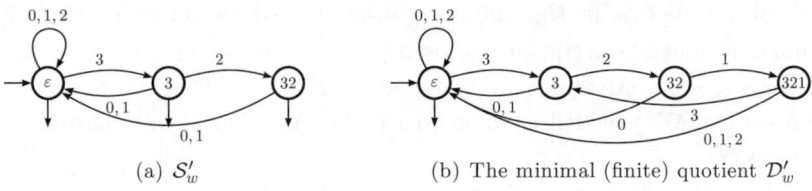

Fig. 2.12 Finite automata for S_w and D_w, $w = (3\,2\,1)^\omega$.

Proposition 2.3.6 *Let v be an lsm-word in $A^\mathbb{N}$. Then, the subshift S_v is of finite type if, and only if, v is purely periodic.*

Proof Suppose that $v = (v_1 v_2 \cdots v_p)^\omega$ and consider the set

$$X'_v = \{v_{[n]} b \in A^* \mid 0 \leq n \leq p-1, \forall b \in A,\ b > v_{n+1}\}.$$

It is easy to check that $S_v = S(X'_v)$. The converse follows from the fact that v is a lsm-word. □

2.3.2 Real base

In this section we consider a base β which is a real number > 1. The reader can consult (Lothaire 2002, Chapter 7) for the proof of some results presented below, and other related results.

Any number x in the interval $[0, 1)$ has a so-called *greedy β-expansion* given by a greedy algorithm (Rényi 1957): let $r_0 = x$, and, for $j \geq 1$, let $x_j = \lfloor \beta r_{j-1} \rfloor$ and $r_j = \{\beta r_{j-1}\}$. Then $x = \sum_{j=1}^\infty x_j \beta^{-j}$, where the x_j's are integer digits in the alphabet $A_\beta = \{0, 1, \ldots, \lceil \beta \rceil - 1\}$. The greedy β-expansion of x is denoted by $\mathsf{d}_\beta(x)$. We also write $x = .x_1 x_2 \cdots$. The same expansion can be obtained by the *β-transformation* on $[0, 1)$: let $\tau_\beta(x) = \{\beta x\}$. Then, for $j \geq 1$, $x_j = \lfloor \beta \tau_\beta^{j-1}(x) \rfloor$.

Note that, when β is an integer, we recover the classical expansion of any x in $[0, 1)$ defined in Section 2.2.

The same algorithm can be applied to $x = 1$, and we obtain the so-called *β-expansion* of 1, $\mathsf{d}_\beta(1)$. Note that, if β is not an integer, then $\mathsf{d}_\beta(1)$ is an infinite word on A_β, but if β is an integer then $\mathsf{d}_\beta(1) = \beta 0^\omega$.

If $x > 1$, there exists $k \geq 0$ such that x/β^k belongs to the interval $[0, 1)$. If $\mathsf{d}_\beta(x/\beta^k) = .x_1 x_2 \cdots$, then $x = x_1 \cdots x_k \boldsymbol{\cdot} x_{k+1} x_{k+2} \cdots$. The greedy β-expansion of x is also denoted $\langle x \rangle_\beta$. The following lemma is an immediate consequence of the greedy algorithm.

Lemma 2.3.7 *An infinite sequence of non-negative integers $(x_i)_{i \geq 1}$ is the* greedy β-expansion of a real number x of $[0,1)$ (resp. of 1) if, and only if, for every $i \geq 1$ (resp. $i \geq 2$), $x_i\beta^{-i} + x_{i+1}\beta^{-i-1} + \cdots < \beta^{-i+1}$.

As in the usual numeration systems, the order between real numbers is given by the lexicographic order on greedy β-expansions.

Proposition 2.3.8 *Let x and y be two real numbers from $[0,1)$. Then $x < y$ if, and only if, $\mathsf{d}_\beta(x) < \mathsf{d}_\beta(y)$.*

Proof Let $\mathsf{d}_\beta(x) = (x_i)_{i \geq 1}$ and let $\mathsf{d}_\beta(y) = (y_i)_{i \geq 1}$, and suppose that $\mathsf{d}_\beta(x) < \mathsf{d}_\beta(y)$. There exists $k \geq 1$ such that $x_k < y_k$ and $x_1 \cdots x_{k-1} = y_1 \cdots y_{k-1}$. Hence $x \leq y_1 \beta^{-1} + \cdots + y_{k-1}\beta^{-k+1} + (y_k - 1)\beta^{-k} + x_{k+1}\beta^{-k-1} + x_{k+2}\beta^{-k-2} + \cdots < y$ since $x_{k+1}\beta^{-k-1} + x_{k+2}\beta^{-k-2} + \cdots < \beta^{-k}$ by Lemma 2.3.7. The converse is immediate. $\square$

A number may have several different writings in base β, which we call β-representations. The greedy β-expansion is characterised by the following property.

Proposition 2.3.9 *The greedy β-expansion of a real number x of $[0,1)$ is the greatest of all the β-representations of x with respect to the lexicographic order.*

Example 2.3.10 Let φ be the Golden Ratio $\frac{1+\sqrt{5}}{2}$. The greedy φ-expansion of $x = 3 - \sqrt{5}$ is equal to 10010^ω. Different φ-representations of x are 01110^ω, or $100(01)^\omega$ for instance.

If a representation ends in infinitely many zeroes, like $u0^\omega$, the trailing zeroes are omitted and the representation is said to be *finite*.

The greedy β-expansion of $x \in [0,1]$ is finite if, and only if, $\tau_\beta^i(x) = 0$ for some i, and it is eventually periodic if, and only if, the set $\{\tau_\beta^i(x) \mid i \geq 1\}$ is finite.

2.3.2.1 The β-shift

Denote by D_β the set of greedy β-expansions of numbers of $[0,1)$. It is a shift-invariant subset of $A_\beta^\mathbb{N}$. The β-shift S_β is the closure of D_β. Note that D_β and S_β have the same set of finite factors. When β is an integer, S_β is the full β-shift $A_\beta^\mathbb{N}$.

A finite (resp. infinite) word is said to be β-*admissible* if it is a factor of an element of D_β (resp. an element of D_β).

The greedy β-expansion of 1 plays a special role in this theory. Let

$d_\beta(1) = (t_n)_{n \geq 1}$ be the greedy β-expansion of 1. We define also the *quasi-greedy expansion* $d^*_\beta(1)$ of 1 by: if $d_\beta(1) = t_1 \cdots t_m$ is finite, then $d^*_\beta(1) = (t_1 \cdots t_{m-1}(t_m - 1))^\omega$, $d^*_\beta(1) = d_\beta(1)$ otherwise.

Theorem 2.3.11 (Parry 1960) *Let $\beta > 1$ be a real number, and let s be an infinite sequence of non-negative integers. The sequence s belongs to D_β if and only if for all $k \geq 0$*
$$\sigma^k(s) < d^*_\beta(1)$$
and s belongs to S_β if, and only if, for all $k \geq 0$
$$\sigma^k(s) \leq d^*_\beta(1).$$

Definition 2.3.12 A number β such that $d_\beta(1)$ is eventually periodic is called a *Parry number*. If $d_\beta(1)$ is finite then β is called a *simple Parry number*.

Example 2.3.13 1. Let φ be the Golden Ratio $\frac{1+\sqrt{5}}{2}$. The expansion of 1 is finite, equal to 11.
2. Let $\theta = \frac{3+\sqrt{5}}{2}$. The expansion of 1 is eventually periodic, equal to $d_\theta(1) = 21^\omega$.
3. Let $\beta = \frac{3}{2}$. Then $d_\beta(1) = 101000001\cdots$ is aperiodic.

Remark 2.3.14 Note that the greedy β-expansion of 1 is never purely periodic.

As a corollary of Theorem 2.3.11 it follows that $d^*_\beta(1)$ is an lsm-word and $S_\beta = S_{d^*_\beta(1)}$ with the notation of Definition 2.3.2. By Propositions 2.3.3, 2.3.4 and 2.3.6 follow then the well-known properties of the β-shift (established in (Ito and Takahashi 1974), (Bertrand-Mathis 1986), (Blanchard 1989)).

Theorem 2.3.15 *The β-shift S_β is a coded symbolic dynamical system which is*

(i) *sofic if, and only if, $d_\beta(1)$ is eventually periodic, i.e., β is a Parry number*
(ii) *of finite type if, and only if, $d_\beta(1)$ is finite, i.e., β is a simple Parry number.*

Remark 2.3.16 Since a sofic symbolic dynamical system is of finite type if, and only if, it can be recognised by a local automaton, see (Béal 1993), it follows that, when β is a simple Parry number the automaton recognising the β-shift can be chosen to be local.

Example 2.3.17 1. Let φ be the Golden Ratio $\frac{1+\sqrt{5}}{2}$. The automaton of Figure 2.14 (a) below recognising S_φ is local, because every admissible word with last letter 0 (resp. 1) arrives in state 0 (resp. 1).
2. Let $\theta = \frac{3+\sqrt{5}}{2}$. Then $\mathsf{d}_\theta(1) = 21^\omega$. The automaton of Figure 2.13 recognising S_θ is not local, since there are two different loops labelled by 1.

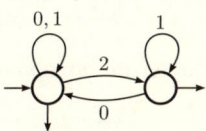

Fig. 2.13 Finite automaton for the θ-shift, $\theta = \frac{3+\sqrt{5}}{2}$.

The following result is a reformulation of Proposition 2.3.4.

Proposition 2.3.18 *The set D_β is recognisable by a finite Büchi automaton if, and only if, $\mathsf{d}_\beta(1)$ is eventually periodic.*

Example 2.3.19 Since $\mathsf{d}_\varphi(1) = 11$, the φ-shift is a system of finite type, recognised by the finite automaton of Figure 2.14 (a). The set D_φ is recognised by the finite Büchi automaton of Figure 2.14 (b).

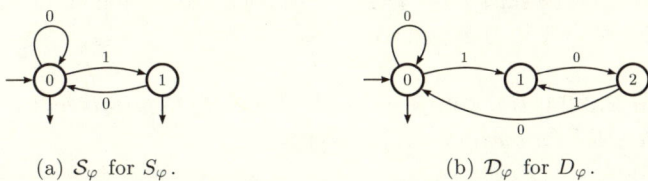

(a) $\mathcal{S}_\varphi$ for S_φ.

(b) $\mathcal{D}_\varphi$ for D_φ.

Fig. 2.14 Finite automata for S_φ and D_φ, $\varphi = \frac{1+\sqrt{5}}{2}$.

There is an important case where the β-expansion of 1 is eventually periodic. A *Pisot number* is an algebraic integer greater than 1 such that all its Galois conjugates have modulus less than one. The natural integers and the Golden Ratio are Pisot numbers.

Theorem 2.3.20 (Schmidt 1980a) *If β is a Pisot number, then every number of $\mathbb{Q}(\beta) \cap [0,1]$ has an eventually periodic β-expansion.*

As a consequence we obtain the important result, see also (Bertrand 1977).

Theorem 2.3.21 *If β is a Pisot number, then the β-shift is a sofic system.*

The *topological entropy* of a subshift $S \subseteq A^{\mathbb{N}}$ is defined as

$$h(S) = \lim_{n \to \infty} \frac{1}{n} \log(L_n(S))$$

where $L_n(S)$ denotes the number of factors of length n in S. One proof of the following well-known result using the fact that the β-shift is a coded system can be found in (Lothaire 2002, Chapter 1).

Proposition 2.3.22 *The topological entropy of the β-shift is equal to $\log \beta$.*

2.3.2.2 The (F) Property

If β is an integer, then every positive integer has a *finite* β-expansion, but this is not true in general when β is not an integer. However, it is easy to see that for the Golden Ratio φ, every positive integer has a finite expansion, for instance, $\langle 2 \rangle_\varphi = 10.01$.

More generally, it is interesting to find numbers having this property. We recall some definitions and results from (Frougny and Solomyak 1992).

Definition 2.3.23 A number β is said to *satisfy the (F) Property* if every element of $\mathbb{Z}[\beta^{-1}] \cap [0,1)$ has a finite greedy β-expansion.
A number β is said to *satisfy the (PF) Property* if every element of $\mathbb{N}[\beta^{-1}] \cap [0,1)$ has a finite greedy β-expansion.

Proposition 2.3.24 *If β satisfies the (F) Property then β is a Pisot number. Moreover, the following are equivalent:*
- *β satisfies the (F) Property*
- *β satisfies the (PF) Property and $\mathsf{d}_\beta(1)$ is finite.*

There are Pisot numbers β with $\mathsf{d}_\beta(1)$ finite that do not satisfy the (F) Property, for instance the Pisot number with minimal polynomial $X^4 - 2X^3 - X - 1$. Here $\mathsf{d}_\beta(1) = 2011$ and $\langle 3 \rangle_\beta = 10.111(00012)^\omega$.

The problem of characterising Pisot numbers satisfying the (F) Property is still open. Up to now, the only families satisfying this property are the following ones.

Theorem 2.3.25 *Let $\beta > 1$ be a root of a polynomial in $\mathbb{Z}[X]$ of the form $M(X) = X^g - b_1 X^{g-1} - b_2 X^{g-2} - \cdots - b_g$. If one of the following properties holds, then β satisfies the (F) Property:*
(i) $b_1 \geq b_2 \geq \cdots \geq b_g > 0$,
(ii) $b_i \geq 0$ for $1 \leq i \leq g$ and $b_1 > \sum_{i=2}^{g} b_i$.

Part (i) is from (Frougny and Solomyak 1992) and Part (ii) from (Hollander 1996).

Cubic Pisot units satisfying (F) are characterised by the following.

Theorem 2.3.26 (Akiyama 2000) *A cubic Pisot unit β satisfies the (F) Property if, and only if, it is a root of the polynomial $M(X) = X^3 - aX^2 - bX - 1$ of $\mathbb{Z}[X]$ with $a \geq 0$ and $-1 \leq b \leq a+1$.*

A family of Pisot numbers satisfying (PF) is the following one.

Theorem 2.3.27 *Let β be such that $\mathsf{d}_\beta(1) = t_1 t_2 \cdots t_m (t_{m+1})^\omega$ with $t_1 \geq t_2 \geq \cdots \geq t_m > t_{m+1} > 0$. Then β is a Pisot number which satisfies the (PF) Property.*

Corollary 2.3.28 *Every quadratic Pisot number satisfies the (PF) Property.*

Example 2.3.29 The number $\theta = \frac{3+\sqrt{5}}{2}$, with $\mathsf{d}_\theta = 21^\omega$ satisfies the (PF) Property, but not the (F) Property, since $\langle \theta - 1 \rangle_\theta = 1.1^\omega$.

2.3.2.3 Digit-set conversion and normalisation

Let C be an arbitrary alphabet of digits. The *normalisation* $\nu_{\beta,C}$ in base β on C is the partial function which maps any β-representation on C of a given number of $[0,1)$ onto the greedy β-expansion of that number:

$$\nu_{\beta,C} : C^\mathbb{N} \to A_\beta^\mathbb{N} \qquad (c_i)_{i \geq 1} \mapsto \mathsf{d}_\beta(\sum_{i \geq 1} c_i \beta^{-i}).$$

The function $\nu_{\beta,C}$ is partial since as C may contain negative digits, a word of C^* may represent a negative number, which has no β-expansion. Note that, as for the integer bases, addition and multiplication by a positive integer constant K are particular instances of normalisation. Addition consists in normalising on the alphabet $\{0, \ldots, 2(\lceil \beta \rceil - 1)\}$, and multiplication by K on the alphabet $\{0, \ldots, K(\lceil \beta \rceil - 1)\}$.

We first adapt the notions of zero automaton and digit-conversion transducers given in Section 2.2.3 for integer base to the non-integer base β.

Zero automaton The evaluator $\mathcal{Z}_\beta$ in base β is defined as in integer base but for the set of states which is $\mathbb{Z}[\beta]$. The initial state is 0 and the transitions are of the form:

$$\forall s, t \in \mathbb{Z}[\beta] \qquad \forall a \in \mathbb{Z} \qquad s \xrightarrow[\mathcal{Z}_\beta]{a} t \quad \text{if, and only if,} \quad t = \beta s + a. \qquad (2.11)$$

Let $B_d = \{-d, \ldots, d\}$ where d is a positive integer, $d \geq \lfloor \beta \rfloor$.

Proposition 2.3.30 *An infinite word z in $B_d^{\mathbb{N}}$ has value 0 in base β if, and only if, it is accepted by the Büchi automaton $\mathcal{Z}_{\beta,d} = (Q_d, B_d, E, \{0\}, Q_d)$ where the transitions in E are those defined by (2.11) and $Q_d = \mathbb{Z}[\beta] \cap [-\frac{d}{\beta-1}, \frac{d}{\beta-1}]$.*

Proof By the definition of $\mathcal{Z}_{\beta,d}$, every infinite word z that labels an infinite computation in $\mathcal{Z}_{\beta,d}$ is accepted by $\mathcal{Z}_{\beta,d}$. For every $n \geq 1$, $|\pi_p(z_1 \cdots z_n)| \leq \frac{d}{\beta-1}$ and then $\pi_\beta(.z) = \lim_{n \to +\infty} \frac{1}{\beta^n} \pi_p(z_1 z_2 \cdots z_n) = 0$.

Conversely, let z in $B_d^{\mathbb{N}}$ which does not label a computation in $\mathcal{Z}_{\beta,d}$, that is, there exists a prefix w of z such that

$$0 \xrightarrow[\mathcal{Z}_{\beta,d}]{w} t \quad \text{with} \quad t > d/(\beta-1).$$

We have

$$\pi_p(.z) \geq \pi_p(.w\overrightarrow{d}^\omega) = \frac{1}{\beta^{|w|}}\left(t - \frac{d}{\beta-1}\right) > 0.$$

(The case $t < -d/(\beta-1)$ is identical.) □

This automaton is called the *zero automaton* in base β over the alphabet B_d. It is not finite in general. Our aim is now to prove the following result.

Theorem 2.3.31 *The following conditions are equivalent:*
(i) *the zero automaton $\mathcal{Z}_{\beta,d}$ is finite for every $d \geq \lfloor \beta \rfloor$*
(ii) *the zero automaton $\mathcal{Z}_{\beta,d}$ is finite for one $d \geq \lfloor \beta \rfloor + 1$*
(iii) *β is a Pisot number.*

The proof relies on the following statements.

Lemma 2.3.32 *If $\mathcal{Z}_{\beta,d}$ is finite, then β is an algebraic integer.*

Proof Let $\mathsf{d}_\beta(1) = (t_i)_{i \geq 1}$. Then $(-1)t_1 t_2 \cdots$ is the label of a path in $\mathcal{Z}_{\beta,d}$, and there exist n and p such that the states $\pi_\beta((-1)t_1 t_2 \cdots t_n)$ and $\pi_\beta((-1)t_1 t_2 \cdots t_n \cdots t_{n+p})$ are the same. □

We now suppose that β is an algebraic integer with minimal polynomial M_β of degree g. Denote $\beta_1 = \beta, \beta_2, \ldots, \beta_g$ the roots of M_β. On the discrete lattice of rank g, $\mathbb{Z}[X]/(M_\beta) \simeq \mathbb{Z}[\beta]$, a norm is defined as

$$\|P(X)\| = \max_{1 \leq i \leq g} |P(\beta_i)|. \tag{2.12}$$

Proposition 2.3.33 *If β is a Pisot number, then $\mathcal{Z}_{\beta,d}$ is finite for every $d \geq \lfloor \beta \rfloor$.*

Proof Every state s in Q_d is associated with the label of the shortest path $z_1 z_2 \cdots z_n$ from 0 to s in the automaton. Thus $s = s(\beta) = z_1 \beta^{n-1} + \cdots + z_n$, with $s(X)$ in $\mathbb{Z}[X]/(M_\beta)$ and $|s| = |s(\beta)| \leq \frac{d}{\beta-1}$. For every conjugate β_i with $|\beta_i| < 1$, we have $|s(\beta_i)| \leq \frac{d}{1-|\beta_i|}$. Since β is Pisot, this is true for $2 \leq i \leq g$. Thus every state of Q_d is bounded in norm, and so there is only a finite number of them. □

Example 2.3.34 The zero automaton on $\{-1, 0, 1\}$ for $\varphi = \frac{1+\sqrt{5}}{2}$ is drawn in Figure 2.15.

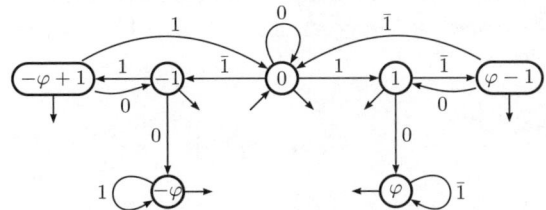

Fig. 2.15 Finite zero automaton $\mathcal{Z}_{\varphi,1}$, $\varphi = \frac{1+\sqrt{5}}{2}$.

See (Berend and Frougny 1994) for a proof that Part (i) implies (iii) of Theorem 2.3.31.

Proposition 2.3.35 *If the zero automaton $\mathcal{Z}_{\beta,d}$ is finite for every $d \geq \lfloor \beta \rfloor$, then β is a Pisot number.*

The core of Proposition 2.3.35 consists in using, with techniques of complex analysis, the following lemma *for every integer d*.

Lemma 2.3.36 *If the automaton $\mathcal{Z}_{\beta,d}$ is finite, then for every conjugate β_i with $|\beta_i| > 1$, if $s = s(\beta)$ belongs to Q_d then $|s(\beta_i)| \leq \frac{d}{|\beta_i|-1}$.*

Proof Let $z_1 z_2 \cdots$ be the label of a path recognised by $\mathcal{Z}_{\beta,d}$ with origin 0. Since Q_d is finite there exist n and p such that $s = s(\beta) = z_1 \beta^{n-1} + \cdots + z_n = z_1 \beta^{n+p-1} + \cdots + z_{n+p}$. Thus for every conjugate β_i with $|\beta_i| > 1$, $z_1 \beta_i^{n-1} + \cdots + z_n = z_1 \beta_i^{n+p-1} + \cdots + z_{n+p} = \beta_i^p (z_1 \beta_i^{n-1} + \cdots + z_n) + z_{n+1} \beta_i^{p-1} + \cdots + z_{n+p}$, thus

$$|z_1 \beta_i^{n-1} + \cdots + z_n| \leq \frac{d}{|\beta_i|^p - 1} \frac{|\beta_i|^p - 1}{|\beta_i| - 1} = \frac{d}{|\beta_i| - 1}.$$

□

Example 2.3.37 Take β the root > 1 of the polynomial $X^4 - 2X^3 - 2X^2 - 2$. Then $\mathsf{d}_\beta(1) = 2202$ and β is a simple Parry number, but it is not a Pisot number, since there is another root $\alpha \approx -1.134186$. By a direct computation it can be shown that the path of label $\bar{1}221\bar{1}1\bar{2}201$ in $\mathcal{Z}_{\beta,2}$ with origin 0 leads to a state $s = s(\beta)$ such that $s(\alpha) > 2/(|\alpha|-1)$. Lemma 2.3.36 implies that, for every $d \geq 2$, the set of words of $B_d^{\mathbb{N}}$ having value 0 is not recognisable by a finite automaton.

Normalisation Take two alphabets of integers C and A. Let $d = \max(|c - a|)$ for c in C and a in A, and let $B_d = \{-d, \ldots, d\}$ as above. As in Section 2.2.2.1, one constructs from the zero automaton $\mathcal{Z}_{\beta,d}$ a *digit-conversion transducer* or *converter* $\mathcal{C}_\beta(C \times A)$. The transitions are defined by

$$s \xrightarrow[\mathcal{C}_\beta(C \times A)]{c|a} t \quad \text{if, and only if,} \quad s \xrightarrow[\mathcal{Z}_{\beta,d}]{c-a} t.$$

Thus one obtains the following proposition.

Proposition 2.3.38 *The converter* $\mathcal{C}_\beta(C \times A)$ *recognises the set*

$$\{(x, y) \in C^{\mathbb{N}} \times A^{\mathbb{N}} \mid \pi_\beta(x) = \pi_\beta(y)\}.$$

If β is a Pisot number, then $\mathcal{C}_\beta(C \times A)$ is finite.

Theorem 2.3.39 (Frougny 1992) *If β is a Pisot number, then normalisation in base β on any alphabet C is realisable by a finite letter-to-letter transducer.*

Proof Since β is a Pisot number the automaton $\mathcal{D}_\beta$ recognising D_β is finite by Proposition 2.3.18. The normaliser $\mathcal{N}_\beta(C)$ is obtained as the composition of $\mathcal{C}_\beta(C \times A_\beta)$ with the transducer which realises the intersection with D_β. $\square$

It is easy to check that for any fixed digit alphabet C, normalisation in base β on C is a *bounded-length discrepancy function* (see Section 2.6.3). It follows then, that if normalisation in base β on an alphabet C is realisable by a finite transducer, it is realisable by a finite *letter-to-letter* transducer, and then that the *zero automaton* $\mathcal{Z}_{\beta,d}$ is finite for $d = \max(|c - a|)$, for c in C and a in A_β.

The following result allows us to prove that (ii) implies (i) in Theorem 2.3.31.

Proposition 2.3.40 (Frougny and Sakarovitch 1999) *If normalisation in base β on the alphabet $A'_\beta = \{0, \ldots, \lfloor\beta\rfloor, \lfloor\beta\rfloor + 1\}$ is realisable by a finite transducer, then normalisation in base β is realisable by a finite transducer on any alphabet.*

In view of Example 2.3.37, we set the following conjecture.

Conjecture 2.3.41 If the zero automaton $\mathcal{Z}_{\beta,d}$ is finite for $d = \lfloor\beta\rfloor$ then β is a Pisot number.

2.3.3 U-systems

We now consider another generalisation of the integer base numeration systems which only allows us to represent natural integers. The base is replaced by a *basis* which is an infinite sequence of positive integers (also called *scale*) and which plays the role of the sequence of the powers of the integer base. The classical example is the Fibonacci numeration system. These systems have been first defined and studied in full generality in (Fraenkel 1985).

We shall see that, under mild and natural hypotheses, the basis is associated with a real number β, as the Fibonacci numeration system is associated with the Golden Ratio. Then, many of the properties established for numeration in base β transfer to the U-system, but the situation is far more intricate. In fact, even in the simple case where the β is an integer, the language of the numeration system may or may not be a rational language according to the initial conditions (see Example 2.3.58).

2.3.3.1 Rationality of U-expansions

A *basis* is a strictly increasing sequence of integers $U = (u_n)_{n \geq 0}$ with $u_0 = 1$. A *representation in the system U* – or a *U-representation* – of a non-negative integer N is a finite sequence of integers $(d_i)_{k \geq i \geq 0}$ such that

$$N = \sum_{i=0}^{k} d_i u_i.$$

Such a representation will be written $d_k \cdots d_0$, most significant digit first.

Among all possible U-representations of a given non-negative integer N, one is distinguished and called the *U-expansion* of N. It is also called the *greedy U-representation*, since it can be obtained by the following greedy algorithm: given integers m and p let us denote by $q(m,p)$ and $r(m,p)$ the quotient and the remainder of the Euclidean division of m by p. Let $k \geq 0$ such that $u_k \leq N < u_{k+1}$ and let $d_k = q(N, u_k)$ and $r_k = r(N, u_k)$, and, for $i = k-1, \ldots, 0$, $d_i = q(r_{i+1}, u_i)$ and $r_i = r(r_{i+1}, u_i)$. Then $N = d_k u_k + \cdots + d_0 u_0$. The U-expansion of N is denoted by $\langle N \rangle_U$.

By convention the U-expansion of 0 is the empty word ε. Under the hypothesis that the ratio u_{n+1}/u_n is bounded by a constant as n tends to infinity, the digits of the U-expansion of any positive integer N are bounded and contained in a *canonical* finite alphabet A_U associated with U.

Example 2.3.42 Let $F = (F_n)_{n \geq 0}$ be the sequence of Fibonacci numbers, $F = \{1, 2, 3, 5, \ldots\}$. The canonical alphabet is equal to $\{0, 1\}$. The F-expansion of the number 11 is 10100, another F-representation is 10011.

The U-expansions are characterised by the following.

Lemma 2.3.43 *The word $d_k \cdots d_0$, where each d_i, for $k \geq i \geq 0$, is a non-negative integer and $d_k \neq 0$, is the U-expansion of some positive integer if, and only if, for each i, $d_i u_i + \cdots + d_0 u_0 < u_{i+1}$.*

Proposition 2.3.44 *The U-expansion of an integer is the greatest in the radix order of all the U-representations of that integer.*

Proof Let $v = d_k \cdots d_0$ be the greedy U-representation of N, and let $w = w_j \cdots w_0$ be another representation. Since $u_k \leq N < u_{k+1}$, then $k \geq j$. If $k > j$, then $v \succ w$. If $k = j$, suppose $v \prec w$. Thus there exist i, $k \geq i \geq 0$, such that $d_i < w_i$ and $d_k \cdots d_{i+1} = w_k \cdots w_{i+1}$. Hence $d_i u_i + \cdots + d_0 u_0 = w_i u_i + \cdots + w_0 u_0$, but $d_i u_i + \cdots + d_0 u_0 \leq (w_i - 1) u_i + d_{i-1} u_{i-1} + \cdots + d_0 u_0$, so $u_i + w_{i-1} u_{i-1} + \cdots + w_0 u_0 \leq d_{i-1} u_{i-1} + \cdots + d_0 u_0 < u_i$ since v is greedy, a contradiction. □

As for the beta-expansions, the order between integers is given by the radix order on their U-expansions.

Proposition 2.3.45 *Let M and N be two positive integers. Then $M < N$ if, and only if, $\langle M \rangle_U \prec \langle N \rangle_U$.*

The set of U-expansions of all the non-negative integers is denoted by $L(U)$.

Example 2.3.46 Let F be the sequence of Fibonacci numbers. Then $L(F)$ is the set of words without the factor 11, and not beginning with a 0:

$$L(F) = 1\{0,1\}^* \setminus \{0,1\}^* 11 \{0,1\}^* \cup \{\varepsilon\}.$$

When the sequence U satisfies a linear recurrence with integral coefficients, that is, when U is a linear recurrent sequence, we say that U defines a *linear numeration system* or that U is a *linear recurrent basis*.

Proposition 2.3.47 (Shallit 1994) *Let U be a basis. If $L(U)$ is a rational language, then U is a linear recurrent sequence.*

Proof (Loraud 1995) Let ℓ_n (resp. k_n) be the number of words of length n in $L(U)$ (resp. in $0^*L(U)$). Since a word in $L(U)$ does not begin with a 0, we have $k_n = \ell_0 + \ell_1 + \cdots + \ell_n$ for every n and then $k_n = u_n$ by Lemma 2.3.43. If $L(U)$ is a rational language, so is $0^*L(U)$ and $U = (u_n)_{n \geq 0}$ is a linear recurrent sequence, a classical result in automata theory (see Theorem 2.6.2). $\square$

The results on β-expansions transfer to the U-expansions when U satisfies some conditions. The results below were established in (Hollander 1998). A linear recurrent basis $U = (u_n)_{n \geq 0}$ is said to satisfy the *dominant root condition* if $\lim_{n \to \infty} u_{n+1}/u_n = \beta$ for some $\beta > 1$.

Lemma 2.3.48 *Let U be a linear recurrent basis, with characteristic polynomial $C_U(X)$. Assume that $C_U(X)$ has a unique root β, possibly with multiplicity, of maximum modulus, and assume that β is real. Then U satisfies the dominant root condition for β.*

For a language L, we denote by $\mathsf{Maxlg}\,(L)$ the set of words of L which have no greater word of the same length in L in the radix order. It is known that if L is rational, so is $\mathsf{Maxlg}\,(L)$ (Proposition 2.6.4). The following is also a classical result of automata theory (see Proposition 2.6.3).

Lemma 2.3.49 *Let M be a language which contains exactly one word of every length. If M is rational, then there exist an integer p, a finite family of words x_i, y_i and z_i, with $|y_i| = p$, and a finite set of words M_0 such that*

$$M = \bigcup_{i=1}^{i=p} x_i y_i^* z_i \cup M_0 \qquad (2.13)$$

where the union is disjoint.

For every n in $\mathbb{N}$, let m_n be the word of length n of $L(U)$ which is maximum in the radix order : $m_n = \langle u_n - 1 \rangle_U$, and $\mathsf{Maxlg}\,(L(U)) = \cup_{n \geq 0} m_n$. Note that the empty word $\varepsilon = m_0$ belongs to M. The following result is similar to the lexicographical characterisation of the β-shift given by Parry, see Theorem 2.3.11.

Proposition 2.3.50 *The following holds:*

$$L(U) = \cup_{n \geq 0} \{v \in A_U^n \mid \text{every suffix of length } i \leq n \text{ of } v \text{ is } \preceq m_i\}.$$

Using the previous result, one can construct a finite automaton similar to the one defined for the β-shift.

Proposition 2.3.51 *If* $\mathsf{Maxlg}\,(L(U))$ *is rational, so is* $L(U)$.

The following lemma shows that the β-expansion of 1 governs the U-expansions when β is the dominant root of U.

Lemma 2.3.52 *Suppose that U has a dominant root β, and let $\mathsf{d}_\beta(1) = (t_n)_{n \geq 1}$. Then for each j there exist n and a word w_j of length $n-j$ such that $m_n = \langle u_n - 1 \rangle_U = t_1 \cdots t_j w_j$.*

Proposition 2.3.53 (Hollander 1998) *Let U be a linear recurrent basis with dominant root β. If $L(U)$ is rational then β is a Parry number.*

Proof [Sketch] If $L(U)$ is rational, then $\mathsf{Maxlg}\,(L(U))$ is of the form (2.13). By Lemma 2.3.52, for each j, there exist an n and a word w_j of length $n-j$ such that $m_n = t_1 \cdots t_j w_j$. Combining the two properties, it follows that $\mathsf{d}_\beta(1)$ must be finite or eventually periodic. □

From now on β is a Parry number. In this case, there is a polynomial satisfied by β which arises from the greedy expansion of 1. If $\mathsf{d}_\beta(1)$ is finite, $\mathsf{d}_\beta(1) = t_1 \cdots t_m$, then set

$$G_\beta(X) = X^m - \sum_{i=1}^m t_i X^{m-i}.$$

If $\mathsf{d}_\beta(1)$ is infinite eventually periodic, $\mathsf{d}_\beta(1) = t_1 \cdots t_m (t_{m+1} \cdots t_{m+p})^\omega$, with m and p minimal, then set

$$G_\beta(X) = X^{m+p} - \sum_{i=1}^{m+p} t_i X^{m+p-i} - X^m + \sum_{i=1}^m t_i X^{m-i}.$$

Such a polynomial is called the *canonical beta-polynomial* for β. Note that in general G_β is not equal to the minimal polynomial of β but is a multiple of it.

Example 2.3.54 Let η be the root > 1 of the polynomial $M_\eta = X^3 - X - 1$. This number is the smallest Pisot number. Since $\mathsf{d}_\eta(1) = 10001$, the canonical beta-polynomial is $G_\eta = X^5 - X^4 - 1$.

Number representation and finite automata

We will need a slightly more general definition. If $\mathsf{d}_\beta(1)$ is infinite eventually periodic, $\mathsf{d}_\beta(1) = t_1 \cdots t_m (t_{m+1} \cdots t_{m+p})^\omega$, with m and p minimal, set $r = p$. If $\mathsf{d}_\beta(1)$ is finite, $\mathsf{d}_\beta(1) = t_1 \cdots t_m$, then set $r = m$. An *extended beta-polynomial* is a polynomial of the form

$$H_\beta(X) = G_\beta(X)(1 + X^r + \cdots + X^{rk})X^n$$

for k in $\mathbb{N}$ and n in $\mathbb{N}$.

When $\mathsf{d}_\beta(1)$ is infinite an extended beta-polynomial corresponds to taking m and p not minimal. When $\mathsf{d}_\beta(1)$ is finite an extended beta-polynomial corresponds to taking improper expansions of 1 of the form $(t_1 \cdots t_{m-1}(t_m - 1))^k t_1 \cdots t_m$, and to any writing of $\mathsf{d}_\beta^*(1)$ as uv^ω.

Example 2.3.55 The canonical beta-polynomial for the Golden Ratio is $G_\varphi = X^2 - X - 1$. The polynomial $X^4 - X^3 - X - 1 = G_\varphi(1 + X^2)$ is an extended beta-polynomial corresponding to the improper expansion 1011 of 1.

Lemma 2.3.56 *Let $H_\beta(X)$ be an extended polynomial for $\beta > 1$, and assume that $H_\beta(X) = C_U(X)$. Then U satisfies the dominant root condition for β, and β is a simple root of $H_\beta(X)$.*

The following theorem shows that the situation for linear numeration systems is much more complicated than for the β-shift.

Theorem 2.3.57 (Hollander 1998) *Let U be a linear recurrent basis whose dominant root β is a Parry number.*

- *If $\mathsf{d}_\beta(1)$ is infinite eventually periodic, then $L(U)$ is rational if, and only if, U satisfies an extended beta-polynomial for β.*
- *If $\mathsf{d}_\beta(1)$ is finite, of length m, then: if U satisfies an extended beta-polynomial for β then $L(U)$ is rational; and conversely if $L(U)$ is rational, then U satisfies either an extended beta-polynomial for β, $H_\beta(X)$, or a polynomial of the form $(X^m - 1)H_\beta(X)$.*

In the finite case, rationality indeed depends on initial conditions.

Example 2.3.58 (Hollander 1998) Take $u_n = 4u_{n-1} - 3u_{n-2}$ with $C_U(X) = (X-1)(X-3)$. The dominant root is $\beta = 3$.
Take $u_0 = 1$ and $u_1 = 4$. Then $u_n = 3u_{n-1} + 1$, and so the language of maximal words is $M = 30^*$, and $L(U)$ is rational.
Take $u_0 = 1$ and $u_1 = 2$. Then $u_n = 3u_{n-1} - 1 = (3^n + 1)/2$, and

$A_U = \{0,1,2\}$. Let k be the largest integer such that m_n begins with k digits 2. Thus k is the largest integer such that

$$\frac{3^n+1}{2} > 2(\frac{3^{n-1}+1}{2} + \cdots + \frac{3^{n-k}+1}{2})$$

that is, $3^{n-k}+1 > 2n$, and $3^{n-k}+1+2(n-k) > 2n$. As $n \to \infty$, both $k \to \infty$ and $n-k \to \infty$, and $L(U)$ is not rational.

We now define a numeration system canonically associated with a real number β in a way that gives the numeration system the same dynamical properties as the β-shift.

Definition 2.3.59 The *numeration system associated with β* is defined by the basis $U_\beta = (u_n)_{n \geq 0}$ as follows:
If $\mathsf{d}_\beta(1)$ is finite, $\mathsf{d}_\beta(1) = t_1 \cdots t_m$, set

$$u_n = t_1 u_{n-1} + \cdots + t_m u_{n-m} \text{ for } n \geq m,$$

$u_0 = 1$, and for $1 \leq i \leq m-1$, $u_i = t_1 u_{i-1} + \cdots + t_i u_0 + 1$.
If $\mathsf{d}_\beta(1) = (t_i)_{i \geq 1}$ is infinite, set

$$u_n = t_1 u_{n-1} + t_2 u_{n-2} + \cdots + t_n u_0 + 1, \text{ for } n \geq 1, \ u_0 = 1.$$

If $\mathsf{d}_\beta(1)$ is finite or eventually periodic, the sequence U_β is linearly recurrent, and its characteristic polynomial is thus the canonical beta-polynomial of β.

Example 2.3.60 The linear numeration system associated with the Golden Ratio is the Fibonacci numeration system.

Proposition 2.3.61 (Bertrand-Mathis 1989) *Let $\beta > 1$ be a real number. Then $L(U_\beta) = L(S_\beta)$.*

Example 2.3.62 Take the Pisot number $\theta = \frac{3+\sqrt{5}}{2}$, then $\mathsf{d}_\theta(1) = 21^\omega$, and $U_\theta = \{1,3,8,21,55,144,377,\ldots\}$ is the sequence of Fibonacci numbers of even index. The beta-polynomial $G_\theta(X) = X^2 - 3X + 1$ is equal to the minimal polynomial of θ. The set $L(U_\theta)$ is recognisable by the finite automaton of Figure 2.13 above, which recognises the θ-shift.

On the other hand, consider the linear recurrent basis $R_\theta = \{1,2,6,17,46,122,321,\ldots\}$ defined by $r_n = 4r_{n-1} - 4r_{n-2} + r_{n-3}$ for $n \geq 3$, $r_0 = 1$, $r_1 = 2$, $r_2 = 6$. Then θ is the dominant root of R_θ; the characteristic polynomial of R_θ is equal to $(X-1)(X^2-3X+1)$. By showing that R_θ does not satisfy an extended beta-polynomial, Theorem 2.3.57 implies that the set $L(R_\theta)$ is not recognisable by a finite automaton. A direct combinatorial proof can be found in (Frougny 2002).

2.3.3.2 Normalisation

Let C be an arbitrary alphabet of digits. The *normalisation* $\nu_{U,C}$ in basis U on C is the partial function which maps any U-representation on C of any positive integer n onto the U-expansion of n:

$$\nu_{U,C}: C^* \to A_U^* \qquad c_k \cdots c_0 \mapsto \langle \sum_{i=0}^{k} c_i u_i \rangle_U.$$

As for beta-expansions, one can define the zero automaton and the converter for a U-system. Let us say that U is a *Pisot basis* if U is a linear recurrent basis whose characteristic polynomial is the minimal polynomial of a Pisot number. It follows from Theorem 2.3.57 that if U is a Pisot basis, then $L(U)$ is rational.

Proposition 2.3.63 *Let U be a Pisot basis. Then, for any alphabet of digits, the zero automaton and the converter in the system U are finite.*

Example 2.3.64 The zero automaton in the Fibonacci numeration system and for the alphabet $\{-1, 0, 1\}$ is the automaton of Figure 2.15, without the states labelled φ and $-\varphi$, and with 0 as unique final state.

By a similar construction to the one exposed in Section 2.3.2.3, we obtain the following result.

Proposition 2.3.65 (Frougny and Solomyak 1996) *Let U be a Pisot basis. For any digit alphabet C, the normalisation $\nu_{U,C}$ is realisable by a finite letter-to-letter transducer.*

Example 2.3.66 Normalisation on $\{0, 1\}$ in the Fibonacci numeration system consists in replacing every factor 011 by 100. The finite transducer realising normalisation is shown in Figure 2.16. For sake of simplicity, this normaliser does not accept words which begin with 11.

2.3.3.3 Successor function

The successor function is usually and canonically defined on $\mathbb{N}$: $n \mapsto n+1$. What we call 'successor function' here is of course the same function, but lifted at the level of expansions in the system we consider. Successor function is a special case of addition and thus of normalisation. When the latter

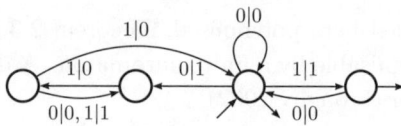

Fig. 2.16 Finite normaliser on $\{0,1\}$ for the Fibonacci numeration system.

is a rational function, or even realised by a letter-to-letter (right sequential) transducer, so is the successor function, without any ado. But this successor function is such a special case that we can give statements under weaker hypotheses than the ones that assure the rationality of normalisation.

Let U be a basis and, as above, $L = L(U)$ the set of U-expansions. The successor function in the basis U is thus the function Succ_L which maps every word of L onto its successor in L in the radix order. If L is a rational language, it is thus known that Succ_L is a synchronous relation, even a (left and right) letter-to-letter rational relation, even a piecewise right sequential function (see Proposition 2.6.7, Corollary 2.6.11, Proposition 2.6.14 in Section 2.6). From Proposition 2.3.61 above, we then have the following consequence of these results.

Proposition 2.3.67 *Let β be a Parry number and U_β the linear numeration system associated with β. The successor function in the numeration system U_β is realisable by a letter-to-letter transducer.*

In general, the successor function in a linear numeration system is *not co-sequential*, as shown by the next example.

Example 2.3.68 Take the Pisot number $\theta = \frac{3+\sqrt{5}}{2}$, see Example 2.3.62. By the foregoing, $L(U_\theta)$ is rational, and $\mathrm{Succ}_{L(U_\theta)}$ is realisable by a finite transducer. For every n, the words $v_n = 021^n$ and $w_n = 01^{n+1}$ are in $L(U_\theta)$. We have $\mathrm{Succ}_{L(U_\theta)}(v_n) = 10^{n+1}$ and $\mathrm{Succ}_{L(U_\theta)}(w_n) = 01^n 2$.

The *suffix distance* $\mathrm{d}_s(x,y)$ of two words x and y is

$$\mathrm{d}_s(x,y) = |x| + |y| - 2\,|x \wedge_s y|$$

where $x \wedge_s y$ is the *longest common suffix* of x and y. Hence we have $\mathrm{d}_s(v_n, w_n) = 4$ and $\mathrm{d}_s\bigl(\mathrm{Succ}_{L(U_\theta)}(v_n), \mathrm{Succ}_{L(U_\theta)}(w_n)\bigr) = 2(n+2)$. By the characterisation of co-sequential functions due to Choffrut (see Theorem 2.6.13), $\mathrm{Succ}_{L(U_\theta)}$ is not co-sequential.

The conditions under which the successor function in a linear numeration system is co-sequential are indeed completely determined.

Theorem 2.3.69 (Frougny 1997) *Let U be a numeration system such that $L(U)$ is rational. The successor function in the system U is co-sequential if, and only if, the set $\mathsf{Maxlg}\,(L(U))$ is of the form*

$$\mathsf{Maxlg}\,(L(U)) = \bigcup_{i=1}^{i=p} y_i^* z_i \cup M_0 \qquad (2.14)$$

where M_0 is finite, $|y_i| = p$ and the union is disjoint.

In the case of linear numeration system with dominant root, the previous result can be refined.

Theorem 2.3.70 *Let U be a linear recurrent basis whose dominant root β is a Parry number. Then the successor function in the system U is co-sequential if, and only if, the following conditions hold:*
(i) *β is a simple Parry number;*
(ii) *U satisfies the canonical beta-polynomial for β.*

Example 2.3.71 The successor function in the Fibonacci numeration system is realised by a finite right sequential transducer, see Figure 2.17.

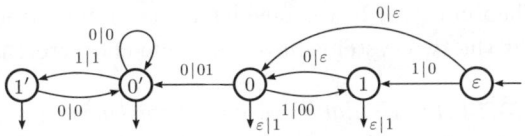

Fig. 2.17 Finite right sequential transducer for the successor function in the Fibonacci numeration system.

2.3.4 Base changing

As far as comparison, or conversion, of the expansions of numbers in different real bases is concerned, the situation is very similar to the one with integer bases. In the background, and for the negative part, stand the generalisations of Cobham's Theorem – we choose the one due to Bès. If U is a basis, a set of natural integers is said to be U-*recognisable* if the set of the U-expansions of its elements is a rational set.

Theorem 2.3.72 (Bès 2000) *Let U and V be two Pisot basis, associated with two multiplicatively independent Pisot numbers. A set X of positive integers is both U- and V-recognisable if, and only if, it is recognisable.*

From this result follows that the conversion between the expansions in two such linear numeration systems U and V cannot be realised by a finite transducer.

2.3.4.1 Multiplicatively dependent bases

We now consider the case where the bases β and γ are multiplicatively dependent. When β and γ are integers, then the conversion from base β to base γ is realisable by a finite right sequential transducer (Proposition 2.2.17).

Proposition 2.3.73 *Let β and γ be two multiplicatively dependent Pisot numbers. The conversion from base γ to base β is realisable by a finite transducer.*

Proof Set $\delta = \beta^k = \gamma^\ell$ and let $(x_i)_{i \geq 1} = \mathsf{d}_\delta(x)$ where x is in $[0, 1)$. Then $0^{k-1} x_1 0^{k-1} x_2 0^{k-1} \cdots$ is a β-representation of x on the alphabet A_δ. Since β is Pisot, normalisation in base β on the alphabet A_δ is realisable by a finite transducer. Similarly the conversion from base δ to base γ is realisable by a finite transducer. By composition and inversion of relations realised by finite transducers, the result follows (see Section 2.6). □

Now, as in Theorem 2.3.72, we consider linear numeration systems. We first suppose that the two systems have the same characteristic polynomial.

Proposition 2.3.74 *Let U and V be two Pisot basis, associated with the same Pisot number, but defined by different initial conditions. The conversion from a V-representation of a positive integer to the U-expansion of that integer is computable by a finite transducer.*

Proof By Proposition 2.3.65 normalisation in the system U is computable by a finite transducer on any alphabet. Suppose that M_β, the minimal polynomial of β, has degree g. The family $\{u_n, u_{n+1}, \ldots, u_{n+g-1} \mid n \geq 0\}$ is free, because its annihilator polynomial is M_β. Since U and V have the same characteristic polynomial, it is known from standard results of linear algebra that there exist rational constants λ_i such that, for each $n \geq 0$, $v_n = \lambda_1 u_{n+g-1} + \cdots + \lambda_g u_n$. One can assume that the λ_i's are all of the form p_i/q where the p_i's belong to $\mathbb{Z}$ and q belongs to $\mathbb{N}$, $q \neq 0$. Let N be a positive integer and consider a V-representation $c_j \cdots c_0$ of N, where the c_i's are in an alphabet of digits $B \supseteq A_V$. Then $qN = c_j q v_j + \cdots + c_0 q v_0$. Since for $n \geq 0$, $qv_n = p_1 u_{n+g-1} + \cdots + p_g u_n$, we get that qN is of the form $qN = d_{j+g-1} u_{j+g-1} + \cdots + d_0 u_0$. Since each digit d_i, for $0 \leq i \leq j+g-1$, is a linear combination of q, p_1, $\ldots$, p_g, and the c_i's, we get that d_i is an element of a finite alphabet of digits $D \supset A_U$. By assumption, $\nu_{U,D}$ is computable by a finite automaton. It remains to show that the function which maps $\nu_{U,D}(d_{j+g-1}\cdots d_0) = <qN>_U$ onto $<N>_U$ is computable by a finite automaton, and this is due to the fact that it is the inverse of the multiplication by the natural q, which is computable by a finite automaton in the system U. □

Definition 2.3.75 Let β be a Pisot number of degree g, and denote $\beta_1 = \beta, \ldots, \beta_g$ the roots of the minimal polynomial M_β. The *Lucas-like numeration system* associated with β is the system defined by the basis $V_\beta = (v_n)_{n \geq 0}$ where

$$v_0 = 1, \text{ and for } n \geq 1, \ v_n = \beta_1^n + \cdots + \beta_g^n.$$

The characteristic polynomial of V_β is equal to M_β.

This terminology comes from the fact that for the Golden Ratio φ, V_φ is the sequence of Lucas numbers. On the other hand, the numeration system U_β associated with β in Definition 2.3.59 is a *Fibonacci-like* numeration system, since, for the Golden Ratio φ, U_φ is the sequence of Fibonacci numbers.

Proposition 2.3.76 *Let β be a Pisot number, and let $\delta = \beta^k$. The conversion from the Lucas-like numeration system V_δ to the the Lucas-like numeration system V_β is realisable by a finite transducer.*

Proof The conjugates of δ are of the form $\delta_i = \beta_i^k$, for $2 \leq i \leq g$. Set $V_\delta = (w_n)_{n \geq 0}$ with $w_n = \delta_1^n + \cdots + \delta_m^n$ for $n \geq 1$. For $n \geq 1$, $w_n = v_{kn}$. Thus any V_δ-representation of an integer N of the form $d_j \cdots d_0$ gives a V_β-representation of N of the form $d_j 0^{k-1} d_{j-1} 0^{k-1} \cdots d_1 0^{k-1} d_0$. Since the normalisation in the system V_β is computable by a finite transducer on any alphabet by Proposition 2.3.65, the result follows. □

Theorem 2.3.77 (Frougny 2002) *Let U and V be two Pisot basis, associated with two multiplicatively dependent Pisot numbers. Then the conversion from a V-representation of a positive integer to the U-expansion of that integer is computable by a finite transducer.*

Proof Set $\delta = \beta^k = \gamma^\ell$. As in Proposition 2.3.76, the conversion from the Lucas-like numeration system V_δ to the the Lucas-like numeration system V_γ is realisable by a finite transducer. By Proposition 2.3.74, the conversion from V to V_γ and the conversion from V_β to U are realisable by a finite transducer, and the result follows. □

Corollary 2.3.78 *Let U and V be two Pisot basis, associated with two multiplicatively dependent Pisot numbers. Then the U-recognisable sets and V-recognisable sets of natural integers coincide.*

2.3.4.2 Base β and U_β numeration system

When β is an integer, β-expansions and U_β-expansions of the positive integers are the same. There is a particular case of Pisot numbers for which the conversion from base β to the U_β numeration system is realisable by means of a finite transducer.

Let us take the example of the Golden Ratio φ and the Fibonacci numeration system. By Theorem 2.3.25, φ satisfies the (F) Property, so the greedy φ-expansion of every positive integer is finite. In this section we want to answer the following questions. Does there exist a characterisation of the greedy φ-expansions of the positive integers? Is there any relation between the greedy φ-expansion of a positive integer and its greedy representation in the Fibonacci system? Table 2.1 below gives the φ-expansion of the first 10 integers together with their Fibonacci greedy representation.

N	Fibonacci representations	φ-expansions	Folded φ-expansions
1	1	1.	$\begin{smallmatrix}1\\0\end{smallmatrix}$
2	10	10.01	$\begin{smallmatrix}1&0\\1&0\end{smallmatrix}$
3	100	100.01	$\begin{smallmatrix}1&0&0\\0&1&0\end{smallmatrix}$
4	101	101.01	$\begin{smallmatrix}1&0&1\\0&1&0\end{smallmatrix}$
5	1000	1000.1001	$\begin{smallmatrix}1&0&0&0\\1&0&0&1\end{smallmatrix}$
6	1001	1010.0001	$\begin{smallmatrix}1&0&1&0\\1&0&0&0\end{smallmatrix}$
7	1010	10000.0001	$\begin{smallmatrix}1&0&0&0&0\\0&1&0&0&0\end{smallmatrix}$
8	10000	10001.0001	$\begin{smallmatrix}1&0&0&0&1\\0&1&0&0&0\end{smallmatrix}$
9	10001	10010.0101	$\begin{smallmatrix}1&0&0&1&0\\0&1&0&1&0\end{smallmatrix}$
10	10010	10100.0101	$\begin{smallmatrix}1&0&1&0&0\\0&1&0&1&0\end{smallmatrix}$

Table 2.1. *Fibonacci expansions, φ-expansions, and folded φ-expansions of the 10 first integers.*

In fact the results are not only valid for the Golden Ratio, but for the larger class of *quadratic Pisot units*. A quadratic Pisot unit is an algebraic number whose minimal polynomial is of the form $X^2 - rX - 1$ with $r \geq 1$ or $X^2 - rX + 1$ with $r \geq 3$. By Corollary 2.3.28, every quadratic Pisot number satisfies the (PF) Property, and thus the expansion of every positive integer is finite. If the β-expansion of a positive integer n is of the form $u.v$, by padding the shortest word by 0's one can suppose that they have the same length. The *folded β-expansion* of n is the couple $\binom{u}{\tilde{v}}$, where $\tilde{v}$ is the mirror image of v.

Theorem 2.3.79 (Frougny and Sakarovitch 1999) *Let β be a quadratic Pisot unit. There exists a letter-to-letter finite transducer that maps the U_β-representation of any positive integer onto its folded β-expansion.*

Since the image of a function computable by a finite letter-to-letter transducer is a rational language, it then follows immediately from Theorem 2.3.79 that we have:

Corollary 2.3.80 *Let β be a quadratic Pisot unit. The set of folded β-expansions of all the non-negative integers is a rational language.*

By a result of (Rosenberg 1967) it follows thus that the set of β-expansions of all the non-negative integers is a *linear context-free language*. The following result shows that only quadratic Pisot units enjoy this property.

Theorem 2.3.81 (Frougny and Solomyak 1999) *Let $\beta > 1$ be a non-integral real number such that the β-expansion of every non-negative integer is finite. Let $R_\beta \subset A_\beta^* . A_\beta^*$ be the set of β-expansions of all the non-negative integers. If R_β is a context-free language, then β must be a quadratic Pisot unit.*

2.4 Canonical numeration systems

In this section we present another generalisation of the integer base number system, in which the expansion of a number is given by a right-to-left algorithm. The canonical numeration systems have been extensively studied, and we refer the reader to (Scheicher and Thuswaldner 2004), (Akiyama and Rao 2005), (Brunotte, Huszti, and Pethő 2006) for some recent contributions, and (Barat, Berthé, Liardet, et al. 2006) for a survey.

We also present briefly a new concept, the shift radix systems, which is a generalisation of both the Pisot base and the canonical numeration systems.

2.4.1 Canonical numeration systems in algebraic number fields

The elements of this section are taken in particular from (Gilbert 1981, Gilbert 1991, Kátai and Kovács 1981).

Let β be an algebraic integer of modulus > 1, and let A be a finite set of elements of $\mathbb{Z}[\beta]$ containing zero.

Definition 2.4.1 The pair (β, A) is a *canonical numeration system* (CNS for short) if every element z of $\mathbb{Z}[\beta]$ has a unique integer representation $d_k \cdots d_0$ with d_j in A, $d_k \neq 0$, that we denote $\langle z \rangle_\beta = d_k \cdots d_0$, and such that $z = \pi_\beta(d_k \cdots d_0) = \sum_{j=0}^{k} d_j \beta^j$.

Example 2.4.2
- The negative integer base $\beta = -b$, with $b \geq 2$, forms a CNS with the alphabet $\{0, \ldots, b-1\}$, see (Grünwald 1885).
- Base $\beta = 3$ with the alphabet $\{-1, 0, 1\}$ forms a CNS, see (Knuth 1998).
- The Penney numeration system with base $\beta = -1 \pm i$ and digit set $\{0, 1\}$ forms a CNS, see (Penney 1964).

Let $M_\beta(X) = X^g + b_{g-1}X^{g-1} + \cdots + b_0$ be the minimal polynomial of β. The *norm* of β is $N(\beta) = |b_0|$. A set $R \subset \mathbb{Z}[\beta]$ is a *complete residue system* for $\mathbb{Z}[\beta]$ modulo β if every element of $\mathbb{Z}[\beta]$ is congruent modulo β to a unique element of R.

It is classical (Theorem of Sylvester) that a complete residue system of elements of $\mathbb{Z}[\beta]$ modulo β contains $N(\beta)$ elements, for instance the set $A_\beta = \{0, \ldots, N(\beta) - 1\}$.

Proposition 2.4.3 *Suppose that every element of $\mathbb{Z}[\beta]$ has a finite integer representation in the CNS (β, A). Then this representation is unique if, and only if, A is a complete residue system for $\mathbb{Z}[\beta]$ modulo β, that contains zero.*

Proof Suppose that the representation of $z \in \mathbb{Z}[\beta]$ is $d_k \cdots d_0$. Then $z \sim d_0$ $(\bmod\ \beta)$, thus A must contain a complete residue system modulo β.

Now suppose that two digits c and d of A are congruent modulo β. Then $c - d = e\beta$ for some e in $\mathbb{Z}[\beta]$. Let $\langle e \rangle_\beta = e_k \cdots e_0$. Then $c = e\beta + d$, so c has two representations, c itself, and $e_k \cdots e_0 d$.

Conversely, suppose that there exists $z \in \mathbb{Z}[\beta]$ with two different representations, $d_k \cdots d_0$ and $c_\ell \cdots c_0$. One can suppose that $k \geq \ell$, and set $c_j = 0$ for $\ell + 1 \leq j \leq k$. Then the polynomial $(d_k - c_k)X^k + \cdots + (d_0 - c_0)$ vanishes at $X = \beta$, and it is thus divisible by the minimal polynomial $M_\beta(X)$. Contradiction, since $|d_0 - c_0| < N(\beta)$. □

Given β and A a complete residue system, a word $d_k \cdots d_0$ with d_j in

A is a representation of $z \in \mathbb{Z}[\beta]$ if $d_0 \sim z \pmod{\beta}$ and $d_k \cdots d_1$ is the representation of $(z - d_0)/\beta$. Thus we define

$$\Phi_\beta : \mathbb{Z}[\beta] \to \mathbb{Z}[\beta] \qquad (2.15)$$
$$z \mapsto \frac{z-d}{\beta} \text{ with } d \sim z \pmod{\beta}.$$

The digits d_j in the representation of z are given by $d_j = \Phi_\beta^j(z) \pmod{\beta}$. Thus the representation of z in the system (β, A) is finite if, and only if, the iterates $\Phi_\beta^j(z)$, $j \geq 0$, eventually reach 0.

Remark that all words of A^* are admissible.

Proposition 2.4.4 *If (β, A) is a canonical numeration system then*
(i) *β and all its conjugates have moduli greater than 1*
(ii) *β has no positive real conjugate.*

Proof (i) Suppose that there is a conjugate β_i with $|\beta_i| < 1$. Let z be in $\mathbb{Z}[\beta]$ with $\langle z \rangle_\beta = d_k \cdots d_0$, d_j in A. Let $z_i = \sum_{j=0}^{k} d_j \beta_i^j$. Set $m_A = \max(|a|, a \in A)$. Then $|z_i| < m_A/(1 - |\beta_i|)$, and so there exist elements in $\mathbb{Z}[\beta]$ with no representation in (β, A).
(ii) Let β_i be a conjugate of β which is real and positive. Suppose -1 could be represented in the system as $-1 = \sum_{j=0}^{k} d_j \beta^j$. Then $-1 = \sum_{j=0}^{k} d_j \beta_i^j$, which is impossible. $\square$

Note that (ii) implies that if (β, A) is a CNS then the constant term of the minimal polynomial is positive.

An element z in $\mathbb{Q}(\beta)$ has a representation $\langle z \rangle_\beta = d_k \cdots d_0 . d_{-1} d_{-2} \cdots$ in the CNS (β, A) if $z = \sum_{i=-\infty}^{k} d_i \beta^i$ with d_i in A. The following result is similar to the results in integer and non-integer real base, see (Gilbert 1981, Gilbert 1991).

Proposition 2.4.5 *If (β, A) is a canonical numeration system then every element of the field $\mathbb{Q}(\beta)$ has an eventually periodic representation in (β, A).*

2.4.2 Normalisation in canonical numeration systems

The results presented in this section primarily appeared in (Grabner, Kirschenhofer, and Prodinger 1998), (Thuswaldner 1998) and (Safer 1998), (Scheicher and Thuswaldner 2004).

Let (β, A) be a canonical numeration system. Let $C \supset A$ be a finite alphabet of digits in $\mathbb{Z}[\beta]$. The normalization on C in the system (β, A) is the function

$$\nu_{\beta,C} : C^* \longrightarrow A^* \qquad c_k \cdots c_0 \longmapsto \langle \sum_{j=0}^{k} c_j \beta^j \rangle_\beta.$$

As in the previous sections, we define the zero automaton, on a finite symmetric alphabet D of digits in $\mathbb{Z}[\beta]$, that contains A. The *zero automaton* $\mathcal{Z}_{\beta,D}$ on D is defined as follows: $\mathcal{Z}_{\beta,D} = (\mathbb{Z}[\beta], D, E, \{0\}, \{0\})$ where the transitions in E are defined by

$$\forall s,t \in \mathbb{Z}[\beta], \qquad \forall a \in D, \quad s \xrightarrow[\mathcal{Z}_{\beta,D}]{a} t \quad \text{if, and only if,} \quad t = \beta s + a. \quad (2.16)$$

This automaton accepts the writings of 0 in base β on the alphabet D. Let $m_D = \max\{|a| \mid a \in D\}$ and let $Q_D = \{s \in \mathbb{Z}[\beta] \mid |s| \leq \frac{m_D}{|\beta|-1}\}$.

Proposition 2.4.6 *The* trim *part of $\mathcal{Z}_{\beta,D}$ contains only states belonging to Q_D.*

Proof As D contains A and is symmetrical, every element of $\mathbb{Z}[\beta]$ is accessible in $\mathcal{Z}_{\beta,D}$.

Suppose that $e_k \cdots e_0$ is a word of D^* such that $\sum_{j=0}^{k} e_j \beta^j = 0$. Then, for $1 \leq j \leq k$, $s_j = \beta^{j-1} e_k + \cdots + e_{k-j+1} = -\beta^{-j+1}(\beta^{j-2} e_{j-2} + \cdots + e_0)$, thus $|s_j| < \frac{m_D}{|\beta|-1}$, and $e_k \cdots e_0$ is the label of a path

$$0 \xrightarrow{e_k} s_1 \xrightarrow{e_{k-1}} \cdots s_k \xrightarrow{e_0} s_{k+1} = 0$$

in $\mathcal{Z}_{\beta,D}$ with all the states in Q_D. □

Lemma 2.4.7 *If β and all its conjugates have moduli greater than 1 then for every finite alphabet D the zero automaton $\mathcal{Z}_{\beta,D}$ is finite.*

Proof Recall that the norm defined on $\mathbb{Z}[\beta] \simeq \mathbb{Z}[X]/(M_\beta)$ is defined by $||P(X)|| = \max_{1 \leq i \leq g} |P(\beta_i)|$, see (2.12). Let $s = s(\beta)$ be in Q_D. Then for $1 \leq i \leq g$, $|s(\beta_i)| < \frac{m_D}{|\beta_i|-1}$. Since the elements of Q_D are bounded in norm in the discrete lattice $\mathbb{Z}[\beta]$, Q_D is finite and the automaton $\mathcal{Z}_{\beta,D}$ is finite. □

We now consider the normalisation from an alphabet C in the CNS (β, A). Let D be a symmetrized alphabet of digits in $\mathbb{Z}[\beta]$ containing the set $\{c-a \mid c \in C, a \in A\}$. As explained in the integer base case, one can associate with the zero automaton $\mathcal{Z}_{\beta,D}$ a converter $\mathcal{C}_\beta(C \times A)$. The transitions are defined by

$$s \xrightarrow[\mathcal{C}_\beta(C\times A)]{c|a} t \quad \text{if, and only if,} \quad s \xrightarrow[\mathcal{Z}_{\beta,D}]{c-a} t.$$

Lemma 2.4.8 *If A is a complete residue system modulo β then the converter $\mathcal{C}_\beta(C\times A)$ is input co-deterministic.*

Proof By definition there is an edge $s \xrightarrow{c|a} t$ in $\mathcal{C}_\beta(C\times A)$ if, and only if, $\beta s + c = t + a$. If there is another edge $s' \xrightarrow{c|a'} t$ in $\mathcal{C}_\beta(C\times A)$, then $\beta(s-s') = a - a'$, which is impossible since A is a complete residue system. □

It is thus more natural to define a right sequential letter-to-letter transducer, the normaliser $\mathcal{N}_\beta(C)$, with

$$t \xrightarrow[\mathcal{N}_\beta(C)]{c|a} s \quad \text{if, and only if,} \quad (-s) \xrightarrow[\mathcal{C}_\beta(C\times A)]{c|a} (-t).$$

Let $c_k \cdots c_0 \in C^*$. Setting $s_0 = 0$, there is a unique path in $\mathcal{N}_\beta(C)$

$$s_{k+1} \xleftarrow{c_k|d_k} s_k \xleftarrow{c_{k-1}|d_{k-1}} s_{k-1} \cdots \xleftarrow{c_1|d_1} s_1 \xleftarrow{c_0|d_0} s_0$$

and

$$\sum_{j=0}^{k} c_j \beta^j = \left(\sum_{j=0}^{k} d_j \beta^j\right) + s_{k+1}\beta^{k+1}. \tag{2.17}$$

Remark 2.4.9 If any element of Q_D has a finite integer representation in the system (β, A) (with A a complete residue system modulo β) then the normaliser $\mathcal{N}_\beta(C)$ converts any element z in $\mathbb{Z}[\beta]$ with a representation in C^* into its (β, A) integer representation.

Proof If $z = \sum_{j=0}^{k} c_j \beta^j$, then there exists a path in $\mathcal{N}_\beta(C)$ satisfying (2.17), and $\langle z \rangle_\beta = \langle s_{k+1}\rangle_\beta d_k \cdots d_0$. □

Remark 2.4.10 The normaliser $\mathcal{N}_\beta(C)$ can be used as an algorithm to represent any $z \in \mathbb{Z}[\beta]$ in the system (β, A) (with A a complete residue system modulo β). In fact, given z, there exists a C such that z belongs to C. Feed the transducer with z as input. There exists a unique path

$$\Phi_\beta^{k+1}(z) \xleftarrow{0|d_k} \Phi_\beta^{k}(z) \xleftarrow{0|d_{k-1}} \Phi_\beta^{k-1}(z) \cdots \xleftarrow{0|d_1} \Phi_\beta(z) \xleftarrow{z|d_0} 0$$

and $\langle z \rangle_\beta = d_k \cdots d_0$ if, and only if, $\Phi_\beta^{k+1}(z) = 0$.

From Proposition 2.4.4, Lemma 2.4.7 and Lemma 2.4.8 follows the following result.

Proposition 2.4.11 *If the system (β, A) is a canonical numeration system then the right sequential normaliser $\mathcal{N}_\beta(C)$ is finite for every alphabet C.*

2.4.3 Bases for canonical numeration systems

In general, it is difficult to determine which numbers are suitable bases for a CNS. However, several results are known. In the particular case where β is a Gaussian integer and A is an alphabet of natural integers there is a nice characterisation due to (Kátai and Szabó 1975).

Theorem 2.4.12 *Let β be a Gaussian integer of norm N, and let $A = \{0, \ldots, N-1\}$. Then (β, A) is a canonical numeration system for the complex numbers if, and only if, $\beta = -n \pm i$, for some $n \geq 1$ (and $N = n^2$).*

It is noteworthy that any complex number has a representation – not necessarily unique – in this system.

Quadratic CNS have been characterised in (Kátai and Kovács 1981) and in (Gilbert 1981). In (Brunotte 2001, Brunotte 2002) are characterised all CNS whose bases are roots of trinomials. In the general case (Akiyama and Pethő 2002) have given an algorithm to decide whether a number β is the base of a CNS.

Theorem 2.4.13 *Let β be an algebraic integer with minimal polynomial $M_\beta(X) = X^g + b_{g-1}X^{g-1} + \cdots + b_0$. If one of the following properties is satisfied then β is a base for a CNS:*

(i) $b_0 \geq 2$ and $b_0 \geq b_1 \geq \cdots \geq b_{g-1} \geq 1$
(ii) $b_2 \geq 0, \ldots, b_{g-1} \geq 0$, $1 + \sum_{i=0}^{g-1} b_i \geq 0$ and $b_0 > 1 + \sum_{i=1}^{g-1} |b_i|$.

Part (i) is due to (Kovács 1981), and Part (ii) has been obtained by (Scheicher and Thuswaldner 2004) using automata.

2.4.4 Shift radix systems

In (Akiyama, Borbély, Brunotte, et al. 2005) the concept of shift radix system was introduced to unify canonical numeration systems and β-expansions. Although these two numeration systems are quite different, they are close relatively to some finiteness properties, which means that all numbers of a certain set admit finite expansions.

Definition 2.4.14 Let $\mathbf{r} = (r_1, \ldots, r_d)$ be an element of $\mathbb{R}^d$. Define a mapping $\mu_{\mathbf{r}} : \mathbb{Z}^d \to \mathbb{Z}^d$ by

$$\mu_{\mathbf{r}}((z_1, \ldots, z_d)) = (z_2, \ldots, z_d, -\lfloor r_1 z_1 + \cdots + r_d z_d \rfloor).$$

We say that $\mu_{\mathbf{r}}$ has the *finiteness property* if for every $\mathbf{z}$ in $\mathbb{Z}^d$ there exists a k such that $\mu_{\mathbf{r}}^k(\mathbf{z}) = 0$. In that case $(\mathbb{Z}, \mu_{\mathbf{r}})$ is called a *shift radix system* or SRS.

2.4.4.1 Connection with Pisot numbers and the (F) property

In (Akiyama and Scheicher 2005) it is indicated that the origin of SRS can be found in (Hollander 1996).

Theorem 2.4.15 (Akiyama, Borbély, Brunotte, Pethő, and Thuswaldner 2005) *Let $\beta > 1$ be an algebraic integer with minimal polynomial*

$$M_\beta(X) = X^g + b_{g-1} X^{g-1} + \cdots + b_0 \in \mathbb{Z}[X].$$

Write $M_\beta(X) = (X - \beta)(X^{g-1} + r_{g-1} X^{g-2} + \cdots + r_1)$ and let $\mathbf{r} = (r_1, \ldots, r_{g-1})$. Then β satisfies the (F) property if, and only if, $\mathbf{r}$ gives a $(g-1)$-dimensional SRS.

Proof It is easy to see that β satisfies the (F) property if, and only if, each element of $\mathbb{Z}[\beta] \cap [0, \infty)$ has a finite greedy β-expansion. For $1 \le i \le g-1$, $r_i = -(\frac{b_{i-1}}{\beta} + \cdots + \frac{b_0}{\beta^i})$ and $r_g = 1$. The ring $\mathbb{Z}[\beta]$ is generated by $\{1, \beta, \ldots, \beta^{g-1}\}$ as a $\mathbb{Z}$-module; the same is true for $\{r_1, \ldots, r_g\}$. Thus every element z of $\mathbb{Z}[\beta] \cap [0, 1)$ can be expressed as $z = \sum_{i=1}^{g} z_i r_i$. The β-transformation of z can be written $\tau_\beta(z) = \sum_{i=1}^{g} z_{i+1} r_i$ with z_{g+1} such that $0 \le z_2 r_1 + \cdots + z_{g+1} r_g < 1$, more precisely

$$z_{g+1} = -\lfloor z_2 r_1 + \cdots + z_g r_{g-1} \rfloor.$$

Then $\mu_{\mathbf{r}}(z_1, \ldots, z_{g-1}) = (z_2, \ldots, z_g)$. □

The roots of the polynomial $X^{g-1} + r_{g-1} X^{g-2} + \cdots + r_1$ have modulus less than one, and it can be proved that the SRS algorithm associated with $(r_1, \ldots, r_{g-1})$ always leads to a periodic orbit, and thus that every positive element of $\mathbb{Z}[\beta]$ has an eventually periodic greedy β-expansion. The same can be proved for every positive element of $\mathbb{Q}[\beta]$, which reproves Theorem 2.3.20.

2.4.4.2 Connection with canonical numeration systems

Theorem 2.4.16 (Akiyama, Borbély, Brunotte, Pethő, and Thuswaldner 2005) *The polynomial* $X^g + b_{g-1}X^{g-1} + \cdots + b_0$ *gives a CNS if, and only if,*

$$\mathbf{r} = (\frac{1}{b_0}, \frac{b_{g-1}}{b_0}, \ldots, \frac{b_1}{b_0})$$

gives a g–dimensional SRS.

Proof Take z in $\mathbb{Z}[\beta]$. Then z can be written as $z = \sum_{i=0}^{g-1} z_i \beta^i$ with z_i in $\mathbb{Z}$. The mapping Φ_β (see (2.15)) can be extended as a mapping $\widetilde{\Phi}_\beta : \mathbb{Z}^g \to \mathbb{Z}^g$ defined as

$$\widetilde{\Phi}_\beta((z_0, \ldots, z_{g-2}, z_{g-1})) = (z_1 - qb_1, \ldots, z_{g-1} - qb_{g-1}, -q))$$

with $q = \lfloor z_0/b_0 \rfloor$.

For easier notation, set $b_g = 1$. The basis $\{1, \beta, \ldots, \beta^{g-1}\}$ can be replaced by the basis $\{w_1, \ldots, w_g\}$ with $w_j = \sum_{i=g-j+1}^{g} b_j \beta^{i+j-g-1}$ for $1 \leq j \leq g$. Now, if $z = \sum_{i=1}^{g} y_j w_j$, we can define a map Ψ_β playing the same role as Φ_β by

$$\Psi_\beta(z) = (\sum_{i=1}^{g-1} y_{j+1} w_j) - w_g \lfloor \frac{b_1 y_g + \cdots + b_g y_1}{b_0} \rfloor.$$

This maps is extended as a mapping $\widetilde{\Psi}_\beta : \mathbb{Z}^g \to \mathbb{Z}^g$ defined by

$$\widetilde{\Psi}_\beta((y_1, \ldots, y_{g-1}, y_g)) = (y_2, \ldots, y_g, -\lfloor \frac{b_1 y_g + \cdots + b_g y_1}{b_0} \rfloor)$$

and $\widetilde{\Psi}_\beta$ is just the SRS mapping $\mu_\mathbf{r}$. □

2.5 Representation in rational base

We now turn to the problem of the representation of numbers, integers or reals, again in a base *which is not an integer* but a rational number – and thus certainly *not a Pisot number*, as it has been the case in most of the preceding sections. The greedy algorithm which was ubiquitous there and underlying almost every construction is now inappropriate or, to tell the truth, one cannot tell anything of its outcome. We shall make use instead of an algorithm which is reminiscent of the division algorithm defined with integer base and which produces, as the division algorithm, the digits of the representations *from right to left*.

All the results of this section are taken, and their presentation is adapted, from (Akiyama, Frougny, and Sakarovitch 2008).

2.5.1 Representation of integers

Let p and q be two co-prime integers, $p > q \geqslant 1$. The *definition* of the numeration system in base $\frac{p}{q}$ itself, and thus the evaluation map, will follow from the algorithm which computes the representation of the integers.

2.5.1.1 The modified division algorithm

Let N be any positive integer; let us write $N_0 = N$ and, for $i \geqslant 0$, write

$$q N_i = p N_{i+1} + a_i \tag{2.18}$$

where a_i is the remainder of the division of $q N_i$ by p, and thus belongs to $A_p = \{0, \ldots, p-1\}$. Since N_{i+1} is strictly smaller than N_i, the division (2.18) can be repeated only a finite number of times, until eventually $N_{k+1} = 0$ for some k. The sequence of successive divisions (2.18) for $i = 0$ to $i = k$ is thus an *algorithm* – that in the sequel is referred to as the *Modified Division*, or *MD*, *algorithm* – which given N produces the digits a_0, a_1, ..., a_k, and it holds that:

$$N = \sum_{i=0}^{k} \frac{a_i}{q} \left(\frac{p}{q}\right)^i. \tag{2.19}$$

We will say that the word $a_k \cdots a_0$, computed from N from right to left, that is to say *least significant digit first*, is a $\frac{p}{q}$-*representation* of N.

Let $U_{\frac{p}{q}}$ be the sequence defined by:

$$U_{\frac{p}{q}} = \{u_i = \frac{1}{q}\left(\frac{p}{q}\right)^i \mid i \in \mathbb{Z}\}.$$

We will say that $U_{\frac{p}{q}}$, together with the digit alphabet A_p is the numeration system in base $\frac{p}{q}$ or the $\frac{p}{q}$ numeration system. If $q = 1$, it is exactly the classical numeration system in base p. But, on the other hand, this definition *does not* match the one we have given for the numeration system in base β in Section 2.3: $U_{\frac{p}{q}}$ *is not* the sequence of powers of $\frac{p}{q}$ but rather these powers *divided by q* and the digits *are not* the integers smaller than $\frac{p}{q}$ but rather the integers *whose quotient by q* is smaller than $\frac{p}{q}$. The *evaluation map* $\pi_{\frac{p}{q}} : A_p^* \to \mathbb{Q}$ is defined accordingly: for every word w of A_p^*, we have

$$w = a_k a_{k-1} \cdots a_1 a_0 \quad \longmapsto \quad \pi_{\frac{p}{q}}(w) = \sum_{i=0}^{k} a_i u_i = \sum_{i=0}^{k} \frac{a_i}{q}\left(\frac{p}{q}\right)^i. \tag{2.20}$$

With the same proof as for an integer base system (*cf.* Lemma 2.2.1), we have:

Lemma 2.5.1 *The restriction of $\pi_{\frac{p}{q}}$ to A_p^k is injective, for every k.*

As for integer base, $\pi_{\frac{p}{q}}$ is not injective on the whole A_p^* since for any u in A_p^* and any integer h it holds: $\pi_{\frac{p}{q}}(0^h u) = \pi_{\frac{p}{q}}(u)$. On the other hand, Lemma 2.5.1 implies that this is the only possibility and we have:

$$\pi_{\frac{p}{q}}(u) = \pi_{\frac{p}{q}}(v) \text{ and } |u| > |v| \implies u = 0^h v \text{ with } h = |u| - |v|. \tag{2.21}$$

Theorem 2.5.2 *Every non-negative integer N has a $\frac{p}{q}$-representation which is an integer representation. It is the unique finite $\frac{p}{q}$-representation of N.*

Proof Let $a_k \cdots a_0$ be the $\frac{p}{q}$-representation given to N by the MD algorithm, and suppose that there exists another *finite* representation of N in the system $U_{\frac{p}{q}}$, of the form $e_\ell e_{\ell-1} \cdots e_0 . e_{-1} \cdots e_{-m}$ with $e_{-m} \neq 0$. Then

$$q\left(\frac{p}{q}\right)^m N = \sum_{i=-m}^{\ell} e_i \left(\frac{p}{q}\right)^{m+i} = \sum_{i=0}^{k} a_i \left(\frac{p}{q}\right)^{m+i}$$

and therefore $\pi_{\frac{p}{q}}(e_\ell \cdots e_0 e_{-1} e_{-2} \cdots e_{-m}) = \pi_{\frac{p}{q}}(a_k a_{k-2} \cdots a_0 0^m)$. Contradiction between (2.21) and $e_{-m} \neq 0$. □

This unique finite $\frac{p}{q}$-representation of N (under the condition that the leading digit is not 0) will be called the $\frac{p}{q}$-*expansion* of N and written $\langle N \rangle_{\frac{p}{q}}$. By convention and as in the three preceding sections, the $\frac{p}{q}$-expansion of 0 is the empty word ε.

Example 2.5.3 Let $p = 3$ and $q = 2$, then $A_3 = \{0, 1, 2\}$ – this will be our main running example in this section. Table 2.2 gives the $\frac{3}{2}$-expansions of the first twelve non-negative integers.

ε	0	2120	6
2	1	2122	7
21	2	21011	8
210	3	21200	9
212	4	21202	10
2101	5	21221	11

Table 2.2. The $\frac{3}{2}$-expansion of the first twelve integers.

We let $L_{\frac{p}{q}}$ denote the set of $\frac{p}{q}$-*expansions* of the non-negative integers:

$$L_{\frac{p}{q}} = \{\langle N \rangle_{\frac{p}{q}} \mid N \in \mathbb{N}\}.$$

In contrast with the three preceding sections, and as we shall see below, $L_{\frac{p}{q}}$ is not a rational set. Before getting to this point, let us note that the same order properties as for integer base systems hold for the $\frac{p}{q}$ numeration system, provided only the words in $L_{\frac{p}{q}}$ are considered.

Proposition 2.5.4 *Let v and w be in $L_{\frac{p}{q}}$. Then $v \preceq w$ if, and only if, $\pi_{\frac{p}{q}}(v) \leqslant \pi_{\frac{p}{q}}(w)$.*

Proof Let $v = a_k \cdots a_0$ and $w = b_\ell \cdots b_0$ be the $\frac{p}{q}$-expansions of the integers $m = \pi_{\frac{p}{q}}(v)$ and $n = \pi_{\frac{p}{q}}(w)$ respectively. By Theorem 2.5.2, we already know that $v = w$ if, and only if, $\pi_{\frac{p}{q}}(v) = \pi_{\frac{p}{q}}(w)$. The proof goes by induction on ℓ, which is (by hypothesis) greater than or equal to k. The proposition holds for $\ell = 0$.

Let us write $v' = a_k \cdots a_1$ and $w' = b_\ell \cdots b_1$, and $m' = \pi_{\frac{p}{q}}(v')$ and $n' = \pi_{\frac{p}{q}}(w')$ are integers. It holds:

$$n - m = \frac{p}{q}(n' - m') + \frac{1}{q}(b_0 - a_0).$$

Now $v \prec w$ implies that either $v' \prec w'$ or $v' = w'$ and $a_0 < b_0$. If $v' \prec w'$, then $n' - m' \geqslant 1$ by induction hypothesis and thus $n - m > 0$ since $b_0 - a_0 \geqslant -(p-1)$. If $v' = w'$, then $n - m = \frac{1}{q}(b_0 - a_0) > 0$. □

Corollary 2.5.5 *Let v and w be in $0^* L_{\frac{p}{q}}$ and of equal length. Then $v \leq w$ if, and only if, $\pi_{\frac{p}{q}}(v) \leqslant \pi_{\frac{p}{q}}(w)$.*

It is to be noted also that these statements do not hold without the hypothesis that v and w belong to $L_{\frac{p}{q}}$ (to $0^* L_{\frac{p}{q}}$ respectively). For instance, $\pi_{\frac{3}{2}}(1\,0) = 3/4 < \pi_{\frac{3}{2}}(2) = 1$ and $\pi_{\frac{3}{2}}(2\,0\,0\,0) = 27/16 < \pi_{\frac{3}{2}}(0\,2\,1\,2) = 4$.

2.5.1.2 The set of $\frac{p}{q}$-expansions of the integers

It is a very intriguing, and totally open, question to characterise the set $L_{\frac{p}{q}}$. As far as now, we can only make basic observations.

By construction, $L_{\frac{p}{q}}$ is *prefix-closed*, that is, any prefix of any word of $L_{\frac{p}{q}}$ is in $L_{\frac{p}{q}}$. A simple look at Table 2.2 shows that it is *not suffix-closed*. In fact, *every* word of A_p^* is a suffix of some words in $L_{\frac{p}{q}}$. More precisely, we have the following statement.

Proposition 2.5.6 *For every integer k and every word w in A_p^k, there exists a unique integer n, $0 \leqslant n < p^k$ such that w is the suffix of length k of the $\frac{p}{q}$-expansion of all integers m congruent to n modulo p^k.*

Proof Given any integer $n = n_0$, the division (2.18) repeated k times yields:

$$q^k n_0 = p^k n_k + q^k \pi_{\frac{p}{q}}(a_{k-1}a_{k-2}\cdots a_0). \tag{2.22}$$

If we do the same for another integer $m = m_0$ and perform the subtraction on the two sides of Equation (2.22), then:

$$q^k(n_0 - m_0) = p^k(n_k - m_k)$$
$$+ q^k\left(\pi_{\frac{p}{q}}(a_{k-1}a_{k-2}\cdots a_0) - \pi_{\frac{p}{q}}(b_{k-1}b_{k-2}\cdots b_0)\right).$$

As q^k is prime with p^k, and using Lemma 2.5.1, then:

$$n - m \equiv 0 \pmod{p^k} \iff a_{k-1}a_{k-2}\cdots a_0 = b_{k-1}b_{k-2}\cdots b_0. \tag{2.23}$$

Since there are exactly p^k words in A_p^k, each of them must appear once and only once when n ranges from 0 to $p^k - 1$ and (2.23) gives the second part of the statement. $\square$

It follows that a word w of length k is a *right context* for the $\frac{p}{q}$-expansions $\langle n \rangle_{\frac{p}{q}}$ and $\langle m \rangle_{\frac{p}{q}}$ of two integers n and m for $L_{\frac{p}{q}}$, that is, both $\langle n \rangle_{\frac{p}{q}} w$ and $\langle m \rangle_{\frac{p}{q}} w$ are in $L_{\frac{p}{q}}$, if, and only if, n and m are congruent modulo q^k. This implies immediately that the *coarsest right regular equivalence* that saturates $L_{\frac{p}{q}}$ is the identity, hence in particular is not of finite index. A classical statement in formal language theory (see for instance (Hopcroft, Motwani, and Ullman 2006)) then implies:

Corollary 2.5.7 *If $q \neq 1$, then $L_{\frac{p}{q}}$ is not a regular language.*

Along the same lines it is easy to give a more precise statement on suffixes that are powers of a given word.

Lemma 2.5.8 *Let w be in $L_{\frac{p}{q}}$ and $w = uv$ be a proper factorization of w. Then uv^k belongs to $L_{\frac{p}{q}}$ only if $q^{(k-1)|v|}$ divides $\pi_{\frac{p}{q}}(w) - \pi_{\frac{p}{q}}(u)$.*

Proof The word uv^k belongs to $L_{\frac{p}{q}}$ only if

$$\pi_{\frac{p}{q}}(uv^k) - \pi_{\frac{p}{q}}(uv^{k-1}) = \left(\frac{p}{q}\right)^{|v|}(\pi_{\frac{p}{q}}(uv^{k-1}) - \pi_{\frac{p}{q}}(uv^{k-2})) = \cdots$$
$$= \left(\frac{p}{q}\right)^{(k-1)|v|}(\pi_{\frac{p}{q}}(uv) - \pi_{\frac{p}{q}}(u))$$

is in $\mathbb{Z}$. And this is possible only if $q^{(k-1)|v|}$ divides $\pi_{\frac{p}{q}}(uv) - \pi_{\frac{p}{q}}(u)$. $\square$

Lemma 2.5.8 will be used in the sequel to show that the closure of $L_{\frac{p}{q}}$ does not contain eventually periodic infinite words; combined with the classical 'pumping lemma' (see (Hopcroft, Motwani, and Ullman 2006) and Lemma 1.3.14), it implies another statement related to formal language theory:

Corollary 2.5.9 *If $q \neq 1$, then $L_{\frac{p}{q}}$ is not a context-free language.*

2.5.1.3 The evaluator and the converters

We build an evaluator and zero automata in a similar way as the one we followed for integer base. Let $\frac{p}{q}$ be the base fixed as before but the digits be a priori *any integer*, positive or negative. The evaluator $\mathcal{Z}_{\frac{p}{q}}$ has the set of q-decimal numbers, that is, $\mathbb{Z}[\frac{1}{q}]$, as set of states, it reads (from left to right) the numbers (written on the 'alphabet' $\mathbb{Z}$), and runs in such a way that, at every step of the reading, the reached state indicates the value of the portion of the number read so far. The initial state of $\mathcal{Z}_{\frac{p}{q}}$ is thus 0 and its transitions are of the form:

$$\forall s,t \in \mathbb{Z}[\frac{1}{q}], \ \forall a \in \mathbb{Z} \qquad s \xrightarrow[\mathcal{Z}_{\frac{p}{q}}]{a} t \quad \text{if, and only if,} \quad qt = ps + a, \tag{2.24}$$

from which we get the expected behaviour:

$$\forall w \in \{\mathbb{Z}\}^* \qquad 0 \xrightarrow[\mathcal{Z}_{\frac{p}{q}}]{w} \pi_{\frac{p}{q}}(w).$$

It follows from (2.24) that $\mathcal{Z}_{\frac{p}{q}}$ is both *deterministic* and *co-deterministic*.

As above, we shall make use of finite parts of $\mathcal{Z}_{\frac{p}{q}}$. First, we restrict the alphabet to be a finite subset of $\mathbb{Z}$: $B_d = \{-d, \ldots, d\}$ with $d \geq p-1$ and thus $A_p \subset B_d$. Second, we choose 0 as unique final state and we get a zero automaton $\mathcal{Z}_{\frac{p}{q},d} = \left(\mathbb{Z}[\frac{1}{q}], B_d, E, \{0\}, \{0\} \right)$ where the transitions in E are those defined by (2.24). This automaton accepts thus the writings of 0 (in base $\frac{p}{q}$ and on the alphabet B_d). It is still infinite but we have the following.

Proposition 2.5.10 *The trim part of $\mathcal{Z}_{\frac{p}{q},d}$ is finite and its set of states is $H = \{-h, \ldots, h\}$ where $h = \lfloor \frac{d-q}{p-q} \rfloor$.*

Proof As B_d contains A_p and is symmetrical, every z in $\mathbb{Z}$ is accessible in $\mathcal{Z}_{\frac{p}{q},d}$. On the other hand, no state in $\mathbb{Z}[\frac{1}{q}] \setminus \mathbb{Z}$ is co-accessible to 0 in $\mathcal{Z}_{\frac{p}{q},d}$.

If m is a positive integer strictly larger than $(d-q)/(p-q)$, the 'smallest' reachable state from m, that is, the smallest integer which is larger than, or

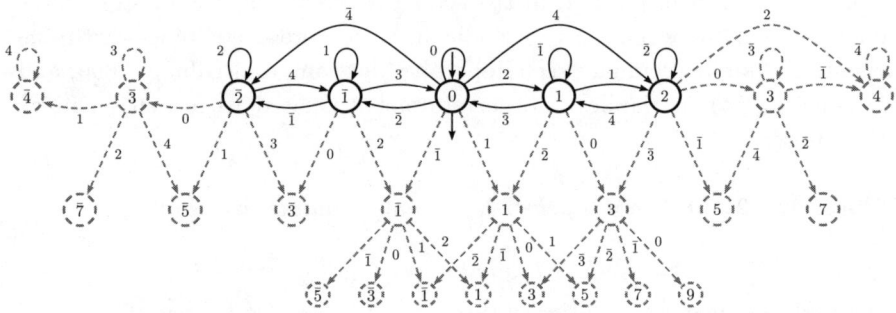

Fig. 2.18 A partial view of $\mathcal{Z}_{\frac{3}{2},4}$.
The upper row consists of the states whose labels are integers; the row below of the states whose labels are of the form $n/2$, with odd n; the next row of those whose labels are of the form $n/4$, with odd n; etc. For the readability of the figure, not all transitions labelled in B_4 are drawn.

equal to, $\frac{1}{q}(mp-d)$, is also larger than, or equal to, m: m is not co-accessible in $\mathcal{Z}_{\frac{p}{q},d}$ and the same is true if m is strictly smaller than $-(d-q)/(p-1)$.

Conversely, if m is a positive integer smaller than $(d-q)/(p-q)$, then the integer $k = p + (m-1)(p-q)$ is smaller than, or equal to, d and $m \xrightarrow{\bar{k}} (m-1)$ is a transition in $\mathcal{Z}_{\frac{p}{q},d}$. Hence, by induction, a path from m to 0 in $\mathcal{Z}_{\frac{p}{q},d}$. $\square$

By definition, the trim part of $\mathcal{Z}_{\frac{p}{q},d}$ is the *strongly connected component* of 0. Figure 2.18 shows $\mathcal{Z}_{\frac{3}{2},4}$.

Let $\mathcal{Z}_{\frac{p}{q},d}$ denote the automaton reduced to its trim part only, with set of states H. And as above again, the automaton $\mathcal{Z}_{\frac{p}{q},d}$ will serve as the base for the construction of a series of converters and normalisers exactly as in the case of integer base.

Figure 2.19 (a) shows the right sequential converter that realises addition in the $\frac{3}{2}$ numeration system; Figure 2.19 (b) shows the right sequential converter on the alphabet $\{\bar{1},0,1,2\}$ in the $\frac{3}{2}$ numeration system.

Remark 2.5.11 A converter reads words on a digit alphabet C, and outputs an equivalent $\frac{p}{q}$-representation on another alphabet A, even for words v such that $\pi_{\frac{p}{q}}(v)$ is not an integer.

As a corollary to the construction of the converter, it is easy to build a *letter-to-letter right sequential transducer* that realises the successor function for the $\frac{p}{q}$ numeration system.

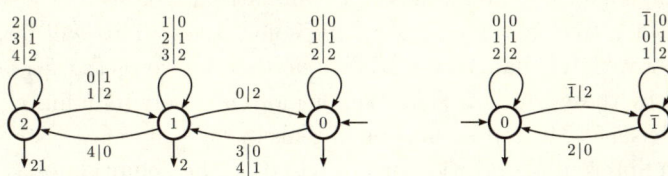

(a) The converter for the addition. (b) The converter on $\{-1, 0, 1, 2\}$.

Fig. 2.19 Two converters for the $\frac{3}{2}$ numeration system.

2.5.2 Representation of the reals

Every infinite word $u = (a_i)_{i \geq 1}$ in $A_p^{\mathbb{N}}$ is given a real value x by the evaluation map $\pi_{\frac{p}{q}}$:

$$u = a_1 a_2 \cdots \longmapsto x = \pi_{\frac{p}{q}}(u) = \sum_{i=1}^{\infty} \frac{a_i}{q} \left(\frac{p}{q}\right)^{-i}$$

and u is called a $\frac{p}{q}$-*representation* of x. We use the same conventions as in the preceding sections and we have:

$$\forall u = (a_i)_{i \geq 1} \in A_p^{\mathbb{N}} \quad \pi_{\frac{p}{q}}(.u) = \lim_{n \to +\infty} \left(\frac{q}{p}\right)^n \pi_{\frac{p}{q}}(a_1 a_2 \cdots a_n), \quad (2.25)$$

$$\forall u \in A_p^{\mathbb{N}}, \forall w \in A_p^{*} \quad \pi_p(.wu) = \left(\frac{q}{p}\right)^{|w|} (\pi_p(w) + \pi_p(.u)).$$

Proposition 2.5.12 *The map $\pi_{\frac{p}{q}} : A_p^{\mathbb{N}} \to \mathbb{R}$ is continuous.*

Our purpose here is to associate with every real number a $\frac{p}{q}$-representation which will be as canonical as possible. In contrast with what is done in integer or Pisot base numeration systems, where the canonical representation – the *greedy expansion* – is defined by an algorithm which *computes* it for every real, we set *a priori* what are these canonical $\frac{p}{q}$-expansions.

2.5.2.1 Construction of the tree $T_{\frac{p}{q}}$

The free monoid A_p^{*} is classically represented as the nodes of the (infinite) full p-ary tree: every node is labelled by a word in A_p^{*} and has p children, every edge between a node and its children is labelled by one of the letters of A_p and the label of a node is precisely the label of the (unique) path that goes from the root to that node.

As the language $L_{\frac{p}{q}}$ is prefix-closed, it can naturally be seen as a subtree of the full p-ary tree, obtained by cutting some edges. This will form the tree $T_{\frac{p}{q}}$ (after we have changed the label of nodes from words to the numbers represented by these words). This tree, or more precisely its infinite paths, will be the basis for the representation of reals in the $\frac{p}{q}$ number system. We give now an 'internal' description of $T_{\frac{p}{q}}$, based on the definition of a family of maps from $\mathbb{N}$ to $\mathbb{N}$, which will be proved to be effective for the study of infinite paths.

Definition 2.5.13 (i) For each a in A_p, let $\psi_a : \mathbb{N} \to \mathbb{N}$ be the *partial map* defined by:

$$\forall n \in \mathbb{N} \quad \psi_a(n) = \begin{cases} \frac{1}{q}(pn+a) & \text{if } \frac{1}{q}(pn+a) \in \mathbb{N} \\ \text{undefined} & \text{otherwise} \end{cases}$$

We write $\mathsf{e}(n) = \{a \in A_p \mid \psi_a(n) \text{ is defined}\}$, $\mathsf{Me}(n) = \max\{\mathsf{e}(n)\}$ for the largest digit for which $\psi_a(n)$ is defined, and $\mathsf{me}(n) = \min\{\mathsf{e}(n)\}$ for the smallest digit with the same property.

(ii) The tree $T_{\frac{p}{q}}$ is the labelled infinite tree (*where both nodes and edges are labelled*) constructed as follows. The nodes are labelled in $\mathbb{N}$, the edges in A, and the root is labelled by 0. The children of a node labelled by n are nodes labelled by $\psi_a(n)$ for a in $\mathsf{e}(n)$, and the edge from n to $\psi_a(n)$ is labelled by a.

(iii) We call *path label* of a node s of $T_{\frac{p}{q}}$, and write $\mathsf{p}(s)$, the label of *the path* from the root of $T_{\frac{p}{q}}$ to s. We denote by $I_{\frac{p}{q}}$ the subtree of $T_{\frac{p}{q}}$ made of nodes whose path label does not begin with a 0.

The very way $T_{\frac{p}{q}}$ is defined implies that if two nodes have the same label, they are the root of two isomorphic subtrees of $T_{\frac{p}{q}}$ and it follows from Proposition 2.5.6 that the converse is true, that is two nodes which hold distinct labels are the root of two distinct subtrees of $T_{\frac{p}{q}}$. As no two nodes of $I_{\frac{p}{q}}$ have the same label, hence we have:

Proposition 2.5.14 *If $q \neq 1$ no two subtrees of $I_{\frac{p}{q}}$ are isomorphic.*

Definition 2.5.13 and the MD algorithm imply directly the following facts.

Lemma 2.5.15 *For every n in $\mathbb{N}$, it holds:*

(i) $\mathsf{me}(n) = \mathsf{e}(n) \cap \{0, 1, \ldots, q-1\}$ *and* $\mathsf{Me}(n) = \mathsf{e}(n) \cap \{p-q, \ldots, p-1\}$.

(ii) $a \in \mathsf{e}(n)$ *and* $a + q \in A_p \implies a + q \in \mathsf{e}(n)$.

(iii) $a, a+q \in \mathsf{e}(n) \implies \psi_{a+q}(n) = \psi_a(n) + 1$.

(iv) $\mathsf{me}(n+1) = \mathsf{Me}(n) + q - p$ *and* $\psi_{\mathsf{me}(n+1)}(n+1) = \psi_{\mathsf{Me}(n)}(n) + 1$.

And finally:

(v) *The label of every node s of $T_{\frac{p}{q}}$ is $\pi_{\frac{p}{q}}(p(s))$.*

We denote by $W(n)$ (resp. by $w(n)$) the label of the infinite path that starts from a node with label n and that follows always the edges with the maximal (resp. minimal) digit label. Such a word is said to be a *maximal* word (resp. a *minimal* word) in $T_{\frac{p}{q}}$. We note: $\mathbf{t}_{\frac{p}{q}} = W(0)$ and $\boldsymbol{\omega}_{\frac{p}{q}} = \pi_{\frac{p}{q}}(.\mathbf{t}_{\frac{p}{q}})$. (It holds $\boldsymbol{\omega}_{\frac{p}{q}} < \frac{p-1}{p-q}$.)

The infinite word $\mathbf{t}_{\frac{p}{q}}$ is the maximal element with respect to the lexicographic order of the label of all infinite paths of $T_{\frac{p}{q}}$ that start from the root. Notice that, for any rational $\frac{p}{q}$, 0^ω is the minimal element with respect to the lexicographic order of the label of all infinite paths of $T_{\frac{p}{q}}$ and that, if $q = 1$, that is, in an integer base, $W(n) = (p-1)^\omega$, and $w(n) = 0^\omega$ for every n in $\mathbb{N}$.

Example 2.5.16 For $\frac{p}{q} = \frac{3}{2}$, $\mathbf{t}_{\frac{3}{2}} = 2122111221211221211\cdots$.

We call *branching* a node v of $T_{\frac{p}{q}}$ if it has at least two children, that is, if $e(\pi_{\frac{p}{q}}(p(v)))$ has at least two elements. Direct computations yields the following.

Lemma 2.5.17 *Let v be any branching node in $T_{\frac{p}{q}}$, and $n = \pi_{\frac{p}{q}}(p(v))$ its label. Let a_1 and $b_1 = a_1 + q$ be in $e(n)$ and let $m_1 = \psi_{a_1}(n)$ and $m_2 = \psi_{b_1}(n) = m_1 + 1$. Write $W(m_1) = a_2 a_3 \cdots$ and $w(m_2) = b_2 b_3 \cdots$. It then holds:*

$$\pi_{\frac{p}{q}}(.a_1 a_2 a_3 \cdots) = \pi_{\frac{p}{q}}(.b_1 b_2 b_3 \cdots). \tag{2.26}$$

2.5.2.2 The $\frac{p}{q}$-expansions of real numbers

Notation 2.5.18 Let us denote by $W_{\frac{p}{q}}$ the subset of $A_p^{\mathbb{N}}$ that consists of the labels of infinite paths starting from the root of $T_{\frac{p}{q}}$.

Note that the finite prefixes of the elements of $W_{\frac{p}{q}}$ are the words in $0^* L_{\frac{p}{q}}$. A direct consequence of Lemma 2.5.8 is the following.

Proposition 2.5.19 *If $q > 1$, then no element of $W_{\frac{p}{q}}$ is eventually periodic, but 0^ω.*

As announced, the set of $\frac{p}{q}$-expansions is defined *a priori* and not algorithmically.

Definition 2.5.20 *The set of expansions in the $\frac{p}{q}$ numeration system is $W_{\frac{p}{q}}$.*

In other words, an element u of $W_{\frac{p}{q}}$ is a $\frac{p}{q}$-expansion of the real $x = \pi_{\frac{p}{q}}(u)$ and conversely any element of $A_p^{\mathbb{N}}$ which does not belong to $W_{\frac{p}{q}}$ is not a $\frac{p}{q}$-expansion. The following Lemma 2.5.21 and Theorem 2.5.23 tell that $\frac{p}{q}$-expansions are not too many nor too few respectively and vindicate the definition.

Lemma 2.5.21 *The map* $\pi_{\frac{p}{q}}: W_{\frac{p}{q}} \to \mathbb{R}$ *is order preserving.*

Proof Let $u = (a_i)_{i \geq 1}$ and $v = (b_i)_{i \geq 1}$ be in $W_{\frac{p}{q}}$. If $u \leq v$ then, for every k in $\mathbb{N}$, $a_1 a_2 \cdots a_k \leq b_1 b_2 \cdots b_k$ and then, by Corollary 2.5.5, $\pi_{\frac{p}{q}}(a_1 a_2 \cdots a_k) \leqslant \pi_{\frac{p}{q}}(b_1 b_2 \cdots b_k)$. By (2.25), $\pi_{\frac{p}{q}}(.u) \leqslant \pi_{\frac{p}{q}}(.v)$. □

By contrast, it follows from the examples given after Corollary 2.5.5 that the map $\pi_{\frac{p}{q}}: A_p^{\mathbb{N}} \to \mathbb{R}$ is not order preserving.

Notation 2.5.22 Let $X_{\frac{p}{q}} = \pi_{\frac{p}{q}}\left(W_{\frac{p}{q}}\right)$. The elements of $X_{\frac{p}{q}}$ are non-negative real numbers less than or equal to $\omega_{\frac{p}{q}}$:
$$X_{\frac{p}{q}} \subseteq [0, \omega_{\frac{p}{q}}].$$

Theorem 2.5.23 *Every real in* $[0, \omega_{\frac{p}{q}}]$ *has at least one $\frac{p}{q}$-expansion, that is,* $X_{\frac{p}{q}} = [0, \omega_{\frac{p}{q}}]$.

Proof By definition, the set $W_{\frac{p}{q}}$ is the set of infinite words w in $A_p^{\mathbb{N}}$ such that any prefix of w is in $0^* L_{\frac{p}{q}}$. As $0^* L_{\frac{p}{q}}$ is *prefix-closed* – since $L_{\frac{p}{q}}$ is prefix-closed *and* the empty word belongs to $L_{\frac{p}{q}}$ – $W_{\frac{p}{q}}$ is closed (see (Perrin and Pin 2003)) in the compact set $A_p^{\mathbb{N}}$, hence compact. Since $\pi_{\frac{p}{q}}$ is continuous, $X_{\frac{p}{q}}$ is closed.

Suppose that $[0, \omega_{\frac{p}{q}}] \setminus X_{\frac{p}{q}}$ is a non-empty open set, containing a real t. Let $y = \sup\{x \in X_{\frac{p}{q}} \mid x < t\}$ and $z = \inf\{x \in X_{\frac{p}{q}} \mid x > t\}$. Since $X_{\frac{p}{q}}$ is closed, y and z both belong to $X_{\frac{p}{q}}$. Let $u = a_1 a_2 \cdots$ be the largest $\frac{p}{q}$-expansion of y and $v = b_1 b_2 \cdots$ the smallest $\frac{p}{q}$-expansion of z (in the lexicographic order). Of course, $u < v$ since $u \neq v$. Let $a_1 \cdots a_N$ be the *longest common prefix* of u and v (with the convention that N can be 0). Set $m = \pi_{\frac{p}{q}}(a_1 \cdots a_N .)$, $n = \pi_{\frac{p}{q}}(a_1 \cdots a_N a_{N+1} .)$ and $r = \pi_{\frac{p}{q}}(a_1 \cdots a_N b_{N+1} .)$. Then
$$u \leq a_1 \cdots a_N a_{N+1} \mathsf{W}(n) < a_1 \cdots a_N b_{N+1} \mathsf{w}(r) \leq v.$$

By the choice of v, $\pi_{\frac{p}{q}}(.a_1 \cdots a_N a_{N+1} \mathsf{W}(n)) < z$, and by the choice of u, $u = a_1 \cdots a_N a_{N+1} \mathsf{W}(n)$. Symmetrically, $v = a_1 \cdots a_N b_{N+1} \mathsf{w}(r)$.

If $a_{N+1} + q < b_{N+1}$, then there exists a digit c in $\mathsf{e}(m)$ such that $a_{N+1} +

$q \leqslant c < b_{N+1}$. For any w' in $A_p^{\mathbb{N}}$ such that $w = a_1 \cdots a_N c w'$ is in $W_{\frac{p}{q}}$ (and there exist some), we have

$$u < w < v.$$

Whatever the value of $\pi_{\frac{p}{q}}(.w)$, y or z, we have a contradiction with the extremal choice of u and v.

If $a_{N+1} + q = b_{N+1}$, then $r = n+1$ and $z = y$ by Lemma 2.5.17, hence a contradiction. And thus $X_{\frac{p}{q}} = [0, \omega_{\frac{p}{q}}]$. □

A word in $W_{\frac{p}{q}}$ is said to be *eventually maximal* (resp. *eventually minimal*) if it has a suffix which is a maximal (resp. minimal) word.

The following statement shows that in spite of the non-rationality of $W_{\frac{p}{q}}$ the $\frac{p}{q}$-expansions of reals behave very much as the expansions obtained by a greedy algorithm in an integer or in a real base.

Theorem 2.5.24 *The set of reals in $X_{\frac{p}{q}}$ that have more than one $\frac{p}{q}$-expansion is countably infinite in bijection with the set of branching nodes in $T_{\frac{p}{q}}$. The $\frac{p}{q}$-expansions of such reals are eventually maximal or eventually minimal. If $p \geqslant 2q - 1$, then no real number has more than two $\frac{p}{q}$-expansions.*

Remark 2.5.25 In contrast with the classical representations of reals, the finite prefixes of a $\frac{p}{q}$-expansion of a real number, completed by zeroes, *are not* $\frac{p}{q}$-expansions of real numbers (though they can be given a value by the function π of course), that is to say, if a non-empty word w is in $L_{\frac{p}{q}}$, then the word $w 0^\omega$ does not belong to $W_{\frac{p}{q}}$.

It is an open problem, a challenging one, to prove that the hypothesis $p \geqslant 2q - 1$ in Theorem 2.5.24 is not necessary and that a real has never more than two $\frac{p}{q}$-expansions (with the meaning we have given to it) for any rational $\frac{p}{q}$.

2.6 A primer on finite automata and transducers

The matter developed in this chapter calls for definitions and results on finite automata and transducers that go beyond those given in Chapter 1 and we have gathered them in this section.

The notation follows the one adopted in (Sakarovitch 2003), where the proofs of the statements can be found as well – unless otherwise stated. The definitions are sometimes made simpler for their intended scope is the content of this chapter only.

2.6.1 Automata

Let us first complete the definitions and results on finite automata given in Chapter 1. We call *recognisable* or *rational* the languages of A^* recognised by a finite automaton – that were rather called *regular* in Section 1.3 – and we denote this family by Rat A^*. Since every finite automaton is equivalent to a deterministic one, we have:

Theorem 2.6.1 Rat A^* *is an effective Boolean algebra of languages.*

The *generating function* of a language L of A^* is the series

$$\Psi_L(X) = \sum_{n \in \mathbb{N}} \ell_n X^n$$

where ℓ_n is the number of words of L of length n: $\ell_n = \mathrm{Card}\,(L \cap A^n)$.

A series $\Phi(X)$ is called a *rational function* if it is the quotient of two polynomials $P(X)$ and $Q(X)$ of $\mathbb{Z}[X]$: $\Phi(X) = \frac{P(X)}{Q(X)}$. A classical result in algebra states that a series $\Phi(X) = \sum_{n \in \mathbb{N}} a_n X^n$ is rational if, and only if, its coefficients a_n satisfy a linear recurrence relation with coefficients in $\mathbb{Z}$. The following result is not for nothing in the choice of *rational* rather than *regular* for languages recognised by finite automata.

Theorem 2.6.2 (Chomsky and Miller 1958) *The generating function of a rational language is a rational function.*

Proof Let $\mathcal{A}$ be a finite *deterministic* automaton (of dimension Q) which recognises the language L and let $\mathbf{M}$ be the *adjacency matrix* of $\mathcal{A}$. Write $\mathbf{l}(n)$ for the vector of dimension Q whose pth entry is the number of words of length n which label paths from state p to a final state in $\mathcal{A}$: $\ell_n = \mathbf{l}_i(n)$ for the initial state i of $\mathcal{A}$. As $\mathcal{A}$ is deterministic, it holds

$$\forall n \in \mathbb{N}\;\; \mathbf{l}(n+1) = \mathbf{M}\,\mathbf{l}(n). \tag{2.27}$$

By the Cayley–Hamilton Theorem, $\mathbf{M}$ is a zero of its *characteristic polynomial*, that is:

$$\mathbf{M}^k - z_1\,\mathbf{M}^{k-1} - \cdots - z_{k-1}\,\mathbf{M} - z_k\,\mathbf{I} = 0,$$

which by (2.27) yields a linear recurrence relation for the $\ell(n)$ and thus for their ith entries. $\square$

A language L of A^* is said to have *bounded growth* if the coefficients of its generating function are uniformly bounded, that is, if for every n there are less than k words of length n in L, for a fixed integer k. If x, y and z are words in A^*, the language $x\,y^*z$ is called a *ray language*. A ray language, or any finite union of ray languages, is rational and has bounded growth. The following converse is folklore (see (Sakarovitch 2003) and see also in Section 3.3.2, the proof of Theorem 3.3.16).

Proposition 2.6.3 *A rational language L has bounded growth if and only if it is a finite union of ray languages.*

An automaton is said to be *k-local* if the end of any computation of length k depends on its label only, and not on its origin. Remark that a 1-local automaton is deterministic.

2.6.2 Transducers

As defined in Chapter 1, a transducer $\mathcal{T}$ (from A^* to B^*) is an automaton whose transitions are labelled by pairs of words (elements of $A^* \times B^*$). We write $\mathcal{T} = (Q, A, B, E, I, T)$ where $E \subseteq Q \times A^* \times B^* \times Q$ is the set of transitions and where I and T are subsets of Q which we consider as functions from Q into $\mathbb{B}$ in view of forthcoming generalisations. The transducer $\mathcal{T}$ is *finite* if E, and thus the useful part of Q, is finite.

The set of labels of successful computations, which we denote by $|\mathcal{T}|$, is a subset of $A^* \times B^*$, that is, the graph of a relation from A^* to B^*, the *relation realised by $\mathcal{T}$*. If $\mathcal{T}$ is finite, $|\mathcal{T}|$ is a *rational subset* of $A^* \times B^*$, hence realises a *rational relation*. If the labels of the transitions of a (finite) transducer $\mathcal{T}$ are projected on the first (resp. the second) component, we get a (finite) automaton, which we call the *(underlying) input automaton* (resp. the *(underlying) output automaton*) which recognises the domain (resp. the image) of the relation $|\mathcal{T}|$: both are rational languages of A^* (resp. of B^*). Remark also that *morphisms* (from a free monoid into another) are realised by one state transducers.

In contrast with Theorem 2.6.1, rational relations *are not closed under intersection*, and thus the set of rational relations is not a Boolean algebra. Moreover, as the Post Correspondence Problem may easily be described as the intersection of the graph of two morphisms, it is not decidable whether the intersection of two rational relations is empty, from which one deduces that equivalence of rational relations is not decidable.

On the positive side, rational relations from a free monoid into another one are *closed under composition*, and the image of a rational language by a rational relation is rational. From the definition itself follows that the inverse of a rational relation is a rational relation (it suffices to exchange the first and the second components of the labels).

The model of finite transducers may be transformed, without changing the class of realised relations, in order to allow various proofs. In particular, the initial and final functions (from Q to $\mathbb{B}$) may be generalised to functions from Q into $(\varepsilon \times B^*)$ – or, by abuse, from Q into B^* – together with the adequate, and obvious, modification of the definition of the label of a computation.

Figure 2.20 shows three transducers: one for the identity ι, one for ι_K the identity restricted to the rational set $K = a^*b^*$, that is, the intersection with K, and one for the relation γ' which maps every word w onto the set of words of the same length as w and greater in the lexicographic order (assuming that $a < b$).

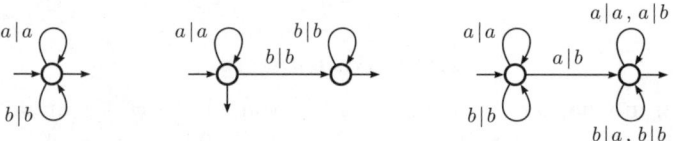

Fig. 2.20 Three transducers.

As an example of the usefulness of rational relations in the study of rational languages, let us give a simple and short proof of a classical property often credited to (Shallit 1994) and that appears several times in this chapter. If L is a language, we denote by $\mathsf{minlg}\,(L)$ (resp. $\mathsf{Maxlg}\,(L)$) the set of words of L which have no lesser (resp. greater) word of the same length in L in the lexicographic, or radix order (they coincide on words of the same length).

Proposition 2.6.4 *If L is a rational language, then so are $\mathsf{minlg}\,(L)$ and $\mathsf{Maxlg}\,(L)$.*

Proof Any word v of L which is greater (in the lexicographic order) than another word u of L of the same length belongs to $\iota_L(\gamma'(\iota_L(u)))$. Thus $\mathsf{minlg}\,(L) = L \setminus \mathsf{Im}[\iota_L \circ \gamma' \circ \iota_L]$, and is rational when L is.

An analogous equality holds for $\mathsf{Maxlg}\,(L)$. □

2.6.3 Synchronous transducers and relations

The three transducers of Figure 2.20 have the property that the label of every transition is a pair of letters, which immediately implies that they realise *length preserving* relations. Being length preserving however is somewhat too strong a restriction and this constraint is relaxed by allowing the replacement, in either component, of a letter by a padding symbol which does not belong to any alphabet – traditionally denoted by a $ – under the 'padding condition', that is, no letter can appear after the padding symbol on the same component. Such transducers are called *synchronous transducers*. They realise *synchronous relations*,† obtained by the projection which erases the padding symbol, and the family of synchronous relations (from A^* into B^*) is denoted by Syn $A^* \times B^*$.

The introduction of the padding symbol is more than a technical trick since in particular it is not decidable whether a given rational relation is synchronous or not (see (Frougny and Sakarovitch 1993)). However synchronous relations are a very natural subfamily of rational relations and they have been given a logical characterisation in (Eilenberg, Elgot, and Shepherdson 1969). Most of the rational relations that are considered in this chapter are synchronous. Figure 2.21 shows a synchronous transducer for the complement of the identity ι, and one for the relation γ which maps every word w onto the set of words which are greater than w in the radix order.

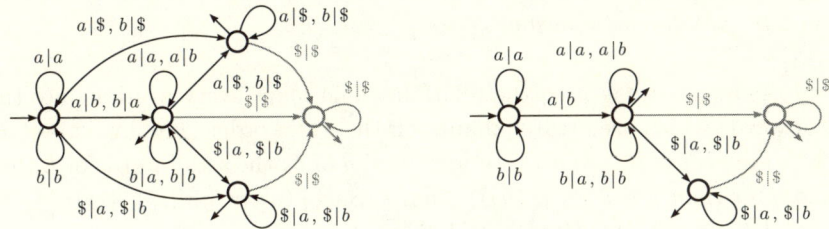

Fig. 2.21 Two synchronous transducers.

Thanks to the following two properties, Syn $A^* \times B^*$ provides a family of rational relations which can be fruitfully used in constructions and proofs. The first one follows from the fact that the pairs of letters from two alphabets can be considered as letters from the product alphabet and thus synchronous tranducers as finite automata.

Theorem 2.6.5 Syn $A^* \times B^*$ *is an effective Boolean algebra of rational relations.*

Theorem 2.6.6 Syn $A^* \times B^*$ *is closed under composition.*

Let $\mathcal{T} = (Q, A_\$, B_\$, E, I, T)$ and $\mathcal{U} = (R, B_\$, C_\$, F, J, U)$ be two synchronous transducers which realise the two relations $|\mathcal{T}|: A^* \to B^*$ and

† In (Sakarovitch 2003), synchronous relations are defined as relations realised by letter-to-letter transducers whose final functions maps states into (Rat $A^* \times \varepsilon) \cup (\varepsilon \times $Rat $B^*)$. Hopefully, the two definitions are equivalent; the present one is preferred as it makes Theorem 2.6.5 and Theorem 2.6.6 more evident.

$|\mathcal{U}|\colon B^* \to C^*$ respectively. Let then $\mathcal{T} \circ \mathcal{U}$ be the synchronous transducer $\mathcal{T} \circ \mathcal{U} = (\,Q \times R, A_\$, C_\$, G, I \times J, T \times U\,)$ defined by

$$G = \{((p,r),(a,c),(q,s)) \mid \exists b \in B_\$ \quad (p,(a,b),q) \in E, (r,(b,c),s) \in F\}.$$

Without loss of generality, we can assume that both $\mathcal{T}$ and $\mathcal{U}$ are completed by transitions labelled by $(\$,\$)$ and that go from every final state to a sink state equipped with a loop labelled in the same way (the grey part in the transducers of Figure 2.21). Under this assumption, it is a formality to check that $\mathcal{T} \circ \mathcal{U}$ realises the relation $|\mathcal{U}| \circ |\mathcal{T}|\colon A^* \to C^*$. This construction is used in Section 2.2.2.2 for the construction of $\mathcal{W}_2''$.

The fruitfulness of the notion is visible in establishing the following property. If A is a totally ordered alphabet, the *radix order* is a well-ordering on A^* and thus on any of its subset L; we denote by Succ_L the function which maps every word of L onto its successor in L in the radix order.

Proposition 2.6.7 *If L is a rational language, then Succ_L is a synchronous (functional) relation.*

Proof As above, we write γ the relation which maps every word w onto the set of words which are greater than w in the radix order. For any subset K of A^*, $\min(K) = K \setminus \gamma(K)$. For any word u of L, the set of words of L that are greater than u is $\iota_L(\gamma(\iota_L(u)))$. Hence $\mathrm{Succ}_L(u) = \min(\iota_L(\gamma(\iota_L(u)))) = \iota_L(\gamma(\iota_L(u))) \setminus \gamma(\iota_L(\gamma(\iota_L(u))))$ and $\mathrm{Succ}_L = \iota_L \circ \gamma \circ \iota_L \setminus \gamma \circ \iota_L \circ \gamma \circ \iota_L$ is a synchronous relation by Theorem 2.6.5 and Theorem 2.6.6. □

Remark 2.6.8 If we take a slightly more general definition for the successor function, namely, a function ω whose restriction on L realises Succ_L, we may find *non-rational languages* whose successor function is realised by a finite (letter-to-letter right) transducer. In this case, L is strictly contained in $\mathrm{Dom}\,\omega$. Such an example is given by the numeration system in rational base (see Section 2.5).

2.6.4 The left-right duality

Before studying further specialised classes of transducers, let us recall and precise the conventions and terminology relative to the duality between the *left-to-right* and *right-to-left* reading.

The *transpose* of a word of A^*, $w = a_1 a_2 \cdots a_n$, with the a_i's in A, is the word $w^t = a_n a_{n-1} \cdots a_1$, that is, the sequence of letters obtained by reading w *from right to left*. Transposition is additively extended to subsets of A^*: $L^t = \bigcup_{w \in L} w^t$.

The *transpose* of an automaton $\mathcal{A} = (Q, A, E, I, T)$, is the automaton $\mathcal{A}^t = (Q, A, E^t, T, I)$ where $E^t = \{(q, a, p) \mid (p, a, q) \in E\}$. Obviously, $L(\mathcal{A}^t) = [L(\mathcal{A})]^t$.

A number of properties of automata are *directed*, that is, correspond to properties of the reading of words from left to right; *e.g.* being deterministic. If $\mathcal{A}^t$ has such a property P, $\mathcal{A}$ is said to have the property *co-P*. For instance, $\mathcal{A}$ is *co-deterministic* if $\mathcal{A}^t$ is deterministic.

Another way to bring the left-right duality into play is to consider *right automata*, that is, automata that read words *from right to left* (a procedure that can prove to be natural when reading numbers: from least to most significant digit). It amounts to the same thing to say that w is accepted by a right automaton $\mathcal{A}$ or that it is accepted by the (left) automaton $\mathcal{A}^t$.

These notions go over to transducers. The *transpose* of a transducer $\mathcal{T} = (Q, A, B, E, I, T)$, is the transducer $\mathcal{T}^t = (Q, A, B, E^t, T, I)$ where $E^t = \{(q, (f^t, g^t), p) \mid (p, (f, g), q) \in E\}$ and $|\mathcal{T}^t| = |\mathcal{T}|^t$. A *right transducer* reads the input word, and 'writes' the output word from right to left. As above, the relation realised by a right transducer $\mathcal{T}$ is the same as the one realised by the (left) transducer $\mathcal{T}^t$.

Being synchronous is a *directed* notion, because of the 'padding condition' or, to state it in another way, because the padding symbols are written at the *right end* of words, and we have (implicitly) defined it for (left) transducer, thus we have defined the *left synchronous relations*. A relation is *co-synchronous* – we also say *right synchronous* if it is realised by a synchronous right transducer, or by the transpose of a synchronous transducer. In general, a left synchronous relation is not a right synchronous one. Relations that are both left and right synchronous have been characterised recently (Carton 2009). We consider below an important particular case of such relations.

2.6.5 Letter-to-letter transducers and bld-relations

We call *letter-to-letter transducer* (with a slight abuse of words) a transducer whose transitions are labelled by pairs of letters *and* whose initial and final functions map states into $(A^* \times \varepsilon) \cup (\varepsilon \times B^*)$. In a relation realised by such a transducer, the lengths of a word and its images are not necessarily equal but their difference is bounded. More important, the converse of this simple observation is true.

Let $\theta \colon A^* \to B^*$ be a relation with the property that there exists an integer k such that, for every f in A^* and every g in $\theta(f)$, then $||f| - |g|| \leqslant k$. If $k = 0$, θ is a *length preserving relation*; for an arbitrary k, θ has been called a *bounded length difference relation* (Frougny and Sakarovitch 1993) or *bounded length discrepancy relation* (Sakarovitch 2003), *bld-relation* for short in any case.

It is not difficult to verify that a rational relation is bld if, and only if, any transducer $\mathcal{T}$ (without padding symbol!) which realises θ has the property that the label of every circuit in $\mathcal{T}$ is such that the length of the 'input' is equal to the length of the 'output', a property which is thus decidable. The following result is essentially due to Eilenberg who proves it for length-preserving relations (Eilenberg 1974); it has been extended to bld-relations in (Frougny and Sakarovitch 1993). It relates a property of the graph of a rational relation (being bld) to the way this relation may be realised (being synchronous).

Proposition 2.6.9 *A bld-rational relation is both left and right synchronous.*

The next characterisation of bld relations within synchronous ones goes back to (Elgot and Mezei 1965).

Proposition 2.6.10 *A left (or right) synchronous relation with finite image and finite co-image is a bld-rational relation.*

Corollary 2.6.11 *If L is a rational language, then Succ_L is realised by a finite letter-to-letter right transducer.*

2.6.6 Sequential transducers and functions

A transducer (from A^* to B^*) is said to be *sequential* (resp. *co-sequential*) if its underlying input automaton is *deterministic* (resp. *co-deterministic*) and the initial and final functions map its states into $\varepsilon \times B^*$. With this definition, a sequential, or co-sequential, transducer realises a *functional* relation. A function (from A^* to B^*) is said to be *sequential* (resp. *co-sequential*) if it is realised by a *sequential* (resp. *co-sequential*) transducer. Of course, a co-sequential function is realised by a sequential right transducer.

Sequential functions are characterised within rational functions by a topological criterion in the following way: the *prefix distance* d of two words u and v is defined as $\mathrm{d}(u, v) = |u| + |v| - 2|u \wedge v|$, where $u \wedge v$ is the *longest common prefix* of u and v.

Definition 2.6.12 *A function φ is said to be k-Lipschitz (for the prefix distance) if:*

$$\forall u, v \in \mathsf{Dom}\,\varphi, \quad \mathrm{d}(\varphi(u), \varphi(v)) \leq k\,\mathrm{d}(u, v).$$

The function φ is Lipschitz if there exists a k such that φ is k-Lipschitz.

Theorem 2.6.13 (Choffrut 1977) *A rational function is sequential if, and only if, it is Lipschitz.*

By the left-right duality, we define the *suffix distance* d_s on A^*: $d_s(u, v) = |u| + |v| - 2|u \wedge_s v|$, where $u \wedge_s v$ is the *longest common suffix* of u and v. A rational function is co-sequential if, and only if, it is Lipschitz for the suffix distance. At this point, it cannot be skipped that sequentiality is a decidable property for functions realised by finite transducers (see (Choffrut 1977)) although this does not play any role in this chapter.

We call *piecewise (co-)sequential* a function that is a finite union of (co-)sequential functions (thus with disjont domains). And we have the following.

Proposition 2.6.14 (Angrand and Sakarovitch) *If L is a rational language, then Succ_L is a piecewise co-sequential function.*

2.7 Notes

As we have already mentioned, we consider only *positional* numeration systems. However, we indicate a pioneer work on the relations between numeration and finite automata, which is the paper (Raney 1973), in which it is proved that the continued fractions expansion of a real number can be coded by an infinite word on a two-letter alphabet, and that homographic transformations can be realised by finite transducers.

2.7.1 Representation in integer base

On the links between numeration, logic and finite automata there is a survey (Bruyère, Hansel, Michaux, et al. 1994).

A generalisation of Cobham's Theorem to real numbers has been established in a series of papers (Boigelot and Brusten 2009, Boigelot, Brusten, and Bruyère 2008). It is proved in particular that, if a set S of positive real numbers is recognised by a finite weak deterministic automaton in two integer bases that are multiplicatively independent, then S is definable in $\langle \mathbb{R}, \mathbb{Z}, +, < \rangle$, which means that S is a finite union of intervals with rational endpoints.

2.7.2 Representation in real base

Symbolic dynamical systems defined by a particular order on the set of infinite words on a finite alphabet have been studied from an ergodic point of view in (Takahashi 1980).

An algebraic integer $\beta > 1$ is a *Salem number* if all its Galois conjugates have modulus ≤ 1, with at least one conjugate with modulus 1. It has been proved in (Boyd 1989) that every Salem number of degree 4 is a Parry number. Boyd also conjectured that it is still true in degree 6, but false for degree ≥ 8. An algebraic integer $\beta > 1$ is a *Perron number* if all its Galois conjugates have modulus $< \beta$. Perron numbers are introduced in (Lind 1984). Every Parry number is a Perron number. In (Solomyak 1994) and in (Flatto, Lagarias, and Poonen 1994) is proved that all the Galois conjugates of a Parry number have modulus strictly less than the Golden Ratio. Beta-expansions also appear in the mathematical description of quasicrystals, see (Gazeau, Nešetřil, and B. Rovan, eds 2007).

The study of the β-shift from the point of view of the Chomsky hierarchy has been done by K. Johnson. A symbolic dynamical system is said to be *context-free* if the set of its finite admissible factors is a context-free language. It is proved in (Johnson 1999) that the β-shift is context-free if, and only if, it is sofic.

In Section 2.2.2.3 we have presented expansions of minimal weight in base 2. Recently, the investigation of minimal weight expansions has been extended to the Fibonacci numeration system in (Heuberger 2004), and an equivalent to the NAF has been defined. When β is a Pisot number the set of β-expansions of minimal weight, where the weight is the absolute sum of the digits, is recognisable by a finite automaton, (Frougny and Steiner 2008). For the Golden Ratio φ the average weight of φ-expansions on the alphabet $\{-1, 0, 1\}$ of the numbers of absolute value less than M is $\frac{1}{5} \log_\varphi M$, which means that typically only every fifth digit is non-zero. Note that the corresponding value for 2-expansions of minimal weight is $\frac{1}{3} \log_2 M$, see (Arno and Wheeler 1993, Bosma 2001), and that $\frac{1}{5} \log_\varphi M \approx 0.288 \log_2 M$.

Fractals and tilings are the subject of Chapter 5 of this book. Let us just mention some works using finite automata associated with numeration in an irrational base. The celebrated Rauzy fractal is associated with numeration in base the *Tribonacci number* which is the root > 1 of the polynomial $X^3 - X^2 - X - 1$. The boundary of the Rauzy fractal (and of more general fractals associated with Pisot numbers) has been described by a finite automaton in (Messaoudi 1998, Messaoudi 2000) and (Durand and Messaoudi 2009).

Finite automata and substitutions are treated in (Pytheas Fogg 2002,

Chapter 7). (Canterini and Siegel 2001a, Canterini and Siegel 2001b) have defined the prefix-suffix automaton associated with a substitution of Pisot type.

Beta-expansions have been extended to the case of finite fields by (Hbaib and Mkaouar 2006) and (Scheicher 2007). Here β is an element of the field of formal Laurent series $\mathbb{F}((X^{-1}))$, with $|\beta| > 1$. The main difference with the classical real base is that all the expansions are admissible. Moreover the (F) Property is satisfied if and only if β is a Pisot element of $\mathbb{F}((X^{-1}))$, that is to say, β is an algebraic integer over $\mathbb{F}[X]$ such that for all Galois conjugates $|\beta_i| < 1$ (Scheicher 2007).

2.7.3 Canonical numeration systems

In the case where the alphabet associated with a number β is $A_\beta = \{0, 1, \ldots, N(\beta) - 1\}$, the 'clearing algorithm' of (Gilbert 1981) gives an easy way of computing the expansion of an integer in the system (β, A_β).

Tilings generated by a canonical numeration system have been investigated by many authors. The first one is probably the *twin dragon* tiling, linked to the Penney CNS defined by the base $-1 + i$, which was obtained by Knuth as the set $\{z \in \mathbb{C} \mid z = \sum_{j \geq 0} d_j (-1 + i)^{-j}, d_j \in \{0, 1\}\}$, see (Knuth 1998).

There are contributions on fractals and tilings in (Gilbert 1991), (Scheicher and Thuswaldner 2002) and (Akiyama and Thuswaldner 2005).

There have been a number of generalisations of CNS. Let us mention that the case where β is not an algebraic integer but an algebraic number has been considered in particular in (Gilbert 1991). It is mentioned that for any rational $p/q > 1$, $\beta = -p/q$ with digit set $\{0, 1, \ldots, p-1\}$ forms a CNS in which any number of $\mathbb{Z}[1/q]$ has a finite representation.

Scheicher and Thuswaldner investigated number systems in polynomial rings over finite fields (Scheicher and Thuswaldner 2003).

2.7.4 Representation in rational base

Expansions in rational base are linked to the problem of the distribution of the fractional part of the powers of rational numbers.

The distribution modulo 1 of the powers of a rational number, indeed the problem of proving whether they form a dense set or not, is an old problem. Pisot, Vijayaraghavan and André Weil have shown that there are infinitely many limit points. With this problem as a background, Mahler asked in (Mahler 1968) whether there exists a non-zero real z such that the fractional part of $z(3/2)^n$ for $n = 0, 1, \ldots$ fall into $[0, 1/2[$. It is not known whether such a real – called a Z-number – does exist but Mahler showed that the set of Z-numbers is at most countable. His proof is based on the fact that the fractional part of a Z-number (if it exists) has an expansion in base 3/2 which is entirely determined by its integral part.

Koksma proved that for almost every real number $\theta > 1$ the sequence $(\{\theta^n\})_n$ is uniformly distributed in $[0,1]$, but very few results are known for specific values of θ. One of these is that *if θ is a Pisot number*, then the above sequence converges to 0 if we identify $[0,1[$ with $\mathbb{R}/\mathbb{Z}$.

The next step in attacking this problem has been *to fix the rational $\frac{p}{q}$* and to study the distribution of the sequence

$$f_n(z) = \left\{ z \left(\frac{p}{q}\right)^n \right\}$$

according to the value of the real number z. Once again, the sequence $f_n(z)$ is uniformly distributed for almost all $z > 0$, but nothing is known for specific values of z.

In the search for z's for which the sequence $f_n(z)$ is *not uniformly distributed*, Mahler considered those for which the sequence is eventually contained in $[0, \frac{1}{2}[$. Mahler's notation is generalised as follow: let I be a (strict) subset of $[0,1[$ and let

$$\mathbf{Z}_{\frac{p}{q}}(I) = \{z \in \mathbb{R} \mid \left\{ z \left(\frac{p}{q}\right)^n \right\} \text{ stays eventually in } I \}.$$

Mahler's problem is to ask whether $\mathbf{Z}_{\frac{3}{2}}\left([0, \frac{1}{2}[\right)$ is empty or not.

Mahler's work has been developed in two directions: the search for subsets I as large as possible such that $\mathbf{Z}_{\frac{p}{q}}(I)$ is empty and conversely the search for subsets I as small as possible such that $\mathbf{Z}_{\frac{p}{q}}(I)$ is non-empty.

Along the first line, remarkable progress has been made by Flatto et al. (Flatto, Lagarias, and Pollington 1995) who proved that the set of reals s such that $\mathbf{Z}_{\frac{p}{q}}\left([s, s + \frac{1}{p}[\right)$ is empty is *dense* in $[0, 1 - \frac{1}{p}]$, and Bugeaud (Bugeaud 2004b) proved that its complement is of Lebesgue measure 0. Along the other line, Pollington (Pollington 1981) showed that $\mathbf{Z}_{\frac{3}{2}}\left([\frac{4}{65}, \frac{61}{65}[\right)$ is non-empty.

It is proved in (Akiyama, Frougny, and Sakarovitch 2008) that if $p \geqslant 2q - 1$, there exists a subset $Y_{\frac{p}{q}}$ of $[0,1[$, of Lebesgue measure $\frac{q}{p}$, such that $\mathbf{Z}_{\frac{p}{q}}\left(Y_{\frac{p}{q}}\right)$ is countably infinite. The elements of $\mathbf{Z}_{\frac{p}{q}}\left(Y_{\frac{p}{q}}\right)$ are indeed the reals which have two $\frac{p}{q}$-expansions. Coming back to the historical 3/2 case, we have that the set of positive numbers z such that $\left\{ z \left(\frac{3}{2}\right)^n \right\} \in [0, 1/3[\cup [2/3, 1[$ for $n = 0, 1, 2, \ldots$ is countably infinite. It is noteworthy that the expansion 'computed' by Mahler for his Z-numbers happens to be exactly one of the $\frac{3}{2}$-expansions presented in Section 2.5 – if it exists.

3
Abstract numeration systems

Pierre Lecomte,

Michel Rigo

The primary motivation for the introduction of the abstract numeration systems stems from the celebrated theorem of Cobham dating back to 1969 about the so-called recognisable sets of integers in any integer base numeration system. Representations of numbers are words over a finite alphabet. There is a one-to-one correspondence between the sets of numbers and the languages made of the corresponding representations. Hence it is natural to consider questions related to formal language theory. In particular, we study sets of integers corresponding to regular languages. The different sections of this chapter are largely independent. However, Section 3.2 presents basic concepts and notation used in all later sections. The main focus is on the representation of integers. Extension to abstract numeration systems of the notion of recognisable sets of integers is studied in Section 3.3. In particular, we present some results about the stability of recognisability after multiplication by a constant. This requires us to discuss the complexity (or counting) function of regular languages. Section 3.4 is about the extension – to any substitutive sequence – of Cobham's theorem from 1972 about the equality of the set of infinite k-automatic words and the set of images under codings of the fixed points of substitutions of constant length k. The notion of an $\mathcal{S}$-automatic sequence is then introduced and various applications to $\mathcal{S}$-recognisability are considered. This chapter ends with a discussion about the representation of real numbers using abstract numeration systems.

3.1 Motivations

The primary role of a numeration system is to replace numbers which by essence are abstract objects by their representations which are words over suitable alphabets. As an example, the k-ary system replaces integers by

Combinatorics, Automata and Number Theory, ed. Valérie Berthé and Michel Rigo.
Published by Cambridge University Press. ©Cambridge University Press 2010.

their representations in base k. Denote by B_k the set

$$\{0,\ldots,k-1\}^* \setminus (0\{0,\ldots,k-1\}^*)$$

of words over $\{0,\ldots,k-1\}$ not starting with 0. The one-to-one correspondence mapping a non-negative integer n onto its k-ary representation $\mathrm{rep}_k(n) \in B_k$, also denoted $\langle n \rangle_k$ in Chapter 2, can be extended to a one-to-one correspondence between $2^{\mathbb{N}}$ and 2^{B_k}: any set $X \subseteq \mathbb{N}$ is associated with the language $\mathrm{rep}_k(X)$ made up from the k-ary representations of the numbers in X. It is therefore natural to study the relationship existing between the arithmetic properties of integers and the syntactical properties of the corresponding representations in a given numeration system. From the point of view of formal language theory, one can focus on those sets $X \subseteq \mathbb{N}$ for which a (deterministic) finite automaton can be used to decide for any given word w over $\{0,\ldots,k-1\}$ whether or not w belongs to $\mathrm{rep}_k(X)$. Sets having such a property are called k-*recognisable* sets. In some sense, a k-recognisable set can be considered as particularly simple because through the k-ary numeration system it has a simple algorithmic description. Recall that in the *Chomsky hierarchy*, see for instance (Sudkamp 2005), (Shallit 2008), deterministic finite automata accepting regular languages are the simplest model of computation. However, dealing with k-recognisable sets has a price. As observed by A. Cobham, see Theorem 1.5.5: k-recognisability depends heavily on the choice of the base and sets which are k-recognisable for all $k \geq 2$ are exactly the eventually periodic sets. For that matter, also see Chapter 2 and in particular Subsection 2.2.4.

First we recall the notion of representation with respect to a U-system. Also see Section 2.3.3.

Definition 3.1.1 Let us extend the notion of k-ary numeration system by replacing the sequence $(k^n)_{n \geq 0}$ with some increasing sequence $U = (U_n)_{n \geq 0}$ of integers such that $U_0 = 1$. Using successive Euclidean divisions, we define the U-*representation* of any positive integer n. Let ℓ be such that $U_\ell \leq n < U_{\ell+1}$. We can greedily decompose n in a unique way as

$$n = \sum_{k=0}^{\ell} c_k U_k \text{ with } c_\ell \neq 0 \text{ and } \sum_{k=0}^{i} c_k U_k < U_{i+1},\ \forall i \in \{0,\ldots,\ell\}. \quad (3.1)$$

This latter greedy condition implies that, for all $i \in \{0,\ldots,\ell\}$, we have

$$c_i \in \{0,\ldots,\lceil U_{i+1}/U_i \rceil - 1\}.$$

The U-*representation* of n is $c_\ell \cdots c_0$ and is denoted by $\mathrm{rep}_U(n)$. We set

$\mathrm{rep}_U(0) := \varepsilon$. Since we are interested in language theoretic properties related to U-representations, we assume moreover that the set $\{U_{i+1}/U_i \mid i \geq 0\}$ is bounded from above to ensure that $\mathrm{rep}_U(\mathbb{N}) = \{\mathrm{rep}_U(n) \mid n \in \mathbb{N}\}$ is a language over a finite alphabet. We set A_U to be the minimal (or canonical) alphabet of this language, i.e., $A_U = \mathrm{alph}(\mathrm{rep}_U(\mathbb{N}))$. We can similarly to the integer base systems define the notion of U-recognisable sets. A set $X \subseteq \mathbb{N}$ is said to be *U-recognisable*, if $\mathrm{rep}_U(X)$ is accepted by a DFA.

These systems can be referred to as *positional numeration systems* and the corresponding sequence U is usually called the *scale* or the *basis* of the system. The greediness of the U-representations implies the next proposition. We recall Definition 1.2.15 for the definition of the genealogical ordering $\prec$.

Proposition 3.1.2 *For all $m, n \in \mathbb{N}$, we have*

$$m < n \Leftrightarrow \mathrm{rep}_U(m) \prec \mathrm{rep}_U(n)$$

where the genealogical ordering $\prec$ is induced by the natural ordering of the alphabet $A_U \subset \mathbb{N}$.

Definition 3.1.3 In what follows, in particular for Propositions 3.1.5 and 3.1.9, when speaking of a *numeration system* $U = (U_n)_{n \geq 0}$ we assume that U is increasing, that $U_0 = 1$ and that the set $\{U_{i+1}/U_i \mid i \geq 0\}$ is bounded.

Amongst the possibly U-recognisable subsets of $\mathbb{N}$, the whole set $\mathbb{N}$ is of special interest. It seems natural to consider numeration systems $U = (U_n)_{n \geq 0}$ for which $\mathrm{rep}_U(\mathbb{N})$ is regular, i.e., for which $\mathbb{N}$ is U-recognisable. In that case, we have an algorithm using a constant amount of memory and working in time proportional to the length of the input – namely a DFA – to check whether or not any given word over A_U is a valid U-representation. Let us investigate a little bit further what is implied by the U-recognisability of $\mathbb{N}$. We recall that a multi-graph is a graph which is permitted to have multiple edges, that is, edges that connect the same pair of vertices. First we start with the following lemma whose proof can be compared with the proof of Proposition 2.6.2.

Lemma 3.1.4 *Let $G = (V, E)$ be a directed finite multi-graph where V is the set of vertices of G, E is its multi-set of arcs in $V \times V$ and let $q, r \in V$. The map $\mathcal{U}_{q,r} : \mathbb{N} \to \mathbb{N}$ counting the number $\mathcal{U}_{q,r}(n)$ of directed paths of length n from q to r satisfies a linear recurrence relation with (constant) integer coefficients.*

Proof Consider the adjacency matrix $\mathbf{M} \in \mathbb{N}^{V \times V}$ of G: for all vertices

$x, y \in V$, $\mathbf{M}_{x,y}$ is the number of arcs from x to y, i.e., paths of length 1. A simple induction shows that, for all $x, y \in V$ and all $n \in \mathbb{N}$, $[\mathbf{M}^n]_{x,y}$ is the number of paths of length n from x to y. By the Cayley-Hamilton theorem, if $C(X) = \det(\mathbf{M} - X\mathbf{I}) = c_k X^k + \cdots + c_1 X + c_0 \in \mathbb{Z}[X]$ is the characteristic polynomial of $\mathbf{M}$ where $\mathbf{I}$ is the identity matrix of size $k = \mathrm{Card}(V)$, then $C(\mathbf{M}) = \mathbf{0}$. Multiplying by $\mathbf{M}^n$, $n \geq 0$, gives $c_k \mathbf{M}^{n+k} + \cdots + c_1 \mathbf{M}^{n+1} + c_0 \mathbf{M}^n = \mathbf{0}$. To conclude the proof, observe that this latter relation between matrices holds component-wise. □

The next result is a reformulation of Proposition 2.3.47.

Proposition 3.1.5 *Let $U = (U_n)_{n \geq 0}$ be a numeration system as given in Definition 3.1.3. If $\mathbb{N}$ is U-recognisable, then the sequence $(U_n)_{n \geq 0}$ satisfies a linear recurrence relation with (constant) integer coefficients.*

Proof Note that $\mathrm{rep}_U(U_\ell) = 10^\ell$ for all $\ell \geq 0$. Amongst the words of length $\ell + 1$ in $\mathrm{rep}_U(\mathbb{N})$, the smallest one for the genealogical ordering is 10^ℓ. Consequently, for all $\ell \geq 0$, $U_{\ell+1} - U_\ell$ is exactly the number of words of length $\ell + 1$ in $\mathrm{rep}_U(\mathbb{N})$. Since this latter language is regular, it is accepted by a DFA and the number of words of length n in $\mathrm{rep}_U(\mathbb{N})$ is equal to the number of paths of length n from the initial state to the final ones. Using Lemma 3.1.4 we deduce that the sequence $(\mathrm{Card}(\mathrm{rep}_U(\mathbb{N}) \cap A_U^n))_{n \geq 0}$ satisfies a linear recurrence relation with integer coefficients and the conclusion follows easily. □

As sketched by the next two examples, the converse of Proposition 3.1.5 does not hold in general. Sufficient conditions for $\mathbb{N}$ to be U-recognisable are considered in (Loraud 1995), (Hollander 1998). See Theorem 2.3.57. Also see Example 3.1.

Example 3.1.6 (Shallit 1994) Such a counter-example is given by the sequence $(U_n)_{n \geq 0}$ defined by $U_n = (n+1)^2$. Then we have $U_0 = 1$, $U_1 = 4$, $U_2 = 9$ and $U_{n+3} = 3 U_{n+2} - 3 U_{n+1} + U_n$. In that case, $\mathrm{rep}_U(\mathbb{N}) \cap 10^* 10^* = \{10^a 10^b \mid b^2 < 2a + 4\}$ showing with the pumping lemma that $\mathbb{N}$ is not U-recognisable.

Example 3.1.7 (Frougny 2002) We sketch another counter-example related to β-expansions, see Example 2.3.62 for details. Let $\beta = (3 + \sqrt{5})/2$. The β-expansion of 1 is 21^ω. Consider the sequence $(U_n)_{n \geq 0}$ satisfying the recurrence relation $U_{n+3} = 4 U_{n+2} - 4 U_{n+1} + U_n$, for all $n \geq 0$, with $U_0 = 1$, $U_1 = 2$ and $U_2 = 6$. Proceed by contradiction and assume that $\mathbb{N}$ is U-recognisable. Using the postponed Lemma 3.3.5, the set $X = \{U_n - 1 \mid$

$n \geq 0\}$ is U-recognisable because $\operatorname{rep}_U(X) = \mathsf{Maxlg}\,(\operatorname{rep}_U(\mathbb{N}))$. Due to the β-expansion of 1, one can show that all but a finite number of words in $\operatorname{rep}_U(X)$ are of the kind $21^{i_n} 2w_n$ where $i_n \to \infty$ and $|w_n| \to \infty$ as $n \to \infty$. Therefore the pumping lemma shows that $\operatorname{rep}_U(X)$ is not regular.

It is probably worth recalling here a standard result about the general form of linear recurrence sequences, see any standard textbook like (Graham, Knuth, and Patashnik 1989). We assume that all the coefficients and the initial conditions belong to some field extension $\mathbb{K}$ of characteristic zero where the characteristic polynomial of the recurrence factorises as linear factors.

Theorem 3.1.8 *Let $k \geq 1$ and $r_0, \ldots, r_{k-1} \in \mathbb{K}$. Let $(U_n)_{n\geq 0}$ be a sequence satisfying, for all $n \geq 0$,*

$$U_{n+k} = r_{k-1} U_{n+k-1} + \cdots + r_0 U_n.$$

If $\alpha_1, \ldots, \alpha_t$ are the roots of the characteristic polynomial $X^k - r_{k-1} X^{k-1} - \cdots - r_0$ of the recurrence with respective multiplicities $m_1, \ldots, m_t$, then there exist polynomials $P_1, \ldots, P_t \in \mathbb{K}[X]$ of degree respectively less than $m_1, \ldots, m_t$ and depending only on the initial conditions $U_0, \ldots, U_{k-1} \in \mathbb{K}$ such that

$$\forall n \geq 0,\ U_n = P_1(n)\,\alpha_1^n + \cdots + P_t(n)\,\alpha_t^n.$$

Let $B \subset \mathbb{Z}$ be an alphabet. The function $\operatorname{val}_{B,U} : B^* \to \mathbb{Z}$ maps any word $w = c_\ell \cdots c_0 \in B^*$ onto $\operatorname{val}_{B,U}(w) = \sum_{k=0}^{\ell} c_k U_k$. It is clear that, for all $n \in \mathbb{N}$, $\operatorname{val}_{B,U}(\operatorname{rep}_U(n)) = n$. On the other hand, for all $w \in B^*$, such that $\operatorname{val}_{B,U}(w) \geq 0$, the so-called *normalisation* maps w onto $\operatorname{rep}_U(\operatorname{val}_{B,U}(w))$ which is not necessarily equal to w. Indeed, to apply $\operatorname{val}_{B,U}$, it is not required that w is a greedy U-representation. For example, considering the Fibonacci numeration system $F = (1, 2, 3, 5, \ldots)$, $\operatorname{rep}_F(\operatorname{val}_{\{0,1\},F}(11)) = 100$. Note that in general, if the alphabet B contains negative elements, then the normalisation is a partial function whose domain is a strict subset of B^*.

We already know that eventually periodic sets are k-recognisable for all $k \geq 2$. What can be said in a wider framework?

Proposition 3.1.9 *Let $p, r \geq 0$. If $(U_n)_{n\geq 0}$ is a numeration system given as in Definition 3.1.3 and satisfying a linear recurrence relation with integer coefficients, then*

$$\operatorname{val}_{A_U,U}^{-1}(p\mathbb{N} + r) = \left\{ c_\ell \cdots c_0 \in A_U^* \ \Big|\ \sum_{k=0}^{\ell} c_k U_k \in p\mathbb{N} + r \right\}$$

is accepted by a DFA that can be effectively constructed. In particular, if $\mathbb{N}$ is U-recognisable, then any eventually periodic set is U-recognisable.

Prior to the proof, notice that for any integer $n \geq 0$, $\mathrm{val}_{A_U,U}^{-1}(n) \setminus 0 A_U^*$ is a finite set of words $\{x_1, \ldots, x_{t_n}\}$ over A_U such that $\mathrm{val}_{A_U,U}(x_i) = n$ for all $i = 1, \ldots, t_n$. This non-empty set contains in particular $\mathrm{rep}_U(n)$.

Proof Since regular sets are stable under finite modifications, *i.e.*, adding and/or removing a finite number of words to a regular language gives a regular language, we can assume that $p > r \geq 0$. The sequence $(U_n \bmod p)_{n \geq 0}$ is eventually periodic say, with preperiod m and period q, that is, for all $i \geq m$, $U_i \equiv U_{i+q} \pmod{p}$. We build a deterministic finite automaton $\mathcal{A}$ accepting reversal of the words in $\{w \in A_U^* \mid \mathrm{val}_U(w) \in p\mathbb{N} + r\}$. The alphabet of the automaton is A_U. States are ordered pairs (t, s) where $0 \leq t < p$ and $0 \leq s < m + q$. The first component of a state handles the value modulo p of the digits that have been read and the second component takes care of the periodicity of $(U_n \bmod p)_{n \geq 0}$. The initial state is $(0, 0)$. Final states are the ones with the first component equal to r. Transitions are defined as follows

$$\forall s < m + q - 1 : \ (t, s) \xrightarrow{j} (jU_s + t \bmod p, \ s + 1)$$

and

$$(t, m + q - 1) \xrightarrow{j} (jU_{m+q-1} + t \bmod p, \ m)$$

for all $j \in A_U$. Note that $\mathcal{A}$ does not check the greediness of the accepted words, the construction only relies on the U-numerical value of the words modulo p. For the particular case, one has to consider the intersection of two regular languages $\mathrm{rep}_U(\mathbb{N}) \cap \mathrm{val}_U^{-1}(p\mathbb{N} + r)$. □

Taking into account this latter result, Cobham's theorem and also the above discussion about deciding whether a word is a valid U-representation or not, the recognisability of $\mathbb{N}$ is desirable and can be considered as a natural expectation for any numeration system. In particular in view of the above proposition, $\mathbb{N}$ is U-recognisable if, and only if, all eventually periodic sets are U-recognisable. If this becomes our basic requirement, we can *consider the problem the other way round*. Instead of taking a sequence U of integers and looking for conditions that guarantee the U-recognisability of $\mathbb{N}$, we take an arbitrary infinite regular language L over an alphabet A to build a numeration system, this language L being viewed as the set of valid representations of all the integers. Indeed a numeration system U is characterised by the language of all the representations and its monotonicity. In view of Proposition 3.1.2 about order-preserving representations, if the alphabet A is totally ordered, say $(A, <)$, we order the words of L by the increasing genealogical order induced by the ordering of A, say $w_0 \prec w_1 \prec w_2 \prec \cdots$. This ordering of L gives a one-to-one correspondence between $\mathbb{N}$ and L: with $n \in \mathbb{N}$ is associated $w_n \in L$. Such a bijection is the essence of a numeration system: associating a representation with any integer. We have therefore the following formal definition.

Definition 3.1.10 An *abstract numeration system* (or *ANS* for short) is a triple $\mathcal{S} = (L, A, <)$ where L is an infinite regular language over a totally ordered alphabet $(A, <)$. The map $\operatorname{rep}_\mathcal{S} : \mathbb{N} \to L$ is the one-to-one correspondence mapping $n \in \mathbb{N}$ onto the $(n+1)$th word in the genealogically ordered language L, which is called the $\mathcal{S}$-*representation* of n. The $\mathcal{S}$-representation of 0 is the first word in L. The inverse map is denoted by $\operatorname{val}_\mathcal{S} : L \to \mathbb{N}$. If w is a word in L, $\operatorname{val}_\mathcal{S}(w)$ is its $\mathcal{S}$-*numerical value*.

Note that one could relax the assumption about the regularity of L in the definition of an ANS $\mathcal{S} = (L, A, <)$. In that case, we still have to consider words of L in ascending genealogical order. This would give a wider framework to work with, but then we lose the recognisability of $\mathbb{N}$.

Now let us present four examples. Some of them can be related to a suitably chosen sequence $(U_n)_{n \geq 0}$, others can not, showing that the class of ANS is strictly larger than the usual class of numeration systems given by Definition 3.1.1 and for which $\mathbb{N}$ is U-recognisable.

Example 3.1.11 Let U be a numeration system in the sense of Definition 3.1.1 such that $\mathbb{N}$ is U-recognisable. In view of Proposition 3.1.2, this numeration system can be considered as an ANS by enumerating the words of $\operatorname{rep}_U(\mathbb{N})$ by the genealogical order induced by the natural ordering of the digits. As an example, taking the language $1\{0,01\}^* \cup \{\varepsilon\}$ with the natural ordering $0 < 1$ gives back the Fibonacci system and the language $B_k = \{0, \ldots, k-1\}^* \setminus 0\{0, \ldots, k-1\}^*$ gives the k-ary system.

Example 3.1.12 Consider $L = a^*b^*$ with $a < b$ and the ANS $\mathcal{S} = (L, \{a, b\}, <)$. The first few words in L in ascending genealogical order are

$$\varepsilon \prec a \prec b \prec aa \prec ab \prec bb \prec aaa \prec aab \prec abb \prec bbb \prec \cdots.$$

For example, $\operatorname{val}_\mathcal{S}(abb) = 8$ and $\operatorname{rep}_\mathcal{S}(3) = aa$. If we consider the bijection from L to $\mathbb{N}^2$ mapping the word $a^i b^j$ onto the pair (i, j), $i, j \geq 0$, it is not difficult to see that the genealogical ordering of L corresponds to the primitive recursive Peano enumeration of $\mathbb{N}^2$, that is

$$\text{val}_{\mathcal{S}}(a^i b^j) = \frac{1}{2}(i+j)(i+j+1) + j. \tag{3.2}$$

Let us pursue this example a little bit further. Assume that we have a map $v : \{a, b\} \to \mathbb{N}$ which assigns some weight to a and b. We show that there exists *no* sequence $U = (U_n)_{n \geq 0}$ defining a numeration system in the sense of Definition 3.1.1 such that, for all words $w_\ell \cdots w_0 \in L$,

$$\text{val}_{\mathcal{S}}(w_\ell \cdots w_0) = \sum_{k=0}^{\ell} v(w_k) U_k.$$

We proceed by contradiction and we assume that such a sequence exists. Since $U_0 = 1$ and $\text{val}_{\mathcal{S}}(a) = 1$, $\text{val}_{\mathcal{S}}(b) = 2$, we must have $v(a) = 1$ and $v(b) = 2$. Notice that $\text{val}_{\mathcal{S}}(aa) = 3$ and this quantity should be equal to $v(a)U_1 + v(a)U_0$. Consequently, $U_1 = 2$. Therefore $v(b)U_1 + v(b)U_0 = 6$ but $\text{val}_{\mathcal{S}}(bb) = 5$, which gives a contradiction.

This example shows that the family of ANS contains more numeration systems that those of Definition 3.1.1 for which $\mathbb{N}$ is U-recognisable. To contrast with ANS which only depend on the genealogical ordering, recall that the systems associated with Definition 3.1.1 are referred as positional numeration systems. As we shall soon see in Lemma 3.2.2, the general expression of $\text{val}_{\mathcal{S}}(w)$ for an ANS $\mathcal{S} = (L, A, <)$ and a word $w \in L$ involves usually more than a single linear recurrence sequence.

Example 3.1.13 (Allowing leading zeroes) The reader may have noticed that we have defined greedy U-representations as words not starting with zero. It not only makes the definition unambiguous but this choice was made on purpose because in the context of abstract numeration systems, adding leading zeroes to a word changes its length and therefore its position in the genealogically ordered language. As an example, consider the language $\{0,1\}^*$. The first few words in this language are ε, 0, 1, 00, 01, 10, 11, 000. So for the ANS $\mathcal{S} = (\{0,1\}^*, \{0,1\}, 0 < 1)$, we get $\text{val}_{\mathcal{S}}(0) = 1$, $\text{val}_{\mathcal{S}}(00) = 3$ and so on. Actually, if one considers the map v defined as $v(0) = 1$ and $v(1) = 2$, it is not difficult to see that $\text{val}_{\mathcal{S}}(w_\ell \cdots w_0) = \sum_{k=0}^{\ell} v(w_k) 2^k$ which corresponds to the so-called 2-*adic numeration system*: any non-negative integer is uniquely represented as a word over $\{1,2\}$ with the sequence $(2^n)_{n \geq 0}$ being the underlying scale.

Example 3.1.14 (Pisot numeration system) Recall from Chapter 2 that a *Pisot number* is an algebraic integer $\alpha > 1$ whose conjugates have modulus less than 1. Consider a linear recurrence sequence $(U_n)_{n \geq 0}$ whose characteristic polynomial is the minimal polynomial of a Pisot number α of degree k. If the integer initial conditions are $1 = U_0 < U_1 < \cdots < U_{k-1}$, then there exists some $c > 0$ such that $U_n \sim c\alpha^n$ and moreover $|U_n - c\alpha^n| \to 0$, as n tends to infinity, because we can apply Theorem 3.1.8 about the general solution of a linear recurrence† and for any other root $\beta \neq \alpha$ of the characteristic polynomial of the recurrence, since $|\beta| < 1$, we have $\beta^n \to 0$ as $n \to \infty$. This sequence can be used to define a numeration system in the sense of Definition 3.1.1. It is well-known that for such a system, $\mathbb{N}$ is U-recognisable. Moreover, all the nice properties of the integer base numeration systems still hold: logical or substitutive characterisations of the U-recognisable sets, stability of U-recognisability under addition and multiplication by a constant, normalisation is computable by finite automata, ...see (Bruyère and Hansel 1997), (Frougny 1992). Since $\mathbb{N}$ is U-recognisable, these 'state-of-the-art' positional numeration systems are all special cases of ANS.

Example 3.1.15 (Prefix-closed language) In the case of an ANS based on a prefix-closed language, we propose a useful picture of the map val_S. This is simply another expression of the genealogical ordering of L. As an example consider the language $L = \{a, ba\}^*\{\varepsilon, b\}$ and $a < b$. In Figure 3.1 we represent the first three levels of the corresponding *trie*, *i.e.*, a rooted tree where the edges are labelled by letters from A, and the nodes are labelled by prefixes of words in the considered language L. Let $u \in A^*$, $a \in A$. If ua is (a prefix of) a word in L, then there is an edge between u and ua. Note that for a prefix-closed language L, all prefixes of words in L belong to L. In the nodes, we have written the S-numerical value of the corresponding words in L. The root is associated with ε. When considering a prefix-closed

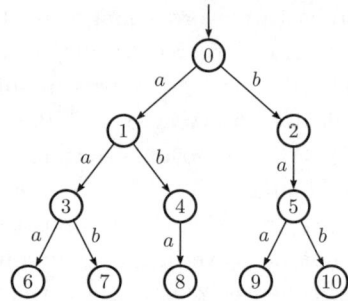

Fig. 3.1 A trie for words of length ≤ 3 in L.

---

† Remember that all the roots of the minimal polynomial of an algebraic number are simple.

language ordered by genealogical order, the nth level of the trie contains all words of L of length n in lexicographic order from left to right assuming that the sons of a node are also ordered with respect to the ordering of the alphabet.

Definition 3.1.16 For a given ANS $\mathcal{S} = (L, A, <)$, any integer n is mapped onto a word $\text{rep}_\mathcal{S}(n)$ and any subset X of $\mathbb{N}$ is mapped onto a language $\text{rep}_\mathcal{S}(X) \subseteq L$. We have therefore a one-to-one correspondence between $2^\mathbb{N}$ and 2^L. In this general framework of abstract numeration systems, we are interested in sets X of integers such that $\text{rep}_\mathcal{S}(X)$ is regular. These sets are called $\mathcal{S}$-recognisable sets.

Example 3.1.17 Considering the ANS $\mathcal{S} = (a^*b^*, \{a, b\}, a < b)$ from Example 3.1.12, the set X of triangular numbers

$$X = \{0, 1, 3, 6, 10, \ldots\} = \{n(n+1)/2 \mid n \geq 0\}$$

is $\mathcal{S}$-recognisable. Indeed, it is easy to check that $\text{rep}_\mathcal{S}(X) = a^*$ because the number of words of length $n \geq 0$ in a^*b^* is exactly $n + 1$. If we consider the ANS $\mathcal{R} = (a^*b^*, \{a, b\}, b < a)$ where the ordering of the alphabet has been reversed, then $\text{rep}_\mathcal{R}(X) = b^*$.

3.2 Computing numerical values and $\mathcal{S}$-representations

Let $\mathcal{S} = (L, A, <)$ be an abstract numeration system. Since L is a regular language, we can consider a complete DFA $\mathcal{A} = (Q, A, E, \{q_0\}, T)$ having $\delta_\mathcal{A} : Q \times A^* \to Q$ as (extended) transition function. We write $q.w$ as a shorthand for $\delta_\mathcal{A}(q, w)$ if the context is clear, $q \in Q$, $w \in A^*$. First we show, as a consequence of the genealogical ordering of L, that the function $\text{val}_\mathcal{S}$ can be computed recursively and we obtain a decomposition of any integer using functions $\mathcal{U}$ and $\mathcal{V}$ counting the number of words accepted from the different states of $\mathcal{A}$ and defined below.

For all $q \in Q$, $L_q = \{w \in A^* \mid q.w \in F\}$ is the regular language of words accepted in $\mathcal{A}$ starting from state q. We set

$$\mathcal{U}_q(n) := \text{Card}(L_q \cap A^n) \quad \text{and} \quad \mathcal{V}_q(n) := \sum_{k=0}^{n} \mathcal{U}_q(k) \qquad (3.3)$$

being respectively the number of words of length n and, at most n, accepted from q. From Lemma 3.1.4, all the sequences $(\mathcal{U}_q(n))_{n \geq 0}$, $q \in Q$, satisfy the same linear recurrence relation. Indeed, $\mathcal{U}_q(n)$ is the sum over all the final states $f \in T$ of the number of paths of length n from q to f. Moreover, $(\mathcal{V}_q(n))_{n \geq 0}$ satisfies a linear recurrence relation that can be derived from the one satisfied by $(\mathcal{U}_q(n))_{n\geq 0}$, simply by observing that, for all $n \geq 0$, we have $\mathcal{V}_q(n+1) - \mathcal{V}_q(n) = \mathcal{U}_q(n+1)$. Also we write

$$\mathcal{U}(n) := \mathcal{U}_{q_0}(n) = \mathrm{Card}(L \cap A^n) \quad \text{and} \quad \mathcal{V}(n) := \mathcal{V}_{q_0}(n) = \mathrm{Card}(L \cap A^{\leq n}).$$

Note that these two maps $\mathcal{U}(n)$ and $\mathcal{V}(n)$ are independent of the choice of the DFA accepting L. They only depend on the language L, so if emphasis on L is needed, we also use notation like $\mathcal{U}_L(n)$ and $\mathcal{V}_L(n)$. The map $\mathcal{U} : \mathbb{N} \to \mathbb{N}$ is often called the *counting function* or (*combinatorial*) *complexity function* of L (compare with Definition 1.2.12).

Since, for all $q \in Q$, the language L_q is regular, we can consider the ANS $\mathcal{S}_q = (L_q, A, <)$. The corresponding maps $\mathrm{val}_{\mathcal{S}_q}$ and $\mathrm{rep}_{\mathcal{S}_q}$ are respectively denoted by val_q and rep_q. For some $q \in Q$, L_q can possibly be finite. If this is the case, we extend the definition of an ANS to allow this situation but the domain of rep_q is therefore $\{0, \ldots, \mathrm{Card}\, L_q - 1\}$.

Example 3.2.1 Consider the regular language L accepted by the DFA depicted in Figure 3.2 having states q_0, q_1 and q_2. With notation introduced

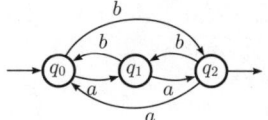

Fig. 3.2 A DFA accepting the language $L \subset \{a, b\}^*$.

above, the first few words in $L = L_{q_0}$, L_{q_1} and L_{q_2} are respectively

$$\begin{aligned}
L_{q_0} &= \{b, aa, abb, bab, bba, aaab, aaba, abaa, baaa, bbbb, aaaaa, \ldots\} \\
L_{q_1} &= \{a, bb, aab, aba, baa, aaaa, abbb, babb, bbab, bbba, aaabb, \ldots\} \\
L_{q_2} &= \{\varepsilon, ab, ba, aaa, bbb, aabb, abab, abba, baab, baba, bbaa, aaaab, \ldots\}
\end{aligned}$$

and the adjacency matrix of the automaton is

$$\mathbf{M} = \begin{pmatrix} 0 & 1 & 1 \\ 1 & 0 & 1 \\ 1 & 1 & 0 \end{pmatrix}.$$

Therefore, using the same technique as in the proof of Lemma 3.1.4 (simply compute the characteristic polynomial of $\mathbf{M}$), we get that the sequences $(\mathcal{U}_{q_i}(n))_{n\geq 0}$ satisfy $\mathcal{U}_{q_i}(n+3) = 3\mathcal{U}_{q_i}(n+1) + 2\mathcal{U}_{q_i}(n)$ for all $n \geq 0$. We have computed the first few values of these sequences:

	0	1	2	3	4	5	6	7	8	9	10
$\mathcal{U}_{q_0}(n) = \mathcal{U}_{q_1}(n)$	0	1	1	3	5	11	21	43	85	171	341
$\mathcal{U}_{q_2}(n)$	1	0	2	2	6	10	22	42	86	170	342

Abstract numeration systems 119

For instance, $\text{rep}_{\mathcal{S}}(0) = \text{rep}_{q_0}(0) = b$, $\text{rep}_{q_1}(0) = a$ and $\text{rep}_{q_2}(0) = \varepsilon$. In the same way, $\text{val}_{\mathcal{S}}(abb) = \text{val}_{q_0}(abb) = 2$, $\text{val}_{q_1}(aab) = 2 = \text{val}_{q_2}(ba)$.

Now that we have a good knowledge of the different maps val_q, $\mathcal{U}_q$ and $\mathcal{V}_q$, we present a lemma used to compute recursively the $\mathcal{S}$-numerical value of any word in L.

Lemma 3.2.2 *Let $\mathcal{S} = (L, A, <)$ be an ANS where L is accepted by a DFA $\mathcal{A} = (Q, A, E, \{q_0\}, T)$. Let $q \in Q$. If the word xy belongs to L_q where the factor y is non-empty, then*

$$\text{val}_q(xy) = \text{val}_{q.x}(y) + \mathcal{V}_q(|xy| - 1) - \mathcal{V}_{q.x}(|y| - 1) + \sum_{\substack{w < x \\ |w| = |x|}} \mathcal{U}_{q.w}(|y|).$$

Proof We have to compute the number of words belonging to L_q and genealogically less than xy. There are three kinds of such words. The first ones are the words of length less than $|xy|$. We have $\mathcal{V}_q(|xy| - 1)$ such words. Then we have to take into account words in L_q of length $|xy|$ having a prefix w such that $|w| = |x|$ and $w < x$. It is clear that there are

$$\text{Card}\{wz \in L_q \mid w < x, |w| = |x|, |z| = |y|\} = \sum_{\substack{w < x \\ |w| = |x|}} \mathcal{U}_{q.w}(|y|)$$

words of this kind. Finally, we have words in L_q of length $|xy|$ having x as prefix and lexicographically less than xy. We have to count the number of words in $L_{q.x}$ of length $|y|$ lexicographically less than y. We get $\text{val}_{q.x}(y) - \mathcal{V}_{q.x}(|y| - 1)$ such words because $\text{val}_{q.x}(y)$ is the total number of words less than y in $L_{q.x}$ and we have to subtract words of length less than $|y|$. □

For ANS we have a 'multi-scale' analogue to the decomposition (3.1) occurring in positional numeration systems. Let $\mathcal{S} = (L, A, <)$ be an ANS where L is accepted by a DFA $\mathcal{A}$. Instead of having a unique sequence $(U_n)_{n \geq 0}$ to express the numerical value of a word $c_\ell \cdots c_0$ as $\sum_{k=0}^{\ell} c_k U_k$, we are considering the several sequences $(\mathcal{U}_q(n))_{n \geq 0}$, in fact, as many sequences as states in $\mathcal{A}$.

Theorem 3.2.3 *Let $\mathcal{S} = (L, A, <)$ be an ANS where L is accepted by the DFA $\mathcal{A} = (Q, A, E, \{q_0\}, T)$. Let $w = w_1 \cdots w_n \in L$. Then we have*

$$\text{val}_{\mathcal{S}}(w) = \sum_{q \in Q} \sum_{i=1}^{|w|} b_{q,i}(w) \, \mathcal{U}_q(|w| - i) \qquad (3.4)$$

where for $i = 1, \ldots, |w|$,

$$b_{q,i}(w) = \mathrm{Card}\{a \in A \mid a < w_i,\ q_0.w_1 \cdots w_{i-1}a = q\} + \mathbf{I}_{q,q_0} \qquad (3.5)$$

where $\mathbf{I}$ is the identity matrix in $\{0,1\}^{Q \times Q}$, so $\mathbf{I}_{q,q_0} = 1$ if, and only if, $q = q_0$. Moreover, these coefficients are bounded:

$$0 \le \sum_{q \in Q} b_{q,i}(w) \le \mathrm{Card}\, A.$$

Proof Formula (3.4) can be proved using Lemma 3.2.2 inductively. Also it can be proved by observing that the summand for $(q,i) \in Q \times \{1, \ldots, |w|\}$ with $q \ne q_0$ is the number of words $v = v_1 \cdots v_n$ of length $|w|$ which have prefix $w_1 \cdots w_{i-1}a$ with $a < w_i$, which means that $v \prec w$, the state q is reached after reading the first i letters of v, and the suffix $v_{i+1} \cdots v_n$ is accepted from state q. For $q = q_0$ the summand for (q_0, i) equals the number with the same descriptions as above plus the number of words of length $|w| - i$ which are accepted by the automaton starting from q_0. Summing over all possible pairs (q, i) first gives the number of words $v \prec w$ with $|v| = |w|$, the extra summand for $q = q_0$ equals the number of words v in L with $|v| < |w|$. Altogether this equals $\mathrm{val}_{\mathcal{S}}(w)$. $\square$

The following proposition asserts that the coefficients of the decomposition (3.4) can be obtained almost *automatically*. This is merely the translation of (3.5) but this fact will play an important role when dealing with the representation of real numbers in Section 3.5.

Proposition 3.2.4 *Let $\mathcal{S} = (L, A, <)$ be an ANS where L is accepted by a DFA $\mathcal{A} = (Q, A, E, \{q_0\}, T)$. For any $q \in Q$, one can efficiently build a sequential letter-to-letter transducer $\mathcal{T}_q$ computing, for all i, the coefficients $b_{q,i}(w)$ occurring in (3.4), i.e., to any input $w_1 \cdots w_n \in L$ is associated the output $b_{q,1}(w) \cdots b_{q,n}(w)$ of $\mathcal{T}_q$.*

Proof It is a direct consequence of (3.5). Let $q \in Q$. We build a transducer $\mathcal{T}_q$ having the same set of states and the same initial state and final states as $\mathcal{A}$. The input and output alphabets of $\mathcal{T}_q$ are respectively A and $\{0, \ldots, \mathrm{Card}\, A\}$. For any transition $(r, c, s) \in E$ appearing in $\mathcal{A}$, we take for the transducer $\mathcal{T}_q$ the transition $(r, (c, x_{r,c}), s)$ where $x_{r,c} = \mathrm{Card}\{a \in A \mid a < c, r.a = q\} + \mathbf{I}_{q_0, q}$. $\square$

Example 3.2.5 Consider the ANS from Example 3.2.1. The corresponding three transducers are given in Figure 3.3 (one for each state of the DFA). Notice that the output associated with a is always 0 except for q_0 where it

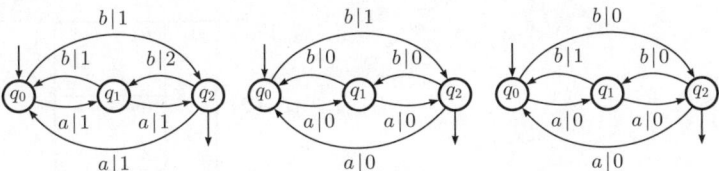

Fig. 3.3 The three transducers $\mathcal{T}_{q_0}$, $\mathcal{T}_{q_1}$ and $\mathcal{T}_{q_2}$.

is 1. Consider for instance the word $abaa$ which is such that $\text{val}_{\mathcal{S}}(abaa) = 7$. Feeding the transducers with this word gives respectively the words 1111, 0000 and 0100. Therefore, we find the expected decomposition of 7:

$$\begin{array}{llllll}
& \mathbf{1}\mathcal{U}_{q_0}(3) & +\mathbf{1}\mathcal{U}_{q_0}(2) & +\mathbf{1}\mathcal{U}_{q_0}(1) & +\mathbf{1}\mathcal{U}_{q_0}(0) & \\
+ & \mathbf{0}\mathcal{U}_{q_1}(3) & +\mathbf{0}\mathcal{U}_{q_1}(2) & +\mathbf{0}\mathcal{U}_{q_1}(1) & +\mathbf{0}\mathcal{U}_{q_1}(0) & \\
+ & \mathbf{0}\mathcal{U}_{q_2}(3) & +\mathbf{1}\mathcal{U}_{q_2}(2) & +\mathbf{0}\mathcal{U}_{q_2}(1) & +\mathbf{0}\mathcal{U}_{q_2}(0) & = 3+1+1+0+2.
\end{array}$$

Now let us turn our attention to the computation of $\text{rep}_{\mathcal{S}}(n)$ where $\mathcal{S} = (L, A, <)$ and $A = \{a_1 < \cdots < a_t\}$. We assume that we have at our disposal a DFA $\mathcal{M}$ having q_0 as initial state and accepting L. In particular, $\mathcal{U}_q(n)$ and $\mathcal{V}_q(n)$ can be obtained using the linear recurrence relations derived from $\mathcal{M}$ and its adjacency matrix. As usual, we simply write $q.w$ for the action of w in A^* on q in the set of states of $\mathcal{M}$.

Observe that $|\text{rep}_{\mathcal{S}}(n)| = \ell > 0$ if, and only if, $\mathcal{V}(\ell-1) \leq n < \mathcal{V}(\ell)$. Indeed, if L contains some words of length ℓ, then the first word of length ℓ has position $\mathcal{V}(\ell-1)$ in the genealogically ordered language L and $\mathcal{U}(\ell) > 0$. So, for all $n \geq 0$, we get

$$|\text{rep}_{\mathcal{S}}(n)| = \inf\{m \in \mathbb{N} \mid n < \mathcal{V}(m)\}.$$

Let $n \geq 0$ and $\ell = |\text{rep}_{\mathcal{S}}(n)|$. To determine the first letter of the $\mathcal{S}$-representation of n, we compute, for all $s \in \{1, \ldots, t\}$, the number $N[\ell, a_s]$ of words of length ℓ belonging to L and beginning with $a_1, a_2, \ldots$ or a_s. It is given by $N[\ell, a_s] := \sum_{i=1}^{s} \mathcal{U}_{q_0.a_i}(\ell-1)$. For convenience, we set $N[\ell, a_0] = 0$. There exists a unique r such that $N[\ell, a_{r-1}] \leq n - \mathcal{V}(\ell-1) < N[\ell, a_r]$ and the first letter of the $\mathcal{S}$-representation of n is therefore a_r. We proceed in the same way to determine the other letters of the $\mathcal{S}$-representation. Table 3.1 sketches the structure of the genealogically ordered language L for words of length ℓ with their corresponding position in L. The pseudocode algorithm presented in Table 3.2 computes the $\mathcal{S}$-representation w of n. In the last line of this algorithm, wa_j represents the concatenation of the word w and the letter a_j.

$\mathcal{V}_{\ell-1}$	a_1	a_1	$\cdots$
	$\vdots$	$\vdots$	
$\mathcal{V}_{\ell-1} + \mathcal{U}_{q_0.a_1a_1}(\ell-2)$		a_2	$\cdots$
		$\vdots$	
	$\vdots$		
	a_1	a_p	$\cdots$
$\mathcal{V}_{\ell-1} + \mathcal{U}_{q_0.a_1}(\ell-1)$	a_2	a_1	$\cdots$
	$\vdots$	$\vdots$	
$\mathcal{V}_{\ell-1} + \mathcal{U}_{q_0.a_1}(\ell-1) + \mathcal{U}_{q_0.a_2a_1}(\ell-2)$		a_2	$\cdots$
		$\vdots$	
	a_2	a_p	$\cdots$
	$\vdots$		
$\mathcal{V}_{\ell-1} + \sum_{i=1}^{p-1} \mathcal{U}_{q_0.a_i}(\ell-1)$	a_p	a_1	$\cdots$
	$\vdots$	$\vdots$	
$\mathcal{V}_{\ell-1} + \sum_{i=1}^{p-1} \mathcal{U}_{q_0.a_i}(\ell-1) + \mathcal{U}_{q_0.a_pa_1}(\ell-2)$		a_2	$\cdots$
		$\vdots$	
	$\vdots$		
$\mathcal{V}_\ell - 1$	a_p	a_p	$\cdots$

Table 3.1. *Words in L of length ℓ in increasing genealogical order and their corresponding $\mathcal{S}$-numerical values.*

3.3 $\mathcal{S}$-recognisable sets

The aim of this section is to present some properties of $\mathcal{S}$-recognisable sets of integers. We know that eventually periodic sets are k-recognisable, for all $k \geq 2$, and by Proposition 3.1.9 also U-recognisable for numeration systems such that $\mathbb{N}$ is U-recognisable. Interestingly this property† still holds for ANS which is somehow encouraging if one thinks about a possible analogue of the Cobham theorem.

Theorem 3.3.1 *Let $\mathcal{S} = (L, A, <)$ be an ANS. Any eventually periodic set is $\mathcal{S}$-recognisable.*

Due to the importance of this result, we provide two different proofs. The first one is direct: we show that the minimal automaton of the set

† It was the very first result we were looking for. Getting it was a true motivation for the study of ANS.

```
Find the unique ℓ be such that 𝒱(ℓ − 1) ≤ n < 𝒱(ℓ)
q ← q₀
m ← n − 𝒱(ℓ − 1)
w ← ε
FOR i = 1 TO ℓ DO
    s ← 1
    WHILE m ≥ 𝒰_{q.a_s}(ℓ − i) DO
        m ← m − 𝒰_{q.a_s}(ℓ − i)
        s ← s + 1
    END-WHILE
    q ← q.a_s
    w ← wa_s
END-FOR
```

Table 3.2. *An algorithm for computing* $\mathrm{rep}_\mathcal{S}(n)$.

of representations of any eventually periodic set is finite. It presents some sharp argument but it does not provide any 'constructive feeling' about the machinery behind as does the second proof.

Prior to these proofs we can make the following observation. It is well-known that taking in a regular language the smallest (respectively largest) word of every length for the genealogical ordering gives again a regular language, see Proposition 2.6.4 and Lemma 3.3.5. We can reformulate Theorem 3.3.1 to obtain some *decimation* operation preserving the regularity of languages.

Theorem 3.3.2 *Let* $(A, <)$ *be a totally ordered alphabet. If we order the words of a regular language* $L \subseteq A^*$ *in the genealogical order induced by* $<$, *say* $w_0 \prec w_1 \prec w_2 \prec \cdots$, *then for all* $p > r \geq 0$ *the language* $\{w_{np+r} \in L \mid n \geq 0\}$ *is regular.*

Let us present a first proof of Theorem 3.3.1 or equivalently of the above theorem.

Proof It is well-known that a language $M \subseteq A^*$ is regular if, and only if, its minimal automaton $\mathcal{A}_M$ is finite. The set of states of $\mathcal{A}_M$ is $\{w^{-1}M \mid w \in A^*\}$ where $w^{-1}M = \{u \mid wu \in M\}$. See any standard textbook about automata theory like (Eilenberg 1974) or (Sakarovitch 2003).

Since a finite union of regular languages is regular and since adding or removing a finite number of words in a regular language does not change its regularity, it is enough to show that the minimal automaton $\mathcal{A}_P$ of the language $P = \mathrm{rep}_\mathcal{S}(p\mathbb{N} + r) \subseteq A^*$ is finite, with $p > r \geq 0$. The states of $\mathcal{A}_P$ are the sets

$$w^{-1}P = \{x \in A^* \mid \mathrm{val}_\mathcal{S}(wx) \equiv r \pmod{p}\}, w \in A^*.$$

Consider the regular language L on which the ANS $\mathcal{S}$ is built and its corresponding minimal automaton $\mathcal{A}_L$. In fact we could consider any DFA accepting L, the arguments remain unchanged. The reader should be careful, we are considering two different minimal automata: $\mathcal{A}_L$ which we know is finite and $\mathcal{A}_P$ which we would like to prove to be finite. We know that, for all states q of $\mathcal{A}_L$, the sequences $(\mathcal{U}_q(n))_{n\geq 0}$ and $(\mathcal{V}_q(n))_{n\geq 0}$ introduced in (3.3) satisfy a linear recurrence equation and are therefore eventually periodic mod p. Let q_0 be the initial state of $\mathcal{A}_L$. Assume that the period of the sequence $(\mathcal{V}_{q_0}(n) \bmod p)_{n\geq 0}$ is t and its preperiod is s. By Lemma 3.2.2, we have

$$\mathrm{val}_{\mathcal{S}}(wx) = \mathrm{val}_{q_0.w}(x) + \mathcal{V}_{q_0}(|wx|-1) - \mathcal{V}_{q_0.w}(|x|-1) + \sum_{\substack{v<w \\ |v|=|w|}} \mathcal{U}_{q_0.v}(|x|).$$

Since $\mathcal{A}_L$ is finite, $q_0.w$ can only take a finite number of values in Q, the set of states of $\mathcal{A}_L$. Working modulo p, for $|w|>s$, the term $\mathcal{V}_{q_0}(|wx|-1)$ can be written as $\mathcal{V}_{q_0}(|x|+i)$ for some $i \in \{0,\ldots,t-1\}$ because $(\mathcal{V}_{q_0}(n) \bmod p)_{n\geq 0}$ is eventually periodic. Modulo p, for all $w \in A^*$, there exist coefficients $j_q \in \{0,\ldots,p-1\}$ such that

$$\sum_{\substack{v<w \\ |w|=|v|}} \mathcal{U}_{q_0.v}(|x|) \equiv \sum_{q\in Q} j_q \mathcal{U}_q(|x|) \pmod{p}.$$

Note that the number of maps $n \mapsto \sum_{q\in Q} j_q \mathcal{U}_q(n)$ is finite and bounded by $p^{\mathrm{Card}\, Q}$. Consequently, for any $w \in A^*$ such that $|w|>s$, the set $w^{-1}P$ is of the form

$$\{x \mid \mathrm{val}_k(x) + \mathcal{V}_{q_0}(|x|+i) - \mathcal{V}_k(|x|-1) + \sum_{q\in Q} j_q \mathcal{U}_q(|x|) \equiv r \pmod{p}\}$$

for some $k \in Q$, $j_q \in \{0,\ldots,p-1\}$ and $i \in \{0,\ldots,t-1\}$. So, there are finitely many sets of this kind and the set $\{w^{-1}P \mid w \in A^*\}$ of states of the minimal automaton of P is finite. □

Now let us consider an alternative proof followed by an example.

Idea of the proof. We notice that all the sequences occurring in Lemma 3.2.2 are eventually periodic modulo p with some common period M and they are all periodic after at most K terms. We build an NFA which reads entries from the left, say leading letter first, and which computes the numerical value of entries modulo p. In order to apply Lemma 3.2.2, we have to keep track of the state the DFA accepting L is in when reading such an entry. Also we have to deal with the common period M: when we enter a new word w, the value of $|w| \bmod M$ is guessed non-deterministically. Only a correct guess can lead to the unique final state. The last K letters are treated separately because we cannot rely any more on the periodic structure.

Proof Let $\mathcal{A} = (Q, A, E, \{q_0\}, T)$ be a DFA accepting the language L with $\delta_{\mathcal{A}} : Q \times A \to Q$ as the transition function. We give a method to construct an NFA accepting $\text{rep}_S(p\mathbb{N} + r)$. The key argument is again that, for all $q \in Q$, the sequences $(\mathcal{U}_q(n) \bmod p)_{n \geq 0}$ and $(\mathcal{V}_q(n) \bmod p)_{n \geq 0}$ are eventually periodic. Therefore, for each $q \in Q$, there exist g_q, h_q, s_q and t_q belonging to $\mathbb{N}$ such that $h_q, t_q \geq 1$,

$$\forall n \geq g_q, \ \mathcal{U}_n(q) \equiv \mathcal{U}_{n+h_q}(q) \pmod{p}$$

and

$$\forall n \geq s_q, \ \mathcal{V}_n(q) \equiv \mathcal{V}_{n+t_q}(q) \pmod{p}.$$

Set M to be the least common multiple of the constants h_q and t_q and

$$K = \max\left\{\sup_{q \in Q} g_q, \sup_{q \in Q} s_q + 1\right\}.$$

Taking $s_q + 1$ instead of s_q is due to the term $\mathcal{V}_{q.a}(|y| - 1)$ in the expression of $\text{val}_q(ay)$ given by Lemma 3.2.2: for all $a \in A$ and all $y \in A^+$ such that $ay \in L_q$, we have

$$\text{val}_q(ay) = \text{val}_{q.a}(y) + \overbrace{\mathcal{V}_q(|y|) - \mathcal{V}_{q.a}(|y|-1) + \sum_{\substack{b < a \\ b \in A}} \mathcal{U}_{q.b}(|y|)}^{=:R(q,a,|y|)}. \tag{3.6}$$

This shows that for $|y| \geq K$, $\text{val}_q(ay)$ is congruent to $\text{val}_{q.a}(y)$ modulo p but a quantity $R(q, a, |y|) \bmod p$ depending only on q, a and $|y| \bmod M$ has to be added. Hence, for $n \geq 1$ and letters $a_1, \ldots, a_{K+n} \in A$, we obtain inductively that $\text{val}_q(a_1 \cdots a_n a_{n+1} \cdots a_{n+K})$ is equal to

$$\begin{aligned} & R(q, a_1, K+n-1) + R(q.a_1, a_2, K+n-2) + \cdots \\ + & R(q.a_1 \cdots a_{n-1}, a_n, K) + \text{val}_{q.a_1 \cdots a_n}(a_{n+1} \cdots a_{n+K}). \end{aligned}$$

We will mimic this latter decomposition using the following NFA. Consider the NFA $\mathcal{B} = (Q' \cup \{f\}, A, E', I, F)$ where $Q' = Q \times \{0, \ldots, p-1\} \times \{0, \ldots, M-1\}$, $I = \{(q_0, 0, j) \mid j = 0, \ldots, M-1\}$ and $F = \{f\}$ where $f \notin Q'$. Let us show that this NFA accepts the language $\text{rep}_S(p\mathbb{N} + r) \cap A^{\geq K}$. The language $\text{rep}_S(p\mathbb{N} + r) \cap A^{<K}$ is finite and can be handled separately. The first component of any state of $\mathcal{B}$ is used to store and mimic the behaviour of $\mathcal{A}$. The estimated numerical value modulo p resulting from the letters that have already been read, is stored in the second component (starting from zero, first we add $R(q, a_1, K+n-1)$, then

$R(q.a_1, a_2, K+n-2)$, etc.). The length modulo M of the remaining part of the word to be read is stored in the last component of the state, this length is unknown at the beginning and will non-deterministically be guessed by $\mathcal{B}$. Now we will explain the details. The transition relation of $\mathcal{B}$ is such that

$$\begin{cases} ((q,i,j), a, (\delta_A(q,a), k, j-1)) \in E', & \text{if } j \in \{1, \ldots, M-1\}; \\ ((q,i,j), a, (\delta_A(q,a), k, M-1)) \in E', & \text{if } j = 0 \end{cases} \quad (3.7)$$

where the unique k, depending on q, a, i and j, is easily computed using (3.6). Actually $k = i + R(q, a, j) \bmod p$. If $x \in L_q \cap A^K$ and $i \in \{0, \ldots, p-1\}$ are such that $\mathrm{val}_q(x) + i \equiv r \pmod{p}$ then we also add

$$((q, i, K \bmod M), x, f) \in E'$$

and note that these are the only relations leading to the final state. The reading of a word w of length at least K could *a priori* be started from any of the M initial states of $\mathcal{B}$. But note that only one of these states has to be chosen with respect to $|w|$ to reach the unique final state f at the end of the reading of w. $\square$

Example 3.3.3 We apply the above construction to obtain an NFA recognising $\mathrm{rep}_\mathcal{S}(3\mathbb{N}+1)$ where $\mathcal{S}$ is the ANS based on the language L of the words over $\{a,b\}$ having an even number of b. We assume that $a < b$. The minimal automaton of L is depicted in Figure 3.4. We have

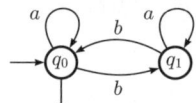

Fig. 3.4 DFA accepting words with an even number of b.

$$\begin{cases} \mathcal{U}_{q_0}(n) = 2^{n-1}, \forall n \geq 1 \\ \mathcal{U}_{q_0}(0) = 1 \end{cases} \quad \text{and} \quad \begin{cases} \mathcal{U}_{q_1}(n) = 2^{n-1}, \forall n \geq 1 \\ \mathcal{U}_{q_1}(0) = 0. \end{cases}$$

For all $n \geq 1$, $\mathcal{U}_{q_0}(n) = \mathcal{U}_{q_1}(n) \equiv (-1)^{n-1} \pmod{3}$ and, for all $n \in \mathbb{N}$, $\mathcal{V}_{q_0}(n) \equiv (-1)^n \pmod{3}$ and $\mathcal{V}_{q_1}(n) \equiv (-1)^n - 1 \pmod{3}$. Using the notation given in the previous proof, we set $K = 1$ and $M = 2$. From (3.6), we get the following relations modulo 3. If $|w| \geq 1$,

$$\begin{aligned} \mathrm{val}_{q_0}(aw) &\equiv \mathrm{val}_{q_0}(w) + (-1)^{|w|+1} &\pmod{3} \\ \mathrm{val}_{q_0}(bw) &\equiv \mathrm{val}_{q_1}(w) + (-1)^{|w|} + 1 &\pmod{3} \\ \mathrm{val}_{q_1}(aw) &\equiv \mathrm{val}_{q_1}(w) + (-1)^{|w|+1} &\pmod{3} \\ \mathrm{val}_{q_1}(bw) &\equiv \mathrm{val}_{q_0}(w) + (-1)^{|w|} - 1 &\pmod{3} \end{aligned}$$

where we can notice that the last term depends only on $|w| \pmod{2}$. Taking these relations into account, we define as in (3.7) the main part of the transition relation:

	$(q_0,0,0)$	$(q_0,1,0)$	$(q_0,2,0)$	$(q_0,0,1)$	$(q_0,1,1)$	$(q_0,2,1)$
a	$(q_0,2,1)$	$(q_0,0,1)$	$(q_0,1,1)$	$(q_0,1,0)$	$(q_0,2,0)$	$(q_0,0,0)$
b	$(q_1,2,1)$	$(q_1,0,1)$	$(q_1,1,1)$	$(q_1,0,0)$	$(q_1,1,0)$	$(q_1,2,0)$

	$(q_1,0,0)$	$(q_1,1,0)$	$(q_1,2,0)$	$(q_1,0,1)$	$(q_1,1,1)$	$(q_1,2,1)$
a	$(q_1,2,1)$	$(q_1,0,1)$	$(q_1,1,1)$	$(q_1,1,0)$	$(q_1,2,0)$	$(q_1,0,0)$
b	$(q_0,0,1)$	$(q_0,1,1)$	$(q_0,2,1)$	$(q_0,1,0)$	$(q_0,2,0)$	$(q_0,0,0)$

For instance, $((q_0,1,0),b,(q_1,0,1)) \in E$ because in the minimal automaton of L, $q_0.b = q_1$ and if $|w| \equiv 0 \pmod 2$, then $1+(-1)^{|w|}+1 \equiv 0 \pmod 3$. To conclude, observe that $\mathrm{val}_{q_0}(a) = 1$, $b \notin L_{q_0}$, $a \notin L_{q_1}$ and $\mathrm{val}_{q_1}(b) = 0$. So, $((q_0,0,1),a,f)$ and $((q_1,1,1),b,f)$ also belong to the relation defining the NFA.

To reach the final state, the words of even, respectively odd, length have to be read starting from the initial state $(q_0,0,0)$, respectively the second initial state $(q_0,0,1)$. If the reading of a word begins in the wrong initial state with respect to the parity of its length, then no path can reach the final state.

In the general framework of abstract numeration systems, we can consider several kinds of questions about $\mathcal{S}$-recognisability of sets of integers. They are natural extensions of those considered in the classical context of positional numeration systems.

- For a given set $X \subseteq \mathbb{N}$, can we build an ANS $\mathcal{S}$ such that X is $\mathcal{S}$-recognisable?
- For a given ANS $\mathcal{S}$, what kind of arithmetic operations on sets of integers do preserve $\mathcal{S}$-recognisability?
- For a given ANS $\mathcal{S}$, what can be said about the $\mathcal{S}$-recognisable subsets of $\mathbb{N}$?
- In particular, can we obtain some characterisation (logical, arithmetic, whatever ...) of the $\mathcal{S}$-recognisable subsets of $\mathbb{N}$?
- How is $\mathcal{S}$-recognisability dependent on the ANS?

As the reader may observe many challenging questions can be considered in this context, also see the bibliographic notes at the end of the chapter for other related questions. We are far from being able to answer all of them but in the next pages, we will develop some of these topics. Also the use of ANS casts some new light on well-known results occurring in the classical context. Let us start with the following result.

Proposition 3.3.4 (Translation by a constant) *Let $\mathcal{S} = (L, A, <)$ be an ANS. If $X \subseteq \mathbb{N}$ is $\mathcal{S}$-recognisable, then also $X + t$ is $\mathcal{S}$-recognisable for all $t \in \mathbb{N}$.*

Proof See for instance (Lecomte and Rigo 2001). Taking into account the theory of synchronised relations (Frougny and Sakarovitch 1993), the *successor* map defined on L by $w \mapsto \text{rep}_\mathcal{S}(\text{val}_\mathcal{S}(w) + 1)$ is shown to be realised by a left letter-to-letter finite transducer (see Corollary 2.6.11) and the conclusion follows. Also the paper (Angrand and Sakarovitch) is relevant in that context, see Proposition 2.6.14. □

The following result is proved in (Shallit 1994), also see Proposition 2.6.4.

Lemma 3.3.5 *Let L be a regular language over the totally ordered alphabet $(A, <)$. The following languages are regular:*

$$\text{minlg}(L) = \{u \in L \mid w \in L, w \neq u, |w| = |u| \Rightarrow u \prec w\},$$

$$\text{Maxlg}(L) = \{u \in L \mid w \in L, w \neq u, |w| = |u| \Rightarrow w \prec u\}.$$

The following observation is an immediate reformulation of the above result. Because it will be used quite often we state it as a lemma.

Lemma 3.3.6 *Let $\mathcal{S} = (L, A, <)$ be an ANS. The set $\{\mathcal{V}_L(n) \mid n \geq 0\} = \{\text{Card}(L \cap A^{\leq n}) \mid n \geq 0\}$ is $\mathcal{S}$-recognisable.*

Example 3.1.17 about the set of triangular numbers $\{P(n) \mid n \geq 0\}$ where $P(n) = n(n+1)/2$ can be revisited in light of this result.

3.3.1 Building ANS to recognise specific sets

Considering an infinite set $X \subseteq \mathbb{N}$ we can under particular circumstances look for an ANS $\mathcal{S} = (L, A, <)$ such that $X = \{\mathcal{V}_L(n) \mid n \geq 0\}$. This is the case when X has the form given in the following result whose proof is the main goal of this subsection.

Theorem 3.3.7 *Let $m \geq 1$. For $i = 1, \ldots, m$, let P_i be polynomials belonging to $\mathbb{Q}[X]$ such that $P_i(\mathbb{N}) \subseteq \mathbb{N}$ and let c_i be non-negative integers. Set*

$$f : \mathbb{N} \to \mathbb{N}, \; n \mapsto \sum_{i=1}^{m} P_i(n) \, c_i^n.$$

The range $f(\mathbb{N})$ is $\mathcal{S}$-recognisable, for some ANS $\mathcal{S}$ which can be effectively constructed.

The idea of the proof is to build a suitable regular language L having the 'right' counting function i.e., such that $f(\mathbb{N}) = \{\mathcal{V}_L(n) \mid n \geq 0\}$ or $\mathcal{U}_L(n) = f(n+1) - f(n)$ for large enough n. In view of Lemma 3.1.4 and Theorem 3.1.8, it seems reasonable to build such a regular language. Then the conclusion will trivially follow from Lemma 3.3.6. Note that this result can also be related to the work of (Carton and Thomas 2002) and this connection will be discussed in Section 3.4.1.

Example 3.3.8 It is a classical result that, for all integer bases $k \geq 2$, the set of squares is never k-recognisable, see again (Eilenberg 1974). See Example 1.3.16 for the base 10 case. Nevertheless, one can observe that $(n+1)^2 - n^2 = 2n+1$ and the language $L = a^*b^* \cup a^*c^*$ has exactly $2n+1$ words of length n for all $n \geq 0$. Hence $\{n^2 \mid n \geq 0\}$ is $\mathcal{S}$-recognisable for any ANS based on L whatever is the total ordering on $\{a, b, c\}$.

Remark 3.3.9 In the above discussion, the $\mathcal{S}$-recognisability of the considered set does not depend on the ordering of the alphabet. What only matters is to apply Lemma 3.3.5 to the function $\mathcal{V}_L$ which remains unaffected when reordering the alphabet.

Definition 3.3.10 Let x and y be two words in A^*. The *shuffle* of x and y is the finite language $x \shuffle y$ defined by

$$\{x_1 y_1 \cdots x_n y_n \mid x = x_1 \cdots x_n, y = y_1 \cdots y_n, n \geq 1, x_i, y_i \in A^*\}.$$

The *shuffle* of two languages $L_1, L_2 \subseteq A^*$ is the language

$$L_1 \shuffle L_2 = \{w \mid \exists x \in L_1, y \in L_2 : w \in x \shuffle y\} = \bigcup_{\substack{x \in L_1, \\ y \in L_2}} x \shuffle y.$$

If L_1, L_2 are regular then also $L_1 \shuffle L_2$ is regular, see for instance (Eilenberg 1974, Proposition 3.5).

For each $k \in \mathbb{N}$, we build recursively a regular language $L[n \mapsto n^k]$ such that $\mathcal{U}_{L[n \mapsto n^k]}(n) = n^k$ for all $n \in \mathbb{N}$. The first two languages $L[n \mapsto 1]$ and $L[n \mapsto n]$ are defined by $L[n \mapsto 1] = a^*$ and $L[n \mapsto n] = a^+b^*$. Let $k \geq 2$ and assume that we have $L[n \mapsto n^0], \ldots, L[n \mapsto n^{k-1}]$ at our disposal. The induction step relies on the fact that if, for all $n \geq 0$, $\mathcal{U}_M(n) = (n+1)^{k-1}$ then $\mathcal{U}_{M \shuffle \{c\}}(n) = n^k$ provided that c is not a letter in $\mathrm{alph}(M)$. Indeed, for each of the $(n+1)^{k-1}$ words w of length n in M, $w \shuffle c$ contains $n+1$ words of length $n+1$. So there are exactly $(n+1)^k$ words of length $n+1$

in $M \sqcup \{c\}$. Due to

$$(n+1)^{k-1} = \sum_{j=0}^{k-1} \binom{k-1}{j} n^j$$

we build M as a finite union of the languages $L[n \mapsto n^0], \ldots, L[n \mapsto n^{k-1}]$ written *over pairwise disjoint alphabets* $A_{i,j}$, i.e., if $(i,j) \neq (i',j')$, then $A_{i,j} \cap A_{i',j'} = \emptyset$:

$$M = \bigcup_{j=0}^{k-1} \bigcup_{i=1}^{\binom{k-1}{j}} L_{i,j}$$

where $L_{i,j} \subseteq A_{i,j}^*$ is a copy of $L[n \mapsto n^j]$.

Proposition 3.3.11 *Let $P \in \mathbb{N}[X]$. There exists an ANS $\mathcal{S} = (L, A, <)$ such that $P(\mathbb{N})$ is $\mathcal{S}$-recognisable.*

Proof The case where P is constant, is trivial. By Proposition 3.3.4 we may assume that $P(0) = 0$. Since the polynomial $P(n+1) - P(n)$ only contains powers of n with non-negative integer coefficients, by a union of copies of languages $L[n \mapsto n^k]$ over disjoint alphabets we can build a regular language $L \subseteq A^*$ such that $\mathcal{U}_L(n) = P(n+1) - P(n)$. Fix a total ordering $<$ on A and let $\mathcal{S} = (L, A, <)$.

To conclude the proof, we still need to find some integer ℓ such that the first word of length ℓ in L has $P(\ell)$ as numerical value. From the above discussion, this will imply that, for all $n \geq \ell$, $P(n)$ is the numerical value of the first word of length n in L. This is the aim of the next paragraph.

We can assume that $\varepsilon \in L$ and that the first word w of length 2 in the genealogically ordered language L is such that $\text{val}_\mathcal{S}(w) = P(2)$. Indeed, adding or removing a finite number of words of length 1 in a regular language does not alter its regularity. We can add new letters to the alphabet to increase at will the number of words of length 1. Note that we have to consider words of length 2 and not words of length 1 because $P(1)$ is not necessarily equal to one and therefore cannot possibly be represented by the first word of length 1. Contrarily to words of length 1, there is a single word of length 0 so we have no freedom to modify the number of words of length 0.

Let $n \geq 2$. Since $\mathcal{U}_L(n) = P(n+1) - P(n)$, if the numerical value of the first word of length n is $P(n)$ then the numerical value of the first word of length $n+1$ is $P(n+1)$. Consequently, we have

$$\text{rep}_\mathcal{S}(P(\mathbb{N}) \setminus \{P(1)\}) = \text{Min}_\prec(L \setminus A).$$

By Lemma 3.3.5, $P(\mathbb{N})$ is $\mathcal{S}$-recognisable. A single word should possibly be added to take into account the $\mathcal{S}$-representation of $P(1)$. □

Lemma 3.3.12 *Let k and t be two positive integers. There exists a regular language $L[n \mapsto n^k - t\,n^{k-1}]$ such that*

$$\mathcal{U}_{L[n \mapsto n^k - t\,n^{k-1}]}(n) = \begin{cases} n^k - t\,n^{k-1}, & \text{if } n \geq t, \\ 0, & \text{otherwise.} \end{cases}$$

Proof For $k = 1$ take the language $L[n \mapsto n^1 - t\,n^0] = a^{t+1}a^*b^*$. Now assume that $k \geq 2$. From the above discussion, we have $L[n \mapsto n^k] = M \sqcup \{a\}$ where $L[n \mapsto n^k] \subseteq A^*$ and a not belonging to $\text{alph}(M)$. Let $n \geq 1$. For $i = 1, \ldots, n$, $L[n \mapsto n^k]$ has exactly n^{k-1} words of length n with a occurring at position i (say, counted from the right). The language

$$L[n \mapsto n^k - t\,n^{k-1}] = L[n \mapsto n^k] \setminus \bigcup_{i=0}^{t-1} A^* a\, A^i \quad (3.8)$$

has $n^k - t\,n^{k-1}$ words of length n for $n \geq t$. □

Proposition 3.3.13 *Let $P \in \mathbb{Z}[X]$ be such that $P(\mathbb{N}) \subseteq \mathbb{N}$. There exists an ANS $\mathcal{S} = (L, A, <)$ such that $P(\mathbb{N})$ is $\mathcal{S}$-recognisable.*

Proof Without loss of generality, we may assume that $\deg(P) = d + 1 \geq 1$. We proceed as in the proof of Proposition 3.3.11 and consider the polynomial $Q(n) = P(n+1) - P(n)$. Since $P(\mathbb{N}) \subseteq \mathbb{N}$, the leading coefficients of P and Q are positive. By possibly adding extra terms of the form $X^j - X^j$, if $\deg(Q) = d$ then to take advantage of the previous lemma, $Q(X)$ can be written as

$$\sum_{\ell=0}^{d} c_\ell X^\ell + X^{i_1+1} - t_1 X^{i_1} + \cdots + X^{i_r+1} - t_r X^{i_r} \quad (3.9)$$

for some $c_0, \ldots, c_d \in \mathbb{N}$, $i_1, \ldots, i_r \in \{0, \ldots, d-1\}$ and $t_1, \ldots, t_r \in \mathbb{N} \setminus \{0\}$. Let $t = \sup\{t_1, \ldots, t_r, 2, m\}$ where m is the least integer such that $P(n) < P(n+1)$ for all $n \geq m$. Making the union of regular languages over disjoint alphabets of the kind $L[n \mapsto n^\ell]$ and $L[n \mapsto n^{i_j} - t_j\, n^{i_j - 1}]$ given by Lemma 3.3.12, we get a regular language L satisfying, for all $n \geq t$, $\mathcal{U}_L(n) = Q(n)$.

Since $t \geq 2$, we can assume that L contains exactly $P(t)$ words of length at most $t - 1$. This can be achieved by adding or removing a finite number of words from the language L. Let $\mathcal{S}$ be an ANS based on the ordered regular language L. The first word of length t has a numerical value equal to $P(t)$ and, for all $n \geq t$, $\mathcal{U}_L(n) = P(n+1) - P(n)$. Then we get

$$\operatorname{rep}_\mathcal{S}(P(\mathbb{N})) = (\operatorname{minlg}(L) \cap A^{\geq t}) \cup \{\operatorname{rep}_\mathcal{S}(P(0)), \ldots, \operatorname{rep}_\mathcal{S}(P(t-1))\}$$

where $A^{\geq t}$ denotes the set of all words of length at least t. By Lemma 3.3.5, $\operatorname{rep}_\mathcal{S}(P(\mathbb{N}))$ is regular. □

As the third step we get a theorem of recognisability in the general case of polynomials with rational coefficients. Interestingly, the proof relies on the $\mathcal{S}$-recognisability of arithmetic progressions.

Proposition 3.3.14 *Let $P \in \mathbb{Q}[X]$ be such that $P(\mathbb{N}) \subseteq \mathbb{N}$. There exists an ANS $\mathcal{S} = (L, A, <)$ such that $P(\mathbb{N})$ is $\mathcal{S}$-recognisable.*

Proof Assume that $\deg(P) = d \geq 1$. Let $s_0, \ldots, s_d, c_d \in \mathbb{N} \setminus \{0\}$ and $c_0, \ldots, c_{d-1} \in \mathbb{Z}$ be such that

$$P(X) = \frac{c_d}{s_d} X^d + \frac{c_{d-1}}{s_{d-1}} X^{d-1} + \cdots + \frac{c_0}{s_0}.$$

Let s be the least common multiple of $s_0, \ldots, s_d$. One has $sP = Q$ with $Q \in \mathbb{Z}[X]$. Since $P(\mathbb{N}) \subseteq \mathbb{N}$, then $Q(\mathbb{N}) \subseteq s\mathbb{N}$. As in the proof of Proposition 3.3.13, there exist $t = \sup\{2, m\}$ where m is the least integer such that $P(n) < P(n+1)$, for all $n \geq m$, and a regular language M over a totally ordered alphabet $(A, <)$ such that, for all $n \geq t$,

$$\mathcal{U}_M(n) = Q(n+1) - Q(n) = s\left(P(n+1) - P(n)\right).$$

We modify M by possibly adding or removing a finite number of words to get $\mathcal{V}_M(t-1) = sP(t) = Q(t)$. Otherwise stated, if we set $\mathcal{R} = (M, A, <)$ and w is the first word of length t in M, then $\operatorname{val}_\mathcal{R}(w) = Q(t)$. By Theorem 3.3.1, the arithmetic progression $s\mathbb{N}$ is $\mathcal{R}$-recognisable. Consequently, $L = \operatorname{rep}_\mathcal{R}(s\mathbb{N})$ is a regular language such that

$$\mathcal{V}_L(t-1) = P(t) \text{ and, } \forall n \geq t, \ \mathcal{U}_L(n) = P(n+1) - P(n).$$

Indeed L is obtained by taking in the genealogically ordered language M the words at position $is+1$, $i \in \mathbb{N}$. Since the first word of length t in M is the first word of length t in L and its position in the genealogically ordered language L is $P(t)$, the conclusion follows from Lemma 3.3.5. □

Proposition 3.3.15 *Let $c \in \mathbb{N} \setminus \{0, 1\}$ and P be a polynomial in $\mathbb{Q}[X]$ such that $P(\mathbb{N}) \subseteq \mathbb{N}$. There exists a numeration system $\mathcal{S}$ such that the set $\{P(n) c^n \mid n \in \mathbb{N}\}$ is $\mathcal{S}$-recognisable.*

Proof First assume that $P \in \mathbb{Z}[X]$ and that it is non-constant. We show how to construct a regular language L such that for all large enough n,

$$\mathcal{U}_L(n) = P(n+1)\,c^{n+1} - P(n)\,c^n = [c\,P(n+1) - P(n)]\,c^n.$$

The assumption $P(\mathbb{N}) \subseteq \mathbb{N}$ implies that the polynomial $c\,P(n+1) - P(n) \in \mathbb{Z}[X]$ has a positive leading coefficient. We can apply the same decomposition as in (3.9) and therefore proceed as in the proof of Proposition 3.3.13.

To get such a language L, it is enough to show how to construct, for all $k \geq 0$, a regular language $L[n \mapsto n^k c^n]$ having $n^k\,c^n$ words of length $n \geq 0$ and, for all $t \geq 1$, a regular language with $(n^k - t\,n^{k-1})\,c^n$ words of length $n \geq t$. As an intermediate step, also we construct, for all $k > i \geq 0$, regular languages $M_{k,i}$ having $n^i c^n$ words of length $n - k + i$ for all $n > k$.

If $\mathrm{Card}(A) = c$, note that $L[n \mapsto n^0 c^n] = A^*$. So we build $L[n \mapsto n^1 c^n]$ and $M_{1,0}$ first. Let $A_1, \ldots, A_c$ be c pairwise disjoint alphabets of cardinality c. The language $M_{1,0} = A_1^* \cup \cdots \cup A_c^*$ is such that $\mathcal{U}_{M_{1,0}}(n-1) = c^n$ for all $n > 1$. Let a_1 be a letter not in $\mathrm{alph}(M_{1,0})$. To obtain $L[n \mapsto n^1 c^n]$ we take the words of length at least 2 in $M_{1,0} \sqcup \{a_1\}$ and add c distinct words of length 1.

Let $k \geq 2$. Assume that we have $M_{k-1,0}, \ldots, M_{k-1,k-2}$ at our disposal. We have to construct languages $M_{k,0}, \ldots, M_{k,k-1}$ and $L[n \mapsto n^k c^n]$. Let $A_1, \ldots, A_{c^k}$ be c^k pairwise disjoint alphabets of cardinality c. The language $M_{k,0} = A_1^* \cup \cdots \cup A_{c^k}^*$ is such that $\mathcal{U}_{M_{k,0}}(n-k) = c^n$ for all $n > k$. Now assume that we have $M_{k,i}$ for some $i < k - 1$. Let a_{i+1} be a letter not in $\mathrm{alph}(M_{k,i})$. Then for $n > k$, $M_{k,i} \sqcup \{a_{i+1}\}$ has $n^i(n-k+i+1)c^n$ words of length $n-k+i+1$ because for each word of length $n-k+i$ in $M_{k,i}$ we can put the extra letter a_{i+1} in $n-k+i+1$ positions. To get $M_{k,i+1}$ we make the union of $M_{k,i} \sqcup \{a_{i+1}\}$ and $k-i-1$ copies over disjoint alphabets of languages of the kind $M_{k,i}\,a_{i+1}$. The extra letter concatenated at the end of each words in $M_{k,i}$ ensures that $M_{k,i}\,a_{i+1}$ has $n^i c^n$ words of length $n-k+i+1$. Now $M_{k,k-1}$ has, for all $n > k$, $n^{k-1} c^n$ words of length $n-1$. So if a_k does not belong to $\mathrm{alph}(M_{k,k-1})$, we consider the words of length at least $k+1$ in $M_{k,k-1} \sqcup \{a_k\}$ and we add a suitable number of words of shorter length to get $L[n \mapsto n^k c^n]$. Since a shuffle operation is involved in this latter construction, using the same argument as in (3.8) we can build a regular language having a complexity function $(n^k - t\,n^{k-1})c^n$ for all $n \geq t$.

To conclude the proof, if $P \in \mathbb{Q}[X] \setminus \mathbb{Z}[X]$, then apply the same trick as in the proof of Proposition 3.3.14. $\square$

Repeating the construction given in this latter proof to get several regular languages over distinct alphabets, the reader should be convinced that Theorem 3.3.7 stated at the beginning of this section is derived easily.

3.3.2 ANS based on slender languages

A language L is said to be *slender*, if there exists d such that $\mathcal{U}_L(n) \leq d$ for all $n \geq 0$. For ANS based on slender languages, $\mathcal{S}$-recognisable sets are completely characterised.

Theorem 3.3.16 *Let $L \subseteq A^*$ be a slender regular language and $\mathcal{S} = (L, A, <)$. A set $X \subseteq \mathbb{N}$ is $\mathcal{S}$-recognisable if, and only if, X is a finite union of arithmetic progressions.*

Proof By a well-known characterisation of slender languages†, there exist $k \geq 1$ and words x_i, y_i, z_i, $1 \leq i \leq k$, such that

$$L = \bigcup_{i=1}^{k} x_i \, y_i^* z_i \cup F, \ x_i, z_i \in A^*, y_i \in A^+$$

where the sets $x_i \, y_i^* z_i$ are pairwise disjoint and F is a finite set. The sequence $(\mathcal{U}_L(n))_{n \in \mathbb{N}}$ is eventually periodic of period $p = \text{lcm}_i |y_i|$. Moreover, for n large enough, if $x_i \, y_i^n \, z_i$ is the mth word of length $|x_i \, z_i| + n |y_i|$ then $x_i \, y_i^{n+p/|y_i|} z_i$ is the mth word of length $|x_i \, z_i| + n |y_i| + p$. Roughly speaking, for sufficiently large n, the structures of the ordered sets of words of length n and $n+p$ are the same. The regular subsets of L are of the form

$$\bigcup_{j \in J} x_{i_j} \, (y_{i_j}^{t_j})^* z_{i_j} \cup F' \tag{3.10}$$

where J is a finite set, $i_j \in \{1, \ldots, k\}$, $t_j \in \mathbb{N}$ and F' is a finite subset of L. Now we can conclude. If X is $\mathcal{S}$-recognisable, then $\text{rep}_\mathcal{S}(X)$ is a regular subset of L of the form (3.10). In view of the first part of the proof, it is clear that X is eventually periodic with period $\text{lcm}(p, \text{lcm}_j |y_{i_j}^{t_j}|)$. The converse follows from Theorem 3.3.1. □

Example 3.3.17 Consider the language $L = ab^*c \cup b(aa)^*c$. It contains exactly two words of each positive even length: $ab^{2i}c \prec ba^{2i}c$ and one word for each odd length larger than 2: $ab^{2i+1}c$. The sequence $\mathcal{U}_L(n)$ is eventually periodic of period two: $0, 0, 2, 1, 2, 1, \ldots$.

Corollary 3.3.18 *Let $\mathcal{S}$ be a numeration system based on a slender language. If $X, Y \subseteq \mathbb{N}$ are $\mathcal{S}$-recognisable, then $X + Y$ and tX are $\mathcal{S}$-recognisable for all $t \in \mathbb{N}$.*

---

† See for instance (Păun and Salomaa 1995) or independently (Shallit 1994). Compare this result with the one given in Theorem 3.3.21.

Proof It is clear that if $X, Y \subseteq \mathbb{N}$ are eventually periodic, then $X + Y$ and tX are also eventually periodic. □

3.3.3 Multiplication by a constant

From Corollary 3.3.18 given above, if $\mathcal{S} = (L, A, <)$ is an ANS based on a slender language and if $X \subseteq \mathbb{N}$ is $\mathcal{S}$-recognisable, then for any given non-negative integer t, the set tX is again $\mathcal{S}$-recognisable. Such a property is also well-known for the usual k-ary numeration systems and more generally for the Pisot numeration systems sketched in Example 3.1.14. One can therefore consider the following general problem of *characterising ANS $\mathcal{S}$ such that multiplication by a constant is $\mathcal{S}$-recognisability-preserving* that is, for all $\mathcal{S}$-recognisable sets $X \subseteq \mathbb{N}$ and for all $t \in \mathbb{N}$, the set $tX = \{tx \mid x \in X\}$ is still $\mathcal{S}$-recognisable. This is the most basic arithmetic operation to consider. A more ambitious task is to consider addition of two $\mathcal{S}$-recognisable sets $X + Y = \{x + y \mid x \in X, y \in Y\}$ and look for ANS such that the resulting set is again $\mathcal{S}$-recognisable. Note that even for the usual integer base systems, if X and Y are two k-recognisable sets of integers, then in general the set $X.Y = \{xy \mid x \in X, y \in Y\}$ is not k-recognisable.

Definition 3.3.19 If $w_1, \ldots, w_n$ are words over A of arbitrary, and not necessarily equal, lengths then the padding $(w_1, \ldots, w_n)^\#$ is defined as

$$(\#^{m-|w_1|} w_1, \ldots, \#^{m-|w_n|} w_n)$$

where $m = \max\{|w_1|, \ldots, |w_n|\}$ and $\#$ is a new padding symbol. Such an n-tuple can be considered as a single word over the alphabet $(A \cup \{\#\})^n$ obtained as the Cartesian product of n copies of $A \cup \{\#\}$. The concatenation of two words $(u_1, \ldots, u_n)$ and $(v_1, \ldots, v_n)$ over $(A \cup \{\#\})^n$ is $(u_1 v_1, \ldots, u_n v_n)$ where the usual concatenation product is considered component-wise.

Remark 3.3.20 Assume that addition in an ANS $\mathcal{S} = (L, A, <)$ is computable by finite automaton, *i.e.*, its graph

$$G = \{(\text{rep}_\mathcal{S}(x), \text{rep}_\mathcal{S}(y), \text{rep}_\mathcal{S}(x + y))^\# \mid x, y \geq 0\}$$

is regular where the shortest words are padded with an extra symbol $\#$ to get three components of same length, that is, we get words over the Cartesian product $(A \cup \{\#\}) \times (A \cup \{\#\}) \times (A \cup \{\#\})$ and the corresponding DFA reads simultaneously one symbol from the three components. By considering $G \cap \{(v, v, w)^\# \mid v \in L, w \in A^*\}$ then multiplication by 2 is also computable by finite automaton and in particular, multiplication by 2 is therefore $\mathcal{S}$-recognisability-preserving. By iterating this kind of argument, if addition is computable by finite automaton, then it is the same for multiplication by any constant $t \in \mathbb{N}$.

First let us recap some well-known facts about the complexity function of regular languages. A language is said to be *polynomial* (or *sparse*) if there exists some non-negative integer k such that $\mathcal{U}_L(n) \in \mathcal{O}(n^k)$. If the regular language is infinite, one can show that there exist a constant $C > 0$ and an infinite sequence of integers $n_1 < n_2 < \cdots$ such that $\mathcal{U}_L(n_j) \geq C n_j^k$ for all $j \geq 1$. Infinite slender languages are in particular polynomial. Deterministic finite automata accepting polynomial regular languages have some specific properties. A well-known description of the polynomial regular languages and the dichotomy existing with exponential languages are for instance given in (Szilard, Yu, Zhang, et al. 1994). We recall these two statements below. Obviously only states which are both accessible and co-accessible have an impact on the complexity function (see Chapter 1 for definition of accessibility).

Theorem 3.3.21 *Let* $\mathcal{A} = (Q, A, E, \{q_0\}, T)$ *be an accessible and co-accessible DFA. The language L accepted by $\mathcal{A}$ is polynomial if, and only if, all states $q \in Q$ belong to at most one cycle in $\mathcal{A}$. In particular, L is polynomial if, and only if, it is a finite union of languages of the form* $u_1 v_1^* u_2 v_2^* \cdots u_t v_t^* u_{t+1}$.

Note that it is algorithmically decidable whether or not the language accepted by a DFA given as an input is polynomial. This question is for instance considered in Theorem 11.1.27.

Theorem 3.3.22 *A regular language L is either polynomial or there exist $C > 0$ and an infinite sequence $n_1 < n_2 < \cdots$ such that the complexity function of L satisfies, for all $j \geq 1$, $\mathcal{U}_L(n_j) = 2^{f(n_j)}$ where $f(n_j) \geq C n_j$.*

If a regular language is not polynomial, then we shall say that it is *exponential*. Note that in general, we cannot give a suitable lower bound on $\mathcal{U}_L(n)$ for every large enough n. It is the reason why in the above theorem, we have not written $f \in \Omega(n)$ but instead have given a lower bound for infinitely many n. Consider for instance the regular language $L = \{aa, ab, ba, bb\}^*$. We have, for all $n \geq 0$, $\mathcal{U}_L(2n) = 2^{2n}$ and $\mathcal{U}_L(2n+1) = 0$. Therefore this language is exponential but even in this case, for infinitely many m, $\mathcal{U}_L(m) = 0$.

Next we sketch a picture of the main results about preservation of $\mathcal{S}$-recognisability after multiplication by a constant.

Proposition 3.3.23 Let $S = (a^*b^*, \{a,b\}, a < b)$. If $t \in \mathbb{N}$ is an odd square, then for every S-recognisable set $X \subseteq \mathbb{N}$, tX is again S-recognisable. Otherwise, there exists an S-recognisable set $Y \subseteq \mathbb{N}$ such that tY is not S-recognisable.

The proof is given in (Lecomte and Rigo 2001) and involves Pell equations. It is due to the expression (3.2) of $\mathrm{val}_S(a^i b^j)$ which is a polynomial of degree 2 in i and j. Let us point out that the arguments developed in this proof are also useful to give a counter-example showing that S-automaticity is not preserved after periodic deletion, see (Rigo and Maes 2002). The fact revealed in the previous proposition when t is not a square is a special case of some general phenomenom (Rigo 2002) about ANS on polynomial regular languages.

Theorem 3.3.24 Let S be an ANS based on a polynomial regular language L. Suppose that there exist $C > 0$ and an integer $k \geq 1$ such that $\mathcal{U}_L(n) \in \mathcal{O}(n^k)$ and for infinitely many n, $\mathcal{U}_L(n) \geq Cn^k$. If t is not the $(k+1)$th power of an integer, then there exists an S-recognisable set $Y \subseteq \mathbb{N}$ such that tY is not S-recognisable.

Having this theorem at hand, it seems natural to determine which suitable powers may preserve recognisability. For specific kind of polynomial languages of any degree, we have the following result. Notice that for $n = 1$, we have a slender language and the case $n = 2$ is exactly the one considered in Proposition 3.3.23.

Proposition 3.3.25 (Charlier, Rigo, and Steiner 2008) Let $n \geq 3$. Let A be the ordered alphabet $\{a_1 < \cdots < a_n\}$ and $S = (a_1^* \cdots a_n^*, A, <)$ be an ANS. For all $t \geq 2$, there exists an S-recognisable set $Y \subseteq \mathbb{N}$ such that tY is not S-recognisable.

Also details are given in (Charlier 2009). In general, exponential languages with a polynomial complement do not preserve recognisability after multiplication by a constant.

Proposition 3.3.26 Let A be an alphabet such that $\mathrm{Card}(A) \geq 2$. Let $L \subset A^*$ be an infinite polynomial regular language and S be an ANS based on its complement $A^* \setminus L$. There exists an S-recognisable set $Y \subseteq \mathbb{N}$ and an integer t such that tY is not S-recognisable.

Arguments appearing in Example 3.1.7 and in the proof of this proposition as given in (Rigo 2002) are based on similar techniques involving the pumping lemma.

In view of these results and except for very special cases like the slender languages or A^*, we can conclude that the only regular languages that are possibly suited to define ANS for which recognisability is preserved after arithmetic operations, are necessarily exponential languages with exponential complement.

3.4 Automatic sequences

In (Cobham 1972), it is shown that an infinite word $x = x_0 x_1 x_2 \cdots$ over an alphabet B is obtained as the image under a coding $\tau : A \to B$ of the fixed point of a morphism $\sigma : A \to A^*$ of *constant length* k if, and only if, there exists a DFAO $\mathcal{A} = (Q, A, \delta, \{q_0\}, B, \mu)$ such that $x_n = \mu(\delta(q_0, \mathrm{rep}_k(n)))$ for all $n \geq 0$. This result closely relates the constant length of the morphism to the base k numeration system and also explains the established terminology of k-automatic sequences †, *i.e.*, the nth term of the sequence is generated by an automaton with output 'fed' with the k-ary representation of n.

A natural generalisation of this iterative process used to define infinite words is to relax the hypothesis about the constant length of σ and to consider an arbitrary non-erasing prolongable morphism $\sigma : A \to A^*$ and an extra coding. Hence what kind of numeration system should replace the usual k-ary one? In this section, we prove the following analogue of Cobham's theorem from 1972 where ANS come into play and we discuss some of its applications to $\mathcal{S}$-recognisable sets.

Theorem 3.4.1 *Let $x = x_0 x_1 x_2 \cdots$ be an infinite word over an alphabet B. This word is substitutive if, and only if, there exists an ANS $\mathcal{S} = (L, A, <)$ and a DFAO $(Q, A, \delta, \{q_0\}, B, \mu)$ such that for all $n \geq 0$, $x_n = \mu(\delta(q_0, \mathrm{rep}_{\mathcal{S}}(n)))$.*

This result splits into Propositions 3.4.12 and 3.4.16. The next definition naturally extends the classical generation process of k-automatic sequences.

Definition 3.4.2 Let $\mathcal{S} = (L, A, <)$ be an ANS. We say that an infinite word $x = x_0 x_1 x_2 \cdots \in B^{\mathbb{N}}$ is $\mathcal{S}$-*automatic*, if there exists a DFAO $(Q, A, \delta, \{q_0\}, B, \mu)$ such that $x_n = \mu(\delta(q_0, \mathrm{rep}_{\mathcal{S}}(n)))$ for all $n \geq 0$.

---

† Originally A. Cobham was using the terminology of *tag sequences* referring to the generation process of the infinite word. Let us quote (Cobham 1972): 'Adding a feedback feature which permits symbols produced at early stages of the generating process to be re-examined at later stages increases flexibility and the variety of sequences generable by devices so augmented is substantially richer.... Suppose we have generated symbols with index 0 through $2k - 1$ and that our left hand points at the k-th of these, our right hand at the last. We observe the symbol at which our left hand is pointing and write with our right the $2k$-th and $(2k+1)$-st as prescribed. Moving our left hand one symbol to the right, we are in position to repeat the procedure.'

Example 3.4.3 We consider the alphabets $A = \{a, b\}$, $B = \{0, 1, 2, 3\}$, the ANS $\mathcal{S} = (a^*b^*, A, a < b)$ of Example 3.1.12 and the DFAO depicted in Figure 3.5 We obtain the first few terms of the corresponding $\mathcal{S}$-automatic

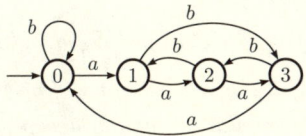

Fig. 3.5 A DFAO with output alphabet $\{0, 1, 2, 3\}$.

sequence $x = 01023031200231010123023031203120231002310212\cdots$.

Notice that taking the ANS $\mathcal{R} = (\{a, ba\}^*\{\varepsilon, b\}, \{a, b\}, a < b)$ of Example 3.1.15 we obtain another infinite word $y = 01023\underline{1}31023\cdots$ which is $\mathcal{R}$-automatic (underlined letters indicate the differences between x and y). This stresses the fact that an $\mathcal{S}$-automatic sequence really depends on two ingredients: an ANS and a DFAO.

Example 3.4.4 Let $\mathcal{S} = (B_k, \{0, \ldots, k-1\}, <)$ be the ANS corresponding to the usual k-ary numeration system where the language $B_k = \{0, \ldots, k-1\}^* \setminus 0\{0, \ldots, k-1\}^*$ is as in Example 3.1.11. By considering a DFAO $(Q, \{0, \ldots, k-1\}, \delta, \{q_0\}, B, \mu)$, the sequence defined, for all $n \geq 0$, by $x_n = \mu(\delta(q_0, \text{rep}_\mathcal{S}(n)))$ is $\mathcal{S}$-automatic. So k-automatic sequences are special cases of $\mathcal{S}$-automatic sequences.

The next lemma is a powerful result that allows us to get rid of the erasing behaviour that can appear in the two morphisms used for generating a substitutive word and restricts the second one to a coding. A proof† of this result is given in (Allouche and Shallit 2003). This result is also expressed by Theorem 4.6.1.

Lemma 3.4.5 (Cobham 1968) *Let A, B, C be three alphabets. Consider two arbitrary morphisms $\sigma : A \to A^*$ and $\tau : A \to B^*$ such that $\tau(\sigma^\omega(a))$ is an infinite word. There exist a non-erasing morphism $\alpha : C \to C^+$ prolongable on a letter $c \in C$ and a coding $\beta : C \to B$ such that*

$$\tau(\sigma^\omega(a)) = \beta(\alpha^\omega(c)).$$

The idea of the following result is to consider the end point, whether or not it is a final state, of all paths that can be achieved in a DFA. These paths are naturally genealogically ordered with respect to their label. Recall that S is the shift operator introduced in Chapter 1.

† Have a look, one needs to define *dead* and *moribund* letters.

Lemma 3.4.6 Let $A = \{a_1 < \cdots < a_n\}$ be a totally ordered alphabet, $\mathcal{A} = (Q, A, E, \{q_0\}, T)$ be a DFA where E defines a partial function $\delta_{\mathcal{A}} : Q \times A \to Q$ and let $z \notin Q$. Define the morphism $\psi_{\mathcal{A}} : Q \cup \{z\} \to (Q \cup \{z\})^*$ by $\psi_{\mathcal{A}}(z) = z\, q_0$ and, for all $q \in Q$,

$$\psi_{\mathcal{A}}(q) = \delta_{\mathcal{A}}(q, a_1) \cdots \delta_{\mathcal{A}}(q, a_n).$$

In this latter expression, if $\delta_{\mathcal{A}}(q, a_i)$ is not defined for some i, then it is replaced by ε. Let L be the regular language accepted by $(Q, A, E, \{q_0\}, Q)$ where all states of $\mathcal{A}$ are final. Then the shifted sequence $S(\psi_{\mathcal{A}}^\omega(z))$ is the sequence $(x_n)_{n \in \mathbb{N}}$ of the states reached in $\mathcal{A}$ by the words of L in genealogical order, i.e., for all $n \in \mathbb{N}$,

$$x_n = \delta_{\mathcal{A}}(q_0, w_n)$$

where w_n is the $(n+1)$th word of the genealogically ordered language L.

Prior to the proof, let us give a short example to set properly the framework.

Example 3.4.7 Consider the DFA given in Figure 3.6. Note that the automaton is not complete, the transition function δ is partial: from q_1 one cannot read b. Assume $a < b$. The sequence of the ordered words in the

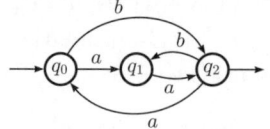

Fig. 3.6 A DFA.

language accepted by the automaton where all states are considered as final states are

$$(w_n)_{n \geq 0} = \varepsilon,\ a,\ b,\ aa,\ ba,\ bb,\ aaa,\ aab,\ baa,\ bab,\ bba,\ aaaa, \ldots.$$

The corresponding sequence of states is

$$(\delta(q_0, w_n))_{n \geq 0} = q_0, q_1, q_2, q_2, q_0, q_1, q_0, q_1, q_1, q_2, q_2, q_1, \ldots.$$

For instance, the second q_2 in the sequence, *i.e.*, its fourth element, is the state reached by the DFA when reading aa, *i.e.*, the fourth word w_3, from q_0. Now consider the morphism

$$\psi : \begin{cases} z & \mapsto z\, q_0 \\ q_0 & \mapsto q_1\, q_2 \\ q_1 & \mapsto q_2 \\ q_2 & \mapsto q_0\, q_1. \end{cases}$$

One can observe that the introduction of the extra letter z gives a prolongable morphism: in this example only $\psi(z)$ begins with z, for $x \in \{q_0, q_1, q_2\}$, $\psi(x)$ does not start with x. Now, one can compute the prefix of $\psi^\omega(z)$,

$$\psi^\omega(z) = z\, q_0\, q_1\, q_2\, q_2\, q_0\, q_1\, q_0\, q_1\, q_1\, q_2\, q_2\, q_1\, q_2\, q_2\, q_2\, \cdots.$$

Now let us give the proof of Lemma 3.4.6.

Proof First observe that we have the following factorisation:

$$\psi_{\mathcal{A}}^\omega(z) = z x_0 x_1 x_2 \cdots = z\, q_0\, \psi_{\mathcal{A}}(q_0)\, \psi_{\mathcal{A}}^2(q_0) \cdots$$

and $x_0 = q_0 = \delta_{\mathcal{A}}(q_0, \varepsilon)$. Then by definition of $\psi_{\mathcal{A}}$, if $x_n = \delta_{\mathcal{A}}(q_0, w_n)$, $n \geq 0$, then the factor

$$u_n = \psi(x_n) = \delta_{\mathcal{A}}(q_0, w_n a_1) \cdots \delta_{\mathcal{A}}(q_0, w_n a_n) \tag{3.11}$$

appears in $\psi_{\mathcal{A}}^\omega(z)$ with the usual convention of replacing with ε the undefined transitions. Indeed, $z x_0 x_1 x_2 \cdots$ is a fixed point of $\psi_{\mathcal{A}}$ and each x_n produces a factor $\psi_{\mathcal{A}}(x_n) = u_n$ appearing later on in the infinite word. Moreover this factor is preceded by $\delta_{\mathcal{A}}(q_0, w_{n-1} a_1) \cdots \delta_{\mathcal{A}}(q_0, w_{n-1} a_n)$ and followed by $\delta_{\mathcal{A}}(q_0, w_{n+1} a_1) \cdots \delta_{\mathcal{A}}(q_0, w_{n+1} a_n)$. It is therefore clear that we get all states reached from the initial state when considering in increasing genealogical order the labels of all the paths in $\mathcal{A}$. $\square$

Remark 3.4.8 We use the notation of Lemma 3.4.6. Note that the morphism $\psi_{\mathcal{A}}$ given in the previous statement depends only on Q, A and E but not on the set of final states T. In the literature, one can find the terminology *transition structure* when final states are unspecified or unimportant. In particular, there is no relation between the language recognised by $\mathcal{A}$ and the infinite word $S(\psi_{\mathcal{A}}^\omega(z))$.

If $\mathcal{A}$ contains no cycle, *i.e.*, if the language recognised by $\mathcal{A}$ is finite, then the fixed point of $\psi_{\mathcal{A}}$ starting with z is also finite.

Example 3.4.9 Also the reader could have the feeling that the introduction of the extra letter z such that $\psi_{\mathcal{A}}(z) = z q_0$ is artificial. In particular, if the initial state has a loop. Let us consider the following example given by the DFA depicted in Figure 3.7. Let us compare the infinite words generated by the morphism

$$\mu : \begin{cases} q_0 & \mapsto & q_0\, q_2 \\ q_1 & \mapsto & q_2 \\ q_2 & \mapsto & q_0\, q_1 \end{cases}$$

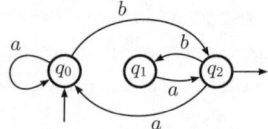

Fig. 3.7 Another DFA.

and by the morphism ψ_A given by Lemma 3.4.6 and defined by $\psi_A(z) = zq_0$ and $\psi_A(x) = \mu(x)$ for $x \in \{q_0, q_1, q_2\}$. We get

$$\psi_A^\omega(z) = zq_0 q_0 q_2 q_0 q_2 q_0 q_1 q_0 q_2 q_0 q_1 q_0 q_2 q_2 q_0 q_2 \cdots$$

but

$$\mu^\omega(q_0) = q_0 q_2 q_0 q_1 q_0 q_2 q_2 q_0 q_2 q_0 q_1 q_0 q_1 q_0 q_2 \cdots .$$

One can show that the sequence $\mu^\omega(q_0)$ is the sequence of states reached from q_0 by considering only words in the DFA starting with b instead of taking into account all the possible paths. Of course, one cannot simply remove the loops of label a from q_0 because it may be used not only by paths starting with a.

In Lemma 3.4.6, we have in a canonical way associated with any DFA, even with any labelled directed graph, a morphism. Now we present some kind of converse construction. Note that this construction is very close to the prefix-suffix graph introduced in Definition 5.2.4.

Definition 3.4.10 Let us adapt a classical construction encountered in the case of k-automatic sequences. Any pair given by a morphism $\sigma : A \to A^*$ and a letter $a \in A$ can be canonically associated with a DFA denoted $\mathcal{A}_{\sigma,a}$ and defined as follows. Let $\|\sigma\| = \max_{b \in A} |\sigma(b)|$. The alphabet of $\mathcal{A}_{\sigma,a}$ is $\{1, \ldots, \|\sigma\|\}$, its set of states is A. The initial state is a. For all $b \in A$ and $i \in \{1, \ldots, |\sigma(b)|\}$, we set $\delta(b, i) = \sigma(b)[i, i]$ to define the partial transition function of $\mathcal{A}_{\sigma,a}$. There is usually no need to specify the final states. One can for instance set $T = A$ as the set of final states.

Moreover, if an extra morphism $\tau : A \to B^*$ is given, then we extend $\mathcal{A}_{\sigma,a}$ to define a DFAO $\mathcal{A}_{\sigma,a,\tau}$ where the output function is given precisely by τ.

Example 3.4.11 With the notation of the previous definition, consider the alphabets $A = \{a, b, c\}$, $B = \{d, e\}$ and the morphisms

$$\sigma : A \to A^+, \begin{cases} a & \mapsto & abc \\ b & \mapsto & bc \\ c & \mapsto & aac \end{cases} \text{ and } \tau : A \to B, \begin{cases} a & \mapsto & d \\ b, c & \mapsto & e. \end{cases}$$

The corresponding automaton $\mathcal{A}_{\sigma,a,\tau}$ is given in Figure 3.8 and the output function is represented on the outgoing arrows.

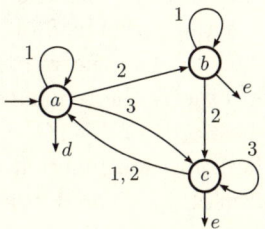

Fig. 3.8 The automaton $\mathcal{A}_{\sigma,a,\tau}$.

Proposition 3.4.12 *Let $\sigma : A \to A^*$ be a morphism prolongable on the letter $a \in A$ and $\tau : A \to B^*$ be a morphism such that $x = \tau(\sigma^\omega(a))$ is infinite. There exists an abstract numeration system $\mathcal{S}$ such that $x \in B^\mathbb{N}$ is an $\mathcal{S}$-automatic sequence.*

Proof Thanks to Lemma 3.4.5, we can assume that σ is non-erasing and that τ is a coding. Let $C = \{1, \ldots, \|\sigma\|\} \subset \mathbb{N}$ and consider the automaton $\mathcal{A}_{\sigma,a} = (A, C, E, \{a\}, T)$ as given in Definition 3.4.10 where $T = A$, i.e., all states are final.

Let $L \subseteq C^*$ be the language recognised by $\mathcal{A}_{\sigma,a}$. This language will be used to build an abstract numeration system $\mathcal{S}$ to show that x is $\mathcal{S}$-automatic. The alphabet C being a subset of $\mathbb{N}$, we will consider the natural ordering of C. Since $\sigma(a) \in aA^+$, it is clear that if $w \in L$ then $1w \in L$. Indeed by definition of $\mathcal{A}_{\sigma,a}$, its initial state a has a loop labelled by 1, the first letter in C. If we apply Lemma 3.4.6 to this automaton $\mathcal{A}_{\sigma,a}$, we obtain a morphism $\psi_{\mathcal{A}_{\sigma,a}}$ generating the sequence of the states reached by the words of L. This morphism is defined as follows. Let $z \notin A$. We have $\psi_{\mathcal{A}_{\sigma,a}}(z) = za$ and, for all $b \in A$, $\psi_{\mathcal{A}_{\sigma,a}}(b) = \sigma(b)$.

The main point leading to the conclusion is to compare $\psi^\omega_{\mathcal{A}_{\sigma,a}}(z)$ and $\sigma^\omega(a)$. There exists $u \in A^+$ such that $\sigma(a) = au$. We have the following factorisations

$$\sigma^\omega(a) = au\,\sigma(u)\,\sigma^2(u)\,\sigma^3(u)\cdots$$

and

$$\psi^\omega_{\mathcal{A}_{\sigma,a}}(z) = za\,a\,u\,\sigma(a)\,\sigma(u)\,\sigma^2(a)\,\sigma^2(u)\,\sigma^3(a)\,\sigma^3(u)\cdots.$$

If we erase the factors z, a, $\sigma(a)$, $\sigma^2(a), \ldots$ occurring in that order in the above factorisation of $\psi^\omega_{\mathcal{A}_{\sigma,a}}(z)$, we recover $\sigma^\omega(a)$. Recall that $\psi^\omega_{\mathcal{A}_{\sigma,a}}(z)$ is, except for z, the sequence of states reached in $\mathcal{A}_{\sigma,a}$ by considering all the possible paths in genealogical order. The second occurrence of a in $\psi_{\mathcal{A}_{\sigma,a}}^{\omega}(z)$ is the state reached in $\mathcal{A}_{\sigma,a}$ when reading $1 \in L$. By the property (3.11) of $\psi_{\mathcal{A}_{\sigma,a}}$, the factor $\sigma^n(a)$ in the above factorisation corresponds to the states reached in $\mathcal{A}_{\sigma,a}$ when reading the words in L of length $n+1$ starting with 1. Consequently, when giving to $\mathcal{A}_{\sigma,a}$ the words of $L \setminus 1C^*$ in increasing genealogical order, we build exactly the sequence $\sigma^\omega(a) = (y_n)_{n \geq 0}$, i.e., if $w_0 \prec w_1 \prec w_2 \prec \cdots$ are the words of $L \setminus 1C^*$ in genealogical order, then $y_n = \delta(a, w_n)$ where δ is the transition function of $\mathcal{A}_{\sigma,a}$. To conclude, one has to consider the automaton $\mathcal{A}_{\sigma,a,\tau}$ as a DFAO with the ANS $\mathcal{S}$ built over $L \setminus 1C^*$ to see that the sequence $\tau(\sigma^\omega(a))$ is $\mathcal{S}$-automatic. $\square$

Example 3.4.13 We illustrate the previous proof by considering the morphisms of Example 3.4.11 and the automaton $\mathcal{A}_{\sigma,a,\tau}$ given in Figure 3.8. We thus have a morphism $\psi_{\mathcal{A}_{\sigma,a}}$

$$\psi_{\mathcal{A}_{\sigma,a}} : A \cup \{z\} \to (A \cup \{z\})^+ : \begin{cases} z & \mapsto za \\ a & \mapsto \sigma(a) = abc \\ b & \mapsto \sigma(b) = bc \\ c & \mapsto \sigma(c) = aac. \end{cases}$$

Let $u = bc$ and $\sigma(a) = au$. If we underline the factors $z, a, \sigma(a), \sigma^2(a), \ldots$ we have

$$\psi_{\mathcal{A}_{\sigma,a}}^\omega(z) = \underline{z}a\underline{ab}c\underline{bc}bcaac\underline{abc}bc\underline{aac}bcaacabcabcaac\cdots.$$

Erasing the underlined factors, we get $abcbcaacbcaacabcabcaac\cdots$ which is exactly $\sigma^\omega(a)$.

The statement of the next result explicitly introduces the language that was built in the proof of Proposition 3.4.12. We can say that the language $L \setminus 1C^*$ is the *directive language* of σ: if the letters in $\sigma^\omega(a)$ are indexed by the words in $L \setminus 1C^*$, then we know precisely which letter is producing which factor through the morphism.

Corollary 3.4.14 *Let $\sigma : A \to A^*$ be a non-erasing morphism prolongable on the letter $a \in A$ such that $x = (x_n)_{n \geq 0} = \sigma^\omega(a)$ is infinite. Consider the ANS $\mathcal{S}$ built over $L \setminus 1C^*$ where $C = \{1, \ldots, \max_{b \in A} |\sigma(b)|\}$ and L is the language accepted by $\mathcal{A}_{\sigma,a}$. Let $w \in L$ be such that $|\sigma(x_{\mathrm{val}_\mathcal{S}(w)})| = \ell$. Then*

$$\sigma(x_{\mathrm{val}_\mathcal{S}(w)}) = x_{\mathrm{val}_\mathcal{S}(w1)} \cdots x_{\mathrm{val}_\mathcal{S}(w\ell)}.$$

In the above formula, for $i \in \{1, \ldots, \ell\}$, wi has to be understood as the concatenation of $w \in L \subseteq C^*$ and $i \in C$.

Proof It is a direct consequence of the proofs of Lemma 3.4.6 and Proposition 3.4.12.

An independent proof is the following one. We even get another proof that $x_n = \delta_\sigma(a, \mathrm{rep}_S(n))$ where δ_σ is the partial transition function of $\mathcal{A}_{\sigma,a}$.

Consider the adjacency matrix $\mathbf{M} \in \mathbb{N}^{A \times A}$ of $\mathcal{A}_{\sigma,a}$, see Section 1.4 for the definition. For all $s > 0$ and $b, c \in A$, $[\mathbf{M}^s]_{b,c}$ is the number of paths of length s from b to c in $\mathcal{A}_{\sigma,a}$. Since all states of this latter automaton are final, the number N_s of words of length s accepted by $\mathcal{A}_{\sigma,a}$ is obtained by summing up all the entries of $\mathbf{M}^s$ in the row corresponding to the initial state a. Because $\mathcal{A}_{\sigma,a}$ has a loop of label 1 in a, the number of words of length s accepted by $\mathcal{A}_{\sigma,a}$ and starting with 1 is equal to the number N_{s-1} of words of length $s-1$ accepted by $\mathcal{A}_{\sigma,a}$. Consequently, the number of words of length s in the language $L \setminus 1C^*$ is exactly $N_s - N_{s-1}$. From the definition of $\mathcal{A}_{\sigma,a}$, the matrix $\mathbf{M}$ can also be related to the morphism σ and $\mathbf{M}_{b,c}$ is the number of occurrences of c in $\sigma(b)$. Summing up all entries in the row of $\mathbf{M}^s$ corresponding to a gives $|\sigma^s(a)|$. Therefore, the number of words of length s in $L \setminus 1C^*$ is $|\sigma^s(a)| - |\sigma^{s-1}(a)|$ and we get that

$$|\mathrm{rep}_S(n)| = s \Leftrightarrow n \in \{|\sigma^{s-1}(a)|, \ldots, |\sigma^s(a)| - 1\}. \tag{3.12}$$

In particular, if $0 < n < |\sigma(a)|$, we have $|\mathrm{rep}_S(n)| = 1$ and in this case† $\mathrm{rep}_S(n) = n + 1$. Since we have $\mathrm{rep}_S(0) = \varepsilon$ and $\sigma(a) = au$, for some $u \in \Sigma^*$, we get $x_0 = a = \delta_\sigma(a, \mathrm{rep}_S(0))$. Hence, by the definition of $\mathcal{A}_{\sigma,a}$, we have that $x_n = \delta_\sigma(a, \mathrm{rep}_S(n))$ for $n < |\sigma(a)|$. Now let $s > 0$ and assume that $x_n = \delta_\sigma(a, \mathrm{rep}_S(n))$ for all $n < |\sigma^s(a)|$. Let $|\sigma^s(a)| \le n < |\sigma^{s+1}(a)|$. There exists a unique $|\sigma^{s-1}(a)| \le m < |\sigma^s(a)|$ such that

$$\sigma^{s+1}(a) = \underbrace{\sigma^{s-1}(a)\,u\,x_m\,v}_{\sigma^s(a)}\,\sigma(u)\,\underbrace{y\,x_n\,z}_{\sigma(x_m)}\,\sigma(v),$$

for some words u, v, y, z. Therefore $x_n = (\sigma(x_m))_{i-1}$ for some $i \in \{1, \ldots, |\sigma(x_m)|\}$. Then by the definition of $\mathcal{A}_{\sigma,a}$, we have

$$x_n = \delta_\sigma(x_m, i) = \delta_\sigma(\delta_\sigma(a, \mathrm{rep}_S(m)), i) = \delta_\sigma(a, \mathrm{rep}_S(m)i)$$

and in view of condition (3.12) and again by the definition of $\mathcal{A}_{\sigma,a}$, we get

$$\mathrm{val}_S(\mathrm{rep}_S(m)i) = |\sigma^s(a)| + |\sigma(x_{|\sigma^{s-1}(a)|})| + \cdots + |\sigma(x_{m-1})| + i = n.$$

Hence, $\mathrm{rep}_S(n) = \mathrm{rep}_S(m)i$ and the result follows. □

Example 3.4.15 The infinite word generated by the morphism σ given in Example 3.4.11 is $(x_n)_{n \ge 0} = abc\underline{b}caac\underline{b}caacabcabcaac\cdots$. The first few

---

† In order to have $\mathrm{rep}_S(n) = n$, one could work instead with the alphabet $C' = \{0, \ldots, \max_{b \in A} |\sigma(b)| - 1\}$.

words without leading 1 accepted by the automaton given in Figure 3.8 where all states are final are $\varepsilon, 2, 3, 21, 22, 31, 32, 33, 211, \ldots$. This provides us with an ANS $\mathcal{S}$.

For instance, we consider the element $x_3 = b$. This is why it has been underlined. We know that $\sigma(b) = bc$. So this latter factor should appear later on in the infinite word and the previous corollary permits us to find where it occurs. The $\mathcal{S}$-representation of 3 is 21. So we have to consider the words 211 and 212 – only these two words because $|\sigma(b)| = 2$ – and $\mathrm{val}_{\mathcal{S}}(211) = 8$, $\mathrm{val}_{\mathcal{S}}(212) = 9$. Therefore, one can check that $x_8 x_9 = \sigma(b) = bc$.

Now we turn to the converse of Proposition 3.4.12.

Proposition 3.4.16 *Every $\mathcal{S}$-automatic sequence is substitutive.*

Proof Let $\mathcal{S} = (L, A, <)$ be an ANS. Let $\mathcal{A} = (Q, A, E, \{q_0\}, T)$ be a complete DFA with transition function $\delta_{\mathcal{A}} : Q \times A^* \to Q$ recognising L and $\mathcal{B} = (R, A, \delta_{\mathcal{B}}, \{r_0\}, B, \mu)$ be a DFAO generating an $\mathcal{S}$-automatic sequence $x = (x_n)_{n \geq 0}$ over B, i.e., for all $n \geq 0$, $x_n = \mu(\delta_{\mathcal{B}}(r_0, \mathrm{rep}_{\mathcal{S}}(n)))$.

Consider the Cartesian product automaton $\mathcal{P} = \mathcal{A} \times \mathcal{B}$ defined as follows. The set of states of $\mathcal{P}$ is $Q \times R$. The initial state is (q_0, r_0) and the alphabet is A. For any word $w \in A^*$, the transition function $\Delta : (Q \times R) \times A^* \to Q \times R$ is given by

$$\Delta((q, r), w) = (\delta_{\mathcal{A}}(q, w), \delta_{\mathcal{B}}(r, w)).$$

This means that the product automaton mimics in a single automaton, the behaviours of both $\mathcal{A}$ and $\mathcal{B}$. In particular, after reading w in $\mathcal{P}$, $\Delta((q_0, r_0), w)$ belongs to $F \times R$ if, and only if, w belongs to L. Moreover, if $\mathrm{rep}_{\mathcal{S}}(n) = w$ and $\Delta((q_0, r_0), w) = (q, r)$, then $x_n = \mu(r)$.

Now we can apply Lemma 3.4.6 to $\mathcal{P}$ and define a morphism $\psi_{\mathcal{P}}$ prolongable on a letter z which does not belong to $Q \times R$. In view of the previous paragraph, we define $\nu : (Q \times R) \cup \{z\} \to B^*$ by

$$\nu(q, r) = \begin{cases} \mu(r), & \text{if } q \in F, \\ \varepsilon, & \text{otherwise} \end{cases}$$

and $\nu(z) = \varepsilon$. As Lemma 3.4.6 can be used to describe the sequence of reached states, $\nu(\psi_{\mathcal{P}}(z))$ is exactly the sequence $(x_n)_{n \geq 0}$. $\square$

Note that the morphisms obtained at the end of this proof are erasing. Again, if needed, Lemma 3.4.5 can be used.

3.4.1 Some properties of $\mathcal{S}$-automatic sequences

Here we give a characterisation of $\mathcal{S}$-automatic sequences in terms of finiteness of its $\mathcal{S}$-kernel and then study the relationship between $\mathcal{S}$-automaticity and $\mathcal{S}$-recognisability. We conclude this subsection with some discussion about the theorem of Cobham.

Definition 3.4.17 Let $\mathcal{S} = (L, A, <)$ be an ANS. For each word w in A^*, we define an, possibly finite or empty, ordered set of integers:

$$\mathcal{I}_\mathcal{S}(w) = \{n \mid \mathrm{rep}_\mathcal{S}(n) \in A^*w\} = \mathrm{val}_\mathcal{S}(L \cap A^*w) = \{i_{w,0} < i_{w,1} < \cdots\}.$$

In particular if s is a suffix of w, then $\mathcal{I}_\mathcal{S}(w) \subseteq \mathcal{I}_\mathcal{S}(s)$. Also we define a partial function $\alpha_\mathcal{S}$ mapping $(w, n) \in A^* \times \mathbb{N}$ onto $i_{w,n}$. For all $w \in A^*$, defining the ANS $\mathcal{S}_w = (L \cap A^*w, A, <)$, we get

$$\alpha_\mathcal{S}(w, n) = \mathrm{val}_\mathcal{S}(\mathrm{rep}_{\mathcal{S}_w}(n)), \text{ whenever defined.}$$

Example 3.4.18 Consider the language L accepted by the DFA depicted in Figure 3.2 from Example 3.2.1. Since the first few words in L are

$\mathrm{rep}_\mathcal{S}(n)$	b	aa	abb	bab	bba	$aaab$	$aaba$	$abaa$	$baaa$	$bbbb$	$\ldots$
n	0	1	2	3	4	5	6	7	8	9	$\ldots$

we get $\mathcal{I}_\mathcal{S}(\varepsilon) = \mathbb{N}$, $\mathcal{I}_\mathcal{S}(a) = \{1, 4, 6, 7, 8, \ldots\}$, $\mathcal{I}_\mathcal{S}(b) = \{0, 2, 3, 5, 9, \ldots\}$, $\mathcal{I}_\mathcal{S}(aa) = \{1, 7, 8, \ldots\}$, etc. So $\alpha_\mathcal{S}(\varepsilon, n) = n$ for all $n \geq 0$, $\alpha_\mathcal{S}(a, 0) = 1$, $\alpha_\mathcal{S}(a, 1) = 4$, $\alpha_\mathcal{S}(a, 2) = 6$, $\alpha_\mathcal{S}(b, 0) = 0$, $\alpha_\mathcal{S}(b, 1) = 2$, $\alpha_\mathcal{S}(b, 2) = 3$, etc.

Recall that, for $k \geq 2$, the *k-kernel* (also see Definition 9.1.1 where it is used to define the concept of automatic sequence, as recalled below) of a sequence $(x_n)_{n \geq 0}$ is the set of subsequences defined as

$$\{(x_{k^j n + r})_{n \geq 0} \mid j \geq 0, 0 \leq r < j\}.$$

Otherwise stated, we consider all the subsequences obtained by taking all the indices that are congruent modulo a power of k. It is well-known that *a sequence is k-automatic if, and only if, its k-kernel is finite*, see for instance (Allouche and Shallit 2003). With the definition of the map $\alpha_\mathcal{S}$ introduced above, but by writing α_k when dealing with the usual k-ary numeration system on $B_k = \{0, \ldots, k-1\}^* \setminus 0\{0, \ldots, k-1\}^*$, the usual k-kernel of a sequence $(x_n)_{n \geq 0}$ can be rewritten as

$$\{(x_{\alpha_k(w,n)})_{n \geq 0} \mid w \in \{0, \ldots, k-1\}^*\}$$

because a word u over $\{0, \ldots, k-1\}$ ends with a suffix w of length j if, and only if, $\mathrm{val}_k(u) \mod k^j = \mathrm{val}_k(w)$.

Definition 3.4.19 Let $S = (L, A, <)$ be an ANS. The S-*kernel* of the sequence $(x_n)_{n \geq 0}$ is the set of subsequences $\{(x_{\alpha_S(w,n)})_{n \geq 0} \mid w \in A^*\}$.

Theorem 3.4.20 (Rigo and Maes 2002) *A sequence $x = (x_n)_{n \geq 0} \in A^{\mathbb{N}}$ is S-automatic if, and only if, its S-kernel is finite.*

The proof is similar to the classical one.

Remark 3.4.21 It is obvious that a set of integers is S-recognisable if, and only if, its characteristic word is S-automatic. Therefore a sequence $x = (x_n)_{n \geq 0} \in A^{\mathbb{N}}$ is S-automatic if, and only if, for all $a \in A$, the a-*fiber*, i.e., the set $\{n \mid x_n = a\}$, is S-automatic.

Properties of the complexity function p_w counting the number of factors of a substitutive word w are well-known, see Chapter 4. These facts can be taken into account to show that some sets are S-recognisable for *no* ANS S. Take the *Champernowne word* over $\{0,1\}$ $c = 0110111001011101111000\cdots$ obtained as the ordered juxtaposition of the binary representations of the integers. It is the characteristic word of a set of integers $\{1, 2, 4, 5, 6, \ldots\}$ which is never S-recognisable because the complexity function of c is $p_c(n) = 2^n$ which is not an admissible behaviour for a substitutive word. Also it can be shown that the set of primes is never S-recognisable (Mauduit 1988), (Mauduit 1992), (Rigo 2000).

Remark 3.4.22 It is not difficult to prove that the characteristic sequence of the set of squares can be generated using the morphism $\sigma : a \mapsto abcd$, $b \mapsto b$, $c \mapsto cdd$, $d \mapsto d$ iterated on a and a coding $\tau : a, b \mapsto 1, c, d \mapsto 0$ (also see the morphism and the coding given in Example 1.2.23). We can compare this result with Example 3.3.8 and observe that the construction developed in Section 3.3.1 can also be presented in the context of substitutive words. In particular, one can notice that the same kind of results have been obtained independently in (Carton and Thomas 2002) where some decidability of the logical structure $\langle \mathbb{N}, + \rangle$ extended with a substitutive predicate is sought.

It is time to come back to the Cobham theorem (Theorem 1.5.5) and its generalisation to ANS. Indeed, now we hope that thanks to Theorem 3.4.1 the reader is convinced that both formalisms of substitution or ANS are well suited to define and study a relevant notion of recognisable sets of integers. Let x, y be infinite fixed points of two morphisms μ, ν: $\mu(x) = x$, $\nu(y) = y$ and α, β be two codings. Roughly speaking, we would like to have a result of the kind: if μ and ν are 'independent' in a sense to be defined and if $\alpha(x) = \beta(y)$, then the word $\alpha(x)$ is eventually periodic. Following G. Hansel's work about syndeticity (Hansel 1982, Hansel 1998), F. Durand has made a lot of progress in that direction. For instance, if μ and ν are primitive and if the corresponding dominating eigenvalues are multiplicatively independent then the theorem of Cobham still holds, see (Durand 1998a). Later on more cases can be taken into account, see (Durand 1998c), (Durand 2002) and also (Durand and Rigo 2009) where the situation of two ANS, one defined on a polynomial language and the other on an exponential one, is considered. To obtain full generality, only a few cases remain unsolved.

Remark 3.4.23 Up to now there is no proof of a Cobham-like theorem for a substitution having no main sub-substitution having the same dominating eigenvalue like $a \mapsto aa0$, $0 \mapsto 01$ and $1 \mapsto 0$. In this latter example, the dominating eigenvalue is 2 but the substitution restricted to $\{0,1\}$ has $(1+\sqrt{5})/2$ as dominating eigenvalue.

3.4.2 The HD0L ω-equivalence and periodicity problems

We recall some definitions about the so-called *Lindenmayer systems*. For an account of these systems we refer to (Kari, Rozenberg, and Salomaa 1997), also see Section 10.1. A *D0L system* is a triple $G = (A, \sigma, u)$ where A is a finite alphabet, u is a word over A and $\sigma : A^* \to A^*$ is a morphism, the acronym D stands for 'deterministic' and 0 stands for 'zero-sided'. An *HD0L system*, where H stands for 'homomorphism', is a 5-tuple $G = (A, B, \sigma, \tau, u)$ where (A, σ, u) is a D0L system, B is a finite alphabet and $\tau : A^* \to B^*$ is a morphism. If u is a prefix of $\sigma(u)$ and the set $\{\sigma^n(u) \mid n \geq 0\}$ is infinite, we denote $\sigma^\omega(u) = \lim_{n \to \infty} \sigma^n(u)$. Similarly, if $G = (A, B, \sigma, \tau, u)$ is an HD0L system, u is prefix of $\sigma(u)$ and the set $\{\tau(\sigma^n(u)) \mid n \geq 0\}$ is infinite, we denote $\omega(G) = \lim_{n \to \infty} \tau(\sigma^n(u))$. The *HD0L ω-equivalence problem* is stated as follows. Let $G_i = (A_i, B_i, \sigma_i, \tau_i, u_i)$, $i = 1, 2$, be two HD0L systems such that $\omega(G_1)$ and $\omega(G_2)$ exist. If $\omega(G_1) = \omega(G_2)$, then the two HD0L systems G_1 and G_2 are said to be ω-*equivalent*. Is it possible to decide whether or not G_1 and G_2 are ω-equivalent? In fact, HD0L systems are closely related to substitutive words.

Lemma 3.4.24 *Let $G_1 = (A, B, \mu, \nu, w)$ be an HD0L system such that $\omega(G_1)$ exists and $|w| > 1$. Then there exists an HD0L system $G_2 = (C, B, \sigma, \tau, c)$ ω-equivalent to G_1 where the letter $c \in C$ is prefix of $\sigma(c)$.*

Proof Assume that $\mu(w) = wu$ for some $u \in A^+$ and $w = w_1 \cdots w_\ell$, $\ell \geq 2$, with $w_i \in A$ for $1 \leq i \leq \ell$. We have $\mu^n(w) = w\,u\,\mu(u) \cdots \mu^{n-1}(u)$ for all $n \geq 1$. Let us introduce $\ell + 1$ new letters $c, \overline{w}_1, \ldots, \overline{w}_\ell$ which do not belong to A. The alphabet C is defined by $C = A \cup \{c, \overline{w}_1, \ldots, \overline{w}_\ell\}$. The morphism $\sigma : C^* \to C^*$ is defined as follows, $\sigma : c \mapsto c\overline{w}_1$, $\overline{w}_1 \mapsto \overline{w}_2, \ldots, \overline{w}_{\ell-1} \mapsto \overline{w}_\ell$, $\overline{w}_\ell \mapsto u$ and for $a \in A$, $\sigma(a) = \mu(a)$. We get

$$\lim_{n \to \infty} \sigma^n(c) = c\overline{w}_1 \cdots \overline{w}_\ell \, u \, \mu(u) \, \mu^2(u) \, \mu^3(u) \cdots.$$

To conclude the proof, we define τ by $\tau(c) = \varepsilon$, $\tau(\overline{w}_i) = \nu(w_i)$ for $1 \leq i \leq \ell$ and $\tau(a) = \nu(a)$ for $a \in A$. It is obvious that $\tau(\sigma^\omega(c)) = \nu(\mu^\omega(w))$. □

Remark 3.4.25 With the above lemma and Theorem 3.4.1, many classical open decision problems about HD0L systems can be restated in the framework of ANS. See in particular Chapter 10. The HD0L ω-equivalence problem is equivalent to the following problem expressed in terms of ANS. Let $\mathcal{S}_i = (L_i, A_i, <_i)$, $i = 1, 2$, be two abstract numeration systems. Is it decidable, given regular languages $K_i \subseteq L_i$, $i = 1, 2$, whether or not $\text{val}_{\mathcal{S}_1}(K_1) = \text{val}_{\mathcal{S}_2}(K_2)$? In the same way, the problem of deciding whether or not a given infinite HD0L word $\omega(G)$ is eventually periodic is equivalent to the following problem. Let $\mathcal{S} = (L, A, <)$ be an ANS. *Is it decidable, given a regular language $K \subseteq L$, whether or not $\text{val}_\mathcal{S}(K)$ is eventually periodic?*

3.4.3 Multidimensional setting

If one goes to the multidimensional case, it is not difficult to mimic as follows the construction of (Salon 1987), where images of letters are finite multidimensional words with square or cube shapes of same dimension. Let $d \geq 2$, $\mathcal{S} = (L, A, <)$ and $\#$ be a symbol not in A. The idea to define an $\mathcal{S}$-automatic d-dimensional sequence $\mathbf{x} = (x_{i_1, \ldots, i_d})_{i_1, \ldots, i_d \geq 0}$ over an alphabet B (*i.e.*, a map from $\mathbb{N}^d$ onto B) is to consider a DFAO $\mathcal{B} = (Q, (A \cup \{\#\})^d, \delta_\mathcal{B}, \{q_0\}, B, \mu)$ over the alphabet $(A \cup \{\#\})^d$ and to define

$$x_{i_1, \ldots, i_d} = \mu(\delta_\mathcal{B}(q_0, (\text{rep}_\mathcal{S}(i_1), \ldots, \text{rep}_\mathcal{S}(i_d))^\#)).$$

The padding operator $\#$ has been given in Definition 3.3.19.

One can therefore ask if Theorem 3.4.1 can be extended to this setting. We mention the following result without giving much detail, merely some informal description. Also see (Charlier 2009).

Theorem 3.4.26 (Charlier, Kärki, and Rigo 2010) *Let $d \geq 1$. The d-dimensional infinite word $\mathbf{x} = (x_{i_1, \ldots, i_d})_{i_1, \ldots, i_d \geq 0}$ is $\mathcal{S}$-automatic for some abstract numeration system $\mathcal{S} = (L, A, <)$ where $\varepsilon \in L$ if, and only if, $\mathbf{x}$ is the image under a coding of a morphic shape-symmetric infinite d-dimensional word.*

Abstract numeration systems 151

Observe that the proof of Proposition 3.4.12 makes use at the very beginning of Lemma 3.4.5. So one of the main difficulties occurring in the proof of the above theorem is that Lemma 3.4.5 has to be generalised to a multi-dimensional setting. This is some technical business that we do not want to present here. Nevertheless, we briefly describe using an example what is the idea of the shape-symmetry introduced in (Maes 1999). Indeed, this notion can be defined with plenty of detail and indices but a glimpse should be enough to have a good idea of the result above, also see (Maes 1998), (Maes 2000).

Example 3.4.27 Consider a map μ defined on the alphabet $\{a, \ldots, h\}$ and whose images are finite rectangular arrays. We can use the same formalism as in the definition of words and also define the concatenation of words in any of the two directions provided that they have compatible shapes. For instance, $\mu(a)$ and $\mu(b)$ can be concatenated horizontally but not vertically.

$$\mu(a) = \mu(f) = \begin{array}{|c|c|} \hline a & b \\ \hline c & d \\ \hline \end{array},\ \mu(b) = \begin{array}{|c|} \hline e \\ \hline c \\ \hline \end{array},\ \mu(c) = \begin{array}{|c|c|} \hline e & b \\ \hline \end{array},\ \mu(d) = \begin{array}{|c|} \hline f \\ \hline \end{array},$$

$$\mu(e) = \begin{array}{|c|c|} \hline e & b \\ \hline g & d \\ \hline \end{array},\ \mu(g) = \begin{array}{|c|c|} \hline h & b \\ \hline \end{array},\ \mu(h) = \begin{array}{|c|c|} \hline h & b \\ \hline c & d \\ \hline \end{array}.$$

In Figure 3.9 we have represented the first iterations of μ on the letter a. As for prolongable morphisms, one can expect that this process will lead to some bidimensional fixed point $(x_{i,j})_{i,j\geq 0}$. The shape-symmetry mainly refers to the fact that, for all i, j, if the image by μ of $x_{i,j}$ is a rectangle of size $\ell \times m$, then the image by μ of $x_{j,i}$ is a rectangle of size $m \times \ell$. Also one must ensure that the images of all the letters in a given column (respectively row) have images which are rectangles with same length (respectively height). An equivalent formulation is that the image by μ of any diagonal element $x_{i,i}$ is a square. Some details are omitted, see (Maes 1999) for a complete description.

$$\mu(a) = \begin{array}{|c|c|} \hline a & b \\ \hline c & d \\ \hline \end{array},\ \mu^2(a) = \begin{array}{|c|c|c|} \hline a & b & e \\ \hline c & d & c \\ \hline e & b & f \\ \hline \end{array},\ \mu^3(a) = \begin{array}{|c|c|c|c|c|} \hline a & b & e & e & b \\ \hline c & d & c & g & d \\ \hline e & b & f & e & b \\ \hline e & b & e & a & b \\ \hline g & d & c & c & d \\ \hline \end{array}$$

Fig. 3.9 The first few iterations of μ.

3.5 Representing real numbers

The basic aim of this section is to introduce and summarise the material found in (Lecomte and Rigo 2002), (Lecomte and Rigo 2004). The concern is to extend the use of an ANS $\mathcal{S} = (L, A, <)$ to represent real numbers and to study the properties of the proposed extension.

Roughly, the problem is reduced to the representation of numbers belonging to some subinterval of $[0, 1]$. The idea is to associate with a real number x an infinite word $w \in A^\omega$ that plays a role similar to its decimal expansion: w will be the limit of a sequence of words $w^{(n)}$ in L used to produce more and more accurate rational approximations $x^{(n)}$ of x that eventually converge to it. Let us explain how they mimic the decimal system or more generally, β-numeration systems, see Chapter 2. The rational approximations provided by the decimal expansion $.d_1 d_2 \cdots d_\ell \cdots$ of a real number in $(1/10, 1)$ are $\frac{d_1}{10}, \frac{d_1 d_2}{100}, \ldots, \frac{d_1 \cdots d_\ell}{10^\ell}, \ldots$. They all take the form of a fraction whose numerator is a prefix of some length ℓ of the expansion and the denominator is the number of integers whose decimal representation has *length at most* ℓ. With that scheme in mind, for a sequence $(w^{(n)})_{n \geq 0}$ of words in L converging to an infinite word w, we set

$$x^{(n)} = \frac{\mathrm{val}_\mathcal{S}(w^{(n)})}{\mathcal{V}_L(|w^{(n)}|)}. \tag{3.13}$$

Under some suitable assumptions, $(x^{(n)})_{n \geq 0}$ is a converging numerical sequence and its limit x belongs to some interval canonically associated with the language L. The prefixes $w^{(n)}$ of the representation w can be used to approximate x. In the sequel, we assume that $\mathcal{A}$ is the minimal automaton of the regular language L.

3.5.1 Extending $\mathrm{val}_\mathcal{S}$

The set of the representations of the real numbers that we will describe is

$$\mathrm{Adh}(L) := \{ w \in A^\omega \mid \exists (w^{(n)})_{n \geq 0} \in L^\mathbb{N}, \lim_{n \to \infty} w^{(n)} = w \}.$$

This notion of adherence appears in (Nivat 1978) and is studied in (Boasson and Nivat 1980).

Proposition 3.5.1 *The set $\mathrm{Adh}(L)$ is uncountable if, and only if, there exist two cycles $\mathcal{C}$ and $\mathcal{C}'$ in any DFA accepting L such that $\mathcal{C} \cap \mathcal{C}' \neq \emptyset$ and $\mathcal{C} \cup \mathcal{C}'$ contains an accessible state and a co-accessible state.*

In the sequel, $\mathrm{Adh}(L)$ is obviously supposed to be uncountable. Also we make some additional assumptions in order that the sequences (3.13) converge when w belongs to $\mathrm{Adh}(L)$.

Hypothesis 3.5.2 For each $q \in Q$, either

(i) there exists $N_q \in \mathbb{N}$ such that $\mathcal{U}_q(n) = 0$, for all $n > N_q$, or
(ii) there exist $\theta_q \geq 1$, $P_q(x) \in \mathbb{R}[x]$ and $c_q > 0$ such that

$$\lim_{n \to \infty} \frac{\mathcal{U}_q(n)}{P_q(n)\theta_q^n} = c_q.$$

Since $\mathrm{Adh}(L)$ is uncountable, it follows from the above proposition that the language L has an exponential growth and therefore that $\theta := \theta_{q_0} > 1$. Replacing P_{q_0} by P_{q_0}/c_{q_0}, we may assume in what follows that

$$\lim_{n \to \infty} \frac{\mathcal{U}_{q_0}(n)}{P_{q_0}(n)\theta^n} = 1 =: a_{q_0} \quad \text{and, for all } q \in Q, \quad a_q := \lim_{n \to \infty} \frac{\mathcal{U}_q(n)}{P_{q_0}(n)\theta^n} \geq 0.$$

Proposition 3.5.3 *Let $\mathcal{S}$ be an ANS based on a language satisfying Hypothesis 3.5.2. If $w = w_0 w_1 \cdots \in \mathrm{Adh}(L)$ is the limit of a sequence $(w^{(n)})_{n \geq 0}$ of words in L, then*

$$x := \lim_{n \to \infty} \frac{\mathrm{val}_\mathcal{S}(w^{(n)})}{\mathcal{V}_L(|w^{(n)}|)} = \frac{\theta - 1}{\theta^2} \sum_{q \in Q} a_q \sum_{j=0}^{\infty} b_{q,j} \theta^{-j}$$

with the coefficients $b_{q,j}$ defined in (3.5). In particular, x is independent of the sequence in $L^\mathbb{N}$ converging to w. Moreover, it belongs to $[1/\theta, 1]$. Conversely every element in $[1/\theta, 1]$ is the limit of a sequence of the form (3.13) for some $w \in \mathrm{Adh}(L)$.

In this latter proposition, x is said to be the *numerical value* $\mathrm{val}_\mathcal{S}(w)$ of w. In the same way, the infinite word w is said to be an $\mathcal{S}$-*representation* of the real number x.

Proposition 3.5.4 *Let $\mathcal{S}$ be an ANS based on a language satisfying Hypothesis 3.5.2. The map $\mathrm{val}_\mathcal{S} : \mathrm{Adh}(L) \to [1/\theta, 1]$ is increasing and uniformly continuous.*

Some elements in $[1/\theta, 1]$ may have more than one $\mathcal{S}$-representation in $\mathrm{Adh}(L)$ and possibly infinitely many. This problem will be discussed in the next subsection.

Recall that each Pisot number β defines a unique positional and linear Bertrand numeration system $U_\beta = (U_n)_{n \in \mathbb{N}}$. Recall Example 3.1.14 and see (Bruyère and Hansel 1997). One can show (Frougny and Solomyak 1996) that the language L_β of all the normalised representations computed by the greedy algorithm satisfies Hypothesis 3.5.2, with $\theta = \beta$. In (Lecomte and Rigo 2004), also it is shown that the $\mathcal{S}$-representations of the elements of $[1/\beta, 1]$ in the ANS based upon L_β and the classical β-developments of these numbers coincide. In particular, $\mathrm{Adh}(L_\beta)$ is the set of these developments.

Example 3.5.5 Consider the classical Fibonacci system (for integers) or the β-numeration system related to the Golden Ratio φ. The language of all the representations of integers not starting with 0 is accepted by the DFA depicted in Figure 3.10. Let us consider the ANS based on this language and show that we get back the usual φ-development. If we set $\lambda = \frac{5+\sqrt{5}}{10}$,

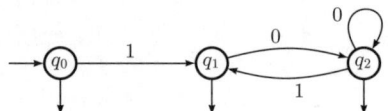

Fig. 3.10 A Fibonacci ANS.

then an easy computation shows that $\mathcal{U}_{q_0}(n) \sim \lambda \varphi^{n-1}$, $\mathcal{U}_{q_1}(n) \sim \lambda \varphi^n$ and $\mathcal{U}_{q_2}(n) \sim \lambda \varphi^{n+1}$. Setting P_{q_0} to the constant λ/φ and dividing $\mathcal{U}_q(n)$ by $P_{q_0} \varphi^n$, we get $a_{q_0} = 1$, $a_{q_1} = \varphi$ and $a_{q_2} = \varphi^2$. For any infinite word $w_0 w_1 \cdots \in \mathrm{Adh}(L)$, the formula of Proposition 3.5.3 becomes

$$\frac{\varphi-1}{\varphi^2} \sum_{j=0}^{\infty} \varphi^{-j} + (\varphi-1) \sum_{j=0}^{\infty} b_{q_2,j} \varphi^{-j} = \varphi^{-1} + \sum_{j \geq 2} w_j \varphi^{-j-1}$$

because for all $j \geq 0$ we have $b_{q_0,j} = 1$, $b_{q_1,j} = 0$, $b_{q_2,0} = 0$ and for $j > 0$ $b_{q_2,j} = w_j$. To obtain the last equality we used the fact that $\varphi - 1 = \varphi^{-1}$.

3.5.2 The intervals I_u

Let us have a closer look at the approximations of the elements in $[1/\theta, 1]$ by finite words. Let u be a word of length ℓ and denote by I_u the set of real numbers $x \in [1/\theta, 1]$ having an $\mathcal{S}$-representation starting with u. If I_u is non-empty, i.e., if u is a prefix of some element in $\mathrm{Adh}(L)$, then it is a closed interval. In particular, $I_\varepsilon = [1/\theta, 1]$. Moreover, if u is a prefix of v, then $I_u \supset I_v$ and if $w = w_0 w_1 \cdots \in \mathrm{Adh}(L)$ is an $\mathcal{S}$-representation of x, then x belongs to $I_{w_0 \cdots w_{\ell-1}}$ for all ℓ.

The set $\mathcal{I}_\ell$ of non-empty intervals I_u such that $|u| = \ell$ defines a covering of $[1/\theta, 1]$ made of closed subintervals with disjoint interiors. Otherwise stated, there exist $k(\ell)$ and real numbers $\kappa_1^\ell = 1/\theta \leq \cdots \leq \kappa_{k(\ell)+1}^\ell = 1$ such that

$$\mathcal{I}_\ell = \{[\kappa_j^\ell, \kappa_{j+1}^\ell] \mid j = 1, ..., k(\ell)\}.$$

Each κ_j^ℓ, $1 < j \leq k(\ell)$, has at least two representations, as it is the upper bound of some I_u and the lower bound of some I_v. It may well occur that Adh(L) contains infinitely many words having prefix u although I_u contains exactly one element, i.e., $\kappa_j^\ell = \kappa_{j+1}^\ell$ for some j. This one has then infinitely many representations. Obviously, vanishing constants a_p are of no use to compute val$_\mathcal{S}(w)$ and this causes that phenomenon, see Proposition 3.5.3. It follows easily from Hypothesis 3.5.2 that if $a_q = 0$ and $p = q.u$ for some word u, then $a_p = 0$. Thanks to Proposition 3.5.3, we may delete from $\mathcal{A}$ the states q such that $a_q = 0$ and the corresponding edges without changing the representations of real numbers, up to the fact that we replace $\mathcal{A}$ and L by the simplified automaton and its language. Now, a real number in $(1/\theta, 1)$ has exactly one representation if it is not the endpoint of some I_u and exactly two representations otherwise.

Proposition 3.5.6 *Let $\mathcal{S}$ be an ANS based on a language satisfying Hypothesis 3.5.2 and let $\mathbf{M} \in \mathbb{N}^{Q \times Q}$ be the adjacency matrix of $\mathcal{A}$. For all states p, we have*

$$a_p = \frac{1}{\theta} \sum_{q \in Q} \mathbf{M}_{pq} a_q.$$

If u is a prefix of length ℓ of some element of Adh(L), then

$$I_u = \left[\frac{1}{\theta} + \frac{\theta-1}{\theta^{\ell+1}} \sum_{\substack{|t|=\ell \\ t<u}} a_{q_0.t}, \frac{1}{\theta} + \frac{\theta-1}{\theta^{\ell+1}} \sum_{\substack{|t|=\ell \\ t \leq u}} a_{q_0.t} \right].$$

In particular, θ and the numbers κ_j^ℓ are algebraic.

3.5.3 A dynamical point of view

In order to understand the structure of the set of intervals I_u, it is useful, say, to normalise them in some way. This will allow us to design an algorithm to compute representations of real numbers in $[1/\theta, 1]$ and to study the set of these which have an eventually periodic expansion. To that end, for any interval $I = [s, t]$, $s < t$, we use the increasing bijection

$$f_I : I \to [0,1], \quad x \mapsto \frac{x-s}{t-s}$$

and we say that $f_I(x)$ is the *relative position* of $x \in I$ (inside I). More generally, the relative position of a subset E of I inside I will be $f_I(E)$.

Proposition 3.5.7 *Let u and v be two words such that $q_0.u = q_0.v$. For each $a \in A$, $I_{ua} \neq \emptyset$ if, and only if, $I_{va} \neq \emptyset$. Moreover, if $I_{ua} \neq \emptyset$, then the relative positions of I_{ua} inside I_u is equal to that of I_{va} inside I_v.*

The above proposition is a key point in our study as it tells that the interval I_u only depends upon the states $q_0.u$, and not specifically upon u. We use it to construct a dynamical system $(Q \times [0,1], T)$ which encodes the relationship between the sets $\mathcal{I}_\ell$.

Let q be any given state of $\mathcal{A}$. As the latter is accessible, $q = q_0.u$ for some u of length say ℓ. Then

$$I_u = [\kappa_i^{\ell+1}, \kappa_j^{\ell+1}]$$

for some $i < j$ and we get a partition

$$[0,1] = [\kappa_i', \kappa_{i+1}') \cup \cdots \cup [\kappa_r', \kappa_{r+1}') \cup \cdots \cup [\kappa_{j-1}', \kappa_j']$$

where, for simplicity, κ_r' denotes the relative position of $\kappa_r^{\ell+1}$ inside I_u. Of course, the elements of that partition are nothing but the non-empty intervals among the I_{ua}, $a \in A$. We let $R_{q,a}$ denote the relative position of such an I_{ua} inside I_u and we define the function T by

$$T: Q \times [0,1] \to Q \times [0,1], \ (q,x) \mapsto (q.a, f_{R_{q,a}}(x))$$

where a is the unique letter such that $x \in R_{q,a}$.

The algorithm in Table 3.3 computes prefixes of the representation of a real number $x \in [1/\theta, 1]$ by applying iteratively T to the initial data (q_0, x). The length ℓ of the prefixes is determined by some halting condition.

```
INPUT : x ∈ [1/θ, 1]
q ← q₀
u ← ε
I ← Iε
y ← f_I(x)
REPEAT
    DETERMINE a ∈ A such that y ∈ R_{q,a}
    q ← q.a
    u ← CONCATENATE(u, a)
    I ← R_{q,a}
    y ← f_I(x)
UNTIL |u| = ℓ
```

Table 3.3. *An algorithm computing a prefix of length ℓ of an $\mathcal{S}$-representation of the real x.*

Example 3.5.8 Let us continue Example 3.5.5. Clearly, the state q_0 occurs only once and a representation always starts with 1. From q_1, one can only reach q_2 reading 0. So a discussion has to be made only for state q_2. The interval $I_{10} =$ splits into I_{100} and I_{101}. The relative position of $\varphi^{-1} + \varphi^{-3}$ inside $[\varphi^{-1}, 1]$ is φ^{-1}. So we get the partition $[0, \varphi^{-1}) \cup [\varphi^{-1}, 1]$. A scheme of application of Algorithm 3.3 is given in Figure 3.11.

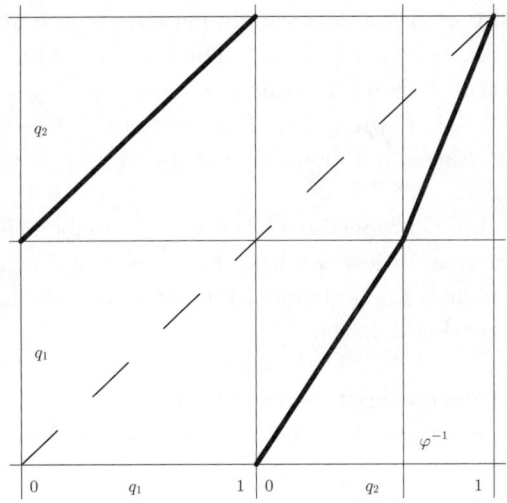

Fig. 3.11 The map T restricted to $\{q_1, q_2\} \times [0,1]$.

3.5.4 Real numbers with eventually periodic representations

In the k-ary numeration system, the set of real numbers having an eventually periodic representations is $\mathbb{Q}$. In particular, it is dense in $\mathbb{R}$ and has strong algebraic properties (it is a field). It is of course tempting to investigate the properties of the set of eventually periodic words belonging to $\mathrm{Adh}(L)$. This is a hard problem and only a few results are known up to now in that direction.

We begin with some interesting facts. For any set of (infinite) words X, we let $\mathrm{eper}(X)$ denote the set of eventually periodic elements in X, by $\mathrm{per}(X)$ the set of their periods and by $\mathrm{preper}(X)$ the set of their preperiods, i.e., if uv^ω belongs to $\mathrm{eper}(X)$, then v belongs to $\mathrm{per}(X)$ and u belongs to $\mathrm{preper}(X)$.

Proposition 3.5.9 *The sets $\mathrm{per}(\mathrm{Adh}(L))$ and $\mathrm{preper}(\mathrm{Adh}(L))$ are regular. Moreover, $\mathrm{eper}(\mathrm{Adh}(L))$ is dense in $\mathrm{Adh}(L)$.*

The real numbers having an eventually periodic representation can be characterised in terms of T.

Theorem 3.5.10 *A number $x \in [1/\theta, 1]$ has an eventually periodic representation if, and only if, there exists $r < s$ such that*
$$T^r(q_0, x) = T^s(q_0, x).$$
In particular, each number κ_i^ℓ has an eventually periodic representation.

Let us explain why κ_i^ℓ has an eventually periodic representation. There is a word m such that $I_m = [\kappa_i^\ell, \kappa_{i+1}^\ell]$. The representation of κ_i^ℓ given by Algorithm 3.3 starts with m. In other words, $T^\ell(q_0, \kappa_i^\ell) = (q_0.m, 0)$. But then, clearly, for $r > \ell$, $T^r(q_0, \kappa_i^\ell)$ is of the form $(q, 0)$ for some q. As there are finitely many states, it follows that $T^r(q_0, \kappa_i^\ell) = T^s(q_0, \kappa_i^\ell)$ for some $r < s$.

As for the algebraic properties of the set of numbers having an eventually periodic representation, we have the following result similar to the one given independently in (Bertrand 1977), (Schmidt 1980a) (also see Theorem 2.3.20 and Section 2.3.2.1).

Theorem 3.5.11 (Rigo and Steiner 2005) *Let $\mathcal{S}$ be an ANS based on a language satisfying Hypothesis 3.5.2. If the corresponding real number θ is a Pisot number, then*

$$\mathrm{val}_\mathcal{S}(eper(Adh(L))) = \mathbb{Q}(\theta) \cap [1/\theta, 1]$$

but if θ is neither a Pisot number nor a Salem number, then

$$\mathbb{Q}(\theta) \cap [1/\theta, 1] \not\subseteq \mathrm{val}_\mathcal{S}(eper(Adh(L))).$$

3.6 Exercises and open problems

Exercise 3.1 (Charlier 2009) Consider the sequence $U = (U_n)_{n \geq 0}$ given by $U_i = i + 1$ for $i = 0, 1, 2, 3$ and $U_n = 2U_{n-1}$ for all $n \geq 4$. Show that $\mathbb{N}$ is U-recognisable. Show that for all $k \geq 1$, there exist no $a_{k-1}, \ldots, a_0 \in \mathbb{C}$ with $a_0 \neq 0$ such that, for all $n \geq 0$,

$$U_{n+k} = a_{k-1} U_{n+k-1} + \cdots + a_0 U_n.$$

In the terminology of (Berstel and Reutenauer 1988), U does not satisfy any *strict* linear recurrence relation. Hint: $a_0 \neq 0$ is invertible.

Exercise 3.2 Give a proof of (3.4) given in page 119 using Lemma 3.2.2.

Exercise 3.3 Consider the two abstract numeration systems based on a^*b^* obtained by changing the ordering on the alphabet, $\mathcal{S} = (a^*b^*, \{a,b\}, a < b)$ and $\mathcal{R} = (a^*b^*, \{a,b\}, b < a)$. Study the function $f_{\mathcal{S},\mathcal{R}} : \mathbb{N}^2 \to \mathbb{N}^2, (i,j) \mapsto (x,y)$ such that $\mathrm{rep}_\mathcal{R}(\mathrm{val}_\mathcal{S}(a^i b^j)) = a^x b^y$.

Exercise 3.4 Show that, in general, changing the ordering of the alphabet is not a recognisability-preserving operation. A counter-example is given in (Lecomte and Rigo 2001).

Exercise 3.5 Find a closed formula for the expression of $\mathrm{val}_\mathcal{S}(a^i b^j c^k)$ for the ANS $\mathcal{S} = (a^*b^*c^*, \{a,b,c\}, a<b<c\}$. In this system which set of integers is represented respectively by a^*, b^* and c^*?

Exercise 3.6 Generalise ANS on a^*b^* or $a^*b^*c^*$ by considering the ANS $\mathcal{S} = (a_1^* \cdots a_t^*, \{a_1, \ldots, a_t\}, a_1 < \cdots < a_t)$ where $a_1, \ldots, a_t$ are t distinct letters. Show that this system is equivalent to the so-called *binomial numeration system* defined as follows, see (Fraenkel 1985). Any integer $n \geq 0$ can be uniquely written as

$$n = \binom{z_t}{t} + \binom{z_{t-1}}{t-1} + \cdots + \binom{z_1}{1}$$

with $z_t > z_{t-1} > \cdots > z_1 \geq 0$. Indeed, show that we have

$$\mathrm{val}_\mathcal{S}(a_1^{n_1} \cdots a_t^{n_t}) = \sum_{i=1}^{t} \binom{n_i + \cdots + n_t + t - i}{t - i + 1}.$$

For details see (Charlier, Rigo, and Steiner 2008). Also see connection with (Lew, Morales, and Sánchez-Flores 1996).

Exercise 3.7 Consider the ANS given in Example 3.1.15. This system seems to be related to the Fibonacci numeration system. Is it possible to assign weights $v(a)$ and $v(b)$ to a and b to recover the usual Fibonacci system, i.e., such that, for all $w_\ell \cdots w_0 \in L$, $\mathrm{val}_\mathcal{S}(w) = \sum_{k=0}^{\ell} v(w_i) F_i$?

Exercise 3.8 Let $\mathcal{S} = (a^*b^*, \{a,b\}, a<b\}$. Show that the formal series $\sum_{w \in L} \mathrm{val}_\mathcal{S}(w)\, w$ is rational in the sense of (Berstel and Reutenauer 1988) (also see the definition given in Section 2.6.1). In particular, we get the linear representation (λ, μ, γ) where $\mu : \{a,b\}^* \to \mathbb{N}^{3\times 3}$ is a morphism of monoids defined by

$$\mu(a) = \begin{pmatrix} 1 & 1 & 0 \\ 0 & 1 & 1 \\ 0 & 0 & 1 \end{pmatrix},\ \mu(b) = \begin{pmatrix} 1 & 1 & 1 \\ 0 & 1 & 1 \\ 0 & 0 & 1 \end{pmatrix},$$

$\lambda = \begin{pmatrix} 1 & 0 & 0 \end{pmatrix}$ and $\gamma = \begin{pmatrix} 0 & 1 & 1 \end{pmatrix}$ such that $\mathrm{val}_\mathcal{S}(w) = \lambda \mu(w)\,^t\gamma$. This result holds for any ANS, see (Rigo 2002) and independently (Choffrut and Goldwurm 1995) where $\mathrm{val}_\mathcal{S}$ is called *ranking*.

Exercise 3.9 Consider the ANS $\mathcal{S} = (\{a,b\}^* \setminus a^*, \{a,b\}, a<b)$ and the set Y such that $\mathrm{rep}_\mathcal{S}(Y) = a^*b$. Is the set $\mathrm{val}_\mathcal{S}(2\,\mathrm{rep}_\mathcal{S}(Y)) = Z$ still recognisable? We suggest to write a small computer program to list the first 100 elements in $\mathrm{rep}_\mathcal{S}(Z)$.

Exercise 3.10 Let $L \subset A^*$ be a cofinite language not equal to A^*. Study the preservation of $\mathcal{S}$-recognisability after multiplication by a constant for ANS based on L. Notice that if $L = A^*$, then ANS defined on L is equivalent to the usual integer base (Card A)-ary system.

Exercise 3.11 (Open problem) For the usual k-ary numeration system, a logical characterisation of the k-recognisable sets by first order logical formula from $\langle \mathbb{N}, +, V_k \rangle$ is well-known. Could one imagine a logical characterisation of the $\mathcal{S}$-recognisable sets in a suitable logical structure?

Exercise 3.12 (Open problem) Assume that $P \in \mathbb{Q}[X]$ is such that $P(\mathbb{N}) \subseteq \mathbb{N}$. If $P(\mathbb{N})$ is $\mathcal{S}$-recognisable for $\mathcal{S} = (L, A, <)$, what information on L can be obtained? For instance, is L polynomial?

Exercise 3.13 (Open problem) Let $\mathcal{S}$ be an ANS. Find necessary and/or sufficient conditions for the existence of an increasing sequence $(U_n)_{n \geq 0}$ of integers such that $U_0 = 1$ and a map $v : A \to \mathbb{N}$ such that $\mathrm{val}_{\mathcal{S}}(w) = \sum_{i=0}^{\ell} v(w_i) U_i$ for all $w = w_\ell \cdots w_0 \in L$.

Exercise 3.14 (Open problem) This problem is also discussed in the bibliographic notes and in Remark 3.4.25. Does there exist an algorithm, given an ANS $\mathcal{S} = (L, A, <)$ and any $\mathcal{S}$-recognisable set X of integers given by a DFA, which can be used to decide whether or not X is eventually periodic?

Exercise 3.15 (Open problem) Obtain a general Cobham-like theorem for ANS. See in particular, Remark 3.4.23.

3.7 Notes

Properties of k-recognisable sets are well-known. For a survey, see for instance (Bruyère, Hansel, Michaux, et al. 1994). This paper explains in particular the logical characterisation of the k-recognisable sets in terms of first order logical formulas in an extension of the Presburger arithmetic $\langle \mathbb{N}, + \rangle$ with a valuation V_k. Most of the characterisations encountered for k-ary systems can be extended to the Pisot systems of Example 3.1.14. See (Bruyère and Hansel 1997) which partially relies on (Frougny 1992) about the *normalisation* function computable by a finite automaton. Also it is probably worth having a look at (Shallit 1994) which has been cited many times in this chapter.

Slender languages have been considered in several contexts and particularly in some decision problems. See (Honkala 1997), (Honkala 1998),

(Honkala 2001b). For a study of morphisms and/or languages with polynomial growth, also see (Mauduit 1986).

It is interesting to note that Lemma 3.3.5 about the regularity of the set of minimal words of each length in a regular language has been extended to context-free languages. If L is context-free, then $\mathsf{minlg}\,(L)$ is again context-free (Berstel and Boasson 1997). It is also shown that if $\mathrm{Pref}(\mathrm{Adh}(L)) = L$, then $\mathsf{minlg}\,(L)$ is regular.

State complexity issues about *decimations* treated in Theorem 3.3.2 are considered in (Krieger, Miller, Rampersad, et al. 2009). In this paper, the authors moreover provide an example of an ANS $\mathcal{G}$ based on a non-regular context-free language such that $\mathrm{rep}_{\mathcal{G}}(2\mathbb{N})$ is not context-free. In (Berstel, Boasson, Carton, et al. 2006), some operations preserving regular languages are discussed.

The idea of Definition 3.4.10 associating with a morphism some canonical automaton already appears in the seminal paper (Cobham 1972). Of course, when considering a uniform morphism, the resulting automaton is complete.

More on extension of $\mathcal{S}$-automaticity to the multidimensional case can be found in (Rigo and Maes 2002) and (Nicolay and Rigo 2007). Following the work of A. Fraenkel, applications of ANS, and in particular the use of Corollary 3.4.14, to combinatorial game theory appear in (Duchêne and Rigo 2008a) and (Duchêne and Rigo 2008b). See (Duchêne, Fraenkel, Nowakowski, et al. 2009) for an application of shape-symmetric bidimensional morphisms to Wythoff's game. In this paper, it is proved that the set of losing positions defines a shape-symmetric morphic array.

Consider the following decision problem. Let $\mathcal{S} = (L, A, <)$ be an ANS. For any $\mathcal{S}$-recognisable set of integers given by a DFA, decide whether or not this set is eventually periodic. As explained in Remark 3.4.25 this problem can also be stated in terms of HD0L systems. The purely substitutive case is settled positively in (Harju and Linna 1986), (Pansiot 1986). For k-automatic sequences, the problem is solved in (Honkala 1986). See (Leroux 2005) where a polynomial time general procedure for d-dimensional subsets of $\langle \mathbb{Z}, + \rangle$ is given. An elegant and simple approach based on the construction of an NFA can be found in (Allouche, Rampersad, and Shallit 2009). In (Honkala and Rigo 2004), some special cases expressed in terms of ANS are treated. Recently, (Bell, Charlier, Fraenkel, et al. 2009) covers a large class of ANS for which the problem is decidable, also see (Charlier 2009). The general problem is still open.

Several other topics related to ANS and in particular to the representation of real numbers are the following ones. For most of the situations described below some extra assumptions on the language are often required like having a DFA with a dominating eigenvalue. The introduction in the sense of (Grabner, Liardet, and Tichy 1995) of the *odometer* for ANS is made in (Berthé and Rigo 2007b). The idea is to define in a proper way a map sending an infinite word onto its 'successor', also see Section 6.5. As an example, for positional numeration systems like the Fibonacci numeration system, this map sends $010100(10)^\omega$ onto $0000(10)^\omega$ and one has to study carry propagation. The definition of the odometer for ANS acts on a pair made of an infinite word and the infinite sequence of states corresponding to the path followed in the automaton when reading this word. The idea is to replace a non-maximal prefix of length ℓ read from some state q by the next word of same length ℓ accepted from q. Continuing the Fibonacci example, the successor of 1010 is 10000 which explains what we get above by taking mirror images. Note that considering pairs of letters and states is equivalent to considering local automata. Some *tilings* given in the framework of ANS have been presented in (Berthé and Rigo 2007a), see connection with Chapters 2 and 5. An analogue to the classical *sum-of-digits* function (see Chapter 9) can be defined as follows. Consider a map $f : A \to \mathbb{R}$ to define a completely *additive function*, i.e., for all $w = w_1 \cdots w_\ell \in A^*$, $f(w) = \sum_{i=1}^{\ell} f(w_i)$. The behaviour and distribution of the corresponding summatory function $\sum_{w \in L} f(w)$ is studied in (Grabner and Rigo 2003) and (Grabner and Rigo 2007). Extensions of β-expansions and of the map $T_\beta : x \mapsto \{\beta x\}$, see Section 2.3.2, is presented in (Rigo and Steiner 2005). As in Lemma 3.2.2, it involves for ANS as many maps as states in the minimal automaton of L.

The framework of Section 3.5 extends to a larger class of numeration systems in (Charlier, Le Gonidec, and Rigo) including numeration systems based on a non-regular language such as the one coming from rational base numeration systems (see Section 2.5).

4
Factor complexity

Julien Cassaigne,

François Nicolas

4.1 Introduction

Given an infinite word, we can study the language of its finite factors. Intuitively, we expect that 'simple' words (because they are generated by simple devices, or have some regularity property) will also have a 'simple' language of factors. One way to quantify this is to just count factors of each length. In doing so, we associate a function with the infinite word considered: its *factor complexity function*.

This function was introduced in 1938 by Gustav A. Hedlund and Marston Morse (Morse and Hedlund 1938), under the name *block growth*, as a tool to study symbolic dynamical systems. The name *subword complexity* was given in 1975 by Andrzej Ehrenfeucht, Kwok Pun Lee and Grzegorz Rozenberg (Ehrenfeucht, Lee, and Rozenberg 1975). Here we use *factor complexity* for consistency with the use of *factor* (see Section 1.2.3).

Factor complexity should not be confused with other notions of complexity, like algorithmic complexity or Kolmogorov complexity. We shall not use them here, and by 'complexity' we always mean 'factor complexity'.

Factor complexity will be used in Chapter 7 in combinatorial criteria for unique ergodicity, see Section 7.3.1. In Chapter 8 it is applied to the expansion of real numbers, see Theorem 8.1.6.

4.2 Definitions, basic properties and first examples

4.2.1 Factor complexity

Let A be a finite alphabet, of cardinality d. We consider an infinite word $u = u_0 u_1 u_2 \cdots \in A^{\mathbb{N}}$. (We could also work with bi-infinite words in $A^{\mathbb{Z}}$;

Combinatorics, Automata and Number Theory, ed. Valérie Berthé and Michel Rigo.
Published by Cambridge University Press. ©Cambridge University Press 2010.

most of the results that we will mention for infinite words remain valid for bi-infinite words.)

We recall Definitions 1.2.8 and 1.2.12.

For each $n \in \mathbb{N}$, $L_n(u)$ is the set of factors of length n in u, while $L(u)$ is the set of all factors in u, and is called the *language* of u.

For each $n \in \mathbb{N}$, let $p_u(n)$ be the cardinality of $L_n(u)$. Then p_u is a function from $\mathbb{N}$ to $\mathbb{N}$ which is called the *complexity function* of u. Often, when no confusion is possible, u will be kept implicit and we shall just write p.

Note that one can also define, in exactly the same terms, the complexity function p_w of a finite word $w \in A^*$, as well as the complexity function p_X of a language $X \subseteq A^*$ or of an infinitary language $X \subseteq A^\mathbb{N}$, a factor of X being a factor of any element of X. Here we shall focus on the complexity function of infinite words.

4.2.2 Periodic words

Example 4.2.1 Let $u = (aab)^\omega = aabaabaab\cdots$. Then $L(u) = (aab)^*\{\varepsilon, a, aa\} \cup (aba)^*\{\varepsilon, a, ab\} \cup (baa)^*\{\varepsilon, b, ba\}$, and the complexity function of u is as follows:

- $p(0) = 1$, as $L_0(u) = \{\varepsilon\}$,
- $p(1) = 2$, as $L_1(u) = \{a, b\}$ and
- $p(n) = 3$ for $n \geqslant 2$, as $L_n(u)$ contains exactly one word beginning with aa, one word beginning with ab and one word beginning with ba.

More generally, let $z \in A^+$ be a *primitive word*, i.e., a non-empty word which is not a power of a shorter word. Then $u = z^\omega = zzz\cdots$ is a periodic infinite word (see Definition 1.2.11), with minimal period $|z|$; the word z is called the *period cycle* of u. The complexity function of u satisfies $p(n) = |z|$ for each $n \geqslant |z|$, as there is exactly one factor of length n beginning with each conjugate of z (see also Exercise 4.1).

This can be extended to *eventually periodic words*, of the form $u = tz^\omega$, where t is a finite word called *transient part*. The length of t is called *preperiod*.

Proposition 4.2.2 *Let $u = tz^\omega$ be an eventually periodic word. Then its complexity is bounded by $|tz|$. Moreover, if z is primitive and t is either empty, or ends with a letter different from the last letter of z (any eventually periodic word can be uniquely written in this form), then $p(n) = |tz|$ for all $n \geqslant |tz|$.*

Proof If $i \geq |tz|$ and $n \in \mathbb{N}$, then the factor of length n of u occurring at position i also occurs at position $i - |z|$. Therefore all factors have an occurrence at a position less than $|tz|$, and as there are $|tz|$ such positions, there are at most $|tz|$ factors of each length.

Assume that z is primitive, t minimal and $n \geq |tz|$. If $p(n) < |tz|$, then there are two positions $i < j < |tz|$ such that the same factor $w = w_0 w_1 \cdots w_{n-1}$ of length n occurs at positions i and j. The suffix of length $|z|$ of w occurs in u at positions $i+n-|z|$ and $j+n-|z|$. Since both positions are at least $|t|$ and z is primitive, this is possible only if $(j-i)$ is a multiple of $|z|$. Let $j - i = m|z|$ with $m \geq 1$. As a consequence, $i \leq j - |z| < |t|$ and t is not empty. Therefore $w_{|t|-i-1} = u_{|t|-1}$ is the last letter of t. But $w_{|t|-i-1}$ is also $u_{|t|-1+m|z|}$, the last letter of z, a contradiction. □

4.2.3 Maximal complexity

Let $d = \operatorname{Card} A$ be the cardinality of A. As there are d^n words in A^n, obviously $p(n) \leq d^n$. This bound can be attained, as shown by the following construction.

Example 4.2.3 The free monoid A^* is a countable set. Let $A^* = \{w_0, w_1, w_2, \ldots\}$ be an enumeration of A^*, for instance in radix order: $A^* = \{\varepsilon, a, b, aa, ab, ba, bb, aaa, aab, \ldots\}$ (assuming a binary alphabet). Let $u = w_0 w_1 w_2 \cdots$ be the infinite word obtained by concatenating the w_i:

$$u = abaaabbabbaaaaababaabbbaababbbabbbaaaaaaabaaba \cdots.$$

It is called the *Champernowne word* (Barbier 1887, Champernowne 1933) and its complexity is $p(n) = d^n$, since by construction every word in A^* occurs in it.

Variants of this construction can be used to get other complexity functions, see Exercise 4.2.

4.2.4 Monotonicity

Not every function f such that $1 \leq f(n) \leq (\operatorname{Card} A)^n$ is the complexity function of an infinite word over A; characterising such functions is currently an open problem. However, some necessary conditions can be given, and we shall see a few of them. We start with a simple one.

Proposition 4.2.4 *The complexity function is non-decreasing: for all n, $p(n) \leq p(n+1)$.*

Proof For each factor $w \in L_n(u)$, we can choose a position $i(w)$ such that w occurs at that position in u. Then let $e(w)$ be the factor of length $n+1$ that occurs at position $i(w)$. Obviously, w is a prefix of $e(w)$. We have thus defined a map e from $L_n(u)$ to $L_{n+1}(u)$. This map is clearly injective: two different words of length n cannot be prefixes of the same word of length $n+1$. Therefore the cardinality of $L_{n+1}(u)$ is larger than or equal to that of $L_n(u)$: $p(n+1) \geqslant p(n)$. □

4.3 The theorem of Morse and Hedlund

4.3.1 Statement and corollaries

We have seen that eventually periodic words have bounded complexity. Our first theorem states that not only the converse holds, but actually the complexity of a non-eventually-periodic word is far from being bounded.

Theorem 4.3.1 (Theorem of Morse and Hedlund) *Let u be an infinite word and p its factor complexity function. Then either u is eventually periodic, or p is strictly increasing.*

Corollary 4.3.2 *Assume that for some integer n, $p(n) \leqslant n$. Then u is eventually periodic.*

Proof [Proof of Corollary 4.3.2] If u is not eventually periodic, then by Theorem 4.3.1, p is strictly increasing, *i.e.*, $p(n+1) \geqslant p(n)+1$. As $p(0) = 1$, we deduce that $p(n) \geqslant n+1$ for all n. □

Another corollary is that functions asymptotically equivalent to $\sqrt{n}$ or to $n/2$, for instance, are not complexity functions.

To conclude the section, let us state the variant of Theorem 4.3.1 that deals with an arbitrary language L and its factor complexity function p: if there exists $n \in \mathbb{N}$ such that $p(n) \leqslant n$ then the complexity p is bounded (Ehrenfeucht and Rozenberg 1982, Balogh and Bollobás 2005, Cassaigne and Nicolas 2009) and there exists a finite language X such that

$$L \subseteq \{xy^k z \mid (x, y, z, k) \in X \times X \times X \times \mathbb{N}\}$$

(Ehrenfeucht and Rozenberg 1982). The result is as similar as possible to the theorem of Morse and Hedlund because:

- The complexity function of an arbitrary language might be bounded but not eventually constant: if $L = \{10^{2n}1 \mid n \in \mathbb{N}\}$ then, for $n \geqslant 2$, $p(n) = 3$ if n is odd and $p(n) = 4$ if n is even.

- The complexity function of an arbitrary language might be unbounded but not eventually increasing: if

$$L = \{0^n 1^n \mid n \in \mathbb{N}\} \cup \{01^{2n}0 \mid n \in \mathbb{N}\} \cup \{10^{2n}1 \mid n \in \mathbb{N}\}$$

then, for $n \geq 3$, $p(n) = n+3$ if n is odd and $p(n) = n+5$ if n is even.

4.3.2 Proof of the theorem of Morse and Hedlund

Proof [Proof of Theorem 4.3.1] By Proposition 4.2.4, $p(n)$ is always non-decreasing. Assume that it is not strictly increasing; then there is some n such that $p(n) = p(n+1)$. Let $M = p(n)$.

Consider then the map e defined in the proof of Proposition 4.2.4. It is an injective map between two sets of equal cardinality, hence it is a bijection. Define now $f(w)$ as the suffix of length n of $e(w)$: f is a map (not necessarily injective) in $L_n(u)$. A map in a finite set always has a cycle: there exist $x \in L_n(u)$ and $m \in [\![1, M]\!]$ such that $f^m(x) = x$ (start from an arbitrary y and consider the $M+1$ elements y, $f(y)$, $f^2(y) = f(f(y))$, ..., $f^M(y)$; since the set $L_n(u)$ has M elements, two of them must be equal).

Assume now that some word $w \in L_n(u)$ occurs at position i in u. Let z be the factor of length $n+1$ that occurs at the same position i in u. As e is a bijection, $z = e(w')$ for some $w' \in L_n(u)$; but then w' is a prefix of z, and so is w, so that $w' = w$ and $z = e(w)$. Now the suffix of length n of z is $f(w)$, and it occurs at position $i+1$ in u. If we iterate this, we conclude that $f^j(w)$ occurs at position $i+j$ in u.

In particular, let $w = x$. Then $f^{j+m}(x) = f^j(x)$ occurs at positions $i+j+m$ and $i+j$ in u, so the letters at those positions are the same: $u_{i+j+m} = u_{i+j}$, for all $j \in \mathbb{N}$. We have just shown that u is eventually periodic with period m and preperiod at most i. □

The main ingredient in the above proof is to discuss how factors can be extended. In Section 4.5 we will develop tools for this, and with them the proof will become much shorter.

4.3.3 Sturmian words

Corollary 4.3.2 is sharp: there exist infinite words with a complexity equal to $n+1$ for all n. Such words are called *Sturmian words*, see Definition 1.2.13. The theory of Sturmian words is quite developed, and we shall not study it in detail here. The reader is referred to (Lothaire 2002, Chapter 2).

Example 4.3.3 Define a sequence $(f_i)_{i \geqslant 0}$ of finite words by: $f_0 = a$, $f_1 = ab$, $f_{i+1} = f_i f_{i-1}$. As f_i is a proper prefix of f_{i+1}, there exists a unique infinite word u such that all f_i are prefixes of u: the *Fibonacci infinite word*.

$$u = abaababaabaababaabab aabab aabaabab aabab \cdots.$$

We shall prove that $p_u(n) = n+1$ later, once we have introduced appropriate tools in Section 4.5.

4.4 High complexity

4.4.1 Subadditivity

Let $u \in A^{\mathbb{N}}$ be an infinite word. We have seen in Proposition 4.2.4 that for all $n \in \mathbb{N}$, $p_u(n) \leqslant p_u(n+1)$. On the other hand, if w is a factor of length n occurring in u, then there are at most Card A factors of length $n+1$ occurring in u and having w as a prefix, since they can be written as wa for some $a \in A$. Therefore, for all $n \geqslant 0$, one has $p_u(n+1) \leqslant (\text{Card } A) p_u(n)$. More generally:

Proposition 4.4.1 *Let p be the complexity function of an infinite word u. Then, for all non-negative integers m and n, it satisfies $p(m+n) \leqslant p(m)p(n)$.*

In other words, the real-valued function $\log p$ is subadditive.

Proof Given $w \in L_{m+n}(u)$, let w_1 be its prefix of length m and w_2 its suffix of length n: then $w = w_1 w_2$, with $w_1 \in L_m(u)$ and $w_2 \in L_n(u)$. We have just defined an injective map from $L_{m+n}(u)$ into $L_m(u) \times L_n(u)$, hence $p(m+n) \leqslant p(m)p(n)$. □

Corollary 4.4.2 *Assume that $L(u) \neq A^*$. Then there exists a real number α, with $1 \leqslant \alpha < \text{Card } A$, such that $p(n) = \mathcal{O}(\alpha^n)$.*

Proof Let $d = \text{Card } A$. There exists some m such that $L_m(u) \neq A^m$. Then $p(m) < d^m$. Let $\alpha = p(m)^{1/m}$: we have $1 \leqslant \alpha < d$.

Let now n be any integer, and $q = \lceil n/m \rceil$, so that $n \leqslant mq < m+n$. Iterating Proposition 4.4.1, we get that $p(mq) \leqslant p(m)^q$. And by Proposition 4.2.4, $p(n) \leqslant p(mq)$. We thus have

$$p(n) \leqslant p(m)^q = \alpha^{mq} \leqslant \alpha^{m+n} = p(m)\alpha^n$$

therefore $p(n) = \mathcal{O}(\alpha^n)$. □

A consequence of Corollary 4.4.2 is that a function equivalent to $2^n/n$, for instance, cannot be the complexity function of a binary infinite word. See also Exercise 4.5.

4.4.2 Topological entropy

The following proposition allows to associate with any infinite word a single real number, that can then be used to classify infinite words.

Proposition 4.4.3 *Let p be the complexity function of an infinite word u. Then the sequence $\left(\frac{\log p(n)}{n}\right)_{n \geqslant 1}$ converges to a real number h, with $0 \leqslant h \leqslant \log \operatorname{Card} \operatorname{alph} u$.*

This number h is called the *topological entropy* of u.

The convergence of $\frac{\log p(n)}{n}$ is a direct consequence of the subadditivity of $\log p(n)$ by Fekete's lemma (Fekete 1923), see Lemma 11.1.1. We include a direct proof for completeness.

Proof Let $h = \inf_{n \geqslant 1} \frac{\log p(n)}{n}$. We shall prove that $\frac{\log p(n)}{n}$ converges to h. Fix $\delta > 0$. Then, by definition of the infimum, there exists m such that $\frac{\log p(m)}{m} < h + \delta$; let $\alpha = p(m)^{1/m}$, so that $\log \alpha < h + \delta$. Proceed now as in the proof of Corollary 4.4.2, to conclude that for any n, $p(n) \leqslant p(m)\alpha^n$. Then $\frac{\log p(n)}{n} \leqslant \frac{\log p(m)}{n} + \log \alpha < \frac{\log p(m)}{n} + h + \delta$. But $\frac{\log p(m)}{n} < \delta$ for n large enough, and we then have $h \leqslant \frac{\log p(n)}{n} < h + 2\delta$. As δ can be taken arbitrarily small, this shows that the sequence converges to h. It is clear that $0 \leqslant h \leqslant \log \operatorname{Card} \operatorname{alph} u$. □

4.4.3 Words with arbitrary entropy

We have already encountered, in Sections 4.2.2 and 4.2.3, words with topological entropy 0 and $\log \operatorname{Card} A$, respectively. The following theorem shows that topological entropy can take any real value between 0 and $\log \operatorname{Card} A$.

Theorem 4.4.4 (Grillenberger 1972) *Let A be an alphabet with $d \geqslant 2$ letters. For any $h \in [0, \log d)$, there exists a uniformly recurrent infinite word in $A^{\mathbb{N}}$ with topological entropy h.*

Actually, Grillenberger's theorem is stronger, as the constructed words are also uniquely ergodic (see Proposition 7.4.3 in Chapter 7). Removing this requirement allows for a simpler construction, that we present here.

Note that it is not possible to extend the theorem to $h = \log d$. When Card $A \geqslant 2$, no word in $A^{\mathbb{N}}$ with topological entropy log Card A is uniformly recurrent, since every finite word occurs in it by Corollary 4.4.2.

Proof Let $h \in [0, \log d)$ be fixed. Since uniformly recurrent words with entropy 0 and $\log d$ are easy to construct, assume that $0 < h < \log d$.

Let (q_k) be a sequence of positive integers, which will be specified later. We construct a family of finite languages D_k recursively as follows. First set $D_0 = A$ and choose an order on A. Then, for all $k \in \mathbb{N}$, let u_k be the concatenation of all words in D_k in lexicographic order, and set $D_{k+1} = u_k D_k^{q_k}$. It is clear that elements of D_k have the same length n_k satisfying $n_{k+1} = (r_k + q_k)n_k$, where

$$r_k = \text{Card}\, D_k = d^{\prod_{i=0}^{k-1} q_i}.$$

It is also clear that u_k is a prefix of u_{k+1}, so that the infinite word $u = \lim_{k \to \infty} u_k$ is well-defined. Moreover u is uniformly recurrent since u_k occurs in u with gaps bounded by n_{k+1}. Actually, u belongs to the class of Toeplitz words (Jacobs and Keane 1969), which are uniformly recurrent.

Let us now prove by induction that, if x is a prefix of a word in D_k and y a suffix of another word in D_k, with $|xy| = n_k$, then $xy \in D_k$. This obviously holds for $k = 0$, as $n_k = 1$. Assume that the assertion holds for some k, and let now x be a prefix of a word in D_{k+1} and y a suffix of a word w in D_{k+1}. If $|x| \leqslant |u_k|$, then x is a common prefix of all words in D_{k+1} so that $xy = w \in D_{k+1}$. Otherwise, let $|x| = |u_k| + hn_k + j$, with $j \in [\![1, n_k]\!]$, and write $x = x''x'$, $y = y'y''$, with $|x'| = j$ and $|y'| = n_k - j$. By construction, $x'' \in u_k D_k^h$, x' is a prefix of a word in D_k, y' is a suffix of a word in D_k, $y'' \in D_k^{q_k - h - 1}$, and $|x'y'| = n_k$. It follows from the induction hypothesis that $x'y' \in D_k$, and thus $xy = x''(x'y')y'' \in u_k D_k^{q_k} = D_{k+1}$.

We deduce that $L_{n_k}(u) \subseteq \bigcup_{i=0}^{n_k-1} D_k^{(i)}$, where $D_k^{(i)} = \{yx \mid x \in A^i, y \in A^*, xy \in D_k\}$ is obtained by iterating i times the circular shift on D_k. More precisely, if $w \in L_{n_k}(u)$ occurs at position s in u, with $s \equiv i \pmod{n_k}$ for some $i \in [\![0, n_k - 1]\!]$, then $w \in D_k^{(i)}$.

It follows that $r_k \leqslant p(n_k) \leqslant n_k r_k$. The topological entropy of u is then

$$h' = \lim_{k \to \infty} \log(r_k)/n_k$$
$$= (\log d) \prod_{k=0}^{\infty} \frac{q_k}{q_k + r_k}.$$

We can then inductively construct q_k in order to get $h' = h$. Start with

$r_0 = d$ and $z_0 = \frac{\log d}{h} > 1$. Given r_k and z_k, let $q_k = 1 + \lfloor \frac{r_k}{z_k - 1} \rfloor$, $r_{k+1} = r_k^{q_k}$, and $z_{k+1} = z_k \frac{q_k}{q_k + r_k}$. If $z_k > 1$, then $q_k > \frac{r_k}{z_k - 1} > 0$, so that $1 + \frac{r_k}{q_k} < z_k$ and $z_{k+1} > 1$. By induction, it follows that $z_k > 1$ for all $k \in \mathbb{N}$, so that q_k is well-defined and $q_k \in \mathbb{N}_{>0}$ for all $k \in \mathbb{N}$.

As z_k is decreasing and bounded from below, it has a limit $\zeta \geqslant 1$. If $\zeta > 1$, we have $q_k \leqslant 1 + \frac{r_k}{\zeta - 1} \leqslant \frac{\zeta}{\zeta - 1} r_k$, for all $k \in \mathbb{N}$ (as $r_k \geqslant 1$). It follows that $\frac{z_{k+1}}{z_k} = \frac{q_k}{q_k + r_k} \leqslant \frac{\zeta}{2\zeta - 1} < 1$, a contradiction with the convergence of (z_k). So $\zeta = 1$.

The above construction can thus be applied to (q_k) and yields a uniformly recurrent word with topological entropy $h' = h\zeta = h$. □

4.5 Tools for low complexity

4.5.1 Special and bispecial factors

In the proof of the Theorem of Morse and Hedlund we had to discuss how factors could be extended into longer factors. Let us now formalise this.

Throughout this section, $u \in A^{\mathbb{N}}$ is a fixed infinite word. As with the complexity function, we make the convention that, when only one infinite word is considered, it is often kept implicit to avoid heavy notation. If there is a need to consider several infinite words at the same time, then a notation like $d^+(w)$ (defined below) can be replaced with $d_u^+(w)$ to specify that it applies to the infinite word u.

Let w be a factor of u. A letter that occurs in u immediately after an occurrence of w is called a *right extension* of w. The set of right extensions of w in u is denoted $E^+(w)$. The *right valence* of w (in u) is the number of different right extensions of w:

$$d^+(w) = \text{Card}\{x \in A \mid wx \in L(u)\}.$$

Left extensions $E^-(w)$ and the *left valence* $d^-(w)$ are defined similarly.

A factor having right valence at least 2 is called *right special*; a factor having left valence at least 2 is called *left special*; and a factor that is both right special and left special is called *bispecial*. We denote respectively by $RS_n(u)$, $LS_n(u)$ and $BS_n(u)$ the sets of right special, left special and bispecial factors of length n in u.

In the proof of Proposition 4.2.4, we used the fact that $d^+(w)$ is always at least one, so w is right special if, and only if, $d^+(w) \neq 1$. On the other hand, $d^-(w)$ can be 0, but only under particular circumstances: when w is a *unioccurrent prefix* of u, i.e., a prefix that occurs nowhere else in u. In most of the examples we will encounter, this does not happen because the infinite word u is *recurrent*, i.e., all of its factors occur infinitely often (see Definition 1.2.9). When u is not recurrent, another prefix will play a particular role: the longest prefix which is not unioccurrent. We name it the *exceptional prefix*. We denote by $LS'_n(u)$ the set containing all left special factors of length n as well as the unioccurrent prefix of length n, if it exists, so that $LS'_n(u) = \{w \in L_n(u) \mid d^-(w) \neq 1\}$, and by $BS'_n(u)$ the set containing all bispecial factors of length n as well as the exceptional prefix, if it exists and its length is n.

When u is recurrent, then right and left valences behave in a symmetrical way, and we have $LS'_n(u) = LS_n(u)$ and $BS'_n(u) = BS_n(u)$.

The *extension type* of a factor w in u is the set $E(w)$ of pairs (a,b) in $A \times A$ such that w can be extended in both directions as awb:

$$E(w) = \{(a,b) \in A^2 \mid awb \in L(u)\}.$$

Note that $E(w) \subseteq E^-(w) \times E^+(w)$, but equality does not always hold. The *(bilateral) multiplicity* of a factor w is the number

$$m(w) = \operatorname{Card} E(w) - d^-(w) - d^+(w) + 1.$$

Proposition 4.5.1 *Let w be a factor of an infinite word u such that $m(w) \neq 0$. Then either w is bispecial, or w is the exceptional prefix.*

Proof Assume that w is not bispecial. Then either $d^+(w) = 1$ or $d^-(w) \leqslant 1$. If $d^+(w) = 1$, then let $E^+(w) = \{b\}$. Every occurrence of w in u is followed by b, and we have $E(w) = E^-(w) \times \{b\}$, so $\operatorname{Card} E(w) = d^-(w)$ and $m(w) = 0$. If $d^-(w) = 1$ and w is not a prefix of u, then $m(w) = 0$ by a symmetric argument. If $d^-(w) = 0$, then w is a unioccurrent prefix and $d^+(w) = 1$, so $m(w) = 0$.

It remains to consider the case where $d^+(w) \geqslant 2$, $d^-(w) = 1$ and w is a prefix of u. Let $E^-(w) = \{a\}$ and $b \in A$ be such that wb is a prefix of u. If wb has at least one other occurrence, which is preceded by a, then $E(w) = \{a\} \times E^+(w)$, so $m(w) = 0$. Otherwise wb is unioccurrent, therefore w is the exceptional prefix (then $E(w) = \{a\} \times (E^+(w) \setminus \{b\})$ and $m(w) = -1$). □

In particular, if u is recurrent, there is no exceptional prefix and only bispecial factors can have non-zero multiplicity. Note also that, when w is neither a unioccurrent prefix nor the exceptional prefix, $E(w)$ contains information about left and right extensions of w, so the multiplicity can be deduced from the knowledge of $E(w)$ only.

A bispecial factor is said to be *strong* if $m(w) > 0$, *weak* if $m(w) < 0$

and *neutral* if $m(w) = 0$. A bispecial factor whose extension type satisfies $E(w) \subseteq \{a\} \times A \cup A \times \{b\}$ for some $(a, b) \in E(w)$ is said to be *ordinary*. Every ordinary bispecial factor is neutral. Note that the converse does not hold if Card $A \geqslant 3$.

Example 4.5.2 Let $u = abc(abbbbc)^\omega$: it is an eventually periodic, non-recurrent infinite word. Prefixes of length 3 or more are unioccurrent and the exceptional prefix is ab. Bispecial factors are ε, b, bb, and bbb. Their multiplicities are $m(\varepsilon) = -1$, $m(b) = 1$, $m(bb) = 0$, $m(bbb) = -1$ and $m(ab) = -1$.

See Exercise 4.7 for an application of bispecial factors.

4.5.2 Finite differences of the complexity function

We now define new functions, derived from the complexity function. Let $s(n)$ be the *first finite difference* of $p(n)$, i.e., $s(n) = p(n+1) - p(n)$; let $b(n)$ be its *second finite difference*, i.e., $b(n) = s(n+1) - s(n) = p(n+2) - 2p(n+1) + p(n)$. (The choice of the names s and b will become clear later.)

Proposition 4.5.3 *Knowing either the function s, or $p(1)$ and the function b, one can recover the function p using the following formulas:*

$$p(n) = 1 + \sum_{\ell=0}^{n-1} s(\ell) = 1 + (p(1) - 1)n + \sum_{m=0}^{n-1} (n - 1 - m)b(m).$$

Proof The formula $p(n) = 1 + \sum_{\ell=0}^{n-1} s(\ell)$ is an immediate consequence of the definition of $s(n)$, as $p(0) = 1$. Similarly, from the definition of $b(n)$ we get $s(n) = s(0) + \sum_{m=0}^{n-1} b(m)$. If we substitute the latter formula in the former, noting that $s(0) = p(1) - 1$, we get $p(n) = 1 + (p(1) - 1)n + \sum_{\ell=0}^{n-1} \sum_{m=0}^{\ell-1} b(m)$. Inverting the two sums, and computing $\sum_{\ell=m+1}^{n-1} b(m) = (n - 1 - m)b(m)$, we get the desired formula. □

Proposition 4.5.3 is particularly useful when p grows slowly, as then usually s and b will take small values and will be easier to evaluate. For instance, an infinite word is Sturmian ($p(n) = n + 1$) if, and only if, $s(n)$ is identically 1, or equivalently if exactly two different letters occur ($p(1) = 2$) and $b(n)$ is identically 0.

4.5.3 Special factors and complexity

To evaluate s or b, we can use special and bispecial factors.

Theorem 4.5.4 *Let $u \in A^{\mathbb{N}}$ be an infinite word, p its complexity function, and s and b its first two finite differences. Then we have, for any $n \in \mathbb{N}$:*

(i) $s(n) = \sum_{w \in RS_n(u)} (d^+(w) - 1)$,

(ii) $s(n) = \sum_{w \in LS'_n(u)} (d^-(w) - 1)$,

(iii) $b(n) = \sum_{w \in BS'_n(u)} m(w)$.

In other words, $s(n)$ is the number of right special factors w of u of length n, counted with multiplicity $d^+(w) - 1$. It is also, if u is recurrent, the number of left special factors w of u of length n, counted with multiplicity $d^-(w) - 1$. Finally, $b(n)$ is the number of bispecial factors w of u of length n (as well as, if u is not recurrent, the exceptional prefix), counted with multiplicity $m(w)$, the bilateral multiplicity. Note that, unlike $d^+(w) - 1$ which is always non-negative, $d^-(w) - 1$ may take the value -1 if w is a unioccurrent prefix, and $m(w)$ may be negative, even when u is recurrent. Note also that the sums can always be taken over the whole set $L_n(u)$, as the multiplicity is zero outside the considered set.

If u is a bi-infinite word instead of an infinite word, the conclusions of Theorem 4.5.4 also hold, and there are no more unioccurrent nor exceptional prefixes, so $LS'_n(u) = LS_n(u)$ and $BS'_n(u) = BS_n(u)$.

Proof As factors of length $n + 1$ can be enumerated by first enumerating their prefix (resp. suffix) of length n, then for each of them their right (resp. left) extensions, we get

$$p(n+1) = \sum_{w \in L_n(u)} d^+(w) = \sum_{w \in L_n(u)} d^-(w).$$

Therefore

$$s(n) = \sum_{w \in L_n(u)} (d^+(w) - 1) = \sum_{w \in L_n(u)} (d^-(w) - 1).$$

But $d^+(w) - 1$ is non-zero if, and only if, w is right special; and $d^-(w) - 1$ is non-zero if, and only if, w is left special or a unioccurrent prefix; hence (i) and (ii).

To get (iii), observe first that $m(w)$ can be defined for any factor w, but is non-zero only if w is bispecial or the exceptional prefix. Summing $m(w)$ over $L_n(u)$, and separating the four terms, we get

$$\sum_{w \in L_n(u)} m(w) = p(n+2) - p(n+1) - p(n+1) + p(n) = b(n).$$

□

In the case of a binary alphabet, valences of special factors are always 2 so we do not need to use multiplicities for them; as for bispecial factors, their multiplicity $m(w)$ can only be -1, 0, or 1. Theorem 4.5.4 becomes:

Corollary 4.5.5 *Let u be a binary infinite word. Then $s(n)$ is the number of its right special factors of length n. If u is recurrent, then $s(n)$ is also the number of its left special factors of length n, and $b(n) = sb(n) - wb(n)$, where $sb(n)$ is the number of strong bispecial factors of length n, and $wb(n)$ is the number of weak bispecial factors of length n.*

Proof Let $A = \{a, b\}$. The only non-trivial argument is that $m(w) \in \{-1, 0, 1\}$. Indeed, assume that u is recurrent and w is bispecial. Then wa, wb, aw, bw are all factors of u. The words aw and bw may not be right special, but each has at least one right extension: we get two factors aws and bwt with $s, t \in A$. Thus the extension type $E(w)$ contains at least the two elements (a, s) and (b, t), and at most four elements, so $2 \leqslant \operatorname{Card} E(w) \leqslant 4$, and we get $-1 \leqslant m(w) = \operatorname{Card} E(w) - 3 \leqslant 1$. □

The multiplicity $m(w)$ somehow measures the correlation between left and right extensions of w. In the binary case, weak bispecial factors have completely correlated extensions (once the left extension is chosen, the right extension is imposed) while strong bispecial factors have non-correlated extensions (the left and right extensions can be chosen independently).

4.5.4 Another proof of the theorem of Morse and Hedlund

Using special factors, we get a new proof of the theorem of Morse and Hedlund:

Proof [Proof of Theorem 4.3.1] Let u be a non-eventually-periodic infinite word, and $n \in \mathbb{N}$. As $L_n(u)$ is finite, there is at least one factor $w \in L_n(u)$ that occurs at least twice, say at positions i and j, $i < j$. There exists $m \in \mathbb{N}$ such that $u_{i+m} \neq u_{j+m}$, otherwise $j - i$ would be a period; and $m \geqslant n$ since the same word w of length n occurs at positions i and j. Let $w' = u_i u_{i+1} \cdots u_{i+m-1}$. Then w' is right special, since it can be extended both by u_{i+m} and u_{j+m}, and so is its suffix of length n. Therefore there is at least one special factor of each length, and by Theorem 4.5.4, $s(n) = p(n+1) - p(n) \geqslant 1$. So p is strictly increasing. □

4.5.5 Rauzy graphs

Factor extensions, valences, special and bispecial factors, etc., can all be visualised using a graph representation of $L_n(u)$, introduced by Gérard Rauzy (Rauzy 1983).

The *Rauzy graph* (or *factor graph*) of order n of u is the directed graph G_n defined as follows: G_n has $p(n)$ vertices labelled with elements of $L_n(u)$ (and identified with their labels), and there is an edge from v to w if, and only if, there exist two letters $a, b \in A$ such that $vb = aw \in L_{n+1}(u)$. We decide to label this edge with the letter a (sometimes it is more convenient to label it with b, or with the word vb). The graph G_n has therefore $p(n+1)$ edges.

The indegree of a vertex w is the left valence $d^-(w)$, and the outdegree is the right valence $d^+(w)$. Therefore, a factor is right special if, and only if, the corresponding vertex has at least two outgoing edges, and similarly left special and bispecial factors are easily recognised.

If we read the factors of length n of u in the order in which they occur, i.e., starting with $u_0 u_1 \cdots u_{n-1}$, then $u_1 u_2 \cdots u_n$, etc., and consider the corresponding vertices in G_n, we obtain an infinite path in G_n, as two vertices visited consecutively are always linked by an edge, and this infinite path is labelled by u. Similarly, from a factor z of length at least n, we obtain a finite path in G_n of length $|z| - n$, starting at the vertex labelled with the prefix of length n of z and ending at the one labelled with its suffix s of length n. If the label of this path is w, then $z = ws$.

We can view G_n as a non-deterministic finite automaton, in which all states are initial and final, see Definition 1.3.6. Let $L(G_n)$ be the regular language recognised by G_n.

Proposition 4.5.6 *The sequence of regular languages $L(G_n)$ approaches $L(u)$ from above, i.e., $L(u) = \bigcap_{n \in \mathbb{N}} L(G_n)$. More precisely, $L(G_{n+1}) \subseteq L(G_n)$, and if $m \leqslant n$, $L(G_n) \cap A^m = L_m(u)$.*

Proof Let $w \in L(u)$. Then w can be extended to some $z = ws \in L_{|w|+n}(u)$, and z defines a path in G_n labelled by w, so $L(u) \subseteq L(G_n)$.

Let $w \in L(G_{n+1})$. Then w is the label of a path $(v_0, v_1, \ldots, v_{|w|})$ in G_{n+1}. Let v'_i be the prefix of length n of v_i: then $(v'_0, v'_1, \ldots, v'_{|w|})$ is a path in G_n which is also labelled by w, so $L(G_{n+1}) \subseteq L(G_n)$.

Let $m \leqslant n$ and consider a word $w \in L(G_n) \cap A^m$. It is the label of a path of length m in G_n, starting at some vertex v. Then w is the prefix of length m of v, so $w \in L_m(u)$. As we already know that $L_m(u) \subseteq L(G_n)$, we have $L(G_n) \cap A^m = L_m(u)$.

Let $L = \bigcap_{n \in \mathbb{N}} L(G_n)$. Then for all m, $L \cap A^m = L_m(u)$, and $L = L(u)$.

□

Note that actually, $L(G_n) \cap A^m = L_m(u)$ holds also when $m = n + 1$. For other properties of Rauzy graphs, see Exercises 4.8 and 4.9 and Section 7.3.2.

4.5.6 Application to Sturmian words

We conclude this section with a description of Rauzy graphs for Sturmian words.

Lemma 4.5.7 *Sturmian words are recurrent.*

Proof Let u be a non-recurrent infinite word. Then there is a word w occurring only finitely many times in u (for instance, a unioccurrent prefix): let $n = |w|$ and v be an infinite word obtained by deleting a sufficiently long prefix from u so that $w \notin L(v)$. Then $p_v(n) < p_u(n)$. If v is eventually periodic, then so is u, therefore p_u is bounded and u is not Sturmian. Otherwise, by Theorem 4.3.1, p_v is strictly increasing so $p_v(n) \geqslant n+1$, and then $p_u(n) \geqslant n+2$ so u is not Sturmian either. □

Proposition 4.5.8 *Let u be a Sturmian word and $n \in \mathbb{N}$. The word u has exactly one left special factor of length n and one right special factor of length n.*

Denote by v the left special factor of length n of u, and by w its right special factor of length n. Then the Rauzy graph G_n has one of the following two shapes:

- *If $v = w$, then G_n consists of two disjoint loops around the vertex labelled by w.*
- *If $v \neq w$, then G_n has three branches: one linking v to w, and two linking w to v.*

Proof By Lemma 4.5.7, u is recurrent. We can therefore apply Corollary 4.5.5: as $p(n) = n+1$ for all $n \in \mathbb{N}$, then $s(n) = 1$, so there are exactly one right special factor and one left special factor of each length. There is at most one bispecial factor (when the right special factor and the left special factor coincide), and then it is necessarily neutral as $b(n) = 0$.

If $v = w$, then all vertices of G_n other than w have indegree 1 (as u is recurrent, it cannot be 0) and outdegree 1, while w being bispecial has indegree and outdegree 2. The only way to connect these vertices is with two disjoint loops around w.

If $v \neq w$, then G_n has two distinguished vertices v (indegree 2, outdegree 1) and w (indegree 1, outdegree 2). There are two ways to connect them, one as described and the other one with a loop around v, a loop around w, and a branch connecting w to v. But this latter graph cannot be the Rauzy graph of a recurrent word, since it is not strongly connected, so it has to be excluded. $\square$

The technique used to describe the Rauzy graphs of Sturmian words can be generalised to other words with known factor complexity: once $s(n)$ is known, there are finitely many possible shapes for G_n. This number can be further reduced if the word is recurrent. For instance, if u is a binary recurrent word with $s(n) = 2$, then G_n can have 10 different shapes (Rote 1994).

The same technique can also be used to characterise classes of words for which the complexity is not specified but satisfies certain conditions, for instance $p(n) \sim n$ (Aberkane 2003). See also Exercise 4.31.

4.5.7 Sparse words

We study here a particular class of words, useful to construct words with specified low complexity. First, we introduce *return words* (see also Chapter 6):

Definition 4.5.9 Let $u \in A^\mathbb{N}$ be an infinite word, and $w \in L(u)$. A word $v \in A^+$ is a *return word* of w in u if the following three conditions hold:

- vw is a factor of u,
- w is a prefix of vw,
- w is not an inner factor of vw.

The set of return words of w in u is denoted $\mathcal{R}_u(w)$.

Remark 4.5.10 Return words are always primitive. Indeed, if w is a prefix of $z^k w$ with $k \geqslant 2$, then w is also a prefix of zw so it is an inner factor of $z^k w$. Note also that $\mathcal{R}_u(w)$ is always a code.

Remark 4.5.11 In the Rauzy graph $G_{|w|}$, the return words of w label certain paths from w to itself. The converse is usually not true (indeed, labels of paths in $G_{|w|}$ need not even be factors of u). However, if u is recurrent, for any edge e in $G_{|w|}$ there is a path from w to w through e labelled with a return word of w.

Definition 4.5.12 An infinite word $u \in A^{\mathbb{N}}$ is called *sparse* if there exists a letter $0 \in A$ such that for all $n \in \mathbb{N}$, $0^n \in L(u)$ and $\operatorname{Card} \mathcal{R}_u(0^n) = 2$.

Observe that since $\mathcal{R}_u(\varepsilon) = \operatorname{alph} u$, sparse words are always binary, so we may assume $A = \{0, 1\}$.

Example 4.5.13 Let $(e_k)_{k \geqslant 0}$ be an increasing sequence of non-negative integers. The word $u = 10^{e_0} 10^{e_1} 10^{e_2} 1 \cdots$ is sparse. Indeed, $\mathcal{R}_u(0^n) = \{0, 0^n 1\}$.

But sparse words may also be recurrent, as the next example, sometimes known as the *universal counter-example*, shows.

Example 4.5.14 Let $(e_k)_{k \geqslant 0}$ be an increasing sequence of non-negative integers. Let (u_k) be the sequence of binary words defined by $u_0 = 1$ and $u_{k+1} = u_k 0^{e_k} u_k$. Then u_k converges to an infinite word $u = \lim_{k \to \infty} u_k$ which is recurrent and sparse. Indeed, $\mathcal{R}_u(0^n) = \{0, 0^n u_j\}$ where $j = \min\{k \in \mathbb{N} \mid e_k \geqslant n\}$.

Using special words, the factor complexity of sparse words can easily be evaluated.

Lemma 4.5.15 *Let u be a sparse word. There exists a sequence of finite words $(z_n)_{n \geqslant 0}$ such that for all $n \in \mathbb{N}$, $\mathcal{R}_u(0^n) = \{0, 0^n z_n\}$. Moreover, $z_0 = 1$ and $z_{n+1} \in z_n(0^n z_n)^*$.*

Proof By definition of sparse words, $\mathcal{R}_u(0^n)$ has exactly two elements. As 0^{n+1} occurs in u, one of these elements must be 0. The other one cannot be a prefix of 0^n, as a return word is primitive by Remark 4.5.10, so it must be of the form $0^n z_n$. Clearly $z_0 = 1$, as $\mathcal{R}_u(\varepsilon) = \operatorname{alph} u$. As 0^n is a prefix of 0^{n+1}, one has $\mathcal{R}_u(0^{n+1}) \subseteq \mathcal{R}_u(0^n)^+$. But as $0^{n+1} z_{n+1} 0^{n+1}$ must not contain 0^{n+1} as an inner factor, necessarily $0^{n+1} z_{n+1} \in 0(0^n z_n)^+$. □

Proposition 4.5.16 *Let u be a sparse word. Then u has at most $n+1$ right special factors of length n, and therefore $p_u(n) \leqslant \frac{n^2+n+2}{2}$.*

Proof Let w be a right special factor of u, let k be the largest integer such that 0^k occurs in w, and write $w = w'0^k w''$ with $|w''|$ minimal. By Lemma 4.5.15, every occurrence of 0^k in u is followed by either 0 or $z_k 0^k$. In particular, as w is special, both $0^k w''0$ and $0^k w''1$ are factors of u, hence both $w''0$ and $w''1$ are prefixes of words in $\{0, z_k 0^k\}^*$. If $w'' \neq \varepsilon$, this is impossible: by minimality, neither 0 nor $z_k 0^k$ are prefixes of w'', so both $w''0$ and $w''1$ have to be prefixes of $z_k 0^k$, which is absurd.

As w is special, it occurs at least twice in u, so there is an occurrence of 0^k starting before the second occurrence of w. Then w is a suffix of a word in $\{0, z_k 0^k\}^*$, and as 0^{k+1} does not occur in w, it is actually a suffix of the periodic left-infinite word ${}^\omega(z_k 0^k)$.

We have just proved that

$$RS_n(u) \subseteq \left\{ \operatorname{suff}_n {}^\omega(z_k 0^k) \mid k \in [\![0,n]\!] \right\},$$

which implies that $s(n) \leqslant n+1$. Summing according to Proposition 4.5.3, we obtain

$$p(n) \leqslant 1 + \sum_{m=0}^{n-1}(m+1) = \frac{n^2 + n + 2}{2}.$$

$\square$

For particular sparse words, complexity can be estimated more precisely:

Example 4.5.17 We continue Example 4.5.13. Right special factors of u are the suffixes of $0^{e_k-1}10^{e_k}$ for all $k \in \mathbb{N}$, with the convention that $e_{-1} = 0$. Indeed, a word containing two 1's occurs only once so it cannot be special; a word of the form $0^m 10^n$ is right special if, and only if, $n = e_k$ for some k and $m \leqslant e_{k-1}$, otherwise it cannot be followed by 1; and a word of the form 0^n is always right special. It follows that

$$s(n) = 1 + \operatorname{Card}\{k \in \mathbb{N} \mid e_k \leqslant n-1 \leqslant e_k + e_{k-1}\}$$

since $\operatorname{suff}_n 0^{e_k-1}10^{e_k}$ is either too short or equal to 0^n if $n \notin [\![e_k+1, e_k + e_{k-1}+1]\!]$.

Techniques to evaluate the cardinality of such *sandwich sets* are developed in Section 4.7.2.3. For instance, we have:

(i) If $e_k = \Theta(\beta^k)$ for some $\beta \in \mathbb{R}_{>0}$, then $s(n) = O(1)$ and $p(n) = \Theta(n)$. Indeed, if $C_1 \beta^k \leqslant e_k \leqslant C_2 \beta^k$, then $e_k \leqslant n-1 \leqslant e_k + e_{k-1}$ is possible only if

$$k \in \left[\log_\beta\left(\frac{n-1}{C_2(1+1/\beta)}\right), \log_\beta\left(\frac{n-1}{C_1}\right) \right],$$

which is an interval of bounded length.

(ii) If $e_k \sim Ck^r$ for some $C \in \mathbb{R}_{>0}$ and $r \in \mathbb{N}_{>0}$, then $s(n) = \Theta(\sqrt[r]{n})$ and $p(n) = \Theta(n\sqrt[r]{n})$. Indeed, if $(3C/4)k^r \leqslant e_k \leqslant (5C/4)k^r$ for k large enough, then for n large enough $e_k \leqslant n-1 \leqslant e_k + e_{k-1}$ is possible only if $k \leqslant \left(\frac{n-1}{3C/4}\right)^{1/r}$, so that $s(n) = \mathcal{O}(\sqrt[r]{n})$, and it holds as soon as
$$k \in \left[\left(\frac{n-1}{3C/2}\right)^{1/r}, \left(\frac{n-1}{5C/4}\right)^{1/r}\right],$$
so that $s(n) = \Omega(\sqrt[r]{n})$. Then $p(n) = \Theta(n\sqrt[r]{n})$ follows by summation. Note that $p(n) = \Theta(n\sqrt[r]{n})$ still holds if the assumption $e_k \sim Ck^r$ is replaced with $e_k = \Theta(k^r)$, but then it is no longer true that $s(n) = \Theta(\sqrt[r]{n})$.

See also Exercise 4.10.

4.6 Morphisms and complexity

4.6.1 Complements on morphisms and infinite words

Morphisms of free monoids are introduced in Section 1.2.4. We recall that, if $\sigma\colon A^* \to B^*$ is a non-erasing morphism, it can be extended to a map $\sigma\colon A^\mathbb{N} \to B^\mathbb{N}$. More generally, any morphism $\sigma\colon A^* \to B^*$ can be extended to a map $\sigma\colon A^\mathbb{N} \to B^\mathbb{N} \cup B^*$. The image of an infinite word $u = u_0 u_1 u_2 \cdots$ is the infinite concatenation of images of its letters, $\sigma(u) = \sigma(u_0)\sigma(u_1)\sigma(u_2)\cdots$. Most of the time this is an infinite word in $B^\mathbb{N}$, but it could also be a finite word, if all $\sigma(u_n)$ are empty words after some point.

Morphisms are one of the most useful tools to construct and transform non-periodic infinite words with interesting properties related to factor complexity. In particular, they allow us to define *morphic* and *purely morphic* infinite words, see Definition 1.2.18.

The following result, due to Alan Cobham (Cobham 1968) will be useful to simplify the use of morphic words. Its proof can be found in (Allouche and Shallit 2003, Theorem 7.5.1 and Corollary 7.7.5).

Theorem 4.6.1 (Cobham's theorem on morphic words)

(i) *An infinite word $u \in A^\mathbb{N}$ is morphic if, and only if, it can be written as $u = \tau(\sigma^\omega(x))$, where B is an alphabet, $x \in B$, $\sigma\colon B^* \to B^*$ is a non-erasing endomorphism, and $\tau\colon B^* \to A^*$ is a letter-to-letter morphism.*

(ii) *The image of a morphic word under a morphism is either finite or again morphic.*

Note that it is not always the case that a purely morphic word can be written as $\sigma^\omega(x)$ with σ non-erasing, see Exercise 4.11.

Remark 4.6.2 Every periodic infinite word is purely morphic: for every $z \in A^+$, z^ω is generated by the endomorphism of A^* that maps each letter in A to z^2.

Remark 4.6.3 Every eventually periodic infinite word $u = tz^\omega$ is morphic: indeed, u is the image of 01^ω under the morphism that maps 0 to t and 1 to z. Moreover, 01^ω is purely morphic because it is generated by the endomorphism of $\{0,1\}^*$ that maps 0 to 01 and 1 to 1.

Example 4.6.4 The infinite word 001^ω is morphic by Remark 4.6.3 but it is not purely morphic: indeed, if a purely morphic word $u = \sigma^\omega(a)$ starts with two consecutive occurrences of the letter a, then aa must be a prefix of $\sigma(a)$ and thus a occurs infinitely many times in u. See also Example 10.1.2.

Example 4.6.5 The paperfolding word is morphic but not purely morphic. It is $\tau(\sigma^\omega(a))$, where $\sigma(a) = aa'$, $\sigma(b) = ab'$, $\sigma(a') = ba'$, $\sigma(b') = bb'$, and τ removes the primes. See (Allouche 1992) for the computation of its complexity.

See Exercises 4.12 and 4.13 for other examples of morphic words that are not purely morphic. See also Remark 4.7.3.

Theorem 4.6.1 allows us to rewrite a morphic word $\tau(\sigma^\omega(a))$ as $\tilde{\tau}(\tilde{\sigma}^\omega(a))$, where $\tilde{\sigma}$ has better properties than σ, namely, it is non-erasing.

4.6.2 Action of a morphism on factor complexity

Let $u \in A^\mathbb{N}$ be an infinite word, and $\tau\colon A^* \to B^*$ a morphism. Knowing the factor complexity of u, we want to estimate the complexity of $v = \tau(u)$.

If τ is a letter-to-letter morphism, this is easy:

Lemma 4.6.6 *For any infinite word $u \in A^\mathbb{N}$ and any letter-to-letter morphism $\tau\colon A^* \to B^*$, the complexity functions of u and $\tau(u)$ satisfy $p_{\tau(u)}(n) \leqslant p_u(n)$ for every $n \in \mathbb{N}$.*

Proof Since τ is letter-to-letter, it preserves lengths. Therefore $L_n(\tau(u)) = \tau(L_n(u))$, hence $p_{\tau(u)}(n) \leqslant p_u(n)$. □

Recall that $\|\tau\|$ denotes the width of τ (see Definition 1.2.20). Lemma 4.6.6 can be generalised as follows:

Lemma 4.6.7 *For any infinite word $u \in A^{\mathbb{N}}$ and any non-erasing morphism $\tau \colon A^* \to B^*$, the complexity functions of u and $\tau(u)$ satisfy $p_{\tau(u)}(n) \leqslant \|\tau\| p_u(n)$ for every $n \in \mathbb{N}$.*

Proof Let $v = \tau(u)$, and consider a word $w' \in L_n(v)$.

Let $x' \in B^*$ be such that $x'w'$ is a prefix of v, $x \in A^*$ be the longest prefix of u such that $\tau(x)$ is a prefix of x', and $w \in L_n(u)$ be the word of length n such that xw is a prefix of u. Write $w = ay$, with $a \in A$ and $|y| = n - 1$.

Let $k = |x'| - |\tau(x)|$. The maximality of $|x|$ implies that x' is a proper prefix of $\tau(xa)$, so that $0 \leqslant k < |\tau(a)| \leqslant \|\tau\| = \max_{b \in A} |\tau(b)|$. Moreover, $|\tau(xw)| = |\tau(xa)| + |\tau(y)|$, with $|\tau(xa)| \geqslant |x'| + 1$ and $|\tau(y)| \geqslant |y| = n - 1$ as τ is non-erasing, so that $|\tau(xw)| \geqslant |x'| + n = |x'w'|$. It follows that w' occurs at position k in $\tau(w)$.

The function that maps $w' \in L_n(v)$ to $(w, k) \in L_n(u) \times [\![0, \|\tau\| - 1]\!]$ is therefore injective, and the result follows. $\square$

Lemma 4.6.7 previously appeared in (Allouche and Shallit 2003, Theorem 10.2.4), and in a more general form, but with a larger constant, in (Ehrenfeucht and Rozenberg 1982, Theorem 3). See also (Pansiot 1984, Lemma 4.2). No such incquality holds when τ is erasing, see Exercise 4.14.

In the other direction, to get a useful lower bound on $p_v(n)$ we need to exclude some morphisms, such as the morphisms that map all letters of B to powers of the same word, and thus map any infinite word to a periodic word. The right notion is that of *injective* morphism, *i.e.*, a morphism that never maps two distinct words to the same image. Recall that a morphism of free monoids $\tau \colon A^* \to B^*$ is injective if, and only if, letters of A are mapped to distinct words and the language $\tau(A)$ is a *code*, see Definition 1.2.3.

Lemma 4.6.8 *For any infinite word $u \in A^{\mathbb{N}}$ and any injective morphism $\tau \colon A^* \to B^*$, the complexity functions of u and $\tau(u)$ satisfy for every $n \in \mathbb{N}$*

$$p_{\tau(u)}(n) \geqslant \frac{1}{\|\tau\|} p_u\left(\left\lfloor \frac{n}{\|\tau\|} \right\rfloor\right).$$

Proof Let $v = \tau(u)$, and consider a word $w \in L_n(u)$.

Let E be the set of words $x \in L(u)$ such that w is a prefix of x and $|\tau(x)| \leqslant \|\tau\| n$. Note that $w \in E$ so that E is non-empty. Also, since τ is injective, it is also non-erasing, and thus $x \in E$ implies $|x| \leqslant \|\tau\| n$, so that E is finite. We can therefore choose an element y of E of maximal length.

Since u is an infinite word, there is a letter $a \in A$ such that $ya \in L(u)$. The maximality of y then ensures that $|\tau(ya)| > \|\tau\| n$. Let w' be the prefix of $\tau(ya)$ of length $\|\tau\|n$ and $k = \|\tau\|n - |\tau(y)|$. We have $0 \leqslant k < |\tau(a)| \leqslant \|\tau\|$.

The function that maps $w \in L_n(u)$ to $(w', k) \in L_{\|\tau\|n}(v) \times [\![0, \|\tau\| - 1]\!]$ is injective. Indeed, given (w', k), $\tau(y)$ can be recovered by deleting the last k letters of w', then the injectivity of τ guarantees that y is unique, and w is obtained as the prefix of length n of y. We deduce that $p_u(n) \leqslant \|\tau\| p_v(\|\tau\|n)$ for all $n \in \mathbb{N}$.

We then have
$$p_v(n) \geqslant p_v\left(\|\tau\|\left\lfloor\frac{n}{\|\tau\|}\right\rfloor\right) \geqslant \frac{1}{\|\tau\|}p_u\left(\left\lfloor\frac{n}{\|\tau\|}\right\rfloor\right).$$
□

Lemma 4.6.8 above is a sligthly improved version of Proposition 7 in (Cassaigne and Nicolas 2003).

Lemma 4.6.9 *Let $u \in A^\mathbb{N}$ and let $B \subseteq A$ be such that $L(u) \cap B^*$ is finite. Let $\chi\colon A^* \to (A \setminus B)^*$ denote the morphism that erases the letters in B: $\chi(b) = \varepsilon$ for every $b \in B$ and $\chi(c) = c$ for every $c \in A \setminus B$.*

Then, $\chi(u)$ is an infinite word and its complexity function satisfies
$$p_u(n) \geqslant p_{\chi(u)}\left(\left\lfloor\tfrac{1}{M+1}n\right\rfloor\right)$$
for every nwhere M denotes the maximum length of a word in $L(u) \cap B^*$.

Proof Let $f_n\colon A^* \to (A \setminus B)^{\leqslant n}$ be the function defined by: $f_n(x) = \text{pref}_n \chi(x)$ for every $x \in A^*$. It is clear that $f_n\bigl(L_{(M+1)n}(u)\bigr) = L_n(\chi(u))$, so $p_{\chi(u)}(n) \leqslant p_u((M+1)n)$. Substituting n with $\left\lfloor\tfrac{1}{M+1}n\right\rfloor$ in the previous inequality, we get
$$p_{\chi(u)}\left(\left\lfloor\tfrac{1}{M+1}n\right\rfloor\right) \leqslant p_u\left((M+1)\left\lfloor\tfrac{1}{M+1}n\right\rfloor\right) \leqslant p_u(n).$$
□

Note that the inequalities that we have obtained in this section between p_u and $p_{\tau(u)}$ do not imply similar inequalities between the first differences s_u and $s_{\tau(u)}$, even in the case where τ is a letter-to-letter morphism. There does not seem to be an easy way to obtain such an inequality.

Remark 4.6.10 *For every $u \in A^\mathbb{N}$, every $w \in A^*$, and every $n \in \mathbb{N}$,*
$$p_u(n) \leqslant p_{wu}(n) \leqslant p_u(n) + |w|.$$

4.7 The theorem of Pansiot

The aim of this section is to present a comprehensive proof of a surprising result from J.-J. Pansiot (Pansiot 1984): the factor complexities of purely morphic words can only adopt five different asymptotic behaviours.

4.7.1 Statement and discussion

Recall that $f(n) = \Theta(g(n))$ means that there exist two positive real constants λ and μ such that, for all n large enough, $\lambda g(n) \leqslant f(n) \leqslant \mu g(n)$.

Theorem 4.7.1 (J.-J. Pansiot) *Let u be a purely morphic word and p its complexity function. Then one of the following holds:*

(i) $p(n) = \Theta(1)$,
(ii) $p(n) = \Theta(n)$,
(iii) $p(n) = \Theta(n \log \log n)$,
(iv) $p(n) = \Theta(n \log n)$ *or*
(v) $p(n) = \Theta(n^2)$.

Theorem 4.7.1 was stated in (Pansiot 1984). Pansiot later generalised his classification theorem to the complexity functions of D0L-languages (Pansiot 1985). He also remarked that the complexity function of general morphic words may behave differently:

Proposition 4.7.2 (J.-J. Pansiot) *For each integer $r \geqslant 1$, there exists a binary morphic word u such that $p_u(n) = \Theta(n\sqrt[r]{n})$.*

Proof Let $A = \{a, b_0, b_1, b_2, \ldots, b_r\}$ be an alphabet with cardinality $r+2$. Let $\sigma \colon A^* \to A^*$ be the endomorphism defined by: $\sigma(a) = ab_r$, $\sigma(b_0) = b_0$, and $\sigma(b_i) = b_i b_{i-1}$ for every $i \in [\![1, r]\!]$. Let $\chi \colon A^* \to \{0, 1\}^*$ be the morphism defined by: $\chi(a) = \varepsilon$, $\chi(b_i) = 0$ for every $i \in [\![0, r-1]\!]$, and $\chi(b_r) = 1$.

Let $u = \chi(\sigma^\omega(a))$. By Theorem 4.6.1.(ii), as u is infinite, it is a morphic word. Let $e_k = |\sigma^k(b_r)| - 1$ for each $k \in \mathbb{N}$. It is clear that $\chi(\sigma^k(b_r)) = 10^{e_k}$, and thus

$$u = \chi\left(ab_r \sigma(b_r) \sigma^2(b_r) \sigma^3(b_r) \cdots \right) = 10^{e_0} 10^{e_1} 10^{e_2} 10^{e_3} \cdots.$$

Since the sequence $(e_k)_{k \geqslant 0}$ is increasing, u is a sparse word as in Example 4.5.17, with $e_k \sim k^r/r!$ (in fact, e_k is a polynomial function of k with degree r) so that $p_u(n) = \Theta(n\sqrt[r]{n})$. $\square$

Remark 4.7.3 Proposition 4.7.2 provides new examples of words that are morphic but not purely morphic. See Section 4.6.1. Indeed, when $r \geqslant 2$, Theorem 4.7.1 ensures that the complexity function of a purely morphic words does not satisfy $p(n) = \Theta(n\sqrt[r]{n})$.

In fact, a stronger statement than Proposition 4.7.2 holds: for each integer $r \geqslant 1$, the complexity function of some *recurrent* binary morphic word is in $\Theta(n\sqrt[r]{n})$. The reader is referred to Exercise 4.15.

It is not known whether other complexity classes are possible for morphic words. However, R. Devyatov recently proved (Devyatov 2008) that the factor complexity of a morphic word is either in $\Theta(n\sqrt[r]{n})$ for some positive integer r, or in $\mathcal{O}(n \log n)$.

4.7.2 Asymptotic analysis results

No combinatorics on words is involved in this section. Its aim is to prove several auxiliary results that will be used in the proof of Theorem 4.7.1.

4.7.2.1 An asymptotic scale

Given two ordered pairs (λ_1, μ_1), $(\lambda_2, \mu_2) \in \mathbb{R} \times \mathbb{R}$, (λ_1, μ_1) is said to be *lexicographically smaller* than (λ_2, μ_2) if the following holds:

$$(\lambda_1 < \lambda_2) \text{ or } (\lambda_1 = \lambda_2 \text{ and } \mu_1 < \mu_2).$$

Claim 4.7.4 *For any (β_1, α_1), $(\beta_2, \alpha_2) \in (\mathbb{R}_{>0} \times \mathbb{R}) \cup \{(0,0)\}$, $k^{\alpha_1}\beta_1^k = o(k^{\alpha_2}\beta_2^k)$ as k tends to ∞ if, and only if, (β_1, α_1) is lexicographically smaller than (β_2, α_2).*

The function mapping each ordered pair $(\beta, \alpha) \in (\mathbb{R}_{>0} \times \mathbb{R}) \cup \{(0,0)\}$ to the sequence $(k^\alpha \beta^k)_{k \geqslant 1}$ is one-to-one, but if $\beta = 0$ then $(k^\alpha \beta^k)_{k \geqslant 1}$ is identically zero for every $\alpha \in \mathbb{R}$.

Since the lexicographic order is total, it follows from Claim 4.7.4 that the sequences of the form $(k^\alpha \beta^k)_{k \geqslant 1}$ with $(\beta, \alpha) \in (\mathbb{R}_{>0} \times \mathbb{R}) \cup \{(0,0)\}$ are linearly ordered by their asymptotic orders of growth: for any (β_1, α_1), $(\beta_2, \alpha_2) \in (\mathbb{R}_{>0} \times \mathbb{R}) \cup \{(0,0)\}$ with $(\beta_1, \alpha_1) \neq (\beta_2, \alpha_2)$, either $k^{\alpha_1}\beta_1^k = o(k^{\alpha_2}\beta_2^k)$ or $k^{\alpha_2}\beta_2^k = o(k^{\alpha_1}\beta_1^k)$ as k tends to ∞.

4.7.2.2 Matrix iteration

Definition 4.7.5 (Vector norm) Let V be a real or complex vector space. A *norm* on V is a mapping $\|\cdot\|$ from V to $\mathbb{R}$ such that the following three properties hold for all vectors $x, y \in V$, and all scalars $\lambda \in \mathbb{R}$:

(i) $\|x\| = 0$ if, and only if, x is the zero vector,

(ii) $\|\lambda x\| = |\lambda|\, \|x\|$ and
(iii) $\|x + y\| \leqslant \|x\| + \|y\|$.

Throughout this section, d denotes a positive integer. A fundamental theorem from I. Gelfand states that for every norm $\|\cdot\|$ on $\mathbb{C}^{d\times d}$ and every matrix $\mathbf{M} \in \mathbb{C}^{d\times d}$, $\sqrt[k]{\|\mathbf{M}^k\|}$ converges to the spectral radius of $\mathbf{M}$ as k tends to ∞. The aim of this section is to prove the following variant of Gelfand's result:

Theorem 4.7.6 *Let $\|\cdot\|$ be a norm on $\mathbb{C}^{d\times d}$. For each $\mathbf{M} \in \mathbb{C}^{d\times d}$, there exists $(\beta, \alpha) \in (\mathbb{R}_{>0} \times \mathbb{N}) \cup \{(0,0)\}$ such that $\|\mathbf{M}^k\| = \Theta(k^\alpha \beta^k)$ as k tends to ∞.*

Let us first comment the statement of Theorem 4.7.6. The theorem of Gelfand ensures that β equals the spectral radius of $\mathbf{M}$. Moreover, if $\mathbf{M}$ is not nilpotent then $\alpha+1$ equals the maximum size of the Jordan blocks of $\mathbf{M}$ with spectral radius β (Varga 2000), and if $\mathbf{M}$ is nilpotent then $\beta = \alpha = 0$.

Let us now turn to the proof of Theorem 4.7.6. It relies on the following basic result, that can be found for instance in (Lang 1983):

Theorem 4.7.7 (Equivalence of norms) *Let V be real or complex vector space. If V is finite-dimensional then all norms on V are equivalent: for any norms $\|\cdot\|_A$ and $\|\cdot\|_B$ on V, there exist positive real numbers λ and μ such that $\lambda \|x\|_A \leqslant \|x\|_B \leqslant \mu \|x\|_A$ for every $x \in V$.*

Corollary 4.7.8 *For any norms $\|\cdot\|_A$ and $\|\cdot\|_B$ on $\mathbb{C}^{d\times d}$ and for any $\mathbf{M} \in \mathbb{C}^{d\times d}$, $\|\mathbf{M}^k\|_A = \Theta\left(\|\mathbf{M}^k\|_B\right)$ as k tends to ∞.*

Proof [Proof of Theorem 4.7.6] Let $\mathbf{Q} \in \mathbb{C}^{d\times d}$ be a non-singular matrix such that $\mathbf{QMQ}^{-1}$ is in Jordan normal form: there exist $\mathbf{D}, \mathbf{N} \in \mathbb{C}^{d\times d}$ such that $\mathbf{D}$ is diagonal, $\mathbf{N}$ is nilpotent, $\mathbf{QMQ}^{-1} = \mathbf{D} + \mathbf{N}$, and $\mathbf{DN} = \mathbf{ND}$. For all $i, j \in [\![1, d]\!]$, let $m_{i,j} : \mathbb{N} \to \mathbb{C}$ be the function mapping each $k \in \mathbb{N}$ to the $(i,j)^{\text{th}}$ entry of $\mathbf{QM}^k\mathbf{Q}^{-1}$. By Corollary 4.7.8, we may assume that $\|\mathbf{X}\|$ equals the maximum magnitude of the entries of $\mathbf{QXQ}^{-1}$ for every $\mathbf{X} \in \mathbb{C}^{d\times d}$: $\|\mathbf{M}^k\| = \max_{i,j \in [\![1,d]\!]} |m_{i,j}(k)|$ for every $k \in \mathbb{N}$.

Let us study the asymptotic behaviours of the $m_{i,j}$'s. For every $k \in \mathbb{N}$, the binomial theorem yields

$$\mathbf{QM}^k\mathbf{Q}^{-1} = (\mathbf{D} + \mathbf{N})^k = \sum_{h=0}^{k} \binom{k}{h} \mathbf{D}^{k-h} \mathbf{N}^h.$$

Besides, $\mathbf{N}^h$ is a zero matrix for every $h \geqslant d$, and thus

$$\mathbf{Q}\mathbf{M}^k\mathbf{Q}^{-1} = \sum_{h=0}^{d-1} \binom{k}{h} \mathbf{D}^{k-h}\mathbf{N}^h$$

for every $k \geqslant d-1$. Hence, there exist an eigenvalue λ_i of $\mathbf{D}$ and a complex polynomial $f_{i,j}$ such that $m_{i,j}(k) = f_{i,j}(k)\lambda_i^k$ for every $k \geqslant d$. It follows that there exists $(\beta_i, \alpha_{i,j}) \in (\mathbb{R}_{>0} \times \mathbb{N}) \cup \{(0,0)\}$ such that $m_{i,j}(k) = \Theta\left(k^{\alpha_{i,j}}\beta_i^k\right)$:

- If $f_{i,j}$ is not identically zero then $\beta_i = |\lambda_i|$ and $\alpha_{i,j}$ equals the degree of $f_{i,j}$.
- If $f_{i,j}$ is identically zero then $(\beta_i, \alpha_{i,j}) = (0,0)$.

Let (β, α) be the greatest element of $\{(\beta_i, \alpha_{i,j}) \mid i,j \in [\![1,d]\!]\}$ according to the lexicographic order. It follows from Claim 4.7.4 that $\|\mathbf{M}^k\| = \Theta(k^\alpha \beta^k)$. $\square$

Example 4.7.9 Consider the 2-by-2 integer matrix

$$\mathbf{M} = \begin{bmatrix} 4 & -3 \\ 3 & 4 \end{bmatrix}.$$

It is easy to check that

$$\mathbf{M}^k = 5^k \begin{bmatrix} \cos(k\theta) & -\sin(k\theta) \\ \sin(k\theta) & \cos(k\theta) \end{bmatrix}$$

for every $k \in \mathbb{N}$, where θ is an argument of $4 + 3\sqrt{-1}$. Therefore, $\|\mathbf{M}^k\| = \Theta(5^k)$ as k tends to ∞. For instance, let $\|\cdot\|_\infty$ denote the *Chebyshev norm* on $\mathbb{C}^{2\times 2}$: for each $\mathbf{X} \in \mathbb{C}^{2\times 2}$, $\|\mathbf{X}\|_\infty$ equals the maximum magnitude of the entries of $\mathbf{X}$. For every $k \in \mathbb{N}$, $\|\mathbf{M}^k\|_\infty$ obeys the inequalities $\frac{\sqrt{2}}{2}5^k \leqslant \|\mathbf{M}^k\|_\infty \leqslant 5^k$. However, no entry of $\mathbf{M}^k$ behaves in the same way as $\|\mathbf{M}^k\|$ as k tends to ∞. Indeed both sets $\{\cos(k\theta) \mid k \in \mathbb{N}\}$ and $\{\sin(k\theta) \mid k \in \mathbb{N}\}$ are dense subsets of $[-1, +1]$.

4.7.2.3 The sandwich set theorem

Definition 4.7.10 Let f_1 and f_2 be real-valued functions defined on a superset of $\mathbb{N}$. For each $y \in \mathbb{R}$, $E_y(f_1, f_2)$ denotes the set of all $k \in \mathbb{N}$ such that $f_1(k) \leqslant y \leqslant f_2(k)$.

The aim of this section is to prove the following theorem:

Theorem 4.7.11 *For each* $i \in \{1,2\}$, *let* $(\beta_i, \alpha_i) \in \mathbb{R}_{>1} \times \mathbb{R}$ *and let*

$f_i \colon \mathbb{N} \to \mathbb{R}_{\geqslant 0}$ be a function such that $f_i(k) = \Theta(k^{\alpha_i} \beta_i^k)$ as k tends to ∞. Let $\Delta \colon \mathbb{R}_{>1} \to \mathbb{R}$ be the function defined by:

$$\Delta(y) = -\left(\frac{1}{\log \beta_2} - \frac{1}{\log \beta_1}\right) \log y + \left(\frac{\alpha_2}{\log \beta_2} - \frac{\alpha_1}{\log \beta_1}\right) \log \log y$$

for every real $y > 1$. If (β_1, α_1) is lexicographically smaller than or equal to (β_2, α_2) then $\operatorname{Card} E_y(f_1, f_2) = \Delta(y) + \mathcal{O}(1)$ as y tends to ∞.

The following corollary of Theorem 4.7.11 plays a crucial role in the proof of the theorem of Pansiot:

Corollary 4.7.12 *In the notation of Theorem 4.7.11,*

(i) *if $\beta_1 = \beta_2$ and $\alpha_1 = \alpha_2$ then $\operatorname{Card} E_y(f_1, f_2) = \mathcal{O}(1)$,*
(ii) *if $\beta_1 = \beta_2$ and $\alpha_1 < \alpha_2$ then $\operatorname{Card} E_y(f_1, f_2) = \Theta(\log \log y)$ and*
(iii) *if $\beta_1 < \beta_2$ then $\operatorname{Card} E_y(f_1, f_2) = \Theta(\log y)$.*

Let us briefly study the case where (β_2, α_2) is lexicographically smaller than (β_1, α_1). In this case $\Delta(y)$ tends to $-\infty$ as y tends to ∞. However, $f_2 = o(f_1)$ by Claim 4.7.4, so $F = \{k \in \mathbb{N} \mid f_1(k) \leqslant f_2(k)\}$ is finite, and thus $E_y(f_1, f_2)$ is empty for every $y > \max f_2(F)$.

Lemma 4.7.13 *For any increasing bijections $f_1, f_2 \colon \mathbb{R}_{\geqslant 0} \to \mathbb{R}_{\geqslant 0}$,*

$$\operatorname{Card} E_y(f_1, f_2) = \max\left\{f_1^{-1}(y) - f_2^{-1}(y), 0\right\} + \mathcal{O}(1)$$

as y tends to ∞.

If f_1 and f_2 are bijections from $\mathbb{R}_{\geqslant 0}$ onto itself such that $f_1(x) \leqslant f_2(x)$ for all sufficiently large real $x \geqslant 0$ then $\operatorname{Card} E_y(f_1, f_2) = f_1^{-1}(y) - f_2^{-1}(y) + \mathcal{O}(1)$, and thus estimating the cardinality of $E_y(f_1, f_2)$ reduces to computing asymptotic expansions for $f_1^{-1}(y)$ and $f_2^{-1}(y)$.

Proof [Proof of Lemma 4.7.13] Remark that

$$E_y(f_1, f_2) = \left[f_2^{-1}(y), f_1^{-1}(y)\right] \cap \mathbb{N}$$

for every real $y \geqslant 0$, and that

$$\max\{x_2 - x_1, 0\} \leqslant \operatorname{Card}\left([x_1, x_2] \cap \mathbb{N}\right) \leqslant \max\{x_2 - x_1, 0\} + 1$$

for all reals $x_1, x_2 \geqslant 0$. $\square$

Lemma 4.7.14 *Let $\beta, \alpha \in \mathbb{R}$ with $\beta > 1$, and let $f \colon \mathbb{R}_{\geqslant 0} \to \mathbb{R}_{\geqslant 0}$ be an increasing bijection. If $f(x) = \Theta(x^\alpha \beta^x)$ as x tends to ∞ then*

$$f^{-1}(y) = \frac{1}{\log \beta} \log y - \frac{\alpha}{\log \beta} \log \log y + \mathcal{O}(1)$$

as y tends to ∞.

Proof Let x tend to ∞. Note that $f_1(x) = \Theta(f_2(x))$ if, and only if, $\log f_1(x) = \log f_2(x) + \mathcal{O}(1)$ for any eventually positive functions f_1, $f_2 \colon \mathbb{R}_{\geq 0} \to \mathbb{R}_{\geq 0}$. It follows:

$$\log f(x) = x \log \beta + \alpha \log x + \mathcal{O}(1) \qquad (4.1)$$
$$= \Theta(x)$$

and

$$\log \log f(x) = \log x + \mathcal{O}(1). \qquad (4.2)$$

Combining Equations (4.1) and (4.2) yields

$$x \log \beta - \log f(x) + \alpha \log \log f(x) = \mathcal{O}(1). \qquad (4.3)$$

Now remark that $\lim_{y \to \infty} f^{-1}(y) = \infty$. Substituting x with $f^{-1}(y)$ in Equation (4.3), we get

$$f^{-1}(y) \log \beta - \log y + \alpha \log \log y = \mathcal{O}(1)$$

as y tends to ∞. The desired asymptotic expansion of $f^{-1}(y)$ follows. $\square$

Proof [Proof of Theorem 4.7.11] For each $i \in \{1, 2\}$, let $g_i \colon \mathbb{R}_{\geq 0} \to \mathbb{R}_{\geq 0}$ be an increasing bijection such that $g_i(x) = x^{\alpha_i} \beta_i^x$ for all sufficiently large real $x \geq 0$. There exists $(\lambda_i, \mu_i, k_i) \in \mathbb{R}_{>0} \times \mathbb{R}_{>0} \times \mathbb{N}$ such that

$$\lambda_i g_i(k) \leq f_i(k) \leq \mu_i g_i(k)$$

for every integer $k \geq k_i$. Let $k_0 = \max\{k_1, k_2\}$. It is clear that

$$E_y(\mu_1 g_1, \lambda_2 g_2) \setminus [\![0, k_0 - 1]\!] \subseteq E_y(f_1, f_2) \subseteq E_y(\lambda_1 g_1, \mu_2 g_2) \cup [\![0, k_0 - 1]\!],$$

so

$$\operatorname{Card} E_y(\mu_1 g_1, \lambda_2 g_2) - k_0 \leq \operatorname{Card} E_y(f_1, f_2) \leq \operatorname{Card} E_y(\lambda_1 g_1, \mu_2 g_2) + k_0.$$

Therefore, to obtain $\operatorname{Card} E_y(f_1, f_2) = \Delta(y) + \mathcal{O}(1)$ as desired, it suffices to prove that

$$\operatorname{Card} E_y(h_1, h_2) = \Delta(y) + \mathcal{O}(1) \qquad (4.4)$$

for each $(h_1, h_2) \in \{(\mu_1 g_1, \lambda_2 g_2), (\lambda_1 g_1, \mu_2 g_2)\}$. The benefit in replacing $E_y(f_1, f_2)$ with $E_y(h_1, h_2)$ is that both h_1 and h_2 are bijections from $\mathbb{R}_{\geq 0}$ onto itself, so Lemmas 4.7.14 and 4.7.13 apply: $h_1^{-1}(y) - h_2^{-1}(y) = \Delta(y) + \mathcal{O}(1)$, and thus

$$\operatorname{Card} E_y(h_1, h_2) = \max\{\Delta(y) + \mathcal{O}(1), 0\} + \mathcal{O}(1). \qquad (4.5)$$

Now remark that either Δ is identically zero or $\lim_{y \to \infty} \Delta(y) = \infty$: Δ is identically zero if $(\beta_1, \alpha_1) = (\beta_2, \alpha_2)$, and $\lim_{y \to \infty} \Delta(y) = \infty$ if (β_1, α_1) is lexicographically smaller than (β_2, α_2). In both cases, we have

$$\max\{\Delta(y) + \mathcal{O}(1), 0\} = \Delta(y) + \mathcal{O}(1). \tag{4.6}$$

Combining Equations (4.5) and (4.6), we get Equation (4.4). □

4.7.3 Around the theorem of Salomaa and Soittola

This section is dedicated to the following unsurprising but non-trivial theorem:

Theorem 4.7.15 (A. Salomaa and M. Soittola) *For every endomorphism* $\sigma \colon A^* \to A^*$ *and every* $x \in A^*$, *there exists* $(\beta, \alpha) \in (\mathbb{R}_{\geqslant 1} \times \mathbb{N}) \cup \{(0,0)\}$ *such that* $|\sigma^k(x)| = \Theta(k^\alpha \beta^k)$ *as k tends to* ∞.

Note that $\beta = \alpha = 0$ only occurs in the degenerated case where x is erased by a power of σ. Note also that the sequence $(|\sigma^k(x)|)_{k \geqslant 0}$ is completely determined by the incidence matrix of σ (denoted $\mathbf{M}_\sigma$) and the Parikh vector of x (denoted $\mathbf{P}(x)$): $|\sigma^k(x)| = \mathbf{U} \mathbf{M}_\sigma^k \mathbf{P}(x)$ for every $k \in \mathbb{N}$, where $\mathbf{U} \in \mathbb{R}^{1 \times A}$ is the row vector whose entries are all ones.

4.7.3.1 Proof of the theorem of Salomaa and Soittola

The original result from Salomaa and Soittola (Salomaa and Soittola 1978) is, in fact, stronger than Theorem 4.7.15: see Exercise 4.16. However, Theorem 4.7.15 is both easier to prove and sufficient for our purpose. The proof of Theorem 4.7.15 presented below first appeared in (Cassaigne, Mauduit, and Nicolas 2008).

Lemma 4.7.16 *Let* $\sigma \colon A^* \to A^*$ *be an endomorphism and let* $x, y \in A^*$. *If y is a factor of* $\sigma^{k_0}(x)$ *for some* $k_0 \in \mathbb{N}$ *then* $|\sigma^k(y)| = \mathcal{O}(|\sigma^k(x)|)$ *as k tends to* ∞.

Proof If y is a factor of $\sigma^{k_0}(x)$ then for every $k \in \mathbb{N}$, $\sigma^k(y)$ is a factor of $\sigma^{k+k_0}(x)$, and thus $|\sigma^k(y)| \leqslant |\sigma^{k+k_0}(x)| \leqslant \|\sigma^{k_0}\| |\sigma^k(x)|$. □

Proof [Proof of Theorem 4.7.15] Without loss of generality, we may assume that for each $a \in A$, there exists $j \in \mathbb{N}$ such that a occurs in $\sigma^j(x)$. Let $\|\cdot\|_1$ denote the *Manhattan norm* on $\mathbb{C}^{A \times A}$: for each $\mathbf{X} \in \mathbb{C}^{A \times A}$, $\|\mathbf{X}\|_1$ equals the sum of the magnitudes of the entries of $\mathbf{X}$. Let $S_k = \|\mathbf{M}_\sigma^k\|_1$ for every $k \in \mathbb{N}$.

By Theorem 4.7.6, there exists $(\beta, \alpha) \in (\mathbb{R}_{>0} \times \mathbb{N}) \cup \{(0,0)\}$ such that $S_k = \Theta(k^\alpha \beta^k)$. Note that β belongs to $\{0\} \cup \mathbb{R}_{\geqslant 1}$ because $S_k \in \mathbb{N}$ for every $k \in \mathbb{N}$. It now suffices to prove that $|\sigma^k(x)| = \Theta(S_k)$.

First of all, remark that

$$S_k = \|\mathbf{M}_{\sigma^k}\|_1 = \sum_{a \in A} \sum_{b \in A} |\sigma^k(a)|_b = \sum_{a \in A} |\sigma^k(a)|$$

because $\sum_{b \in A} |w|_b = |w|$ for every $w \in A^*$. On the one hand, we have

$$|\sigma^k(x)| = \sum_{a \in A} |x|_a |\sigma^k(a)| \leqslant |x| S_k,$$

because $|x|_a \leqslant |x|$ for every $a \in A$. It follows that $|\sigma^k(x)| = \mathcal{O}(S_k)$. On the other hand, Lemma 4.7.16 ensures that for each $a \in A$, $|\sigma^k(a)| = \mathcal{O}(|\sigma^k(x)|)$, so $S_k = \mathcal{O}(|\sigma^k(x)|)$. □

4.7.3.2 Some consequences of the theorem of Salomaa and Soittola

Combining Claim 4.7.4 and Theorem 4.7.15, we get:

Lemma 4.7.17 *For each $i \in \{1, 2\}$, let A_i be an alphabet, let $\sigma_i \colon A_i^* \to A_i^*$ be an endomorphism, and let $x_i \in A_i^*$. At least one of the following three relations holds as k tends to ∞: $|\sigma_1^k(x_1)| = o(|\sigma_2^k(x_2)|)$, or $|\sigma_1^k(x_1)| = \Theta(|\sigma_2^k(x_2)|)$, or $|\sigma_2^k(x_2)| = o(|\sigma_1^k(x_1)|)$.*

In particular, Lemma 4.7.17 implies that for any endomorphism $\sigma \colon A^* \to A^*$, the letters in A are linearly pre-ordered by their orders of growth under σ.

Lemma 4.7.18 *For any endomorphism $\sigma \colon A^* \to A^*$, any $x \in A^*$ and any $k_0 \in \mathbb{N}$, $|\sigma^{k+k_0}(x)| = \Theta(|\sigma^k(x)|)$ as k tends to ∞.*

Proof Without loss of generality, we may assume $k_0 = 1$, i.e., we only need to check

$$|\sigma^{k+1}(x)| = \Theta(|\sigma^k(x)|). \tag{4.7}$$

For any $\beta, \alpha \in \mathbb{C}$ with $\beta \neq 0$, it is clear that $(k+1)^\alpha \beta^{k+1} \sim \beta k^\alpha \beta^k = \Theta(k^\alpha \beta^k)$, so by Theorem 4.7.15, Equation (4.7) holds if $\sigma^k(x) \neq \varepsilon$ for every $k \in \mathbb{N}$. If there exists $d \in \mathbb{N}$ such that $\sigma^d(x) = \varepsilon$ then $|\sigma^{k+1}(x)| = |\sigma^k(x)| = 0$ for every $k \geqslant d$, and thus Equation (4.7) trivially holds in this case. □

A more explicit result than Lemma 4.7.18 can be proven without the help of Theorem 4.7.15: see Exercise 4.17.

Proposition 4.7.19 *Let $\sigma\colon A^* \to A^*$ be an endomorphism, let $x \in A^*$, and let $k_0 \in \mathbb{N}$ with $\sigma^{k_0}(x) \neq \varepsilon$.*

(i) *For each $a \in \mathrm{alph}(\sigma^{k_0}(x))$, $|\sigma^k(a)| = \mathcal{O}(|\sigma^k(x)|)$ as k tends to ∞.*
(ii) *There exists $b \in \mathrm{alph}(\sigma^{k_0}(x))$ such that $|\sigma^k(b)| = \Theta(|\sigma^k(x)|)$ as k tends to ∞.*

Proof (i) follows from Lemma 4.7.16, so we only need to prove (ii).

Let us first assume that $k_0 = 0$. Lemma 4.7.17 ensures that there exists $b \in \mathrm{alph}(x)$ such that for every $a \in \mathrm{alph}(x)$, $|\sigma^k(a)| = \mathcal{O}(|\sigma^k(b)|)$. It is easy to see that $|\sigma^k(b)| = \Theta(|\sigma^k(x)|)$.

Let us now consider the general case. Substituting x with $\sigma^{k_0}(x)$ in the previous discussion, we get that there exists $b \in \mathrm{alph}(\sigma^{k_0}(x))$ such that $|\sigma^k(b)| = \Theta(|\sigma^{k+k_0}(x)|)$, and thus $|\sigma^k(b)| = \Theta(|\sigma^k(x)|)$ by Lemma 4.7.18. □

Proposition 4.7.20 *Let $\sigma\colon A^* \to A^*$ be an endomorphism and let $\beta \in \mathbb{R}_{\geq 1}$. If there exists $(x, \alpha) \in A^* \times \mathbb{N}$ such that $|\sigma^k(x)| = \Theta(k^\alpha \beta^k)$ as k tends to ∞ then there exists $b \in A$ such that $|\sigma^k(b)| = \mathcal{O}(\beta^k)$ as k tends to ∞.*

Proof For every morphism $\tau\colon A^* \to B^*$ and every $w \in A^+$, we have
$$\min_{a \in A} |\tau(a)| \leq \frac{|\tau(w)|}{|w|}.$$

So, letting $\tau = \sigma^k$ and $w = \sigma^k(x)$, we get
$$\min_{a \in A} |\sigma^k(a)| \leq \frac{|\sigma^{2k}(x)|}{|\sigma^k(x)|} \tag{4.8}$$

for every $k \in \mathbb{N}$. On the one hand, the asymptotic behaviour of the right-hand side of Equation (4.8) is clear:
$$\frac{|\sigma^{2k}(x)|}{|\sigma^k(x)|} = \Theta\left(\frac{(2k)^\alpha \beta^{2k}}{k^\alpha \beta^k}\right) = \Theta(\beta^k). \tag{4.9}$$

On the other hand, Lemma 4.7.17 ensures that there exists $b \in A$ such that for every $a \in A$, $|\sigma^k(b)| = \mathcal{O}(|\sigma^k(a)|)$. It is easy to see that b satisfies
$$|\sigma^k(b)| = \Theta\left(\min_{a \in A} |\sigma^k(a)|\right). \tag{4.10}$$

Combining Equations (4.10), (4.8), and (4.9) yields $|\sigma^k(b)| = \mathcal{O}(\beta^k)$. □

Although Proposition 4.7.20 is sufficient for our purpose, its conclusion can be significantly strengthened (Salomaa and Soittola 1978): see Exercise 4.18.

Definition 4.7.21 Let $\sigma\colon A^* \to A^*$ be an endomorphism and let $x \in A^*$.

- We say that x is *exponentially growing* under σ if there exists a real $\beta > 1$ such that $|\sigma^k(x)| = \Omega(\beta^k)$ as k tends to ∞.
- We say that x is *polynomially bounded* under σ if there exists $\alpha \in \mathbb{N}$ such that $|\sigma^k(x)| = \mathcal{O}(k^\alpha)$ as k tends to ∞.

It follows from Theorem 4.7.15 that x is either exponentially growing or polynomially bounded under σ.

Proposition 4.7.22 *Let $\sigma\colon A^* \to A^*$ be an endomorphism, let $x \in A^*$ be such that x grows exponentially under σ, and let $(\beta, \alpha) \in \mathbb{R}_{>1} \times \mathbb{N}$ be such that $|\sigma^k(x)| = \Theta(k^\alpha \beta^k)$ as k tends to ∞. Let C denote the set of letters $c \in A$ such that $|\sigma^k(c)| = \Theta(k^\alpha \beta^k)$ as k tends to ∞. As k tends to ∞, $|\sigma^k(x)|_C = \Theta(\beta^k)$.*

Note that Proposition 4.7.19.(ii) ensures that $|\sigma^k(x)|_C$ is positive for every $k \in \mathbb{N}$. Proposition 4.7.22 shows that the exponential growth of x under σ compels the exponential growth of $|\sigma^k(x)|_C$ as k tends to ∞.

Proof [Proof of Proposition 4.7.22] For each $k \geq 1$, let $f_k = k^\alpha \beta^k$. We make the convention that $f_0 = 1$ (even if $\alpha > 0$), so that f_k is positive for every $k \in \mathbb{N}$. Let $\lambda, \mu \in \mathbb{R}_{>0}$ be such that $\lambda f_k \leq |\sigma^k(c)| \leq \mu f_k$ for every $(c, k) \in C \times \mathbb{N}$. Let $\lambda', \mu' \in \mathbb{R}_{>0}$ be such that $\lambda' f_k \leq |\sigma^k(x)| \leq \mu' f_k$ for every $k \in \mathbb{N}$. Let B denote the set of letters $b \in A$ such that $|\sigma^k(b)| = o(f_k)$. For each $k \in \mathbb{N}$, let $g_k = \max_{b \in B} |\sigma^k(b)|$.

By Lemma 4.7.17, $B \cup C$ equals the set of letters $a \in A$ such that $|\sigma^k(a)| = \mathcal{O}(f_k)$. Let $j, k \in \mathbb{N}$. Proposition 4.7.19.(i) ensures that $\sigma^k(x) \in (B \cup C)^*$, so we can write

$$|\sigma^{j+k}(x)| = \sum_{b \in B} |\sigma^j(b)| \, |\sigma^k(x)|_b + \sum_{c \in C} |\sigma^j(c)| \, |\sigma^k(x)|_c. \qquad (4.11)$$

Let us estimate both of the two terms on the right-hand side of Equation (4.11):

$$0 \leq \sum_{b \in B} |\sigma^j(b)| \, |\sigma^k(x)|_b \leq g_j \, |\sigma^k(x)|_B$$

and

$$\lambda f_j \, |\sigma^k(x)|_C \leq \sum_{c \in C} |\sigma^j(c)| \, |\sigma^k(x)|_c \leq \mu f_j \, |\sigma^k(x)|_C.$$

It follows on the one hand

$$\lambda' f_{j+k} \leq \left|\sigma^{j+k}(x)\right| \leq g_j \left|\sigma^k(x)\right|_B + \mu f_j \left|\sigma^k(x)\right|_C$$

and on the other hand

$$\lambda f_j \left|\sigma^k(x)\right|_C \leq \left|\sigma^{j+k}(x)\right| \leq \mu' f_{j+k}$$

therefore

$$\frac{\lambda' f_{j+k}}{\mu f_j} - \frac{g_j \left|\sigma^k(x)\right|_B}{\mu f_j} \leq \left|\sigma^k(x)\right|_C \leq \frac{\mu' f_{j+k}}{\lambda f_j}.$$

For fixed k, as j tends to ∞, $\dfrac{f_{j+k}}{f_j}$ tends to β^k and $\dfrac{g_j}{f_j}$ tends to 0, so

$$\frac{\lambda'}{\mu} \beta^k \leq \left|\sigma^k(x)\right|_C \leq \frac{\mu'}{\lambda} \beta^k.$$

□

The following example illustrates Proposition 4.7.22 (see also Section 4.10.5).

Example 4.7.23 Let $\sigma \colon \{0,1\}^* \to \{0,1\}^*$ be the endomorphism defined by: $\sigma(0) = 010$ and $\sigma(1) = 11$. For every $k \in \mathbb{N}$, it is clear that $\sigma^k(1) = 1^{2^k}$, $\left|\sigma^k(0)\right| = (\frac{1}{2}k + 1)2^k$, and $\left|\sigma^k(0)\right|_0 = 2^k$.

Lemma 4.7.24 *Let $(f_k)_{k \geq 0}$ be a sequence of positive real numbers such that the ratio $f_{k+1} f_k^{-1}$ converges to a real limit, denoted β, as k tends to ∞. If $\beta > 1$ then*

$$\sum_{j=0}^{k} f_j \sim \frac{\beta}{\beta - 1} f_k$$

as k tends to ∞.

Proof For each $k \in \mathbb{N}$, let

$$F_k = \sum_{j=0}^{k} f_j.$$

Let r, s be real numbers such that $0 \leq r < \beta^{-1} < s < 1$. Since the ratio $f_i f_{i+1}^{-1}$ tends to β^{-1} as i tends to ∞, there exists $k_0 \in \mathbb{N}$ such that

$$r \leq \frac{f_i}{f_{i+1}} \leq s \qquad (4.12)$$

for every $i > k_0$. Multiplying Equation (4.12) over all $i \in [\![j, k-1]\!]$ yields

$$r^{k-j} \leqslant \frac{f_j}{f_k} \leqslant s^{k-j} \tag{4.13}$$

for any j and k satisfying $k_0 < j \leqslant k$. Then, summing Equation (4.13) over all $j \in [\![k_0+1, k]\!]$, we obtain

$$\frac{1-r^{k-k_0}}{1-r} \leqslant \frac{F_k}{f_k} - \frac{F_{k_0}}{f_k} \leqslant \frac{1-s^{k-k_0}}{1-s} \tag{4.14}$$

for any $k \geqslant k_0$. As k tends to ∞, r^{k-k_0}, s^{k-k_0}, and $\frac{F_{k_0}}{f_k}$ tend to 0: indeed, substituting j with $k_0 + 1$ in Equation (4.13), we get $f_k^{-1} \leqslant C^{-1} s^k$ for every $k > k_0$, where $C = f_{k_0+1} s^{k_0+1}$ is a constant. Hence, Equation (4.14) yields

$$\frac{1}{1-r} \leqslant \liminf_{k \to \infty} \frac{F_k}{f_k} \leqslant \limsup_{k \to \infty} \frac{F_k}{f_k} \leqslant \frac{1}{1-s}.$$

Since r and s can be chosen arbitrarily close to β^{-1}, it follows that

$$\lim_{k \to \infty} \frac{F_k}{f_k} = \frac{1}{1-\beta^{-1}} = \frac{\beta}{\beta-1}.$$

□

Lemma 4.7.24 is related to the d'Alembert ratio test. The case where $\beta < 1$ is studied in Exercise 4.19.

Lemma 4.7.25 *Let $\sigma \colon A^* \to A^*$ be an endomorphism and let $x \in A^*$. If x grows exponentially under σ then*

$$\sum_{j=0}^{k-1} |\sigma^j(x)| = \Theta(|\sigma^k(x)|) \tag{4.15}$$

as k tends to ∞.

Proof Let $(\beta, \alpha) \in \mathbb{R}_{>1} \times \mathbb{R}$ and let $f_k = k^\alpha \beta^k$ for every integer $k \geqslant 1$. We make the convention that $f_0 = 1$. It follows from Lemma 4.7.24 that

$$\sum_{j=0}^{k-1} f_j \sim \tfrac{\beta}{\beta-1} f_{k-1} \sim \tfrac{1}{\beta-1} f_k = \Theta(f_k).$$

By Theorem 4.7.15, the pair (β, α) can be chosen in such a way that $|\sigma^k(x)| = \Theta(f_k)$, so Equation (4.15) holds. □

If x is polynomially growing under σ then Equation (4.15) does not hold:
$$\sum_{j=0}^{k-1} |\sigma^j(x)| = \Theta(k |\sigma^k(x)|)$$
as k tends to ∞.

4.7.4 Centric factors

Definition 4.7.26 Let (x, y, z) be a triple of words. We say that a word w is a *centric factor* of (x, y, z) if there exist a non-empty suffix x' of x and a non-empty prefix z' of z such that $w = x'yz'$.

If w is a centric factor of (x, y, z) then w is a factor of xyz and y is an inner factor of w. The converse is false in general:

Example 4.7.27 Let $(x, y, z) = (010, 00, 10)$. The word $w = 1000$ is such that w is a factor of xyz and y is an inner factor of w, but w is not a centric factor of (x, y, z). The centric factors of (x, y, z) are 0001, 10001, 00010, 010001, 100010 and xyz.

Remark 4.7.28 Let n be a positive integer. No triple of words admits more than $n - 1$ distinct centric factors with length n.

Definition 4.7.29 Let $\sigma \colon A^* \to A^*$ be an endomorphism and let x, y, $z \in A^*$. Define $L^\sigma(x, y, z)$ as the set of all $w \in A^*$ such that w is a centric factor of $(\sigma^k(x), \sigma^k(y), \sigma^k(z))$ for some $k \in \mathbb{N}$. For each $n \in \mathbb{N}$, define $L_n^\sigma(x, y, z) = L^\sigma(x, y, z) \cap A^n$.

Remark 4.7.30 Let x, y_1, y_2, $z \in A^*$.

(i) Every centric factor of $(x, y_1 y_2, z)$ is a centric factor of $(x, y_1, y_2 z)$ and of (xy_1, y_2, z).
(ii) $L^\sigma(x, y_1 y_2, z)$ is a subset of $L^\sigma(x, y_1, y_2 z) \cap L^\sigma(xy_1, y_2, z)$.

The next result plays a crucial role all through the proof of the theorem of Pansiot.

Theorem 4.7.31 Let $u \in A^\mathbb{N}$ be a purely morphic word and let $\sigma \colon A^* \to A^*$ be an endomorphism that generates u. There exists a finite subset $G \subseteq A \times A^+ \times A$ such that
$$L_n(u) = \bigcup_{(a, w, a') \in G} L_n^\sigma(a, w, a')$$
for every integer $n \geqslant 3$.

The property of G can be restated as follows: for each $x \in A^*$ with $|x| \geq 3$, x is a factor of u if, and only if, there exists $(a, w, a') \in G$ such that $x \in L^\sigma(a, w, a')$.

Proof [Proof of Theorem 4.7.31] Let G be the set of all $(a, w, a') \in A \times A^+ \times A$ such that $|w| \leq 2 \|\sigma\| - 2$ and awa' is a factor of u.

For every $(a, w, a') \in G$ and every $k \in \mathbb{N}$, $\sigma^k(a)\sigma^k(w)\sigma^k(a') = \sigma^k(awa')$ is a factor of u, so all centric factors of $(\sigma^k(a), \sigma^k(w), \sigma^k(a'))$ are factors of u.

Conversely, let x be a factor of u with $|x| \geq 3$. Let u_0 denote the first letter of u and let $r \in \mathbb{N}$ be such that x is a factor of $\sigma^r(u_0)$. Let $(x_i)_{0 \leq i \leq r}$ be a sequence with terms in $L(u)$ satisfying: $x_r = x$, and for each $i \in [\![0, r-1]\!]$, x_i is a shortest factor of $\sigma^i(u_0)$ such that x_{i+1} is a factor of $\sigma(x_i)$. Let $q = 1 + \max\{i \in [\![0, r]\!] \mid |x_i| \leq 2\}$ (note that q is well-defined and $1 \leq q \leq r$). For each $i \in [\![q, r]\!]$, write x_i in the form $x_i = a_i w_i a_i'$ with $(a_i, w_i, a_i') \in A \times A^+ \times A$. Clearly, x_{i+1} is a centric factor of $(\sigma(a_i), \sigma(w_i), \sigma(a_i'))$ for each $i \in [\![q, r-1]\!]$. More generally, x_j is a centric factor of $(\sigma^{j-i}(a_i), \sigma^{j-i}(w_i), \sigma^{j-i}(a_i'))$ for all $i, j \in [\![q, r]\!]$ with $i \leq j$. In particular, x is a centric factor of $(\sigma^k(a_q), \sigma^k(w_q), \sigma^k(a_q'))$ with $k = r - q$. Moreover, (a_q, w_q, a_q') belongs to G. Indeed, x_q is a factor of $\sigma(x_{q-1})$, and thus $|w_q| = |x_q| - 2 \leq |\sigma(x_{q-1})| - 2 \leq 2\|\sigma\| - 2$. □

With a slight modification, we can also account for words of length 2:

Corollary 4.7.32 *In the notation of Theorem 4.7.31, there exists a finite subset $H \subseteq A \times A^* \times A$ such that*

$$L(u) \setminus (A \cup \{\varepsilon\}) = \bigcup_{(a,w,a') \in H} L^\sigma(a, w, a').$$

Proof Let G_0 be the set of $(a, w, a') \in A \times A^* \times A$ such that $w = \varepsilon$ and $aa' \in L_2(u)$. The set $H = G \cup G_0$ satisfies the desired property. □

4.7.5 Everywhere-growing endomorphisms

4.7.5.1 Definitions and basic properties

Definition 4.7.33 Let $\sigma \colon A^* \to A^*$ be an endomorphism and let $x \in A^*$.

- We say that x is *growing* under σ if $|\sigma^k(x)|$ tends to ∞ as k tends to ∞.
- We say that x is *bounded* under σ if $|\sigma^k(x)|$ is bounded. Then the sequence of words $(\sigma^k(x))_{k \geq 0}$ is eventually periodic.

Every word over A is either growing or bounded under σ.

Example 4.7.34 If an endomorphism σ is prolongable on a letter u_0 then u_0 is growing under σ.

Definition 4.7.35 An endomorphism $\sigma\colon A^* \to A^*$ is called *everywhere-growing* if every letter in A is growing under σ.

Remark 4.7.36 Let $\sigma\colon A^* \to A^*$ be an everywhere-growing endomorphism. Any fixed point of σ is a purely morphic infinite word generated by σ.

The aim of Section 4.7.5 is to prove that the complexity function of any fixed point of any everywhere-growing endomorphism belongs to one of the following three classes: $\mathcal{O}(n)$, $\Theta(n \log \log n)$, or $\Theta(n \log n)$. Note that if the complexity function p of an infinite word satisfies $p(n) = \mathcal{O}(n)$ then the theorem of Morse and Hedlund ensures that either $p(n) = \Theta(1)$ or $p(n) = \Theta(n)$.

Definition 4.7.37 A morphism $\tau\colon A^* \to B^*$ is called *expansive* if $|\tau(a)| \geqslant 2$ for every $a \in A$.

Every expansive endomorphism is everywhere-growing. However, the Fibonacci morphism (see Example 1.2.22) is everywhere-growing but it is not expansive. In the previous literature, everywhere-growing endomorphisms are defined as expansive endomorphisms. Since each everywhere-growing endomorphism admits an expansive power, everywhere-growing endomorphisms generate the same purely morphic words as expansive endomorphisms.

Applying Proposition 4.7.20 with $\beta = 1$, we get that if an endomorphism admits a polynomially bounded word x then it also admits a bounded letter b. Hence, we can state:

Proposition 4.7.38 Let $\sigma\colon A^* \to A^*$ be an everywhere-growing endomorphism. Every non-empty word over A is exponentially growing under σ.

Proposition 4.7.38 can be proven directly without the help of Proposition 4.7.20 or Theorem 4.7.15: see Exercise 4.20.

4.7.5.2 Three classes of everywhere-growing endomorphisms

Definition 4.7.39 (Quasi-uniform) We say that an endomorphism $\sigma\colon A^* \to A^*$ is *quasi-uniform* if there exists $\beta \in \mathbb{R}_{\geqslant 1}$ such that for each $a \in A$, $|\sigma^k(a)| = \Theta(\beta^k)$ as k tends to ∞.

In fact, each $x \in A^+$ satisfies $|\sigma^k(x)| = \Theta(\beta^k)$ as k tends to ∞. For every integer $\beta \geqslant 1$, all β-uniform endomorphisms are quasi-uniform. Note that quasi-uniform endomorphisms are not necessarily everywhere-growing since β may equal 1. This convention ensures that all irreducible endomorphisms are quasi-uniform: see Exercise 4.22.

Definition 4.7.40 (Polynomially diverging) We say that an everywhere-growing endomorphism $\sigma \colon A^* \to A^*$ is *polynomially diverging* if there exist $\beta \in \mathbb{R}_{>1}$ and a non-identically-zero function $\alpha \colon A \to \mathbb{N}$ such that for each $a \in A$, $|\sigma^k(a)| = \Theta(k^{\alpha(a)} \beta^k)$ as k tends to ∞.

Note that:

- By Proposition 4.7.20, there exists $a \in A$ such that $\alpha(a) = 0$. In fact, there exists an integer $r \geqslant 1$ such that $\alpha(A) = [\![0, r]\!]$: see Exercise 4.18.
- The function α is extendable to A^+: for each $w \in A^+$, $|\sigma^k(w)| = \Theta(k^{\alpha(w)} \beta^k)$ as k tends to ∞, where $\alpha(w) = \max\{\alpha(a) \mid a \in \mathrm{alph}(w)\}$.

Definition 4.7.41 (Exponentially diverging) We say that an everywhere-growing endomorphism $\sigma \colon A^* \to A^*$ is *exponentially diverging* if there exist $a_1, a_2 \in A$, $\alpha_1, \alpha_2 \in \mathbb{N}$, and $\beta_1, \beta_2 \in \mathbb{R}_{>1}$ with $\beta_1 \neq \beta_2$ such that for each $i \in \{1, 2\}$, $|\sigma^k(a_i)| = \Theta(k^{\alpha_i} \beta_i^k)$ as k tends to ∞.

Example 4.7.42 For each $r \in \mathbb{N}$, consider the expansive endomorphism $\sigma_r \colon \{0, 1\}^* \to \{0, 1\}^*$ defined by $\sigma_r(0) = 010^r$ and $\sigma_r(1) = 11$. Clearly, σ_0 is uniform. The case of σ_1 is studied in Example 4.7.23: σ_1 is polynomially diverging. For each integer $r \geqslant 2$, σ_r is exponentially diverging: for all $k \in \mathbb{N}$, $|\sigma_r^k(0)| = \frac{r}{r-1}(r+1)^k - \frac{1}{r-1}2^k$.

Three classes of endomorphisms are introduced in Definitions 4.7.39, 4.7.40, and 4.7.41. Clearly, those classes are pairwise disjoint. Moreover each everywhere-growing endomorphism belongs to one of them: see Theorem 4.7.15 and Proposition 4.7.38. Note also that two everywhere-growing endomorphisms belong to the same class if their incidence matrices are equal. We can now make precise the main result of Section 4.7.5:

Theorem 4.7.43 *Let $u \in A^{\mathbb{N}}$ with $\mathrm{alph}(u) = A$ and let $\sigma \colon A^* \to A^*$ be an everywhere-growing endomorphism such that $\sigma(u) = u$.*

- *If σ is quasi-uniform then $p_u(n) = \mathcal{O}(n)$.*
- *If σ is polynomially diverging then $p_u(n) = \Theta(n \log \log n)$.*
- *If σ is exponentially diverging then $p_u(n) = \Theta(n \log n)$.*

Note that the hypothesis $\text{alph}(u) = A$ is not disposable because any everywhere-growing endomorphism can be obtained as a restriction of some exponentially diverging endomorphism. Indeed, consider an arbitrary everywhere-growing endomorphism $\sigma \colon A^* \to A^*$. Let b be a letter such that $b \notin A$, let $B = A \cup \{b\}$, and let $\tau \colon B^* \to B^*$ be the endomorphism defined by: $\tau(a) = \sigma(a)$ for every $a \in A$ and $\tau(b) = b^{\|\sigma\|+1}$. It is clear that τ is exponentially diverging and that σ is a restriction of τ. It follows that any fixed point of σ is a fixed point of τ.

The proof of Theorem 4.7.43 can be found in Sections 4.7.5.3 and 4.7.5.4. More precisely, Theorem 4.7.43 is obtained as the conjunction of Theorems 4.7.44, 4.7.47, and 4.7.48 below.

4.7.5.3 Upper bounds

The first important result of this section is:

Theorem 4.7.44 (A. Ehrenfeucht, K. P. Lee and G. Rozenberg) *The complexity function of an infinite word generated by an everywhere-growing endomorphism is in $\mathcal{O}(n \log n)$.*

Ehrenfeucht, Lee and Rozenberg proved that for any D0L-system (A, σ, w) such that σ is expansive, the complexity function of the D0L-language $\{\sigma^k(w) \mid k \in \mathbb{N}\}$ is in $\mathcal{O}(n \log n)$ (Ehrenfeucht, Lee, and Rozenberg 1975). This result implies Theorem 4.7.44.

Definition 4.7.45 For every endomorphism $\sigma \colon A^* \to A^*$, every $x, y, z \in A^*$, and every $n \in \mathbb{N}$, define $K_n^\sigma(x, y, z)$ as the set of all $k \in \mathbb{N}$ such that $|\sigma^k(y)| + 2 \leqslant n \leqslant |\sigma^k(xyz)|$.

Note that $K_n^\sigma(x, y, z) = E_n(f_1, f_2)$, where functions $f_1, f_2 \colon \mathbb{N} \to \mathbb{N}$ are given by: $f_1(k) = |\sigma^k(y)| + 2$ and $f_2(k) = |\sigma^k(xyz)|$ for every $k \in \mathbb{N}$. Hence, if y is exponentially growing under σ then it follows from Theorem 4.7.15 and Corollary 4.7.12 that $\text{Card}\, K_n^\sigma(x, y, z) = \mathcal{O}(1)$, or $\text{Card}\, K_n^\sigma(x, y, z) = \Theta(\log \log n)$, or $\text{Card}\, K_n^\sigma(x, y, z) = \Theta(\log n)$.

Lemma 4.7.46 *Let $u \in A^{\mathbb{N}}$ and let $\sigma \colon A^* \to A^*$ be an everywhere-growing endomorphism such that $\sigma(u) = u$. There exists a finite subset $G \subseteq A \times A^+ \times A$ such that*

$$p_u(n) \leqslant (n-1) \sum_{(a, w, a') \in G} \text{Card}\, K_n^\sigma(a, w, a')$$

for every integer $n \geqslant 3$.

Proof By Theorem 4.7.31, it suffices to check that for any x, y, $z \in A^*$, and any $n \geqslant 1$,

$$\operatorname{Card} L_n^\sigma(x,y,z) \leqslant (n-1) \operatorname{Card} K_n^\sigma(x,y,z).$$

Let $k \in \mathbb{N}$. If $k \notin K_n^\sigma(x,y,z)$ then $(\sigma^k(x), \sigma^k(y), \sigma^k(z))$ does not admit any centric factor with length n: indeed, a triple of words (x', y', z') admits at least one centric factor with length n only if $|y'| + 2 \leqslant n \leqslant |x'y'z'|$. If $k \in K_n^\sigma(x,y,z)$ then $(\sigma^k(x), \sigma^k(y), \sigma^k(z))$ admit at most $n-1$ centric factors with length n (see Remark 4.7.28). □

Proof [Proof of Theorem 4.7.44] Let $\sigma\colon A^* \to A^*$ be an everywhere-growing endomorphism. Let x, y, $z \in A^*$ with $y \neq \varepsilon$. The integer set $K_n^\sigma(x,y,z)$ is a subset of $\{k \in \mathbb{N} \mid |\sigma^k(y)| + 2 \leqslant n\}$ and the cardinality of the latter integer set is in $\mathcal{O}(\log n)$ because y grows exponentially under σ by Proposition 4.7.38. It now follows from Lemma 4.7.46 that the complexity function of any infinite word generated by σ is in $\mathcal{O}(n \log n)$. □

The next step is to prove:

Theorem 4.7.47 *Let $u \in A^\mathbb{N}$ and let $\sigma\colon A^* \to A^*$ be an everywhere-growing endomorphism such that $\sigma(u) = u$.*

- *If σ is quasi-uniform then $p_u(n) = \mathcal{O}(n)$.*
- *If σ is polynomially diverging then $p_u(n) = \mathcal{O}(n \log \log n)$.*

Proof Assume that σ is quasi-uniform. Corollary 4.7.12 ensures that for any x, y, $z \in A^*$ with $y \neq \varepsilon$, $\operatorname{Card} K_n^\sigma(x,y,z) = \mathcal{O}(1)$. It now follows from Lemma 4.7.46 that $p_u(n) = \mathcal{O}(n)$.

Assume that σ is polynomially diverging. There exist $\beta \in \mathbb{R}_{>1}$ and a function $\alpha\colon A^+ \to \mathbb{N}$ such that for each $w \in A^+$, $|\sigma^k(w)| = \Theta(k^{\alpha(w)} \beta^k)$. Theorem 4.7.11 ensures that for any x, y, $z \in A^*$ with $y \neq \varepsilon$,

$$\operatorname{Card} K_n^\sigma(x,y,z) = \frac{\alpha(xyz) - \alpha(y)}{\beta} \log \log n + \mathcal{O}(1) = \mathcal{O}(\log \log n).$$

It now follows from Lemma 4.7.46 that $p_u(n) = \mathcal{O}(n \log \log n)$. □

4.7.5.4 Lower bounds

The aim of this section is to prove:

Theorem 4.7.48 *Let $u \in A^\mathbb{N}$ with $\operatorname{alph}(u) = A$ and let $\sigma\colon A^* \to A^*$ be an everywhere-growing endomorphism such that $\sigma(u) = u$.*

- *If σ is polynomially diverging then $p_u(n) = \Omega(n \log \log n)$.*

- If σ is exponentially diverging then $p_u(n) = \Omega(n \log n)$.

In fact, Pansiot proved a stronger result than Theorem 4.7.48 (Pansiot 1984): see Exercise 4.24.(i).

Lemma 4.7.49 *Let $v \in \{0,1\}^{\mathbb{N}}$, let $(w_k)_{k \geqslant 0} \in (\{0,1\}^*)^{\mathbb{N}}$, and let $(e_k)_{k \geqslant 0} \in \mathbb{N}^{\mathbb{N}}$ be such that for every $k \in \mathbb{N}$, w_k is a suffix of w_{k+1}, $e_k < e_{k+1}$, and $w_k 10^{e_k} 1$ is a factor of v. For each $n \in \mathbb{N}$, $s_v(n)$ is greater than or equal to the number of $k \in \mathbb{N}$ satisfying*

$$1 + e_k \leqslant n \leqslant 1 + e_k + |w_k|. \tag{4.16}$$

Proof Let K denote the set of all $k \in \mathbb{N}$ satisfying Equation (4.16). For each $k \in K$, define $x_k = \text{suff}_n w_k 10^{e_k}$ and remark that x_k is a right special factor of v:
- $x_k 1$ is a factor of v because $x_k 1$ is a suffix of $w_k 10^{e_k} 1$,
- $x_k 0$ is a factor of v because $x_k 0$ is a factor of $w_{k+1} 10^{e_{k+1}}$,

It follows $s_v(n) \geqslant \text{Card}\{x_k \mid k \in K\}$. Besides, for every $i, j \in K$ with $i < j$, the letter of x_i ocurring at position $n - e_i + 1$ is a 1 while the letter of x_j occuring at the same position is a 0. Therefore, the x_k's are pairwise distinct: $\text{Card}\{x_k \mid k \in K\} = \text{Card } K$. □

The conclusion of Lemma 4.7.49 states that $s_v(n)$ is greater than or equal to $\text{Card } E_n(f_1, f_2)$ for every $n \in \mathbb{N}$, where functions $f_1, f_2 \colon \mathbb{N} \to \mathbb{N}$ are given by: $f_1(k) = 1 + e_k$ and $f_2(k) = 1 + e_k + |w_k|$ for every $k \in \mathbb{N}$.

Lemma 4.7.50 *Let $u \in A^{\mathbb{N}}$ and let $\sigma \colon A^* \to A^*$ be an everywhere-growing endomorphism such that $\sigma(u) = u$. For each $a \in \text{alph}(u)$, there exists $a' \in \text{alph}(u)$ such that a' occurs in u infinitely many times and $|\sigma^k(a')| = \Theta(|\sigma^k(a)|)$ as k tends to ∞.*

Proof Let C denote the set of letters $c \in A$ such that $|\sigma^k(c)| = \Theta(|\sigma^k(a)|)$. It follows from Propositions 4.7.22 and 4.7.38 that $\lim_{k \to \infty} |\sigma^k(a)|_C = \infty$, so there exists $a' \in C$ such that the sequence $(|\sigma^k(a)|_{a'})_{k \geqslant 0}$ is unbounded. Since $\sigma^k(a)$ is a factor of u for every $k \in \mathbb{N}$, a' occurs in u infinitely many times. □

A stronger version of Lemma 4.7.50 can be found in Exercise 4.23.

Let $\tau \colon A^* \to A^*$ be an endomorphism, let $c \in A$ be such that c occurs in $\tau(c)$, and let $y, z \in A^*$ be such that $\tau(c) = ycz$. For each $k \in \mathbb{N}$, define:

$$y_k = \tau^{k-1}(y) \cdots \tau^3(y) \tau^2(y) \tau(y) y,$$
$$z_k = z\tau(z)\tau^2(z)\tau^3(z) \cdots \tau^{k-1}(z).$$

It is clear that $\tau^k(c) = y_k c z_k$ for every integer $k \in \mathbb{N}$. Note that y_k is a suffix of y_{k+1} and that z_k is a prefix of z_{k+1}: $y_{k+1} = \tau^k(y) y_k$ and $z_{k+1} = z_k \tau^k(z)$. Hence, iterating τ on c makes a word 'grow' on each side of c.

Claim 4.7.51 *Let X be a finite set and let N denote the factorial of $\operatorname{Card} X$. Every function $f \colon X \to X$ satisfies $f^N = f^{2N}$.*

Note that $f^N = f^{2N}$ implies $f^N = f^{kN}$ for every integer $k \geqslant 1$.

Lemma 4.7.52 *Let $\sigma \colon A^* \to A^*$ be an endomorphism and let $B, C \subseteq A$ be such that $A = B \cup C$, $B \cap C = \emptyset$, $\sigma(B) \subseteq B^*$ and $\sigma(C) \subseteq A^* C A^*$. Let N denote the factorial of $\operatorname{Card} C$. For each $c \in C$, there exist $c_1, c_2 \in C$ such that*

(i) $\sigma^N(c) \in B^* c_1 A^* \cap A^* c_2 B^*$,
(ii) $\sigma^N(c_1) \in B^* c_1 A^*$ and
(iii) $\sigma^N(c_2) \in A^* c_2 B^*$.

For instance, B and C can be taken as the sets of bounded and growing letters under σ, respectively.

Proof [Proof of Lemma 4.7.52] Let $f_1 \colon C \to C$ be the function mapping each $c \in C$ to the leftmost letter of $\sigma(c)$ that belongs to C: $\sigma(c) \in B^* f_1(c) A^*$ for every $c \in C$. The property $\sigma(C) \subseteq A^* C A^*$ ensures that f_1 is well-defined. In the same way, let $f_2 \colon C \to C$ be the function mapping each $c \in C$ to the rightmost letter of $\sigma(c)$ that belongs to C: $\sigma(c) \in A^* f_2(c) B^*$ for every $c \in C$. Claim 4.7.51 ensures that $f_i^N = f_i^{2N}$ for each $i \in \{1, 2\}$. Besides, from $\sigma(B) \subseteq B^*$, we deduce that $\sigma^k(c) \in B^* f_1^k(c) A^* \cap A^* f_2^k(c) B^*$ for every $k \geqslant 1$ and every $c \in C$. Therefore, for any given letter $c \in C$, $c_1 = f_1^N(c)$ and $c_2 = f_2^N(c)$ satisfy the desired properties (i), (ii) and (iii). □

Remark 4.7.53 In the notation of Lemma 4.7.52, for every $(c, x, c') \in C \times B^* \times C$ and every $k \in \mathbb{N}$, there exists a unique $(\check{c}, \check{x}, \check{c}') \in C \times B^* \times C$ such that $\check{c}\check{x}\check{c}'$ is a centric factor of $(\sigma^k(c), \sigma^k(x), \sigma^k(c'))$: $\check{c}$ is the rightmost letter of $\sigma^k(c)$ that belongs to C and $\check{c}'$ is the leftmost letter of $\sigma^k(c')$ that belongs to C.

Proof [Proof of Theorem 4.7.48] Let us first study the case where σ is polynomially diverging. There exist $\beta \in \mathbb{R}_{>1}$ and a function $\alpha \colon A^+ \to \mathbb{N}$ such that for each $w \in A^+$, $|\sigma^k(w)| = \Theta(k^{\alpha(w)} \beta^k)$. Let $B = \{b \in A \mid \alpha(b) = 0\}$ and $C = \{c \in A \mid \alpha(c) \geqslant 1\}$. The key idea of the proof is to iterate a suitable power of σ on a suitable element of $L(u) \cap C B^+ C$.

The following properties hold:

(i) $A = B \cup C$ and $B \cap C = \emptyset$.

(ii) $L(u) \cap CB^+C \neq \emptyset$.

(iii) $\sigma(B) \subseteq B^+$ and $\sigma(C) \subseteq A^*CA^*$.

Property (i) and the fact that C is non-empty are trivial. Moreover, B is also non-empty by Proposition 4.7.20, so property (ii) follows from Lemma 4.7.50. Property (iii) holds because for each $a \in A$, $\alpha(a) = \max\{\alpha(a') \mid a' \in \text{alph}(\sigma(a))\}$ (apply Proposition 4.7.19 with $k_0 = 1$).

Let $\chi \colon A^* \to \{0,1\}^*$ be the letter-to-letter morphism defined by: $\chi(b) = 0$ for every $b \in B$ and $\chi(c) = 1$ for every $c \in C$. The definition of χ is consistent by property (i). Let $v = \chi(u)$. Let us prove that $s_v(n) = \Omega(\log \log n)$. By Proposition 4.5.3 and Lemma 4.6.6, such a lower bound on $s_v(n)$ ensures that the complexity functions of v and u are in $\Omega(n \log \log n)$. It is obtained by the mean of Lemma 4.7.49 as explained below.

Let $M \geqslant 1$ be such that σ^M is expansive, let N denote the factorial of $\text{Card } C$, and let $\tau = \sigma^{MN}$: τ is expansive, $\tau(u) = u$, and for every $w \in A^+$, $|\tau^k(w)| = \Theta(k^{\alpha(w)}\beta^{kMN})$. By property (ii), there exist $c, c' \in C$, and $x \in B^+$ such that cxc' is a factor of u. Moreover, Lemma 4.7.52 applies by properties (i) and (iii), so we may assume that $\tau(c) \in A^*cB^*$ and $\tau(c') \in B^*c'A^*$: if needed, replace cxc' with the unique centric factor of $(\tau(c), \tau(x), \tau(c'))$ that belongs to CB^+C (see Remark 4.7.53). Let z, $z' \in B^*$, and let $y \in A^*$ be such that $\tau(c) = ycz$ and $\tau(c') \in z'c'A^*$.

For each $k \in \mathbb{N}$, let

$$w_k = \chi\left(\tau^{k-1}(y) \cdots \tau^3(y)\tau^2(y)\tau(y)y\right)$$

and

$$e_k = \left(\sum_{j=0}^{k-1} |\tau^j(z)|\right) + |\tau^k(x)| + \left(\sum_{j=0}^{k-1} |\tau^j(z')|\right).$$

It is clear that w_k is a suffix of w_{k+1}: $w_{k+1} = \chi(\tau^k(y))w_k$; since τ is expansive, $|\tau^k(x)|$ is smaller than $|\tau^{k+1}(x)|$ and thus e_k is smaller than e_{k+1}; as a prefix of $\chi(\tau^k(cxc'))$, $w_k 10^{e_k} 1$ is a factor of v. Hence, Lemma 4.7.49 applies, so it suffices to check that the cardinality of

$$K_n = \{k \in \mathbb{N} \mid 1 + e_k \leqslant n \leqslant 1 + e_k + |w_k|\}$$

is in $\Omega(\log \log n)$. By Lemma 4.7.25,

$$e_k = \Theta(|\tau^k(z)|) + |\tau^k(x)| + \Theta(|\tau^k(z')|) = \Theta(\beta^{kMN})$$

and

$$|w_k| = |\tau^k(c)| - 1 - \sum_{j=0}^{k-1}|\tau^j(z)|$$
$$= |\tau^k(c)| - 1 - \Theta(|\tau^k(z)|)$$
$$= \Theta(k^{\alpha(c)}\beta^{kMN}) - 1 - \Theta(\beta^{kMN})$$
$$= \Theta(k^{\alpha(c)}\beta^{kMN}),$$

so Theorem 4.7.11 ensures

$$\operatorname{Card} K_n = \frac{\alpha(c)}{\beta^{MN}} \log\log n + \mathcal{O}(1) = \Theta(\log\log n).$$

This concludes the proof for the case where σ is polynomially diverging.

The case where σ is exponentially diverging can basically be handled in the same way. The main difference lies in the choice of B and C. Let u_0 denote the first letter of u and let $(\beta, \alpha) \in \mathbb{R}_{>1} \times \mathbb{N}$ be such that $|\sigma^k(u_0)| = \Theta(k^\alpha \beta^k)$. Define B as the set of $b \in A$ such that $|\sigma^k(b)| = o(\beta^k)$ and define C as the set of $c \in A$ such that $|\sigma^k(c)| = \Omega(\beta^k)$. The rest of the proof is left to the reader. $\square$

4.7.6 Bounded letters

In order to conclude the proof of the theorem of Pansiot, we now turn to endomorphisms for which some letters are bounded. Pansiot's theorem is obtained as the conjunction of Theorems 4.7.55 and 4.7.66 below.

Throughout this section,

- σ denotes an endomorphism of A^*,
- u denotes a purely morphic word generated by σ,
- u_0 denotes the first letter of u,
- B denotes the set of all bounded letters under σ and
- C denotes the set of all growing letters under σ.

It is clear that $A = B \cup C$, $B \cap C = \emptyset$, B^* is the set of all bounded words under σ, $\sigma(B) \subseteq B^*$, A^*CA^* is the set of all growing words under σ, and $\sigma(C) \subseteq A^*CA^*$. Note that the sets B and C are not completely determined by u:

Example 4.7.54 Let $\check{\sigma}, \hat{\sigma}: \{0,1,2,3\}^* \to \{0,1,2,3\}^*$ be the endomorphisms defined by: $\check{\sigma}(0) = \hat{\sigma}(0) = 023$, $\check{\sigma}(1) = \hat{\sigma}(1) = 1$, $\check{\sigma}(2) = 21$,

$\check{\sigma}(3) = 3$, $\hat{\sigma}(2) = 2$, and $\hat{\sigma}(3) = 13$. For every $k \in \mathbb{N}$, it is easy to check that $\check{\sigma}^k(23) = 21^k 3 = \hat{\sigma}^k(23)$ so

$$\check{\sigma}^\omega(0) = 0\,23\,213\,2113\,21113\,211113\,2111113\,21111113 \cdots = \hat{\sigma}^\omega(0).$$

However, 2 is growing under $\check{\sigma}$ whereas 2 is bounded under $\hat{\sigma}$, and 3 is growing under $\hat{\sigma}$ whereas 3 is bounded under $\check{\sigma}$.

4.7.6.1 Subquadratic complexities

Theorem 4.7.55 *Let $u \in A^\mathbb{N}$ be a purely morphic word and let $\sigma \colon A^* \to A^*$ be an endomorphism that generates u. If at most finitely many distinct factors of u are bounded under σ then one of the following holds:*

(i) $p(n) = \Theta(1)$,
(ii) $p(n) = \Theta(n)$,
(iii) $p(n) = \Theta(n \log \log n)$ *or*
(iv) $p(n) = \Theta(n \log n)$.

Proof The proof relies on Theorem 4.7.43: the method is to bound the complexity function of u from above and from below with the complexity functions of two infinite words generated by everywhere-growing endomorphisms that belong to the same class. Without loss of generality, we may assume that $A = \mathrm{alph}(u)$. Let u_0 denote the first letter of u.

Upper bound. Let $X = B^*CB^* \cap L(u)$: X is finite by hypothesis. Let $\hat{A}$ be an alphabet with the same cardinality as X and let $\hat{\tau} \colon \hat{A}^* \to A^*$ be a morphism that induces a bijection from $\hat{A}$ onto X. (Note that $\hat{\tau}$ is injective only if $B = \emptyset$ and $X = A = C$.) Remark that $\sigma(X) \subseteq X^+$. Hence, there exists an endomorphism $\hat{\sigma} \colon \hat{A}^* \to \hat{A}^*$ such that $\hat{\tau} \circ \hat{\sigma} = \sigma \circ \hat{\tau}$: to construct it, factorise arbitrarily $\sigma(x)$ over X for each $x \in X$. Remark that $u_0 \in C \subseteq X$ and $\sigma(u_0) \in u_0 X^+$: if $\sigma(u_0)$ did not belong to $u_0 X^+$ then $\sigma(u_0)$ would belong to $u_0 B^*$ and thus u would belong to $u_0 B^\mathbb{N}$. Therefore, we may assume that $\hat{u}_0$ is the first letter of $\hat{\sigma}(\hat{u}_0)$, where $\hat{u}_0$ is the unique element of $\hat{A}$ that satisfies $\hat{\tau}(\hat{u}_0) = u_0$.

Let us prove that $\hat{\sigma}$ is everywhere-growing. Let $\hat{c} \in \hat{A}$. For every $k \in \mathbb{N}$, it holds

$$\hat{\tau} \circ \hat{\sigma}^k = \sigma^k \circ \hat{\tau}, \qquad (4.17)$$

so we get $\|\hat{\tau}\| \, |\hat{\sigma}^k(\hat{c})| \geqslant |\hat{\tau}(\hat{\sigma}^k(\hat{c}))| = |\sigma^k(\hat{\tau}(\hat{c}))|$. Besides, $\hat{\tau}(\hat{c})$ is growing under σ as an element of B^*CB^*. It follows that $\hat{c}$ is growing under $\hat{\sigma}$. Thus $\hat{\sigma}$ is everywhere-growing: this implies that $\hat{\sigma}$ is prolongable on $\hat{u}_0$. Let $\hat{u} = \hat{\sigma}^\omega(\hat{u}_0)$.

Another consequence of Equation (4.17) is that $\hat{\tau}(\hat{\sigma}^k(\hat{u}_0)) = \sigma^k(u_0)$ for every $k \in \mathbb{N}$. Therefore, $\hat{u}$ satisfies $\hat{\tau}(\hat{u}) = u$, and thus Lemma 4.6.7 ensures

$$p_u(n) = \mathcal{O}(p_{\hat{u}}(n)). \qquad (4.18)$$

At this point, it follows from Theorem 4.7.44 that the complexity functions of u and $\hat{u}$ are in $\mathcal{O}(n \log n)$.

Lower bound. Let $\check{\tau}\colon A^* \to C^*$ be the morphism defined by: $\check{\tau}(b) = \varepsilon$ for every $b \in B$ and $\check{\tau}(c) = c$ for every $c \in C$. Let $\check{\sigma}$ denote the restriction of $\check{\tau} \circ \sigma$ to C^*: $\check{\sigma}$ is an endomorphism of C^* and u_0 is the first letter of $\check{\sigma}(u_0)$. For each $k \in \mathbb{N}$, it holds that

$$\check{\tau} \circ \sigma^k = \check{\sigma}^k \circ \check{\tau},$$

so we get $\check{\tau}(\sigma^k(u_0)) = \check{\sigma}^k(u_0)$. It follows that $\check{\sigma}$ is prolongable on u_0 and that $\check{u} = \check{\sigma}^\omega(u_0)$ is the image of u under $\check{\tau}$. Therefore, Lemma 4.6.9 ensures

$$p_u(n) \geqslant p_{\check{u}}\left(\left\lfloor \tfrac{1}{M+1} n \right\rfloor\right) \qquad (4.19)$$

for all $n \in \mathbb{N}$, where M denotes the maximum length of a word in $L(u) \cap B^*$.

Matching the bounds. Consider a pair $(c, \hat{c}) \in C \times \hat{A}$ such that $\hat{\tau}(\hat{c}) \in B^*cB^*$. Observe that

$$\left|\hat{\sigma}^k(\hat{c})\right| = \left|\sigma^k(c)\right|_C = \left|\check{\sigma}^k(c)\right|$$

for every $k \in \mathbb{N}$. Therefore, $\hat{\sigma}$ and $\check{\sigma}$ are everywhere-growing endomorphisms that belong to the same class: $\hat{\sigma}$ and $\check{\sigma}$ are both quasi-uniform, or both polynomially diverging, or both exponentially diverging. Let us first consider the case where $\hat{\sigma}$ is quasi-uniform. Then, according to Theorem 4.7.43, the complexity function of $\hat{u}$ is in $\mathcal{O}(n)$. By Equation (4.18), it follows that the complexity function of u is also in $\mathcal{O}(n)$. Hence, the theorem of Morse and Hedlund ensures that $p_u = \Theta(1)$ or $p_u(n) = \Theta(n)$. However, it may happen that $p_u(n) = \Theta(1)$ while $p_{\hat{u}}(n) = \Theta(n)$:

Remark 4.7.56 Consider the case where $A = \{0, 1, 2\}$, $\sigma(0) = 01$, $\sigma(1) = 1212$, $\sigma(2) = \varepsilon$, and $u = \sigma^\omega(0) = 01(12)^\omega$. We then have $B = \{2\}$, $C = \{0, 1\}$, and $X = \{0, 1, 12, 21, 212\}$. Without loss of generality, we may assume that $\hat{A} = \{[0], [1], [12], [21], [212]\}$ and $\hat{\tau}([x]) = x$ for every $x \in X$. The endomorphism $\hat{\sigma}$ can be chosen as follows: $\hat{\sigma}([0]) = [0][1]$, $\hat{\sigma}([1]) = [1][212]$, and $\hat{\sigma}([212]) = \hat{\sigma}([12]) = [12][12]$. Although $\sigma^\omega(u_0)$ is eventually periodic, it is not the case of $\hat{u} = \hat{\sigma}^\omega([0])$:

$$\hat{u} = [0][1]\,[1][212][12]^{e_0}\,[1][212][12]^{e_1}\,[1][212][12]^{e_2}\,[1][212][12]^{e_3}\cdots$$

where $e_k = 2^{k+1} - 2$ for every $k \in \mathbb{N}$.

If $\hat{\sigma}$ and $\check{\sigma}$ are polynomially diverging then according to Theorem 4.7.43, the complexity functions $\check{u}$ and $\hat{u}$ are both in $\Theta(n \log \log n)$, and thus $p_u(n) = \Theta(n \log \log n)$ by Equations (4.18) and (4.19). The case where $\hat{\sigma}$ and $\check{\sigma}$ are polynomially diverging is handled in the same way. $\square$

4.7.6.2 Quadratic complexities

Theorem 4.7.57 (A. Ehrenfeucht, K. P. Lee and G. Rozenberg) *The complexity function of any purely morphic word is in $\mathcal{O}(n^2)$.*

An explicit bound for the hidden constant with respect to the cardinality of the alphabet and to the width of the generating morphism can be found in (Allouche and Shallit 2003, Theorem 10.4.7).

Corollary 4.7.58 *The complexity function of any morphic word is in $\mathcal{O}(n^2)$.*

Proof This follows immediately from Theorem 4.7.57 and Lemma 4.6.6. $\square$

In 1975, A. Ehrenfeucht, K. P. Lee, and G. Rozenberg proved that the complexity function of any D0L-language is in $\mathcal{O}(n^2)$ (Ehrenfeucht, Lee, and Rozenberg 1975) which implies Theorem 4.7.57. See also (Rozenberg and Salomaa 1980) and (Allouche and Shallit 2003). Subsequently, Ehrenfeucht and Rozenberg generalised the result: they proved that the complexity function of any HD0L-language is in $\mathcal{O}(n^2)$ (Ehrenfeucht and Rozenberg 1982), which implies Corollary 4.7.58.

Lemma 4.7.59 *For any endomorphism $\sigma \colon A^* \to A^*$, there exists an integer $M \geqslant 1$ such that for every $x \in A^*$, x is bounded under σ if, and only if, $\sigma^{2M}(x) = \sigma^M(x)$.*

Proof Let $X = \{\sigma^k(b) \mid (b, k) \in B \times \mathbb{N}\}$. Clearly, X is finite and $\sigma(X) \subseteq X$. It follows from Claim 4.7.51 that the factorial of Card X is a suitable choice for M. $\square$

Exercise 4.26 is related to the problem of the effective computation of M.

Lemma 4.7.60 *Let W and X be sets, let $\sigma \colon W \to W$ and $f \colon W \to X$ be functions, let T be a positive integer, and let n be a non-negative integer. The following two assertions are equivalent:*

(i) *For every $w \in W$, the sequence $\bigl(f(\sigma^k(w))\bigr)_{k \geq 0}$ is eventually periodic with period T and preperiod nT.*
(ii) *For every $w \in W$, the sequence $\bigl(f(\sigma^{kT}(w))\bigr)_{k \geq n}$ is constant.*

Proof For any fixed element $w \in W$, the following two assertions are equivalent:

(i) The sequence $\bigl(f(\sigma^k(w))\bigr)_{k \geq 0}$ is eventually periodic with period T and preperiod nT.
(ii) For each $w' \in \{w, \sigma(w), \sigma^2(w), \sigma^3(w), \ldots, \sigma^{T-1}(w)\}$, the sequence $\bigl(f(\sigma^{kT}(w'))\bigr)_{k \geq n}$ is constant.

□

Lemma 4.7.61 *For each endomorphism $\sigma \colon A^* \to A^*$, there exists an integer $T \geq 1$ such that for every $(w,n) \in A^* \times \mathbb{N}$, the sequences $\bigl(\mathrm{pref}_n \sigma^k(w)\bigr)_{k \geq 0}$ and $\bigl(\mathrm{suff}_n \sigma^k(w)\bigr)_{k \geq 0}$ are eventually periodic with period T and preperiod nT.*

Let us note that Lemma 4.7.61 originally appeared in (Ehrenfeucht, Lee, and Rozenberg 1975), see also Lemma 10.4.6 in (Allouche and Shallit 2003).

Proof [Proof of Lemma 4.7.61] By Lemma 4.7.59, there exists $M \geq 1$ such that $\sigma^{2M}(w) = \sigma^M(w)$ for every $w \in B^*$. Let N denote the factorial of $\mathrm{Card}\, C$, let $\tau = \sigma^{MN}$, and $T = 3MN$. Our task is to check that T satisfies the desired properties. By Lemma 4.7.60, it suffices to prove that for any given $(w,n) \in A^* \times \mathbb{N}$, the sequence $\bigl(\mathrm{pref}_n \tau^{3k}(w)\bigr)_{k \geq n}$ is constant: the case of $\bigl(\mathrm{suff}_n \tau^{3k}(w)\bigr)_{k \geq n}$ is symmetric. This is obvious for $n = 0$, so we assume $n \geq 1$. Our method is to prove that the sequence $\bigl(\mathrm{pref}_n \tau^k(w)\bigr)_{k \geq n+2}$ is constant.

If $w \in B^*$ then the sequence $\bigl(\tau^k(w)\bigr)_{k \geq 1}$ is constant, and thus the desired property holds. Conversely, assume that $w \in A^*CA^*$. Let $(x,c) \in B^* \times C$ be such that xc is a prefix of $\tau(w)$. By Lemma 4.7.52, there exists $y \in B^*$ such that yc is a prefix of $\tau(c)$. Since, in addition, $\tau(\tau(x)) = \tau(x)$, we get that $\tau(x)\tau^{k-1}(c)$ is a prefix $\tau^k(w)$ for every $k \geq 2$. To conclude the proof, we distinguish the two cases: in the first case, iterating τ on c 'pushes' powers of $\tau(y)$ on the left of c, and in the second case, iterating τ on c generates an infinite word as explained below.

First, assume $\tau(y) \neq \varepsilon$. For every $k \geq 1$, $(\tau(y))^{k-1} yc$ is a prefix of $\tau^k(c)$ because $\tau(\tau(y)) = \tau(y)$. Therefore, $\tau(x)(\tau(y))^{k-2}$ is a prefix of $\tau^k(w)$ for every $k \geq 2$. For every $k \geq n+2$, $\tau(x)(\tau(y))^n$ is thus a prefix of $\tau^k(w)$, and since the length of $\tau(x)\,(\tau(y))^n$ is at least n, we get

$$\operatorname{pref}_n \tau^k(w) = \operatorname{pref}_n \left(\tau(x)\,(\tau(y))^n\right).$$

As the right-hand side is independent from k, the case $\tau(y) \neq \varepsilon$ is treated.

Second, assume $\tau(y) = \varepsilon$. Then, $\tau^k(c)$ is a prefix of $\tau^{k+1}(c)$ for every $k \geqslant 1$, and thus $\tau(x)\tau^n(c)$ is a prefix of $\tau^k(w)$ for every $k \geqslant n+1$. Besides, c is growing under σ and τ, so $\tau^k(c)$ is in fact a proper prefix of $\tau^{k+1}(c)$ for every $k \geqslant 1$. It follows that the length of $\tau^n(c)$ is at least n, and thus

$$\operatorname{pref}_n \tau^k(w) = \operatorname{pref}_n \left(\tau(x)\tau^n(c)\right)$$

for every $k \geqslant n+1$. □

Proof [Proof of Theorem 4.7.57] Let $u \in A^{\mathbb{N}}$ be a purely morphic word and let $\sigma \colon A^* \to A^*$ be an endomorphism that generates u. Let us prove that $p_u(n) = \mathcal{O}(n^2)$. By Corollary 4.7.32, it suffices to prove that for any x, y, $z \in A^*$, $\operatorname{Card} L_n^\sigma(x,y,z) = \mathcal{O}(n^2)$. Moreover, $L^\sigma(x,y,z)$ is a subset of $L^\sigma(x,\varepsilon,yz)$ (see Remark 4.7.30.(ii)), so our task reduces to proving that $\operatorname{Card} L_n^\sigma(x,\varepsilon,yz) = \mathcal{O}(n^2)$.

For all n, $k \in \mathbb{N}$, let $x_{n,k} = \operatorname{suff}_n \sigma^k(x)$ and $y_{n,k} = \operatorname{pref}_n \sigma^k(yz)$, and for all $n \in \mathbb{N}$, let $X_n = \{(x_{n,k}, \varepsilon, y_{n,k}) \mid k \in \mathbb{N}\}$. Lemma 4.7.61 ensures that X_n is finite and that there exists a constant $T \geqslant 1$ such that for all $n \in \mathbb{N}$, $\operatorname{Card}(X_n) \leqslant (n+1)T$. Besides, the centric factors of $(\sigma^k(x), \varepsilon, \sigma^k(yz))$ with length n are exactly the centric factors of $(x_{n,k}, \varepsilon, y_{n,k})$ with length n. Therefore, $L_n^\sigma(x,\varepsilon,yz)$ is the set of all words with length n that are centric factors of some triples in X_n. The inequality $\operatorname{Card} L_n^\sigma(x,\varepsilon,yz) \leqslant n \operatorname{Card}(X_n)$ follows (see Remark 4.7.28).

Hence, we get

$$\operatorname{Card} L_n^\sigma(x,y,z) \leqslant \operatorname{Card} L_n^\sigma(x,\varepsilon,yz) \leqslant n \operatorname{Card}(X_n) \leqslant Tn(n+1).$$

□

Proposition 4.7.62 *Let $u \in A^{\mathbb{N}}$ be a purely morphic word and let $\sigma \colon A^* \to A^*$ be an endomorphism that generates u. Let B denote the set of bounded letters under σ and let C denote the set of growing letters under σ. There exists a finite subset $Q \subseteq C \times B^* \times B^* \times B^* \times B^* \times B^* \times C$ such that $L(u) \cap CB^*C$ equals the set of all words of the form $cy_1 z_1^k x z_2^k y_2 c'$ with $(c, y_1, z_1, x, z_2, y_2, c') \in Q$ and $k \in \mathbb{N}$.*

Proof By Lemma 4.7.59, there exists $M \geqslant 1$ such that $\sigma^{2M}(w) = \sigma^M(w)$ for every $w \in B^*$. Let N denote the factorial of $\operatorname{Card} C$ and let $\tau = \sigma^{MN}$.

Since τ generates u, Corollary 4.7.32 ensures that there exists a finite subset $H \subseteq A \times A^* \times A$ such that
$$L(u) \cap CB^*C = \bigcup_{(a,w,a') \in H} (L^\tau(a,w,a') \cap CB^*C).$$

Let us prove that for each $(a, w, a') \in A \times A^* \times A$, there exists a finite subset $Q_{a,w,a'} \subseteq C \times B^* \times B^* \times B^* \times B^* \times B^* \times C$ such that $L^\tau(a, w, a') \cap CB^*C$ equals the set of all words of the form $cy_1 z_1^k x z_2^k y_2 c'$ with $(c, y_1, z_1, x, z_2, y_2, c') \in Q_{a,w,a'}$ and $k \in \mathbb{N}$.

If $(a, w, a') \notin C \times B^* \times C$ then $L^\tau(a, w, a') \cap CB^*C$ is empty: if $a \in B$ then $L^\tau(a, w, a') \subseteq BA^+$, if $w \notin B^*$ then $L^\tau(a, w, a') \subseteq A^+CA^+$, and if $a' \in B$ then $L^\tau(a, w, a') \subseteq A^+B$. A suitable choice for $Q_{a,w,a'}$ is thus the empty set.

Let us now study the case where $(a, w, a') \in C \times B^* \times C$. For each $k \in \mathbb{N}$, let us compute the unique centric factor of $\bigl(\tau^k(a), \tau^k(w), \tau^k(a')\bigr)$ that belongs to CB^*C (see Remark 4.7.53). The replacement of σ with τ simplifies the problem. Let $c, c' \in C$, and $x_1, x_2 \in B^*$ be such that cx_1 is a suffix of $\tau(a)$ and $x_2 c'$ is a prefix of $\tau(a')$. It is clear that $cx_1 \tau(w) x_2 c'$ is the unique centric factor of $(\tau(a), \tau(w), \tau(a'))$ that belongs to CB^*C. By Lemma 4.7.52, there exist $y_1, y_2 \in B^*$ such that cy_1 is a suffix of $\tau(c)$ and $y_2 c'$ is a prefix of $\tau(c')$. Let us repeatedly use the fact that $\tau(\tau(b)) = \tau(b)$ for every $b \in B$. Let $z_i = \tau(y_i)$ for $i \in \{1, 2\}$. For every $k \geqslant 1$, $cy_1 z_1^{k-1}$ is a suffix of $\tau^k(c)$ and $z_2^{k-1} y_2 c'$ is a prefix of $\tau^k(c')$. For every $k \geqslant 2$, $cy_1 z_1^{k-2} \tau(x_1)$ is a suffix of $\tau^k(a)$ and $\tau(x_2) z_2^{k-2} y_2 c'$ is a prefix of $\tau^k(a')$, so letting $x = \tau(x_1 w x_2)$, $cy_1 z_1^{k-2} x z_2^{k-2} y_2 c'$ is the unique centric factor of $\bigl(\tau^k(a), \tau^k(w), \tau^k(a')\bigr)$ that belongs to CB^*C. The following formula summarises the previous discussion:
$$L^\tau(a, w, a') \cap CB^*C = \{awa', cx_1 \tau(w) x_2 c'\} \cup \{cy_1 z_1^k x z_2^k y_2 c' \mid k \in \mathbb{N}\}.$$

Therefore,
$$\{(a, \varepsilon, \varepsilon, w, \varepsilon, \varepsilon, a'), (c, x_1, \varepsilon, \tau(w), \varepsilon, x_2, c'), (c, y_1, z_1, x, z_2, y_2, c')\}$$
is a suitable choice for $Q_{a,w,a'}$.

Clearly,
$$Q = \bigcup_{(a,w,a') \in H} Q_{a,w,a'}$$
satisfies the desired property. $\square$

Claim 4.7.63 *Let $w \in \{0,1\}^n$, and for each $i \in \{1, 2\}$, let $q_i, e_i \in \mathbb{N}$ be such that $10^{e_i} 1$ occurs in w at position q_i and $e_i \geqslant \frac{1}{2}n - 1$. Then $(q_1, e_1) = (q_2, e_2)$.*

The bound $\frac{1}{2}n - 1$ is sharp: if one of e_1 or e_2 is smaller than this, then w may contain for instance $10^{e_1} 10^{e_2} 1$.

Proposition 4.7.64 *Let $v \in \{0,1\}^{\mathbb{N}}$, let $r \in \mathbb{R}_{\geqslant 1}$, and let $(e_k)_{k \geqslant 0} \in \mathbb{N}^{\mathbb{N}}$ satisfying the following properties:*

(i) *For all $k \in \mathbb{N}$, $10^{e_k} 1$ is a factor of v.*
(ii) *For all $h, k \in \mathbb{N}$, $h \neq k$ implies $e_h \neq e_k$.*
(iii) *As k tends to ∞, $\dfrac{e_k}{k^r}$ converges to a positive, finite limit.*

Then the complexity function of v satisfies $p_v(n) = \Omega\big(n^{1+1/r}\big)$.

Proof Let $v_n = 0^n v$, and let $K_n = \big\{k \in \mathbb{N} \mid \frac{1}{2}n - 1 \leqslant e_k \leqslant \frac{3}{4}n - 2\big\}$. First of all, note that

$$p_v(n) \geqslant p_{v_n}(n) - n \tag{4.20}$$

(see Remark 4.6.10), and that Condition (iii) yields the following estimate of the cardinality of K_n (see Example 4.5.17.(ii)):

$$\operatorname{Card} K_n = \Omega\big(n^{1/r}\big). \tag{4.21}$$

Let us now construct an injection from $[\![0, \lfloor \frac{1}{4}n \rfloor]\!] \times K_n$ into $L_n(v_n)$. Every $(q,k) \in [\![0, \lfloor \frac{1}{4}n \rfloor]\!] \times K_n$ satisfies $q + |10^{e_k} 1| \leqslant n$, so there exists $w_{q,k,n} \in L_n(v_n)$ such that $10^{e_k} 1$ occurs in $w_{q,k,n}$ at position q. It follows from Claim 4.7.63 that the function mapping each $(q,k) \in [\![0, \lfloor \frac{1}{4}n \rfloor]\!] \times K_n$ to $w_{q,k,n} \in L_n(v_n)$ is injective. (At this point, the introduction of v_n is easy to understand. Indeed, $w_{q,k,n}$ may not be a factor of v: $w_{q,k,n}$ is a factor of v only if $10^{e_k} 1$ occurs in v at some position greater than or equal to q.) Hence, we get

$$p_{v_n}(n) \geqslant \operatorname{Card}\left([\![0, \lfloor \tfrac{1}{4}n \rfloor]\!] \times K_n\right) \geqslant \tfrac{1}{4}n \operatorname{Card} K_n. \tag{4.22}$$

Combining Equations (4.20), (4.22), and (4.21) yields $p_v(n) = \Omega\big(n^{1+1/r}\big)$. □

Lemma 4.7.65 *Let $u \in A^{\mathbb{N}}$ be a purely morphic word and let $\sigma \colon A^* \to A^*$ be an endomorphism that generates u. If u is not eventually periodic then there exists $c \in A$ such that c is growing under σ and c occurs in u infinitely many times.*

Proof Let u_0 denote the first letter of u, and let $x \in A^+$ be such that $\sigma(u_0) = u_0 x$.

First, consider the case where $x \in B^*$. By Lemma 4.7.59, there exists $M \geq 1$ such that $\sigma^{2M}(w) = \sigma^M(w)$ for every $w \in B^*$; let $\tau = \sigma^M$ and let $y \in A^+$ be such that $\tau(u_0) = u_0 y$. Remark that $y = x\sigma(x)\sigma^2(x)\sigma^3(x)\cdots\sigma^{M-1}(x) \in B^*$. Hence, $\tau^k(u_0) = \tau(u_0)(\tau(y))^{k-1}$ for every $k \geq 1$, so u can be written in the form $u = \tau(u_0)(\tau(y))^\omega$, and thus u is eventually periodic.

Second, assume that $x \in A^*CA^*$. Then, for each $k \in \mathbb{N}$, there exists $c_k \in C$ such that c_k occurs in $\sigma^k(x)$. Remark that for every $c \in A$, c occurs in u infinitely many times if, and only if, c occurs in $\sigma^k(x)$ for infinitely many $k \in \mathbb{N}$. Therefore, any c satisfying $c = c_k$ for infinitely many $k \in \mathbb{N}$ satisfies the desired properties. $\square$

Theorem 4.7.66 *Let $u \in A^\mathbb{N}$ be a purely morphic word and let $\sigma\colon A^* \to A^*$ be an endomorphism that generates u. If u is not eventually periodic and if infinitely many distinct factors of u are bounded under σ then $p_u(n) = \Theta(n^2)$.*

Proof Assume that u is not eventually periodic and that $L(u) \cap B^*$ is infinite. Theorem 4.7.57 reduces our task to proving that $p_u(n) = \Omega(n^2)$. Then, $L(u) \cap CB^*C$ is infinite by Lemma 4.7.65, and thus Proposition 4.7.62 ensures that there exist $c, c' \in C$, and $y_1, z_1, x, z_2, y_2 \in B^*$ with $z_1 z_2 \neq \varepsilon$ such that $cy_1 z_1^k x z_2^k y_2 c'$ is a factor of u for every $k \in \mathbb{N}$. Let $\chi\colon A^* \to \{0,1\}^*$ be the letter-to-letter morphism defined by: $\chi(b) = 0$ for every $b \in B$ and $\chi(c) = 1$ for every $c \in C$. Let $v = \chi(u)$ and for each $k \in \mathbb{N}$, let $e_k = |z_1 z_2|k + |y_1 x y_2|$. Since $\chi(cy_1 z_1^k x z_2^k y_2 c') = 10^{e_k} 1$ is a factor of v for every $k \in \mathbb{N}$, Proposition 4.7.64 ensures that $p_v(n) = \Omega(n^2)$, and thus the desired asymptotic lower bound for p_u follows by Lemma 4.6.6. $\square$

Example 4.7.67 Let $\sigma\colon \{0,1\}^* \to \{0,1\}^*$ be the endomorphism defined by: $\sigma(0) = 001$ and $\sigma(1) = 1$. For every $n \in \mathbb{N}$, $01^n 0$ is a factor of $\sigma^\omega(0)$. Therefore, $\sigma^\omega(0)$ is not eventually periodic. Moreover, 1 is bounded under σ, so Theorem 4.7.66 ensures that the complexity of $\sigma^\omega(0)$ is in $\Theta(n^2)$. Note that $\sigma^\omega(0)$ is a sparse word; the complexity of such a word is computed more precisely in Exercise 4.10.

4.8 Complexity of automatic words

Let $\sigma\colon A^* \to A^*$ be a q-uniform morphism, $q \geq 2$, and $\tau\colon A^* \to B^*$ be a letter-to-letter morphism. Then the infinite word $u = \tau(\sigma^\omega(a)) \in B^\mathbb{N}$ is a *q-automatic word*. The alphabet A is known as the *internal alphabet* of u. It follows from Theorem 4.7.47 and Lemma 4.6.6 that the factor complexities of $\sigma^\omega(a)$ and u are in $\mathcal{O}(n)$. However, this can be proved in a much simpler way using the fact that σ is uniform. An explicit upper-bound on the hidden constant was given by A. Cobham (Cobham 1972), see also (Allouche and Shallit 2003, Theorem 10.3.1): $p_u(n) \leqslant qd^2 n$ for every integer $n \geqslant 1$, where d denotes the cardinality of A. We slightly sharpen his result:

Theorem 4.8.1 *Let u be a q-automatic word, with internal alphabet A. Then, for all $n \geqslant 2$, $p_u(n) \leqslant (d^2 + \frac{q-2}{2}d)(n-1)$, where $d = \operatorname{Card} A$. In particular, $p_u(n) = \mathcal{O}(n)$.*

Proof Let $u = \tau(v)$, with $v = \sigma^\omega(a)$, where σ is a q-uniform endomorphism of A^*, τ is letter-to-letter, and $a \in A$. By Lemma 4.6.6, it is sufficient to prove the desired inequality for p_v. In the rest of the proof, $p = p_v$.

Let us first prove that, for all $m \in \mathbb{N}$ and $i \in [\![0, q]\!]$,

$$p(qm + i + 1) \leqslant (q - i)p(m + 1) + ip(m + 2). \tag{4.23}$$

Let w be a factor of length $qm + i + 1$ of v. Choose an occurrence of w in v and write its position as $qn + j$, with $0 \leqslant j < q$. If $i + j + 1 \leqslant q$, let $z = v_n v_{n+1} \cdots v_{n+m}$; if $i + j + 1 > q$, let $z = v_n v_{n+1} \cdots v_{n+m+1}$. In both cases, $\sigma(z)$ occurs at position qn in v, and this occurrence completely covers the occurrence of w at position $qn + j$. Consequently, w occurs at position j in $\sigma(z)$. The word w is entirely defined by (j, v), and there are exactly $(q - i)p(m + 1) + ip(m + 2)$ such pairs.

Let $C = d^2 + \frac{q-2}{2}d$. Obviously $p(2) \leqslant d^2 \leqslant C$. For $m = 0$ and $2 \leqslant i \leqslant q$, Equation (4.23) yields $p(i + 1) \leqslant (q - i)d + id^2 \leqslant Ci$. Hence $p(n) \leqslant C(n - 1)$ holds for $2 \leqslant n \leqslant q + 1$. For larger n, the result follows then from Equation (4.23) by induction: let $n = qm + i + 1$ with $m \geqslant 1$ and $1 \leqslant i \leqslant q$, and observe that $n > m + 2$, so that the induction hypothesis can be used for $p(m + 1)$ and $p(m + 2)$. We then have

$$\begin{aligned} p(qm + i + 1) &\leqslant (q - i)p(m + 1) + ip(m + 2) \\ &\leqslant (q - i)Cm + iC(m + 1) = C(qm + i), \end{aligned}$$

i.e., $p(n) \leqslant C(n - 1)$. □

4.9 Control of bispecial factors

4.9.1 A technical lemma

We present here a powerful technical lemma that allows us to control the number of strong bispecial factors in an infinite word. We then show two applications of this lemma.

This lemma appeared in a similar form, but in the binary case only, in (Cassaigne and Chekhova 2006), and in a different form in (Cassaigne 1996). Its proof is based on Rauzy graphs.

Lemma 4.9.1 *Let $u \in A^{\mathbb{N}}$ be a recurrent infinite word, or $u \in A^{\mathbb{Z}}$ be a bi-infinite word. Let p be the complexity function of u and s its first finite difference. Let $0 < n_1 < n_2$ be integers. Let $Z^+ = \{w \in L(u) \mid n_1 \leqslant |w| < n_2$ and $m(w) > 0\}$ be the set of strong bispecial factors of u with length between n_1 and $n_2 - 1$. Then*

$$\sum_{w \in Z^+} m(w) \leqslant s(n_1)\left(s(n_1) + s(n_2) + (p(n_2+1) - p(n_1+1))\frac{s(n_1)}{n_1}\right).$$

Note that if A is a binary alphabet, the left-hand side reduces to Card Z^+.

Proof Let $G = G_{n_1}(u)$ be the Rauzy graph of order n_1 of u. Vertices of G are identified with factors in $L_{n_1}(u)$, and edges of G with factors in $L_{n_1+1}(u)$: if e is an edge from v to w, then v is a prefix and w a suffix of e. Let $Y = \bigcup_{n_1 < n \leqslant n_2} LS_n(u)$ be the set of left special factors of u with length between $n_1 + 1$ and n_2, and let $Z = \bigcup_{n_1 \leqslant n < n_2} BS_n(u)$ be the set of bispecial factors of u with length between n_1 and $n_2 - 1$.

We define a weight function $\pi \colon L_{n_1+1}(u) \to \mathbb{N}$ on edges of G as follows: $\pi(e) = \sum_{w \in Y \cap A^*e} (d^-(w) - 1)$. In other words, $\pi(e)$ is the number of left special factors $w \in Y$ ending with e, counted with multiplicity $d^-(w) - 1$. Observe that $\pi(e) = \text{Card}(L_{n_2+1}(u) \cap A^*e) - 1$ (by an argument similar to the proof of Theorem 4.5.4, with a telescopic sum). We also define, for any vertex v of G, $\Delta\pi(v) = \sum_{a \in E^+(v)} \pi(va) - \sum_{a \in E^-(v)} \pi(av)$, the variation between the total weight of incoming edges and that of outgoing edges at v.

We first evaluate $\Delta\pi(v)$:

$$\Delta\pi(v) = \sum_{a \in E^+(v)} (\text{Card}(L_{n_2+1}(u) \cap A^*va) - 1)$$

$$- \sum_{a \in E^-(v)} (\text{Card}(L_{n_2+1}(u) \cap A^*av) - 1)$$

$$= \sum_{w \in L_{n_2}(u) \cap A^*v} d^+(w) - d^+(v) - \text{Card}(L_{n_2+1}(u) \cap A^*v) + d^-(v)$$

$$= \sum_{w \in L_{n_2}(u) \cap A^*v} (d^+(w) - d^-(w)) - (d^+(v) - d^-(v)). \quad (4.24)$$

If $G' = (V', E')$ is a subgraph of G, let $\pi(G')$ be the sum of $\pi(e)$ over all edges $e \in E'$, and $\Delta \pi(G')$ be the sum of $\Delta \pi(v)$ over all vertices $v \in V'$. In particular, $\pi(G) = \sum_{w \in Y}(d^-(w) - 1) = p(n_2 + 1) - p(n_1 + 1)$ by summation of Theorem 4.5.4.(ii), and $\Delta \pi(G) = 0$.

From Equation (4.24) we get a lower bound on $\Delta \pi(G')$:

$$\Delta \pi(G') = \sum_{w \in L_{n_2}(u) \cap A^*V'} (d^+(w) - d^-(w)) - \sum_{v \in V'} (d^+(v) - d^-(v))$$

$$\geqslant - \sum_{w \in L_{n_2}(u) \cap A^*V'} (d^-(w) - 1) - \sum_{v \in V'} (d^+(v) - 1)$$

$$\geqslant -s(n_2) - s(n_1) \qquad (4.25)$$

the last inequality coming from Theorem 4.5.4.(i) and (ii).

For every $v \in L_{n_1}(u)$, let $a_v \in E^+(v)$ be such that va_v has maximal weight among all edges of G starting in v (if several edges have the same weight, choose one arbitrarily). Let r_v be the maximal weight of edges starting in v other than va_v. If $w \in Z^+ \cap A^*v$ is a strong bispecial factor ending with v, then

$$m(w) = \sum_{a \in E^+(w)} (d^-(wa) - 1) - d^-(w) + 1$$

$$\leqslant \sum_{a \in E^+(w) \setminus \{a_v\}} (d^-(wa) - 1). \qquad (4.26)$$

Indeed, if $a_v \in E^+(w)$, then $(d^-(wa_v) - 1) - d^-(w) + 1 \leqslant 0$; and otherwise, $-d^-(w) + 1 \leqslant 0$. Let $M = \sum_{w \in Z^+} m(w)$. Summing Equation (4.26) over all $w \in Z^+$, we get

$$M = \sum_{w \in Z^+} m(w)$$

$$\leqslant \sum_{v \in L_{n_1}(u)} \sum_{w \in Z^+ \cap A^*v} \sum_{a \in E^+(w) \setminus \{a_v\}} (d^-(wa) - 1)$$

$$\leqslant \sum_{v \in L_{n_1}(u)} \sum_{a \in E^+(v) \setminus \{a_v\}} \pi(va)$$

$$\leqslant \sum_{v \in RS_{n_1}(u)} \sum_{a \in E^+(v) \setminus \{a_v\}} r_v$$

$$\leqslant s(n_1) \max_{v \in L_{n_1}(u)} r_v.$$

Therefore we can find one vertex v of G such that $r_v \geqslant M/s(n_1)$, i.e., with two outgoing edges of weight at least $M/s(n_1)$. Let e_1 and e_2 be those edges, and v_1 and v_2 be their respective destinations.

Let now $f = M/s(n_1)^2 - s(n_2)/s(n_1) - 1$. For each $i \in \{1,2\}$, let $G'_i = (V'_i, E'_i)$ be the subgraph of G containing v_i and all vertices that can be reached from v_i using only edges of weight at least f, along with those edges. We prove by contradiction that v is in G'_i. Assume the contrary. Denote by I_i the set of edges of G that end in a vertex of G'_i but do not belong to G'_i themselves, and by O_i the set of edges of G that start in a vertex of G'_i but do not belong to G'_i. Note that $I_i \cap O_i$ is not necessarily empty. By construction, $\pi(e) < f$ for all $e \in O_i$. Also, $0 < \operatorname{Card} O_i \leqslant 1 + \sum_{w \in V'_i}(d^+(w) - 1)$, with equality when G'_i is a tree, hence $\operatorname{Card} O_i \leqslant 1 + s(n_1) - (d^+(v) - 1) \leqslant s(n_1)$ (note that $d^+(v) \geqslant 2$). As v is not in G'_i, we have $e_i \in I_i$. Then

$$\Delta\pi(G'_i) = \sum_{e \in O_i} \pi(e) - \sum_{e \in I_i} \pi(e)$$
$$< s(n_1)f - \pi(e_i)$$
$$< -s(n_1) - s(n_2)$$

contradicting Equation (4.25).

Therefore G'_1 and G'_2 both contain v, and in fact $G'_1 = G'_2$. This subgraph of G contains two loops around v, starting with e_1 and e_2. These loops may not be disjoint, but we may choose them so that there is a vertex v', two paths γ_1 and γ_2 from v to v', and a (possibly empty) path γ_3 from v' to v, so that the paths are disjoint apart from their endpoints, and the loops are respectively $\gamma_1 \cup \gamma_3$ and $\gamma_2 \cup \gamma_3$.

Let x_i be the label of γ_i, so that $vx_1 \in A^*v'$, $vx_2 \in A^*v'$ and $v'x_3 \in A^*v$. Then $vx_1x_3x_2$ and $vx_2x_3x_1$ are two distinct words (since x_1 and x_2 start with different letters) of the same length in A^*v'. This is possible only if $|x_1x_3x_2| > n_1$. Hence G'_1 contains more than n_1 edges. We thus have $\pi(G) \geqslant \pi(G'_1) \geqslant n_1 f$. Replacing f with its value and $\pi(G)$ with $p(n_2 + 1) - p(n_1 + 1)$, we get the desired inequality. □

The assumption that u is recurrent or bi-infinite in Lemma 4.9.1 is needed so that every factor has at least one left extension. If u is a non-recurrent infinite word, then a workaround is to consider the bi-infinite word $v = {}^\omega zu$, where z is a letter that does not occur in $\operatorname{alph} u$. Then $p_v(n) = p_u(n) + n$. If $w \in L(u)$, then the bilateral multiplicity of w in v is the same as its bilateral multiplicity in u (if w is a prefix of u, then in v it has one extra left extension, z, and one extra element in its extension type; otherwise equality is obvious). If $w \in L(v) \setminus L(u)$, then $E^-(w) = \{z\}$ and $m(w) = 0$. So the strong bispecial factors of v are the same as those of u, with the same multiplicity. Applying Lemma 4.9.1 to v, we get:

Corollary 4.9.2 *Let* $u \in A^{\mathbb{N}}$ *be an infinite word, and* p, s, n_1, n_2 *and* Z^+ *as in Lemma 4.9.1. Then*

$$\sum_{w \in Z^+} m(w) \leqslant s(n_1)\bigg(s(n_1) + s(n_2) \\ + (p(n_2+1) - p(n_1+1) + n_2 - n_1)\frac{s(n_1)}{n_1}\bigg).$$

4.9.2 A bound on $s(n)$

The first application of Lemma 4.9.1 is the following result (Cassaigne 1996) that characterises infinite words with a linearly growing complexity.

Theorem 4.9.3 *Let* $u \in A^{\mathbb{N}}$ *be an infinite word, p its complexity function, and s its first finite difference. Then $p(n) = \mathcal{O}(n)$ if, and only if, s is bounded.*

Proof It is clear that if s is bounded, then $p(n) = \mathcal{O}(n)$. Indeed, if $s(n) \leqslant M$ for all $n \in \mathbb{N}$, then $p(n) \leqslant Mn + 1$ for all $n \in \mathbb{N}$.

Conversely, assume that $p(n) \leqslant an + 1$ for all n. Then there are infinitely many n such that $s(n) \leqslant 2a$, and if n_1 and n_2 are two large enough consecutive such n, then $n_2 < 2n_1$.

If u is eventually periodic, the result obviously holds. Otherwise, $p(n_1 + 1) \geqslant n_1 + 2$ and we have $p(n_2+1) - p(n_1+1) \leqslant p(2n_1) - n_1 - 2 \leqslant 2an_1 - n_1$.

Let $Z^+ = \{w \in L(u) \mid n_1 \leqslant |w| < n_2 \text{ and } m(w) > 0\}$ and apply Corollary 4.9.2:

$$\sum_{w \in Z^+} m(w) \leqslant s(n_1)\bigg(s(n_1) + s(n_2) \\ + (p(n_2+1) - p(n_1+1) + n_2 - n_1)\frac{s(n_1)}{n_1}\bigg)$$
$$\leqslant 2a\bigg(4a + (2an_1 - n_1 + n_2 - n_1)\frac{2a}{n_1}\bigg)$$
$$\leqslant 8a^2(1+a).$$

Then, for all $n \in [\![n_1, n_2]\!]$, summing Theorem 4.5.4.(iii) we get

$$s(n) = s(n_1) + \sum_{k=n_1}^{n-1} \sum_{w \in BS_k(u)} m(w)$$
$$\leqslant 2a + 8a^2(1+a).$$

□

For a slight generalisation of Theorem 4.9.3, see Exercise 4.28. The bound $8a^3 + 8a^2 + 2a$ in the proof is obviously not optimal, and it would be interesting to investigate what the optimal bound is.

4.9.3 The limit of $p(n)/n$

The second application is a result that was conjectured by A. Heinis, who proved it in the case $\alpha \leqslant 2$ (Heinis 2002), then later extended his method to $\alpha \leqslant 3$ (private communication).

Theorem 4.9.4 *Let $u \in A^{\mathbb{N}}$ be an infinite word. Assume that $\alpha = \lim_{n \to \infty} \frac{p_u(n)}{n}$ exists. Then $\alpha \in \mathbb{N}$.*

Proof We may assume that $\alpha > 0$, so that u is not eventually periodic. Since $p(n) = \mathcal{O}(n)$, by Theorem 4.9.3 there exists some constant S such that $s(n) \leqslant S$ for all $n \in \mathbb{N}$.

Fix $n_1 \in \mathbb{N}$. Let $Z^- = \{w \in L(u) \mid n_1 \leqslant |w| < 2n_1 \text{ and } m(w) < 0\}$ and $Z^+ = \{w \in L(u) \mid n_1 \leqslant |w| < 2n_1 \text{ and } m(w) > 0\}$, and apply Corollary 4.9.2 with $n_2 = 2n_1$, noting that $p(2n_1+1) - p(n_1+1) = \sum_{n=n_1+1}^{2n_1} s(n) \leqslant n_1 S$:

$$\sum_{w \in Z^+} m(w) \leqslant s(n_1)\bigg(s(n_1) + s(2n_1)$$
$$+ (p(2n_1+1) - p(n_1+1) + 2n_1 - n_1)\frac{s(n_1)}{n_1}\bigg)$$
$$\leqslant S(2S + (n_1 S + n_1)\frac{S}{n_1})$$
$$\leqslant S^2(S+3).$$

As $\sum_{w \in Z^- \cup Z^+} m(w) = s(2n_1) - s(n_1) \geqslant -S$, we deduce that $\sum_{w \in Z^-} |m(w)| \leqslant S^2(S+3) + S$, and finally that $\operatorname{Card}(Z^- \cup Z^+) \leqslant \sum_{w \in Z^- \cup Z^+} |m(w)| \leqslant 2S^3 + 6S^2 + S$: The number of non-neutral (*i.e.*, strong or weak) bispecial factors of length between n_1 and $2n_1 - 1$ is bounded.

Therefore there exists $\varepsilon \in \mathbb{R}_{>0}$ (*e.g.*, $\varepsilon = 1/20S^3$) such that, for every large enough n_1, there is an interval $[\![q, q'-1]\!] \subseteq [\![n_1, 2n_1-1]\!]$ with $q' \geqslant (1+\varepsilon)q$ containing no non-neutral bispecial factor. Then $s(n) = s(q)$ for $q \leqslant n \leqslant q'$, and $p(q') - p(q) = (q'-q)s(q)$. But $p(n) = \alpha n + o(n)$, so that $\alpha(q'-q) + o(q') = (q'-q)s(q)$, and $\alpha = s(q) + o(q'/(q'-q)) = s(q) + o(1)$, hence $\alpha = s(q)$ is an integer. □

4.10 Examples of complexity computations for morphic words

4.10.1 General ideas

We now turn to the practical problem of finding an expression for $p(n)$ for a given infinite word. We shall study a few examples, all of which are morphic words, using the tools developed in Section 4.5. Other examples can be found for instance in (Allouche 1994).

A consequence of Theorem 4.5.4 is that to compute the complexity function of an infinite word, it is sufficient to describe all its bispecial factors, including their length and their multiplicity. The exact method to produce this description depends on how u is defined. When u is purely morphic, generated by an endomorphism σ, it is usually possible (with some technical assumptions on σ, including a *synchronisation lemma* such as Lemmas 4.10.2, 4.10.6, and 4.10.8 below) to express the set of bispecial factors in u as a finite union of families satisfying a recurrence relation of the form $w_{j+1} = \hat{\sigma}(w_j)$, where $\hat{\sigma}$ is a map closely related to σ.

4.10.2 The characteristic word of powers of 2

Let u be defined by $u_n = b$ if $n = 2^i$ for some i, $u_n = a$ otherwise:

$$u = abbabaaabaaaaaaabaaaaaaaaaaaaaaab\cdots.$$

This word is not recurrent. Any factor that contains at least two occurrences of b occurs only once, as the distance between consecutive occurrences of b never takes the same value. It is a morphic word, see Example 1.2.24 and Exercise 4.12.

To compute the complexity, we shall count factors according to the number of b they contain. Fix $n \in \mathbb{N}$.

First case: factors that contain no b. There is one such word for each length, a^n, and it is a factor of u.

Second case: factors that contain exactly one b. There are n such words of length n (characterised by the position of the b), and all of them are factors of u.

Third case: words containing at least two occurrences of b. Such factors occur only once, so it is sufficient to count the positions at which they occur. Let w be the factor of length n starting at position j.

First subcase: $n \leqslant 2$. The only possibility is $j = 2$, $w = bb$.

Second subcase: $3 \cdot 2^{i-1} \leqslant n \leqslant 2^{i+1}$ for some $i \geqslant 1$. If $0 \leqslant j \leqslant 2^{i-1}$, then w contains a b at positions $2^{i-1} - j$ and $2^i - j$. If $2^{i-1} + 1 \leqslant j \leqslant 2^i$, then w contains a b at positions $2^i - j$ and $2^{i+1} - j$. If $j > 2^i$, then w contains at most one b, since $n < 2^{i+2} - 2^{i+1} + 1$. Altogether we have $2^i + 1$ possibilities.

Third subcase: $2^{i+1}+1 \leqslant n \leqslant 3 \cdot 2^i - 1$ for some $i \geqslant 1$. If $0 \leqslant j \leqslant 2^i$, then w contains at least two occurrences of b as above. If $2^i + 1 \leqslant j \leqslant 2^{i+2} - n$, then w contains only one b, at position $2^{i+1} - j$. If $2^{i+2} + 1 - n \leqslant j \leqslant 2^{i+1}$, then w contains a b at positions $2^{i+1} - j$ and $2^{i+2} - j$. If $j > 2^{i+1}$, then w contains at most one b, since $n < 2^{i+3} - 2^{i+2} + 1$. Altogether we have $n - 2^i + 1$ possibilities.

Summing up the different cases, we find that:

- $p(0) = 1$, $p(1) = 2$, $p(2) = 4$;
- if $3 \cdot 2^{i-1} \leqslant n \leqslant 2^{i+1}$ for some $i \geqslant 1$, then $p(n) = n + 2^i + 2$;
- if $2^{i+1} \leqslant n \leqslant 3 \cdot 2^i$ for some $i \geqslant 1$, then $p(n) = 2n - 2^i + 2$.

We could have obtained the same result with less computations by counting only right special factors. Indeed, a special factor occurs at least twice, so it cannot contain two occurrences of b. Only the first two cases remain.

First case: factors that contain no b. The factor a^n is indeed right special.

Second case: factors that contain one b. Assume that $w = a^k b a^j$ is right special. Then $wb = a^k b a^j b$ occurs, so $j = 2^i - 1$ for some i. If $i = 0$, we find that $w = b$ or $w = ab$. Otherwise, wb has to be a factor of $ba^{2^{i-1}-1}ba^{2^i-1}b$, so $k \leqslant 2^{i-1} - 1$. As $k + 1 + j = n$, this is the case if, and only if, $0 \leqslant n - 2^i \leqslant 2^{i-1} - 1$, i.e., $2^i \leqslant n \leqslant 3 \cdot 2^{i-1} - 1$.

Summing up both cases, we find that $s(n) = 1$ if $n = 0$ or $3 \cdot 2^{i-1} \leqslant n \leqslant 2^{i+1} - 1$ for some $i \geqslant 1$, $s(n) = 2$ if $n = 1$ or $2^{i+1} \leqslant n \leqslant 3 \cdot 2^i - 1$ for some $i \geqslant 1$, and $s(2) = 3$. Then Proposition 4.5.3 allows us to compute $p(n)$.

4.10.3 The Fibonacci word

Recall (see Example 1.2.22) that the Fibonacci word is $u = \varphi^\omega(a)$ where $\varphi(a) = ab$ and $\varphi(b) = a$.

Lemma 4.10.1 *The language of the Fibonacci word is closed under mirror.*

Proof Let $\tilde{\varphi}$ be the morphism defined by $\tilde{\varphi}(a) = ba$ and $\tilde{\varphi}(b) = a$. Observe first that, for any $w \in \{a, b\}^*$, the mirror image of $\varphi(w)$ is $\widetilde{\varphi(w)} = \tilde{\varphi}(\tilde{w})$. Second, for any $w \in \{a, b\}^*$, $a\tilde{\varphi}(w) = \varphi(w)a$.

We now prove by induction on i that the mirror image of $f_i = \varphi^i(a)$ is a factor of u. This holds obviously true for $i = 0$. Assume that $\tilde{f}_i \in L(u)$. Then $\tilde{f}_{i+1} = \widetilde{\varphi(f_i)} = \tilde{\varphi}(\tilde{f}_i)$. We know that $\tilde{f}_i$ is a factor of u; some extension $\tilde{f}_i x$, with $x \in \{a, b\}$, is also a factor of u. Then $\varphi(\tilde{f}_i x)$ is also a factor of u, as well as its prefix $\varphi(\tilde{f}_i)a = a\tilde{\varphi}(\tilde{f}_i) = a\tilde{f}_{i+1}$, so $\tilde{f}_{i+1}$ is a factor of u.

As any factor of u is a factor of some f_i, we conclude that $L(u)$ is closed under mirror. □

Lemma 4.10.2 *If $\varphi(w)a$ is a factor of u, then w is a factor of u.*

Proof If $w = \varepsilon$, obviously w is a factor of u. We now assume $w \neq \varepsilon$. As $u = \varphi(u)$, there exists $v \in L(u)$ such that $\varphi(w)a$ is a factor of $\varphi(v)$, and we can choose v of minimal length. Then $\varphi(w)a = t_1\varphi(v')t_2$, where $v = x_1 v' x_2$, $t_1 \in \{a, b, ab\}$ is a non-empty suffix of $\varphi(x_1)$, and $t_2 \in \{a, ab\}$ is a non-empty prefix of $\varphi(x_2)$. But $t_1 = b$ is impossible, as $\varphi(w)a$ starts with a. So t_1 is either a or ab, and in both cases $t_1 = \varphi(x_1)$. Similarly, t_2 cannot be ab, as $\varphi(w)a$ ends with a, so $t_2 = a$. We then have $\varphi(w)a = \varphi(x_1 v')a$, and as $\{\varphi(a), \varphi(b)\}$ is a suffix code, φ is injective and we conclude that $w = x_1 v'$, a factor of u. □

Proposition 4.10.3 *The left special factors in the Fibonacci word are its prefixes. Its right special factors are the mirror images of its prefixes.*

Proof If w is left special, then $\varphi(w)$ is also left special. Indeed, if aw and bw are both factors of u, then so are $ab\varphi(w)$ and $a\varphi(w)$. Iterating this, we get that all f_i are left special, and therefore all prefixes of u are left special (since a prefix of a left special factor is left special).

Conversely, assume that w is left special. Necessarily w begins with a. If w ends with b, then it is always followed by a so that $w' = wa$ is also left special; otherwise we let $w' = w$. Then w' ends with a and contains no bb, so it can be factored over $\{ab, a\}$, the last factor being a. There exists then some v such that $w' = \varphi(v)a$. By Lemma 4.10.2, this implies that v is a factor of u. Moreover, we know that w' is left special, so its extensions aw' and bw' are factors of u, and the latter is always preceded by a. We can apply Lemma 4.10.2 again to $aw' = \varphi(bv)a$ and to $abw' = \varphi(av)a$ to conclude that bv and av are factors of u, i.e., v is left special. If $v \neq \varepsilon$, we have $|v| < |\varphi(v)|$ (since v starts with a) and $|\varphi(v)| = |w'| - 1 \leq |w|$. We can then proceed by induction: assume that v is a prefix of u. Then vx is a prefix of u for some letter x, and then w is a prefix of $w' = \varphi(v)a$ which is a prefix of $\varphi(vx)$ which is a prefix of u.

Lemma 4.10.1 implies that right special factors are the mirror images of left special factors. □

Corollary 4.10.4 *The Fibonacci word is Sturmian, i.e., its complexity is $p(n) = n + 1$.*

Proof Proposition 4.10.3 implies that there is exactly one right special factor of each length. By Corollary 4.5.5, we conclude that $s(n) = 1$ so $p(n) = n + 1$. □

The main argument in the proof is Lemma 4.10.2. It allows us to 'desubstitute' φ, and to relate every special factor to a shorter one, and eventually to the empty word. A similar argument is always used when computing the complexity for a morphic word.

4.10.4 The Thue–Morse word

The Thue–Morse word is the purely morphic word

$$t = abbabaabbaababbabaababbaabbabaabbaababbaabbabaababbabaabbaab\cdots,$$

fixed point of the morphism θ with $\theta(a) = ab$ and $\theta(b) = ba$.

It is recurrent, so we may use bispecial factors and Corollary 4.5.5.

Proposition 4.10.5 *The bispecial factors in t are*

- *strong bispecial factors:* ε, $\theta^m(ab)$, $\theta^m(ba)$ *for* $m \geqslant 0$;
- *neutral bispecial factors:* a, b;
- *weak bispecial factors:* $\theta^m(aba)$, $\theta^m(bab)$ *for* $m \geqslant 0$.

The proof relies on the following lemma, that plays the same role as Lemma 4.10.2 for the Fibonacci word:

Lemma 4.10.6 *Every factor of t is of the form $w = r_1 \cdot \theta(v) \cdot r_2$ with $v \in L(t)$ and $r_i \in \{\varepsilon, a, b\}$. If $|w| \geqslant 5$, then this decomposition is unique.*

The proof of Lemma 4.10.6 is omitted here.

Proof [Proof of Proposition 4.10.5] Let w be a factor such that $|w| \geqslant 5$. According to the lemma, w can be uniquely written as $w = r_1 \cdot \theta(v) \cdot r_2$. Assume $r_2 = a$. If wa were a factor of t, the lemma could also be applied to it: $wa = r'_1 \cdot \theta(v') \cdot r'_2$. Since $\theta(v')$ cannot end with aa, the only possibility is $r'_2 = a$, and then $w = r'_1 \cdot \theta(v') \cdot \varepsilon$, contradicting uniqueness. So $r_2 = a$ implies that w is always followed by b. Similarly, if $r_2 = b$, then w is always followed by a. If w is right special, necessarily $r_2 = \varepsilon$. Then wa can be written as $wa = r_1 \cdot \theta(v) \cdot a$, hence (by the above argument) it is always followed by b, and the decomposition of wab is $wab = r_1 \cdot \theta(va) \cdot \varepsilon$. Therefore va is a factor of t. For the same reason vb is also a factor of t, so v is right special.

We have shown that right special factors of length at least 5 are all of the form $w = r_1\theta(v)$, where v itself is right special. Left special factors have a symmetrical property, and consequently bispecial factors of length at least 5 are all images under θ of a shorter bispecial factor. Moreover, the extension type of $\theta(v)$ is determined by that of v and has the same cardinality, so that $m(\theta(v)) = m(v)$.

It now suffices to list short bispecial factors to conclude: ε, ab, ba, $abba$, $baab$ are strong, a and b are neutral, and aba and bab are weak. $\square$

We deduce from Proposition 4.10.5 formulas for $s(n)$ and $p(n)$ (Brlek 1989).

Corollary 4.10.7 *For the Thue–Morse word, we have*

$$s(n) = \begin{cases} 1 & \text{if } n = 0 \\ 2 & \text{if } 0 < n \leqslant 2 \\ 4 & \text{if } 2 \cdot 2^m < n \leqslant 3 \cdot 2^m \\ 2 & \text{if } 3 \cdot 2^m < n \leqslant 4 \cdot 2^m \end{cases} \quad \text{and}$$

$$p(n) = \begin{cases} 1 & \text{if } n = 0 \\ 2 & \text{if } n = 1 \\ 4 & \text{if } n = 2 \\ 4n - 2 \cdot 2^m - 4 & \text{if } 2 \cdot 2^m < n \leqslant 3 \cdot 2^m \\ 2n + 4 \cdot 2^m - 2 & \text{if } 3 \cdot 2^m < n \leqslant 4 \cdot 2^m \end{cases},$$

for all $m \in \mathbb{N}$.

Observe that $\limsup_{n \to \infty} \frac{p(n)}{n} = \frac{10}{3}$ while $\liminf_{n \to \infty} \frac{p(n)}{n} = 3$.

For more on the complexity of the Thue–Morse word, see Exercise 4.31. See also Exercise 4.30.

4.10.5 An example with slightly higher complexity

Let $u = \sigma^\omega(a)$ with $\sigma(a) = aba$ and $\sigma(b) = bb$:

$$u = ababbabab bbbabab baba bbbbbbbbab babab abb bbb babab bababbbbbbbbb \cdots.$$

Here the key lemma is

Lemma 4.10.8 *If $w \in L(u)$, then $w = r_1 \cdot \sigma(v) \cdot r_2$ with $r_1 \in \{\varepsilon, a, b, ba\}$, $v \in L(u)$, and $r_2 \in \{\varepsilon, a, b, ab\}$. If $|w| \geqslant 4$ and $w \notin b^*$, then this decomposition is unique.*

The only bispecial factor of length less than 4 which is not a power of b is bab, and it generates a family $\sigma^m(bab)$ of weak bispecial factors, of length $2^m(3 + m/2)$. Powers of b are all bispecial, strong if the length is a power of two, neutral otherwise.

We have: $sb(n) = 1$ if $n = 2^m$, $sb(n) = 0$ otherwise; $wb(n) = 1$ if $n = 2^m(3 + m/2)$, $wb(n) = 0$ otherwise. Then $s(n) = 1 + q - r$, where $q = \lceil \log_2 n \rceil$ is the number of strong bispecial factors of length less than n, and $r = \left\lceil \frac{W(128n \log 2)}{\log 2} - 6 \right\rceil$ is the number of weak bispecial factors of length less than n, where W is the unique analytic real function on $(-1/e, +\infty)$ such that $W(x)e^{W(x)} = x$.

Then finally

$$\begin{aligned} p(n) &= 1 + \sum_{m=0}^{n-1} s(m) \\ &= 1 + n + \sum_{m=0}^{q-1} (n - 1 - 2^m) - \sum_{m=0}^{r-1} (n - 1 - 2^m(3 + m/2)) \\ &= 1 + n + (n-1)q - (2^q - 1) - (n-1)r + (2^{r+1} + 2^{r-1}r - 2). \end{aligned}$$

In this expression, dominant terms are nq and nr. Then

$$p(n) = n \frac{\log n - W(128n \log 2)}{\log 2} + \mathcal{O}(n).$$

Using the fact that $W(x) = \log x - \log \log x + \mathcal{O}\left(\frac{\log \log x}{\log x}\right)$ we obtain an equivalent of the complexity function: $p(n) \sim n \log_2 \log_2 n$.

See also Exercise 4.10 for a more general approach that includes this example.

4.11 Complexity computation for an s-adic family of words

4.11.1 An s-adic family

The method used in Section 4.10 to compute the complexity of morphic words can actually be used for other infinite words defined with morphisms.

Definition 4.11.1 Let $(A_i)_{i \geqslant 0}$ be a sequence of alphabets, $(\sigma_i)_{i \geqslant 0}$ a sequence of morphisms such that $\sigma_i \colon A_{i+1}^* \to A_i^*$, and $(a_i)_{i \geqslant 0}$ a sequence of letters with $a_i \in A_i$. Assume that the limit

$$u = \lim_{i \to \infty} (\sigma_0 \circ \sigma_1 \circ \cdots \circ \sigma_{i-1})(a_i)$$

exists and is an infinite word $u \in A_0^{\mathbb{N}}$. Then $(\sigma_i, a_i)_{i \geqslant 0}$ is called an *s-adic construction* of u.

In this section, we shall study the complexity of a particular family of infinite words with an s-adic construction. This family is used in (Bruin and Troubetzkoy 2003) for coding certain interval translation maps. Another example of s-adic construction is given in Exercise 4.32. See also Section 7.5.3 and (Durand 2000, Durand 2003) for other approaches to s-adic constructions.

Let A be the alphabet $A = \{1, 2, 3\}$. For all $h \in \mathbb{N}_{>0}$, define the endomorphism χ_h of A^* by

$$\chi_h : 1 \longmapsto 2$$
$$2 \longmapsto 31^h$$
$$3 \longmapsto 31^{h-1}.$$

Given a sequence $(h_i)_{i \geqslant 0}$ of positive integers, we consider the s-adic construction $(\chi_{h_i}, 3)$: it clearly defines an infinite word

$$u = \lim_{i \to \infty} (\chi_{h_0} \circ \chi_{h_1} \circ \cdots \circ \chi_{h_{i-1}})(3).$$

We assume that the sequence $(h_i)_{i \geqslant 0}$ satisfies the following condition:

$$\begin{cases} h_i \neq 1 \text{ for infinitely many even } i \text{ and} \\ h_i \neq 1 \text{ for infinitely many odd } i. \end{cases} \quad (4.27)$$

This condition eliminates degenerate cases, see Remark 4.11.6. We shall prove:

Theorem 4.11.2 *Let u be the infinite word defined above. Assume that Condition (4.27) holds. Then the complexity function $p(n)$ of u and its first difference $s(n)$ have the following properties:*

(i) *For all $n \geqslant 0$, $s(n) \in \{2, 3\}$, the value 2 being assumed infinitely often.*
(ii) *For all $n \geqslant 1$, $p(n) \leqslant 3n$, and $\lim_{n \to \infty} p(n) - 3n = -\infty$.*

One of the motivation for studying this family of words in (Bruin and Troubetzkoy 2003) was to discuss the unique ergodicity (see Definition 7.2.8) of the associated symbolic dynamical systems X_u, and thus of the interval translation maps that they encode. As Theorem 4.11.2 implies that u is 2-disconnectable (see Definition 7.3.6), by Theorem 7.3.7 it follows that X_u has at most two ergodic measures. This can also be obtained more directly using Theorem 1.1 from (Boshernitzan 1984).

In the case where the numbers h_i are bounded, this can be improved as the set of positive integers n such that $s(n) = 2$ becomes *logarithmically syndetic* (*i.e.*, the ratio between consecutive such n is bounded):

Proposition 4.11.3 *Let u be as in Theorem 4.11.2. If there is a constant C such that for all $i \in \mathbb{N}$, $h_i \leqslant C$, then in every interval of the form $[M, 2CM - 1]$, $M \in \mathbb{N}_{>0}$, there exists at least one n such that $s(n) = 2$.*

By Theorem 3.2 from (Boshernitzan 1984), this implies that X_u is uniquely ergodic.

4.11.2 Reformulation

For technical reasons, we need to slightly change the definition of u.

Let $\chi_{jk} = \chi_k \circ \chi_j$, so that

$$\chi_{jk}: 1 \longmapsto 31^k$$
$$2 \longmapsto 31^{k-1}2^j$$
$$3 \longmapsto 31^{k-1}2^{j-1}.$$

For $i \in \mathbb{N}_{>0}$, let $k_i = h_{2i-2}$ and $j_i = h_{2i-1}$. For any $m \in \mathbb{N}_{>0}$, define the infinite word u_m as the limit

$$u_m = \lim_{i \to \infty} \chi_{j_m k_m} \circ \chi_{j_{m+1} k_{m+1}} \circ \cdots \circ \chi_{j_{m+i} k_{m+i}}(3).$$

Clearly, the word we are interested in is $u = u_1$.

If we consider simultaneously all infinite words u_m, we have the relation

$$u_m = \chi_{j_m k_m}(u_{m+1})$$

which is formally very similar to the relation $v = \sigma(v)$ satisfied by a purely morphic word. It is now possible to adapt the method designed for purely morphic words to get a recursive description of bispecial factors of u_m.

Condition (4.27) now becomes

$$\begin{cases} j_i \neq 1 \text{ for infinitely many } i, \text{ and} \\ k_i \neq 1 \text{ for infinitely many } i. \end{cases} \quad (4.28)$$

We first observe that u_m is uniformly recurrent, and that all three letters actually occur in u_m.

Lemma 4.11.4 *For all $m \in \mathbb{N}$, u_m is uniformly recurrent.*

Proof Let w be any factor of u_m. By construction, it occurs in $\chi_{j_m k_m} \circ \chi_{j_{m+1} k_{m+1}} \circ \cdots \circ \chi_{j_{m+i-1} k_{m+i-1}}(3)$ for some $i \in \mathbb{N}$. It is clear that 3 occurs with bounded gaps in u_{m+i}, since $u_{m+i} = \chi_{j_{m+i} k_{m+i}}(u_{m+i+1})$ and 3 occurs in $\chi_{j_{m+i} k_{m+i}}(a)$ for all $a \in A$. Therefore w occurs with bounded gaps in u_m. □

Lemma 4.11.5 *Assume that Condition (4.28) holds. Then, for all $m \in \mathbb{N}$, $\mathrm{alph}\, u_m = \{1, 2, 3\}$.*

Proof Obviously, $3 \in \mathrm{alph}\, u_m$. By Condition (4.28), there exists $m' \geqslant m$ such that $k_{m'} \geqslant 2$, and then 1 occurs in $\chi_{j_{m'} k_{m'}}(3)$, therefore $1 \in \mathrm{alph}\, u_{m'}$. Similarly, there exists $m'' \geqslant m$ such that $2 \in \mathrm{alph}\, u_{m''}$. Observe that,

Factor complexity 229

for all $a \in A$ and all $i \in \mathbb{N}_{>0}$, the letter a occurs in $\chi_{j_i k_i}(a)$. Consequently, alph $u_{i+1} \subseteq$ alph u_i. Iterating, we get that alph $u_{m'} \subseteq$ alph u_m and alph $u_{m''} \subseteq$ alph u_m, hence the conclusion. □

Remark 4.11.6 If Condition (4.28) is not satisfied, for instance if $j_i = 1$ for all $i \geqslant m_0$, then it is easily seen that 2 does not occur in u_{m_0}, which is then a binary word. In fact, it is a Sturmian word, and it follows that the complexity of u is $p(n) = n + C$ for some constant C and for large enough n.

4.11.3 Synchronisation lemma

The first step is to establish a *synchronisation lemma*.

Lemma 4.11.7 *Let m be a positive integer, and w be a factor of u_m. If w contains at least one of 3, 2^{j_m}, or 1^{k_m}, i.e., if*

$$w \notin N = \{\varepsilon, 1^h, 2^i, 1^h 2^i \mid 1 \leqslant h \leqslant k_m - 1, 1 \leqslant i \leqslant j_m - 1\},$$

then w can be written in a unique way in the form $w = s\chi_{j_m k_m}(v)p$, where v is a factor of u_{m+1},

$$s \in \{\varepsilon, 1^h, 1^{k_m}, 2^i, 2^{j_m}, 1^h 2^{j_m - 1}, 1^h 2^{j_m} \mid 1 \leqslant h \leqslant k_m - 1, 1 \leqslant i \leqslant j_m - 1\},$$

and

$$p \in \{\varepsilon, 3, 31^h, 31^{k_m - 1} 2^i \mid 1 \leqslant h \leqslant k_m - 1, 1 \leqslant i \leqslant j_m - 1\},$$

with the additional condition that v does not end in 3 if $p = \varepsilon$. Moreover, $|v| < |w|$.

Proof As $21 \notin L(u_m)$, the set of factors avoiding 3, 2^{j_m} and 1^{k_m} is indeed N. We observe that 3 occurs only as the initial letter of images of letters under $\chi_{j_m k_m}$, and that 2^{j_m} and 1^{k_m} occur only as suffixes of such images. Any word w containing them can therefore be written in the form $w = s\chi_{j_m k_m}(v)p$, with s a proper suffix of some $\chi_{j_m k_m}(x)$, $x \in A$, p a proper prefix of some $\chi_{j_m k_m}(y)$, $y \in A$ and xvy a factor of u_{m+1}. Since 1^{k_m}, 2^{j_m} and $12^{j_m - 1}3$ occur only in respectively $\chi_{j_m k_m}(1)$, $\chi_{j_m k_m}(2)$ and $\chi_{j_m k_m}(3)3$, they act as recognising words and the decomposition (s, v, p) is unique except in one case: when w ends in $31^{k_m} 2^{j_m - 1}$, this can be the image of 3 or a prefix of the image of 2. To ensure uniqueness, we have to forbid the first situation.

If $k_m \geqslant 2$ or $j_m \geqslant 2$, then $|\chi_{j_m k_m}(v)| > |v|$ as soon as v is non-empty, hence in any case $|v| < |w|$. If $k_m = j_m = 1$, then $|\chi_{j_m k_m}(v)| > |v|$ except when $v = 3^n$, but then $p \neq \varepsilon$ by the additional condition, hence also $|v| < |w|$. □

4.11.4 Bispecial factors

The second step is to establish the list of short (non-synchronised) bispecial factors in u_m, as well as their extension type $E(w)$ and their multiplicity $m(w)$. For this we have to discuss whether j_m or k_m is equal to 1 or not, as this slightly changes the extension possibilities (for instance, $E(\varepsilon)$ is just the set of factors of length 2 in u_m, and new factors occur when blocks 1^{k_m-1} and 2^{j_m-1} vanish).

For simplicity, we shall write elements of $E(w)$ as words of length 2 instead of as pairs of letters.

Lemma 4.11.8 *In u_m, the bispecial factors which do not contain any of the synchronising words 3, 2^{j_m} or 1^{k_m} are as follows*

- *If $j_m \geqslant 2$ and $k_m \geqslant 2$:*
  - ε, *of extension type* $E(\varepsilon) = \{11, 12, 13, 22, 23, 31\}$ *and multiplicity* $m(\varepsilon) = +1$;
  - 1^h, *for* $1 \leqslant h \leqslant k_m - 2$, *of extension type* $E(1^h) = \{11, 12, 13, 31\}$ *and multiplicity* $m(1^h) = 0$;
  - 1^{k_m-1}, *of extension type* $E(1^{k_m-1}) = \{13, 31, 32\}$ *and multiplicity* $m(1^{k_m-1}) = -1$;
  - 2^i, *for* $1 \leqslant i \leqslant j_m - 2$, *of extension type* $E(2^i) = \{12, 22, 23\}$ *and multiplicity* $m(2^i) = 0$;
  - 2^{j_m-1}, *of extension type* $E(2^{j_m-1}) = \{12, 13, 23\}$ *and multiplicity* $m(2^{j_m-1}) = 0$.

- *If $j_m \geqslant 2$ and $k_m = 1$:*
  - ε, *of extension type* $E(\varepsilon) = \{13, 22, 23, 31, 32\}$ *and multiplicity* $m(\varepsilon) = 0$;
  - 2^i, *for* $1 \leqslant i \leqslant j_m - 2$, *of extension type* $E(2^i) = \{22, 23, 32\}$ *and multiplicity* $m(2^i) = 0$;
  - 2^{j_m-1}, *of extension type* $E(2^{j_m-1}) = \{23, 32, 33\}$ *and multiplicity* $m(2^{j_m-1}) = 0$.

- *If $j_m = 1$ and $k_m \geqslant 2$:*
  - ε, *of extension type* $E(\varepsilon) = \{11, 12, 13, 23, 31\}$ *and multiplicity* $m(\varepsilon) = 0$;
  - 1^h, *for* $1 \leqslant h \leqslant k_m - 2$, *of extension type* $E(1^h) = \{11, 12, 13, 31\}$ *and multiplicity* $m(1^h) = 0$;
  - 1^{k_m-1}, *of extension type* $E(1^{k_m-1}) = \{13, 31, 32, 33\}$ *and multiplicity* $m(1^{k_m-1}) = 0$.

- *If $j_m = 1$ and $k_m = 1$:*

– ε, of extension type $E(\varepsilon) = \{13, 23, 31, 32, 33\}$ and multiplicity $m(\varepsilon) = 0$.

Proof This follows from inspection of all elements in N, as defined in Lemma 4.11.7, in each of the four cases. Lemma 4.11.5 ensures that all elements of N actually occur in u_m. □

The third step is to use the synchronisation lemma to express synchronised bispecial factors in terms of shorter bispecial factors.

Lemma 4.11.9 *Let w be a bispecial factor of u_m that contains at least one of the synchronising words 3, 2^{j_m} or 1^{k_m}. Then w can be written in a unique way in the form $w = s\chi_{j_m k_m}(v)p$, where v is a bispecial factor of u_{m+1}, and s and p are as in Lemma 4.11.7.*

Proof Lemma 4.11.7 provides the unique decomposition $w = s\chi_{j_m k_m}(v)p$. If w extends as xwy, with $(x,y) \in A \times A$, then $xs\chi_{j_m k_m}(v)py$ must occur in u_m, which implies (by uniqueness in Lemma 4.11.7) that xs is a suffix of $\chi_{j_m k_m}(x')$ for some letter $x' \in A$ and py is a prefix of $\chi_{j_m k_m}(y')3$ for some letter $y' \in A$, where $x'vy'$ is a factor of u_{m+1} that extends v.

Let us first discuss the right extensions of w (observe that p cannot be shorter than 31^{k_m-1}, or it would have only one right extension); beware that here $E^+(w)$ denotes the right extensions of w in u_m, whereas $E^+(v)$ denotes the right extensions of v in u_{m+1}:

- if $E^+(w) = \{1, 2, 3\}$, then p can only be 31^{k_m-1}, j_m must be 1, and $E^+(v) = \{1, 2, 3\}$;
- if $E^+(w) = \{1, 2\}$, then p can only be 31^{k_m-1}, and $E^+(v)$ contains 1 and at least one of 2 or 3 (necessarily 2 if $j_m = 1$);
- if $E^+(w) = \{1, 3\}$, then p can only be 31^{k_m-1}, j_m must be 1, and $E^+(v)$ contains 1 and 3;
- if $E^+(w) = \{2, 3\}$, then p can only be $31^{k_m-1}2^{j_m-1}$, and $E^+(v)$ contains 2 and 3.

In all the above cases, v is right special. Consider now left extensions:

- if $E^-(w) = \{1, 2, 3\}$, then s can only be empty, j_m and k_m must be 1, and $E^-(v) = \{1, 2, 3\}$;
- if $E^-(w) = \{1, 2\}$, then either s is empty, and $E^-(v)$ contains 1 and at least one of 2 or 3, (necessarily 2 if $j_m = 1$), or $s = 2^{j_m-1}$, $k_m \geqslant 2$, and $E^-(v)$ contains 2 and 3;
- if $E^-(w) = \{1, 3\}$, then s can only be 1^{k_m-1}, j_m must be 1, and $E^-(v)$ contains 1 and 3;

- if $E^-(w) = \{2,3\}$, then s can only be 2^{j_m-1}, k_m must be 1, and $E^-(v)$ contains 2 and 3.

Again, in all the above cases, v is left special. Therefore v is always bispecial. □

Lemma 4.11.9 provides a way to trace any bispecial factor of u_1 back to a non-synchronised bispecial factor of some u_m. Conversely, the bispecial factors in u_m that reduce to a given bispecial factor v in u_{m+1} can be explicitly determined, as well as their extension type knowing the extension type of v:

Lemma 4.11.10 *Let v be a bispecial factor of u_{m+1}. The word $w = s\chi_{j_m k_m}(v)p$, where s and p satisfy the conditions in Lemma 4.11.7, is a bispecial factor of u_m if, and only if, all the following conditions hold (when they apply):*

- *if $E^+(v) = \{1,2,3\}$ and $j_m = 1$, then $p = 31^{k_m-1}$ and $E^+(w) = \{1,2,3\}$;*
- *if $E^+(v) = \{1,2,3\}$ and $j_m \geq 2$, then either $p = 31^{k_m-1}$ and $E^+(w) = \{1,2\}$, or $p = 31^{k_m-1}2^{j_m-1}$ and $E^+(w) = \{2,3\}$;*
- *if $E^+(v) = \{1,2\}$, then $p = 31^{k_m-1}$ and $E^+(w) = \{1,2\}$;*
- *if $E^+(v) = \{1,3\}$ and $j_m = 1$, then $p = 31^{k_m-1}$ and $E^+(w) = \{1,3\}$;*
- *if $E^+(v) = \{1,3\}$ and $j_m \geq 2$, then $p = 31^{k_m-1}$ and $E^+(w) = \{1,2\}$;*
- *if $E^+(v) = \{2,3\}$, then $p = 31^{k_m-1}2^{j_m-1}$ and $E^+(w) = \{2,3\}$;*
- *if $E^-(v) = \{1,2,3\}$ and $j_m = k_m = 1$, then $s = \varepsilon$ and $E^-(w) = \{1,2,3\}$;*
- *if $E^-(v) = \{1,2,3\}$, $j_m = 1$, and $k_m \geq 2$, then either $s = \varepsilon$ and $E^-(w) = \{1,2\}$, or $s = 1^{k_m-1}$ and $E^-(w) = \{1,3\}$;*
- *if $E^-(v) = \{1,2,3\}$, $j_m \geq 2$, and $k_m = 1$, then either $s = \varepsilon$ and $E^-(w) = \{1,2\}$, or $s = 2^{j_m-1}$ and $E^-(w) = \{2,3\}$;*
- *if $E^-(v) = \{1,2,3\}$, $j_m \geq 2$, and $k_m \geq 2$, then either $s = \varepsilon$ and $E^-(w) = \{1,2\}$, or $s = 2^{j_m-1}$ and $E^-(w) = \{1,2\}$;*
- *if $E^-(v) = \{1,2\}$, then $s = \varepsilon$ and $E^-(w) = \{1,2\}$;*
- *if $E^-(v) = \{1,3\}$ and $j_m = 1$, then $s = 1^{k_m-1}$ and $E^-(w) = \{1,3\}$;*
- *if $E^-(v) = \{1,3\}$ and $j_m \geq 2$, then $s = \varepsilon$ and $E^-(w) = \{1,2\}$;*
- *if $E^-(v) = \{2,3\}$ and $k_m = 1$, then $s = 2^{j_m-1}$ and $E^-(w) = \{2,3\}$;*
- *if $E^-(v) = \{2,3\}$ and $k_m \geq 2$, then $s = 2^{j_m-1}$ and $E^-(w) = \{1,2\}$.*

Proof For each of the right or left extension types, one has to compute the images under $\chi_{j_m k_m}$ of each of the possible extensions of v, then compute their longest common prefix or suffix (respectively). When v has only two possible extensions, this prefix or suffix is the only possibility for this side of w; when v has three possible extensions, there is most of the time a second possibility, the longest prefix or suffix of two among the three considered words: if this is the case then both possibilities for w have only two extensions, otherwise the only possibility has three extensions. □

Lemma 4.11.10 deals with each side of the bispecial factors independently. To compute extension types and multiplicities, one has to consider them simultaneously. The number of cases becomes so large that we will not write all of them: the result is best expressed by a picture. For each possible extension type of v, as well as any of the four cases for j_m and k_m (each can be either 1 or larger), there are at most four possibilities for w, each having a determined extension type. We represent this as a finite automaton on the alphabet $B = \{1,2\} \times \{1,2\} \times \{0,1\} \times \{0,1\}$, where the first two components indicate whether j_m and k_m (respectively) are 1 or at least 2, and the last two indicate whether the shorter or longer possibility for w should be chosen.

Define the mappings χ_{jk}^{LR} in A^*, for $j \geq 1$, $k \geq 1$, $L \in \{0,1\}$ and $R \in \{0,1\}$, as follows:

$$\begin{aligned} \chi_{jk}^{00}(v) &= \chi_{jk}(v) 31^{k-1} \\ \chi_{jk}^{01}(v) &= \chi_{jk}(v) 31^{k-1} 2^{j-1} \\ \chi_{1k}^{1R}(v) &= 1^{k-1} \chi_{1k}(v) 31^{k-1} \\ \chi_{jk}^{10}(v) &= 2^{j-1} \chi_{jk}(v) 31^{k-1} \text{ if } j \geq 2 \\ \chi_{jk}^{11}(v) &= 2^{j-1} \chi_{jk}(v) 31^{k-1} 2^{j-1} \text{ if } j \geq 2. \end{aligned}$$

Note that these mappings are not morphisms, that the value of R is unimportant when $j = 1$, and that the value of L is unimportant when $j = k = 1$.

Lemma 4.11.11 *Let $I = \{\varepsilon, 1, 1', 2, 2'\}$ and F be the set of bispecial extension types on A (i.e., subsets $E \subseteq A \times A$ such that the projections $\pi_1(E)$ and $\pi_2(E)$ on each component have at least two elements). There exists a finite automaton $\mathcal{A}$, with states $I \cup F$, where the five states in I are initial and the states in F are final, and alphabet B, that computes the bispecial factors of u_1 in the following sense.*

Let q be a positive integer. Choose symbols L_i and R_i in $\{0,1\}$ for $1 \leq i \leq q-1$, and a non-synchronised bispecial factor w_q in u_q (see Lemma 4.11.8). Let $s \in I$ be $s = \varepsilon$ if $w_q = \varepsilon$, $s = 1$ if $w_q = 1^{k_q - 1}$, $s = 1'$ if $w_q = 1^h$, $1 \leq h \leq k_q - 2$, $s = 2$ if $w_q = 2^{j_q - 1}$, and $s = 2'$ if $w_q = 2^i$, $1 \leq i \leq j_q - 2$. Let $L_q = 0$, $R_q = 0$, $J_i = \max(j_i, 2)$, $K_i = \max(k_i, 2)$, and $b_i = (J_i, K_i, L_i, R_i) \in B$ for $1 \leq i \leq q$. If there is a path in the automaton $\mathcal{A}$, starting in s, labelled with the word $b_q b_{q-1} \cdots b_1 \in B^q$ (note that letters are read from b_q to b_1), and ending in some state labelled with an extension type E, then the word

$$w = \chi_{j_1 k_1}^{L_1 R_1} \circ \chi_{j_2 k_2}^{L_2 R_2} \circ \cdots \circ \chi_{j_{q-1} k_{q-1}}^{L_{q-1} R_{q-1}}(w_q)$$

is a bispecial factor of u_1, with extension type $E(w) = E$. Conversely, all bispecial factors of u_1 are obtained by this procedure.

Proof Let w be a bispecial factor in u_1. Iterating Lemma 4.11.9, we obtain a finite sequence $(w_1, \ldots, w_q)$, where $w_1 = w$, w_i is a bispecial factor of u_i for all $i \in [\![1, q]\!]$, and $w_q \in N$. (Note that the sequence must be finite because Lemma 4.11.7 ensures that the length decreases.) By Lemma 4.11.8, w_q is of one of the five types corresponding to initial states in $\mathcal{A}$, and transitions from I to F are set up in such a way that its extension type is given by the second state in the path. Then Lemma 4.11.10 implies that for each $i \in [\![1, q-1]\!]$, $w_i = \chi_{j_i k_i}^{L_i R_i}(w_{i+1})$ for some choice of L_i and R_i. The extension type of w_i is determined from b_i and $E(w_{i+1})$ according to Lemma 4.11.10, and this defines the other transitions of $\mathcal{A}$. By construction, such a path exists if and only if w is a bispecial factor. □

The automaton $\mathcal{A}$ can be considerably simplified. First, among the 478 states in F, only 23 are reachable from I: they correspond to the extension types of bispecial factors that actually occur in u_1 for some choice of $((j_i, k_i))_{i \geqslant 1}$. As far as complexity is concerned, only bispecial factors with a non-zero bilateral multiplicity are useful. We can therefore also remove from $\mathcal{A}$ all states labelled with extension types that have multiplicity 0 (recall that the bilateral multiplicity can be computed from the extension type alone) and that lead only to other states with multiplicity 0. In particular, all ordinary bispecial factors can be safely ignored as they lead only to other ordinary bispecial factors. (Note that, whereas on a binary alphabet all neutral bispecial factors are ordinary and can be ignored, here some neutral bispecial factors have to be kept as they may lead to bispecial factors of non-zero multiplicity.)

The simplified automaton $\mathcal{A}'$ is represented in Figure 4.1. To represent extension types E graphically, it is convenient to draw a bipartite graph with two copies of A as vertices, and an edge from a to b if, and only if, $(a, b) \in E$. We have also indicated the corresponding bilateral multiplicity.

The automata $\mathcal{A}$ and $\mathcal{A}'$ are deterministic automata on the alphabet B (when one initial state is selected). They can also be viewed as non-deterministic transducers from $(\{1, 2\} \times \{1, 2\})^*$ to $(\{0, 1\} \times \{0, 1\})^*$, which, given $((J_m, K_m), \ldots, (J_1, K_1))$ as input, compute all $((L_{m-1}, R_{m-1}), \ldots, (L_1, R_1))$ that lead to a bispecial factor (of non-zero multiplicity in the case of $\mathcal{A}'$).

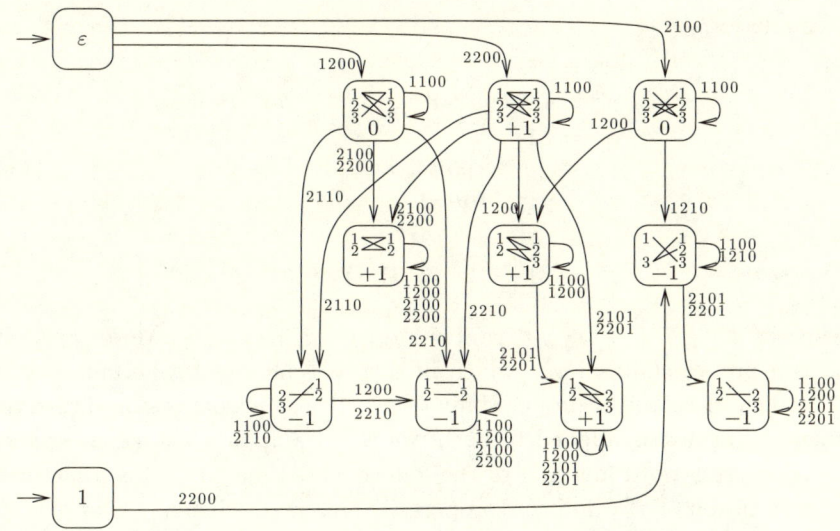

Fig. 4.1 The simplified automaton $\mathcal{A}'$.

4.11.5 The case where $j_i, k_i \geq 2$

Let us first restrict to the case where j_i and k_i are never equal to 1. Then we see in Figure 4.1 that there are only four possible paths in $\mathcal{A}'$, and consequently four families of non-neutral bispecial factors.

Lemma 4.11.12 *Assume that $j_i \geq 2$ and $k_i \geq 2$ for all i. Then the non-neutral bispecial factors of u_1 are the following*

- $x_m^+ = \chi_{j_1 k_1}^{00} \circ \chi_{j_2 k_2}^{00} \circ \cdots \circ \chi_{j_m-1 k_m-1}^{00}(\varepsilon)$, *of multiplicity* $+1$, *for all* $m \geq 1$;
- $x_m^- = \chi_{j_1 k_1}^{00} \circ \chi_{j_2 k_2}^{00} \circ \cdots \circ \chi_{j_{m-2} k_{m-2}}^{00} \circ \chi_{j_{m-1} k_{m-1}}^{10}(\varepsilon)$, *of multiplicity* -1, *for all* $m \geq 2$;
- $y_m^+ = \chi_{j_1 k_1}^{01} \circ \chi_{j_2 k_2}^{01} \circ \cdots \circ \chi_{j_{m-1} k_{m-1}}^{01}(\varepsilon)$, *of multiplicity* $+1$, *for all* $m \geq 2$;
- $y_m^- = \chi_{j_1 k_1}^{01} \circ \chi_{j_2 k_2}^{01} \circ \cdots \circ \chi_{j_{m-1} k_{m-1}}^{01}(1^{k_m-1})$, *of multiplicity* -1, *for all* $m \geq 1$.

Moreover, one has $|x_1^+| < |y_1^-| < |x_2^+|$ *and* $|x_m^+| < |x_m^-| \leq |y_m^+| < |y_m^-| < |x_{m+1}^+|$ *for all* $m \geq 2$.

Proof The expression of the bispecial factors is a direct consequence of Lemma 4.11.11, considering only edges labelled 2200, 2201, 2210 or 2211 in the automaton. We define the families (x_m^-) and (y_m^+) for $m \geq 2$ only so that ε is counted only once.

One has
$$x_1^+ = \varepsilon$$
$$y_1^- = 1^{k_1-1}$$
$$x_2^+ = 31^{k_1-1}$$
$$x_2^- = 2^{j_1-1}31^{k_1-1}$$
$$y_2^+ = 31^{k_1-1}2^{j_1-1}$$
$$y_2^- = (31^{k_1})^{k_2-1}31^{k_1-1}2^{j_1-1}$$
$$x_3^+ = 31^{k_1-1}2^{j_1-1}(31^{k_1})^{k_2-1}31^{k_1-1}$$

hence $|x_1^+| < |y_1^-| < |x_2^+| < |x_2^-| = |y_2^+| < |y_2^-| < |x_3^+|$. More precisely, x_2^+ is a proper suffix of x_2^-, which in turn is a factor of a permutation of y_2^+, which in turn is a proper suffix of y_2^-. These properties are preserved when χ_{jk}^{00} or χ_{jk}^{01} is applied to both words, as well as when χ_{jk}^{00} is applied to the shorter word and χ_{jk}^{01} to the longer one. Finally, $y_1^- 3$ is a factor of a permutation of x_2^+, and this property is preserved when χ_{jk}^{01} is applied to the shorter word and χ_{jk}^{00} to the longer one (observe that $\chi_{jk}^{01}(w)3$ is a prefix of $\chi_{jk}^{00}(w3)$). □

We can now prove a particular case of Theorem 4.11.2:

Proposition 4.11.13 *Assume that $j_i \geqslant 2$ and $k_i \geqslant 2$ for all i. Then for all $n \geqslant 1$, $s(n-1) \in \{2,3\}$ and $p(n) \leqslant 3n$.*

Proof The first part is a consequence of Lemma 4.11.12, as, by summing relation (iii) of Theorem 4.5.4, one has $s(n) = s(0) + \sum_{|w|<n} m(w) \in \{2,3\}$.

The second part is obtained by summing again the first one, as in Proposition 4.5.3, starting from $p(1) = 3$. □

4.11.6 The general case

In the general case, the structure of bispecial factors is much more complicated, as all extension types listed in Figure 4.1 may occur. For any positive integers q and m, let us define *generation q* of bispecial factors of u_m as the set $G_{m,q}$ of bispecial factors obtained from this q in Lemma 4.11.11 applied to u_m. Non-synchronised bispecial factors of u_m form generation 1. Let $G'_{m,q}$ be the set of bispecial factors in $G_{m,q}$ which correspond to paths in automaton $\mathcal{A}'$ from Figure 4.1; this includes all non-neutral bispecial factors. Let also

$$g_{m,q} = \chi_{j_m k_m}^{00} \circ \chi_{j_{m+1} k_{m+1}}^{00} \circ \cdots \circ \chi_{j_{m+q-2} k_{m+q-2}}^{00}(\varepsilon),$$

Factor complexity

so that $g_{m,q} = \chi^{00}_{j_m k_m}(g_{m+1,q-1})$. Note that

$$g_{m,q}1 = \chi_{j_m k_m}(g_{m+1,q-1})31^{k_m-1}1 = \chi_{j_m k_m}(g_{m+1,q-1}1),$$

so $g_{m,q}$ can be alternatively defined by $g_{m,q}1 = \chi_{j_m k_m} \circ \chi_{j_{m+1} k_{m+1}} \circ \cdots \circ \chi_{j_{m+q-2} k_{m+q-2}}(1)$.

A preorder on A^* is defined as follows: $v \leqslant w$ if for every letter $a \in A$, w contains at least as many occurrences of a as v. Obviously, $v \leqslant w$ implies $|v| \leqslant |w|$, but the converse is not true. This preorder is compatible with concatenation as well as with the application of a morphism, in the sense that if $v \leqslant w$ and f is a morphism, then $f(v) \leqslant f(w)$ (whereas the order on lengths may not be preserved).

Lemma 4.11.14 *For any bispecial factor $w \in G'_{m,q}$, one has $|g_{m,q}| \leqslant |w| < |g_{m,q+1}|$.*

Proof We shall prove the stronger inequalities $g_{m,q} \leqslant w$ and $3w \leqslant g_{m,q+1}$. Write

$$w = \chi^{L_m R_m}_{j_m k_m} \circ \chi^{L_{m+1} R_{m+1}}_{j_{m+1} k_{m+1}} \circ \cdots \circ \chi^{L_{m+q-2} R_{m+q-2}}_{j_{m+q-2} k_{m+q-2}}(w_{m+q-1})$$

as in Lemma 4.11.11 (with indices shifted by $m-1$ since we consider the infinite word u_m instead of u_1). Here w_{m+q-1} is a non-synchronised bispecial factor in u_{m+q-1} corresponding to one of the two initial states in $\mathcal{A}'$, i.e., $w_{m+q-1} \in \{\varepsilon, 1^{k_{m+q-1}-1}\}$. Define words w_l for $m \leqslant l < m+q-1$ by $w_l = \chi^{L_l R_l}_{j_l k_l}(w_{l+1})$, so that w_l is a bispecial factor in u_l and $w_m = w$.

The first inequality is easily proved by induction, starting from $g_{m+q-1,1} = \varepsilon \leqslant w_{m+q-1}$, and observing that for any choice of j, k, L, R and any word v, by definition $\chi^{00}_{jk}(v)$ is a factor of $\chi^{LR}_{jk}(v)$.

For the second inequality, we also proceed by induction, starting from

$$3w_{m+q-1} \leqslant 31^{k_{m+q-1}-1} = g_{m+q-1,2}.$$

Assume that $3w_{l+1} \leqslant g_{l+1,m+q-l}$. Then

$$\chi^{00}_{j_l k_l}(3w_{l+1}) \leqslant \chi^{00}_{j_l k_l}(g_{l+1,m+q-l}),$$

i.e.,

$$31^{k_l-1}2^{j_l-1}\chi^{00}_{j_l k_l}(w_{l+1}) \leqslant g_{l,m+q+1-l}.$$

Note that the case where $(L_l, R_l) = (1,1)$ does not occur in $\mathcal{A}'$, so that

$$3w_l = 3\chi^{L_l R_l}_{j_l k_l}(w_{l+1})$$
$$\leqslant 31^{k_l-1}2^{j_l-1}\chi^{00}_{j_l k_l}(w_{l+1})$$
$$\leqslant g_{l,m+q+1-l},$$

concluding the induction step. □

Lemma 4.11.15 *The set $G'_{m,q}$ has at most four elements. More precisely, the following cases are possible:*

(i) $G'_{m,q} = \{x^0_{m,q}\}$ *with* $m(x^0_{m,q}) = 0$;
(ii) $G'_{m,q} = \{x^+_{m,q}, y^-_{m,q}\}$ *with* $m(x^+_{m,q}) = +1$, $m(y^-_{m,q}) = -1$, *and* $x^+_{m,q} \leqslant ay^-_{m,q}$ *for* $a = 1$ *or* $a = 2$;
(iii) $G'_{m,q} = \{x^+_{m,q}, x^-_{m,q}, y^+_{m,q}, y^-_{m,q}\}$ *with* $m(x^+_{m,q}) = +1$, $m(x^-_{m,q}) = -1$, $m(y^+_{m,q}) = +1$, $m(y^-_{m,q}) = -1$, *and* $x^+_{m,q} \leqslant 2x^-_{m,q} \leqslant 2y^+_{m,q} \leqslant 12y^-_{m,q}$.

Proof This is proved by establishing the list of sets of states in $\mathcal{A}'$ that can be reached for a fixed sequence (J_i, K_i) and all possible (L_i, R_i) (or, equivalently, by determinizing $\mathcal{A}'$ viewed as a transducer): these sets have at most four elements. Then inequalities are established by induction on q. □

We have now enough elements to prove Theorem 4.11.2.

Proof [Proof of Theorem 4.11.2] By Lemma 4.11.4, u is recurrent and bispecial factors are sufficient to compute complexity. By Lemma 4.11.14 all bispecial factors in $G'_{1,q}$ lie in the interval $I_q = [\![|g_{1,q}|, |g_{1,q+1}| - 1]\!]$. Lemma 4.11.15 gives the function $b(n)$ on this interval: in case (i) it is constantly 0; in case (ii), as well as in case (iii) when $|x^-_{1,q}| = |y^+_{1,q}|$, the only non-zero values are $b(|x^+_{1,q}|) = +1$ and $b(|y^-_{1,q}|) = -1$; in case (iii) when $|x^-_{1,q}| < |y^+_{1,q}|$, additionally $b(|x^-_{1,q}|) = -1$ and $b(|y^+_{1,q}|) = +1$. In all cases

$$\sum_{|g_{1,q}| \leqslant n < |g_{1,q+1}|} b(n) = 0,$$

which implies that $s(|g_{1,q+1}|) = s(|g_{1,q}|)$, and then by induction this is equal to $s(|g_{1,0}|) = s(0) = p(1) - p(0) = 2$. Then, inside I_q, the values $+1$ and -1 alternate, so that $s(n) \in \{2, 3\}$, and (i) is proved.

Property (ii) is obtained by just summing property (i), according to Proposition 4.5.3, starting from $p(1) = 3$. □

We can then turn to Proposition 4.11.3.

Proof [Proof of Proposition 4.11.3] We have seen in the proof of Theorem 4.11.2 that $s(|g_{1,q}|) = 2$ for all q. It remains to show that at least one $|g_{1,q}|$ occurs in every interval $[\![M, 2CM - 1]\!]$. Recall that

$$g_{1,q} 1 = \chi_{j_1 k_1} \circ \chi_{j_2 k_2} \circ \cdots \circ \chi_{j_{q-1} k_{q-1}}(1),$$

and consequently, for $q \geqslant 2$,

$$g_{1,q}1 = \psi(31^{k_q-1})$$

and

$$g_{1,q+1}1 = \psi(31^{k_q-1-1}2^{j_q-1-1}(31^{k_q-1})^{k_q}),$$

where

$$\psi = \chi_{j_1 k_1} \circ \chi_{j_2 k_2} \circ \cdots \circ \chi_{j_{q-2} k_{q-2}}.$$

Moreover, one can check (by induction on q) that $|\psi(2)| < |\psi(31)|$. Consequently

$$\begin{aligned}|g_{1,q+1}1| &\leqslant |\psi(31^{k_q-1-1}(31)^{j_q-1-1}(31^{k_q-1})^{k_q})| \\ &\leqslant (1+(j_{q-1}-1)+k_q)|\psi(31^{k_q-1})| \\ &\leqslant 2C|g_{1,q}1|\end{aligned}$$

which ensures that at least one $|g_{1,q}|$ occurs in every interval $[\![M, 2CM-1]\!]$, as $|g_{1,2}1| = k_1 \leqslant 2C-1$. □

4.12 Exercises and open problems

We have tried to indicate the difficulty of exercises with stars, although this is necessarily subjective. More difficult exercises are marked (*) or (**), while (***) indicates problems that are (as far as we know) open. Hints for some of the exercises are given at the end of the list.

Section 4.2

Exercise 4.1 Let $u = z^\omega$ with z primitive. By Proposition 4.2.2, $p(n) = |z|$ for $n \geqslant |z|$. Show that $p(n) = |z|$ already for $n = |z|-1$. Give an example with $|z|$ arbitrarily large for which $p(|z|-2) \neq |z|$, and (*) another one for which $p(n) = |z|$ holds for all $n \geqslant \log_2 |z|$.

Exercise 4.2 Let L be the regular language $L = \{a, ba\}^*$. Define an infinite word u by concatenating its elements in any order. Show that the factor complexity of u is independent of the order of enumeration of L, and compute it. (*) Does this independence property hold for arbitrary infinite regular languages ?

Section 4.3

Exercise 4.3 Let $\ell(n)$ be the maximum difference between the positions of two consecutive occurrences of the same factor of length n in u (assuming this difference is bounded). Show that, if u is not eventually periodic, then $\ell(n) \geqslant n+1$ for all n.

Exercise 4.4 Let u be a Sturmian infinite word (so that $p(n) = n+1$). Assume that aa occurs in u, but not aaa. Define a new word v by doubling every letter of u; for instance, starting with the Fibonacci infinite word we get

$$v = aabbaaaabbaabbaaaabbaaaabbaabbaaaabbaabb \cdots .$$

Compute the complexity of v. (*) Generalise to all Sturmian words.

Section 4.4

Exercise 4.5 We know that a function equivalent to $2^n/n$ cannot be the complexity function of a binary word. Can it be the complexity function of an infinite word on a larger alphabet?

Exercise 4.6 (*) Let u be an infinite word with entropy zero. Show that $L(u)$ is a regular language if, and only if, u is eventually periodic.

Section 4.5

Exercise 4.7 Let $\ell(n)$ be defined as in Exercise 4.3. Show that if there is no bispecial factor of length n, then $\ell(n+1) = \ell(n+2)$.

Exercise 4.8 Show that $L(G_n) \cap A^{n+1} = L_{n+1}(u)$, and give an example where $L(G_n) \cap A^{n+2} \neq L_{n+2}(u)$.

Exercise 4.9 Show that $L(G_n) = L(G_{n+1})$ if, and only if, every $w \in BS'_n(u)$ has bilateral multiplicity $m(w) = (d^-(w) - 1)(d^+(w) - 1)$.

Exercise 4.10 Let (e_k), (u_k), and u be as in Example 4.5.14.

(i) Prove that the bispecial factors of u are 0^n and $0^{e_k} u_k 0^{e_k}$, and compute their multiplicity. Deduce that

$$s(n) = 1 + \text{Card}\{k \in \mathbb{N} \mid e_k < n\} - \text{Card}\{k \in \mathbb{N} \mid 2e_k + |u_k| < n\}.$$

(ii) If $e_k = \Theta(\beta^k)$, or $e_k \sim Ck^r$, find the order of growth of $p(n)$.

(iii) (**) Let $f\colon \mathbb{R}_{\geq 0} \to \mathbb{R}$ be a twice derivable real function such that $0 \leq f''(x) \leq 1$ for x large enough and $\lim_{x \to +\infty} f'(x) = +\infty$. Construct a binary infinite word such that $p(n) \sim f(n)$ as n tends to ∞.

Section 4.6

Exercise 4.11 (*) Let $A = \{a, b, c\}$ and $\rho\colon A^* \to A^*$ be the endomorphism defined by $\rho(a) = abc$, $\rho(b) = bac$, $\rho(c) = \varepsilon$. Prove that there is no non-erasing endomorphism of A^* other than the identity fixing $\rho^\omega(a)$. Find an alphabet B, a uniform morphism $\sigma\colon B^* \to B^*$, and a letter-to-letter morphism $\tau\colon B^* \to A^*$ such that $\tau(\sigma^\omega(b)) = \rho^\omega(a)$ for some $b \in B$. (Note that $\rho^\omega(a)$ is therefore automatic.)

Exercise 4.12 The characteristic word of powers of two, defined by $u_n = b$ if $n = 2^i$ for some i, $u_n = a$ otherwise:

$$u = abbabaaabaaaaaaabaaaaaaaaaaaaaaab\cdots$$

is morphic (see Example 1.2.24). Prove that it is not purely morphic.

Exercise 4.13 Let $A = \{a, b, c\}$ and $\sigma_i\colon A^* \to A^*$, for $i \in \mathbb{N}$, be the endomorphism defined by $\sigma_i(a) = b^i c$, $\sigma_i(b) = b$, $\sigma_i(c) = ca$.

(i) Prove that, although the morphism σ_i is not prolongable on the letter a, the sequence of words $(\sigma_i^n(a))_{n \geq 0}$ converges to an infinite word u_i.
(ii) Prove that u_i is morphic and uniformly recurrent.
(iii) (*) Prove that u_i is purely morphic if, and only if, $i \in \{0, 1, 2\}$.

Exercise 4.14 (*) Let $u \in A^\mathbb{N}$ be any infinite word. Prove that there exist an infinite word v, on a possibly different alphabet B, and a morphism $\tau\colon B^* \to A^*$ such that $u = \tau(v)$ and $p_v(n) = \mathcal{O}(n)$. (**) Is it always possible to find such a v with $p_v(n) \sim n\,\mathrm{Card}\,A$?

Section 4.7

Exercise 4.15

(i) Let $A = \{a, b\}$, $B = \{a, b, c\}$, and define the endomorphisms $\sigma_2\colon B^* \to B^*$ and $\tau_2\colon B^* \to A^*$ by $\sigma_2(a) = aab$, $\sigma_2(b) = bcc$, $\sigma_2(c) = c$, $\tau_2(a) = a$, and $\tau_2(b) = \tau_2(c) = b$. Let $u_2 = \tau_2(\sigma_2^\omega(a))$. Prove that u_2 is recurrent and (*) that $p_{u_2}(n) = \Theta(n\sqrt{n})$ as n tends to ∞. Is $p_{u_2}(n) \sim Cn\sqrt{n}$ for some $C \in \mathbb{R}$?

(ii) (*) Prove that for each integer $r \geqslant 1$, there exists a recurrent binary morphic word u_r such that $p_{u_r}(n) = \Theta(n\sqrt[r]{n})$ as n tends to ∞.

Exercise 4.16 (**) Let $\sigma\colon A^* \to A^*$ be an endomorphism and let $x \in A^*$ be such that $\sigma^k(x) \neq \varepsilon$ for every $k \in \mathbb{N}$. Prove that there exist an integer $q \geqslant 1$ and $(\beta, \alpha) \in \mathbb{R}_{\geqslant 1} \times \mathbb{N}$ such that for each $r \in [\![0, q-1]\!]$,
$$\frac{\left|\sigma^{kq+r}(w)\right|}{(kq+r)^\alpha \beta^{kq+r}}$$
converges to a positive, finite limit as k tends to ∞.

Exercise 4.17 Let $\sigma\colon A^* \to A^*$ be an endomorphism and let $k_0 \in \mathbb{N}$. Let d denote the cardinality of A. Assume that $\|\sigma^d\| \neq 0$. Prove that
$$\frac{1}{\|\sigma^d\|}\left|\sigma^k(x)\right| \leqslant \left|\sigma^{k+k_0}(x)\right| \leqslant \|\sigma^{k_0}\|\left|\sigma^k(x)\right|$$
for every $x \in A^*$ and every integer $k \geqslant d$.

Exercise 4.18 (*) Let $\sigma\colon A^* \to A^*$ be an endomorphism, and $(\beta, \alpha) \in \mathbb{R}_{\geqslant 1} \times \mathbb{N}$. Assume that there exists $a \in A$ such that $\left|\sigma^k(a)\right| = \Theta(k^\alpha \beta^k)$ as k tends to ∞. Prove that for each $\alpha' \in [\![0, \alpha]\!]$, there exists $b_{\alpha'} \in A$ such that $\left|\sigma^k(b_{\alpha'})\right| = \Theta(k^{\alpha'} \beta^k)$ as k tends to ∞.

Exercise 4.19 In the notation of Lemma 4.7.24, prove that if $\beta < 1$ then the series $\sum_{k=0}^{\infty} f_k$ converges and $\sum_{k=j}^{\infty} f_k \sim \frac{1}{1-\beta} f_j$ as j tends to ∞.

Exercise 4.20

(i) Find an alphabet A and an everywhere-growing endomorphism $\sigma\colon A^* \to A^*$ such that for every $x \in A^*$, the integer sequence $\left(\left|\sigma^k(x)\right|\right)_{k \geqslant 0}$ is not increasing.
(ii) Let $o\colon A^* \to A^*$ be an everywhere-growing endomorphism. Let d denote the cardinality of A.

  (a) Prove that σ^d is expansive.
  (b) Let $\beta = \sqrt[d]{2}$. Prove that $\left|\sigma^k(x)\right| \geqslant \frac{\beta}{2}\beta^k$ for every $(x, k) \in A^+ \times \mathbb{N}$.

Exercise 4.21 Let $\sigma\colon A^* \to A^*$ be an everywhere-growing endomorphism. Prove that there exist a letter $a \in A$ and an integer $k \geqslant 1$ such that σ^k is prolongable on a.

Exercise 4.22

(i) Let $\sigma\colon A^* \to A^*$ be an endomorphism. Prove that σ is quasi-uniform if, and only if, for every $a, b \in A$, $|\sigma^k(a)| = \Theta(|\sigma^k(b)|)$ as k tends to ∞.

(ii) Prove that any irreducible endomorphism is quasi-uniform.

Exercise 4.23 Let $\sigma\colon A^* \to A^*$ be an endomorphism and let $u_0 \in A$ be such that σ is prolongable on u_0. Prove that the following two assertions are equivalent:

- For each $a \in \mathrm{alph}(\sigma^\omega(u_0))$, there exists $a' \in \mathrm{alph}(\sigma^\omega(u_0))$ such that a' occurs in $\sigma^\omega(u_0)$ infinitely many times and $|\sigma^k(a')| = \Theta(|\sigma^k(a)|)$ as k tends to ∞.
- The letter u_0 is exponentially growing under σ.

Exercise 4.24 Let u be an infinite word.

(i) Prove that if u is generated by some everywhere-growing endomorphism then either u is eventually periodic or $s_u(n) = \Omega\left(\frac{1}{n}p_u(n)\right)$.

(ii) (***) Prove or disprove that if u is purely morphic then either u is eventually periodic or $s_u(n) = \Theta\left(\frac{1}{n}p_u(n)\right)$.

Exercise 4.25 (***) Let u be a purely morphic infinite word generated by an everywhere-growing endomorphism σ, which is not quasi-uniform. Is there always a constant C such that either $s_u(n) \sim C \log\log n$ or $s_u(n) \sim C \log n$?

Exercise 4.26 (**) Let $\sigma\colon A^* \to A^*$ be an endomorphism and let $x \in A^*$ be such that x is bounded under σ. Let d denote the cardinality of A. Let N denote the least common multiple of all positive integers up to d. Prove that $\sigma^d(x) = \sigma^{N+d}(x)$.

Section 4.8

Exercise 4.27 (**) Let $u \in A^\mathbb{N}$ be an automatic word. Prove, without using Theorem 4.9.3, that $s(n)$ is bounded. (***) As the function s takes finitely many values, it can be viewed as an infinite word. Is it always automatic too?

Section 4.9

Exercise 4.28 (*) Let $u \in A^{\mathbb{N}}$ be an infinite word such that $p(n) = \mathcal{O}(n^\alpha)$ for some $\alpha \in \mathbb{R}_{\geq 1}$. Prove that $s(n) = \mathcal{O}(n^{3\alpha-3})$. (***) Is it always true that $s(n) = \mathcal{O}(n^{\alpha-1})$?

Exercise 4.29 Let u be an infinite word such that $p(n) = O(n)$. Define $\alpha = \liminf_{n \to \infty} \frac{p(n)}{n}$ and $\beta = \limsup_{n \to \infty} \frac{p(n)}{n}$.

(i) (**) Assuming that $\alpha \in [1,2]$, prove that $\beta - \alpha \geq \frac{(2-\alpha)(\alpha-1)}{\alpha}$.

(ii) Find a word with $(\alpha, \beta) = (\frac{3}{2}, \frac{5}{3})$.

(iii) (**) Assuming that $\alpha \in [1,2] \setminus \{1, \frac{3}{2}, 2\}$, prove that $\beta - \alpha > \frac{(2-\alpha)(\alpha-1)}{\alpha}$.

(iv) (*) Assuming $\alpha \notin \mathbb{N}$, find a positive lower bound on $\beta - \alpha$.

(v) (***) Describe the set of all possible values of (α, β).

Section 4.10

Exercise 4.30 Compute the complexity of the *period-doubling word*, fixed point of σ with $\sigma(a) = ab$, $\sigma(b) = aa$.

Exercise 4.31 Let t be the Thue–Morse word, and u be the word obtained by doubling every letter of t, i.e., $u = aabbbbaabbaa \cdots$.

(i) Describe the Rauzy graphs of t and u.

(ii) (**) Let v be a recurrent binary word such that $p_v(n) = p_t(n)$ for all $n \in \mathbb{N}$. Prove that either $L(v) = L(t)$ or $L(v) = L(u)$.

Section 4.11

Exercise 4.32 Let $0 < l_i < m_i < n_i$ be integers such that

- l_i tends to infinity,
- n_i/m_i grows fast enough and
- m_i/l_i grows faster enough.

For instance, $l_i = 2^{2 \cdot 2^i + 4}$, $m_i = 2^{8 \cdot 2^i}$, $n_i = 2^{10 \cdot 2^i}$. Let $A = \{0, 1\}$. Define the endomorphisms σ_i in A^* by $\sigma_i(0) = 0^{m_i} 1^{l_i}$ and $\sigma_i(1) = 0^{m_i} 1^{n_i}$.

(i) Prove that $(\sigma_i, 0)_{i \geq 0}$ is an s-adic construction for an infinite word u.

(ii) (*) Prove that the bispecial factors of u form four families of the form $(\hat{\sigma}_0 \circ \cdots \circ \hat{\sigma}_{i-1})(w)$, for $w \in \{\varepsilon, 1, 0^{m_i-1}, 1^{n_i-1}\}$, where $\hat{\sigma}_i(w) = 1^{l_i} \sigma_i(w) 0^{m_i} 1^{l_i}$.

(iii) (**) Prove that $s(n) \in \{2, 3\}$ for $n \geq 1$.

(iv) (**) Prove that $\liminf\limits_{n\to\infty} \frac{p(n)}{n} = 2$ and $\limsup\limits_{n\to\infty} \frac{p(n)}{n} = 3$.

Hints

Exercise 4.1. Two words with the same Parikh vector cannot differ in a single position. For the second example, use de Bruijn cycles.

Exercise 4.2. Observe that $L(u) = \bigcup\limits_{w \in L} L(w)$. The independence property holds for regular languages that are stable under concatenation, but not in general. Consider for instance $L = a^* \cup \{b\}$. It is even possible to construct words with very different complexity functions from the same L: take $L = a^* \cup a^*b$, consider the sparse words $a\,ab\,a^2\,a^2b\,a^3\,a^3b\cdots$ (where a^i and a^ib alternate) and $a\,ab\,a^2\,a^3\,a^2b\,a^4\,a^5\,a^6\,a^7\,a^3b\cdots$ (where a^ib is between a^{2^i-1} and a^{2^i}), and use Example 4.5.17.

Exercise 4.4. Most right (or left) special factors of v can be deduced from those of u. There are two additional special factors, b and aaa.

Exercise 4.5. The answer is no. Use Fekete's lemma (Lemma 11.1.1).

Exercise 4.6. Use Theorem 3.3.21.

Exercise 4.10. For (ii), discuss whether $\beta < 2$, $\beta = 2$, or $\beta > 2$. For (iii), choose appropriate e_k. See (Cassaigne 1997).

Exercise 4.12. Prove that a purely morphic word is recurrent if, and only if, its first letter occurs at least twice.

Exercise 4.13. For (ii), observe that u_i is the image of the Fibonacci word under the morphism τ_i defined by $\tau_i(a) = b^ic$, $\tau_i(b) = a$, and apply Theorem 4.6.1. For (iii), first prove that if $i \in \{0,1,2\}$, the word u_i is generated by ψ_i where $\psi_0(c) = ca$, $\psi_0(a) = c$, $\psi_1(b) = \psi_1(a) = bc$, $\psi_1(c) = a$, $\psi_2(b) = bbcabbc$, $\psi_2(c) = abbc$, and $\psi_2(a) = bbcabbcbbca$. Then use the fact that no prefix of the Fibonacci word is a cube to show that there is no such ψ_i for $i \geq 3$.

Exercise 4.14. Add an extra letter 0 to the alphabet, to be deleted by τ, and insert longer and longer blocks of 0 after each letter of u. A factor of length n then either occurs linearly close to the beginning, or contains only one letter different from 0. See also (Ehrenfeucht and Rozenberg 1982, Theorem 2).

Exercise 4.15. Show that $ab^k a$ occurs in u_2 if, and only if, k is a perfect square. Then use Exercise 4.10.

Exercise 4.16. See (Salomaa and Soittola 1978). Use Berstel's theorem on rational series with positive coefficients (Berstel and Reutenauer 1988).

Exercise 4.18. Let C be the set of letters that have the same growth as a, and B the set of letters with a slower growth. Write

$$\left|\sigma^k(a)\right| = \left|\sigma^k(a)\right|_C + \sum_{i=0}^{k-1}\sum_{b \in B}\sum_{c \in C} \left|\sigma^i(b)\right| \left|\sigma(c)\right|_b \left|\sigma^{k-i-1}(a)\right|_c.$$

Use Proposition 4.7.22 to bound $\left|\sigma^{k-i-1}(a)\right|_C$. Assume that $\left|\sigma^i(b)\right| = \mathcal{O}(i^{\alpha-2}\beta^i)$ for every $b \in B$, and reach a contradiction. See also (Salomaa and Soittola 1978).

Exercise 4.24. For (i), see (Pansiot 1984).

Exercise 4.26. Let B be the set of bounded letters, $B_1 = \bigcap_{k \in \mathbb{N}} \text{alph } \sigma^k(B)$ (*non-transient* bounded letters), $B_2 = \{b \in B_1 \mid \exists k \in \mathbb{N}, \sigma^k(b) = \varepsilon\}$ (*mortal* non-transient bounded letters), and $B_3 = B_1 \setminus B_2$ (*immortal* non-transient bounded letters). Let $d_1 = \text{Card } B - \text{Card } B_1$ and $d_2 = \text{Card } B_2$. Prove that

(i) $\sigma^{d_1}(B) \subseteq B_1^*$,
(ii) $\sigma^{d_2}(B_2) \subseteq \{\varepsilon\}$ and
(iii) for $b \in B_3$, $\sigma^{N_1}(b) \in B_2^* b B_2^*$ where N_1 is the order of some permutation of B_3.

Conclude that $\sigma^{d_1+d_2}(x) = \sigma^{N_1+d_1+d_2}(x)$ for $x \in B^*$.

Exercise 4.27. It is known that s is automatic in a particular case: when the internal alphabet of u is binary (*i.e.*, u is a purely morphic binary word generated by a uniform morphism), see (Tapsoba 1994).

Exercise 4.29. For (i) and (iii), use Rauzy graphs: if $\alpha < 2$, then $s(n) = 1$ infinitely often, and G_n is as in Proposition 4.5.8; see (Heinis 2002). For (ii), use the purely morphic (and 2-automatic) word generated by $\sigma(0) = 0010$ and $\sigma(1) = 1010$, introduced in (Rauzy 1983). For (iv), use the proof of Theorem 4.9.3 to bound $s(n)$, then the proof of Theorem 4.9.4 to find large intervals on which $s(n)$ is constant (for $\alpha < 2$, the bound obtained by this method is much smaller than the bound of (i)). Only partial results are known on (v). For instance, some points like $(\frac{3}{2}, \frac{5}{3})$ are isolated, while others like $(1, 1)$ are not, see (Aberkane 2001).

Exercise 4.30. Use the same method as for the Thue–Morse word.

Exercise 4.31. Observe that the Rauzy graphs of t and u have exactly the same shape, with different labels, except for orders 2 and 3. Discuss the possible Rauzy graphs for v. See (Aberkane and Brlek 2002).

Exercise 4.32. For (iii), determine the multiplicities of bispecial factors and show that, for $i \geqslant 1$, their lengths are all distinct and ordered in such a way that strong and weak bispecial factors alternate. For (iv), use matrices to compute the lengths of bispecial factors more precisely. This construction from (Cassaigne and Kaboré 2009) will be used in Proposition 7.5.12.

5
Substitutions, Rauzy fractals and tilings

Valérie Berthé,

Anne Siegel,

Jörg Thuswaldner

5.1 Introduction

This chapter focuses on multiple tilings associated with substitutive dynamical systems. We recall that a substitutive dynamical system (X_σ, S) is a symbolic dynamical system where the shift S acts on the set X_σ of infinite words having the same language as a given infinite word which is generated by powers of a primitive substitution σ. We restrict to the case where the inflation factor of the substitution σ is a unit Pisot number. With such a substitution σ, we associate a multiple tiling composed of tiles which are given by the unique solution of a set equation expressed in terms of a graph associated with the substitution σ: these tiles are attractors of a graph-directed iterated function system (GIFS). They live in $\mathbb{R}^{n-1}$, where n stands for the cardinality of the alphabet of the substitution. Each of these tiles is compact, it is the closure of its interior, it has non-zero measure and it has a fractal boundary that is also an attractor of a GIFS. These tiles are called *central tiles* or *Rauzy fractals*, according to G. Rauzy who introduced them in (Rauzy 1982).

Central tiles were first introduced in (Rauzy 1982) for the case of the Tribonacci substitution ($1 \mapsto 12$, $2 \mapsto 13$, $3 \mapsto 1$), and then in (Thurston 1989) for the case of the beta-numeration associated with the Tribonacci number (which is the positive root of $X^3 - X^2 - X - 1$). One motivation for Rauzy's construction was to exhibit explicit factors of the substitutive dynamical system (X_σ, S) as translations on compact abelian groups, under the hypothesis that σ is a Pisot substitution.

By extending the seminal construction in (Rauzy 1982), it has been proved that central tiles can be associated with Pisot substitutions (see for instance (Arnoux and Ito 2001) or (Canterini and Siegel 2001b)) as well as

Combinatorics, Automata and Number Theory, ed. Valérie Berthé and Michel Rigo.
Published by Cambridge University Press. ©Cambridge University Press 2010.

with beta-numeration with respect to Pisot numbers (*cf.* (Thurston 1989), (Akiyama 1999) and (Akiyama 2002)). They are conjectured to induce tilings in all these cases. The tiling property is known to be equivalent to the fact that the dynamical system (X_σ, S) has pure discrete spectrum (see (Pytheas Fogg 2002, Chapter 7) and (Barge and Kwapisz 2006)) when σ is a unit Pisot irreducible substitution.

We have chosen here to concentrate on tilings associated with substitutions for the sake of clarity. A similar study can be performed in the framework of beta-numeration, with both viewpoints being intimately connected through the notion of beta-substitution. Indeed, a beta-substitution can be associated with any Parry number β (for more details, see Exercise 5.1 and Section 5.11). In the case where β is a Pisot number, the associated substitution can be Pisot reducible as well as Pisot irreducible. The exposition of the theory of central tiles is much simpler when σ is assumed to be Pisot irreducible, even if it extends to the Pisot reducible case. Hence, we will restrict ourselves to the Pisot irreducible case.

There are several approaches for the definition of central tiles. We detail below a construction for unit Pisot substitutions based on a broken line which is defined in terms of the abelianisation of an infinite word generated by σ. Projecting the vertices of this broken line to the contractive subspace of the incidence matrix of σ along its expanding direction and taking the closure of this set yields the central tile. For more details on different approaches, see the surveys in (Pytheas Fogg 2002, Chapters 7 and 8) and (Berthé and Siegel 2005), as well as the discussion in (Barge and Kwapisz 2006) and (Ito and Rao 2006).

The aim of this chapter is to list a great variety of tiling conditions, by focusing on effectivity issues. These conditions rely on the use of various graphs associated with the substitution σ.

This chapter is organised as follows. Section 5.2 gathers all the introductory material. We assume that we are given a unit Pisot irreducible substitution σ. A suitable decomposition of the space $\mathbb{R}^{n-1}$ is first introduced in Section 5.2.1 with respect to the eigenspaces of the incidence matrix $\mathbf{M}_\sigma$ of σ. A definition of the central tile associated with σ as well as its decomposition into subtiles is then provided in Section 5.2.2. We discuss the graph-directed set equation satisfied by the subtiles in Section 5.2.3. Two (multiple) tilings associated with σ are then introduced in Section 5.3. The first one, introduced in Section 5.3.2, is called tiling of the expanding line. This tiling by intervals tiles the expanding line of the incidence matrix $\mathbf{M}_\sigma$ of σ. The second one is *a priori* not a tiling, but a multiple tiling. It is defined on the contracting space of the incidence matrix $\mathbf{M}_\sigma$, and it is made of translated copies of the subtiles of the central tile. It is called the self-replicating multiple tiling. Note that it is conjectured to be a tiling. It will be the main objective of the present chapter to introduce various graphs that provide conditions for this multiple tiling to be a tiling.

The first series of tiling conditions is expressed in geometric terms directly related to properties of the self-replicating multiple tiling. We start in Section 5.4.1 with a sufficient tiling property inspired by the so-called finiteness property (F) (discussed in Section 2.3.2.2). This leads us to introduce successively several graphs in Section 5.4 and Section 5.5, yielding necessary and sufficient conditions. We then discuss in Section 5.6, 5.7 and 5.8 further formulations for the tiling property expressed in terms of the tiling of the expanding line. They can be considered as dual to the former set of conditions. In particular, a formulation in terms of the so-called *overlap coincidence condition* is provided in Section 5.7, as well as, in Section 5.8, a further effective condition based on the notion of *balanced pairs*.

5.2 Basic definitions

We use the terminology of Section 1.4. Let σ be a substitution over the alphabet $A = \{1, 2, \ldots, n\}$. *In all that follows σ is assumed to be a unit substitution that is Pisot irreducible.* In particular, σ is primitive by Theorem 1.4.9. Let us recall that a primitive substitution always admits a power that is prolongable (see Definition 1.2.18 and (Queffélec 1987, Proposition V.1)), and which thus generates an infinite word. For the sake of simplicity, we assume that σ generates an infinite word according to Definition 1.2.18, that will be denoted as $u = u_0 u_1 \cdots$. We will see later that this causes no loss of generality (see Theorem 5.3.16 and Remark 5.3.17). Let us note that u is uniformly recurrent by Proposition 1.4.6, and that $\sigma(u) = u$, *i.e.*, u is a fixed point of σ.

5.2.1 Space decomposition

We want to give a geometric interpretation of the fixed point $u = u_0 u_1 \cdots$ of the unit Pisot irreducible substitution σ. In the present section we first introduce some algebraic formalism in order to embed u in a subspace of $\mathbb{R}^n$ spanned by the eigenvectors associated with the algebraic conjugates of the Perron–Frobenius eigenvalue of the incidence matrix of σ (see Theorem 1.4.2). Since σ is Pisot irreducible, this subspace turns out to be a hyperplane. We define a suitable projection of $\mathbb{R}^n$ onto this hyperplane. The closure of the projections of the abelianised subwords $\mathbf{P}(u_0 u_1 \cdots u_{N-1})$, for

$N \in \mathbb{N}$, will comprise the so-called *central tile* or *Rauzy fractal* that will be defined in Section 5.2.2.

Eigenvectors and eigenvalues. Let σ be a unit Pisot irreducible substitution. We want to decompose $\mathbb{R}^n$ with respect to certain eigenspaces of the incidence matrix $\mathbf{M}_\sigma$ of σ. Let β be the Perron–Frobenius eigenvalue of $\mathbf{M}_\sigma$. According to our assumptions β is a Pisot unit and n is the algebraic degree of β.

Let $r - 1$ be the number of real conjugates of β (distinct from β). They are denoted by $\beta^{(2)}, \ldots, \beta^{(r)}$. Each corresponding eigenspace has dimension one according to Perron–Frobenius' theorem (Theorem 1.4.2). Let $2s$ be the number of complex conjugates of β. They are denoted by $\beta^{(r+1)}, \overline{\beta^{(r+1)}}, \ldots, \beta^{(r+s)}, \overline{\beta^{(r+s)}}$. Each pair of a complex eigenvector together with its complex conjugate generates a two-dimensional plane. One has $n = r + 2s$ since σ is Pisot irreducible.

Let $\mathbf{v}_\beta$ be a left eigenvector of $\mathbf{M}_\sigma$ (*i.e.*, an eigenvector of $^t\mathbf{M}_\sigma$) associated with the eigenvalue β having positive entries contained in $\mathbb{Z}[\beta]$. Such a vector exists by Perron–Frobenius' theorem. Let $\mathbf{u}_\beta$ be the right eigenvector of $\mathbf{M}_\sigma$ associated with β, and normalised by $\langle \mathbf{v}_\beta, \mathbf{u}_\beta \rangle = 1$. The eigenvector $\mathbf{u}_\beta$ is well defined by the above conditions once $\mathbf{v}_\beta$ is given. Again by Perron–Frobenius' theorem, $\mathbf{u}_\beta$ has positive coordinates in $\mathbb{Q}(\beta)$. We obtain left eigenvectors $\mathbf{v}_{\beta^{(i)}}$ for the algebraic conjugates $\beta^{(i)}$ of β by replacing β by $\beta^{(i)}$ in the coordinates of the vector $\mathbf{v}_\beta$. We similarly obtain the right eigenvectors $\mathbf{u}_{\beta^{(i)}}$. Furthermore, the coordinates of $\mathbf{v}_\beta$ are easily seen to be linearly independent over $\mathbb{Q}$. The same holds for the coordinates of $\mathbf{u}_\beta$.

Remark 5.2.1 Note that this normalisation convention for $\mathbf{u}_\beta$ *a priori* does not correspond to the normalised Perron–Frobenius eigenvector of Theorem 1.4.5 and Proposition 10.4.2 whose coordinates give the frequencies of letters in u (in this latter case, the sum of coordinates equals 1). See also the discussion in Section 5.11.

The right and left eigenvectors are easily seen to satisfy the following relations, for $k \geq 2$, $i \geq 2$, $k \neq i$

$$\langle \mathbf{v}_\beta, \mathbf{u}_{\beta^{(k)}} \rangle = 0, \ \langle \mathbf{v}_{\beta^{(i)}}, \mathbf{u}_{\beta^{(k)}} \rangle = 0, \ \langle \mathbf{v}_{\beta^{(k)}}, \mathbf{u}_{\beta^{(k)}} \rangle = 1. \quad (5.1)$$

For more details see (Canterini and Siegel 2001b, Section 2), (Ei, Ito, and Rao 2006, Lemma 2.5), (Baker, Barge, and Kwapisz 2006) or (Siegel and Thuswaldner 2010).

A suitable decomposition of the space. Using the eigenvectors defined above we can decompose $\mathbb{R}^n$ as follows. The *contracting space* of the matrix

$\mathbf{M}_\sigma$ is the subspace $\mathbb{H}_c$ generated by the eigenvectors $\mathbf{u}_{\beta^{(i)}}$ associated with the $n-1$ conjugates of β (each of which has modulus less than one). The *expanding line* of $\mathbf{M}_\sigma$ is the real line $\mathbb{H}_e$ generated by the eigenvector $\mathbf{u}_\beta$. Note that the subscripts c and e stand here as abbreviations for *contracting* and *expanding*, respectively. The space $\mathbb{H}_c$ has dimension $r+2s-1 = n-1$ so that $\mathbb{H}_c \simeq \mathbb{R}^{n-1}$. Moreover, $\mathbb{H}_c$ is orthogonal to $\mathbf{v}_\beta$, according to (5.1).

We denote by $h_\sigma : \mathbb{H}_c \to \mathbb{H}_c$ the restriction of $\mathbf{M}_\sigma$ to $\mathbb{H}_c$. The mapping h_σ is a uniform contraction whose eigenvalues are the conjugates of β. Note that it scales down the $(n-1)$-dimensional Lebesgue measure by the factor $|\beta^{(2)} \cdots \beta^{(r)}||\beta^{(r+1)}|^2 \cdots |\beta^{(r+s)}|^2 = 1/\beta$, since β is a unit. This contraction mapping will play a prominent role in the sequel.

In order to make the distinction between elements of the n-dimensional space $\mathbb{R}^n$ and elements of the $(n-1)$-dimensional space $\mathbb{H}_c$, we restrict the use of bold symbols for the vectors and linear mappings of $\mathbb{R}^n$.

We denote by μ_k the k-dimensional Lebesgue measure. In particular, we work with μ_{n-1} on $\mathbb{H}_c$, and with μ_1 on $\mathbb{H}_e$.

Projections on the eigenspaces. Let $\pi_c : \mathbb{R}^n \to \mathbb{H}_c$ be the projection of $\mathbb{R}^n$ onto $\mathbb{H}_c$ along $\mathbb{H}_e$, according to the natural decomposition $\mathbb{R}^n = \mathbb{H}_c \oplus \mathbb{H}_e$. We recall that $\mathbf{P}$ denotes the abelianisation mapping defined in Section 1.4. The relation $\mathbf{P}(\sigma(w)) = \mathbf{M}_\sigma \mathbf{P}(w)$ for all $w \in A^*$ implies the commutation relation

$$\forall w \in A^*, \ \pi_c \circ \mathbf{P} \circ \sigma(w) = h_\sigma \circ \pi_c \circ \mathbf{P}(w). \tag{5.2}$$

Relation (5.2) reads as follows: when applying σ to a word w, the abelianisation $\mathbf{P}(w)$ is mapped onto $\mathbf{M}_\sigma \mathbf{P}(w)$, which has *a priori* larger entries since $\mathbf{M}_\sigma$ has non-negative entries, and thus 'moves away' from the origin. However, when considering the projection on $\mathbb{H}_c$ of the abelianisations, the point $\pi_c \circ \mathbf{P}(w)$, which is mapped to the point $h_\sigma \circ \pi_c \circ \mathbf{P}(w)$ when applying σ, gets closer to the origin since h_σ is a uniform contraction. The relation (5.2) will play a key role in the sequel.

We deduce from (5.1) that any element $\mathbf{x} \in \mathbb{R}^d$ admits the decomposition

$$\mathbf{x} = \langle \mathbf{x}, \mathbf{v}_\beta \rangle \mathbf{u}_\beta + \sum_{i=2}^{r+2s} \langle \mathbf{x}, \mathbf{v}_{\beta^{(i)}} \rangle \mathbf{u}_{\beta^{(i)}}. \tag{5.3}$$

For more details, see (Canterini and Siegel 2001b, Section 2.1). This implies that the projection of $\mathbf{x}$ onto $\mathbb{H}_e$ along $\mathbb{H}_c$ is equal to $\langle \mathbf{x}, \mathbf{v}_\beta \rangle \mathbf{u}_\beta$. We thus define

$$\pi_e : \mathbb{R}^n \to \mathbb{R}, \ \mathbf{x} \mapsto \langle \mathbf{x}, \mathbf{v}_\beta \rangle. \tag{5.4}$$

The mapping π_e is the projection of $\mathbb{R}^n$ onto the expanding line along the contracting space $\mathbb{H}_c$ followed by a suitable renormalisation that makes it into a mapping with values in $\mathbb{R}$ and not in $\mathbb{H}_e$. One has $\pi_e(\mathbb{H}_c) = 0$. The mapping π_e measures in some sense the distance to the hyperplane $\mathbb{H}_c$. We thus define the *height* of a vector $\mathbf{x} \in \mathbb{R}^n$ as $\langle \mathbf{x}, \mathbf{v}_\beta \rangle = \pi_e(\mathbf{x})$.

One deduces furthermore from simple algebraic considerations applied to (5.3) that

$$\forall \mathbf{x}, \mathbf{y} \in \mathbb{Q}^n, \ \pi_c(\mathbf{x}) = \pi_c(\mathbf{y}) \iff \langle \mathbf{x}, \mathbf{v}_\beta \rangle = \langle \mathbf{y}, \mathbf{v}_\beta \rangle \iff \mathbf{x} = \mathbf{y}. \quad (5.5)$$

For more details, see (Canterini and Siegel 2001b, Section 2.1).

5.2.2 Central tile

We first introduce the notion of a broken line associated with the fixed point u of σ.

Definition 5.2.2 The *broken line* L_u associated with the fixed point u of the unit Pisot irreducible substitution σ is defined as the broken line in $\mathbb{R}^n$ whose set of vertices is given by $\{\mathbf{P}(u_0 \cdots u_{N-1}) \mid N \in \mathbb{N}\}$.

We can also describe the broken line as a stair made of a union of segments. More precisely, for $\mathbf{x} \in \mathbb{Z}^n$ and $i \in A$, we denote by $[\mathbf{x}, i]_g$ the segment $\{\mathbf{x} + \theta e_i \mid \theta \in [0,1]\}$. We call such a segment a *basic geometric strand*, according to (Barge and Kwapisz 2006). We will use and develop this terminology in Section 5.6.1. The broken line associated with u is thus the union of the basic geometric strands $[\mathbf{P}(u_0 \cdots u_{N-1}), u_N]_g$, for $N \in \mathbb{N}$, i.e.,

$$L_u = \bigcup_{N \in \mathbb{N}} [\mathbf{P}(u_0 \cdots u_{N-1}), u_N]_g.$$

Definition of the central tile. The *central tile* (or *Rauzy fractal*) associated with the unit Pisot irreducible substitution σ is the closure of the projection by π_c onto the contracting space $\mathbb{H}_c$ of the vertices of the broken line L_u associated with the fixed point u of σ, i.e.,

$$\mathcal{T}_\sigma := \overline{\{\pi_c \circ \mathbf{P}(u_0 \cdots u_{N-1}) \mid N \in \mathbb{N}\}}.$$

Subtiles of the central tile $\mathcal{T}_\sigma$ are defined according the the letter u_N occurring after the word $u_0 \cdots u_{N-1}$. Indeed, we set for each $i \in A$

$$\mathcal{T}_\sigma(i) := \overline{\{\pi_c \circ \mathbf{P}(u_0 \cdots u_{N-1}) \mid N \in \mathbb{N}, \ u_N = i\}}.$$

By definition, the central tile $\mathcal{T}_\sigma$ consists of the finite union of its subtiles,

i.e.,
$$\mathcal{T}_\sigma = \bigcup_{i \in A} \mathcal{T}_\sigma(i).$$

We will see later (see Corollary 5.2.8, Theorem 5.3.16 and Remark 5.3.17) that the central tile $\mathcal{T}_\sigma$ and the subtiles $\mathcal{T}_\sigma(i)$ do not depend on the choice of u. They only depend on the substitution σ.

Theorem 5.2.3 *Let σ be a unit Pisot irreducible substitution. The central tile $\mathcal{T}_\sigma$ and the subtiles $\mathcal{T}_\sigma(i)$ associated with σ are compact sets.*

Proof Note that the compactness of the subtiles $\mathcal{T}_\sigma(i)$ is a direct consequence of the compactness of $\mathcal{T}_\sigma$, since they are closed subsets of $\mathcal{T}_\sigma$. To prove the compactness of $\mathcal{T}_\sigma$, it is enough to show that the points $\pi_c \circ \mathbf{P}(u_0 \cdots u_{N-1})$, for $N \in \mathbb{N}$, remain at a uniformly bounded distance of the origin in $\mathbb{H}_c$.

In order to prove this, we use a decomposition of the prefixes $u_0 \cdots u_{N-1}$ into images by powers of σ of a finite number of words. Since $\sigma(u) = u$, there exists a unique $L \leq N$ such that $\sigma(u_0 \cdots u_{L-1})$ is a proper prefix of $u_0 \cdots u_{N-1}$, and $u_0 \cdots u_{N-1}$ is a prefix of $\sigma(u_0 \cdots u_L)$. In other words, there exists a proper prefix p of $\sigma(u_L)$, such that

$$u_0 \cdots u_{N-1} = \sigma(u_0 \cdots u_{L-1}) p \text{ with } \sigma(u_L) = p\, u_N\, s. \qquad (5.6)$$

By iterating this process, one gets for every N an expansion of the form

$$u_0 \cdots u_{N-1} = \sigma^K(p_K) \sigma^{K-1}(p_{K-1}) \cdots \sigma(p_1) p_0,$$

where the p_i belong to a finite set of words that only depends on σ. Note that we have obtained a numeration system on words, the so-called Dumont–Thomas numeration (see Sections 9.4.2 and 5.11 for more details). By (5.2), one has

$$\pi_c \circ \mathbf{P}(u_0 \cdots u_{N-1}) = h_\sigma^K \circ \pi_c \circ \mathbf{P}(p_K) + \cdots + h_\sigma \circ \pi_c \circ \mathbf{P}(p_1) + \pi_c \circ \mathbf{P}(p_0).$$

We know that h_σ is a uniform contraction on $\mathbb{H}_c$. As the $\mathbf{P}(p_i)$ take finitely many values, this implies that the points $\pi_c \circ \mathbf{P}(u_0 \cdots u_{N-1})$, for $N \in \mathbb{N}$, remain at a uniformly bounded distance from the origin, which ends the proof. □

5.2.3 A graph-directed iterated function system

We now discuss a key property of the central tile and its subtiles, namely they satisfy a set equation. By the solution of a set equation we mean

Fig. 5.1 The central tile and its subtiles for the substitution $\sigma(1) = 112$, $\sigma(2) = 113$, $\sigma(3) = 1$ (left), and its decomposition into subtiles (right).

the following. We are given a collection of finitely many compact sets $\{K_1, \ldots, K_q\}$. Each set K_i can be decomposed as a union of contracted copies of itself and the other sets K_j. Associated with such a set equation there is a natural graph: its set of vertices is given by $\{K_i \mid 1 \leq i \leq q\}$ and there is an edge e (labelled $i \xrightarrow{e} j$) from K_i to K_j if K_j appears in the decomposition of K_i.

Let us formalise this concept by introducing the notion of *graph-directed iterated function system*. We consider a finite directed graph G with set of vertices $\{1, \ldots, q\}$ and set of edges E for which each vertex has at least one outgoing edge. With each edge e of the graph, is associated a contractive mapping $\tau_e : \mathbb{R}^n \to \mathbb{R}^n$. We call $(G, \{\tau_e\}_{e \in E})$ a *graph-directed iterated function system* (*GIFS*, for short, see (Mauldin and Williams 1988)).

It can be shown by a fixed point argument that given a GIFS $(G, \{\tau_e\}_{e \in E})$ there exists a unique collection of non-empty compact sets $K_1, \ldots, K_q \subset \mathbb{R}^n$ having the property that

$$K_i = \bigcup_{i \xrightarrow{e} j} \tau_e(K_j),$$

where the union runs over all edges in G leading away from the vertex i. The sets K_i are called *GIFS attractors* or *solutions of the GIFS*. Note that the uniqueness statement does not hold for general sets, but only for non-empty compact sets.

Let us see how to apply this formalism to the subtiles $\mathcal{T}_\sigma(i)$. The graph that will be used is the so-called *prefix-suffix graph*. This graph describes the way images of letters under σ can be decomposed, according to the proof of Theorem 5.2.3. It is the starting point for the construction of several kinds of graphs introduced later in this chapter. For more on this graph, see (Canterini and Siegel 2001a, Canterini and Siegel 2001b).

Definition 5.2.4 (Prefix-suffix graph) Let σ be a substitution over the alphabet A. Let P_σ be the finite set

$$P_\sigma := \{(p, i, s) \in A^* \times A \times A^* \mid \exists j \in A,\ \sigma(j) = pis\}. \qquad (5.7)$$

The set of vertices of the *prefix-suffix graph* $\mathcal{G}_\sigma$ of σ is the alphabet A. There is an edge labelled by $(p, i, s) \in P_\sigma$ from i towards j if, and only if, $pis = \sigma(j)$. We then use the notation $i \xrightarrow{(p,i,s)} j$.

Example 5.2.5 Let us consider as an example the substitution $\sigma(1) = 112$, $\sigma(2) = 113$, $\sigma(3) = 1$, whose central tile is depicted on the left side of Figure 5.1. Its prefix-suffix graph is depicted in Figure 5.2. We recall that ε is the empty word.

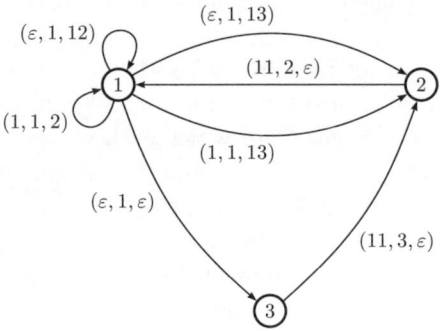

Fig. 5.2 The prefix-suffix graph for $\sigma(1) = 112$, $\sigma(2) = 113$, $\sigma(3) = 1$.

By associating with the edge $e = (p, i, s)$ the contraction mapping

$$\tau_e : \nu \in \mathbb{R}^n \mapsto h_\sigma(\nu) + \pi_c \circ \mathbf{P}(p) \in \mathbb{R}^n,$$

we get the GIFS $(\mathcal{G}_\sigma, \{\tau_e\}_{e \in P_\sigma})$. We now can give explicitly the set equation satisfied by the subtiles of the central tile. This is the content of the following theorem (see (Sirvent and Wang 2002) and also (Ito and Rao 2006)).

Theorem 5.2.6 (Sirvent and Wang 2002) *Let σ be a unit Pisot irreducible substitution over the alphabet A. The subtiles $\mathcal{T}_\sigma(i)$ are the solutions of the GIFS $(\mathcal{G}_\sigma, \{\tau_e\}_{e \in P_\sigma})$, i.e.,*

$$\forall i \in A,\ \mathcal{T}_\sigma(i) = \bigcup_{\substack{j \in A, \\ i \xrightarrow{(p,i,s)} j}} h_\sigma(\mathcal{T}_\sigma(j)) + \pi_c \circ \mathbf{P}(p). \qquad (5.8)$$

Furthermore, the union in (5.8) is a measure disjoint union.

Before giving a proof of this theorem, let us illustrate it on an example.

Example 5.2.7 We continue with the substitution σ of Example 5.2.5. In order to decompose $\mathcal{T}_\sigma(1)$ by (5.16) we look for the outgoing edges for the vertex 1 in the prefix-suffix graph. Equation (5.8) gives

$$\mathcal{T}_\sigma(1) = h_\sigma(\mathcal{T}_\sigma(1)) \cup (h_\sigma(\mathcal{T}_\sigma(1)) + \pi_c(\mathbf{e}_1)) \cup h_\sigma(\mathcal{T}_\sigma(2))$$
$$\cup (h_\sigma(\mathcal{T}_\sigma(2)) + \pi_c(\mathbf{e}_1)) \cup h_\sigma(\mathcal{T}_\sigma(3)),$$
$$\mathcal{T}_\sigma(2) = h_\sigma(\mathcal{T}_\sigma(1)) + 2\pi_c(\mathbf{e}_1),$$
$$\mathcal{T}_\sigma(3) = h_\sigma(\mathcal{T}_\sigma(2)) + 2\pi_c(\mathbf{e}_1).$$

Hence, the largest subtile $\mathcal{T}_\sigma(1)$ can be decomposed into two shrunken copies of $\mathcal{T}_\sigma(1)$, two shrunken copies of $\mathcal{T}_\sigma(2)$ and one shrinked copy of $\mathcal{T}_\sigma(3)$. The subtile $\mathcal{T}_\sigma(2)$ is the geometrically similar image of $\mathcal{T}_\sigma(1)$, and $\mathcal{T}_\sigma(3)$ is the image of $\mathcal{T}_\sigma(2)$. This decomposition is illustrated in Figure 5.1 above. Note that the number of pieces in the decomposition of the subtile $\mathcal{T}_\sigma(i)$ is equal to the number of outgoing edges of the vertex i in the prefix-suffix graph.

Proof of Theorem 5.2.6 We fix $i \in A$ and assume that $u_N = i$. By definition, one has $\pi_c \circ \mathbf{P}(u_0 \cdots u_{N-1}) \in \mathcal{T}_\sigma(i)$. By (5.6), there exist L and a decomposition of $\sigma(u_L)$ as $\sigma(u_L) = pu_N s = pis$ such that $\pi_c \circ \mathbf{P}(u_0 \cdots u_{N-1}) = h_\sigma \circ \pi_c \circ \mathbf{P}(u_0 \cdots u_{L-1}) + \pi_c \circ \mathbf{P}(p)$. We thus get $\pi_c \circ \mathbf{P}(u_0 \cdots u_{N-1}) \in h_\sigma(\mathcal{T}_\sigma(u_L)) + \pi_c \circ \mathbf{P}(p)$. As this is true for each N with $u_N = i$, by grouping by the values of u_L and taking the closure, we obtain the decomposition (5.8) for $\mathcal{T}_\sigma(i)$, i.e.,

$$\mathcal{T}_\sigma(i) = \bigcup_{(p,j,s),\ \sigma(j)=pis} h_\sigma(\mathcal{T}_\sigma(j)) + \pi_c \circ \mathbf{P}(p).$$

Recall that h_σ scales down the $(n-1)$-dimensional Lebesgue measure μ_{n-1} by the factor $1/\beta$. We deduce from (5.8) that

$$\forall i \in A,\ \beta \mu_{n-1}(\mathcal{T}_\sigma(i)) \leq \sum_{j \in A} m_{ij}\, \mu_{n-1}(\mathcal{T}_\sigma(j)), \tag{5.9}$$

where the coefficients m_{ji} denote the entries of the incidence matrix $\mathbf{M}_\sigma$. As β is the Perron–Frobenius eigenvalue of $\mathbf{M}_\sigma$, Lemma 1.4.4 implies the reverse inequality. We thus get equality in (5.9). This implies that no overlap with positive measure occurs in the union in (5.8). □

Note that (5.8) admits the following k-fold iteration for any $k \in \mathbb{N}$ and

$i \in A$

$$\mathcal{T}_\sigma(i) = \bigcup_{(p,j,s),\ \sigma^k(j)=pis} h_\sigma^k(\mathcal{T}_\sigma(j)) + \pi_c \circ \mathbf{P}(p). \tag{5.10}$$

From the uniqueness of the solution of (5.8) for non-empty compact sets we deduce the following result.

Corollary 5.2.8 *Let σ be a unit Pisot irreducible substitution. The central tiles $\mathcal{T}_\sigma$ and the subtiles $\mathcal{T}_\sigma(i)$, for $i \in A$, do not depend on the choice of the fixed point u of σ.*

We know so far that each subtile $\mathcal{T}_\sigma(i)$ can be decomposed into shrunken copies of the subtiles (namely into sets of the form $h_\sigma(\mathcal{T}_\tau(j)) + \pi_c \circ \mathbf{P}(p)$) that are disjoint in measure. To ensure that the subtiles $\mathcal{T}_\sigma(i)$, for $i \in A$, themselves are pairwise disjoint in measure, we introduce the following combinatorial condition on substitutions. For substitutions of constant length this condition goes back to (Dekking 1978), see details in (Pytheas Fogg 2002, Chapter 7).

Definition 5.2.9 (Arnoux and Ito 2001) A substitution σ over the alphabet A satisfies the *combinatorial strong coincidence condition* if for every pair $(j_1, j_2) \in A^2$, there exist $k \in \mathbb{N}$ and $i \in A$ such that $\sigma^k(j_1) = p_1 i s_1$ and $\sigma^k(j_2) = p_2 i s_2$ with $\mathbf{P}(p_1) = \mathbf{P}(p_2)$.

The combinatorial strong coincidence condition is satisfied by every Pisot irreducible substitution over a two-letter alphabet (Barge and Diamond 2002). It is conjectured that every Pisot irreducible substitution satisfies this condition.

The following theorem relates the combinatorial strong coincidence condition to the disjointness of the interiors of the subtiles $\mathcal{T}_\sigma(i)$, $i \in A$.

Theorem 5.2.10 (Arnoux and Ito 2001) *Let σ be a unit Pisot irreducible substitution. If σ satisfies the combinatorial strong coincidence condition, then the subtiles $\mathcal{T}_\sigma(i)$ of the central tile $\mathcal{T}_\sigma$ are measure disjoint.*

Proof The combinatorial strong coincidence condition implies that for every pair of letters (j_1, j_2) there exist a common letter i, a positive integer k and a common abelianised prefix $\mathbf{P}(p)$ such that $h_\sigma^k(\mathcal{T}_\sigma(j_1)) + \pi_c \circ \mathbf{P}(p)$ and $h_\sigma^k(\mathcal{T}_\sigma(j_2)) + \pi_c \circ \mathbf{P}(p)$ both appear in the k-fold iteration (5.10) of the decomposition of $\mathcal{T}_\sigma(i)$ given by (5.8). Theorem 5.2.6 yields that these tiles are disjoint in measure. □

5.3 Tilings

In this section we define a tiling as well as a multiple tiling associated with a unit Pisot irreducible substitution σ. We start with some definitions.

5.3.1 General definitions

Let K_i, $i \in A$, be a finite collection of compact sets of a subspace $\mathbb{H}$ of $\mathbb{R}^n$, with each of the K_i being the closure of its interior. Let p be a positive integer. A *multiple tiling* of degree p of the space $\mathbb{H}$ by the compact sets K_i is a collection of translated copies of the sets K_i of the form $\mathcal{I} := \{K_i + \gamma \mid (\gamma, i) \in \Gamma\}$, where Γ is a subset $\mathbb{H} \times A$, that satisfies the following conditions.

(i) The entire space $\mathbb{H}$ is covered by the elements of $\mathcal{I}$, i.e.,

$$\mathbb{H} = \bigcup_{(\gamma, i) \in \Gamma} K_i + \gamma. \tag{5.11}$$

(ii) Each compact subset of $\mathbb{H}$ intersects a finite number of elements of $\mathcal{I}$.

(iii) Almost every point in $\mathbb{H}$ (with respect to the Lebesgue measure) is covered exactly p times.

The set Γ is called the *translation set*. If the union $\{K_i + \gamma \mid (\gamma, i) \in \Gamma\}$ only satisfies (i), it is said to be a *covering* of $\mathbb{H}$. The sets $K_i + \gamma$ are called *tiles*. In other words, a multiple tiling is a union of tiles $\bigcup_{(\gamma, i) \in \Gamma} K_i + \gamma$ that covers the full space $\mathbb{H}$ with possible overlaps in such a way that almost every point belongs to exactly p tiles. This is illustrated in Figure 5.3 for $p = 2$ with an example obtained in the framework of symmetric beta-expansions taken from (Kalle and Steiner 2009). If $p = 1$, then the multiple tiling is called a *tiling*. See also Figure 5.7 for an example of a tiling.

Condition (ii) means that the first coordinate projection of Γ into $\mathbb{H}$ is a *locally finite* subset of $\mathbb{H}$, i.e., each point in $\mathbb{H}$ has a neighbourhood that intersects only finitely many projected elements of Γ. We also say that Γ is a locally finite set.

Dynamical systems can be associated with tilings in close analogy to dynamical systems associated with substitutions. Indeed, the terminology introduced in Chapter 1 concerning words extends in a natural way to tilings. For more on tiling dynamical systems, see (Solomyak 1997) and (Robinson 2004).

Consider a collection of non-empty compact sets $\{K_i + \gamma \mid (\gamma, i) \in \Gamma\}$ (that is not necessarily a covering or a multiple tiling). A set $K_i + \gamma$ is said to *occur* in $\{K_i + \gamma \mid (\gamma, i) \in \Gamma\}$ if $(\gamma, i) \in \Gamma$. A *patch* is defined as

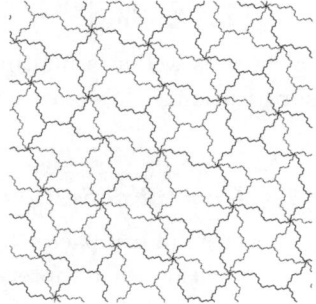

Fig. 5.3 A multiple tiling with $p = 2$.

a finite subset of Γ. It corresponds to a finite union of tiles that occur in $\{K_i + \gamma \mid (\gamma, i) \in \Gamma\}$. The translate of a patch $P = \{(\gamma_1, i_1), \ldots, (\gamma_n, i_n)\}$ by $\nu_0 \in \mathbb{H}$ is defined as $P + \nu_0 := \{(\gamma_1 + \nu_0, i_1), \ldots, (\gamma_n + \nu_0, i_n)\}$. Two patches $P = \{(\gamma_1, i_1), \ldots, (\gamma_n, i_n)\}$ and $P' = \{(\gamma_1', i_1'), \ldots, (\gamma_n', i_n')\}$ are said to be *equivalent* if they coincide up to a translation vector, that is, if there exists $\nu_0 \in \mathbb{H}$ such that $P' = \{(\gamma_1 + \nu_0, i_1), \ldots, (\gamma_n + \nu_0, i_n)\}$.

We now consider a covering of $\mathbb{H}$. We say that a ball $B(\nu, R)$ in $\mathbb{H}$ is contained in a patch $P = \{(\gamma_1, i_1), \ldots, (\gamma_n, i_n)\}$ if $B(\nu, R)$ is a subset of the convex hull of the points γ such that $(\gamma, i) \in P$. We define in a similar way the fact that a patch is contained in a ball. The set Γ is said to be *repetitive* if for any finite patch P, there exists $R > 0$ such that every ball of radius R in $\mathbb{H}$ contains a patch which is equivalent to P. This notion is an analogue of the notion of uniform recurrence for words (see Definition 1.2.9).

A subset of $\mathbb{R}^n$ is said to be a *Delone set* if it is both *uniformly discrete* (there exists $r > 0$ such that any open ball of radius r contains at most one point of this set) and *relatively dense* (there exists $R > 0$ such that every closed ball of radius R contains at least one point of this set). We say by extension that Γ is a *Delone set* if its first coordinate projection on $\mathbb{H}$ is a Delone set. Delone sets have been introduced in the context of point sets and model sets, see *e.g.* (Moody 1997). See also (Lagarias and Pleasants 2002) and (Lagarias and Pleasants 2003) for complexity results on Delone sets that can be compared with analogous results in combinatorics of words on the factor complexity and on the recurrence function.

5.3.2 Tiling of the expanding line

We first associate with σ a tiling by intervals of the expanding half-line $\mathbb{R}_+ \mathbf{u}_\beta \subset \mathbb{H}_e$. It is obtained by projecting the broken line L_u associated with u (see Definition 5.2.2) onto the expanding line $\mathbb{H}_e$ along the contracting

Substitutions, Rauzy fractals and tilings 261

hyperplane $\mathbb{H}_c$ (see Figure 5.4). This induces a tiling of the half-line $\mathbb{R}_+ \mathbf{u}_\beta \subset \mathbb{H}_e$. Using the projection $\pi_e \colon \mathbb{R}^n \to \mathbb{R}$, $\mathbf{x} \mapsto \langle \mathbf{x}, \mathbf{v}_\beta \rangle$, we even get a tiling of $\mathbb{R}_+$ whose tiles are certain translates of the intervals $I_i = [0, \langle \mathbf{e}_i, \mathbf{v}_\beta \rangle]$ for $i = 1, \ldots, n$. In particular, this tiling is obtained by taking the tiles I_{u_0}, $I_{u_1}, \ldots$ adjacent to each other, starting from the origin. The translation set is called the *self-similar translation set* and is equal to

$$\Gamma_e = \{(\pi_e \circ \mathbf{P}(u_0 \cdots u_{N-1}), u_N) \mid N \geq 0\}.$$

Since 0 is an endpoint of a tile and since we have assumed that the coordinates of $\mathbf{v}_\beta$ belong to $\mathbb{Z}[\beta]$, the endpoints of all tiles are contained in $\mathbb{Z}[\beta]$.

We denote the resulting tiling of $\mathbb{R}_+$ by $\mathcal{E}_u$ and refer to it as the *self-similar tiling of the expanding line*. For an illustration, see Figure 5.4. One has

$$\mathcal{E}_u := \{\pi_e [\mathbf{x}, i]_g \mid [\mathbf{x}, i] \in \Gamma_e\}, \tag{5.12}$$

where the basic geometric strand $[\mathbf{x}, i]_g$ is equal to the segment $\{\mathbf{x} + \theta \mathbf{e}_i \mid \theta \in [0, 1]\}$.

The repetitivity of the tiling $\mathcal{E}_u$ is an easy consequence of the fact that u is uniformly recurrent (see Proposition 1.4.6). The terminology 'self-similar' comes from the fact that the set of endpoints of tiles in $\mathcal{E}_u$ is stable by multiplication by β. Sections 5.6, 5.7 and 5.8 rely on this tiling.

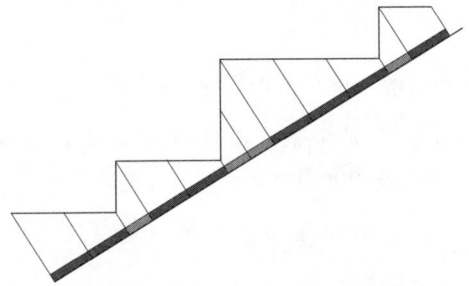

Fig. 5.4 Projecting the broken line L_u. In order to illustrate the relation between the tiling $\mathcal{E}_u$ and the broken line L_u we draw the tiling $\mathcal{E}_u$ parallel to the expanding eigendirection $\mathbf{u}_\beta$ of $\mathbf{M}_\sigma$ and not in the real line, for $\sigma(1) = 112$, $\sigma(2) = 21$.

5.3.3 Self-replicating translation set

We now introduce a multiple tiling associated with the substitution σ. The tiles of this multiple tiling are given by the subtiles $\mathcal{T}_\sigma(i)$, $i \in A$. The corresponding set of translation vectors is obtained by projecting a suitable subset of points of $\mathbb{Z}^n$ on the contracting space $\mathbb{H}_c$. Let us define this set.

Following (Reveillès 1991), we define a notion of discretisation for the hyperplane $\mathbb{H}_c$. The discretised hyperplane is usually called *standard arithmetic discrete hyperplane*. We will use here the shorthand terminology *discrete hyperplane*. We recall that $\mathbb{H}_c$ is the hyperspace orthogonal to the vector $\mathbf{v}_\beta$.

Definition 5.3.1 (Discrete hyperplane) The *discrete hyperplane* associated with $\mathbb{H}_c$ is defined as the set of points $\mathbf{x} \in \mathbb{Z}^n$ that satisfy

$$0 \leq \langle \mathbf{x}, \mathbf{v}_\beta \rangle < \sum_{i \in A} \langle \mathbf{e}_i, \mathbf{v}_\beta \rangle = ||\mathbf{v}_\beta||_1. \tag{5.13}$$

A discrete hyperplane is a discrete set of points. We now introduce a 'continuous' counterpart to this notion.

Definition 5.3.2 (Stepped hyperplane) The *stepped hyperplane* associated with $\mathbb{H}_c$ is defined as the union of faces of unit cubes whose vertices belong to the discrete hyperplane associated with $\mathbb{H}_c$.

We now want to label the faces contained in a stepped hyperplane. For $\mathbf{x} \in \mathbb{Z}^n$ and $i \in A$, the *face of type i located at* $\mathbf{x}$ is defined as the face orthogonal to the ith canonical vector of the translate of the unit cube located at $\mathbf{x}$, i.e.,

$$\mathbf{x} + \{\theta_1 \mathbf{e}_1 + \cdots + \theta_{i-1} \mathbf{e}_{i-1} + \theta_{i+1} \mathbf{e}_{i+1} + \cdots + \theta_n \mathbf{e}_n \mid \theta_j \in [0,1] \text{ for } j \neq i\}.$$

One checks that a face of type i located at $\mathbf{x}$ is a subset of the stepped hyperplane if, and only if, one has

$$0 \leq \langle \mathbf{x}, \mathbf{v}_\beta \rangle < \langle \mathbf{e}_i, \mathbf{v}_\beta \rangle. \tag{5.14}$$

For more details, see for instance the references (Berthé and Vuillon 2000), (Arnoux, Berthé, and Ito 2002) or else (Arnoux, Berthé, and Siegel 2004).

Note that a stepped hyperplane is a hypersurface that lives in $\mathbb{R}^n$, whereas a discrete hyperplane is a subset of $\mathbb{Z}^n$. The discrete hyperplane contains all the vertices of the faces contained in the stepped hyperplane, whereas faces of the stepped hyperplane are labelled by pairs $(\mathbf{x}, i)$ that satisfy (5.14). This labelling thus consists in selecting some vertices among all the vertices of the discrete hyperplane according to the value $\langle \mathbf{x}, \mathbf{v}_\beta \rangle$, hence the difference between the right-hand sides of Inequalities (5.13) and (5.14).

We now project the faces of the stepped hyperplane by π_c.

Proposition 5.3.3 *The collection of projections of the faces of the stepped hyperplane, i.e.,*

$$\{\pi_c([\mathbf{x}, i]_g) \mid \mathbf{x} \in \mathbb{Z}^n, \ i \in A, \ 0 \leq \langle \mathbf{x}, \mathbf{v}_\beta \rangle < \langle \mathbf{e}_i, \mathbf{v}_\beta \rangle \}$$

is a polyhedral tiling of $\mathbb{H}_c$ by n types of projected faces.

For an explicit proof, see (Berthé and Vuillon 2000) or (Arnoux, Berthé, and Ito 2002). A piece of a stepped hyperplane together with its projection by π_c is depicted in Figure 5.5.

Note that in the Pisot reducible case, $\mathbb{H}_c$ is no longer a hyperplane and the projections of faces do overlap. There is no universal construction known to obtain an analogue polyhedral tiling (for special cases where this is possible, see (Ei and Ito 2005, Ei, Ito, and Rao 2006)). Nevertheless, one obtains a polyhedral covering.

 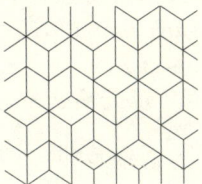

Fig. 5.5 A stepped hyperplane and its projection on $\mathbb{H}_c$ as a polyhedral tiling.

We are now going to replace in this polyhedral tiling projected faces by corresponding subtiles (see Figure 5.7). We will see in Section 5.3.5 that this will yield a multiple tiling (Theorem 5.3.13) which is conjectured to be a tiling. This multiple tiling will be called the *self-replicating multiple tiling*.

With each face of type i located at $\mathbf{x}$ included in the stepped hyperplane, we associate a copy of the tile $\mathcal{T}_\sigma(i)$ located at $\pi_c(\mathbf{x})$ in the contracting space $\mathbb{H}_c$. The *self-replicating translation set* Γ_c is defined as

$$\Gamma_c = \{(\gamma, i) \in \pi_c(\mathbb{Z}^n) \times A \mid \gamma = \pi_c(\mathbf{x}), \mathbf{x} \in \mathbb{Z}^n, \ 0 \leq \langle \mathbf{x}, \mathbf{v}_\beta \rangle < \langle \mathbf{e}_i, \mathbf{v}_\beta \rangle \}. \tag{5.15}$$

An element of the form $(\gamma, i) \in \pi_c(\mathbb{Z}^n) \times A$ is called a *tip*. We denote it by $[\gamma, i]^*$. Tips can be considered as symbolic representations of projections of faces. We denote by $[\gamma, i]_g^*$ the projection by π_c of the face of type i located at $\mathbf{x}$, with $\gamma = \pi_c(\mathbf{x})$, i.e.,

$$[\gamma, i]_g^* := \pi_c([\mathbf{x}, i]^*) \text{ with } \gamma = \pi_c(\mathbf{x}).$$

We thus make the distinction, thanks to the subscript g, between the projected face $[\pi_c(\mathbf{x}), i]_g^*$ and the tip $[\pi_c(\mathbf{x}), i]^*$. The definition of the graphs and the formalism introduced in Sections 5.4 and 5.5 will illustrate the importance of working with symbolic representations.

Note that the discretisation process underlying Definition 5.3.1 is in some sense 'dual' to the notion of broken line (see Definition 5.2.2), hence, the superscript '$*$' in the notation $[\mathbf{x}, i]_g^*$. Projected faces and segments can also be considered as 'dual'. The use of the symbol '$*$' allows us to make the distinction between the notation used for segments and tips. We will develop this duality idea in Section 5.6.1.

Before stating and proving Proposition 5.3.6 below, we need a density result (see Corollary 5.3.5). This density result will be a direct consequence of Kronecker's theorem that we recall here without proof (a proof of this theorem can be found for instance in (Hardy and Wright 1985)).

Theorem 5.3.4 (Kronecker's theorem) *Let $r \geq 1$ and let $\alpha_1, \ldots, \alpha_r$ be real numbers such that $1, \alpha_1, \ldots, \alpha_r$ are rationally independent. For every $\varepsilon > 0$ and for every $(x_1, \ldots, x_r) \in \mathbb{R}^r$, there exist an element $N \in \mathbb{N}$ and $(p_1, \ldots, p_r) \in \mathbb{Z}^r$ such that*

$$\forall i \in \{1, \ldots, r\}, \ |N\alpha_i - p_i - x_i| < \varepsilon.$$

The proof of the following corollary of Kronecker's theorem can be easily adapted from the proof of (Akiyama 1999, Proposition 1) where it is given in the framework of the beta-numeration, by recalling that the coordinates of $\mathbf{v}_\beta$ are rationally independent. A similar argument can be found in (Canterini and Siegel 2001b, Section 3) stated in terms of minimality of a toral addition.

Corollary 5.3.5 *Let σ be a unit Pisot irreducible substitution. The set $\pi_c(\{\mathbf{z} \in \mathbb{Z}^n \mid \langle \mathbf{z}, \mathbf{v}_\beta \rangle \geq 0\})$ is dense in $\mathbb{H}_c$.*

Proposition 5.3.6 *Let σ be a unit Pisot irreducible substitution. Then the following assertions are true.*

(i) *The set Γ_c is a Delone set.*
(ii) *The union $\{\mathcal{T}_\sigma(i) + \gamma \mid [\gamma, i]^* \in \Gamma_c\}$ is a covering of $\mathbb{H}_c$.*

Proof By Proposition 5.3.3, the projections of the faces of the stepped hyperplane by π_c form a polyhedral tiling of $\mathbb{H}_c$ with translation set Γ_c, which implies (i).

Let us prove (ii). Let $\mathbf{z} \in \mathbb{Z}^n$ with $\langle \mathbf{z}, \mathbf{v}_\beta \rangle \geq 0$. There exists $N \in \mathbb{N}$

such that if we set $\mathbf{x} := \mathbf{P}(u_0 \cdots u_{N-1})$, then $\langle \mathbf{x}, \mathbf{v}_\beta \rangle \leq \langle \mathbf{z}, \mathbf{v}_\beta \rangle < \langle \mathbf{x}, \mathbf{v}_\beta \rangle + \langle \mathbf{e}_{u_N}, \mathbf{v}_\beta \rangle$, where $\mathbf{e}_{u_N} = \mathbf{P}(u_N)$. One deduces that $\mathbf{z} - \mathbf{x}$ satisfies (5.14) with $i = u_N$. As $\mathbf{z} = \mathbf{x} + (\mathbf{z} - \mathbf{x})$ this implies that $\pi_c(\mathbf{z}) \in \mathcal{T}_\sigma(i) + \gamma$ for $[\gamma, i]^* \in \Gamma_c$ with $\gamma = \pi_c(\mathbf{z} - \mathbf{x})$ and $i = u_N$.

Let $\nu \in \mathbb{H}_c$. By Corollary 5.3.5 there exists a sequence $(\pi_c(\mathbf{z}_k))_{k \in \mathbb{N}}$ with $\langle \mathbf{z}_k, \mathbf{v}_\beta \rangle \geq 0$ for all k that converges to ν. Furthermore, we have seen that for each k, there exists $[\gamma_k, i_k]^* \in \Gamma_c$ such that $\pi_c(\mathbf{z}_k) \in \mathcal{T}_\sigma(i_k) + \gamma_k$. Since the subtiles $\mathcal{T}_\sigma(i)$, for $i \in A$, are bounded and Γ_c is uniformly discrete, there are infinitely many k for which (γ_k, i_k) takes the same value, say (γ, i). We thus get $\nu \in \mathcal{T}_\sigma(i) + \gamma$, which implies the covering property. This ends the proof of (ii). $\square$

Corollary 5.3.7 *The subtiles $\mathcal{T}_\sigma(i)$, for $i \in A$, have non-empty interior.*

Proof Since the set Γ_c is countable and according to Proposition 5.3.6 (ii), we deduce from Baire's theorem that there exists $i \in A$ such that the interior of $\mathcal{T}_\sigma(i)$ is not empty. We then deduce from the GIFS equation (5.8) and from the primitivity of σ which implies that the prefix-suffix graph $\mathcal{G}_\sigma$ is strongly connected that all subtiles have non-empty interior. $\square$

5.3.4 Tip substitutions

It remains to prove that the collection $\mathcal{I}_\sigma := \{\mathcal{T}_\sigma(i) + \gamma \mid [\gamma, i]^* \in \Gamma_c\}$ yields a multiple tiling of $\mathbb{H}_c$. This will be the content of Theorem 5.3.13 in Section 5.3.5. In order to prove this theorem, we first need to highlight the self-replicating properties of Γ_c. Indeed, Γ_c is stabilised by an inflation mapping acting on $\pi_c(\mathbb{Z}^n) \times A$. This inflation mapping is nothing but a substitution on tips (or, equivalently, on faces of cubes), that is inspired by the GIFS equation (5.8) satisfied by the subtiles. We explain this more precisely in the present section.

Definition 5.3.8 (GIFS substitution) The (n-dimensional) *GIFS substitution on tips* associated with the (one-dimensional) substitution σ, denoted by $\mathbf{E}_1^*$, is defined on patches of tips by

$$\mathbf{E}_1^*\{[\gamma, i]^*\} = \bigcup_{(p,j,s),\ \sigma(j)=pis} \{[h_\sigma^{-1}(\gamma + \pi_c \circ \mathbf{P}(p)), j]^*\}, \qquad (5.16)$$

$$\mathbf{E}_1^*(X_1) \cup \mathbf{E}_1^*(X_2) = \mathbf{E}_1^*(X_1 \cup X_2).$$

We will use the notation $\mathbf{E}_1^*([\gamma, i]^*)$ for $\mathbf{E}_1^*\{[\gamma, i]^*\}$.

Note that we use here the assumption that σ is a unimodular substitution (*i.e.*, its incidence matrix $\mathbf{M}_\sigma$ has determinant ± 1) to ensure that h_σ^{-1} maps

$\pi_c(\mathbb{Z}^n)$ onto $\pi_c(\mathbb{Z}^n)$. Indeed, we use the fact that $h_\sigma \circ \pi_c = \pi_c \circ \mathbf{M}_\sigma$ (see (5.2)), and that the first coordinate γ of a tip belongs to $\pi_c(\mathbb{Z}^n)$.

There is a deep relation between the GIFS substitution $\mathbf{E}_1^*$ and the GIFS equation (5.8), which can indeed be rewritten as

$$\forall [\gamma, i]^* \in \Gamma_c, \; \mathcal{T}_\sigma(i) + \gamma = \bigcup_{[\eta,j]^* \in \mathbf{E}_1^*([\gamma,i]^*)} h_\sigma(\mathcal{T}_\sigma(j) + \eta). \tag{5.17}$$

This formalism will thus be a particularly convenient way to describe the GIFS equation (5.8) in the graph constructions of Section 5.5. It has been introduced by (Arnoux and Ito 2001) and (Sano, Arnoux, and Ito 2001) (under the name *generalised substitutions* with the notation $E_1^*(\sigma)$). We omit here the reference to σ for the sake of simplicity in the notation $\mathbf{E}_1^*$. The subscript of $\mathbf{E}_1^*$ stands for the codimension of faces (in the present chapter, they are codimension one faces of hypercubes), while the superscript of $\mathbf{E}_1^*$ indicates that it is the dual mapping of some mapping $\mathbf{E}_1$, that we will introduce in Section 5.6.1. Examples of generalised substitutions are given in (Pytheas Fogg 2002, Chapter 8). Extensions to more general spaces based on faces of hypercubes having higher codimension have also been provided in (Sano, Arnoux, and Ito 2001).

Example 5.3.9 We continue with the substitution $\sigma(1) = 112$, $\sigma(2) = 113$, $\sigma(3) = 1$ considered in Examples 5.2.5 and 5.2.7.

In order to compute $\mathbf{E}_1^*([0, i]^*)$ by (5.16) we look for the occurrences of the letter 1 in $\sigma(1)$, $\sigma(2)$ and $\sigma(3)$. This yields

$$\mathbf{E}_1^*([0,1]^*) = [0,1]^* \cup [0,2]^* \cup [0,3]^* \cup [h_\sigma^{-1} \circ \pi_c \circ \mathbf{P}(1), 1]^* \cup [h_\sigma^{-1} \circ \pi_c \circ \mathbf{P}(1), 2]^*.$$

We similarly compute

$$\mathbf{E}_1^*([0,2]^*) = [h_\sigma^{-1} \circ \pi_c \circ \mathbf{P}(11),\, 1]^*, \qquad \mathbf{E}_1^*(0,3]^*) = [h_\sigma^{-1} \circ \pi_c \circ \mathbf{P}(11), 2]^*.$$

By applying the commutation relation $\mathbf{h}_c^{-1} \circ \pi_c = \pi_c \circ \mathbf{M}_\sigma^{-1}$ (see (5.2)), one gets $h_\sigma^{-1} \circ \pi_c \circ \mathbf{P}(1) = h_\sigma^{-1} \circ \pi_c(\mathbf{e}_1) = \pi_c(\mathbf{M}_\sigma^{-1}(\mathbf{e}_1)) = \pi_c(\mathbf{e}_3)$. We thus deduce the following relations

$$\begin{aligned}\mathbf{E}_1^*([0,1]^*) &= [0,1]^* \cup [0,2]^* \cup [0,3]^* \cup [\pi_c(\mathbf{e}_3), 1]^* \cup [\pi_c(\mathbf{e}_3), 2]^* \\ \mathbf{E}_1^*([0,2]^*) &= [2\pi_c(\mathbf{e}_3), 1]^* \\ \mathbf{E}_1^*([0,3]^*) &= [2\pi_c(\mathbf{e}_3), 2]^*.\end{aligned}$$

These images are depicted in Figure 5.6, by representing tips as projected faces. Compare with the computation of the decomposition of the subtiles given in Example 5.2.7 which is illustrated in Figure 5.1.

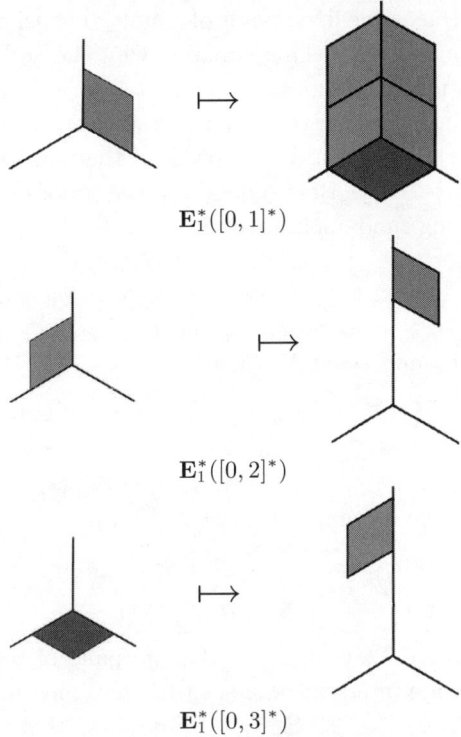

Fig. 5.6 An illustration of the images of the tips $[0,i]^*$, for $i = 1, 2, 3$, under $\mathbf{E}_1^*$ for the substitution $\sigma(1) = 112$, $\sigma(2) = 113$, $\sigma(3) = 1$. We represent here the tip $[\pi_c(\mathbf{x}), i]^*$ by its projection $[\pi_c(\mathbf{x}), i]_g^*$.

Important properties of $\mathbf{E}_1^*$ are subsumed in the following theorem. For a proof, see (Arnoux and Ito 2001) and see also (Arnoux, Berthé, and Siegel 2004).

Theorem 5.3.10 (Arnoux and Ito 2001) *Let σ be a unit Pisot irreducible substitution. Let $\mathbf{E}_1^*$ be its associated GIFS substitution.*

(i) *The images of two different tips in Γ_c under $\mathbf{E}_1^*$ share no tip in common.*
(ii) *The translation set Γ_c is stable under the action of the mapping $\mathbf{E}_1^*$.*
(iii) *The substitution $\mathbf{E}_1^*$ maps Γ_c onto Γ_c, i.e., $\mathbf{E}_1^*(\Gamma_c) = \Gamma_c$.*

According to Assertion (iii) of Theorem 5.3.10, the set of positions of tiles (given by Γ_c) is stable under the action of an inflation rule, namely the mapping $\mathbf{E}_1^*$, which plays the role of the multiplication by β acting on the tiling of the expanding line introduced in Section 5.3.2. In other words, Γ_c can be seen as the fixed point of a multidimensional combinatorial transformation, namely $\mathbf{E}_1^*$. This explains why the set Γ_c is called *self-replicating* translation set. Note that we use the term 'self-replicating' and not 'self-similar' since the mapping h_σ is possibly not a similarity.

This was the first step towards the proof of the multiple tiling property of $\{\mathcal{T}_\sigma(i) + \gamma \mid (\gamma, i) \in \Gamma_c\}$. Before detailing the proof of this property, let us state the following fundamental result.

Proposition 5.3.11 *Let σ be a unit Pisot irreducible substitution and $\mathbf{E}_1^*$ be its associated GIFS substitution. If $[\eta_1, j_1]^*, [\eta_2, j_2]^* \in \mathbf{E}_1^{*N}[\gamma, i]^*$ holds for some $[\gamma, i]^* \in \Gamma_c$ and some N, then*

$$\mu_{n-1}((\mathcal{T}_\sigma(j_1) + \eta_1) \cap (\mathcal{T}_\sigma(j_2) + \eta_2)) = 0.$$

Proof As the GIFS equation (5.17) can be iterated, we obtain the following N-fold iteration of the decomposition of $\mathcal{T}_\sigma(i)$, i.e.,

$$\forall [\gamma, i]^* \in \Gamma_c, \ \mathcal{T}_\sigma(i) + \gamma = \bigcup_{[\eta,j]^* \in \mathbf{E}_1^{*N}([\gamma,i]^*)} h_\sigma^N(\mathcal{T}_\sigma(j) + \eta). \qquad (5.18)$$

According to Theorem 5.2.6 we know that all pairs of pieces in the union on the right-hand side intersect on a set with zero measure.

Thus $[\eta_1, j_1]^*, [\eta_2, j_2]^* \in \mathbf{E}_1^{*N}[\gamma, i]^*$ implies that the intersection $h_\sigma^N(\mathcal{T}_\sigma(j_1) + \eta_1) \cap h_\sigma^N(\mathcal{T}_\sigma(j_2) + \eta_2)$ has zero measure, which yields that the intersection $(\mathcal{T}_\sigma(j_1) + \eta_1) \cap (\mathcal{T}_\sigma(j_2) + \eta_2)$ has measure zero, too. □

This proposition can be read as follows: the GIFS equation (5.17) implies that tiles translated by vectors issued from the tips in $\mathbf{E}_1^{*N}([\gamma, i]^*)$ cannot intersect. This property will be exploited all through this chapter.

5.3.5 Self-replicating multiple tiling

We are now going to prove the multiple tiling property. First we need the following statement on subtiles, whose proof follows the proofs of (Praggastis 1999, Proposition 1.1) and (Sing 2006, Proposition 4.99).

Theorem 5.3.12 *Let σ be a unit Pisot irreducible substitution. The boundary of the central tile $\mathcal{T}_\sigma$ as well as the boundary of each of its subtiles $\mathcal{T}_\sigma(i)$ has zero measure. Moreover, $\mathcal{T}_\sigma$ as well as each of its subtiles is the closure of its interior.*

Proof One has $\tau_e(\partial X) = \partial(\tau_e(X))$, for every $e \in P_\sigma$ and every set X, since in the GIFS equation (5.8) defining the subtiles $\mathcal{T}_\sigma(i)$ the mappings τ_e are homeomorphisms. One has furthermore $\partial(A \cup B) \subseteq \partial A \cup \partial B$. We use the same notation as in the proof of Theorem 5.2.6. From (5.8) we deduce that

$$\beta \mu_{n-1}(\partial \mathcal{T}_\sigma(i)) \leq \sum_{j \in A} m_{ij} \, \mu_{n-1}(\partial \mathcal{T}_\sigma(j)). \tag{5.19}$$

Similarly as in the proof of Theorem 5.2.6, we obtain equality in (5.19). As the union in (5.8) is measure disjoint, the same is true for the sets $\partial \mathcal{T}_\sigma(i)$, for $i \in A$. In particular, they have either all positive measure, or all zero measure. Assume that they have all positive measure. By Corollary 5.3.7, the subtiles $\mathcal{T}_\sigma(i)$ have non-empty interior. Let $i \in A$. Take an open ball B included in the interior of $\mathcal{T}_\sigma(i)$. We consider (5.18) applied to $\mathcal{T}_\sigma(i)$, i.e.,

$$\mathcal{T}_\sigma(i)) = \bigcup_{[\eta,j]^* \in \mathbf{E}_1^{*N}([0,i]^*)} h_\sigma^N(\mathcal{T}_\sigma(j) + \eta). \tag{5.20}$$

We then take N large enough for $\tau_e^N(\mathcal{T}_\sigma(j)) \subseteq B$, for some j such that $\sigma^N(j) = pis$ and $e = (p,i,s) \in P_{\sigma^N}$. Here $e = (p,i,s)$ is an edge of the prefix-suffix graph associated with σ^N. One has $\tau_e(\nu) = h_\sigma^N(\nu) + \pi_c \circ \mathbf{P}(p)$ for $\nu \in \mathbb{H}_c$. Note also that such an integer N exists since the mappings τ_e are contractions. This implies that $\partial(\tau_e^N(\mathcal{T}_\sigma(j))) \cap \partial \mathcal{T}_\sigma(i) = \emptyset$. We also assume N to be large enough for $m_{ij}^N > 0$ (here we use the primitivity of $\mathbf{M}_\sigma$ and m_{ij}^N are the entries of $\mathbf{M}_\sigma^N$). We deduce from (5.20) that

$$\partial \mathcal{T}_\sigma(i) \subseteq \bigcup_{[\eta,k]^* \in \mathbf{E}_1^{*N}([0,i]^*),\, [\eta,k]^* \neq [\pi_c \circ \mathbf{P}(p),j]^*} \partial h_\sigma^N(\mathcal{T}_\sigma(k) + \eta). \tag{5.21}$$

This implies that

$$\mu_{n-1}(\partial \mathcal{T}_\sigma(i)) < \beta^{-N} \sum_{k \in A} m_{ik}^N \, \mu_{n-1}(\partial \mathcal{T}_\sigma(k)),$$

by recalling that the sets in the union on the right-hand side of (5.21) are disjoint in measure. However, this contradicts with the Nth iteration of (5.19) (where the inequality has been proved to be an equality). We thus have proved that the boundary of each subtile has measure zero.

Let us prove now that each subtile is the closure of its interior. Let $i \in A$ and let $\nu \in \mathcal{T}_\sigma(i)$. Let B be an open ball with centre ν. We use as previously the Nth decomposition formula (5.18) for N large enough, and obtain $\nu \in \tau_e^N(\mathcal{T}_\sigma(j)) \subseteq B$, for some j such that $\sigma^N(j) = pis$ and $e = (p,i,s) \in P_{\sigma^N}$. By Corollary 5.3.7, $\mathcal{T}_\sigma(j)$ has non-empty interior, and so does $\tau_e^N(\mathcal{T}_\sigma(j))$. Hence, B contains interior points of $\mathcal{T}_\sigma(i)$. We thus have proved that any open ball centred at ν contains interior points of $\mathcal{T}_\sigma(i)$. Since ν was an arbitrary element of $\mathcal{T}_\sigma$, we conclude that $\mathcal{T}_\sigma(i)$ is the closure of its interior. As $\mathcal{T}_\sigma(i) = \bigcup_{i \in A} \mathcal{T}_\sigma(i)$ the same is true for $\mathcal{T}_\sigma$. □

We now have gathered all prerequisites to be able to prove the following theorem (see (Sirvent and Wang 2002), (Berthé and Siegel 2005) and (Ei, Ito, and Rao 2006)).

Theorem 5.3.13 *Let σ be a unit Pisot irreducible substitution. The collection $\mathcal{I}_\sigma = \{\mathcal{T}_\sigma(i) + \gamma \mid [\gamma, i]^* \in \Gamma_c\}$ is a multiple tiling of $\mathbb{H}_c$. Moreover, Γ_c is repetitive.*

Proof We subdivide the proof into three parts.

The translation set Γ_c is locally finite. As $\mathcal{T}_\sigma(i)$ is compact for each $i \in A$ and as Γ_c is a uniformly discrete set according to Proposition 5.3.6, the collection $\mathcal{I}_\sigma$ is *locally finite*, i.e., there exists a positive integer p such that each point of $\mathbb{H}_c$ is covered at most p times.

The translation set Γ_c is repetitive. We have to prove that for each patch P, there exists $R > 0$ such that each ball of radius R contains a translate of P. Let us fix a finite patch $P = \{[\pi_c(\mathbf{z}_k), i_k]^* \mid 1 \leq k \leq \ell\}$ of Γ_c. Let R_P be chosen in a way that the ball $B(0, R_P)$ contains the patch P.

We introduce the notion of *slice* above $\mathbb{H}_c$. We denote by $L[a, b] = \{\mathbf{x} \in \mathbb{Z}^n \mid a \leq \langle \mathbf{x}, \mathbf{v}_\beta \rangle < b\}$ the set of points whose height is between a and b. Recall that the set Γ_c corresponds to the projection of points in $L[0, ||\mathbf{v}_\beta||_\infty]$.

By (5.15), there exists $\varepsilon_k > 0$ such that $\mathbf{z}_k$ belongs to the slice $L[0, (1 - \varepsilon_k)\langle \mathbf{e}_{i_k}, \mathbf{v}_\beta \rangle]$ for each $k \in \{1, \ldots, \ell\}$. Set $\varepsilon := \frac{1}{2} \min_k \langle \varepsilon_k \mathbf{e}_{i_k}, \mathbf{v}_\beta \rangle$ (note that $\varepsilon > 0$ since P is finite). Still by the definition of Γ_c, we deduce that for every $\mathbf{x} \in \mathbb{Z}^n$, assuming $\mathbf{x} \in L[0, \varepsilon]$ implies that the patch $\pi_c(\mathbf{x}) + P$ belongs to Γ_c.

It now remains to prove that there exists $R > 0$ such that any ball of radius R in $\mathbb{H}_c$ contains a point $\pi_c(\mathbf{x})$ with $\mathbf{x} \in L[0, \varepsilon]$. Recall that the coordinates of $\mathbf{v}_\beta$ are rationally independent. By Kronecker's theorem (Theorem 5.3.4), there exists $\mathbf{x}_0 \in \mathbb{Z}^n$ such that $\mathbf{x}_0 \in L[0, \varepsilon/2]$. Let us divide the slice $L[0, ||\mathbf{v}_\beta||_\infty]$ into $N = \lceil ||\mathbf{v}_\beta||_\infty 2/\varepsilon \rceil$ slices $L[j\varepsilon/2, (j+1)\varepsilon/2]$ of height $\varepsilon/2$. Since $0 < \langle \mathbf{x}_0, \mathbf{v}_\beta \rangle < \varepsilon/2$, each slice can be translated into $L(0, \varepsilon)$: for all $j \leq N$, there exists m_j such that $m_j \mathbf{x}_0 + L[j\varepsilon/2, (j+1)\varepsilon/2] \subset L[0, \varepsilon]$.

Let us fix a point ν in $\mathbb{H}_c$. We use the fact that Γ_c is a Delone set, and in particular, that it is relatively dense (see Proposition 5.3.6). Let $R' > 0$ such that every ball of radius $R' > 0$ contains the image by π_c of a point of the discrete hyperplane (see Definition 5.3.1). In particular, the ball $B(\nu, R')$ contains a point $\pi_c(\mathbf{x})$ with $\mathbf{x} \in L[0, ||\mathbf{v}_\beta||_\infty]$. There exists j such that the point $\mathbf{x}$ belongs to one slice $L[j\varepsilon/2, (j+1)\varepsilon/2]$, hence there exists m_j such that $\mathbf{x} + m_j \mathbf{x}_0 \in L[0, \varepsilon]$. From above, this implies that $\pi_c(\mathbf{x} + m_j \mathbf{x}_0) + P$ occurs in Γ_c.

We deduce that the ball centred at ν with radius $R := R' + \max_k ||m_k \mathbf{x}_0|| + R_P$ contains a copy of the initial patch P up to translation. As $\nu \in \mathbb{H}_c$ was arbitrary this proves the repetitivity of Γ_c.

The collection $\mathcal{I}_\sigma$ is a multiple tiling. Suppose that this is wrong. Since the boundary of each subtile has measure zero by Theorem 5.3.12, the union of the boundaries of all elements of $\mathcal{I}_\sigma$ also has measure zero. Thus there are $\nu_1, \nu_2 \in \mathbb{H}_c$, positive integers $\ell_1 \neq \ell_2$ and $\varepsilon > 0$ such that $B(\nu_j, \varepsilon)$ is covered exactly ℓ_j times by the collection $\mathcal{I}_\sigma$, for $j = 1, 2$, i.e., the points contained in $B(\nu_j, \varepsilon)$ belong to exactly ℓ_j tiles of the collection $\mathcal{I}_\sigma$. More precisely, there are patches $P_1, P_2 \subset \Gamma_c$ of cardinality ℓ_1 and ℓ_2, respectively, such that $B(\nu_j, \varepsilon) \subset \bigcap_{[\gamma, i]^* \in P_j} (\mathcal{I}_\sigma(i) + \gamma)$, for $j = 1, 2$. Moreover, $B(\nu_j, \varepsilon)$ has empty intersection with each tile of $\mathcal{I}_\sigma$ that is not contained in P_j. We assume w.l.o.g. that $\ell_1 < \ell_2$.

Consider now the inflated family $h_\sigma^{-m} \mathcal{I}_\sigma$ (we recall that the inverse of h_σ is an expansive mapping). By the arguments above each point in $h_\sigma^{-m} B(\nu_1, \varepsilon)$ is contained in exactly ℓ_1 tiles of $h_\sigma^{-m} \mathcal{I}_\sigma$. By Theorem 5.3.10 (iii), each tile of $h_\sigma^{-m} \mathcal{I}_\sigma$ has the shape $h_\sigma^{-m}(\mathcal{I}_\sigma(i) + \gamma)$, with $[\gamma, i]^* \in \Gamma_c$. By (5.17) and Proposition 5.3.11, such a tile can be decomposed as a finite union of tiles in $\mathcal{I}_\sigma$ which are pairwise disjoint in measure. Thus almost each point in $h_\sigma^{-m} B(\nu_1, \varepsilon)$ is contained in exactly ℓ_1 tiles of the family $\mathcal{I}_\sigma$.

Since the translation set Γ_c is repetitive, we can choose m so large that $h_\sigma^{-m} B(\nu_1, \varepsilon)$ contains a translated copy $P_2 + \gamma$ of the patch P_2. This means that $B(\nu_2, \varepsilon) + \gamma$ is contained in $h_\sigma^{-m} B(\nu_1, \varepsilon)$ for a large enough m. Recall that $B(\nu_2, \varepsilon)$ is covered exactly ℓ_2 times by $\mathcal{I}_\sigma$. There is *a priori* no reason for $B(\nu_2, \varepsilon) + \gamma$ to be covered exactly ℓ_2 times by $\mathcal{I}_\sigma$. Indeed other tiles might 'invade' $B(\nu_2, \varepsilon) + \gamma$. Nevertheless, it is covered by each element of the patch $P_2 + \gamma$ which implies that $B(\nu_2, \varepsilon) + \gamma$ is covered at least ℓ_2 times by elements of $\mathcal{I}_\sigma$. This yields a contradiction since almost every point in $h_\sigma^{-m} B(\nu_1, \varepsilon)$ is contained in exactly ℓ_1 tiles of $\mathcal{I}_\sigma$, and $\ell_1 < \ell_2$. □

Definition 5.3.14 Let σ be a unit Pisot irreducible substitution. We call the multiple tiling $\mathcal{I}_\sigma$ defined in Theorem 5.3.13 the *self-replicating multiple tiling* associated with σ.

For all known examples of unit Pisot irreducible substitutions the self-replicating multiple tiling is indeed a tiling, as illustrated in Figure 5.7, for instance.

Fig. 5.7 The self-replicating (multiple) tiling for $\sigma(1) = 112$, $\sigma(2) = 113$, $\sigma(3) = 1$. This multiple tiling is indeed a tiling for this substitution.

Definition 5.3.15 (Tiling property) A unit Pisot irreducible substitution σ satisfies the *tiling property* if the self-replicating multiple tiling is a tiling.

The *Pisot conjecture* states that as soon as σ is a unit Pisot irreducible substitution, the tiling property holds. Let us note that the Pisot conjecture has been proved to hold for unit Pisot irreducible substitutions over a two-letter alphabet in (Hollander and Solomyak 2003). The proof strongly relies on the fact that the combinatorial strong coincidence condition is satisfied by every Pisot irreducible substitution over a two-letter alphabet (Barge and Diamond 2002), although the combinatorial strong coincidence condition does not imply the tiling property for a general alphabet.

Note that an immediate reformulation of the tiling property is that $\mu_{n-1}((\mathcal{T}(i) + \gamma) \cap (\mathcal{T}(j) + \eta)) = 0$, for every pair of distinct tiles $\{\mathcal{T}(i) + \gamma, \mathcal{T}(j) + \eta\}$ of the self-replicating multiple tiling.

Note also that in view of the following theorem the assumption that u is generated by σ causes no loss of generality.

Theorem 5.3.16 *Let σ be a unit Pisot irreducible substitution. Let k, ℓ be two positive integers. One has $\mathcal{T}_{\sigma^k} = \mathcal{T}_{\sigma^\ell}$. Furthermore, the substitution σ^k satisfies the tiling property if, and only if, σ^ℓ satisfies the tiling property.*

Proof Let us note that $h_{\sigma^k} = h_\sigma^k$ for all k. According to (5.10), the central tiles $\mathcal{T}_{\sigma^k}$ and $\mathcal{T}_{\sigma^\ell}$ are seen to satisfy

$$\forall i \in A, \ \mathcal{T}_{\sigma^k}(i) = \bigcup_{j,(p,i,s), \ \sigma^{k\ell}(j)=pis} h_\sigma^{k\ell}(\mathcal{T}_{\sigma^k}(j)) + \pi_c \circ \mathbf{P}(p) \quad \text{and}$$

$$\forall i \in A, \ \mathcal{T}_{\sigma^\ell}(i) = \bigcup_{j,(p,i,s), \ \sigma^{k\ell}(j)=pis} h_\sigma^{k\ell}(\mathcal{T}_{\sigma^\ell}(j)) + \pi_c \circ \mathbf{P}(p),$$

respectively. One deduces that $\mathcal{T}_{\sigma^k}(i)$ and $\mathcal{T}_{\sigma^\ell}(i)$ satisfy the same GIFS equation, and thus, that they coincide. Furthermore, the set Γ_c only depends on $\mathbf{v}_\beta$, which is a common left eigenvector for σ^k and σ^ℓ. This concludes the proof. □

Remark 5.3.17 Let σ be a unit Pisot irreducible substitution that is possibly not prolongable. Assume that σ^k is prolongable for some k (such a k always exists by primitivity of σ). Let u be generated by σ^k with $\sigma^k(u) = u$. We define the *central tile* associated with σ as $\mathcal{T}_\sigma := \bigcup_{i \in A} \mathcal{T}_\sigma(i)$, where the non-empty compact sets $\mathcal{T}_\sigma(i)$ are uniquely determined by the following GIFS equation

$$\forall i \in A, \ \mathcal{T}_\sigma(i) = \bigcup_{j,(p,i,s), \ \sigma^j=pis} h_\sigma(\mathcal{T}_\sigma(j)) + \pi_c \circ \mathbf{P}(p).$$

By taking the k-fold iteration of this equation and by uniqueness of its solution, we deduce that $\mathcal{T}_\sigma(i) = \mathcal{T}_{\sigma^k}(i)$, for every $i \in A$.

5.4 Ancestor graphs and tiling conditions

In the remaining part of this chapter we present various conditions for the self-replicating multiple tiling to be a tiling. Recall that the substitution σ satisfies the tiling property if, and only if, each intersection of distinct tiles in the self-replicating multiple tiling has zero measure. In the present section we focus on effective ways to control the measure of intersections of tiles. In Section 5.4.1 we introduce a sufficient condition for the tiling property. In Section 5.4.2 we define a graph that provides an effective way to check this sufficient condition. This leads us to introduce a more intricate graph in Section 5.4.3. This graph yields a necessary and sufficient condition for the tiling property.

5.4.1 Finiteness properties

We have already gained information on intersections of subtiles with zero measure. Indeed, Theorem 5.2.6 states that the shrunken copies of subtiles occurring in the decomposition of each subtile $\mathcal{T}_\sigma(i)$ are disjoint in measure. Moreover, by Theorem 5.2.10, the subtiles $\mathcal{T}_\sigma(i)$, $i \in A$, are disjoint if the substitution σ satisfies the combinatorial strong coincidence condition. We now define a sufficient condition that allows this information to be spread on zero measure intersections throughout the self-replicating multiple tiling, and thus, to exhibit a sufficient condition for the multiple tiling to be indeed a tiling.

Let U denote the patch

$$U := [0,1]^* \cup [0,2]^* \cup \cdots \cup [0,n]^*. \qquad (5.22)$$

It is easy to see that $U \subset \Gamma_c$ by (5.15). Pursuing the analogy between tips and their geometric representations in terms of projected faces (see Section 5.3.3) we call U by slight abuse of language the *lower unit cube*.

One easily checks that U is contained in $\mathbf{E}_1^*(U)$. Indeed, for any $j \in A$, $[0,j^*]$ is contained in $\mathbf{E}_1^*([0,i]^*)$, where i is the first letter of $\sigma(j)$. Hence, we deduce from Theorem 5.3.10 (ii) that the sequence of patches $(\mathbf{E}_1^{*m}(U))_{m \geq 0}$ is an increasing sequence of subsets of Γ_c with respect to inclusion. A specific case occurs when $\mathbf{E}_1^{*m}(U)$ eventually covers the entire self-replicating translation set Γ_c if m tends to infinity. As an illustration, the set $\mathbf{E}_1^{*m}(U)$ is depicted in Figures 5.8 and 5.9, in each case for a specific m, for the substitutions $\sigma(1) = 112$, $\sigma(2) = 113$, $\sigma(3) = 1$ and $\tau(1) = 2$, $\tau(2) = 3$, $\tau(3) = 12$. These pictures indicate that $\mathbf{E}_1^{*m}(U)$ eventually covers the whole self-replicating translation set Γ_c in the case of σ, but not in the case of τ.

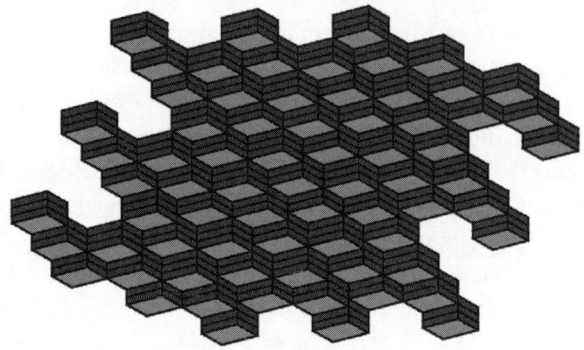

Fig. 5.8 The patch $\mathbf{E}_1^{*5}(U)$ for the substitution $\sigma(1) = 112$, $\sigma(2) = 113$, $\sigma(3) = 1$.

Definition 5.4.1 (Geometric finiteness property) Let σ be a unit Pisot irreducible substitution and $\mathbf{E}_1^*$ be its associated GIFS substitution

Fig. 5.9 The patch $\mathbf{E}_1^{*15}(U)$ for the substitution $\tau(1) = 2$, $\tau(2) = 3$, $\tau(3) = 12$.

on tips. We say that σ satisfies the *geometric finiteness property* if

$$\Gamma_c = \bigcup_{m \in \mathbb{N}} \mathbf{E}_1^{*m}(U). \tag{5.23}$$

Let us see how to propagate information on zero measure intersections inside the subtiles $\mathcal{T}_\sigma(i)$ to all intersections occurring in the self-replicating tiling when σ satisfies the geometric finiteness property.

Theorem 5.4.2 *Let σ be a unit Pisot irreducible substitution. If σ satisfies both the geometric finiteness property and the combinatorial strong coincidence condition, then the self-replicating multiple tiling is a tiling.*

Proof Let us consider two tiles in the self-replicating multiple tiling, namely $\mathcal{T}_\sigma(i_1) + \gamma_1$ and $\mathcal{T}_\sigma(i_2) + \gamma_2$. By the geometric finiteness property and by (5.17), there exist N, j_1, j_2 such that $\mathcal{T}_\sigma(i_1) + \gamma_1$ (respectively $\mathcal{T}_\sigma(i_2) + \gamma_2$) is a piece of the Nth level decomposition (5.18) of a subtile $\mathcal{T}_\sigma(j_1)$ (respectively $\mathcal{T}_\sigma(j_2)$). If $j_1 = j_2$, we are done because we fall into the assumptions of Proposition 5.3.11. If $j_1 \neq j_2$, we know from Theorem 5.2.10 and the combinatorial strong coincidence assumption that $\mathcal{T}_\sigma(j_1)$ and $\mathcal{T}_\sigma(j_2)$ are disjoint up to a set of zero measure. □

A more restrictive condition for tiling is the following *superfiniteness property* (compare with Definition 5.4.1).

Definition 5.4.3 (Geometric superfiniteness property) Let σ be a unit Pisot irreducible substitution and $\mathbf{E}_1^*$ be its associated GIFS substitution on tips. We say that σ satisfies the *geometric superfiniteness property* if there exists $i \in A$ such that

$$\Gamma_c = \bigcup_{m \in \mathbb{N}} \mathbf{E}_1^{*m}([0, i]^*).$$

Note that in this case, the proof of Theorem 5.4.2 applies without requiring the assumption of the combinatorial strong coincidence.

5.4.2 The ancestor graph

We will now discuss how to check in an effective way whether the geometric finiteness property holds. The idea is to prove that the geometric finiteness property is satisfied if, and only if, an explicit finite patch (depending on σ) is eventually covered by the iterations of $\mathbf{E}_1^*$ on the lower unit cube U.

As a consequence of Theorem 5.3.10, every tip has a unique pre-image under the action of $\mathbf{E}_1^*$. We will refer to this pre-image as *ancestor*.

Definition 5.4.4 (Ancestor of a tip) The *ancestor* of $[\eta, j]^* \in \Gamma_c$ is the unique tip $[\gamma, i]^* \in \Gamma_c$ for which $[\eta, j]^* \in \mathbf{E}_1^*([\gamma, i]^*)$.

We have worked so far with the Euclidean norm in $\mathbb{H}_c$. We now introduce a more convenient norm based on (5.3). Let $\|\cdot\|_c$ denote the maximum norm on $\mathbb{H}_c$ with respect to vectors $\mathbf{u}_{\beta^{(i)}}$ for $i \geq 2$, *i.e.*,

$$\forall \nu \in \mathbb{H}_c, \|\nu\|_c = \max\{|\langle \nu, \mathbf{v}_{\beta^{(i)}} \rangle| \mid i = 2, \ldots, r+s\}. \quad (5.24)$$

Let $\beta_{\max} := \max\{|\beta^{(j)}| \mid j \geq 2\}$. One has

$$\|h_\sigma(\nu)\|_c \leq \beta_{\max} \|\nu\|_c \text{ for all } \nu \in \mathbb{H}_c. \quad (5.25)$$

We denote by $B_c(\nu, R)$ the ball centred at ν of radius R with respect to this norm. Let $M_\sigma := \max\{\|\pi_c \circ \mathbf{P}(p)\|_c \mid (p, a, s) \in P_\sigma\}$ (see (5.7) for the definition of P_σ).

Definition 5.4.5 (Seed patch) The *seed patch* V_σ associated with the substitution σ is defined as

$$V_\sigma := \left\{ [\gamma, i]^* \in \Gamma_c \mid \|\gamma\|_c \leq \frac{M_\sigma}{1 - \beta_{\max}} \right\}. \quad (5.26)$$

Remark 5.4.6 Note that $M_\sigma/(1 - \beta_{\max})$ is an upper bound for the diameter of the tiles $\mathcal{T}_\sigma(i)$, according to the proof of Corollary 5.2.8. Thus $0 \in \mathcal{T}_\sigma(j) + \gamma$ implies that $[\gamma, j]^* \in V_\sigma$.

Theorem 5.4.7 *Let σ be a unit Pisot irreducible substitution. One has*

$$\Gamma_c = \bigcup_{m \in \mathbb{N}} \mathbf{E}_1^{*m}(V_\sigma).$$

Proof The definition of $\mathbf{E}_1^*$ yields that if $[\gamma, i]^*$ is the ancestor of the tip $[\eta, j]^*$, then

$$\gamma = h_\sigma(\eta) - \pi_c \circ \mathbf{P}(p), \tag{5.27}$$

where p is a prefix of $\sigma(i)$. We fix $[\eta,j]^* \in \Gamma_c$. Following Definition 5.4.4, let $[\gamma_k, i_k]^*$ be the successive ancestors of $[\eta, j]^*$, i.e., $[\eta, j]^* \in \mathbf{E}_1^*([\gamma_1, i_1]^*)$ and $[\gamma_k, i_k]^* \in \mathbf{E}_1^*([\gamma_{k+1}, i_{k+1}]^*)$ for all $k \geq 1$. By (5.27), one has $\gamma_{k+1} = h_\sigma(\gamma_k) - \pi_c \circ \mathbf{P}(p)$ where p is a prefix of $\sigma(i_k)$. Therefore we have by (5.25)

$$\|\gamma_{k+1}\|_c \leq \beta_{\max} \|\gamma_k\|_c + M_\sigma. \tag{5.28}$$

Let $\alpha \in (\beta_{\max}, 1)$. Then, if $a \geq M_\sigma/(\alpha - \beta_{\max})$, one has $\beta_{\max} a + M_\sigma \leq \alpha a$. This implies

$$\|\gamma_k\|_c \geq \frac{M_\sigma}{\alpha - \beta_{\max}} \implies \|\gamma_{k+1}\|_c \leq \alpha \|\gamma_k\|_c. \tag{5.29}$$

Let $V^{(\alpha)} := \left\{ [\eta, j]^* \in \Gamma_c \mid \|\gamma\|_c < \frac{M_\sigma}{\alpha - \beta_{\max}} \right\}$. All the $V^{(\alpha)}$ are finite patches since Γ_c is uniformly discrete. We also notice that

$$\bigcap_{\beta_{\max} < \alpha < 1} V^{(\alpha)} = V_\sigma.$$

Therefore, there exists $\alpha_0 < 1$ such that $V^{(\alpha_0)} = V_\sigma$ for all $\alpha_0 \leq \alpha < 1$.

By iterating (5.29) we deduce that there is $k \in \mathbb{N}$ such that the kth ancestor $[\gamma_k, i_k]^*$ of $[\eta, j]^*$ satisfies $[\gamma_k, i_k]^* \in V_{\alpha_0}$. As $V^{(\alpha_0)} = V_\sigma$, this implies that $\Gamma_c = \bigcup_{m \in \mathbb{N}} \mathbf{E}_1^{*m}(V_\sigma)$. □

Theorem 5.4.7 is based on the fact that β is a Pisot number. Analogous statements appear in various frameworks, see for instance the references (Akiyama 2000), (Arnoux, Berthé, and Siegel 2004), (Barge and Diamond 2002), (Barge and Kwapisz 2006), (Fernique 2006), (Fuchs and Tijdeman 2006) or (Ito and Rao 2006).

Remark 5.4.8 According to (5.28), one checks that V_σ contains the ancestors of all its elements (but note also that V_σ is not stable under the action of $\mathbf{E}_1^*$). Furthermore, the seed patch V_σ is easily seen to be effectively computable. Note also that $U \subseteq V_\sigma$.

We deduce from Theorem 5.4.7 the following corollary.

Corollary 5.4.9 *If there exists $m \geq 1$ such that $\mathbf{E}_1^{*m}(U)$ contains V_σ, then the geometric finiteness property holds. In this case, we can effectively exhibit such an m.*

The proof of Theorem 5.4.7 mostly relies on the notion of ancestor. In order to obtain an algorithmic way to check the geometric finiteness property, we construct a directed graph based on the notion of ancestor and on the seed patch.

Definition 5.4.10 (Ancestor graph) The vertices of the *ancestor graph* are the tips that occur in the seed patch V_σ introduced in Definition 5.4.5. There is an edge from $[\eta,j]^*$ to $[\gamma,i]^*$ if $[\gamma,i]^*$ is the ancestor of $[\eta,j]^*$, *i.e.*, $[\eta,j]^* \in \mathbf{E}_1^*([\gamma,i]^*)$.

The computation of the ancestor graph is straightforward. First, list the tips that belong to the seed patch V_σ. Then, for every $[\gamma,i]^* \in V_\sigma$, compute the tips $[\eta,j]^* \in \mathbf{E}_1^*([\gamma,i]^*)$, and draw an edge from every $[\eta,j]^*$ to $[\gamma,i]^*$, if $[\eta,j]^* \in V_\sigma$.

Remark 5.4.11 The choice of orientation we have made here for the ancestor graph (which consists in following the ancestor relation) might seem to be counter-intuitive at first sight, and in contradiction with the orientation of edges in the prefix-suffix graph. Note that a converse choice has been made in (Siegel and Thuswaldner 2010) for similar graph constructions. Their purpose was to study the boundary of subtiles and thus, to be able to zoom inside the subtiles. On the opposite, here we want to be able to zoom outside the subtiles in order to cover the self-replicating multiple tiling, hence, to trace back ancestors.

By uniqueness of the ancestor and by stability of V_σ with respect to ancestors, every vertex in the ancestor graph admits exactly one outgoing egde. This implies that every sufficiently long path in this finite graph reaches a cycle and cannot exit from it. By a *cycle*, we mean a closed directed path. Note furthermore that a tip of the form $[0,j]^*$ admits a unique outgoing edge which is also of the form $[0,i]^*$. Indeed, the (unique) ancestor of $[0,j]^*$ is the tip $[0,i]^*$ where i is the first letter of $\sigma(j)$. Hence, once a cycle contains a tip of U, it contains only elements of U. We say that it is *contained* in U. This provides a simple effective condition to check the geometric finiteness property.

Proposition 5.4.12 *Let σ be a unit Pisot irreducible substitution. The geometric finiteness property is satisfied if, and only if, all cycles in the ancestor graph are contained in U.*

Proof If the geometric finiteness property holds, then any sufficiently long path in the ancestor graph contains a tip of U, and thus any cycle in the ancestor graph is contained in U. Conversely, assume that any cycle is contained in U. Any tip in V_σ admits in its sequence of successive ancestors one element that belongs to a cycle, hence to U, which ends the proof in view of Corollary 5.4.9. □

Example 5.4.13 Two examples of ancestor graphs are depicted in Figures 5.10 and 5.11. One can see that the graph corresponding to $\sigma(1) = 112$, $\sigma(2) = 113$, $\sigma(3) = 1$ satisfies the condition of Proposition 5.4.12, whereas the graph corresponding to $\tau(1) = 2$, $\tau(2) = 3$, $\tau(3) = 12$ does not satisfy it. In this second example one can see that the graph, which is made of two connected components, admits two cycles, only one of them made of tips of U.

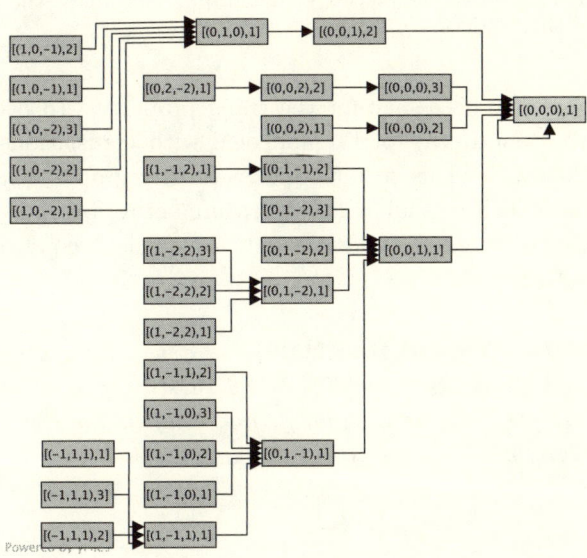

Fig. 5.10 The ancestor graph for the substitution $\sigma(1) = 112$, $\sigma(2) = 113$, $\sigma(3) = 1$. To keep notation simple in the picture of the graph, we omitted the projection π_c in the labels of the vertices.

5.4.3 The two-piece ancestor graph

The geometric finiteness property means that the self-replicating translation set Γ_c can be covered by iterating $\mathbf{E}_1^*$ on the patch U. It turns out that

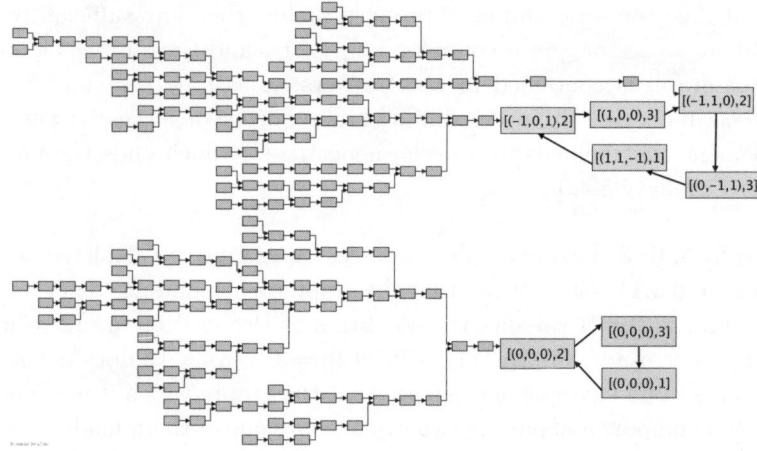

Fig. 5.11 The ancestor graph for the substitution $\tau(1) = 2$, $\tau(2) = 3$, $\tau(3) = 12$. To keep notation simple in the picture of the graph, we omitted the projection π_c in the labels of the vertices.

this condition is not necessary for the tiling property. To overcome this, we will use the repetitivity of Γ_c, and deal with translations of the sets $\mathbf{E}_1^{*m}(U)$. This will lead us to introduce a further graph, inspired by the ancestor graph, which will allow an algorithmic criterion equivalent to the tiling property to be given, and not only a sufficient condition, such as Proposition 5.4.12.

Theorem 5.4.14 (Ito and Rao 2006) *Let σ be a unit Pisot irreducible substitution and $\mathbf{E}_1^*$ be its associated GIFS substitution on tips. The self-replicating multiple tiling is a tiling if, and only if, for every $i \in A$, the radius of the largest ball contained in the union*

$$\bigcup_{[\gamma,j]^* \in \mathbf{E}_1^{*m}([0,i]^*)} [\gamma,j]_g^* \tag{5.30}$$

tends to infinity with m.

Proof Let us note that this statement does not depend on the choice of the norm by the equivalence of norms. Assume that σ satisfies the tiling property. We fix $i \in A$. For $m \in \mathbb{N}$, let $B_c(\delta_m, R_m)$ be the ball (for the norm $\|\cdot\|_c$) with largest radius contained in $h_\sigma^{-m}(\mathcal{T}_\sigma(i))$. Since $\mathcal{T}_\sigma(i)$ has non-empty interior (Corollary 5.3.7) and since h_σ^{-1} is an expansion, we have that $\lim_{m \to +\infty} R_m = \infty$. The GIFS equation (5.17) yields that $h_\sigma^{-m} \mathcal{T}_\sigma(i)$ is covered by the tiles $\mathcal{T}_\sigma(j) + \gamma$ with $[\gamma, j]^* \in \mathbf{E}_1^{*m}([0, i]^*)$. From the tiling assumption, we deduce that for every tip $[\eta, k]^* \notin \mathbf{E}_1^{*m}([0, i]^*)$, the tile $\mathcal{T}_\sigma(k) + \eta$ is measure disjoint from $B_c(\delta_m, R_m)$. Let C denote the diameter of the central tile $\mathcal{T}_\sigma$, i.e., $C = \sup\{||\nu - \nu'||_c \mid \nu, \nu' \in \mathcal{T}_\sigma\}$. Therefore, every tip $[\gamma, j]^*$ with $||\gamma - \delta_m||_c < R_m - C$ has to belong to $\mathbf{E}_1^{*m}([0, i]^*)$. This implies that the radius of the largest ball contained in the union in (5.30) tends to infinity with m.

Conversely, assume that the radius of the largest ball contained in the union in (5.30) tends to infinity with m. Let P be a patch of Γ_c. By repetitivity (Theorem 5.3.13), P is contained, up to a translation vector, in any large enough ball of Γ_c. Therefore there exist $\nu \in \mathbb{H}_c$ and $m > 0$ such that $\nu + P \subset \mathbf{E}_1^{*m}([0, i]^*)$, and thus $P \subset \mathbf{E}_1^{*m}([h_\sigma^m \nu, i]^*)$ by the definition of $\mathbf{E}_1^*$ in (5.16). Proposition 5.3.11 then yields that the tiles $\mathcal{T}_\sigma(j) + \gamma$ with $[\gamma, j] \in P$ have pairwise disjoint interiors. This implies that σ satisfies the tiling property. □

Corollary 5.4.15 *Let σ be a unit Pisot irreducible substitution. The self-replicating multiple tiling is a tiling if, and only if, for every pair of tips $([\eta_1, j_1]^*, [\eta_2, j_2]^*) \in \Gamma_c^2$ there exist $\delta \in \mathbb{H}_c$, $m \geq 0$, and $i \in A$ such that*

$$\delta + \{[\eta_1, j_1]^*, [\eta_2, j_2]^*\} \subset \mathbf{E}_1^{*m}([0, i]^*). \tag{5.31}$$

Proof If this condition is satisfied, Proposition 5.3.11 implies that $\mathcal{T}_\sigma(j_1) + \eta_1$ and $\mathcal{T}_\sigma(j_2) + \eta_2$ do not overlap for arbitrarily chosen $[\eta_1, j_1]^*$ and $[\eta_2, j_2]^*$. Therefore the tiling property is satisfied. Conversely, by Theorem 5.4.14, the tiling property implies that $\mathbf{E}_1^{*m}([0, i]^*)$ contains arbitrarily large balls for large m. Thus, by the repetitivity assertion of Theorem 5.3.13, a translation of each patch $\{[\eta_1, j_1]^*, [\eta_2, j_2]^*\} \subset \Gamma_c$ occurs in $\mathbf{E}_1^{*m}([0, i]^*)$ for some $i \in A$ and some $m \in \mathbb{N}$. □

The formulation of the tiling property given by Corollary 5.4.15 means that every pair of tips belongs to the image of a tip in U up to a common translation vector. In order to check (5.31), we need to trace back ancestors of patches *up to a translation vector*. When dealing with (5.31), we will use the existence of the translation vector δ in order to work only with pairs of tips for which at least one of the elements η_1, η_2 equals 0. We thus introduce the following definition.

Definition 5.4.16 (Two-piece ancestor) Let $\{[\eta_1, j_1]^*, [\eta_2, j_2]^*\}$ be a two-piece patch in Γ_c. A *two-piece ancestor* of this patch is a two-piece patch of the shape $\{[0, i_1]^*, [\gamma, i_2]^*\} \subset \Gamma_c$ for which there exists $\delta \in \mathbb{H}_c$ such that

$$\{[\eta_1, j_1]^*, [\eta_2, j_2]^*\} \subset \delta + \mathbf{E}_1^*\{[0, i_1]^*, [\gamma, i_2]^*\} \tag{5.32}$$

with $\{[\eta_1, j_1]^*, [\eta_2, j_2]^*\} \cap (\delta + \mathbf{E}_1^*[0, i_1]^*) \neq \emptyset$ and $\{[\eta_1, j_1]^*, [\eta_2, j_2]^*\} \cap (\delta + \mathbf{E}_1^*[\gamma, i_2]^*) \neq \emptyset$.

In other words, this means that $\{[\eta_1, j_1]^*, [\eta_2, j_2]^*\}$ appears in the image of the patch $\{[0, i_1]^*, [\gamma, i_2]^*\}$ up to a translated vector, and that the two images of $[0, i_1]^*$ and $[\gamma, i_2]^*$ both have non-empty intersection with $\{[\eta_1, j_1]^*, [\eta_2, j_2]^*\}$. Note that, contrary to the uniquely defined ancestor of a tip (see Definition 5.4.4), a two-piece patch can have several two-piece ancestors. This is due to the freedom given by the translation vector δ. Another important remark is that the tips $[\eta_1, j_1]^*$ and $[\eta_2, j_2]^*$ in Definition 5.4.16 are not required to be different. The same holds for $[0, i_1]^*$ and $[\gamma, i_2]^*$.

In order to check the tiling condition (5.31), we need to recursively check ancestor relations. We thus define a new graph, namely the *two-piece ancestor graph*. To this end we need a new seed patch which is defined as follows.

Definition 5.4.17 (Two-piece seed patch) The *two-piece seed patch* W_σ associated with the substitution σ is defined as

$$W_\sigma := \left\{ [\gamma, i]^* \in \Gamma_c \mid \|\gamma\|_c \leq \frac{2M_\sigma}{1 - \beta_{\max}} \right\}. \qquad (5.33)$$

Remark 5.4.18 Note that $2M_\sigma/(1 - \beta_{\max})$ is at least twice as large as the diameter of the tiles $\mathcal{T}_\sigma(i)$. Thus $\mathcal{T}_\sigma(i) \cap (\mathcal{T}_\sigma(j) + \gamma) \neq \emptyset$ implies that $[\gamma, j]^* \in W_\sigma$.

Moreover, we represent the pair of tips $\{[0, k]^*, [\gamma, \ell]^*\}$ as $[k, \gamma, \ell]^*$, with $k \leq \ell$ if $\gamma = 0$. The condition $k \leq \ell$ if $\gamma = 0$ simply avoids redundancies.

Definition 5.4.19 (Two-piece ancestor graph) The set of vertices of the *two-piece ancestor graph* is equal to

$$\{[k, \gamma, \ell]^* \mid (k, \gamma, \ell) \in A \times \mathbb{H}_c \times A, \ [\gamma, \ell]^* \in W_\sigma, \ k \leq \ell \text{ if } \gamma = 0\}.$$

There is an edge from $[j_1, \eta, j_2]^*$ to $[i_1, \gamma, i_2]^*$ if the patch $\{[0, i_1]^*, [\gamma, i_2]^*\}$ is a two-piece ancestor of the patch $\{[0, j_1]^*, [\eta, j_2]^*\}$.

One checks that each vertex admits at least one outgoing edge. Nevertheless, there might be several outgoing edges.

Following (Siegel and Thuswaldner 2010), the construction of this graph is straightforward when recalling that similar to V_σ, the two-piece seed patch W_σ can be explicitly computed (see Section 5.4.2). The construction can thus be performed in two steps. One first computes the list of vertices

Substitutions, Rauzy fractals and tilings   283

$[i_1, \gamma, i_2]^*$ of the graph, based on the computation of the two-piece seed patch W_σ. Then, by noticing that $[\eta, j]^* \subset \delta + \mathbf{E}_1^*([\gamma, i]^*)$ if, and only if, there exists a prefix p of $\sigma(j)$ such that $\sigma(j) = pis$ and $\delta = \eta - h_\sigma^{-1}(\gamma + \pi_c \circ \mathbf{P}(p))$, one checks whether condition (5.32) is satisfied for each pair of vertices $([j_1, \eta, j_2]^*, [i_1, \gamma, i_2]^*)$.

We need the following easy lemma.

Lemma 5.4.20 *Let $\{[\gamma_1, i_1]^*, [\gamma_2, i_2]^*\}$ be a patch in Γ_c. Then at least one of the sets $\{[0, i_1]^*, [\gamma_2 - \gamma_1, i_2]^*\}$ and $\{[0, i_2]^*, [\gamma_1 - \gamma_2, i_1]^*\}$ is a patch in Γ_c.*

Proof By (5.5), there exists a unique pair vectors $\{\mathbf{x}_1, \mathbf{x}_2\} \subset \mathbb{Z}^n$ such that $\gamma_i = \pi_c(\mathbf{x}_i)$, for $i = 1, 2$. If $\langle \mathbf{x}_1, \mathbf{v}_\beta \rangle \leq \langle \mathbf{x}_2, \mathbf{v}_\beta \rangle$ then $\{[0, i_1]^*, [\gamma_2 - \gamma_1, i_2]^*\}$ is a patch of Γ_c. If the reverse inequality holds, $\{[0, i_2]^*, [\gamma_1 - \gamma_2, i_1]^*\} \subset \Gamma_c$ and we are done. □

We now can state the main result of this section.

Theorem 5.4.21 (Two-piece ancestor graph tiling condition) *Let σ be a unit Pisot irreducible substitution. The substitution σ satisfies the tiling condition if, and only if, from any vertex in the two-piece ancestor graph, there exists a path to a vertex of the shape $[i, 0, i]^*$, for $i \in A$.*

Proof The tiling property is equivalent to (5.31), which is itself equivalent to the following condition: for every two-piece patch $\{[0, j_1]^*, [\eta, j_2]^*\}$, there exist $m \in \mathbb{N}$, $i \in A$ and $\delta \in \mathbb{H}_c$ such that $\{[0, j_1]^*, [\eta, j_2]^*\} \subset \delta + \mathbf{E}_1^{*m}([0, i]^*)$. By the definition of the two-piece ancestor this is equivalent to the fact that for each $\{[0, j_1]^*, [\eta, j_2]^*\}$ there is $m \in \mathbb{N}$ and $i \in A$ such that

$$\{[0, i^*], [0, i^*]\} \text{ is an } m\text{th ancestor of } \{[0, j_1^*], [\eta, j_2^*]\}. \tag{5.34}$$

In order to deduce Theorem 5.4.21, it is sufficient to show that we can assume w.l.o.g. that $[\eta, j_2]^* \in W_\sigma$ holds in (5.34).

Suppose on the contrary that $[\eta, j_2]^* \notin W_\sigma$. Then

$$\|\eta\|_c > \frac{2M_\sigma}{1 - \beta_{\max}}. \tag{5.35}$$

There exists a unique set of two elements $\{[\gamma_1, i_1]^*, [\gamma_2, i_2]^*\} \subset \Gamma_c$ such that

$$[0, j_1]^* \in \mathbf{E}_1^*[\gamma_1, i_1]^* \quad \text{and} \quad [\eta, j_2]^* \in \mathbf{E}_1^*[\gamma_2, i_2]^*. \tag{5.36}$$

By Lemma 5.4.20 one of the sets $\{[0, i_1]^*, [\gamma_2 - \gamma_1, i_2]^*\}$, $\{[0, i_2]^*, [\gamma_1 - \gamma_2, i_1]^*\}$ is contained in Γ_c. Assume that this is true for the first one (the second alternative is handled analogously). Then $\{[0, i_1]^*, [\gamma_2 - \gamma_1, i_2]^*\}$ is a two-piece ancestor of $\{[0, j_1]^*, [\eta, j_2]^*\}$. By (5.27) and (5.36) we have

$$||\gamma_1||_c \leq M_\sigma \quad \text{and} \quad ||\gamma_2||_c \leq \beta_{\max}||\eta||_c + M_\sigma$$

which implies together with (5.35) that $||\gamma_2 - \gamma_1||_c \leq \frac{2M_\sigma}{1-\beta_{\max}} < ||\eta||_c$. Thus, arguing in the same way as in the proof of Theorem 5.4.7 we see that there is a positive integer m' such that $\{[0, j_1]^*, [\eta, j_2]^*\}$ admits an m'th two-piece ancestor $\{[0, k_1]^*, [\gamma', k_2]^*\}$ which satisfies $\gamma' \in W_\sigma$. Thus it suffices to assume $[\eta, j_2]^* \in W_\sigma$ in (5.34) and we are done. □

5.5 Boundary and contact graphs

The tiling condition of Theorem 5.4.21 can be checked in an effective way by constructing the two-piece ancestor graph. However, it turns out that the two-piece ancestor graph is not so easy to handle. Indeed, it can be quite big especially if n is large, and as a second drawback, contrary to the ancestor graph it is not deterministic. We thus introduce two subgraphs of the two-piece ancestor graph, and establish associated tiling conditions, inspired by Proposition 5.4.12 and Theorem 5.4.21.

5.5.1 Boundary graphs

We first state as an immediate consequence of Lemma 5.4.20 the following proposition which shows that we only have to consider intersections between the subtiles $\mathcal{T}(i)$, $i \in A$, and their neighbours in $\mathcal{I}_\sigma$ (see Definition 5.3.14) to check the tiling property.

Proposition 5.5.1 *The tiling property is satisfied if, and only if, $\mathcal{T}_\sigma(i) \cap (\mathcal{T}_\sigma(j) + \gamma)$ has zero measure for every $i \in A$ and every $[\gamma, j]^* \in \Gamma_c$ with $[\gamma, j]^* \neq [0, i]^*$.*

Proof The tiling property holds if, and only if, for any two distinct tips $[\gamma, i]^*, [\eta, j]^* \in \Gamma_c$, we have $\mu_{n-1}((\mathcal{T}_\sigma(i) + \gamma) \cap (\mathcal{T}_\sigma(j) + \eta)) = 0$. But $(\mathcal{T}_\sigma(i)+\gamma) \cap (\mathcal{T}_\sigma(j)+\eta)$ is equal, up to a translation, to $\mathcal{T}_\sigma(i) \cap (\mathcal{T}_\sigma(j)+\eta-\gamma)$, and to $\mathcal{T}_\sigma(j) \cap (\mathcal{T}_\sigma(i)+\gamma-\eta)$. Lemma 5.4.20 implies that either $[\eta-\gamma, j]^* \in \Gamma_c$ or $[\gamma-\eta, i]^* \in \Gamma_c$. This ends the proof. □

As a motivation for the definition of the boundary graph (see Definition 5.5.3 below), let us dwell upon the topological information provided by cycles in the ancestor graph.

Lemma 5.5.2 *A vertex $[\gamma, i]^*$ belongs to a cycle in the ancestor graph if, and only if, $0 \in \mathcal{T}_\sigma(i) + \gamma$.*

Proof We first assume that $[\gamma, i]^*$ belongs to a cycle of the ancestor graph. Thus, there exists $m > 0$ such that $[\gamma, i]^* \in \mathbf{E}_1^{*m}([\gamma, i]^*)$. In view of (5.17) this implies $h_\sigma^m(\mathcal{T}_\sigma(i) + \gamma) \subset \mathcal{T}_\sigma(i) + \gamma$. By iterating this relation and by using the fact that h_σ is a contraction, we deduce that $0 \in \mathcal{T}_\sigma(i) + \gamma$.

Conversely, assume that $0 \in \mathcal{T}_\sigma(i) + \gamma$. We decompose the tile $\mathcal{T}_\sigma(i) + \gamma$ according to (5.17). Since $0 \in \mathcal{T}_\sigma(i) + \gamma$, for every $m \geq 1$ there exists $[\gamma_m, i_m]^* \in \mathbf{E}_1^{*m}([\gamma, i]^*)$ such that $0 \in \mathcal{T}_\sigma(i_m) + \gamma_m$. As, by Remark 5.4.6, we see that $0 \in \mathcal{T}_\sigma(i_m) + \gamma_m$ implies that $\gamma_m \in V_\sigma$, the element $[\gamma_m, i_m]^*$ is a vertex of the ancestor graph. Thus we get the walk

$$[\gamma_m, i_m]^* \to \cdots \to [\gamma_1, i_1]^* \to [\gamma, i]^*$$

in this graph. By the finiteness of V_σ, the sequence $([\gamma_m, i_m]^*)_{m \geq 1}$ takes twice the same value. Let $m_1 < m_2$ be such that $[\gamma_{m_1}, i_{m_1}] = [\gamma_{m_2}, i_{m_2}]$, with $m_1 \neq m_2$. Then, $[\gamma_{m_1}, i_{m_1}]^*$ is contained in a cycle of this graph. Furthermore, there is a walk from $[\gamma_{m_1}, i_{m_1}]^*$ to $[\gamma, i]^*$ in the ancestor graph. Since each vertex in the ancestor graph has a single outgoing edge, this implies that $[\gamma, i]^*$ belongs to the same cycle of the ancestor graph as $[\gamma_{m_1}, i_{m_1}]^*$. □

Lemma 5.5.2 indicates that non-emptyness for solutions of a GIFS equation can be deduced from cycles of the related graph. We now apply this idea together with Proposition 5.5.1 to the two-piece ancestor graph.

Definition 5.5.3 (Boundary graph) Let σ be a unit Pisot irreducible substitution. The *boundary graph* of σ is the subgraph of the two-piece ancestor graph that contains the vertices $[i, \gamma, j]^*$ with $\gamma \neq 0$ or $i \neq j$, for $i, j \in A$, that belong to a cycle, as well as vertices contained in paths leading away from these cycles.

The motivation for this definition will become clearer in the sketch of the proof of next theorem.

Theorem 5.5.4 (Boundary graph tiling condition) *Let σ be a unit Pisot irreducible substitution. The tiling condition is satisfied if, and only if, the spectral radius of the boundary graph (that is, the largest eigenvalue of its adjacency matrix) is strictly smaller than the Perron–Frobenius eigenvalue β of $\mathbf{M}_\sigma$. If this relation holds, the boundary graph provides a GIFS description of the boundary.*

Proof [Sketch] First note that (5.17) implies, assuming the first alternative of Lemma 5.4.20 (the second one can be handled analogously), that

$$\mathcal{T}_\sigma(i_1) \cap (\mathcal{T}_\sigma(i_2) + \gamma) = h_\sigma \left(\bigcup_{\substack{[\eta_1,j_1]^* \in \mathbf{E}_1^*[0,i_1]^* \\ [\eta_2,j_2]^* \in \mathbf{E}_1^*[\gamma,i_2]^*}} ((\mathcal{T}_\sigma(j_1) + \eta_1) \cap (\mathcal{T}_\sigma(j_2) + \eta_2)) \right)$$

$$= h_\sigma \left(\bigcup_{\substack{[0,j_1]^* \in -\eta_1 + \mathbf{E}_1^*[0,i_1]^* \\ [\eta_2-\eta_1,j_2]^* \in -\eta_1 + \mathbf{E}_1^*[\gamma,i_2]^*}} ((\mathcal{T}_\sigma(j_1) \cap (\mathcal{T}_\sigma(j_2) + \eta_2 - \eta_1)) + \eta_1) \right).$$

Since, in view of Remark 5.4.18, the intersections are non-empty only for $\eta_2 - \eta_1 \in W_\sigma$, the equation can be rewritten as

$$\mathcal{T}_\sigma(i_1) \cap (\mathcal{T}_\sigma(i_2) + \gamma) = h_\sigma \left(\bigcup_{[j_1,\eta,j_2]^* \to [i_1,\gamma,i_2]^*} ((\mathcal{T}_\sigma(j_1) \cap (\mathcal{T}_\sigma(j_2) + \eta)) + \delta) \right) \quad (5.37)$$

where the union is taken over all edges of the two-piece ancestor graph which lead to the vertex $[i_1, \gamma, i_2]^*$. Such an intersection is interesting if $[i_1, \gamma, i_2]^*$ is not of the form $[i, 0, i]^*$ for some $i \in A$ since we are only interested in intersections of two different tiles of $\mathcal{I}_\sigma$. If the union on the right-hand side of (5.37) is empty, *i.e.*, if a vertex of the two-piece ancestor graph has no incoming edge, then also the intersection on the left-hand side is empty. Thus we may successively delete all vertices from the two-piece ancestor graph which have no incoming edges. However, as the two-piece ancestor graph is finite these are vertices which either belong to a cycle or to a path that leads away from a cycle. We may also cancel the vertices of the form $[i, 0, i]^*$, for $i \in A$. Indeed, an edge from a vertex of the form $[i, 0, i]^*$ reaches a vertex which is also of the same form $[j, 0, j]^*$ for some $j \in A$. Hence if $[i_1, \gamma, i_2]^*$ is not of the form $[i, 0, i]^*$, then no vertex of this form admits an edge to it. The graph obtained when removing all these vertices is exactly the boundary graph.

Since h_σ is a contraction whose application scales down the $(n-1)$-dimensional Lebesgue measure by the factor $1/\beta$, one deduces that the intersection $\mathcal{T}_\sigma(i_1) \cap (\mathcal{T}_\sigma(i_2) + \gamma)$ has zero measure if the largest eigenvalue of the adjacency matrix of the boundary graph is strictly smaller than β.

The proof of the converse, *i.e.*, that the tiling property implies that the spectral radius of the boundary graph is smaller than β is left as an exercise (see Exercise 5.3). □

Note that (5.37) can be regarded as a GIFS equation for the intersections $\mathcal{T}_\sigma(i_1) \cap (\mathcal{T}_\sigma(i_2) + \gamma)$. In view of Proposition 5.5.5 below, this GIFS can be used to describe $\partial \mathcal{T}_\sigma(i)$ if the tiling condition holds, hence the terminology 'boundary graph'. For more details, see (Siegel and Thuswaldner 2010).

Proposition 5.5.5 *Let σ be a unit Pisot irreducible substitution. One has*

$$\partial \mathcal{T}_\sigma(i) \subseteq \bigcup_{[\gamma,j]^* \neq [0,i]^*,\ [\gamma,j]^* \in \Gamma_c} \mathcal{T}_\sigma(i) \cap (\mathcal{T}_\sigma(j) + \gamma),$$

where equality holds if the tiling property is satisfied.

Proof We fix $i \in A$. Let $\nu \in \partial \mathcal{T}_\sigma(i)$ and assume that ν is not contained in an intersection of the form $\mathcal{T}_\sigma(i) \cap (\mathcal{T}_\sigma(j) + \gamma)$, for $\mathcal{T}_\sigma(j) + \gamma \neq \mathcal{T}_\sigma(i)$. Since the tiles $\mathcal{T}_\sigma(j) + \gamma$ are compact, local finiteness implies that the set $\bigcup_{[\gamma,j]^* \neq [0,i]^*,\ [\gamma,j]^* \in \Gamma_c} (\mathcal{T}_\sigma(j) + \gamma))$ is a closed set in $\mathbb{H}_c$, and thus its complement in $\mathbb{H}_c$ is an open set. Since the complement contains ν, there exists an open neighbourhood B of ν which contains only points of $\mathcal{T}_\sigma(i)$, and no point of other tiles of the self-replicating multiple tiling $\mathcal{I}_\sigma$ of Definition 5.3.14. However, since $\nu \in \partial \mathcal{T}_\sigma(i)$, there is some point $\nu' \in B$ which is not contained in $\mathcal{T}_\sigma(i)$. But then ν' is contained in no tile of $\mathcal{I}_\sigma$, which contradicts the covering property of $\mathcal{I}_\sigma$. We thus have proved the inclusion.

Assume now that the tiling property holds. Moreover, assume that there is $\nu \in \text{int}(\mathcal{T}_\sigma(i))$ which is contained in $\mathcal{T}_\sigma(j) + \gamma$ for some $[\gamma, j]^* \in \Gamma_c$. Then, because $\mathcal{T}_\sigma(i)$ as well as $\mathcal{T}_\sigma(j) + \gamma$ is the closure of its interior (Theorem 5.3.12), there is a $\nu' \in \text{int}(\mathcal{T}_\sigma(i)) \cap \text{int}(\mathcal{T}_\sigma(j) + \gamma)$. Thus, there is a small disk around ν' which is covered by both of these tiles, a contradiction to the tiling property. □

Note that we deduce the following interesting corollary. We only give a proof of the 'if' implication. For a proof of the converse implication, see (Siegel and Thuswaldner 2010).

Corollary 5.5.6 *Let σ be a unit Pisot irreducible substitution. The substitution σ satisfies the geometric finiteness property if, and only if, 0 is an inner point of the central tile $\mathcal{T}_\sigma = \bigcup_{i \in A} \mathcal{T}_\sigma(i)$, and 0 belongs to no other tile of the self-replicating tiling.*

Proof By Proposition 5.4.12 and by Lemma 5.5.2, the geometric finiteness property implies that 0 belongs to no tile of the form $\mathcal{T}_\sigma(i) + \gamma$ with $\gamma \neq 0$ and $i \in A$. Since the tiling property holds and by Proposition 5.5.5, the boundary of $\mathcal{T}_\sigma = \bigcup_{i \in A} \mathcal{T}_\sigma(i)$ is exactly described by the intersections of $\mathcal{T}_\sigma = \bigcup_{i \in A} \mathcal{T}_\sigma(i)$ with the other tiles of the tiling $\mathcal{I}_\sigma$. We deduce that 0 does not belong to the boundary of $\mathcal{T}_\sigma(i)$, and thus that 0 is an inner point of $\mathcal{T}_\sigma = \bigcup_{i \in A} \mathcal{T}_\sigma(i)$. □

Remark 5.5.7 Corollary 5.5.6 was already stated in (Akiyama 2002) in the beta-numeration context. The union of cycles in the ancestor graph is called the *zero-expansion graph* in (Siegel and Thuswaldner 2010). Following the proof of Lemma 5.5.2, labels of edges in cycles allow all Dumont–Thomas expansions (see Sections 5.11 and 9.4.2 for more details) that can be obtained for 0 to be computed explicitly, hence the terminology 'zero-expansion'. Compare also with the notion of *zero automaton* in Chapter 2. The zero-expansion graph allows tiles in $\mathcal{I}_\sigma$ that contain 0 (they are related to cycles in the ancestor graph) to be characterised, and thus, to give a further characterisation of the geometric finiteness property.

Let us end this section with the following statement (given without a proof) that is an analogue of Lemma 5.5.2 and which provides an explicit description of the neighbours of a subtile $\mathcal{T}_\sigma(i)$ with respect to the boundary graph. For a proof, see (Siegel and Thuswaldner 2010).

Proposition 5.5.8 *Let σ be a unit Pisot irreducible substitution. Let $i \in A$ and let $[\gamma, j]^* \in \Gamma_c$ with $\gamma \neq 0$ or $i \neq j$. The intersection $\mathcal{T}_\sigma(i) \cap (\mathcal{T}_\sigma(j) + \gamma)$ is non-empty if, and only if, $[i, \gamma, j]^*$ is a vertex of the boundary graph.*

5.5.2 Approximations of the boundary and contact graphs

Even if the boundary graph is much smaller than the two-piece ancestor graph, its computation relies on the pre-computation of the two-piece seed patch W_σ. However, there is another approach which does not require the pre-computation of this patch. The underlying idea is developed in (Thuswaldner 2006), based on polyhedral approximations of the central tile and its subtiles. Recall that with each tip $[\gamma, i]^*$ we associate the compact polyhedron $[\gamma, i]_g^*$ which is obtained by projecting the corresponding face of type i by π_c (see Section 5.3.3 for details).

Definition 5.5.9 Let σ be a unit Pisot irreducible substitution. The mth *polyhedral approximation* of a subtile $\mathcal{T}_\sigma(i)$ is the union of polyhedra given by

$$\mathcal{T}_\sigma^{(m)}(i) := \bigcup_{[\gamma, i]^* \in \mathbf{E}_1^{*m}[0,i]^*} h_\sigma^m([\gamma, i]_g^*).$$

Proposition 5.5.10 *The Hausdorff limit of the polyhedral approximations $\mathcal{T}_\sigma^{(m)}(i)$ for $m \to \infty$ is the subtile $\mathcal{T}_\sigma(i)$.*

Proof This is a direct consequence of the definition of $\mathbf{E}_1^*$, the GIFS equation (5.8) and the uniqueness of the solution of a GIFS. For more details, see (Arnoux and Ito 2001). □

Remark 5.5.11 Figure 5.9 depicts the image of $\bigcup_{i \in A} \mathcal{T}_\tau^{(15)}(i)$ inflated by h_τ^{-15}, for the substitution $\tau(1) = 2$, $\tau(2) = 3$, $\tau(3) = 12$. We have seen in Example 5.4.13 that σ does not satisfy the geometric finiteness property. Note that Figure 5.9 gives some indication of the fact that 0 (which is dotted) is not an interior point of the central tile.

Definition 5.5.12 (Contact graph) Let σ be a unit Pisot irreducible substitution. A 0 *level vertex* of the two-piece ancestor graph is a vertex of the form $[i, \gamma, j]^*$, such that the $((n-1)$-dimensional) geometric tips $[0, i]_g^*$ and $[\gamma, j]_g^*$ intersect on exactly one face of dimension $n-2$. The *contact graph* is the subgraph of the two-piece ancestor graph whose vertices are the 0 level vertices together with all vertices that can be reached by a path in the two-piece ancestor graph starting at a 0 level vertex.

The contact graph is described in (Thuswaldner 2006). Among the graphs considered for tiling criteria so far, the contact graph is the one which is easiest to compute. Indeed, it can be computed with a recursive procedure that always terminates and does not require the computation of the two-piece seed patch W_σ. We first compute the set of 0 level vertices. For $(i,j) \in A^2$, we take the triple $[i, \pi_c(\mathbf{e}_j), i]^*$ if $\langle \mathbf{v}_\beta, \mathbf{e}_j \rangle < \langle \mathbf{v}_\beta, \mathbf{e}_i \rangle$, or $[i, \pi_c(\mathbf{e}_j - \mathbf{e}_i), j]^*$, otherwise. Note that this proves that the tips coming from a 0 level vertex belong to W_σ. We then construct recursively the contact graph, by computing successors (in the two-piece ancestor graph) of already computed vertices. The choice of the initial set implies that such a vertex always belongs to the two-piece ancestor graph (note that these computations do not require the knowledge of the two-piece ancestor graph; we only use the definition of its edges to compute the successors). The construction ends when no additional edge can be added to the graph, which always happens according to (Thuswaldner 2006).

It turns out that the contact graph provides also a suitable decomposition of the boundary of the central tile, which in turn provides a very simple tiling condition. For a complete proof, see (Thuswaldner 2006) together with (Siegel and Thuswaldner 2010).

Theorem 5.5.13 (Contact graph tiling condition) *Let σ be a unit Pisot irreducible substitution. The tiling property is satisfied if, and only if, the spectral radius of the contact graph is strictly smaller than the Perron–Frobenius eigenvalue β of $\mathbf{M}_\sigma$.*

Proof [Sketch] In (Thuswaldner 2006) it is proved that all overlaps $(\mathcal{T}_\sigma(k_1) + \gamma_1) \cap (\mathcal{T}_\sigma(k_2) + \gamma_2)$ of the self-replicating multiple tiling $\mathcal{I}_\sigma$ are translations of overlaps of the form $\mathcal{T}_\sigma(i_1) \cap (\mathcal{T}_\sigma(i_1) + \gamma)$, where $[i_1, \gamma, i_2]^*$ is a vertex of the contact graph. Note that the proof heavily relies on the polyhedral tiling of Proposition 5.3.3. Moreover, (Thuswaldner 2006) shows that these intersections are the solution of the GIFS equation

$$\mathcal{T}_\sigma(i_1) \cap (\mathcal{T}_\sigma(i_2) + \gamma) = \bigcup_{[j_1, \eta, j_2]^* \to [i_1, \gamma, i_2]^*} ((\mathcal{T}_\sigma(j_1) \cap (\mathcal{T}_\sigma(j_2) + \eta)) + \delta) \quad (5.38)$$

where the union is taken over all edges of the contact graph which lead to the vertex $[i_1, \gamma, i_2]^*$. The proof can now be finished in the same way as the proof of Theorem 5.5.4. □

This tiling condition is definitely the simplest tiling condition that we have considered so far since the contact graph can be computed recursively without the precomputation of the two-piece seed patch W_σ.

It also provides a 'minimal' GIFS for the boundary of the central tile in the sense that it removes from the boundary graph all the intersections that are redundant, *i.e.*, that are included in other intersections. For details and examples of contact graphs for families of substitutions associated with beta-numeration we refer again to (Thuswaldner 2006).

5.6 Geometric coincidences

We consider now a further set of conditions each of which is equivalent to the tiling property. Using the concept of duality, they can be expressed in terms of the tiling of the expanding line.

5.6.1 Strands and duality

Up to now, we have worked with the set Γ_c consisting of tips which correspond to projections of faces by π_c. The notion of tip has allowed us to define the GIFS substitution $\mathbf{E}_1^*$ (see Definition 5.3.8) inspired by the action of the GIFS equations (5.8) that govern the subtiles $\mathcal{T}_\sigma(i)$, for $i \in A$. The tiling property (see Definition 5.3.15) was then expressed in terms of pre-images of tips under $\mathbf{E}_1^*$ (see *e.g.* Theorem 5.4.21).

Substitutions, Rauzy fractals and tilings

We now wish to adopt a dual one-dimensional viewpoint. Instead of working with faces of hypercubes, we will work with line segments, and the projection π_e will play the role of the projection π_c. We consider formal strands to represent stairs joining points in the integer grid $\mathbb{Z}^n$. An element of the form $(\mathbf{x}, i) \in \mathbb{Z}^n \times A$ is called a *basic formal strand*. In the sequel, we use the notation $[\mathbf{x}, i]$ instead of $(\mathbf{x}, i)$ for this object. The *basic geometric strand* $[\mathbf{x}, i]_g$ is defined as the segment connecting $\mathbf{x}$ with $\mathbf{x} + \mathbf{e}_i$. A *formal strand* is then defined as a union of basic formal strands. We similarly define a *geometric strand*.

Note that there is no more '∗' in the notation for strands: $[\mathbf{y}, j]_g$ represents a segment in $\mathbb{R}^n$ while $[\gamma, i]_g^*$ is the projection in $\mathbb{H}_c$ of a face. Faces of hypercubes and segments can be considered as *dual* in the sense of the duality principle of linear algebra (see (Arnoux and Ito 2001)). Note that in (Sano, Arnoux, and Ito 2001) a Poincaré type duality between generalisations of faces and segments is established. As we shall see in the sequel, this duality will allow us to translate properties of the set Γ_c into properties of the self-similar translation set Γ_e, and thus to work with the tiling $\mathcal{E}_u$ of the expanding line.

If we consider a word $w \in A^*$, the point $\mathbf{P}(w) \in \mathbb{Z}^n$ is the abelianisation of w and one builds in a natural way a formal strand and a geometric strand from $\mathbf{0}$ to $\mathbf{P}(w)$ by simply reading the letters in w. Strands allow one to keep track of the combinatorics of a word that would be lost in the abelianisation process.

We now extend the action of σ to unions of basic strands $[\mathbf{x}, i] \in \mathbb{Z}^n \times A$, according to the formalism of (Arnoux and Ito 2001).

Definition 5.6.1 (Geometric realisation of a substitution) Let σ be a unit Pisot irreducible substitution. The *one-dimensional geometric realisation* of σ is defined on the sets of formal strands by

$$\mathbf{E}_1\{[\mathbf{y}, j]\} = \bigcup_{(p,i,s,),\ \sigma(j)=pis} \{[\mathbf{M}_\sigma \mathbf{y} + \mathbf{P}(p), i]\},$$

$$\mathbf{E}_1(Y_1 \cup Y_2) = \mathbf{E}_1(Y_1) \cup \mathbf{E}_1(Y_2).$$

We also use here the notation $\mathbf{E}_1[\mathbf{y}, j]$ for $\mathbf{E}_1\{[\mathbf{y}, j]\}$. With this formalism at hand, the set $\bigcup_{k \geq 0} \mathbf{E}_1^k[\mathbf{0}, u_0]$ generates the broken line L_u, by replacing formal strands by geometric strands. This implies that the broken line L_u is invariant under the action of $\mathbf{E}_1$.

The following lemma enhances the relation between $\mathbf{E}_1$ and the GIFS substitution $\mathbf{E}_1^*$ acting on the translation set Γ_c.

Lemma 5.6.2 (Duality lemma) *Let* $x, y \in \mathbb{Z}^n$. *Then*

$$[\pi_c(\mathbf{y}), j]^* \in \mathbf{E}_1^*([\pi_c(\mathbf{x}), i]^*) \iff [-\mathbf{x}, i] \in \mathbf{E}_1([-\mathbf{y}, j]).$$

Proof Let $\mathbf{x}, \mathbf{y} \in \mathbb{Z}^n$. Then $[\pi_c(\mathbf{y}), j]^* \in \mathbf{E}_1^*([\pi_c(\mathbf{x}), i]^*)$ if, and only if, there exists p such that $\sigma(j) = pis$ and $\pi_c(\mathbf{y}) = h_\sigma^{-1}(\pi_c(\mathbf{x}) + \pi_c(\mathbf{P}(p))) = \pi_c(\mathbf{M}_\sigma^{-1}\mathbf{x} + \mathbf{P}(p))$. Equivalently by (5.5) we have $\mathbf{y} = \mathbf{M}_\sigma^{-1}(\mathbf{x} + \mathbf{P}(p))$, i.e., $-\mathbf{x} = \mathbf{M}_\sigma(-\mathbf{y}) + \mathbf{P}(p)$, which gives $[-\mathbf{x}, i] \in \mathbf{E}_1([-\mathbf{y}, j])$. □

By defining suitable vector spaces on strands and tips, $\mathbf{E}_1^*$ and $\mathbf{E}_1$ are linked to each other by the duality principle of linear algebra (up to a reverse of the orientation of the space that leads to introduce the '−' sign in the statement of Lemma 5.6.2). This is worked out in (Arnoux and Ito 2001). As a geometric interpretation, we can also say that the broken line L_u and the stepped hyperplane are dual to each other.

From Lemma 5.6.2 and the definition of the self-replicating translation set Γ_e, we deduce the following dictionary. The proof is immediate.

Lemma 5.6.3 (Dictionary) *Let σ be a unit Pisot irreducible substitution. The following assertions are true.*

(i) Ancestor. A tip $[\pi_c(\mathbf{x}), i]^*$ is the ancestor of $[\pi_c(\mathbf{y}), j]^*$ if, and only if, $[-\mathbf{x}, i]$ is a segment of the strand $\mathbf{E}_1[-\mathbf{y}, j]$.

(ii) Common ancestor. Two tips $[\pi_c(\mathbf{y}_1), j_1]^*$ and $[\pi_c(\mathbf{y}_2), j_2]^*$ have a common ancestor $[\pi_c(\mathbf{x}), i]^*$ if, and only if, the strands $\mathbf{E}_1[-\mathbf{y}_1, j_1]$ and $\mathbf{E}_1[-\mathbf{y}_2, j_2]$ both contain the segment $[-\mathbf{x}, i]$.

(iii) Two-piece ancestor Equation (5.32). There exists δ such that $\{[\pi_c(\mathbf{y}_1), j_1]^*, [\pi_c(\mathbf{y}_1), j_2]^*\} \subset \delta + \mathbf{E}_1^*\{[0, i_1]^*, [\pi_c(\mathbf{x}), i_2]^*\}$ if, and only if, there exists $\mathbf{z}$ such that $\mathbf{z} + \{[\mathbf{0}, i_1], [-\mathbf{x}, i_2]\} \subset \mathbf{E}_1\{[-\mathbf{y}_1, j_1], \mathbf{E}_1[-\mathbf{y}_2, j_2]\}$.

(iv) Two-piece patch ancestor. There exists an element δ such that $\{[\pi_c(\mathbf{y}_1), j_1]^*, [\pi_c(\mathbf{y}_2), j_2]^*\} \subset \delta + \mathbf{E}_1^*([0, i]^*)$ if, and only if, there exists $\mathbf{z}$ such that $[\mathbf{z}, i] \in \mathbf{E}_1[-\mathbf{y}_1, j_1] \cap \mathbf{E}_1[-\mathbf{y}_2, j_2]$.

5.6.2 Super coincidence condition

The relations established in Section 5.6.1 naturally lead to the following definition which extends the notion of combinatorial strong coincidence introduced in Definition 5.2.9. It can be considered as the dual property of having a common ancestor. This condition was first defined in (Ito and Rao 2006).

Definition 5.6.4 (Geometric strong coincidence) We say that the basic strands $[\mathbf{y}_1, j_1]$ and $[\mathbf{y}_2, j_2]$ have *geometric strong coincidence* if there exists a positive integer such that $\mathbf{E}_1^N[\mathbf{y}_1, j_1]$ and $\mathbf{E}_1^N[\mathbf{y}_2, j_2]$ have at least one basic formal strand in common.

Note that the combinatorial strong coincidence condition of Definition 5.2.9 is equivalent to the fact that $[\mathbf{0}, j_1]$ and $[\mathbf{0}, j_2]$ have geometric strong coincidence for each pair $(j_1, j_2) \in A^2$. Indeed, assume that $\mathbf{E}_1^N[\mathbf{0}, j_1]$ and $\mathbf{E}_1^N[\mathbf{0}, j_2]$ have one basic formal strand in common, say $[\mathbf{x}, i]$. This is equivalent to the existence of the decompositions $\sigma^N(j_1) = p_1 i s_1$ and $\sigma^N(j_2) = p_2 i s_2$, with $\pi_e(\mathbf{x}) = \pi_e \circ \mathbf{P}(p_1) = \pi_e \circ \mathbf{P}(p_2)$. We deduce from (5.5) that $\mathbf{P}(p_1) = \mathbf{P}(p_2)$.

As we shall see, one checks that a two-piece patch $\{[\pi_c(\mathbf{y}_1), j_1]^*, [\pi_c(\mathbf{y}_2), j_2]^*\}$ is a patch of Γ_c up to translation (*i.e.*, there exists δ such that $\{[\pi_c(\mathbf{y}_1) + \delta, j_1]^*, [\pi_c(\mathbf{y}_2) + \delta, j_2]^*\} \subset \Gamma_c$) if, and only if, the projections of the segments $[-\mathbf{y}_1, j_1]_g$ and $[-\mathbf{y}_2, j_2]_g$ along the expanding direction intersect on a non-degenerate interval. Before we make this more precise, we introduce the following definition, due to (Barge and Kwapisz 2006).

Definition 5.6.5 (Height) We say that the basic strands $[\mathbf{x}, i]$ and $[\mathbf{y}, j]$ have *the same height* if

$$\operatorname{int}(\pi_e([\mathbf{x}, i])_g) \cap \operatorname{int}(\pi_e([\mathbf{y}, j]_g)) \neq \emptyset.$$

Lemma 5.6.6 *Let $\mathbf{x}, \mathbf{y} \in \mathbb{Z}^n$ such that $\langle \mathbf{x}, \mathbf{v}_\beta \rangle \leq \langle \mathbf{y}, \mathbf{v}_\beta \rangle$. Then for each $i \in A$ the following assertions are equivalent.*

- *The tip $[\pi_c(-\mathbf{x} + \mathbf{y}), i]^*$ belongs to Γ_c.*
- *For each $j \in A$, the basic strands $[\mathbf{x}, i]$ and $[\mathbf{y}, j]$ have the same height.*

Proof Let $[\mathbf{x}, i]$ and $[\mathbf{y}, j]$ be two formal strands having the same height with $\langle \mathbf{x}, \mathbf{v}_\beta \rangle \leq \langle \mathbf{y}, \mathbf{v}_\beta \rangle$. By (5.4) we have that

$$\langle \mathbf{x}, \mathbf{v}_\beta \rangle \leq \langle \mathbf{y}, \mathbf{v}_\beta \rangle < \langle \mathbf{x} + \mathbf{e}_i, \mathbf{v}_\beta \rangle.$$

This implies that $0 \leq \langle -\mathbf{x} + \mathbf{y}, \mathbf{v}_\beta \rangle < \langle \mathbf{e}_i, \mathbf{v}_\beta \rangle$. We thus deduce that $[\pi_c(-\mathbf{x} + \mathbf{y}), i]^* \in \Gamma_c$. The proof of the converse implication follows along the same lines. $\square$

Definition 5.6.7 (Super coincidence condition) A unit Pisot irreducible substitution σ satisfies the *super coincidence condition* if any two basic strands $[\mathbf{x}, i]$ and $[\mathbf{y}, j]$ have geometric strong coincidence whenever they have the same height.

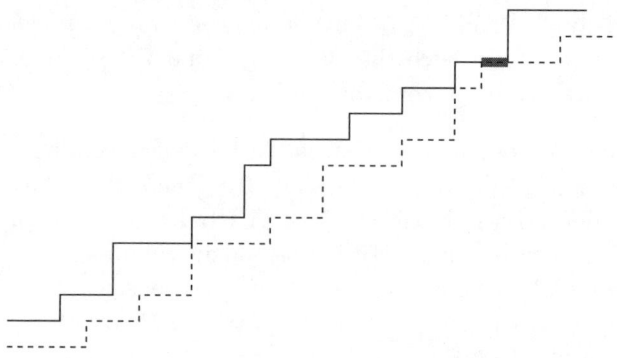

Fig. 5.12 The super coincidence condition.

An illustration of the notion of super coincidence is given in Figure 5.12.

According to (Barge and Kwapisz 2006, Ito and Rao 2006), the tiling condition of Corollary 5.4.15 can be restated as follows.

Theorem 5.6.8 (Ito and Rao 2006) *A unit Pisot irreducible substitution σ satisfies the super coincidence condition if, and only if, the tiling property is satisfied.*

Proof By Corollary 5.4.15, the tiling property is satisfied if, and only if, for any $[\pi_c(\mathbf{y}_1), j_1]^*, [\pi_c(\mathbf{y}_2), j_2]^* \in \Gamma_c$ there is $\delta \in \mathbb{H}_c$ and $i \in A$ such that

$$\{[\pi_c(\mathbf{y}_1), j_1]^*, [\pi_c(\mathbf{y}_2), j_2]^*\} \subset \delta + \mathbf{E}_1^*([0, i]^*).$$

Lemma 5.6.3 (iv) implies that this is equivalent to the fact that there exists $\mathbf{z}$ such that $[\mathbf{z}, i] \in \mathbf{E}_1[-\mathbf{y}_1, j_1] \cap \mathbf{E}_1[-\mathbf{y}_2, j_2]$. Thus, the tiling property is satisfied if, and only if, $[\mathbf{y}_1, j_1]$ and $[\mathbf{y}_2, j_2]$ have geometric strong coincidence for any $[\pi_c(\mathbf{y}_1), j_1]^*, [\pi_c(\mathbf{y}_2), j_2]^* \in \Gamma_c$. In view of Lemma 5.4.20, we may assume w.l.o.g. that $[\pi_c(\mathbf{y}_1) - \pi_c(\mathbf{y}_2), j_1]^* \in \Gamma_c$. Since $[\mathbf{y}_1, j_1]$ and $[\mathbf{y}_2, j_2]$ have geometric strong coincidence if, and only if, $[\mathbf{y}_1 - \mathbf{y}_2, j_1]$ and $[\mathbf{0}, j_2]$ have geometric strong coincidence, the above assertion is equivalent to the fact that $[\mathbf{y}, j_1]$ and $[\mathbf{0}, j_2]$ have geometric strong coincidence if $[\pi_c(\mathbf{y}), j_1]^* \in \Gamma_c$. By Lemma 5.6.6 this is equivalent to the fact that $[\mathbf{y}, j_1]$ and $[\mathbf{0}, j_2]$ have geometric strong coincidence if $[\mathbf{y}, j_1]$ and $[\mathbf{0}, j_2]$ have the same height. However, as $[\mathbf{y}_1, j_1]$ and $[\mathbf{y}_2, j_2]$ have the same height if, and only if, the same is true for $[\mathbf{y}_1 - \mathbf{y}_2, j_1]$ and $[\mathbf{0}, j_2]$, this is equivalent to the super coincidence condition. □

We now define a graph that turns out to be isomorphic to the two-piece ancestor graph. This graph is described in (Barge and Kwapisz 2006).

Similarly as in Section 5.4.3 we introduce the notation $[i_1, \mathbf{x}, i_2]$ for triples $(i_1, \mathbf{x}, i_2) \in A \times \mathbb{Z}^n \times A$ with $[-\pi_c(\mathbf{x}), i_2]^* \in W_\sigma$ where we assume that $i_1 \leq i_2$ if $\mathbf{x} = \mathbf{0}$ to avoid redundancies. The triple $[i_1, \mathbf{x}, i_2]$ represents the pair of tips $[\mathbf{0}, i_1]$ and $[-\mathbf{x}, i_2]$ having the same height by Lemma 5.6.6.

Definition 5.6.9 (Configuration graph) The set of vertices of the *configuration graph* is equal to

$$\{[i_1, \mathbf{x}, i_2] \mid (i_1, \mathbf{x}, i_2) \in A \times \mathbb{Z}^n \times A,\ [-\pi_c(\mathbf{x}), i_2]^* \in W_\sigma,\ i_1 \leq i_2 \text{ if } \mathbf{x} = \mathbf{0}\}.$$

There is an edge from $[j_1, \mathbf{y}, j_2]$ to $[i_1, \mathbf{x}, i_2]$ if there exists $\mathbf{z} \in \mathbb{Z}^n$ such that

$$\mathbf{z} + \{[\mathbf{0}, i_1], [\mathbf{x}, i_2]\} \subset \mathbf{E}_1\{[\mathbf{0}, j_1], [\mathbf{y}, j_2]\}$$

with $\mathbf{z} + [\mathbf{0}, i_1] \cap \mathbf{E}_1\{[\mathbf{0}, j_1], [\mathbf{y}, j_2]\} \neq \emptyset$ and $\mathbf{z} + [\mathbf{x}, i_2] \cap \mathbf{E}_1\{[\mathbf{0}, j_1], [\mathbf{y}, j_2]\} \neq \emptyset$.

Using the duality statements of Lemma 5.6.3 we obtain the following isomorphic graphs.

Proposition 5.6.10 *The configuration graph and the two-piece ancestor graph are isomorphic.*

Proof By definition, $[i_1, \mathbf{x}, i_2]$ is a vertex of the configuration graph if, and only if, $[i_1, -\pi_c(\mathbf{x}), i_2]^*$ is a vertex of the two-piece ancestor graph. By (5.15), we thus get a one-to-one correspondence between the sets of vertices of each graph. Let us consider now edges. According to Lemma 5.6.3 (iii), there exists $\delta \in \mathbb{H}_c$ such that

$$\{[0, j_1]^*, [\pi_c(\mathbf{y}), j_2]^*\} \subset \delta + \mathbf{E}_1^*\{[0, i_1]^*, [\pi_c(\mathbf{x}), i_2]^*\}$$

with

$$\{[0, j_1]^*, [\pi_c(\mathbf{y}), j_2]^*\} \cap (\delta + \mathbf{E}_1^*([0, i_1]^*)) \neq \emptyset \quad \text{and}$$
$$\{[0, j_1]^*, [\pi_c(\mathbf{y}), j_2]^*\} \cap (\delta + \mathbf{E}_1^*([\pi_c(\mathbf{x}), i_2]^*)) \neq \emptyset$$

if, and only if, there exists $\mathbf{z} \in \mathbb{Z}^n$ such that

$$\mathbf{z} + \{[\mathbf{0}, i_1], [-\mathbf{x}, i_2]\} \subset \mathbf{E}_1\{[\mathbf{0}, j_1], [-\mathbf{y}, j_2]\}$$

with

$$(\mathbf{z} + [\mathbf{0}, i_1]) \cap \mathbf{E}_1\{[\mathbf{0}, j_1], [-\mathbf{y}, j_2]\} \neq \emptyset \quad \text{and}$$
$$(\mathbf{z} + [-\mathbf{x}, i_2]) \cap \mathbf{E}_1\{[\mathbf{0}, j_1], [-\mathbf{y}, j_2]\} \neq \emptyset.$$

Hence, also the edges of the configuration graph are in one-to-one correspondence with the edges of the two-piece ancestor graph. $\square$

The following theorem, which has been first proved in (Barge and Kwapisz 2006, Proposition 17.1), is a direct consequence of Proposition 5.6.10 and Theorem 5.4.21.

Theorem 5.6.11 (Barge and Kwapisz 2006, Proposition 17.1) *A unit Pisot irreducible substitution σ satisfies the tiling property if, and only if, from every vertex in the configuration graph, there exists a path to a vertex of the shape $[k, \mathbf{0}, k]$.*

5.7 Overlap coincidences

We are now going to introduce a new viewpoint on configuration graphs and on the super coincidence condition, following the concept of *overlap coincidence*. Overlap coincidence was first used in (Solomyak 1997) for two-dimensional tilings, and later extended to one-dimensional substitution tilings in (Sirvent and Solomyak 2002). As we shall see, this framework allows a graph to be defined that is related to the configuration graph (and the two-piece ancestor graph). Moreover, it provides a simple combinatorial algorithm which allows the tiling property to be decided. In particular, this algorithm avoids having to compute the two-piece seed patch W_σ.

5.7.1 Definitions

So far we have considered pairs of basic strands in $\mathbb{R}^n$ with the same height and checked whether a common basic strand occurs under iterations of $\mathbf{E}_1$. This is the super coincidence condition. By projecting basic strands of the same height by π_e, one recovers intersecting segments, called *overlaps*. The viewpoint used here is to work directly with such intersections. We do not consider all pairs of basic strands with the same height, but we restrict ourselves to basic strands which occur in some translated copies of the broken line L_u.

In order to define suitable translation vectors for the translated copies of L_u, we introduce the set of all possible distances between two tiles of the same type in the self-similar tiling of the expanding line $\mathcal{E}_u = \{\pi_e([\mathbf{x}, i]_g) \mid [\mathbf{x}, i] \in \Gamma_e\}$. Since tiles in the tiling $\mathcal{E}_u$ are ordered according to the fixed point $u = u_0 u_1 u_2 \cdots$, the set of distances between tiles is described by

$$\Xi(\mathcal{E}_u) = \{\lambda \in \mathbb{R}_+ \mid \lambda = \pi_e \circ \mathbf{P}(u_N \cdots u_{N+m-1}),\ N, m \geq 0,\ u_N = u_{N+m}\}$$
$$= \{\lambda \in \mathbb{R}_+ \mid \exists T, T' \in \mathcal{E}_u,\ T' = T + \lambda\}.$$

One has $\Xi(\mathcal{E}_u) \subseteq \mathbb{Z}[\beta]$, since each coordinate of $\mathbf{v}_\beta$ belongs to $\mathbb{Z}[\beta]$. Note that we get $\beta \Xi(\mathcal{E}_u) \subset \Xi(\mathcal{E}_u)$ by using the invariance of u under σ.

Substitutions, Rauzy fractals and tilings

We first introduce the notion of overlap which corresponds to the intersection of the projections of basic geometric strands of the same height, following (Sirvent and Solomyak 2002). Let S, T be two tiles occurring in the self-similar tiling $\mathcal{E}_u$. Note that S and T are intervals. For $\lambda \in \mathbb{R}_+$, the triple (T, S, λ) is called *overlap* if $\mathrm{int}(T \cap (S - \lambda)) \neq \emptyset$. See Figures 5.13 and 5.14 for an illustration.

We now restrict ourselves to sets of pairs of tiles T, S in the self-similar tiling $\mathcal{E}_u$ that are separated by a length belonging to $\Xi(\mathcal{E}_u)$, i.e.,

$$\mathcal{O}_u = \{(T, S, \lambda) \mid \mathrm{int}(T \cap (S - \lambda)) \neq \emptyset,\ T, S \in \mathcal{E}_u,\ \lambda \in \Xi(\mathcal{E}_u)\}.$$

The reason for this restriction for the set of lengths λ will become clear in Section 5.7.2.

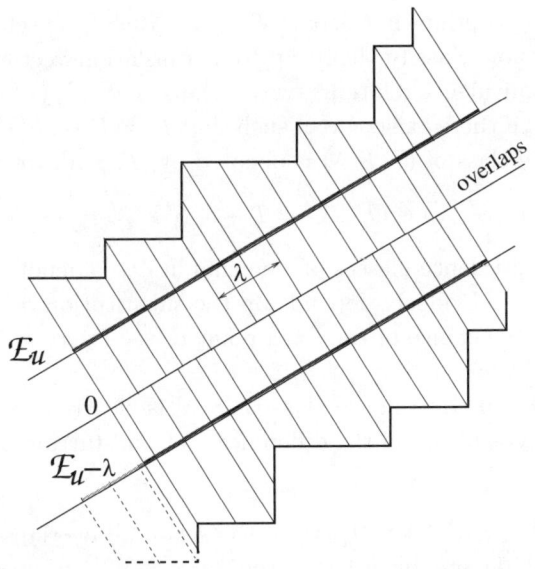

Fig. 5.13 In order to illustrate the relation between the tiling $\mathcal{E}_u$ and the broken lines in $\mathbb{R}^n$ we draw the tilings $\mathcal{E}_u$ and $\mathcal{E}_u - \lambda$ parallel to the expanding eigendirection of M_σ and not in the real line for $\sigma(1) = 112$, $\sigma(2) = 21$.

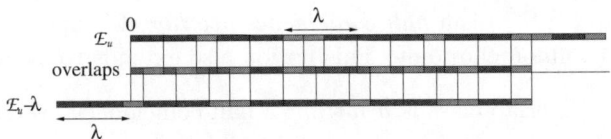

Fig. 5.14 An example of overlaps of a tiling in the real line for $\sigma(1) = 112$, $\sigma(2) = 21$.

The overlaps contained in $\mathcal{O}_u$ can be built in a quite simple geometric way. Indeed, we consider the tiling $\mathcal{E}_u$, and look at the distance λ between two tiles of the same type in $\mathcal{E}_u$. We then shift the tiling $\mathcal{E}_u$ by λ, and take the new tiling by intervals $(\mathcal{E}_u - \lambda) \cap \mathbb{R}_+$, which is not 'synchronised' with $\mathcal{E}_u$, i.e., endpoints of tiles in each tiling do not correspond. We thus define the *synchronised tiling* associated with two tilings by intervals as the tiling by intervals obtained when taking the union of the set of endpoints of both tilings. We use the notation $\mathcal{E} \cap \mathcal{E}'$ for the synchronised tiling associated with $\mathcal{E}$ and $\mathcal{E}'$. We say that we *synchronise* two tilings when we take their synchronised tiling. Synchronising the two tilings $\mathcal{E}_u$ and $\mathcal{E}_u - \lambda$ thus creates new smaller tiles, corresponding to overlaps in $\mathcal{O}_u$. In more geometric terms, we work with the projections of L_u and $L_u - \mathbf{z}$, with $\pi_e(\mathbf{z}) = \lambda$ (see again Figures 5.13 and 5.14).

The number of overlaps in the pair $(\mathcal{E}_u, (\mathcal{E}_u - \lambda) \cap \mathbb{R}^+)$ of self-similar tilings is infinite. We now classify them, up to a translation vector, according to the following equivalence relation: two overlaps (T, S, λ), (T', S', λ') are said to be *equivalent* if there exists $\delta \in \mathbb{R}$ such that $T' = T + \delta$, $S' - \lambda' = S - \lambda + \delta$. The equivalence class of (T, S, λ) is denoted by $[T, S, \lambda]$, i.e,

$$[T, S, \lambda] = \{(T', S', \lambda') \in \mathcal{O}_u \mid T' = T + \delta,\ S' - \lambda' = S - \lambda + \delta,\ \delta \in \mathbb{R}\}.$$

The set of equivalence classes of elements in $\mathcal{O}_u$ is denoted by $[\mathcal{O}_u]$. Let us note that $[\mathcal{O}_u]$ obviously depends on the substitution σ. However, it is independent of the choice of the fixed point u (see Exercise 5.5).

Remark 5.7.1 An equivalence class $[T, S, \lambda]$ is clearly determined by the type of T, the type of S and the difference ν of the starting points of T and $S - \lambda$.

For a fixed $\lambda \in \Xi(\mathcal{E}_u)$ let $\mathcal{O}_u(\lambda)$ be the set of all overlaps (T, S, λ) with $S, T \in \mathcal{E}_u$ and denote by $[\mathcal{O}_u(\lambda)]$ the corresponding set of equivalence classes. We introduce the following terminology.

- A class $[T, S, \lambda] \in [\mathcal{O}_u]$ is called an *overlap class*.
- An overlap (T, S, λ) is a *coincidence overlap* if $T = S - \lambda$. This notion extends to overlap classes.
- An overlap (T, S, λ) is a *half-coincidence overlap* if T and $S - \lambda$ have at least one common endpoint. This notion also extends to overlap classes.

A coincidence overlap class is *a fortiori* a half-coincidence overlap class.

Lemma 5.7.2 *Let σ be a unit Pisot irreducible substitution. The set $[\mathcal{O}_u]$ of overlap classes is finite.*

Note that the proof of this lemma uses similar arguments as, for instance, the proof of Proposition 2.3.33 and Lemma 2.4.7.

Proof Let (T, S, λ) be an overlap. There exist basic geometric strands $[\mathbf{x}, i]_g$, $[\mathbf{y}, j]_g$ in L_u such that $T = \pi_e([\mathbf{x}, i]_g)$, $S = \pi_e([\mathbf{y}, j]_g)$. Let $\nu = \pi_e(\mathbf{y} - \mathbf{x}) - \lambda$ be the difference of the starting points of T and $S - \lambda$. In view of Remark 5.7.1 (note that there are finitely many types of tiles), it suffices to show that only finitely many choices of ν are possible. We have

$$T \cap (S - \lambda) = \pi_e(\mathbf{x}) + (\pi_e[\mathbf{0}, i]_g \cap (\pi_e[\mathbf{0}, j]_g + \nu)).$$

Note that (T, S, λ) is an overlap, if, and only if, $-\pi_e(\mathbf{e}_j) < \nu < \pi_e(\mathbf{e}_i)$.

Furthermore, since $\lambda \in \Xi(\mathcal{E}_u)$, there exists $\mathbf{z} \in \mathbb{Z}^n$ such that $\lambda = \langle \mathbf{z}, \mathbf{v}_\beta \rangle$. Recall that we have assumed that the coordinates of $\mathbf{v}_\beta$ all belong to $\mathbb{Z}[\beta]$. This implies that $\nu = \langle \mathbf{y} - \mathbf{x} - \mathbf{z}, \mathbf{v}_\beta \rangle \in \mathbb{Z}[\beta]$, and that the Galois conjugates of ν are given by $\langle \mathbf{y} - \mathbf{x} - \mathbf{z}, \mathbf{v}_{\beta^{(i)}} \rangle$. We shall prove that these Galois conjugates are uniformly bounded. As they are the coordinates of $\pi_c(\mathbf{y} - \mathbf{x} - \mathbf{z})$ in the basis $(\mathbf{u}_{\beta^{(i)}})_{i \geq 2}$, we have to show the boundedness of $\pi_c(\mathbf{y} - \mathbf{x} - \mathbf{z})$. Since $[\mathbf{x}, i]_g$ and $[\mathbf{y}, j]_g$ are segments L_u, we have $\pi_c(\mathbf{x}), \pi_c(\mathbf{y}) \in \mathcal{T}_\sigma$. Thus Theorem 5.2.3 yields that $\pi_c(\mathbf{x})$ and $\pi_c(\mathbf{y})$ are uniformly bounded. Moreover, as $\pi_e(\mathbf{z}) \in \Xi(\mathcal{E}_u)$, we get that $\pi_c(\mathbf{z}) \in \mathcal{T}_\sigma - \mathcal{T}_\sigma$. Using Theorem 5.2.3 again, this implies that $\pi_c(\mathbf{z})$ is uniformly bounded, too. Therefore ν and all its Galois conjugates are uniformly bounded. Since $\nu \in \mathbb{Z}[\beta]$, there are only finitely many possibilities for ν. □

Remark 5.7.3 This proof yields a strong relation between overlaps and tips: we have proved that with each overlap class $[T, S, \lambda]$ we associate the pair of basic strands $([\mathbf{0}, i], [\mathbf{z}, j])$ that have the same height, with $\mathbf{z} \in \mathbb{Z}^n$ being uniquely determined by $\pi_e(\mathbf{y} - \mathbf{x}) - \lambda = \pi_e(\mathbf{z})$, where $T = \pi_e([\mathbf{x}, i]_g)$, $S = \pi_e([\mathbf{y}, j]_g)$. Note that the position of the basic strand $[\mathbf{z}, j]$ is very specific.

With each overlap (T, S, λ) we associate the intersection $T \cap (S - \lambda)$. Recall that by the invariance of the broken line under $\mathbf{E}_1$ we have the self-similarity equation

$$\beta \pi_e([\mathbf{y}, j]_g) = \bigcup_{[\mathbf{x}, i] \in \mathbf{E}_1 [\mathbf{y}, j]} \pi_e([\mathbf{x}, i]_g).$$

Applying this equation to the intersection $T \cap (S - \lambda)$ yields the decomposition

$$\beta(T \cap (S - \lambda)) = \bigcup_{T' \subset \beta T,\, S' \subset \beta S,\, T' \in \mathcal{E}_u,\, S' \in \mathcal{E}_u} (T' \cap (S' - \beta \lambda)). \quad (5.39)$$

The sets that occur on the right-hand side of (5.39) are tiles of the synchronised tiling associated with the pair $(\mathcal{E}_u, (\mathcal{E}_u - \beta\lambda) \cap \mathbb{R}_+)$.

The following graph will allow us to formulate a new notion of coincidence that will give rise to a further tiling criterion.

Definition 5.7.4 (Graph of overlaps) The *graph of overlaps*, denoted by $\mathcal{G}_\mathcal{O}$, is a directed graph whose set of vertices is the set $[\mathcal{O}_u]$ of overlap classes. There is an edge from $[T, S, \lambda]$ to $[T', S', \lambda']$ if $T' \cap (S' - \lambda')$ is non-empty and appears in the self-similar decomposition (5.39) of $\beta(T \cap (S - \lambda))$.

Note that Lemma 5.7.2 implies that the graph of overlaps $\mathcal{G}_\mathcal{O}$ is finite and does not depend on the choice of u (see Exercise 5.5).

The graph of overlaps is very close to the configuration graph (and thus to the two-piece ancestor graph). The main difference is that the configuration graph and the two-piece ancestor graph are defined in terms of tips or basic strands in the two-piece seed patch W_σ, while the construction of the graph of overlaps does not involve W_σ. Indeed, as seen in the proof of Lemma 5.7.2, the graph of overlaps selects pairs of tips and pairs of basic strands according to the set $\Xi(\mathcal{E}_u)$. This leads to a finite number of pairs of tips or of pairs of basic strands and no reduction to the two-piece seed patch is needed any more.

We introduce a new notion of coincidence which is defined in terms of the graph of overlaps.

Definition 5.7.5 (Strong overlap coincidence condition) The unit Pisot irreducible substitution σ satisfies the *strong overlap coincidence condition* if each vertex in the graph of overlaps $\mathcal{G}_\mathcal{O}$ admits a path leading to an overlap coincidence.

We will prove in Section 5.7.2 that this condition together with the combinatorial strong coincidence condition is equivalent to the tiling property. However, checking the strong overlap coincidence condition is hard since *a priori* it requires to identify *all* non-empty overlaps provided by the expanding tiling, *i.e.*, to consider all parameters $\lambda \in \Xi(\mathcal{E}_u)$. Thus the first problem is to determine $\Xi(\mathcal{E}_u)$. However, fortunately we will see in Section 5.7.2 that we do not need to work with the whole set $\Xi(\mathcal{E}_u)$ in order to check the strong overlap coincidence condition. Indeed, we can restrict ourselves to an arbitrary *single* element $\lambda \in \Xi(\mathcal{E}_u)$. In particular, we will have to consider only synchronisations for the family $\{\mathcal{E}_u, \mathcal{E}_u - \lambda, \mathcal{E}_u - \beta\lambda, \ldots, \mathcal{E}_u - \beta^m \lambda, \ldots\}$.

Definition 5.7.6 (Graph of overlaps of λ) Let $\lambda \in \Xi(\mathcal{E}_u)$. The *graph* *of overlaps of* λ, denoted by $\mathcal{G}_\mathcal{O}(\lambda)$, is the subgraph of the graph of overlaps whose vertices belong to $\bigcup_{i \geq 0} [\mathcal{O}_u(\beta^i \lambda)]$.

As for each $\lambda \in \Xi(\mathcal{E}_u)$ the graph $\mathcal{G}_\mathcal{O}(\lambda)$ is a subgraph of the finite graph of overlaps $\mathcal{G}_\mathcal{O}$, it is itself a finite graph. Moreover, as a finite graph has only finitely many pairwise non-isomorphic subgraphs there are only finitely many different graphs in the class $\{\mathcal{G}_\mathcal{O}(\lambda) \mid \lambda \in \Xi(\mathcal{E}_u)\}$.

To $\mathcal{G}_\mathcal{O}(\lambda)$ we relate the following coincidence condition which will turn out to be equivalent to the strong overlap coincidence condition (see Theorem 5.7.13).

Definition 5.7.7 (Weak overlap coincidence) The unit Pisot irreducible substitution σ satisfies the *weak overlap coincidence condition* if there exists $\lambda \in \Xi(\mathcal{E}_u)$ with $\lambda \neq 0$ such that each vertex in its associated graph of overlaps $\mathcal{G}_\mathcal{O}(\lambda)$ admits a path to an overlap coincidence.

The advantage of this condition is that it can be checked effectively in an easy way.

The following lemma contains a first result on the relation between the strong and the weak overlap coincidence condition.

Lemma 5.7.8 *The strong overlap coincidence condition is true if, and only if, the weak overlap coincidence condition is true for each* $\lambda \in \Xi(\mathcal{E}_u)$.

Proof Assume that the strong coincidence condition is true and choose $\lambda \in \Xi(\mathcal{E}_u)$ arbitrary. Let $[T, S, \lambda]$ be a vertex of $\mathcal{G}_\mathcal{O}(\lambda)$. As $\mathcal{G}_\mathcal{O}(\lambda)$ is a subgraph of $\mathcal{G}_\mathcal{O}$, there is a path in $\mathcal{G}_\mathcal{O}$ from $[T, S, \lambda]$ to a coincidence. However, if a vertex of $\mathcal{G}_\mathcal{O}$ is contained in $\mathcal{G}_\mathcal{O}(\lambda)$, then all its successors in $\mathcal{G}_\mathcal{O}$ are contained in $\mathcal{G}_\mathcal{O}(\lambda)$ in view of (5.39). Thus the path from $[T, S, \lambda]$ to a coincidence which is contained in $\mathcal{G}_\mathcal{O}$ by assumption is also contained in $\mathcal{G}_\mathcal{O}(\lambda)$.

To prove the converse assume that the weak overlap coincidence is true for each $\lambda \in \Xi(\mathcal{E}_u)$. Choose a vertex of $\mathcal{G}_\mathcal{O}$. This vertex is of the form $[T, S, \gamma]$ for some $\gamma \in \Xi(\mathcal{E}_u)$. Thus, in $\mathcal{G}_\mathcal{O}(\gamma)$ there is a path from $[T, S, \gamma]$ to a coincidence. Since $\mathcal{G}_\mathcal{O}(\gamma)$ is a subgraph of $\mathcal{G}_\mathcal{O}$, this path also exists in $\mathcal{G}_\mathcal{O}$. □

5.7.2 Tiling conditions related to overlap graphs

In this section we have two main aims. First we want to prove that the combinatorial strong coincidence condition together with the strong overlap coincidence condition is equivalent to the tiling property. As mentioned above, the strong overlap coincidence condition is hard to check. Thus, in a second step, we show that strong and weak overlap coincidence are equivalent. Summing up we will arrive at a tiling criterion in terms of combinatorial strong coincidence and weak overlap coincidence.

In all what follows we enumerate the tiles of $\mathcal{E}_u$ starting from the tile next to the origin by $T_0, T_1, \ldots$ One has $T_r = \pi_e[\mathbf{P}(u_0 \cdots u_{r-1}), u_r]_g$ for all $r \in \mathbb{N}$. Moreover, for each $\lambda \in \Xi(\mathcal{E}_u)$, we consider the union of tiles $T \in \mathcal{E}_u$ such that $T + \lambda$ also occurs in $\mathcal{E}_u$, i.e.,

$$\mathrm{Occ}(\lambda) := \bigcup_{\{T \in \mathcal{E}_u \mid T + \lambda \in \mathcal{E}_u\}} T.$$

We start with the following lemma (cf. (Solomyak 1997, Proposition 6.7) and (Lee, Moody, and Solomyak 2003, Lemma A.8)), which translates the weak overlap coincidence condition in combinatorial terms (see also the related result (Queffélec 1987, Lemma VI.27)).

Lemma 5.7.9 *Let $\lambda \in \Xi(\mathcal{E}_u)$. The graph of overlaps of λ satisfies the weak overlap coincidence condition if, and only if, there exists some constant $c \in (0, 1)$ such that*

$$\mathrm{Card}\{T_r \in \mathcal{E}_u \mid r \leq N,\ T_r + \beta^m \lambda \notin \mathcal{E}_u\} \ll N c^m \qquad (5.40)$$

holds for all $m \in \mathbb{N}$ when N is large enough in terms of m (the implied constant does not depend on N and m).

Proof We first prove that there is a constant $\tilde{c} \in (0, 1)$ such that

$$\limsup_{N \to \infty} \frac{\mu_1(\{T_r \mid r \leq N,\ T_r + \beta^m \lambda \notin \mathcal{E}_u\})}{\mu_1(\{T_r \mid r \leq N\})} \ll \tilde{c}^m \qquad (5.41)$$

if, and only if, there exists some constant $c \in (0, 1)$ such that (5.40) holds for all $m \in \mathbb{N}$ when N is large enough in terms of m.

Assume that (5.40) holds. Let $L_{\min}$ and $L_{\max}$ be the length of the shortest and longest tile in $\mathcal{E}_u$, respectively. One has for N large enough

$$\mu_1(\{T_r \mid r \leq N,\ T_r + \beta^m \lambda \notin \mathcal{E}_u\}) \leq L_{\max} \mathrm{Card}\{T_r \mid r \leq N,\ T_r + \beta^m \lambda \notin \mathcal{E}_u\}$$

$$\ll c^m N \leq c^m \frac{1}{L_{\min}} \mu_1(\{T_r \mid r \leq N\})$$

$$\ll c^m \mu_1(\{T_r \mid r \leq N\}).$$

Hence,

$$\limsup_{N \to \infty} \frac{\mu_1(\{T_r \mid r \leq N,\ T_r + \beta^m \lambda \notin \mathcal{E}_u\})}{\mu_1(\{T_r \mid r \leq N\})} \ll c^m$$

and (5.41) is true for $\tilde{c} = c$.

Conversely, assume that (5.41) holds for $\tilde{c} \in (0,1)$. Let $c \in (0,1)$ with $\tilde{c} < c$. For N large enough in terms of m, one has

$$\mu_1(\{T_r \mid r \leq N,\ T_r + \beta^m \lambda \notin \mathcal{E}_u\}) \ll c^m \mu_1(\{T_r \mid r \leq N\}).$$

Hence,

$$\mathrm{Card}(\{T_r \mid r \leq N,\ T_r + \beta^m \lambda \notin \mathcal{E}_u\})$$
$$\leq \frac{1}{L_{\min}} \mu_1(\{T_r \mid r \leq N,\ T_r + \beta^m \lambda \notin \mathcal{E}_u\})$$
$$\ll c^m \mu_1(\{T_r \mid r \leq N\}) \leq c^m N L_{\max} \ll c^m N,$$

which ends the proof of the claimed equivalence.

We now prove (5.41). We first assume that the graph of overlaps $\mathcal{G}_\mathcal{O}(\lambda)$ of λ satisfies the weak overlap coincidence condition. By the finiteness of $\mathcal{G}_\mathcal{O}(\lambda)$ there exists a positive integer ℓ such that each vertex admits a path of length bounded by ℓ to an overlap coincidence.

Using (5.39), one checks that $\mathcal{G}_\mathcal{O}(\beta^m \lambda)$ is a subgraph of $\mathcal{G}_\mathcal{O}(\lambda)$ whose paths to overlap coincidences do not get longer as the ones in $\mathcal{G}_\mathcal{O}(\lambda)$. In particular, $\beta^\ell(T \cap (S - \beta^m \lambda))$ contains an overlap coincidence for each overlap $(T, S, \beta^m \lambda)$. One has $\mu_1(\beta^\ell(T \cap (S - \beta^m \lambda))) \leq \beta^\ell L_{\max}$. Moreover, each overlap $(T, S, \beta^m \lambda)$ produces tiles which belong to $\mathrm{Occ}(\beta^{m+\ell} \lambda)$, and whose union has length bounded from below by $L_{\min}$. Let $b := \left(1 - \frac{L_{\min}}{L_{\max} \beta^\ell}\right)$. Since the definition of Occ immediately implies $\beta \mathrm{Occ}(\beta^m \lambda) \subset \mathrm{Occ}(\beta^{m+1} \lambda)$, and, hence, $\beta^\ell \mathrm{Occ}(\beta^m \lambda) \subset \mathrm{Occ}(\beta^{m+\ell} \lambda)$, we deduce that

$$\limsup_{N \to \infty} \frac{\mu_1\{T_r \in \mathcal{E}_u \mid r \leq N,\ T_r + \beta^{m+\ell} \lambda \notin \mathcal{E}_u\}}{\mu_1\{T_r \in \mathcal{E}_u \mid r \leq N\}}$$
$$\leq b \limsup_{N \to \infty} \frac{\mu_1\{T_r \in \mathcal{E}_u \mid r \leq N,\ T_r + \beta^m \lambda \notin \mathcal{E}_u\}}{\mu_1\{T_r \in \mathcal{E}_u \mid r \leq N\}}.$$

Since $b < 1$ does not depend on m and N, by writing $m = k\ell + s$ with $0 \leq s < \ell$ we easily derive (5.41) by iteration.

To prove the converse assume that (5.41) holds and that $\mathcal{G}_\mathcal{O}(\lambda)$ does not satisfy the weak overlap coincidence condition. Then there exists an overlap class $[T, S, \beta^m \gamma]$ such that for every $\ell, m > 0$

$$\beta^\ell(T \cap (S - \beta^m \lambda)) \subset \mathbb{R}_+ \setminus \mathrm{Occ}(\beta^m \lambda).$$

By the repetitivity of $\mathcal{E}_u$ this yields

$$\limsup_{N \to \infty} \frac{\mu_1\{T_r \in \mathcal{E}_u \mid r \leq N,\ T_r + \beta^{m+\ell} \lambda \notin \mathcal{E}_u\}}{\mu_1\{T_r \in \mathcal{E}_u \mid r \leq N\}} \gg 1,$$

uniformly in m, which contradicts (5.41). □

We now establish the relation between the tiling property and the strong overlap coincidence condition. The first part of the following lemma is proved in (Lee 2007) in the general context of substitution Delone sets and Meyer sets. Indeed, it can be derived from the *algebraic coincidence condition* introduced in (Lee 2007). For more details, see the notes at the end of the present chapter.

Lemma 5.7.10 *Let σ be a unit Pisot irreducible substitution. If σ satisfies the strong overlap coincidence condition, then the following assertions are true.*

- *There exists $m \in \mathbb{N}$ such that $\beta^m \left(\Xi(\mathcal{E}_u) - \Xi(\mathcal{E}_u) \right) \cap \mathbb{R}_+ \subset \Xi(\mathcal{E}_u)$.*
- *There exists $m \in \mathbb{N}$ such that $\beta^m \left(\Xi(\mathcal{E}_u) + \Xi(\mathcal{E}_u) \right) \subset \Xi(\mathcal{E}_u)$.*

Proof Let $\lambda_1, \lambda_2 \in \Xi(\mathcal{E}_u)$ with $\lambda_2 - \lambda_1 > 0$ be given. Then Lemma 5.7.9 implies that there exists $m \in \mathbb{N}$ such that for all N large enough and for $i = 1, 2$, one has

$$\operatorname{Card}\{r \leq N \mid T_r + \beta^m \lambda_i \notin \mathcal{E}_u\} < \frac{N}{3}. \tag{5.42}$$

This implies that

$$\operatorname{Card}\{r \leq N \mid T_r + \beta^m \lambda_1 \notin \mathcal{E}_u \text{ or } T_r + \beta^m \lambda_2 \notin \mathcal{E}_u\} < \frac{2N}{3} \tag{5.43}$$

which means that there is some $r \in \mathbb{N}$ such that $T_r + \lambda_1 \beta^m \in \mathcal{E}_u$ and $T_r + \lambda_2 \beta^m \in \mathcal{E}_u$. Thus, $\beta^m (\lambda_2 - \lambda_1) \in \Xi(\mathcal{E}_u)$ which proves the first assertion.

Recall that $L_{\min}$ denotes the length of the shortest tile in $\mathcal{E}_u$. If N is large in terms of m, one has moreover

$$\operatorname{Card}\{\beta^m \lambda_1 / L_{\min} \leq r \leq N \mid T_r - \beta^m \lambda_1 \notin \mathcal{E}_u\} < \frac{N}{3}$$

which implies

$$\operatorname{Card}\{\beta^m \lambda_1 / L_{\min} \leq r \leq N \mid T_r - \beta^m \lambda_1 \notin \mathcal{E}_u \text{ or } T_r + \beta^m \lambda_2 \notin \mathcal{E}_u\} < \frac{2N}{3}.$$

Thus, if N is large in terms of m, there exists $r \leq N$ such that $T_r - \beta^m \lambda_1 \in \mathcal{E}_u$ and $T_r + \lambda_2 \beta^m \in \mathcal{E}_u$. Thus, $\beta^m (\lambda_2 + \lambda_1) \in \Xi(\mathcal{E}_u)$ which proves the second assertion. □

In order to get a relation between the strong overlap coincidence condition and the tiling condition, we need to ensure that all lengths of tiles in $\mathcal{E}_u$ multiplied by some power of β belong to the translation set $\Xi(\mathcal{E}_u)$.

This is realised by assuming the combinatorial strong coincidence condition introduced in Definition 5.2.9.

Lemma 5.7.11 *Let σ be a unit Pisot irreducible substitution that satisfies the combinatorial strong coincidence condition. Then there exists m such that $\beta^m \pi_e \circ \mathbf{P}(i) \in \Xi(\mathcal{E}_u)$ for every letter $i \in A$.*

Proof Let $i \in A$. Let $j \in A$ be such that ij is a factor of the fixed point u. By the combinatorial strong coincidence condition, there exist $m > 0$, a letter k and four words p, q, r, s such that $\sigma^m(i) = pkq$ and $\sigma^m(j) = rks$, with $\mathbf{P}(p) = \mathbf{P}(r)$. Since $\sigma^m(ij)$ is a factor of u, we have that $\pi_e \circ \mathbf{P}(kqr) \in \Xi(\mathcal{E}_u)$. The relation $\mathbf{P}(p) = \mathbf{P}(r)$ yields $\pi_e \circ \mathbf{P}(kqr) = \pi_e \circ \mathbf{P}(pkq) = \pi_e \circ \mathbf{P}(\sigma^m(i)) = \beta^m \pi_e \circ \mathbf{P}(i) \in \Xi(\mathcal{E}_u)$. □

We can now derive the following relation between the strong overlap coincidence condition and the tiling condition.

Theorem 5.7.12 *Let σ be a unit Pisot irreducible substitution. Then σ satisfies the tiling property if, and only if, σ satisfies both the strong overlap coincidence condition and the combinatorial strong coincidence condition.*

Proof We first assume that σ satisfies the tiling property. By Theorem 5.6.8, σ satisfies the super coincidence condition and, in particular, the combinatorial strong coincidence condition. According to Remark 5.7.3, each overlap class $[T, S, \lambda]$ is associated with a pair of basic formal strands $([\mathbf{0}, i], [\mathbf{z}, j])$ having the same height. Since the super coincidence condition holds, there exists m such that $\mathbf{E}_1^m[\mathbf{z}, i]$ and $\mathbf{E}_1^m[0, j]$ contain a common basic formal strand. In other words, one gets an overlap coincidence between $\beta^m T$ and $\beta^m(S - \lambda)$. Therefore the strong overlap coincidence is satisfied.

The converse is slightly more difficult to establish. This is mostly due to the specific positions of the vectors $\mathbf{z}$ associated with overlap classes as noticed in Remark 5.7.3. We assume that σ satisfies both the strong overlap coincidence condition and the combinatorial strong coincidence condition. Let $[\mathbf{x}, i]$ and $[\mathbf{y}, j]$ be a pair of basic formal strands with the same height. Let $T := \pi_e([\mathbf{x}, i]_g)$ and $S := \pi_e([\mathbf{y}, j]_g)$. There exist $\mathbf{z}_1, \mathbf{z}_2 \in \pi_e(\mathbb{Z}^n)$ such that $T \in \mathcal{E}_u - \pi_e(\mathbf{z}_1)$ and $S \in \mathcal{E}_u - \pi_e(\mathbf{z}_2)$. Assume w.l.o.g. that $\pi_e(\mathbf{z}_1) < \pi_e(\mathbf{z}_2)$. Now set $T' := T + \pi_e(\mathbf{z}_1)$ and $S' := S + \pi_e(\mathbf{z}_1)$. Then $T' \in \mathcal{E}_u$ and $S' \in \mathcal{E}_u - \lambda$ with $\lambda := \pi_e(\mathbf{z}_2 - \mathbf{z}_1)$.

By Lemma 5.7.11, the combinatorial strong coincidence condition implies that there exists $m_1 \in \mathbb{N}$ such that $\beta^{m_1} \pi_e \circ \mathbf{P}(k) \in \Xi(\mathcal{E}_u)$ for all $k \in A$. As $\{\mathbf{P}(1), \ldots, \mathbf{P}(n)\} = \{\mathbf{e}(1), \ldots, \mathbf{e}(n)\}$ forms a basis of the lattice $\mathbb{Z}^n$, according to Lemma 5.7.10 for each $\mathbf{z} \in \mathbb{Z}^n$ there exists $m_2 \in \mathbb{N}$ such that $\beta^{m_2}\pi_e(\mathbf{z}) \in \Xi(\mathcal{E}_u)$. Thus there exists $m \in \mathbb{N}$ such that $\lambda \in \Xi(\mathcal{E}_u)$. We thus deduce from the strong overlap coincidence condition that $[\beta^m T, \beta^m S, \beta^m \lambda]$ leads to a coincidence overlap. This implies that the super coincidence condition holds. □

We now turn to the second main result of the present section, the equivalence between the strong and the weak overlap coincidence condition. This is proved in a slightly different context in (Solomyak 1997, Section 6) where it is shown that both conditions are equivalent to the fact that the dynamical system associated with the tiling has pure discrete spectrum (see also (Lee 2007, Section 3) where overlap coincidence is related to pure discrete spectrum). We give a new and direct proof of this result here.

Theorem 5.7.13 *Let σ be a unit Pisot irreducible substitution. Then the following assertions are equivalent.*

(i) *The substitution σ satisfies the weak overlap coincidence condition.*
(ii) *The substitution σ satisfies the strong overlap coincidence condition.*

In view of Lemmas 5.7.8 and 5.7.9 it is clear that Theorem 5.7.13 is a direct consequence of the following lemma. In the proof of this lemma we exploit the fact that weak overlap coincidence implies that the fixed point u of σ is *mean-almost periodic* in the sense of (Queffélec 1987, Definition VI.4).

Lemma 5.7.14 *Let $\lambda_1, \lambda_2 \in \Xi(\mathcal{E}_u)$ with $\lambda_1 \neq 0$. If there exists $c \in (0,1)$ such that*

$$\mathrm{Card}\{T_r \in \mathcal{E}_u \mid r \leq N,\ T_r + \beta^m \lambda_1 \notin \mathcal{E}_u\} \ll N c^m \quad (5.44)$$

for all m when N is large enough in terms of m then

$$\mathrm{Card}\{T_r \in \mathcal{E}_u \mid r \leq N,\ T_r + \beta^m \lambda_2 \notin \mathcal{E}_u\} \ll N c^m$$

for all m when N is large enough in terms of m. The implied constants do not depend on m and N.

Proof Let λ_1 be a non-zero element of $\Xi(\mathcal{E}_u)$ such that (5.44) holds and let $\lambda_2 \in \Xi(\mathcal{E}_u)$.

We consider a tile T that occurs in $\mathrm{Occ}(\lambda_2)$. One has $T + \lambda_2 \in \mathcal{E}_u$. By the repetitivity of $\mathcal{E}_u$, there is a set of tiles $W^{(0)} \subset \mathcal{E}_u$ such that the union of tiles of $W^{(0)}$ is a relatively dense set in $\mathbb{R}_+$ (*i.e.*, there exists $L > 0$ such that every interval of length L in $\mathbb{R}_+$ contains at least one point belonging to one tile of $W^{(0)}$) and $T_r + \beta\lambda_2 \in \mathcal{E}_u$ whenever $T_r \in W^{(0)}$.

We now use the self-similarity of $\mathcal{E}_u$ (or equivalently, the fact that $\sigma(u) = u$). We multiply all tiles of $W^{(0)}$ by β^m and subdivide accordingly to arrive again at $\mathcal{E}_u$. Thus, there exist

$$u_1^{(m)} < v_1^{(m)} < u_2^{(m)} < v_2^{(m)} < u_3^{(m)} < v_3^{(m)} < \cdots$$

with

$$\min_i(v_i^{(m)} - u_i^{(m)}) \gg \beta^m, \ \max_i(u_{i+1}^{(m)} - v_i^{(m)}) \ll \beta^m, \ u_1^{(m)} \ll \beta^m \quad (5.45)$$

such that $T_r + \beta^m \lambda_2 \in \mathcal{E}_u$ whenever $T_r \in W^{(m)}$, where $W^{(m)} := \{T_s \mid s \in \bigcup_{i \geq 1} [\![u_i^{(m)}, v_i^{(m)}]\!]\}$. All bounds in (5.45) follow from the fact that

$$\forall a \in A, \ \beta^m \ll |\sigma^m(a)| \ll \beta^m \quad (5.46)$$

(for more details on (5.46) see Section 4.7.3). To get the upper bound for $\max_i(u_{i+1}^{(m)} - v_i^{(m)})$ also the relative denseness of the union of tiles of $W^{(0)}$ has to be used.

Let $m_0 \in \mathbb{N}$ be fixed in a way that $\beta^{m-m_0}\lambda_1 < \min_i(v_i^{(m)} - u_i^{(m)})$ (such a constant exists in view of (5.45)).

For each $K \in \mathbb{N}$ define the 'exceptional set'

$$S_K := \{T_r \in \mathcal{E}_u \mid T_r - k\beta^{m-m_0}\lambda_1 \notin \mathcal{E}_u \text{ for some } 0 \leq k \leq K\}.$$

Equation (5.44) implies that

$$\mathrm{Card}(S_K \cap \{T_r \mid r \leq N\}) \ll Nc^m \quad (5.47)$$

holds for all m if N large enough (note that the implied constant may depend on K but this is not relevant for us as K will be fixed in a moment). By (5.45) and since $\lambda_1 \neq 0$ we may fix $K \in \mathbb{N}$ in a way that for all $T_r \in \mathcal{E}_u \setminus S_K$ and all $m \in \mathbb{N}$, there exists $k \in [\![0, K]\!]$ such that $T_r - k\beta^{m-m_0}\lambda_1 \in W^{(m)}$.

We set for $k \in [\![0, K]\!]$

$$\mathcal{E}_u^{(k,m)} := \{T_r \in \mathcal{E}_u \mid T_r - k\beta^{m-m_0}\lambda_1 \in W^{(m)}\}.$$

Then

$$\{T_r \mid r \in \mathbb{N}\} = \bigcup_{k=0}^{K} \mathcal{E}_u^{(k,m)} \cup S_K. \quad (5.48)$$

Note that

$$T_r \in \mathcal{E}_u^{(0,m)} \implies T_r + \beta^m \lambda_2 \in \mathcal{E}_u.$$

We now take $k \neq 0$. Let $T_r \in \mathcal{E}_u^{(k,m)}$ such that $T_r + \beta^m \lambda_2 \notin \mathcal{E}_u$. One has $T_r - k\beta^{m-m_0}\lambda_1 \in W^{(m)}$, hence,

$$T_r - k\beta^{m-m_0}\lambda_1 + \beta^m \lambda_2 \in \mathcal{E}_u.$$

We thus have

$$T_r - k\beta^{m-m_0}\lambda_1 + \beta^m \lambda_2 \in \mathcal{E}_u, \ (T_r - k\beta^{m-m_0}\lambda_1 + \beta^m \lambda_2) + k\beta^{m-m_0}\lambda_1 \notin \mathcal{E}_u.$$

Hence, by recalling that $L_{\min}$ is the length of the smallest tile in $\mathcal{E}_u$, we get for $N \in \mathbb{N}$

$$\operatorname{Card}\left\{T_r \in \mathcal{E}_u^{(k,m)} \mid r \leq N, \ T_r + \beta^m \lambda_2 \notin \mathcal{E}_u\right\} \leq$$
$$\operatorname{Card}\left\{T_r \mid r \leq N + \frac{\beta^m \lambda_2}{L_{\min}}, \ T_r + k\beta^{m-m_0}\lambda_1 \notin \mathcal{E}_u\right\}.$$

Putting this together with (5.48), we obtain

$$\operatorname{Card}\{T_r \in \mathcal{E}_u \mid r \leq N, \ T_r + \beta^m \lambda_2 \notin \mathcal{E}_u\}$$
$$\leq \sum_{k=0}^{K} \operatorname{Card}\left\{T_r \in \mathcal{E}_u^{(k,m)} \mid r \leq N, \ T_r + \beta^m \lambda_2 \notin \mathcal{E}_u\right\}$$
$$+ \operatorname{Card}(S_K \cap \{T_r \mid r \leq N\})$$
$$\leq \sum_{k=0}^{K} \operatorname{Card}\{T_r \mid r \leq N + \frac{\beta^m \lambda_2}{L_{\min}}, \ T_r + k\beta^{m-m_0}\lambda_1 \notin \mathcal{E}_u\}$$
$$+ \operatorname{Card}(S_K \cap \{T_r \mid r \leq N\}).$$

We deduce from (5.44), (5.45) and (5.47) that

$$\operatorname{Card}\{T_r \in \mathcal{E}_u \mid r \leq N, \ T_r + \beta^m \lambda_2 \notin \mathcal{E}_u\} \ll \left(N + \frac{K\beta^m \lambda_2}{L_{\min}}\right) c^m + Nc^m.$$

This implies that

$$\operatorname{Card}\{T_r \in \mathcal{E}_u \mid r \leq N, \ T_r + \beta^m \lambda_2 \notin \mathcal{E}_u\} \ll Nc^m$$

holds for all m when N is large in terms of m. □

As mentioned above, Theorem 5.7.13 is an immediate consequence of the previous lemma. According to this theorem it is not necessary to build all of the configuration graph (or of the two-piece ancestor graph) in order to check the tiling property. Indeed, it is enough to build the graph from all pairs of tiles in the broken line that are separated by a fixed vector. We sum this up in the following corollary.

Corollary 5.7.15 *Let σ be a unit Pisot irreducible substitution. Then σ satisfies the tiling property if, and only if, σ satisfies both the weak overlap coincidence condition and the combinatorial strong coincidence condition.*

The computation of the pairs of such tiles still requires the knowledge of the language of the fixed point. In Section 5.8, we will take advantage of the weak overlap property in order to obtain a purely combinatorial characterisation of the tiling property. To this end we will need the following corollary which focuses on gaps between overlaps instead of focusing on density of overlaps. Its proof is an immediate consequence of the proof of Lemma 5.7.9.

Corollary 5.7.16 *Let $\lambda \in \Xi(\mathcal{E}_u)$. The weak overlap coincidence condition is satisfied for a $\lambda \in \Xi(\mathcal{E}_u)$ if, and only if, the distance between two successive coincidence overlaps in $(\mathcal{E}_u, (\mathcal{E}_u - \beta^k \lambda) \cap \mathbb{R}_+)$ is bounded uniformly in k.*

5.8 Balanced pair algorithm

This section is devoted to a further effective condition for the tiling property based on the notion of *balanced pairs* introduced in (Michel 1978) and later used *e.g.* by (Livshits 1987), (Queffélec 1987, Chapter VI), (Sirvent and Solomyak 2002) and (Martensen 2004). The starting point for the balanced pair algorithm is Corollary 5.7.16. It states that the tiling property is strongly related to the uniform boundedness (in k) of the length of gaps between successive coincidence overlaps in pairs of tilings of the form $(\mathcal{E}_u, (\mathcal{E}_u - \beta^k \lambda) \cap \mathbb{R}_+))$. According to Theorem 5.7.13, this property has to be checked only for one suitable $\lambda \in \Xi(\mathcal{E}_u)$. We will choose λ of the form $\lambda = \langle \mathbf{P}(w), \mathbf{v}_\beta \rangle$ for some prefix w of u with $u_0 = u_{|w|}$. For this choice we get that the first tile in $\mathcal{E}_u$ coincides with a tile of $(\mathcal{E}_u - \beta^k \lambda) \cap \mathbb{R}_+$ for each $k \in \mathbb{N}$. Indeed, for each k one has $\beta^k \lambda = \langle \mathbf{M}_\sigma^k \mathbf{P}(w), \mathbf{v}_\beta \rangle = \langle \mathbf{P}(\sigma^k(w)), \mathbf{v}_\beta \rangle$ where $\sigma^k(w)$ is a prefix of u. Note that the first overlap and the last overlaps of the part of the synchronised tiling $\mathcal{E}_u \cap (\mathcal{E}_u - \beta^k \lambda)$ that is located between 0 and $\beta^k \lambda$ are half-coincidences.

In applying the balanced pair algorithm, we will start with gaps between half-coincidence overlaps in order to get the uniform boundedness of gaps between successive coincidence overlaps. A gap between two half-coincidence overlaps can be described as an interval that can be decomposed in two ways: firstly, as a union of consecutive tiles of $\mathcal{E}_u$ and secondly, as a union of consecutive tiles of $(\mathcal{E}_u - \lambda \beta^k) \cap \mathbb{R}_+$. The types of these consecutive tiles correspond to two finite subwords v_1 and v_2 of u, respectively. Since the coordinates of $\mathbf{v}_\beta$ are rationally independent, we have $\mathbf{P}(v_1) = \mathbf{P}(v_2)$. This leads us to introduce the following combinatorial definition.

Definition 5.8.1 (Balanced pair) A pair $(v_1, v_2) \in A^* \times A^*$ is said to be a *combinatorial balanced pair*, or for short, a *balanced pair*, if $\mathbf{P}(v_1) = \mathbf{P}(v_2)$. A *one-letter balanced pair* (also called *coincidence balanced pair*) is a balanced pair of the form (a, a), with $a \in A$.

An *irreducible balanced pair* is a pair (v_1, v_2) with the property that no pair (v_1', v_2'), where v_i' is a proper prefix v_i, $i = 1, 2$, is balanced.

Note that when (v_1, v_2) is a balanced pair, $(\sigma(v_1), \sigma(v_2))$ is balanced as well. Obviously, each balanced pair can be split up uniquely into irreducible balanced pairs. This process is called *reduction*.

The set of irreducible balanced pairs obtained after the reduction of a balanced pair of finite words (v_1, v_2) is denoted by $\mathcal{B}(v_1, v_2)$. We say that a balanced pair (v_1, v_2) *leads to a coincidence* if there exists $N \in \mathbb{N}$ such that $\mathcal{B}(\sigma^N(v_1), \sigma^N(v_2))$ contains a one-letter balanced pair.

The following algorithm will lead to a powerful criterion for the tiling property.

Definition 5.8.2 (Balanced pair algorithm) Let I_0 be a non-empty and finite set of balanced pairs. The *balanced pair algorithm* applied to I_0 successively computes the sets

$$I_k := \bigcup_{(v_1, v_2) \in I_{k-1}} \mathcal{B}(\sigma(v_1), \sigma(v_2)).$$

The algorithm is said to *terminate with rank k*, $k \geq 1$, if $I_{k+1} = I_k$ and each balanced pair $(v_1, v_2) \in I_k$ leads to a coincidence. The algorithm is said to *terminate* if it terminates for some rank $k \geq 1$.

Let $I_0 = \{(v_1, v_2)\}$. If the balanced pair algorithm terminates there exist only finitely many different words w which occur between two consecutive one-letter balanced pairs in the pairs $(\sigma^k(v_1), \sigma^k(v_2))$. Thus the length of such gaps is uniformly bounded in k. We formulate this in a more exact way in the following proposition (see also (Sirvent and Solomyak 2002, Theorem 5.6)).

Proposition 5.8.3 *The balanced pair algorithm applied to a non-empty finite set I_0 of balanced pairs terminates if, and only if, the number of letters between two successive one-letter balanced pairs in $(\sigma^k(v_1), \sigma^k(v_2))$ for any $(v_1, v_2) \in \bigcup_k I_k$, is bounded uniformly in k.*

Example 5.8.4 In this example we want to consider the substitution $\sigma(1) = 112$, $\sigma(2) = 13$ and $\sigma(3) = 1$. It is easy to check that this is a unit Pisot irreducible substitution. We want to perform the balanced pair algorithm for this example, starting from $I_0 = \{(12,21),(13,31),(23,32)\}$. Applying σ and reducing yields

$$(12,21) \to (11213, 13112) \to (1,1)(1213, 3112),$$
$$(13,31) \to (1121, 1112) \to (1,1)(1,1)(21,12),$$
$$(23,32) \to (131, 113) \to (1,1)(31,13).$$

Thus in I_1 we have the one-letter pairs $(2,2)$, $(1,1)$ and the new pair $(1213, 3112)$. Repeating the procedure for I_1 yields the new reduction

$$(1213, 3112) \to (112131121, 111211213) \to (1,1)(1,1)(21,12)(31121, 11213).$$

Now I_2 contains the new pair $(31121, 11213)$. It is treated as follows:

$$(31121, 11213) \to (111211213112, 112112131121)$$
$$\to (1,1)(1,1)(12,21)(1,1)(12,21)(13,31)(1,1)(12,21).$$

This implies that I_3 contains no new pair. Moreover, each of the occurring pairs leads to a one-letter balanced pair. Thus the algorithm terminates.

The notions of balanced pair and reduction also make sense for pairs of infinite words. Let w be a non-empty prefix of the infinite word u. It is not hard to see that the set of irreducible balanced pairs occurring by reducing $(u, S^{|w|}(u))$ (here S denotes the shift) is a finite set (see for instance (Sirvent and Solomyak 2002, Section 3) and Exercise 5.6). Denote this set by $I_0(w)$. We then consider the balanced pairs in $(u, S^{|\sigma(w)|}(u))$, which amounts to applying the reduction process to all balanced pairs in $I_0(w)$. The balanced pair algorithm starting with the set $I_0 = I_0(w)$ is called the *balanced pair algorithm associated with w*.

The balanced pair algorithm is stated in the literature in several different forms. Besides taking the set $I_0(w)$ associated with some prefix w of u as the starting point (Sirvent and Solomyak 2002), the initial set I_0 is defined as $I_0 = \{(ij, ji) \mid i, j \in A, i \neq j\}$ in (Barge and Kwapisz 2006). As we will see, both starting sets lead to the same behaviour. Sometimes, it proves to be more convenient to start with $\{(ij, ji) \mid i \neq j\}$ instead of starting with $I_0(w)$ (even if this latter set can be determined in an effective way).

We now relate the balanced pair algorithm to the overlap coincidence condition.

Theorem 5.8.5 *Let σ be a unit Pisot irreducible substitution. Let w be a prefix of u with $u_{|w|} = u_0$ and set $\lambda = \pi_e \circ \mathbf{P}(w) \in \Xi(\mathcal{E}_u)$. The substitution σ satisfies the weak overlap coincidence condition for λ if, and only if, the balanced pair algorithm associated with w terminates.*

Proof Since w is a prefix of u, $(\mathcal{E}_u - \lambda) \cap \mathbb{R}_+$ is obtained from $\mathcal{E}_u$ by deleting the $|w|$ first tiles. This implies that coincidence overlaps between tiles in $(\mathcal{E}_u, (\mathcal{E}_u - \lambda) \cap \mathbb{R}_+)$ are in one-to-one correspondence with one-letter balanced pairs in $(u, S^{|w|}u)$. The same correspondence holds between $(\mathcal{E}_u, (\mathcal{E}_u - \beta^k \lambda) \cap \mathbb{R}_+)$ and $(u, S^{|\sigma^k(w)|}u)$. Now compare a step from $I_k(w)$ to $I_{k+1}(w)$ in the balanced pair algorithm with the symbolic interpretation of the self-similarity equation (5.39). The result now follows from Proposition 5.8.3 and Corollary 5.7.16. □

Corollary 5.7.15 and Theorem 5.8.5 immediately imply the following result (see also (Sirvent and Solomyak 2002, Section 5)).

Corollary 5.8.6 *A unit Pisot irreducible substitution σ satisfies the tiling property if, and only if, the combinatorial strong coincidence condition is satisfied and there exists a prefix w of u with $u_0 = u_{|w|}$ such that the balanced pair algorithm associated with w terminates.*

An even easier criterion for the tiling property can be obtained by starting the balanced pair algorithm with $I_0 = \{(ij, ji) \mid i \neq j\}$ instead of $I_0(w)$. As will be proved below (see Theorem 5.8.8), applying the balanced pair algorithm with this choice of I_0 allows both the combinatorial strong coincidence condition and the weak overlap condition to be checked at once. This yields the purely combinatorial algorithm for tiling discussed in (Barge and Kwapisz 2006). Before stating and proving Theorem 5.8.8, we establish the following auxiliary result.

Lemma 5.8.7 *Let σ be a unit Pisot irreducible substitution. If the balanced pair algorithm starting with $I_0 = \{(ij, ji) \mid i \neq j\}$ terminates, then the balanced pair algorithm starting with any balanced pair $(w, w') \in A^* \times A^*$ terminates.*

Proof We assume that the balanced pair algorithm starting with $I_0 = \{(ij, ji) \mid i \neq j\}$ terminates. Let $w, w' \in A^*$ be such that $\mathbf{P}(w) = \mathbf{P}(w')$. Let ℓ be equal to the common length $|w| = |w'|$ of w and w'. Since $\mathbf{P}(w) = \mathbf{P}(w')$, there exists a permutation ρ of the set $[\![1, \ell]\!]$ such that $w' = w_\rho$ where w_ρ is a shorthand for the word $w_{\rho(1)} \cdots w_{\rho(\ell)}$. We recall that a *transposition* is a permutation that exchanges two elements and that lets the other elements invariant, and that the symmetric group, i.e., the group of permutations of $[\![1, \ell]\!]$ is generated by the permutations τ_i, for $i \in [\![1, \ell - 1]\!]$, where $\tau_i := (i, i+1)$ is the transposition of $[\![1, \ell]\!]$ that exchanges i and $i+1$. We set $T := \{\tau_i \mid i \in [\![1, \ell - 1]\!]\}$.

We have to show that the balanced pair algorithm starting with (w, w_ρ) terminates for each permutation ρ. We will prove this by induction.

In order to establish the induction start we have to consider the balanced pair algorithm starting with (w, w_τ) for some $\tau \in T$. However, since $\mathcal{B}(w, w_\tau) \subset \{(i,i)\} \cup \{(ij, ji) \mid i \neq j\}$ we only have to consider elements of $I_0 = \{(ij, ji) \mid i \neq j\}$ as the starting set to assure termination of the algorithm. Thus the balanced pair algorithm starting with (w, w_τ) terminates by the assumptions of the lemma.

The induction step is proved if we establish the following 'transitivity property'. Let $\tau, \tau' \in T$. If the balanced pair algorithm starting with (w, w_τ) terminates, and if the balanced pair algorithm starting with $(w_\tau, w_{\tau' \circ \tau})$ terminates, then also the balanced pair algorithm starting with $(w, w_{\tau' \circ \tau})$ terminates. To show this, according to Proposition 5.8.3, it is sufficient to prove that the occurrences between successive one-letter balanced pairs in $(\sigma^k(w), \sigma^k(w_{\tau' \circ \tau}))$ are bounded uniformly in k. As the balanced pair algorithm starting with (w, w_τ) as well as with $(w_\tau, w_{\tau \circ \tau'})$ terminates, Proposition 5.8.3 implies that there exists a constant $C > 0$ such that the occurrences between successive one-letter balanced pairs in $(\sigma^k(w), \sigma^k(w_\tau))$ and in $(\sigma^k(w_\tau), \sigma^k(w_{\tau' \circ \tau}))$ are bounded by C for each k.

Let N be large enough such that $|\sigma^N(a)| \geq C+1$, for every letter $a \in A$. Fix $k \in \mathbb{N}$ and let j be the index of a one-letter balanced pair in the pair of words $(\sigma^k(w), \sigma^k(w_\tau))$, i.e.,

$$\sigma^k(w) = pas, \ \sigma^k(w_\tau) = p'as' \text{ with } \mathbf{P}(p) = \mathbf{P}(p') \text{ and } |p| = |p'| = j-1.$$

Let j' be the index of the first letter of the image $\sigma^N(a)$ of the jth letter in $\sigma^k(w)$ as well as in $\sigma^k(w_\tau)$ under σ^N. This implies that each letter with index $\ell \in \{j', \ldots, j' + C\}$ forms a one-letter balanced pair in the reduction of the pair $(\sigma^{k+N}(w), \sigma^{k+N}(w_\tau))$.

Since by assumption the gaps between one-letter balanced pairs in $(\sigma^{k+N}(w_\tau), \sigma^{k+N}(w_{\tau' \circ \tau}))$ are bounded by C, there is an index $j'' \in \{j', \ldots, j' + C\}$ such that the j''th letters of $\sigma^{k+N}(w_\tau)$ and $\sigma^{k+N}(w_{\tau' \circ \tau})$ coincide. However, as $j'' \in \{j', \ldots, j' + C\}$ also the j''th letters of $\sigma^{k+N}(w)$ and $\sigma^{k+N}(w_\tau)$ coincide. Combining these two assertions we see that the j''th letters of $\sigma^{k+N}(w)$ and $\sigma^{k+N}(w_{\tau' \circ \tau})$ coincide.

Since this argument goes through for every $k \in \mathbb{N}$ and for every index j of a one-letter balanced pair in $(\sigma^k(w), \sigma^k(w_\tau))$, we deduce that the occurrences between successive one-letter balanced pairs in $(\sigma^{k+N}(w), \sigma^{k+N}(w_{\tau' \circ \tau}))$ are bounded uniformly in k by $(C+1)\|\sigma^N\|$ (recall that the width $\|\sigma\|$ of σ is defined as $\|\sigma\| := \max_{a \in A} |\sigma(a)|$ (see Definition 1.2.20)). This establishes the induction step and the lemma is proved. □

Theorem 5.8.8 *Let σ be a unit Pisot irreducible substitution. The balanced pair algorithm starting with $I_0 = \{(ij, ji) \mid i, j \in A, i \neq j\}$ terminates if, and only if, the tiling property is satisfied.*

Proof We first prove that if the balanced pair algorithm starting with $I_0 = \{(ij, ji) \mid i, j \in A, i \neq j\}$ terminates, then σ satisfies the combinatorial strong coincidence condition. Let $i, j \in A$. According to Proposition 5.8.3, the distance between two one-letter balanced pairs in $(\sigma^N(i)\sigma^N(j)), \sigma^N(j)\sigma^N(i))$ is uniformly bounded in N. As $|\sigma^\ell(k)| \to \infty$ for $\ell \to \infty$ for each $k \in A$ this yields the existence of $N \in \mathbb{N}$ such that $\sigma^N(i)$ and $\sigma^N(j)$ can be decomposed as $\sigma^N(i) = par$ and $\sigma^N(j) = qas$, with $\mathbf{P}(p) = \mathbf{P}(q)$. Hence, σ satisfies the combinatorial strong coincidence condition.

We now assume that the balanced pair algorithm starting with $I_0 = \{(ij, ji) \mid i, j \in A, i \neq j\}$ terminates. Let w be a non-empty prefix of u with $u_{|w|} = u_0$. Let $\lambda = \pi_e \circ \mathbf{P}(w) \in \Xi(\mathcal{E}_u)$. Lemma 5.8.7 implies that the balanced pair algorithm associated with w terminates, since the set $I_0(w)$ is finite. Hence, by Corollary 5.8.5, σ satisfies the weak overlap coincidence condition.

Thus we proved that σ satisfies both the combinatorial strong coincidence condition and the weak overlap coincidence condition. Corollary 5.7.15 now implies that the tiling property is satisfied.

Let us now assume that σ satisfies the tiling property. This implies that both the strong overlap and the combinatorial strong coincidence condition hold. Let $(ij, ji) \in I_0$. Let $T_i = \pi_e([0, i]_g)$, $S_j = \pi_e([\mathbf{e}_i, j]_g)$, $T_j = \pi_e([0, j]_g)$, $S_i = \pi_e([\mathbf{e}_j, i]_g)$. By construction T_i and S_j are adjacent intervals, as well as T_j, S_i, and they cover the same interval $I := T_i \cup S_j = T_j \cup S_i$. This interval can be decomposed as the union of three overlap subintervals O_1, O_2, O_3 that do not necessarily belong to $\mathcal{O}_u$. Assume w.l.o.g. that T_i is longer than S_j. Then these intervals are equal to T_j, $T_i \cap S_i$ and S_j. We deduce from Lemma 5.7.11 that there exists M such that $\beta^M O_1, \beta^M O_2, \beta^M O_3$ can be decomposed into overlaps that belong to $\mathcal{O}_u$. Corollary 5.7.16 then implies that gaps between two coincidence overlaps in $\beta^{k+M} O_1$, $\beta^{k+M} O_2$, $\beta^{k+M} O_3$ are uniformly bounded in k, which implies that gaps between one-letter balanced pairs $(\sigma^k(ij), \sigma^k(ji))$ are also uniformly bounded in k. This implies that the balanced pair algorithm starting with I_0 terminates. □

One may define a graph associated with the balanced pair algorithm in an obvious way. The subgraph of this graph consisting of the coincidences (i, i) corresponds to a slightly modified prefix-suffix graph. If one removes this subgraph, and if the balanced pair algorithm ends, the remaining graph is finite, and its dominant eigenvalue corresponds to the asymptotic order of growth of non-coincidences between two different fixed points of powers of σ.

The balanced pair algorithm only terminates if the substitution satisfies the tiling property. Even if the super coincidence conjecture is true, it may well happen that the algorithm takes a long time before it terminates. We refer the reader to (Sirvent and Solomyak 2002, Section 6) where some examples are presented. They show that even quite simple looking substitutions may lead to quite large sets I_k. This suggests that it might be hard to get an analysis of the complexity of the balanced pair algorithm. Nevertheless, it is a useful and easy to implement tool for checking the tiling property for a given example with a purely combinatorial algorithm.

5.9 Conclusion

As a conclusion, let us summarise the different tiling conditions that we have encountered in the present chapter. The main condition can be formulated as follows: every two-piece patch of Γ_c appears (up to a translation vector) in the iterated image of a tip $[0, i]^*$ with $i \in A$. Several variations around this idea have produced the following graphs.

- The *two-piece ancestor graph* traces back ancestors of patches under a generalised substitution that acts on tips that belong to a finite patch (the two-piece seed patch W_σ) of the self-replicating translation set Γ_c. The tiling property is equivalent to the fact that from every vertex, there exists a path to a vertex with the specific shape $[i, 0, i]^*$, for $i \in A$.
- The *boundary graph* is a subset of the *two-piece ancestor graph*. It describes the tiles in the self-replicating multiple tiling that intersect the subtiles of the central tile. The tiling property can be expressed by computing the spectral radius associated with this graph.
- The *contact graph* is a subgraph of the two-piece ancestor graph that can be computed iteratively without the knowledge of the two-piece seed patch. The tiling property can also be expressed in terms of the spectral radius associated to this graph.
- The *configuration graph* is isomorphic to the two-piece ancestor graph. It checks whether every pair of basic strands eventually contains a common basic strand when applying σ iteratively.
- The *graph of overlaps* follows the same type of construction scheme as the the configuration graph. It is restricted to pairs of strands that are related to the fixed point of the substitution. It has to be combined

with the *combinatorial strong coincidence condition* to provide a tiling property.
- The *balanced pair algorithm* is a simple combinatorial process that describes the growth of gaps between coincidence overlaps and checks whether these gaps are uniformly bounded. It terminates whenever the tiling property is satisfied.

Among these conditions, the balanced pair algorithm is definitively the most combinatorial one. However, it does not always terminate. On the contrary, the contact graph can be computed with a purely algorithmic process that does not require additional computations, and provides a complete tiling characterisation. Note that the relations between the contact graph and the balanced pair algorithm remain unclear so far and deserve specific studies. Note also that the contact graph is not defined in the Pisot reducible case.

5.10 Exercises

Exercise 5.1 Let β be a Parry number (see Definition 2.3.12). The β-substitution σ_β associated with β is defined over the alphabet $\{1, \cdots, n\}$, where n stands for the number of states of the automaton $\mathcal{S}_\beta$ defined in the proof of Proposition 2.3.18, as follows: j is the kth letter of $\sigma_\beta(i)$ if, and only if, there is an arrow in $\mathcal{S}_\beta$ from the state i to the state j labelled by $k-1$. Give the β-substitution σ_β explicitly with respect to $\mathsf{d}_\beta(1)$. Compare its prefix-suffix automaton with the automaton $\mathcal{S}_\beta$. Prove that σ_β is primitive. Show that the Perron–Frobenius eigenvalue of its incidence matrix is equal to β. For more details, see (Lothaire 2002, Chapter 7).

Exercise 5.2 Prove that the first coordinate projection of Γ_c on $\mathbb{H}_c$ is not left invariant by any non-zero translation vector.

Exercise 5.3 Prove that the fact that the self-replicating multiple tiling $\mathcal{I}_\sigma$ is actually a tiling implies that the largest eigenvalue of the adjacency matrix of the boundary graph (as well as of the contact graph) is strictly smaller than the largest eigenvalue of $\mathbf{M}_\sigma$.
Hint: Look at the corresponding result for the contact graph of a self-similar lattice tiling in (Gröchenig and Haas 1994, Section 4).

Exercise 5.4 Prove that if two basic strands have geometric strong coincidence, then they have the same height.

Exercise 5.5 Prove that $[\mathcal{O}_u]$ and $[\mathcal{O}_u(\lambda)]$ do not depend on the choice of the fixed point u, but only on σ.
Hint: Use the repetitivity of $\mathcal{E}_u$, or equivalently, the uniform recurrence of the fixed point u.

Exercise 5.6 Prove that the set of irreducible balanced pairs occurring by reducing $(u, S^{|w|}(u))$ (here S denotes the shift) is a finite set.
Hint: Use the uniform recurrence of the fixed point u.

Exercise 5.7 List among the graphs introduced in the present chapter the ones that contain the prefix-suffix graph (or the graph obtained by reversing the direction of its edges) as a subgraph.

5.11 Notes

Section 5.1

As introduced for instance in (Thurston 1989) and in (Fabre 1995), one can associate in a natural way with the β-shift (see Section 2.3.2.1) a substitution σ_β called *β-substitution*, in the case where β is a Parry number. For more details, see Exercise 5.1. Compare also with the ideas underlying Definition 3.4.10. If β is a Pisot number, the associated substitution can be Pisot reducible as well as Pisot irreducible. An example of a Pisot reducible β-substitution is given by the smallest Pisot number β which is the positive root of $X^3 - X - 1$ (see Example 2.3.54).

Fractal geometry is deeply related to the study of numeration systems. One of the most famous examples of fractal tiles that come from numeration systems is the twin dragon fractal related to expansions of Gaussian integers in base $-1 + i$ (see (Knuth 1998, p. 206)). More generally, tilings can be introduced in the framework of canonical numeration systems (see Section 2.4 and the references in the survey (Akiyama and Thuswaldner 2004)), of shift radix systems (see Section 2.4.4 and (Berthé, Siegel, Steiner, et al. 2009)), or of abstract numeration systems (see Chapter 3 and (Berthé and Rigo 2007a)). In particular, the study of the boundary of central tiles has proved to be particularly efficient in order to derive properties of numeration systems. Under the tiling condition and in the cubic case $n = 3$, points lying at the intersection of tiles in the self-replicating tiling have been described as complex numbers with multiple expansions in some numeration system (see e.g. (Messaoudi 1998, Messaoudi 2000, Sadahiro 2006)). For more on the relations between central tiles and numeration systems, see the survey (Barat, Berthé, Liardet, et al. 2006).

The construction of central tiles also has consequences for the effective construction of Markov partitions for toral automorphisms, the main eigenvalue of which is a Pisot number. See, for instance, (Kenyon and Vershik 1998), (Praggastis 1999), (Schmidt 2000), and (Lindenstrauss and Schmidt 2005). For more on connections between beta-numeration, Vershik's adic transformation (see (Vershik and Livshits 1992)) and codings of hyperbolic automorphisms, see the survey (Sidorov 2003), and in the same vein, (Einsiedler and Schmidt 2002).

The study of central tiles has also led to particularly interesting applications in number theory. This was one of the motivations of (Rauzy 1982). Central tiles and their associated tilings are indeed efficient tools to compute best simultaneous Diophantine approximations (see (Chekhova, Hubert, and Messaoudi 2001), (Hubert and Messaoudi 2006) and (Ito, Fujii, Higashino, et al. 2003)), or to characterise points with purely periodic beta-expansions (see (Hama and Imahashi 1997), (Akiyama, Barat, Berthé, et al. 2008) or (Adamczewski, Frougny, Siegel, et al. 2010)).

Section 5.2

In the case of a unit Pisot reducible substitution, besides $\mathbb{H}_c$ and $\mathbb{H}_e$ a third space plays a role. This is the space $\mathbb{H}_s$ generated by the eigenspaces corresponding to the eigenvalues of $\mathbf{M}_\sigma$ that are not conjugate to β. The projection of the broken line L_u on $\mathbb{H}_c$ along $\mathbb{H}_e \oplus \mathbb{H}_s$ still provides a bounded set in $\mathbb{H}_c$ which allows the definition of a central tile also in this case (for more details, see (Ei, Ito, and Rao 2006, Section 3.2) and (Berthé and Siegel 2005)).

The study of the spectrum of Pisot substitutive dynamical systems was one of the main motivations for the introduction of central tiles. Pisot irreducible substitutions are indeed conjectured to have discrete spectrum. For a detailed account of the spectral theory of substitutive dynamical systems, see (Queffélec 1987), (Pytheas Fogg 2002, Chapter 7) and (Barge and Kwapisz 2006). See also Section 6.9.

The notion of *coincidence* (in its various forms) has proved to be an efficient way for proving discrete spectrum. The coincidence condition was first introduced by Dekking (Dekking 1978) for substitutions with constant length. Using this notion, Dekking completely characterised the substitutions with constant length whose associated symbolic dynamical system has discrete spectrum. Later, Arnoux and Ito introduced the notion of combinatorial strong coincidence (under the name *strong coincidence*) in

(Arnoux and Ito 2001), which lead Ito and Rao to define the super coincidence condition (under the name *super coincidence*) in (Ito and Rao 2006). The super coincidence condition has been also introduced independently in (Barge and Kwapisz 2006) (under the name *geometric coincidence condition*). For a complete proof of the equivalence between discrete spectrum and the super coincidence condition for unit Pisot irreducible substitutions, see (Barge and Kwapisz 2006). The notion of coincidence has also been exploited in the framework of substitution tiling spaces and substitution Delone multisets. Lee (see (Lee 2007)) introduced the notion of *algebraic coincidence* in order to characterise substitution Delone multisets that have a pure point diffraction spectrum. Using the notation of Section 5.7 algebraic coincidence can be stated as follows. Let $\Lambda_i := \bigcup_{n \in \mathbb{N},\ u_n = i} T_n$. We say that the substitution σ satisfies the *algebraic coincidence* condition if there exist a positive integer M and $\xi \in \Lambda_i$ for each $i \in A$ such that $\xi + \beta^M \Xi(\mathcal{E}_u) \subseteq \Lambda_i$. For a review of various notions of coincidences that are related to substitutions and substitution Delone multisets, see (Sing 2006) and the discussion in (Lee 2007). See also (Lee, Moody, and Solomyak 2003) and (Fretlöh and Sing 2007) for the related notion of *modular coincidence*.

The numeration system based on words alluded to in the proof of Theorem 5.2.3 is known as the *Dumont–Thomas numeration system* (see e.g. (Dumont and Thomas 1989, Rauzy 1990, Dumont and Thomas 1993) and Section 9.4.2). More precisely one checks that every finite prefix of u can be uniquely expanded as $\sigma^n(p_n)\sigma^{n-1}(p_{n-1})\cdots p_0$, where $p_n \neq \varepsilon$, and $(p_0, a_0, s_0)\cdots(p_n, a_n, s_n)$ is the sequence of labels of a path in the prefix-suffix automaton $\mathcal{G}_\sigma$ (see (Dumont and Thomas 1989, Theorem 1.5)). Hence, we can expand the non-negative integer N as $N = |\sigma^n(p_n)| + \cdots + |p_0|$, where $u_0 \cdots u_{N-1} = \sigma^n(p_n)\sigma^{n-1}(p_{n-1})\cdots p_0$. Let β be the Perron–Frobenius eigenvalue of σ. This numeration system also provides generalised radix expansions of positive real numbers, with digits belonging to a finite subset of the number field $\mathbb{Q}(\beta)$. We first define the mapping $\delta_\sigma : A^* \to \mathbb{Q}(\beta)$, $p \mapsto \langle \mathbf{P}(p), \mathbf{w}_\beta \rangle$, where $\mathbf{w}_\beta$ is a left eigenvector of $\mathbf{M}_\sigma$ with positive entries associated with the Perron–Frobenius eigenvalue β. One has $\delta_\sigma(\sigma^n(p)) = \beta^n \delta_\sigma(p)$, for every n and $p \in A^*$. We then associate with the combinatorial expansion $(p_n, a_n, s_n)\ldots(p_0, a_0, a_0)$ the real number $\delta_\sigma(p_n)\beta^n + \cdots + \delta_\sigma(p_0) \in \mathbb{Q}(\beta)$. To recover the β-numeration in the particular case where σ is a β-substitution, $\mathbf{w}_\beta$ has to be normalised so that its first coordinate is equal to 1: the coordinates of $\mathbf{w}_\beta$ are then of the form $T_\beta^i(1)$, for $0 \le i \le n-1$, with $T_\beta : x \mapsto \{\beta x\}$. We have chosen to work in the present chapter with an eigenvector $\mathbf{v}_\beta$ normalised so that its coordinates belong to $\mathbb{Z}[\beta]$. This choice of normalisation plays a role in particular in the proof of Lemma 5.7.2. One checks that if σ is a unit Pisot irreducible β-substitution, the theory and the results of this chapter also hold by working with $\mathbf{w}_\beta$ normalised in a way that its first coordinate is equal to 1, instead of working with $\mathbf{v}_\beta$. In particular, the proof of Lemma 5.7.2 can easily be adapted, by noticing that there exists a positive integer $D > 0$ such that the coordinates of $\mathbf{w}_\beta$ all belong to $\frac{1}{D}\mathbb{Z}[\beta]$.

Section 5.3

GIFS substitutions, introduced in (Arnoux and Ito 2001), were inspired by the geometric formalism of (Ito and Ohtsuki 1993), whose aim was to provide explicit Markov partitions for hyperbolic automorphisms of the torus associated with particular substitutions produced by Brun's continued fraction algorithm. GIFS substitutions have already proved their efficiency for Diophantine approximation (Ito, Fujii, Higashino, et al. 2003), in word combinatorics (Arnoux, Berthé, and Siegel 2004), and in discrete geometry (Arnoux, Berthé, and Ito 2002), (Fernique 2006) and (Fernique 2009).

There is a second multiple tiling defined on $\mathbb{H}_c$ that plays an important role in the study of the substitutive symbolic dynamical system (X_σ, S). It is obtained by projecting points in $\mathbb{Z}^n$ that lie on the hyperplane with equation $\langle \mathbf{x}, (1, \ldots, 1) \rangle = 0$ by π_c. The corresponding translation set, called the *lattice translation set*, is thus defined as $\{[\gamma, i]^* \in \pi_c(\mathbb{Z}^n) \times A \mid \gamma \in \sum_{k=2}^{n} \mathbb{Z}(\pi_c(\mathbf{e}_k) - \pi_c(\mathbf{e}_1))\}$. It is clearly periodic. According to (Canterini and Siegel 2001b), if σ is a unit Pisot irreducible substitution that satisfies the combinatorial strong coincidence condition, the lattice translation set is a Delone set that also provides a multiple tiling for the subtiles of the central tile. This multiple tiling is called the *lattice multiple tiling*. Rauzy introduced in the seminal paper (Rauzy 1982) the notion of central tile with respect to this tiling. According to (Ito and Rao 2006) (see also (Barge and Kwapisz 2006, Remark 18.5)), we know that the lattice multiple tiling is a tiling if, and only if, the self-replicating multiple tiling is a tiling, if σ is assumed to be a unit Pisot irreducible substitution.

There are two dynamical systems that can be associated in a natural way with a unit Pisot substitution, namely the substitutive dynamical system (X_σ, S) (with its natural $\mathbb{Z}$-action by the shift), and the one-dimensional tiling space associated with the self-similar tiling of the expanding line (described in terms of an $\mathbb{R}$-action by translations). The lattice multiple tiling is intimately connected to the spectral properties of the substitutive dynamical system (X_σ, S) (see (Queffélec 1987) and (Pytheas Fogg 2002, Chapter 7)), whereas the self-replicating multiple tiling is connected to the spectral properties of the one-dimensional tiling space associated with the tiling of the expanding line (see (Barge and Kwapisz 2006)). Note that there exist unit Pisot reducible substitutions for which (X_σ, S) does not have discrete spectrum, as shown by (Baker, Barge, and Kwapisz 2006, Example 5.3). See also (Clark and Sadun 2006) for the study of the spectral impact of deformations of the lengths of tiles for the tiling spaces associated with a substitution. More generally, see (Sadun 2008) for a topological study of tiling spaces with aperiodic order.

Section 5.4

The geometric finiteness property is intimately related to the so-called (F) property (introduced in Section 2.3.2.2 in the beta-numeration framework). It is expressed in (Berthé and Siegel 2005) in terms of the Dumont–Thomas numeration. It also appears in (Fuchs and Tijdeman 2006) in a related context. Note that we can use the vast literature on the (F) property in the beta-numeration framework to exhibit classes of beta-substitutions that satisfy the geometric finiteness property (see *e.g.* (Baker, Barge, and Kwapisz 2006) and (Barge and Kwapisz 2005)).

The so-called (W) or *weak finiteness* property was first introduced in (Hollander 1996). He has proved that the (W) property implies the pure discreteness of the spectrum of the irreducible beta-shift. The (W) property can be stated for a Pisot number β as follows:

$$\forall z \in \mathbb{Z}[\beta^{-1}] \cap [0,1), \ \forall \varepsilon > 0, \ \exists x, y \in \text{Fin}(\beta) \text{ such that } z = x - y \text{ and } y < \varepsilon.$$

The (W) property has been proved in (Akiyama 2002) to be equivalent with the tiling property. An algorithm which can tell whether a given Pisot number β has (F) or (W) property is described in (Akiyama, Rao, and Steiner 2004). The condition of Theorem 5.4.14 is related to the (W) property.

Section 5.5

Similar graphs have appeared in several restricted contexts with different names (see *e.g.* the references in (Akiyama and Thuswaldner 2004) for contact graphs for tiles related to matrix number systems). They are used either to describe beta-expansions for 0 (Akiyama 2002), to describe multiple expansions (Messaoudi 1998, Messaoudi 2000, Durand and Messaoudi 2009), to compute the Hausdorff dimension of the boundary of central tiles (Messaoudi 2000, Feng, Furukado, Ito, et al. 2006,

Thuswaldner 2006), or to obtain pure discrete spectrum conditions for substitutive dynamical systems (by referring to the lattice translation set) (Siegel 2004). The knowledge on intersections between tiles also yields criteria for topological properties of central tiles (connectivity, disklikeness, non-trivial fundamental group) (see (Messaoudi 2000), (Siegel 2004), (Siegel and Thuswaldner 2010)).

Contact graphs are inspired by the contact matrix defined in (Gröchenig and Haas 1994) for self-similar lattice tilings. They have been defined in (Thuswaldner 2006) in the framework of substitutions. The properties of the contact graph are based on the polyhedral tiling generated by the geometric tips. This polyhedral tiling property is very specific to the Pisot irreducible case. This is the main reason why the contact graph can only be defined in the Pisot irreducible case, whereas the two-piece ancestor and the boundary graphs can be defined in the Pisot reducible case with slight modifications (see (Siegel and Thuswaldner 2010)). In the Pisot reducible case, some examples of substitutions have been studied in (Ei and Ito 2005) by using the mth polyhedral approximations of Definition 5.5.9. Unfortunately, no generic algorithm based on this approach exists so far.

Section 5.6

The notion of strand, introduced in (Barge and Diamond 2002), has been very fruitfully developed in the form of the strand space model for one-dimensional substitution tiling spaces in (Barge and Kwapisz 2006), see also (Barge and Kwapisz 2005), (Barge and Diamond 2007) and (Barge, Diamond, and Swanson 2009).

Section 5.7

Lemma 5.7.10 is strongly related to the notion of algebraic coincidence (see (Lee 2007)).

Much more than Lemma 5.7.11 can be said. In fact, the $\mathbb{Z}$-module generated by the lengths $\pi_e \circ \mathbf{P}(i)$, for $i \in A$, is equal to $\Xi(\mathcal{E}_u)$. For a proof, see (Sing 2006, Lemma 6.34) and (Barge and Kwapisz 2006, Section 12). Furthermore, for examples of substitutions for which we have the strict inclusion $\Xi(\mathcal{E}_u) \subset \mathbb{Z}[\beta]$, see (Sing 2006, Remark 6.36).

In (Sirvent and Solomyak 2002) the spectrum of the two dynamical systems associated with a substitution of Pisot type (*i.e.*, the substitutive symbolic space (X_σ, S) and the one-dimensional tiling space), is studied by comparing the balanced pair algorithm (for the $\mathbb{Z}$-action) and the overlap algorithm (for the ℝ-action). It is proved in (Clark and Sadun 2006) and in (Barge and Kwapisz 2006) that for a unit Pisot irreducible substitution, the tiling space has discrete spectrum if, and only if, the substitutive symbolic dynamical system has discrete spectrum.

Section 5.8

In Section 5.8 we have dealt exclusively with Pisot irreducible substitutions. Recently, (Martensen 2004) has generalised the balanced pair algorithm to the Pisot reducible case. In this case one has to identify certain patterns in order to get a proper behaviour of the algorithm, *i.e.*, to show its termination to be equivalent to the fact that the dynamical system associated with the substitution in question is purely discrete.

6
Combinatorics on Bratteli diagrams and dynamical systems

Fabien Durand

The aim of this chapter is to show how Bratteli diagrams are used to study topological dynamical systems. We illustrate their wide range of applications through classical notions: invariant measures, entropy, expansivity, representation theorems, strong orbit equivalence, eigenvalues of the Koopman operator.

6.1 Introduction

In 1972 O. Bratteli (Bratteli 1972) introduced special infinite graphs – subsequently called *Bratteli diagrams* – which conveniently encoded the successive embeddings of an ascending sequence $(A_n)_{n\geq 0}$ of finite-dimensional semi-simple algebras over $\mathbb{C}$ ('multi-matrix algebras'). The sequence $(A_n)_{n\geq 0}$ determines a so-called approximately finite-dimensional (AF) C^*-algebra. Bratteli proved that the equivalence relation on Bratteli diagrams generated by the operation of telescoping is a complete isomorphism invariant for AF-algebras.

From a different direction came the extremely fruitful idea of A. M. Vershik (Vershik 1985) to associate dynamics (*adic transformations*) with Bratteli diagrams (*Markov compacta*) by introducing a lexicographic ordering on the infinite paths of the diagram. By a careful refining of Vershik's construction, R. H. Herman, I. F. Putnam and C. F. Skau (Herman, Putnam, and Skau 1992) succeeded in showing that every minimal Cantor dynamical system is isomorphic to a Bratteli–Vershik dynamical system.

This chapter will give the details of this isomorphism and present some developments.

In this chapter all the dynamical systems (X, T) we consider are such

Combinatorics, Automata and Number Theory, ed. Valérie Berthé and Michel Rigo.
Published by Cambridge University Press. ©Cambridge University Press 2010.

that T is a homeomorphism. We thus work with the two-sided *orbit* of $x \in X$, that is, $\{T^n x \mid n \in \mathbb{Z}\}$.

6.2 Cantor dynamical systems

The notion of dynamical system has been defined (see Section 1.6.2) in two contexts: topological and measure-theoretic. Below we specify a special class of topological dynamical systems with respect to the space where it is defined.

We say that (X, T) is a *Cantor dynamical system* whenever X is a Cantor space, that is, X is non-empty, without isolated points, compact, totally disconnected, and metrisable. We recall that all Cantor spaces are homeomorphic and have a countable basis of their topology consisting of open sets that are also closed. These sets are usually called *clopen*.

Let (X, T) and (Y, S) be two dynamical systems. We say that (Y, S) is a *factor* of (X, T) if there is a continuous and onto map $\varphi : X \to Y$ such that $\varphi \circ T = S \circ \varphi$. The term 'factor' is used here with a completely different meaning as the meaning it has in Chapter 4 *e.g.* Then, φ is called *factor map*. If φ is one-to-one we say that it is an *isomorphism* (it is also called a *conjugacy*), and that (X, T) and (Y, S) are *(topologically) isomorphic*.

Let (X, T) be a minimal Cantor dynamical system and $U \subseteq X$ be a clopen set. Let $T_U : U \to U$ be the map defined by, for $x \in U$,

$$T_U(x) = T^{r_U(x)}(x), \text{ where } r_U(x) = \inf\{n > 0 \mid T^n(x) \in U\}.$$

As (X, T) is minimal, the *first entrance time map to* U, $r_U : X \to \mathbb{Z}$, is well defined and continuous. The pair (U, T_U) is a minimal Cantor dynamical system. We say that (U, T_U) is the *induced dynamical system of* (X, T), and $T_U : U \to U$ the *induced map*, with respect to U.

6.3 Bratteli diagrams

In this section we define the notion of *Bratteli diagram* and of *Bratteli–Vershik dynamical systems*.

6.3.1 Basics on Bratteli diagrams

Definition 6.3.1 A *Bratteli diagram* is an infinite directed graph (V, E) where the *vertex* set V and the *edge* set E can be partitioned into finite sets

$$V = V(0) \cup V(1) \cup V(2) \cup \cdots \text{ and } E = E(1) \cup E(2) \cup \cdots$$

with the following properties:

(i) $V(0) = \{v(0)\}$ is a one-point set,
(ii) $r(E(n)) \subseteq V(n)$, $s(E(n)) \subseteq V(n-1), n = 1, 2, \ldots$,

where $r : E \to V$ is called the *range map* and $s : E \to V$ the *source map*. They satisfy $s^{-1}(v) \neq \emptyset$ for all $v \in V$ and $r^{-1}(v) \neq \emptyset$ for all $v \in V \setminus V(0)$.

It is convenient to give a diagrammatic presentation of the Bratteli diagram with $V(n)$ the vertices at (horizontal) level n, and $E(n)$ the edges (downward directed) connecting the vertices at level $n - 1$ with those at level n. Also, if $\text{Card}(V(n-1)) = t(n-1)$ and $\text{Card}(V(n)) = t(n)$, then $E(n)$ determines a $t(n) \times t(n-1)$ *incidence matrix* $\mathbf{M}(n)$ (see Figure 6.1).

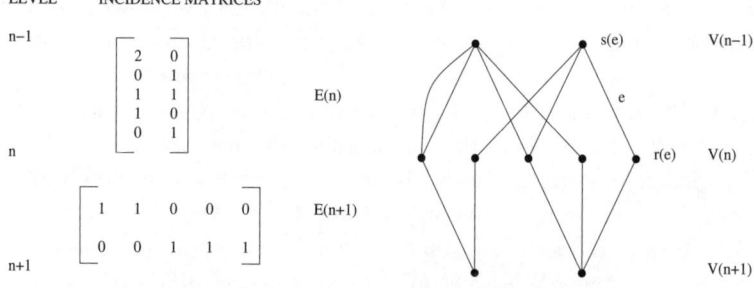

Fig. 6.1 Diagrammatic representation between the levels $n - 1$ and $n + 1$.

We say that two Bratteli diagrams (V, E) and (V', E') are *isomorphic* whenever there exists a pair of bijections $f : V \to V'$, preserving the gradings, and $g : E \to E'$, intertwining the respective source and range maps:

$$s' \circ g = f \circ s \text{ and } r' \circ g = f \circ r.$$

Let $k, l \in \mathbb{Z}_{\geq 0}$ with $k < l$ and let $E(k+1) \circ E(k+2) \circ \cdots \circ E(l)$ denote the set of paths from $V(k)$ to $V(l)$. Specifically,

$$E(k+1) \circ \cdots \circ E(l)$$

$= \{(e_{k+1}, \ldots, e_l) \mid e_i \in E(i), k+1 \leq i \leq l, r(e_i) = s(e_{i+1}), k+1 \leq i \leq l-1\}$.

Remark that the incidence matrix of $E(k+1) \circ \cdots \circ E(l)$ is $\mathbf{M}(l) \cdots \mathbf{M}(k+1)$. We define $r(e_{k+1}, \ldots, e_l) := r(e_l)$ and $s(e_{k+1}, \ldots, e_l) := s(e_{k+1})$.

Given a Bratteli diagram (V, E) and a sequence

$$m_0 = 0 < m_1 < m_2 < \ldots$$

in $\mathbb{Z}^+$, we define the *telescoping* of (V, E) to $\{m_n \mid n \in \mathbb{N}\}$ as the new Bratteli diagram (V', E'), where $V'(n) = V(m_n)$ and $E'(n) = E(m_{n-1} + 1) \circ \cdots \circ E(m_n)$ and the range and source maps are as above (see Figure 6.2).

Combinatorics on Bratteli diagrams and dynamical systems 327

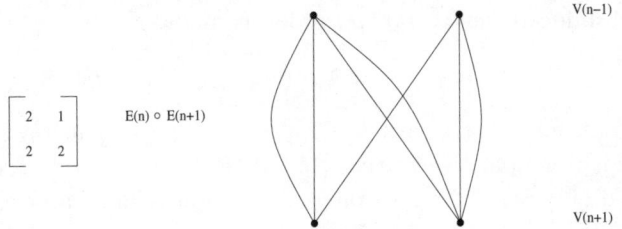

Fig. 6.2 Telescoping between the levels $n-1$ and $n+1$ in the diagram of Figure 6.1.

We say that (V, E) is a *simple Bratteli diagram* if there exists a telescoping (V', E') of (V, E) such that the incidence matrices of (V', E') have only non-zero entries at each level.

We let $\sim$ denote the *equivalence relation on Bratteli diagrams* generated by isomorphism and telescoping. It is not hard to show that $(V^1, E^1) \sim (V^2, E^2)$ if, and only if, there exists a Bratteli diagram (V, E) such that telescoping (V, E) to odd levels $0 < 1 < 3 < \ldots$ yields a telescoping of either (V^1, E^1) or (V^2, E^2), and telescoping (V, E) to even levels $0 < 2 < 4 < \ldots$ yields a telescoping of the other.

We say that a Bratteli diagram B has a *simple hat* whenever it has only simple edges between the top vertex and the first level.

6.3.2 Ordered Bratteli diagrams

Definition 6.3.2 An *ordered Bratteli diagram* $(V, E, \geq)$ is a Bratteli diagram (V, E) together with a partial order $\geq$ on E such that edges e, e' in E are *comparable* if, and only if, $r(e) = r(e')$; in other words, we have a linear order on each set $r^{-1}(\{v\})$, where v belongs to $V \setminus V(0)$ (see Figure 6.3).

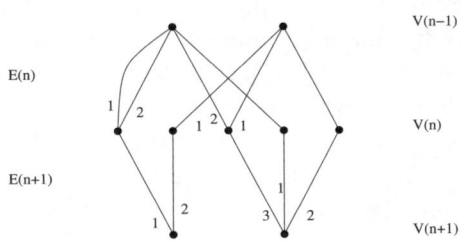

Fig. 6.3 Order on the diagram of Figure 6.1.

Note that if $(V, E, \geq)$ is an ordered Bratteli diagram and $k < l$ in $\mathbb{Z}^+$, then the set $E(k+1) \circ E(k+2) \circ \cdots \circ E(l)$ of paths from $V(k)$ to $V(l)$ may be given an induced (*lexicographic*) order as follows:

$$(e_{k+1}, e_{k+2}, \ldots, e_l) > (f_{k+1}, f_{k+2}, \ldots, f_l)$$

if, and only if, for some i with $k+1 \leq i \leq l$, $e_j = f_j$ for $i < j \leq l$ and $e_i > f_i$. It is a simple observation that if $(V, E, \geq)$ is an ordered Bratteli diagram and (V', E') is a telescoping of (V, E) as defined above, then with the induced order $\geq'$, $(V', E', \geq')$ is again an ordered Bratteli diagram. We say that $(V', E', \geq')$ is a *telescoping* of $(V, E, \geq)$ (see Figure 6.4).

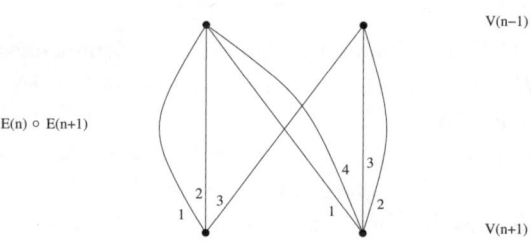

Fig. 6.4 Telescoping of the diagram of Figure 6.3.

Again there is an obvious notion of isomorphism between ordered Bratteli diagrams. Let $\approx$ denote the equivalence relation on ordered Bratteli diagrams generated by isomorphism and by telescoping. One can show that $B^1 \approx B^2$, where $B^1 = (V^1, E^1, \geq^1)$, $B^2 = (V^2, E^2, \geq^2)$, if, and only if, there exists an ordered Bratteli diagram $B = (V, E, \geq)$ such that telescoping B to odd levels $0 < 1 < 3 < \ldots$ yields a telescoping of either B^1 or B^2, and telescoping B to even levels $0 < 2 < 4 < \ldots$ yields a telescoping of the other. This is analogous to the situation for the equivalence relation $\sim$ on Bratteli diagrams as we discussed above. We write $B^1 \sim B^2$ to say $(V^1, E^1) \sim (V^2, E^2)$.

The following notion will be important when we will deal with Bratteli diagrams and subshifts. Fix $n \geq 1$ and let us consider $V(n-1)$ and $V(n)$ as alphabets. For every letter $a \in V(n)$, consider the ordered list $(e_1, \ldots, e_k)$ of edges of $E(n)$ which range at a, and let $(a_1, \ldots, a_k)$ be the ordered list of the labels of the sources of these edges. This defines a morphism $a \mapsto a_1 \cdots a_k$ from $V(n)^*$ to $V(n-1)^*$ we call *the morphism read on $E(n)$*. For example in Figure 6.3 the morphism we read on:

- $E(n)$ is $\tau_n : 0 \mapsto AA$, $1 \mapsto B$, $2 \mapsto BA$, $3 \mapsto A$, $4 \mapsto B$,
- $E(n+1)$ is $\tau_{n+1} : a \mapsto 01$, $b \mapsto 342$,

and on Figure 6.4 the morphism we read on $E(n) \circ E(n+1)$ is $\sigma : a \mapsto AAB$, $b \mapsto ABBA$. We can check we of course that we have $\sigma = \tau_n \circ \tau_{n+1}$.

6.3.3 Dynamics for ordered Bratteli diagrams

Let $B = (V, E, \geq)$ be an ordered Bratteli diagram. Let X_B denote the associated *infinite path space*, i.e.,

$$X_B = \{(e_1, e_2, \ldots) \mid e_i \in E(i), r(e_i) = s(e_{i+1}), i = 1, 2, \ldots\}.$$

We exclude trivial cases and assume henceforth that X_B is an infinite set. Two paths in X_B are said to be *cofinal* if they have the same tails, i.e., the edges agree from a certain level. We endow X_B with a topology by postulating a basis of open sets, namely the family of *cylinder sets*

$$[e_1, e_2, \ldots, e_k]_B = \{(f_1, f_2, \ldots) \in X_B \mid f_i = e_i, 1 \leq i \leq k\}.$$

Each $[e_1, \ldots, e_k]$ is also closed, as is easily seen. When it will be clear from the context we will write $[e_1, \ldots, e_k]$ instead of $[e_1, \ldots, e_k]_B$. Endowed with this topology, we call X_B the *Bratteli compactum* associated with $B = (V, E, \geq)$. Let d_B be the distance on X_B defined by $d_B((e_n)_n, (f_n)_n) = \frac{1}{2^k}$ where $k = \inf\{i \mid e_i \neq f_i\}$. It clearly coincides with the topology of the cylinder sets.

If (V, E) is a simple Bratteli diagram, then X_B has no isolated points, and so is a Cantor space.

Let $x = (e_1, e_2, \ldots)$ be an element of X_B. We will call e_n the nth label of x and denote it by $x(n)$. We let $X_B^{\max}$ denote those elements x of X_B such that $x(n)$ is a maximal edge for all n and $X_B^{\min}$ the analogous set for the minimal edges. It is not difficult to show that $X_B^{\max}$ and $X_B^{\min}$ are non-empty.

Definition 6.3.3 The ordered Bratteli diagram $B = (V, E, \geq)$ is *properly ordered* if it is simple and if $X_B^{\max}$ and $X_B^{\min}$ both are a one point set: $X_B^{\max} = \{x_{\max}\}$ and $X_B^{\min} = \{x_{\min}\}$.

We can now define a map $V_B : X_B \to X_B$, called the *Vershik map* (or the *lexicographic map*), associated with the properly ordered Bratteli diagram $B = (V, E, \geq)$. We call the resulting pair (X_B, V_B) a *Bratteli–Vershik dynamical system*. It is a Cantor dynamical system. Note that B being simple, (X_B, V_B) is minimal.

We let $V_B(x_{\max}) = x_{\min}$. If $x = (e_1, e_2, \ldots) \neq x_{\max}$, let k be the smallest number such that e_k is not a maximal edge. Let f_k be the successor of e_k (and so $r(e_k) = r(f_k)$). Define $V_B(x) = y = (f_1, \ldots, f_{k-1}, f_k, e_{k+1}, e_{k+2}, \ldots)$, where $(f_1, \ldots, f_{k-1})$ is the minimal edge in $E(1) \circ E(2) \circ \cdots \circ E(k-1)$ with range equal to $s(f_k)$.

In the sequel BV will refer to *Bratteli–Vershik*.

6.4 The Bratteli–Vershik model theorem

In what follows (X, T) will always refer to a minimal Cantor dynamical system. We will give all the details of the main result of (Herman, Putnam, and Skau 1992) saying that (X, T) can be topologically realised as a BV-dynamical system. We recall that A. M. Vershik obtained in (Vershik 1985) such a result in a measure-theoretic context.

6.4.1 Existence of Kakutani–Rohlin partitions

A *clopen partition* $\mathcal{P}$ of a set X is a partition whose elements (also called *atoms*) are clopen sets. Note that X being compact these partitions are finite.

Definition 6.4.1 A *Kakutani–Rohlin (KR) partition* of the minimal Cantor dynamical system (X, T) is a clopen partition $\mathcal{P}$ of the form:

$$\mathcal{P} = \{T^j C_k \mid k \in V,\ 0 \leq j < h_k\},$$

where V is a finite set, C_k is a clopen set and h_k is a positive integer. The *tower* k of $\mathcal{P}$ is $\{T^j C_k \mid 0 \leq j < h_k\}$, the set $T^j C_k$ is the jth *level* of the tower k and the *base* of $\mathcal{P}$ is the set $C := \cup_{k \in V} C_k$. The *height* of the tower k is h_k.

Figure 6.6 illustrates the notion of tower. In the sequel we will refer to such a partition as a KR-partition. Let us show such partitions can be chosen to be finer than any given clopen partition. We follow the details given in (Putnam 1989).

Proposition 6.4.2 *Let $\mathcal{Q}$ be a clopen partition of X and C be a clopen set. Then, there exist a clopen partition $C_1, \ldots, C_t$, of C and some integers $(h_i)_{1 \leq i \leq t}$ such that*

$$\mathcal{P} = \{T^j C_i \mid 0 \leq j < h_i, 1 \leq i \leq t\}$$

is finer than $\mathcal{Q}$.

Proof The first entrance time map r_C is continuous and C is compact hence it takes finitely many values: $r_1, r_2, \ldots, r_{t'}$. For all $i \in [1, t']$, we define the clopen set $C_i' = T^{-r_i} C$. Then, the collection

$$\mathcal{P}' = \{T^j C_i' \mid 0 \leq j < r_i, 1 \leq i \leq t'\}$$

is a clopen partition of X but it is not necessarily finer than $\mathcal{Q}$. Let $\mathcal{Q}' = \{P' \cap Q \mid P' \in \mathcal{P}', Q \in \mathcal{Q}\}$. It suffices to find $\mathcal{P}$ finer than $\mathcal{Q}'$. Let $\mathcal{Q}'$

be an atom of $\mathcal{Q}'$. There exists a unique pair (i_0, j_0) with $i_0 \in [1, t']$ and $j_0 \in [0, r_{i_0} - 1]$ such that $Q' \subseteq T^{j_0} C'_{i_0}$. We divide the tower i_0 into two new towers and obtain a new KR-partition $\mathcal{P}''$ with $t' + 1$ towers:

$$\begin{aligned}\mathcal{P}'' = \quad & \{T^j C'_i \mid 0 \leq j < r_i, i \in [1, t'] \setminus \{i_0\}\} \\ \cup \ & \{T^j(Q') \mid -j_0 \leq j < r_{i_0} - j_0\} \\ \cup \ & \{T^j((T^{j_0} C'_{i_0}) \setminus Q') \mid -j_0 \leq j < r_{i_0} - j_0\}.\end{aligned}$$

We repeat this procedure with the new partition $\mathcal{P}''$ and a new atom of $\mathcal{Q}'$. There are finitely many steps and it ends with the KR-partition we are looking for. □

The following theorem is fundamental to represent minimal Cantor dynamical systems by BV-dynamical systems.

Theorem 6.4.3 *Let $x \in X$. There exists a sequence of KR-partitions $(\mathcal{P}(n))_n$ with*

$$\mathcal{P}(n) := \{T^j B_i(n) \mid 0 \leq j < h_i(n),\ 1 \leq i \leq t(n)\}$$

satisfying

(KR1) $\bigcap_n \bigcup_{1 \leq i \leq t(n)} B_i(n) = \{x\}$,
(KR2) $\mathcal{P}(n+1)$ *is finer than* $\mathcal{P}(n)$ *for all* n,
(KR3) $\bigcup_n \mathcal{P}(n)$ *generates the topology of X.*

Proof We start by choosing a decreasing sequence of clopen sets $(C(n))_n$, whose intersection is $\{x\}$, and an increasing sequence of partitions $(\mathcal{P}'(n))_n$ generating the topology. We apply Proposition 6.4.2 to $\mathcal{Q} = \mathcal{P}'(1)$ and $C = C(1)$. We obtain $\mathcal{P}(1)$. We continue applying Proposition 6.4.2 inductively, for $n \geq 2$, to $C = C(n)$ and

$$\mathcal{Q} = \mathcal{P}'(n) \vee \mathcal{P}(n-1) = \{P' \cap P \mid P' \in \mathcal{P}'(n), B \in \mathcal{P}(n-1)\}$$

to obtain $\mathcal{P}(n)$. This achieves the proof. □

6.4.2 From Kakutani–Rohlin partitions to Bratteli–Vershik representations

In the present section we use Theorem 6.4.3 to obtain the BV-representation of minimal Cantor dynamical systems.

Definition 6.4.4 *The properly ordered Bratteli diagram $B = (V, E, \geq)$ is a BV-representation of (X, T) if (X_B, V_B) is isomorphic to (X, T).*

Let $\big(\mathcal{P}(n) = \{T^j B_i(n) \mid 0 \leq j < h_i(n), 1 \leq i \leq t(n)\}\big)_n$ be a sequence of KR-partitions of (X,T) satisfying (KR1), (KR2) and (KR3). We also suppose that $\mathcal{P}(0) = \{X\}$. Hence $t(0) = 1$, $h_1(0) = 1$ and $B_1(0) = X$.

Let $V(n) = \{(n,1), \ldots, (n, t(n))\}$, for $n \geq 0$, and $E(n)$ be the set of quadruples (n, t', t, j) satisfying

$$T^j B_t(n) \subseteq B_{t'}(n-1) \qquad (6.1)$$

for $1 \leq t' \leq t(n-1)$, $1 \leq t \leq t(n)$, $0 \leq j \leq h_t(n) - 1$ and $n \geq 1$. The range and source maps are given by

$$s(n, t', t, j) = (n-1, t') \text{ and } r(n, t', t, j) = (n, t) . \qquad (6.2)$$

Two edges $e_1 = (n_1, t'_1, t_1, j_1)$ and $e_2 = (n_2, t'_2, t_2, j_2)$ are comparable whenever $n_1 = n_2$ and $t_1 = t_2$. In this case we define $e_1 \geq e_2$ if $j_1 \geq j_2$. It is straightforward to verify that $(V, E, \geq)$ is an ordered Bratteli diagram. It is useful to remark, from (6.2), that $((n, t'_n, t_n, j_n))_n$ is an infinite path of $(V, E, \geq)$ if, and only if, $t_{n-1} = t'_n$ for all $n \geq 1$. Hence the paths of the Bratteli diagram have the form $((n, t_{n-1}, t_n, j_n))_n$. Note that (n, t_{n-1}, t_n, j_n) is a minimal edge if, and only if, $j_n = 0$ and is maximal if, and only if, $j_n = h_{t_n}(n) - h_{t_{n-1}}(n-1)$.

For example suppose that $(\mathcal{P}(n))_n$ is a sequence of KR-partitions satisfying (KR1), (KR2) and (KR3) such that

(i) $\mathcal{P}(1) = \{B_1(1), TB_1(1), B_2(1), TB_2(1), T^2 B_2(1)\}$ and
(ii) $\mathcal{P}(2) = \{T^j B_i(2) \mid 0 \leq j < h_i(2), 1 \leq i \leq t(2)\}$ with
  (a) $t(2) = 3$, $h_1(2) = 9$, $h_2(2) = 4$, $h_3(2) = 7$,
  (b) $B_1(2) \subseteq B_1(1)$, $T^2 B_1(2) \subseteq B_1(1)$, $T^4 B_1(2) \subseteq B_2(1)$, $T^7 B_1(2) \subseteq B_1(1)$,
  (c) $B_2(2) \subseteq B_1(1)$, $T^2 B_2(2) \subseteq B_1(1)$,
  (d) $B_3(2) \subseteq B_2(1)$, $T^3 B_3(2) \subseteq B_1(1)$, $T^5 B_3(2) \subseteq B_1(1)$.

This is summarised in Figure 6.5.

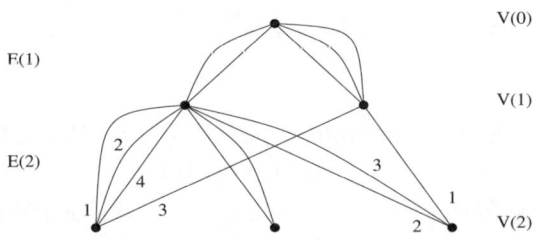

Fig. 6.5 Diagrammatic representation of $\mathcal{P}(0)$, $\mathcal{P}(1)$ and $\mathcal{P}(2)$.

This can be compared to the representation with towers in Figure 6.6.

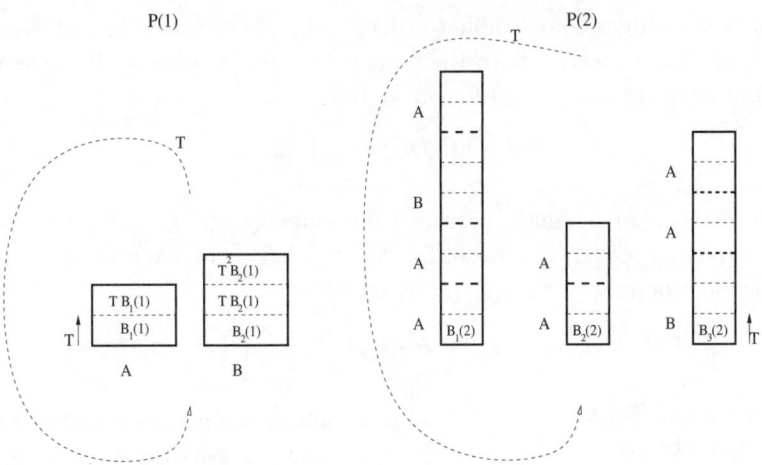

Fig. 6.6 The partition $\mathcal{P}(1)$ consists of two towers called A and B. The dynamics T acts vertically except for the last levels where it goes back to the base. The partition $\mathcal{P}(2)$ can be seen as a concatenation of "vertical pieces" of the towers A and B.

Lemma 6.4.5 *Under the assumptions of this section, the following are equivalent:*

(i) $((n, t_{n-1}, t_n, j_n))_n$ *is an infinite path of* $(V, E, \geq)$,
(ii) $\bigcap_{n \geq 1} T^{\sum_{i=1}^{n} j_i} B_{t_n}(n) \neq \emptyset$ *with* (n, t_{n-1}, t_n, j_n) *satisfying* (6.1),
(iii) $\mathrm{Card}(\bigcap_{n \geq 1} T^{\sum_{i=1}^{n} j_i} B_{t_n}(n)) = 1$ *with* (n, t_{n-1}, t_n, j_n) *satisfying* (6.1).

Moreover, for such an infinite path we have $0 \leq \sum_{i=1}^{n} j_i \leq h_{t_n}(n) - 1$.

Proof [Sketch] From (i) we deduce (ii) using Baire's Theorem. From (ii) we deduce (iii) because the sequence of partitions generates the topology. The last implication is easy to establish. □

We can now state and prove the BV-representation theorem.

Theorem 6.4.6 (Herman, Putnam, and Skau 1992) *There exists a properly ordered Bratteli diagram* $B = (V, E, \geq)$ *such that* (X, T) *is isomorphic to* (X_B, V_B).

Moreover, any contraction of B *yields a BV-representation of* (X, T) *and some of them have incidence matrices with positive entries.*

Proof Let $(\mathcal{P}(n))_n$ and $B = (V, E, \geq)$ be as defined above. Let us show that $X_B^{\min}$ consists of a single path. Let $(n, t_{n-1}, t_n, j_n)_n$ be an infinite path of $X_B^{\min}$. The edges comparable to (n, t_{n-1}, t_n, j_n) are the edges of the form (n, t, t_n, j) and exactly one of them is $(n, t, t_n, 0)$ for some t. It is clearly a minimal edge. Hence $j_n = 0$ for all n. But

$$\emptyset \neq \bigcap_n T^0 B_{t_n}(n) \subseteq \bigcap_n B(n)$$

which consists of a single point. Consequently $((n, t_{n-1}, t_n, 0))_n$ is the unique path of $X_B^{\min}$. The proof for $X_B^{\max}$ is left as an exercise.

Consider the map $\varphi : X_B \to X$ defined by

$$\varphi((n, t_{n-1}, t_n, j_n)_n) = x \text{ where } \{x\} = \bigcap_{n \geq 1} T^{\sum_{i=1}^n j_i} B_{t_n}(n).$$

It is well defined (Lemma 6.4.5) and is a homeomorphism (see Exercise 6.7). Note that $(\mathcal{P}(n))_n$ being a decreasing sequence of partitions, we also have $\{x\} = \bigcap_{n \geq N} T^{\sum_{i=1}^n j_i} B_{t_n}(n)$ for all N.

It remains to show that it commutes with the dynamics. Let $e = (e_n)_n$ be an infinite path of X_B with $e_n = (n, t_{n-1}, t_n, j_n)$. Suppose that e is not the maximal path. Then there exists n_0 such that $V_B(x) = e_1' \cdots e_{n_0-1}' e_{n_0}' e_{n_0+1} e_{n_0+2} \cdots$ where $e_n' = (n, t_{n-1}', t_n', j_n')$, with $j_n' = 0$, $1 \leq n \leq n_0 - 1$ and $e_{n_0}' = (n_0, t_{n_0-1}', t_{n_0}, j_{n_0}')$. Note that, for $1 \leq n \leq n_0 - 1$, the edges e_n being maximal we have $j_n = h_{t_n}(n) - h_{t_{n-1}}(n-1)$. Moreover,

$$T^{h_{t_n}(n) - h_{t_{n-1}}(n-1)} B_{t_n}(n) \subseteq B_{t_{n-1}}(n-1) \text{ for all } 1 \leq n \leq n_0 - 1,$$

$$T^{j_{n_0}} B_{t_{n_0}}(n_0) \subseteq B_{t_{n_0-1}}(n_0 - 1),$$

$$T^{j_{n_0}'} B_{t_{n_0}}(n_0) \subseteq B_{t_{n_0-1}'}(n_0 - 1).$$

From the definition of V_B we deduce $j_{n_0}' = j_{n_0} + h_{t_{n_0-1}}(n_0 - 1)$. Hence $\sum_{1 \leq n \leq n_0} j_n' = j_{n_0} + h_{t_{n_0-1}}(n_0 - 1)$ and

$$\sum_{1 \leq n \leq n_0} j_n = j_{n_0} + \sum_{1 \leq n \leq n_0} h_{t_n}(n) - h_{t_{n-1}}(n-1) = j_{n_0} + h_{t_{n_0-1}}(n_0 - 1) - 1.$$

Suppose now that e is the maximal path. Let $x_{\min}$ be the minimal path of B. Then we have to prove that $\varphi(x_{\min}) = T(\varphi(e))$. But as $\mathcal{P}(0) = \{X\}$, we have $h_{t_0}(0) = 1$ and consequently

$$T(\varphi(\{e\})) = T\left(\bigcap_{n \geq 1} T^{\sum_{i=1}^n h_{t_i}(i) - h_{t_{i-1}}(i-1)} B_{t_n}(n)\right)$$

$$= \bigcap_{n \geq 1} T^{h_{t_n}(n)} B_{t_n}(n) \subseteq \bigcap_{n \geq 1} \bigcup_{1 \leq i \leq t(n)} B_i(n) = \{\varphi(x_{\min})\}.$$

This achieves the main part of the proof. The last part is left as an exercise. $\square$

6.4.3 Kakutani equivalence

Definition 6.4.7 The minimal Cantor dynamical systems (X,T) and (Y,S) are *Kakutani equivalent* if they have (up to isomorphism) a common induced system, i.e., there exist clopen sets $U \subseteq X$ and $V \subseteq Y$ such that the induced systems (X_U, T_U) and (Y_V, S_V) are isomorphic.

Let us relate Kakutani equivalence to Bratteli diagrams.

If $B = (V, E, \geq)$ is a properly ordered Bratteli diagram we may change B into a new properly ordered Bratteli diagram $B' = (V', E', \geq')$ by making a finite change, i.e., by adding and/or removing any finite number of edges (vertices), and then making arbitrary choices of linear orderings of the edges meeting at the same vertex (for a finite number of vertices). So B and B' are cofinally identical, i.e., they only differ on finite initial portions. (Observe that this defines an equivalence relation on the family of properly ordered Bratteli diagrams.) We have the following nice characterisation of the Kakutani equivalence.

Theorem 6.4.8 (Herman, Putnam, and Skau 1992) *Let (X_B, V_B) be the BV-dynamical system associated with the properly ordered Bratteli diagram $B = (V, E, \geq)$. Then the minimal Cantor dynamical system (X, T) is Kakutani equivalent to (X_B, V_B) if, and only if, (X, T) is isomorphic to $(X_{B'}, V_{B'})$, where $B' = (V', E', \geq')$ is obtained from B by a finite change as described above.*

An interesting consequence of this result is the following. Let U be a clopen set of (X_B, V_B). It is a finite union of cylinder sets. We can suppose that they all have the same length, i.e., for some n, $U = \cup_{p \in P} [p]_B$ where P is a set of paths from level n to level 0. To obtain a BV-representation of the induced system on U it suffices to take the properly ordered Bratteli diagram B' which consists of all the paths starting with an element of P endowed with the induced ordering. It is not too much work to prove that the induced system on U is isomorphic to $(X_{B'}, V_{B'})$.

The following result was proved in (Holton and Zamboni 1999). It was also obtained in (Durand 1998b) but not for all cylinders, only for those coming from the prefixes of the fixed points.

Theorem 6.4.9 *Minimal substitution subshifts have finitely many induced subshifts on cylinders up to isomorphism.*

This can be considered as a symbolic counterpart to a theorem of M. Boshernitzan and C. R. Carroll (see (Boshernitzan and Carroll 1997)) which states that an interval exchange transformation defined over a quadratic field has only finitely many induced systems (with respect to some inducing procedure).

According to what has been said before, it is not difficult to show Theorem 6.4.9 does not hold for clopen sets instead of cylinders.

6.5 Examples of BV-models

In this section we give examples of BV-representations for some classical dynamical systems.

6.5.1 Odometers

Let (p_n) be a strictly increasing sequence such that p_n divides p_{n+1} for all n. We endow $\prod_{n=0}^{+\infty} \mathbb{Z}/p_n\mathbb{Z}$ with the product topology of the discrete topologies. The set of (p_n)-adic integers is the inverse limit

$$\mathbb{Z}_{(p_n)} = \left\{ (x_n) \in \prod_{n=0}^{+\infty} \mathbb{Z}/p_n\mathbb{Z} \mid x_n \equiv x_{n+1} \mod p_n \right\}.$$

We endow $\mathbb{Z}_{(p_n)}$ with the induced topology. It is a compact topological ring (see Exercise 6.4). A base of its topology is given by the sets

$$[a_0, a_1, \ldots, a_m] := \{(x_n) \in \mathbb{Z}_{(p_n)} \mid x_i = a_i, 0 \leq i \leq m\}.$$

When $p_n = p^n$ for all n, it defines the classical ring of p-adic integers $\mathbb{Z}_p$. Let $R: \mathbb{Z}_{(p_n)} \to \mathbb{Z}_{(p_n)}$ be the map $x \mapsto x+1$. The pair $(\mathbb{Z}_{(p_n)}, R)$ is called an *odometer in base* (p_n). It is a minimal dynamical system.

For all n, we set $B_1(n) = [0^{n-1}]$, $h(n) = p_n$ and

$$\mathcal{P}(n) := \{R^j B(n) \mid 0 \leq j \leq h(n) - 1\}.$$

Then $(\mathcal{P}(n))_n$ is a sequence of KR-partitions satisfying (KR1), (KR2) and (KR3) (see Exercise 6.9). Remark that $R^j B(n) = [j_0 j_1 \cdots j_{n-1}]$ where $j_i = j \mod p_i$. The edges of the BV-representation of $(\mathbb{Z}_{(p_n)}, R)$ given in Subsection 6.4.2 are of the form $(n, 1, 1, lp_{n-1})$, $0 \leq l \leq q_n - 1 = \frac{p_n}{p_{n-1}} - 1$.

For example if $p_1 = 2$, $p_2 = 10$ and $p_3 = 30$, the first three levels are given in Figure 6.7.

6.5.2 Substitutions

The next result (Theorem 6.5.1 below) was first proven in (Forrest 1997). The proofs given in that paper are mostly of existential nature and do not state a method to compute effectively the BV-representation associated with substitution systems. Let us note that another proof was given in

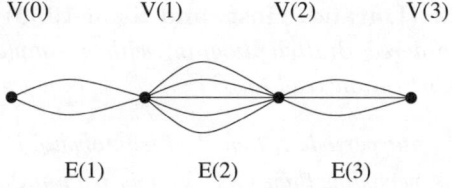

Fig. 6.7 The three first levels of the BV representation of $(\mathbb{Z}_{(p_n)}, R)$.

(Durand, Host, and Skau 1999) that provides such an algorithm. We first need a new definition. Let us note that we work in all this section with bi-infinite words and two-sided shifts.

6.5.2.1 The representation theorem

A Bratteli diagram (V, E) is *stationary* if there exists k such that $k = \text{Card}(V(n))$ for all n, and if (by an appropriate labelling of the vertices) the incidence matrices between level n and $n+1$ are the same $k \times k$ matrix $\mathbf{M}$ for all $n = 1, 2, \ldots$. In other words, beyond level 1 the diagram repeats itself. (Clearly we may label the vertices in $V(n)$ as $v(n, a_1), \ldots, v(n, a_k)$, where $A = \{a_1, \ldots, a_k\}$ is a set of k distinct symbols.)

The ordered Bratteli diagram $B = (V, E, \geq)$ is *stationary* if (V, E) is stationary, and the ordering on the edges with range $v(n, a_i)$ is the same as the ordering on the edges with range $v(m, a_i)$ for $m, n = 2, 3, \ldots$ and $i = 1, \ldots, k$. In other words, beyond level 1 the diagram with the ordering repeats itself.

Theorem 6.5.1 *The family $\mathcal{B}$ of Bratteli–Vershik systems associated with stationary, properly ordered Bratteli diagrams is (up to isomorphism) the disjoint union of the family of substitution minimal systems and the family of stationary odometer systems. Furthermore, the correspondence in question is given by an explicit and algorithmic effective construction.*

In the sequel we describe the algorithm that, starting with a subshift generated by a primitive substitution, gives a stationary BV-representation.

Let B be a stationary ordered Bratteli diagram. The morphism read on $E(n)$ is constant from $n \geq 2$. We call it *the substitution read on B*.

We recall that a subshift is said to be *periodic* if there exist $x \in X$ and an integer k such that $X = \{x, Sx, \ldots, S^k x = x\}$. Otherwise it is said to be *aperiodic*.

Proposition 6.5.2 (Durand, Host, and Skau 1999) *Let B be a stationary, properly ordered Bratteli diagram with a simple hat, and let $\sigma : A^* \to A^*$ be the substitution read on B.*

(i) *If (X_σ, S) is not periodic, then it is isomorphic to (X_B, V_B).*
(ii) *If (X_σ, S) is periodic, then (X_B, V_B) is isomorphic to an odometer in base $(qp^n)_n$, $p, q \in \mathbb{N}$.*

6.5.2.2 Combinatorics on words for Bratteli diagrams

Definition 6.5.3 A substitution σ on the alphabet A is *proper* if there exist an integer $p > 0$ and two letters $r, l \in A$ such that, for every $a \in A$, r is the last letter of $\sigma^p(a)$ and l is the first letter of $\sigma^p(a)$.

A proper substitution has exactly one fixed point (in $A^\mathbb{N}$ and in $A^\mathbb{Z}$). Remark that the substitution read on a stationary properly ordered Bratteli diagram is proper. Consequently, in the light of Proposition 6.5.2, it suffices to find a proper substitution ζ such that (X_ζ, S) is isomorphic to (X_σ, S) to have a stationary BV-representation of a substitution subshift (X_σ, S). In order to find such a substitution we will need the following proposition from (Durand, Host, and Skau 1999) which is a modification of an unpublished result of G. Rauzy. See also Theorem 4.6.1. It has an interest outside of the scope of BV-representations of substitution subshifts. Before stating Proposition 6.5.4, we first need to define the notion of circular code.

Let A be an alphabet, and C a finite subset of A^+. We recall that the set C is a *code* if every word $u \in A^+$ admits at most one decomposition as a concatenation of elements of C. The code C is said to be *circular* if for all words

$$w_1, \ldots, w_j, \ w, \ w'_1, \ldots, w'_k \in C, \ s \in A^+ \text{ and } t \in A^*$$

that satisfy

$$w = ts \text{ and } w_1 \ldots w_j = sw'_1 \ldots w'_k t,$$

then t is the empty word. Note that it follows that $j = k+1$, $w_{i+1} = w'_i$ for $1 \le i \le k$ and $w_1 = s$. One of the interests of circular codes in the present context is that they display a unique decomposition property for sequences. Suppose indeed that C is a circular code on the alphabet A, and that some $x \in A^\mathbb{Z}$ can be decomposed as a concatenation of words $(w_k)_{k \in \mathbb{Z}}$ belonging to C, i.e.,

$$x = \ldots w_{-3} w_{-2} w_{-1} . w_0 w_1 w_2 \ldots$$

Then this decomposition is unique.

Proposition 6.5.4 *Let $y \in R^{\mathbb{N}}$ be a fixed point of a primitive substitution τ on the alphabet R, A an alphabet, $\varphi \colon R^* \to A^*$ a non-erasing morphism, $x = \varphi(y)$ and (X, S) the subshift spanned by x.*

There exist a primitive substitution ζ on an alphabet B, an admissible fixed point z of ζ and a map $\theta \colon B \to A$ such that:

(i) $\theta(z) = x$,
(ii) *if φ is injective and $\varphi(R)$ is a circular code, then θ is an isomorphism from (X_ζ, S) to (X, S),*
(iii) *if τ is proper, then ζ is proper.*

Proof The proof below is very simple, but the notations are, in an unavoidable way, a bit heavy. By substituting τ by a power of itself if needed, we can assume that $|\tau(j)| \geq |\varphi(j)|$ for all $j \in R$. We define:

- an alphabet B by $B = \{(j,p) \mid j \in R, 1 \leq p \leq |\varphi(j)|\}$,
- a map $\theta \colon B \to A$ by $\theta(j,p) = (\varphi(j))_p$,
- a map $\gamma \colon R \to B^+$ by $\gamma(j) = (j,1)(j,2)\cdots(j,|\varphi(j)|)$.

Clearly $\theta \circ \gamma = \varphi$. We define a substitution ζ on B as follows. For j in R and $1 \leq p \leq |\varphi(j)|$, we set

$$\zeta(j,p) = \begin{cases} \gamma\Big((\tau(j))_p\Big) & \text{if} \quad 1 \leq p < |\varphi(j)| \\ \gamma\Big((\tau(j))_{[\![\varphi(j),\tau(j)]\!]}\Big) & \text{if} \quad p = |\varphi(j)|. \end{cases}$$

Hence, for every $j \in R$, $\zeta(\gamma(j)) = \zeta(j,1)\cdots\zeta(j,|\varphi(j)|) = \gamma(\tau(j))$, i.e.,

$$\zeta \circ \gamma = \gamma \circ \tau \tag{6.3}$$

and it follows that

$$\zeta^n \circ \gamma = \gamma \circ \tau^n \text{ for all } n \geq 0.$$

We claim that ζ is primitive. Let n be an integer such that b occurs in $\tau^n(a)$ for all $a, b \in R$. Let (j,p) and (k,q) belong to B. By construction, $\zeta(j,p)$ contains $\gamma(\tau(j)_p)$ as a factor, thus $\zeta^{n+1}(j,p)$ contains $\zeta^n(\gamma(\tau(j)_p)) = \gamma(\tau^n(\tau(j)_p))$ as a factor. By the choice of n, k occurs in $\tau^n(\tau(j)_p)$, thus $\gamma(k)$ is a factor of $\gamma(\tau^n(\tau(j)_p))$, and also of $\zeta^{n+1}(j,p)$. As (k,q) is a letter of $\gamma(k)$, (k,q) occurs in $\zeta^{n+1}(j,p)$ and our claim is proved.

Let $z = \gamma(y)$. By (6.3) we get $\zeta(z) = \gamma(\tau(y)) = \gamma(y) = z$, and z is a fixed point of ζ. By construction, z is uniformly recurrent, thus it is an admissible fixed point of ζ. Moreover, $\theta(z) = \theta(\gamma(y)) = \varphi(y) = x$, and (i) is proved.

Proof of (ii). As θ commutes with the shift and maps z to x, and by minimality of the subshifts, it maps X_ζ onto X. It remains to prove that

$\theta : X_\zeta \to X$ is one-to-one. Let $\alpha \in X$. By definition of X, there exist $t \in X_\tau$ and an integer p, with $0 \le p < |\varphi(t_0)|$, such that $\alpha = S^p\varphi(t)$. Let β be an element of X_ζ with $\theta(\beta) = \alpha$. By definition of γ, there exist some $\delta \in X_\tau$ and some integer q, with $0 \le q < |\gamma(\delta_0)|$, such that $\beta = S^q\gamma(\delta)$. It follows that $S^q\varphi(\delta) = \theta(\beta) = \alpha = S^p\varphi(t)$. As $0 \le q < |\gamma(\delta_0)| = |\varphi(\delta_0)|$ by construction of γ, since $\varphi(R)$ is a circular code and since φ is injective, it follows that $\delta = t$ and $q = p$, thus $\beta = S^p\gamma(t)$: β is uniquely determined by α, and θ is one-to-one.

Proof of (iii). Let $l \in R$ be the letter such that l is the first letter of $\tau(k)$ for every $k \in R$. Let $(j,p) \in B$, and $k = \tau(j)_p$. By definition of ζ, the first letter of $\zeta(j,p)$ is $(k,1)$, and the first letter of $\zeta^2(j,p)$ is the first letter of $\zeta(k,1)$, i.e., $(l,1)$. By the same method, if r is the last letter of $\tau(k)$ for every $k \in R$, then the last letter of $\zeta^2(j,p)$ is $(r, |\varphi(r)|)$ for every $(j,p) \in B$. □

In (Vershik and Livshits 1992) the authors showed that when σ is a primitive substitution then the subshift it generates can be represented (in a measure-theoretic sense) by an ordered Bratteli diagram B where σ is the substitution we read on B. For example, in the case of the *Morse substitution* $a \mapsto ab$, $b \mapsto ba$ the Bratteli diagram they consider is given in Figure 6.8.

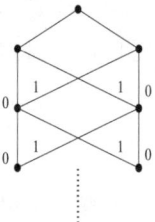

Fig. 6.8 The Bratteli diagram for the Thue–Morse substitution.

It is clear that it has two maximal and two minimal paths. Hence this representation does not fit our settings.

Given a primitive substitution σ on the alphabet A such that (X_σ, S) is aperiodic, let us describe how to construct a primitive proper substitution ζ such that (X_σ, S) is isomorphic to (X_ζ, S). With the techniques used below, we do not need to suppose primitivity. We only need σ to generate a minimal subshift. For example, this includes the Chacon substitution $0 \mapsto 0010$ and $1 \mapsto 1$. An illustration of the construction given below is given in Section 6.5.2.4.

6.5.2.3 Return words

In order to find ζ we need to introduce the notion of return words. For more details, the reader is referred to e.g., (Durand 1998b), (Durand 2000), (Durand 2003). We define an *occurrence of u.v in the bi-infinite word x* to be an integer n such that $x_{[n-|u|,n+|v|-1]} = uv$. A finite word w on A is a *return word to u.v in x* if there exist two consecutive occurrences j, k of $u.v$ in x such that $w = x_{[j,k-1]}$.

Let $x \in A^{\mathbb{Z}}$ be one of the fixed points of σ. We write $r = x_{-1}$ and $l = x_0$, so that r is the last letter of $\sigma(r)$ and l the first letter of $\sigma(l)$.

We denote by $L(\sigma)$ the set of finite words appearing in X_σ and we call it the *language* of σ. Let $\mathcal{R}$ be the set of return words to $r.l$. By minimality, it is a finite set. We set $R = \{1, \ldots, \text{Card}(\mathcal{R})\}$ and let $\varphi : R \to \mathcal{R}$ be the bijection defined as follows: let $\mathcal{R}$ be ordered according to the rank of first occurrence in $x_{[0,+\infty)}$, and $\varphi(k)$ defined to be the kth element of $\mathcal{R}$ for this order. Let z be the unique element of $R^{\mathbb{Z}}$ such that $\varphi(z) = x$.

We define now a substitution τ on the alphabet R. Let $j \in R$, and $w = w_1 \ldots w_k = \varphi(j) \in \mathcal{R}$. We have $rwl \in L(\sigma)$, $w_1 = l$ and $w_k = r$. As $\sigma(x) = x$, the finite word $\sigma(rwl) = \sigma(r)\sigma(w)\sigma(l)$ belongs also to $L(\sigma)$. But r is the last letter of $\sigma(r)$, and l the first of $\sigma(l)$, consequently, $r\sigma(w)l$ belongs to $L(\sigma)$. Note also that the first letter of $\sigma(w)$ is l and its last letter r. Then, rl is a prefix and also a suffix of $r\sigma(w)l$. Therefore, the finite word $\sigma(w)$ appears in x between two occurrences of $r.l$, thus it is a concatenation of return words, i.e., it belongs to $\varphi(R^+)$. Then, there exists a unique finite word $u \in R^+$ such that $\sigma(w) = \varphi(u)$.

We define $\tau(j) = u$. This defines a substitution τ on the alphabet R, characterised by

$$\varphi \circ \tau = \sigma \circ \varphi.$$

It follows that $\varphi \circ \tau^n = \sigma^n \circ \varphi$ for each $n \geq 0$.

The first element in the decomposition of $x_{[0,+\infty)}$ in return words is $\varphi(1)$, i.e., $\varphi(1)l$ is a prefix of $x_{[0,+\infty)}$. Let n be large enough for $|\sigma^n(l)| > |\varphi(1)|$. As $\sigma^n(l)$ and $\varphi(1)l$ are both prefixes of $x_{[0,+\infty)}$, $\varphi(1)l$ is a prefix of $\sigma^n(l)$. Let $j \in R$, and $w = \varphi(j)$. As l is the first letter of w, $\sigma^n(l)$ is a prefix of $\sigma^n(w)$, and so is $\varphi(1)l$. It follows that $\varphi(1)$ is the first element in the decomposition of $\sigma^n(w) = \varphi(\tau^n(j))$ in a concatenation of return words, i.e., 1 is the first letter of $\tau^n(j)$.

Let $m = z_{-1}$. The word $r\varphi(m)$ is a suffix of $x_{(-\infty,-1]}$, and the same argument shows that, for every large enough n and every $j \in R$, m is the last letter of $\tau^n(j)$. This implies that τ is proper.

Let $k > 0$ be an occurrence of $r.l$ large enough for every return word $w \in \mathcal{R}$ to appear in the decomposition of $x_{[0,k)}$, i.e., for every $j \in R$ to occur in the finite word $u \in R^+$ defined by $\varphi(u) = x_{[0,k)}$. Let n be so large that $|\sigma^n(l)| > k$. Let $i, j \in R$. As above, $x_{[0,k)}l$ is a prefix of $\sigma^n(l)$, which is a prefix of $\sigma^n(\varphi(i)) = \varphi(\tau^n(i))$. Thus u is a prefix of $\tau^n(i)$, and j occurs in $\tau^n(i)$, hence τ is primitive. Moreover,

$$\varphi(\tau(z)) = \sigma(\varphi(z)) = \sigma(x) = x = \varphi(z),$$

thus $\tau(z) = z$ by the unique decomposition property, and z is the unique fixed point of τ. As $\varphi(z) = x$ is not periodic, z is not periodic.

Proposition 6.5.4 gives, using Proposition 6.5.2, the substitution ζ we are looking for: (X_σ, S) is isomorphic to (X_B, V_B) where B has a simple hat and ζ is read on B.

6.5.2.4 An example: the Chacon substitution

Let us illustrate this on the non-primitive case given by the Chacon substitution σ. For more details about this substitution, see (Ferenczi 1995).

Let $x = \sigma^\omega(0.0)$. This is the *Chacon word*. It is uniformly recurrent. Using the return words to 0.0 we see that $x = \varphi(y)$ where $\varphi : \{a, b, c\}^* \to \{0, 1\}^*$ is defined by $\varphi(a) = 0$, $\varphi(b) = 010$ and $\varphi(c) = 01010$, and $y = \tau^\omega(b.a)$ where τ is defined by $\tau(a) = ab$, $\tau(b) = acb$ and $\tau(c) = accb$. According to the proof of Proposition 6.5.4 we need to take τ^2 instead of τ and we take

(i) $B = \{(a, 1), (b, 1), (b, 2), (b, 3), (c, 1), (c, 2), (c, 3), (c, 4), (c, 5)\}$,
(ii) $\theta : B \to \{0, 1\}$ given by the following table

α	$(a,1)$	$(b,1)$	$(b,2)$	$(b,3)$	$(c,1)$	$(c,2)$	$(c,3)$	$(c,4)$	$(c,5)$
$\theta(\alpha)$	0	0	1	1	0	1	0	1	0

(iii) $\gamma : \{a, b, c\} \to B^+$ defined by

$$\gamma(a) = (a, 1), \ \gamma(b) = (b, 1)(b, 2)(b, 3), \ \gamma(c) = (c, 1) \cdots (c, 5),$$

(iv) the substitution $\zeta : B^* \to B^*$ defined by the following tables

α	$(a,1)$	$(b,1)$	$(b,2)$	$(b,3)$
$\zeta(\alpha)$	$\gamma(\tau(a))$	$\gamma(a)$	$\gamma(b)$	$\gamma(accbacb)$

α	$(c,1)$	$(c,2)$	$(c,3)$	$(c,4)$	$(c,5)$
$\zeta(\alpha)$	$\gamma(a)$	$\gamma(b)$	$\gamma(a)$	$\gamma(c)$	$\gamma(cbaccbacb)$

A BV-representation of the Chacon subshift (*i.e.*, generated by x) is isomorphic to (X_B, V_B) where B a stationary properly ordered Bratteli diagram with a simple hat such that ζ is the substitution we read on it.

6.5.2.5 Generality in the Chacon example

Let $\sigma : A \to A^*$ be a substitution and consider the sets

$$L(\sigma) = \{w \in A^* \mid w \text{ is a factor of some } \sigma^n(a), a \in A, n \in \mathbb{N}\} \text{ and}$$
$$\Omega(\sigma) = \{x \in A^{\mathbb{Z}} \mid x_i x_{i+1} \cdots x_j \in L(\sigma), i, j \in \mathbb{Z}\}.$$

For $\Omega(\sigma)$ to be non-empty, it is necessary and sufficient that there exists $e \in A$ with

$$\lim_{n \to +\infty} |\sigma^n(e)| = +\infty. \tag{6.4}$$

Without lost of generality we suppose that all letters of A have an occurrence in some element of $\Omega(\sigma)$ and that

$$\forall a \in A, \; \exists n \in \mathbb{N}, \; |\sigma^n(a)|_a \geq 1. \tag{6.5}$$

In the paper (Damanik and Lenz 2006) the authors propose to call *substitution subshift* any pair $(\Omega(\sigma), S)$ where σ satisfies (6.4), (6.5) and $L(\sigma) = L(\Omega(\sigma))$, where $L(\Omega(\sigma))$ is the set of factors of elements of $\Omega(\sigma)$. They prove the following result (the definition of linearly recurrent subshift is given in Section 6.5.3).

Theorem 6.5.5 (Damanik and Lenz 2006) *Let $(\Omega(\sigma), S)$ be a substitution subshift. Then the following are equivalent.*

(i) *There exists $e \in A$ satisfying (6.4) which occurs with bounded gaps and furthermore, σ is assumed to satisfy (6.5).*
(ii) *$(\Omega(\sigma), S)$ is minimal.*
(iii) *$(\Omega(\sigma), S)$ is linearly recurrent.*

A fourth equivalent statement could be added: $(\Omega(\sigma), S)$ is uniquely ergodic.

The longest part of the proof given by D. Damanik and D. Lenz is to prove that (ii) implies (iii). The techniques we used before permit us to give an alternative proof and even to prove a bit more.

Proposition 6.5.6 *Let $(\Omega(\sigma), S)$ be a substitution subshift. Then, it is minimal if, and only if, it is isomorphic to some $(\Omega(\tau), S)$ where τ is a primitive proper substitution.*

6.5.3 Linearly recurrent subshifts

In what follows we show that a subshift is linearly recurrent if, and only if, it has a BV-representation where the incidence matrices have positive entries and belong to a finite set.

Before proceeding we need to introduce some new definitions.

Let x be an element of $A^{\mathbb{Z}}$. As we will manipulate return words, it is important to observe that a finite word $w \in A^+$ is a return word to $u.v$ in x if, and only if,

(i) uwv has an occurrence in x, and
(ii) v is a prefix of wv and u is a suffix of uw, and
(iii) the finite word uwv contains exactly two occurrences of the finite word uv.

We denote by $\mathcal{R}_{x,u.v}$ the set of return words to $u.v$ in x. If u is the empty word ε, then we speak of return words to v instead of the return words to $u.v$ and we set $\mathcal{R}_{x,v} = \mathcal{R}_{x,u.v}$. When it will be clear from the context we will refer to $\mathcal{R}_{u.v}$ in place of $\mathcal{R}_{x,u.v}$.

Definition 6.5.7 We say that $x \in A^{\mathbb{Z}}$ is *linearly recurrent* (LR) (with constant $K \in \mathbb{N}$) if it is uniformly recurrent and if for all u having an occurrence in x and all return words w to u in x we have $|w| \leq K|u|$. A subshift is *linearly recurrent* whenever it is the shift orbit closure of a linearly recurrent word.

From Theorem 6.5.5 we know that all minimal substitution subshifts are linearly recurrent.

Remark that all words of a LR subshift are LR with the same constant. Here are some important properties proved in (Durand, Host, and Skau 1999).

Proposition 6.5.8 *Let x be a non-periodic word and suppose that it is linearly recurrent with constant K. Then:*

(i) *The number of distinct factors of length n of x is less than or equal to Kn.*
(ii) *x is $(K+1)$-power free: u^{K+1} has an occurrence in x if, and only if, u is the empty word.*
(iii) *For all u having an occurrence in x and for all $w \in \mathcal{R}_u$ we have $\frac{1}{K}|u| < |w|$.*
(iv) *If u has an occurrence in x then $|\mathcal{R}_u| \leq K(K+1)^2$.*

We suppose now that x is a uniformly recurrent word. It is easy to see that for all $u, v \in L(x)$ the set $\mathcal{R}_{x,u.v}$ is finite.

It will be convenient to label the return words. We enumerate the elements w of $\mathcal{R}_{x,u.v}$ in the order of the first appearance of uwv in $x_{[-|u|,+\infty)}$. This defines a bijective map $\Theta_{x,u.v} : R_{x,u.v} \to \mathcal{R}_{x,u.v} \subset A^+$ where

$\mathcal{R}_{x,u.v} = \{1, \ldots, \text{Card}(\mathcal{R}_{x,u.v})\}$, and $u\Theta_{x,u.v}(k)v$ is the kth finite word of the type uwv ($w \in \mathcal{R}_{x,u.v}$) that occurs in $x_{[-|u|,+\infty)}$. We consider $\mathcal{R}_{x,u.v}$ as an alphabet. The map $\Theta_{x,u.v}$ defines a morphism from $\mathcal{R}_{x,u.v}$ to A^* and the set $\Theta_{x,u.v}(R^*_{x,u.v})$ consists of all concatenations of return words to $u.v$.

In (Durand, Host, and Skau 1999) is proved the following result that we will use in the sequel.

Proposition 6.5.9 *The map $\Theta_{x,u.v} : R^+_{x,u.v} \to A^+$ is one-to-one.*

The following result characterises linearly recurrent subshifts in terms of BV-representations. We first need to recall that a dynamical system (X, T), endowed with the distance d, is *expansive* if there exists ε such that for all pairs of points (x, y), $x \neq y$, there exists n with $d(T^n x, T^n y) \geq \varepsilon$. We say that ε is a *constant of expansivity* of (X, T). It is easy to show that the subshifts are expansive and that, in fact, every expansive dynamical system is isomorphic to a subshift (see (Kůrka 2003) for example).

Theorem 6.5.10 *A subshift is linearly recurrent if, and only if, it has an expansive BV-representation satisfying:*

(i) *its incidence matrices have positive entries and belong to a finite set of matrices,*

(ii) *for all $n \geq 1$ the substitution read on $E(n)$ is proper.*

Proof Let (X, S) be a LR subshift. The periodic case is trivial hence we suppose that (X, S) is not periodic. It suffices to construct a sequence of KR-partitions having the desired properties.

From Proposition 6.5.8 there exists an integer K such that for all u occurring in some $x \in X$ and all $w \in \mathcal{R}_u$, we have

$$\frac{|u|}{K} \leq |w| \leq K|u|.$$

We set $\alpha = K^2(K + 1)$. Let $x = (x_n)_n$ be an element of X. For each non-negative integer n, we set $u_n = x_{-\alpha^n} \cdots x_{-2}x_{-1}$, $v_n = x_0 x_1 \cdots x_{\alpha^n - 1}$, $\mathcal{R}_n = \mathcal{R}_{x, u_n . v_n}$, $R_n = R_{x, u_n . v_n}$ and $\Theta_n = \Theta_{x, u_n . v_n}$.

Now define for all n

$$\mathcal{P}(n) = \left\{ S^j[u_n.wv_n] \mid w \in \mathcal{R}_n,\ 0 \leq j < |w| \right\}.$$

We claim $(\mathcal{P}(n))_n$ is a sequence of KR-partitions having the desired properties. We leave the details to the reader. They can be found in (Durand 2003).

Now let B be a properly ordered Bratteli diagram satisfying (i) and (ii).

Let ε be a constant of expansivity of X_B. Let n_0 be a level of B such that all cylinders $[e_1, \ldots, e_{n_0-1}]$ are included in a ball of radius $\varepsilon/2$. For any vertex $v \in V(n_0-1)$ let h_v denote the number of edges from v to $V(0)$. Now consider the alphabet

$$A = \{(v,j) \mid v \in V(n_0-1),\ 0 \le j < h_v\},$$

the map

$$\begin{aligned} C:\ X_B &\to A \\ (e(n))_n &\mapsto (r(e_{n_0-1}), j), \end{aligned}$$

if $(e(n))_{1 \le n \le n_0 - 1}$ is the jth finite path of B from $r(e_{n_0-1})$ to $V(0)$ with respect to the order on B, and finally define

$$\begin{aligned} \varphi:\ X_B &\to A^{\mathbb{Z}} \\ x &\mapsto (C \circ V_B^n(x))_{n \in \mathbb{Z}}. \end{aligned}$$

We clearly have that (X_B, V_B) is isomorphic to (Ω, S) where $\Omega = \varphi(X_B)$ and S is the shift on A. It remains to show that (Ω, S) is LR.

Let $K = \sup_{n \ge n_0} \max_{v \in V(n)} \sum_{v' \in V(n-1)} \mathbf{M}(n)_{v,v'}$. Condition (i) implies that K is finite. Let $L = \max_{v,v' \in V(n_0-1)} \frac{h_v}{h_{v'}}$.

Let $v \in V(n_0-1)$ and w be a return word to $u = (v,0)(v,1)\cdots(v,h_v-1)$. Due to Condition (i) and Condition (ii) we have

$$|w| \le 2K \max_{v' \in V(n_0-1)} h_{v'} \le 2KL|u|. \tag{6.6}$$

Now for all $n \ge n_0$, let $\tau_n : V(n) \to V(n-1)^*$ be the substitution read on $E(n)$. We set $W = \{(v,0)(v,1)\cdots(v,h_v-1) \mid v \in V(n_0-1)\}$ and we define the morphism $\sigma : V(n_0-1) \to A^*$ by $\sigma(v) = (v,0)(v,1)\cdots(v,h_v-1)$.

It is clear that all the elements of Ω are concatenations of finite words belonging to W. They are also concatenations of finite words belonging to $\sigma \circ \tau_{n_0}(V(n_0))$, and more generally, of finite words belonging to $\sigma \circ \tau_{n_0} \circ \cdots \circ \tau_n(V(n))$ for all $n \ge n_0$. As for (6.6), we can prove that all return words to some elements of $\sigma \circ \tau_{n_0} \circ \cdots \circ \tau_n(V(n))$ satisfy the same inequality.

Now let u be any non-empty finite word appearing in some word of Ω and w be a return word to u. There exists n such that

$$\max_{v \in V(n)} |\sigma \circ \tau_{n_0} \circ \cdots \circ \tau_n(v)| \le |u| < \max_{v \in V(n+1)} |\sigma \circ \tau_{n_0} \cdots \circ \tau_{n+1}(v)|.$$

Then u is a factor of some $\sigma \circ \tau_{n_0} \circ \cdots \circ \tau_{n+1}(vv')$, v and v' belonging to $V(n+1)$. From Condition (i) and Condition (ii), we deduce vv' is a factor of some $\tau_{n+2} \circ \tau_{n+3}(v'')$, $v'' \in V(n+3)$. Then, u is a factor of $\sigma \circ \tau_{n_0} \circ \cdots \circ \tau_{n+3}(v'')$. Consequently

$$|w| \le 2KL|\sigma \circ \tau_{n_0} \circ \cdots \circ \tau_{n+3}(v'')| \le 2KL^4|u|$$

and (Ω, S) is LR. □

Let us call *linearly recurrent* any Cantor dynamical systems having a BV-representation satisfying (i) and (ii) in Theorem 6.5.10. In fact more can be said about these dynamical systems but we need the following theorem proved in (Downarowicz and Maass 2008). It can be seen as an extension of Proposition 6.5.2. We say that a minimal Cantor dynamical system (X, T) has *topological rank* k if k is the smallest integer such that (X, T) has a BV-representation (X_B, V_B) with the sequence of number of vertices $(\text{Card}(V(n)))_n$ bounded by k. When such a k does not exist, we say that it has *infinite topological rank*. Of course, linearly recurrent BV-dynamical systems have finite topological rank. A topological dynamical system (X, T), endowed with the distance d, is said to be *equicontinuous* whenever

$$\forall \varepsilon, \exists \delta > 0, \sup_{n \in \mathbb{Z}} d(T^n x, T^n y) < \varepsilon \text{ if } d(x, y) < \delta.$$

Theorem 6.5.11 *Let (X, T) be a minimal Cantor dynamical system with topological rank $k \in \mathbb{N}$. Then, (X, T) is expansive if, and only if, $k \geq 2$. Otherwise it is equicontinuous.*

Let us take the notation of the proof of Theorem 6.5.10. We call σ_{n_0} the morphism σ and A_{n_0} the alphabet A. Let X_{n_0} be the subset of $A_{n_0}^{\mathbb{Z}}$ consisting of all the words x such that for all i, j, $x_i x_{i+1} \cdots x_j$ is a factor of $\sigma_{n_0} \circ \tau_{n_0} \cdots \circ \tau_n(v)$ for some $n \in \mathbb{N}$ and $v \in V(n)$. It can be checked that (X_{n_0}, S) is a minimal subshift.

Corollary 6.5.12 *Let (X_B, V_B) be a BV-dynamical system with finite topological rank. Then, (X_B, V_B) is expansive if, and only if, there exists n_0 such that (X_{n_0}, S) is not periodic.*

Moreover, if the cylinders $[e_1, \ldots, e_{n_0-1}]$ of X_B are all included in balls of radius $\frac{\varepsilon}{2}$, ε being a constant of expansivity, then (X_B, V_B) is isomorphic to (X_{n_0}, S).

Note that once some (X_{n_0}, S) is not periodic, then (X_n, S) is aperiodic for all $n \geq n_0$.

6.5.4 Sturmian subshifts

We define the morphisms ρ_n and γ_n, $n \in \mathbb{N} \setminus \{0\}$ from $\{0, 1\}$ to $\{0, 1\}^*$ by

$$\begin{array}{c} \rho_n(0) = 01^{n+1} \\ \rho_n(1) = 01^n \end{array} \text{ and } \begin{array}{c} \gamma_n(0) = 10^{n+1} \\ \gamma_n(1) = 10^n. \end{array}$$

The next theorem is due to G. A. Hedlund and M. Morse (Morse and Hedlund 1940).

Theorem 6.5.13 *Let $x \in \{0,1\}^{\mathbb{N}}$ be a Sturmian word. Then*

(i) *there is $n \geq 1$ such that x is a concatenation of finite words belonging to the set $\{01^{n+1}, 01^n\}$ or to the set $\{10^{n+1}, 10^n\}$,*

(ii) *if $x = \rho_n(z)$ or $x = \gamma_n(z)$, for some $n \geq 1$ and $z \in \{0,1\}^{\mathbb{N}}$, then z is Sturmian.*

Proof Assertion 1 follows from Theorem 7.1 in (Morse and Hedlund 1940) and Assertion 2 is Theorem 8.1 in (Morse and Hedlund 1940). □

Let (X, S) be a Sturmian subshift. For $a \in \{0, 1\}$, we recall that $[a] = \{(x_i)_{i \in \mathbb{Z}} \in X \mid x_0 = a\}$.

Corollary 6.5.14 *Let (X, S) be a Sturmian subshift of $\{0,1\}^{\mathbb{Z}}$, i.e., there exists a Sturmian word x such that all bi-infinite words in X have the same language as x. There exists a sequence $(\zeta_n)_{n \in \mathbb{N}}$ taking values in $\{\rho_1, \gamma_1, \rho_2, \gamma_2, \ldots\}$ such that $(\mathcal{P}(n))_n$ is a sequence of KR-partitions of (X, S) satisfying (KR1), (KR2) and (KR3) where $\mathcal{P}(1) = \{[0], [1]\}$ and, for $n \geq 2$,*

$$\mathcal{P}(n) = \left\{ S^k \zeta_1 \cdots \zeta_n([a]) \mid 0 \leq k < |\zeta_1 \cdots \zeta_{n-1}(a)|, \ a \in \{0, 1\} \right\}.$$

It appears that (KR1) is not clearly satisfied. To be convinced, note that, if for example $\zeta_n = \rho_i$, then

$$\zeta_1 \cdots \zeta_n([a]) \subseteq [\zeta_1 \cdots \zeta_{n-1}(1).\zeta_1 \cdots \zeta_n(a)\zeta_1 \cdots \zeta_{n-1}(0)].$$

Let (X, S) be a Sturmian subshift and $(\mathcal{P}(n))_n$ be the sequence of partitions given by Corollary 6.5.14. With such a sequence is associated an ordered Bratteli–Vershik diagram $B = (V, E, \geq)$ which can be described as follows. For all $n \geq 1$, V_n consists of two vertices, the substitution read on E_{n+1} is ζ_n, with $E(1)$ consisting of a simple hat.

It is clear that a Sturmian subshift is linearly recurrent if, and only if, its BV-representation given by Corollary 6.5.14 is such that $\{\zeta_n \mid n \geq 1\}$ is finite. In (Dartnell, Durand, and Maass 2000) it is proven that it is a substitution subshift if, and only if, $(\zeta_n)_n$ is ultimately periodic. This implies the next result (see also (Kůrka 2003)).

Theorem 6.5.15 *Let (X, S) be a Sturmian subshift generated by α. Then, (X, S) is a substitution subshift if, and only if, α is quadratic.*

The proof uses the notion of dimension groups we introduce in Section 6.6. For similar theorems characterising substitutive Sturmian words, see the references in (Lothaire 2002).

6.5.5 Toeplitz subshifts

A word $(x_n)_{n \in \mathbb{K}}$ ($\mathbb{K} = \mathbb{N}$ or $\mathbb{Z}$) on the alphabet A satisfying

$$\forall n \in \mathbb{Z}, \exists p \in \mathbb{N}, \forall k \in \mathbb{Z}, x_n = x_{n+kp}$$

is called a *Toeplitz word*. The subshifts they generate are called *Toeplitz subshifts*. These words and subshifts have been deeply studied since they were introduced in (Jacobs and Keane 1969). We refer to (Downarowicz 2005) for a nice survey on important dynamical results on these subshifts.

A Bratteli diagram has the *equal path number property* if for all $n \geq 1$ and $u, v \in V(n)$ we have $|r^{-1}(u)| = |r^{-1}(v)|$. This property was defined in (Gjerde and Johansen 2000). Note that the representation of odometers given in Section 6.5.1 shares this property.

Theorem 6.5.16 *A minimal subshift is Toeplitz if, and only if, it has an expansive BV-representation (X_B, V_B) where $B = (V, E, \geq)$ has the equal path number property. Moreover, there exist BV-systems having the equal path number property that are neither expansive nor equicontinuous.*

6.5.6 Interval exchange transformations

Let $\zeta = (\Delta_1, \ldots, \Delta_k)$ be a partition of the segment $[0, 1)$ into $k \geq 2$ disjoint intervals of the form $[a, b)$ numbered from left to right. A *(k)-interval exchange transformation* is an onto map $T : [0, 1) \to [0, 1)$ where $T : \Delta_i \to [0, 1)$ is a translation. We remark there exists a permutation $\pi : \{1, \ldots, k\} \to \{1, \ldots, k\}$ such that the $T\Delta_i$ are positioned on $[0, 1)$, from left to right, as follows: $T\Delta_{\pi^{-1}(1)}, \ldots, T\Delta_{\pi^{-1}(k)}$. For more details on interval exchange transformations we refer to (Cornfeld, Fomin, and Sinaĭ 1982). See also Section 7.4.4.

Let us now give what we will later, in the present section, refer to as the *Cantor version of interval exchange transformations*.

Suppose that T is minimal: all its orbits are dense in $[0, 1)$. Let $\mathcal{D}(T) = \{d_1, \ldots, d_k\}$ be the set of left extremal points of the intervals Δ_i and set $\mathcal{O}(T) = \{T^j d \mid j \in \mathbb{Z}, d \in \mathcal{D}(T)\}$. We define

$$X = ([0, 1) \setminus \mathcal{O}(T)) \bigcup \{x^-, x^+ \mid x \in \mathcal{O}(T)\}$$

where $0^- = 1$. Defining $x^- < x^+$ for all $x \in \mathcal{O}(T)$ with the exception of $0^- \geq x$ for all $x \in X$, this extends the natural order on $[0,1)$ to X. Endowed with the topology of intervals, X is a Cantor space because $\mathcal{O}(T)$ is dense in $[0,1)$.

Let $F : X \to X$ defined by $F(y) = T(y)$ if $y \in [0,1) \setminus \mathcal{O}(T)$ and $F(x^\varepsilon) = T(x)^\varepsilon$ if $x \in \mathcal{O}(T)$ where $\varepsilon \in \{+,-\}$. The pair (X, F) is a minimal Cantor dynamical system, that we will refer to as the *Cantor version of the interval exchange* T (for more details see (Gjerde and Johansen 2002)).

Let $\varphi : X \to [0,1)$ be defined by $\varphi(x^+) = \varphi(x^-) = x$ and $\varphi(x) = x$ when $x \notin \mathcal{O}(T)$. This is an onto continuous map. It is one-to-one everywhere except on a countable set of points. Moreover, $\varphi \circ F = T \circ \varphi$.

Theorem 6.5.17 *Let (X, F) be the Cantor version of a minimal k-interval exchange. Then, (X, F) has a BV-representation (X_B, V_B) where $B = (V, E, \geq)$ is such that*

(i) $\mathrm{Card}(V(1)) = k$ *and* $\mathrm{Card}(V(i)) - \mathrm{Card}(V(i+1)) \in \{0,1\}$ *for all* $i \geq 1$,

(ii) *for all $i \geq 1$ when $V(i-1) = V(i)$ the incidence matrix $\mathbf{M}(i)$ of $E(i)$ has the following form*

$$\mathbf{M}(i) = \begin{bmatrix} 1 & 0 & \cdots & 0 & 0 & \cdots & 0 & s_1 \\ 0 & 1 & \cdots & 0 & 0 & \cdots & 0 & s_2 \\ \vdots & \vdots & & \vdots & \vdots & & \vdots & \vdots \\ 0 & 0 & \cdots & 1 & 0 & \cdots & 0 & s_l \\ 0 & 0 & \cdots & 1 & 0 & \cdots & 0 & s_{l+1} \\ 0 & 0 & \cdots & 0 & 1 & \cdots & 0 & s_{l+2} \\ \vdots & \vdots & & \vdots & \vdots & & \vdots & \vdots \\ 0 & 0 & \cdots & 0 & 0 & \cdots & 1 & s_k \end{bmatrix},$$

where $s_i \in \{0, m, m+1\}$, $s_l = m$ and $s_{l+1} = m+1$ for some $m \geq 0$. When $\mathrm{Card}(V(i)) - \mathrm{Card}(V(i+1)) = 1$, the line $l+1$ does not exist. All the entries of $\mathbf{M}(1)$ are equal to 1.

This theorem has been proved in (Gjerde and Johansen 2002). They also showed that there are BV-dynamical systems satisfying the hypothesis of the theorem that are not isomorphic to a Cantor version of an interval exchange transformation.

6.5.7 Representation of non-minimal Cantor dynamical systems

In (Medynets 2006) the author shows that for Cantor dynamical systems without periodic points, but not necessarily minimal, a BV-representation can also be given. It is applied in (Bezuglyi, Kwiatkowski, and Medynets 2008) to subshifts generated by non-primitive substitutions. The authors show that they have stationary BV-representations as in the minimal case.

6.6 Characterisation of Strong Orbit Equivalence

In this section we give the proof of the Strong Orbit Equivalence theorem proved in (Giordano, Putnam, and Skau 1995). The statement is given in terms of Bratteli diagrams and dimension groups.

6.6.1 Dimension groups as ordered groups

An *ordered group* is a pair (G, G^+), where G^+ is a subset of the group G, such that

$$G^+ + G^+ \subseteq G^+, \ G^+ - G^+ = G, \ G^+ \cap (-G^+) = \{0\}.$$

For example, $(\mathbb{Z}^d, (\mathbb{Z}^d)^+)$, with $(\mathbb{Z}^d)^+ = \{(e_1, \ldots, e_d) \in \mathbb{Z}^d \mid e_i \geq 0, \ 1 \leq i \leq d\}$, is an ordered group.

The set G^+ is the *positive cone* of G and its elements are called non-negative. We set, for all $h, g \in G$, $g \geq h$ if, and only if, $g - h \in G^+$. An *order unit* of (G, G^+) is a non-negative element u of G such that

$$\forall g \in G^+, \exists n > 0, \ g \leq nu.$$

Let (G, G^+, u) and (H, H^+, v) be two ordered groups with order units. When $\varphi : G \to H$ is a homomorphism satisfying $\varphi(G^+) \subseteq H^+$ and $\varphi(u) = v$, we say that φ is a *morphism (of ordered groups with order unit)*. When it is clear from the context we write G instead of (G, G^+, u).

A *dimension group* is an ordered group obtained as a direct limit of a sequence of finitely generated free Abelian groups with standard order and positive group homomorphisms as maps. More precisely, let $\left(M(n) : \mathbb{Z}^{d(n-1)} \to \mathbb{Z}^{d(n)}\right)_{n \geq 1}$ be a sequence of homomorphisms such that $M(n)\left(\mathbb{Z}^{d(n-1)}\right)^+ \subseteq \left(\mathbb{Z}^{d(n)}\right)^+$ for all $n \geq 1$. Consider the following subgroups of $\prod_{n \geq 0} \mathbb{Z}^{d(n)}$:

$$\Delta = \left\{ (v(n))_{n \geq 0} \in \prod_{n \geq 0} \mathbb{Z}^{d(n)} \mid M(n)v(n) = v(n+1) \text{ for all } n \text{ large enough} \right\}$$

and

$$\Delta^0 = \{(v(n))_n \in \Delta \mid v(n) = 0 \text{ for all } n \text{ large enough}\}.$$

Let G be the quotient of Δ by Δ^0 and G^+ be the projection in G of the subset

$$\Delta^+ = \left\{ (v(n))_n \in \Delta \mid v(n) \in \left(\mathbb{Z}^{d(n)}\right)^+ \text{ for all } n \text{ large enough} \right\}.$$

We denote by

$$G := \varinjlim_{M(n)} \mathbb{Z}^{d(n)}.$$

The pair (G, G^+) is an ordered group called a *dimension group*. It can be checked that u is an order unit of (G, G^+) if, and only if, it is the image (by the canonical projection) in G of some $(v(n)) \in \Delta^+ \setminus \Delta^0$. We say that (G, G^+, u) is a *dimension group with order unit*.

6.6.2 Dimension groups and coboundaries

Let (X, T) be a minimal Cantor dynamical system. We denote by $C(X, \mathbb{Z})$ the set of continuous maps from X to $\mathbb{Z}$. Consider the map $\beta : C(X, \mathbb{Z}) \to C(X, \mathbb{Z})$ defined by $\beta f = f \circ T - f$ for all $f \in C(X, \mathbb{Z})$. The images of β are called *coboundaries*. Let $H(X, T)$ be the quotient group $C(X, \mathbb{Z})/\beta C(X, \mathbb{Z})$ and π be the natural projection from $C(X, \mathbb{Z})$ to $C(X, \mathbb{Z})/\beta C(X, \mathbb{Z})$. We call *order unit* the image by π of the constant function equal to 1, and we denote it by u_X. The positive cone, $H^+(X, T, \mathbb{Z})$, is the image by π of the set of non-negative functions $C(X, \mathbb{Z}_{\geq 0})$. Finally, the triple

$$DG(X, T) = (H(X, T, \mathbb{Z}), H^+(X, T, \mathbb{Z}), u_X)$$

is an ordered group with order unit. In the next section we will see that it is the *dimension group* generated by the incidence matrices of any BV-representation of (X, T).

6.6.3 Dimension groups and KR-partitions

Let us show now that, thanks to sequences of KR-partitions, the two previous definitions coincide in the context of minimal Cantor dynamical systems.

Let (X, T) be a minimal Cantor dynamical system and let $(\mathcal{P}(n))_n$ with

$$\mathcal{P}(n) = \{T^j B_i(n) \mid 0 \leq j < h_i(n), 1 \leq i \leq t(n)\}$$

be a sequence of KR-partitions satisfying (KR1), (KR2) and (KR3). We recall that from Theorem 6.4.3, such a sequence always exists.

Let $f \in C(X, \mathbb{Z})$. As $(\mathcal{P}(n))_n$ generates the topology, there exists n_0 such that for all $n \geq n_0$, f is constant on each atom of $\mathcal{P}(n)$. Let $(\mathbf{M}(n))_{n \geq 1}$

be the associated sequence of incidence matrices, set $d(n) = |V(n)|$ for all $n \geq 0$, and consider Δ, Δ^0, Δ^+, G and G^+ as defined in Section 6.6.1. We call χ the canonical projection from Δ to G.

Define the column vector $\tilde{\mathbf{f}}(n) \in \mathbb{Z}^{d(n)}$ by letting $\tilde{\mathbf{f}}_i(n)$ be the sum of the values of f over all the levels of the tower i, $1 \leq i \leq t(n)$. It is clear that

$$\tilde{\mathbf{f}}(n+1) = \mathbf{M}(n+1)\tilde{\mathbf{f}}(n).$$

We set $\tilde{J}(f) = \chi(0, \ldots, 0, \tilde{\mathbf{f}}(n_0), \tilde{\mathbf{f}}(n_0+1), \ldots)$. This defines a homomorphism from $C(X, \mathbb{Z})$ to G.

Let $f \in \beta C(X, \mathbb{Z})$. One has $f = g \circ T - g$ with $g \in C(X, \mathbb{Z})$. Then $\tilde{\mathbf{f}}_i(n) = g \circ T^{h_i(n)}(x) - g(x)$ for all $x \in B_i(n)$. But as the partition satisfies (KR1), $T^{h_i(n)} B_i(n) \subseteq B(n)$ and then $(\tilde{\mathbf{f}}(n))_n$ goes to 0. Moreover the $\tilde{\mathbf{f}}_i(n)$ are integers. Hence there exists n_1 such that for $n \geq n_1$, we have $\tilde{\mathbf{f}}(n) = 0$. Consequently $\beta C(X, \mathbb{Z})$ is included in ker $\tilde{J}$.

Now, let $f \in C(X, \mathbb{Z})$ and suppose $\tilde{J}(f) = 0$. Then, there exists n_0 such that $\tilde{\mathbf{f}}(n) = 0$ for all $n \geq n_0$. Take $n \geq n_0$. Let us find $g \in C(X, \mathbb{Z})$ satisfying $f = g \circ T - g$. We define g as follows:

$$g(x) = -(f(x) + f(Tx) + \cdots + f(T^{h_i(n)-1-j}x)) \text{ if } x \in T^j B_i(n).$$

Let $x \in T^j B_i(n)$ with $j \neq h_i(n) - 1$. Then, $T(x)$ belongs to the level $j + 1$ of the tower i and

$$g(T(x)) = -(f(T(x)) + f(T^2(x)) + \cdots + f(T^{h_i(n)-1-(j+1)}(T(x)))).$$

Hence, $f(x) = g \circ T(x) - g(x)$ for all x that are not in $\cup_{1 \leq i \leq t(n)} T^{h_i(n)-1} B_i(n)$. Let $x \in T^{h_i(n)-1} B_i(n)$. Then $T(x)$ belongs to $B_j(n)$ for some j and consequently $g(T(x)) = \tilde{\mathbf{f}}_j(n) = 0$. Finally $g \circ T(x) - g(x) = 0 - (-f(x)) = f(x)$.

This proves that ker $\tilde{J} = \beta C(X, \mathbb{Z})$ and that $\tilde{J}$ induces a one-to-one homomorphism J from $H(X, T)$ to G. We end this section by showing that J is an isomorphism of dimension groups.

Let $(g(n))_n$ be a representant of an element of G. We define $f \in C(X, \mathbb{Z})$ by $f(x) = g(n)_i$ if $x \in B_i(n)$, for some i, and 0 elsewhere. Then, $\tilde{\mathbf{f}}_i(n) = g(n)_i$ and J is onto. We see that $J(H^+(X, T)) = G^+$ and that $J(u_X) = \chi(\mathbf{M}(n) \cdots \mathbf{M}(2)\mathbf{M}(1))_n) = u_G$. Note that $\mathbf{M}(1)$ is a vector since $V(0)$ is a one-point set.

We thus have proved the following result.

Proposition 6.6.1 *Let (X, T) be a minimal Cantor dynamical system. For every BV-representation $B = (V, E, \geq)$ of (X, T) with incidence matrices $(\mathbf{M}(n))_n$ we have that $DG(X, T)$ is isomorphic, as a dimension group, to*

(G, G^+, u_G). In particular, the dimension group does not depend on the BV-representation.

6.6.4 Strong orbit equivalence

In the sequel we present and prove the Strong Orbit Equivalence (SOE) theorem obtained in (Giordano, Putnam, and Skau 1995). We follow the proof proposed in (Glasner and Weiss 1995) giving more details.

We say that two dynamical systems (X, T) and (Y, S) are *orbit equivalent* whenever there exists a homeomorphism $\varphi : X \to Y$ sending orbits to orbits

$$\varphi(\{T^n x \mid n \in \mathbb{Z}\}) = \{S^n \varphi(x) \mid n \in \mathbb{Z}\},$$

for all $x \in X$. This induces the existence of maps $\alpha : X \to \mathbb{Z}$ and $\beta : X \to \mathbb{Z}$ satisfying for all $x \in X$

$$\varphi \circ T(x) = S^{\alpha(x)} \circ \varphi(x) \text{ and } \varphi \circ T^{\beta(x)}(x) = S \circ \varphi(x).$$

When α and β have at most one point of discontinuity, we say that (X, T) and (Y, S) are *strongly orbit equivalent* (SOE). It is natural to consider such a definition because M. Boyle proved in (Boyle 1983) that if α is continuous then (X, T) is conjugate to (Y, S) or to (Y, S^{-1}). In (Giordano, Putnam, and Skau 1995) the authors characterised SOE by means of Bratteli diagrams and dimension groups.

Theorem 6.6.2 (Giordano, Putnam, and Skau 1995) *Let $(X^{(1)}, T)$ and $(X^{(2)}, S)$ be two minimal Cantor dynamical systems. The following are equivalent:*

(i) *There exist two BV-representations, $B^{(1)} = (V^{(1)}, E^{(1)}, \geq^{(1)})$ of $(X^{(1)}, T)$ and $B^{(2)} = (V^{(2)}, E^{(2)}, \geq^{(2)})$ of $(X^{(2)}, S)$, and an unordered Bratteli diagram $B = (V, E)$ of which $B^{(1)}$ and $B^{(2)}$ are contractions.*

(ii) *There exist two BV-representations, $B^{(1)} = (V^{(1)}, E^{(1)}, \geq^{(1)})$ of $(X^{(1)}, T)$ and $B^{(2)} = (V^{(2)}, E^{(2)}, \geq^{(2)})$ of $(X^{(2)}, S)$, and a homeomorphism $F : X_{B^{(1)}} \to X_{B^{(2)}}$ such that $F(x)(n)$ depends only on $x(1) \ldots x(n)$ and $F\left(x_u^{(1)}\right) = x_u^{(2)}$, $u \in \{\min, \max\}$, and having the property that if x and y are cofinal from level n, then $F(x)$ and $F(y)$ are cofinal from level $n + 1$.*

(iii) $(X^{(1)}, T)$ *and* $(X^{(2)}, S)$ *are SOE.*

(iv) $DG(X^{(1)}, T)$ *and* $DG(X^{(2)}, S)$ *are isomorphic as dimension groups with order units.*

Proof Let us show that (i) implies (ii). Note that a contraction of a BV-representation is itself a BV-representation. Hence, by contracting if needed, we can suppose that $B^{(1)}$ is obtained from B by contracting to odd levels while $B^{(2)}$ is obtained by contracting to even levels. Moreover, from Theorem 6.4.6, we can also suppose that all incidence matrices have entries greater than two. This means that every pair of vertices in consecutive levels has at least two connecting edges.

Let $x_{\min}^{(1)}$ and $x_{\max}^{(1)}$ be the minimal and maximal paths of $B^{(1)}$, and $x_{\min}^{(2)}$ and $x_{\max}^{(2)}$ for $B^{(2)}$. There are unique paths $\tilde{x}_{\min}^{(1)}$ and $\tilde{x}_{\min}^{(2)}$ in B that contract respectively to $x_{\min}^{(1)}$ and $x_{\min}^{(2)}$. Choose a path $z_{\min}$ in B passing through the same vertices as $\tilde{x}_{\min}^{(1)}$ does at odd levels and through the same vertices as $\tilde{x}_{\min}^{(2)}$ at even levels. We similarly construct a path $z_{\max}$ by taking care that it does not share any common edge with $z_{\min}$. This is possible because the incidence matrices have entries greater than two.

Let us define two homeomorphisms $F_1 : X_B \to X_{B^{(1)}}$ and $F_2 : X_B \to X_{B^{(2)}}$. In constructing $z_{\min}$, for each even n, we matched a pair of edges in $E_n \circ E_{n+1}$ with an edge in $E_{n/2}^{(1)}$, namely

$$(z_{\min}(n), z_{\min}(n+1)) \to x_{\min}^{(1)}(n/2),$$

and we match a pair in $E_{n+1} \circ E_{n+2}$ with an edge in $E_{(n+2)/2}^{(2)}$, namely

$$(z_{\min}(n+1), z_{\min}(n+2)) \to x_{\min}^{(2)}((n+2)/2).$$

In the same way $(z_{\max}(n), z_{\max}(n+1))$ is matched with $x_{\max}^{(1)}(n/2)$ and $(z_{\max}(n+1), z_{\max}(n+2))$ with $x_{\max}^{(2)}((n+2)/2)$. Now, for all even n, we extend these matchings in an arbitrary way to bijections respecting the range and source maps from $E(n) \circ E(n+1)$ to $E^{(1)}(n/2)$ and from $E(n+1) \circ E(n+2)$ to $E^{(2)}((n+2)/2)$. This defines two homeomorphisms

$$F_1 : X_B \to X_{B^{(1)}} \text{ and } F_2 : X_B \to X_{B^{(2)}}.$$

The homeomorphism $F = F_2 \circ F_1^{-1}$ has the desired properties.

Let us show that (ii) implies (iii). It suffices to show that $(X_{B^{(1)}}, V_{B^{(1)}})$ and $(X_{B^{(2)}}, V_{B^{(2)}})$ are SOE. In a minimal BV-representation, two points belong to the same orbit if, and only if, they are cofinal, except when it is the orbit of the minimal path. Hence F maps orbits to orbits with the possible exception of the orbit of the minimal paths. But as $F\left(x_u^{(1)}\right) = x_u^{(2)}$, $u \in \{\min, \max\}$, this is also true for the orbit of the minimal paths. Consequently, there are maps $\alpha : X_{B^{(1)}} \to \mathbb{Z}$ and $\beta : X_{B^{(1)}} \to \mathbb{Z}$ uniquely defined by the relations

$$F \circ V_{B^{(1)}}(x) = V_{B^{(2)}}^{\alpha(x)} \circ F(x) \text{ and } F \circ V_{B^{(1)}}^{\beta(x)}(x) = V_{B^{(2)}} \circ F(x)$$

for all $x \in X_{B^{(1)}}$. It remains to prove that α and β are continuous with the possible exception of $x_{\max}^{(1)}$ and $x_{\max}^{(2)}$. We do it for α. It is similar for β.

Let $x = (x_n)_n \in X_{B^{(1)}} \setminus \{x_{\max}^{(1)}\}$ and $k = \alpha(x)$. Let n_0 be such that $(x_1, \ldots, x_{n_0})$ has a non-maximal edge and the minimum number of paths from any vertex in V_{n_0-1} to V_0 is greater than k.

Let y belong to the cylinder $[x_1, \ldots, x_{n_0+1}]$. It suffices to show that $\alpha(y) = k$. The paths $V_{B^{(1)}}(x)$ and $V_{B^{(1)}}(y)$ start with the same $n_0 + 1$ first edges. Thus, from the property of F, $F \circ V_{B^{(1)}}(x)$ and $F \circ V_{B^{(1)}}(y)$ start with the same $n_0 + 1$ first edges $f_1, f_2, \ldots, f_{n_0+1}$:

$$F \circ V_{B^{(1)}}(x) = (f_1, f_2, \ldots, f_{n_0+1}, x'_{n_0+2}, \ldots) \text{ and}$$
$$F \circ V_{B^{(1)}}(y) = (f_1, f_2, \ldots, f_{n_0+1}, y'_{n_0+2}, \ldots).$$

For the same reason, and because x and $V_{B^{(1)}}(x)$, and, y and $V_{B^{(1)}}(y)$ are cofinal from $n_0 + 1$, $F(x)$ and $F(y)$ start with the same edges $g_1, g_2, \ldots, g_{n_0+1}$ and

$$F(x) = (g_1, g_2, \ldots, g_{n_0+1}, x'_{n_0+2}, \ldots) \text{ and}$$
$$F(y) = (g_1, g_2, \ldots, g_{n_0+1}, y'_{n_0+2}, \ldots).$$

But as there are at least k paths from any vertex in V_{n_0-1} to V_0, we deduce that $V_{B^{(1)}}^k([g_1, g_2, \ldots, g_{n_0+1}]) = [f_1, f_2, \ldots, f_{n_0+1}]$ because $F \circ V_{B^{(1)}}(x) = V_{B^{(2)}}^k \circ F(x)$. Therefore $F \circ V_{B^{(1)}}(y) = V_{B^{(2)}}^k \circ F(y)$ and $\alpha(y) = k$.

Let us show that (iii) implies (iv). Let $F : (X^{(1)}, T) \to (X^{(2)}, S)$ be a SOE map. Remark that $(X^{(2)}, S)$ is isomorphic to $(X^{(1)}, F^{-1} \circ S \circ F)$. Hence we can suppose $X^{(2)} = X^{(1)} = X$. Then we have

$$T(x) = S^{\alpha(x)}(x) \text{ and } S(x) = T^{\beta(x)}(x)$$

where α and β are continuous everywhere with y as a possible exception.

Let A be a clopen set not containing y. As α is continuous on A the set $\alpha(A)$ is compact and consequently finite: there exist $n_1, \ldots, n_k$ such that $A = \cup_{1 \leq i \leq k} A \cap \alpha^{-1}(\{n_i\})$. We recall that that the *indicator function* of the set X is denoted by $\mathbb{1}_X$. Hence

$$TA = \bigcup_{1 \leq i \leq k} S^{n_i}(A \cap \alpha^{-1}(\{n_i\}))$$

and

$$\mathbb{1}_A \circ T^{-1} = \sum_{1 \leq i \leq k} \mathbb{1}_{A \cap \alpha^{-1}(\{n_i\})} \circ S^{-n_i}.$$

But as $f - f \circ S^{-n} = (\sum_{1 \leq i \leq n} f \circ S^{-i}) \circ S - (\sum_{1 \leq i \leq n} f \circ S^{-i})$, we deduce that $\mathbb{1}_A \circ T^{-1} - \mathbb{1}_A$ belongs to $\beta_S(C(X, \mathbb{Z}))$.

Now suppose that A contains y. Remark that $\mathbb{1}_A \circ T^{-1} - \mathbb{1}_A = \mathbb{1}_{X \setminus A} -$

$\mathbb{1}_{X\setminus A} \circ T^{-1}$. But as y is not contained in $X\setminus A$ we deduce from the previous case that $\mathbb{1}_A \circ T^{-1} - \mathbb{1}_A$ belongs to $\beta_S(C(X,\mathbb{Z}))$. As T is invertible we proved that for every clopen set E, $\mathbb{1}_E \circ T - \mathbb{1}_E$ belongs to $\beta_S(C(X,\mathbb{Z}))$ and consequently that $\beta_T(C(X,\mathbb{Z})) \subseteq \beta_S(C(X,\mathbb{Z}))$. Proceeding similarly with the equality $S(x) = T^{\beta(x)}(x)$ we obtain $\beta_T(C(X,\mathbb{Z})) = \beta_S(C(X,\mathbb{Z}))$.

Let us show that (iv) implies (i). Let $B^{(1)} = (V^{(1)}, E^{(1)}, \geq^{(1)})$ be a BV-representation of $(X^{(1)}, T)$ and $B^{(2)} = (V^{(2)}, E^{(2)}, \geq^{(2)})$ of $(X^{(2)}, S)$. Let $G^{(1)}$ and $G^{(2)}$ be the dimension groups they induced. We recall that

$$G^{(1)} = \lim_{\overrightarrow{M^{(1)}(n)}} \mathbb{Z}^{|V^{(1)}(n)|} \text{ and } G^{(2)} = \lim_{\overrightarrow{M^{(2)}(n)}} \mathbb{Z}^{|V^{(2)}(n)|}.$$

We know from Proposition 6.6.1 that $G^{(1)}$ is isomorphic to $DG(X^{(1)}, T)$ and $G^{(2)}$ is isomorphic to $DG(X^{(2)}, S)$.

Now we shall construct an unordered Bratteli diagram $B = (V, E)$ that contracts to a contraction of $B^{(1)}$ on odd levels and to a contraction of $B^{(2)}$ on even levels. It suffices to give the sets of vertices $V(n)$ and the incidence matrices $(\mathbf{M}(n))_n$ between consecutive levels.

We set $V(1) = V^{(1)}(1)$ and $\mathbf{M}(1) = M^{(1)}(1)$. Looking at the canonical generators of $\mathbb{Z}^{\mathrm{Card}(V(1))}$ as elements of $G^{(2)}$, we can consider that they are elements of some $\mathbb{Z}^{\mathrm{Card}(V^{(2)}(n_2))}$. We set $V(2) = V^{(2)}(n_2)$, and we call $\mathbf{M}(2)$ the map it defines from $\mathbb{Z}^{\mathrm{Card}(V(1))}$ to $\mathbb{Z}^{\mathrm{Card}(V(2))}$. Again, the elements of $\mathbb{Z}^{\mathrm{Card}(V(2))}$ can be considered as elements of $G^{(1)}$, and thus belong to some $\mathbb{Z}^{\mathrm{Card}(V^{(1)}(n_3))}$. We set $V(3) = V^{(1)}(n_3)$ and we call $\mathbf{M}(3)$ the map that it defines from $\mathbb{Z}^{\mathrm{Card}(V(2))}$ to $\mathbb{Z}^{\mathrm{Card}(V(3))}$. Proceeding like this, we obtain the sequence

$$\mathbb{Z} \xrightarrow{\mathbf{M}(1)} \mathbb{Z}^{\mathrm{Card}(V(1))} \xrightarrow{\mathbf{M}(2)} \mathbb{Z}^{\mathrm{Card}(V(2))} \xrightarrow{\mathbf{M}(3)} \mathbb{Z}^{\mathrm{Card}(V(3))} \ldots$$

that is sufficient to define the Bratteli diagram B we are looking for. $\square$

6.7 Entropy

M. Boyle and D. Handelman have proved in (Boyle and Handelman 1994) that every minimal Cantor dynamical system is SOE to a minimal Cantor dynamical system of entropy zero. In this section we give the details of their proof. It suffices to prove that any Bratteli diagram can be telescoped to a diagram admitting an ordering such that the associated Bratteli–Vershik dynamical system has entropy zero.

We suggest (Walters 1982) as an introduction to topological entropy. Let us recall how it can be defined for subshifts (X, S) and Bratteli–Vershik dynamical systems (X_B, V_B).

The entropy of (X, S), denoted by $h(S)$, is the growth rate of the number of finite words of length n occurring in elements of X:

$$h(S) = \limsup_n \frac{\log |L_n(X)|}{n},$$

where $L_n(X) = \{x_i x_{i+1} \cdots x_{i+n-1} \mid i \in \mathbb{Z},\ x \in X\}$ (see also Section 4.4.2).

To give a convenient way to compute the entropy $h(V_B)$ of (X_B, V_B) we need some notations.

For $n \geq 1$, let $P(n)$ be the set of paths from $V(0)$ to $V(n)$. We define π_n on X_B by $\pi_n((e_k)_{k \geq 1}) = (e_1, \ldots, e_n)$. We will consider the set $A_n = \pi_n(X_B)$ as an alphabet. We call S_n the shift on $A_n^{\mathbb{Z}}$. The set

$$X_n = \left\{ \left(\pi_n\left(V_B^k(x)\right)\right)_{k \in \mathbb{Z}} \mid x \in X_B \right\}$$

is included in $A_n^{\mathbb{Z}}$, S_n-invariant and compact. Hence (X_n, S_n) is a subshift. As a consequence of the Theorem 7.6 in (Walters 1982) we have

$$h(V_B) = \lim_{n \to +\infty} h(S_n).$$

Hence, we need first to compute $h(S_n)$. To this end we will need the following subshifts. When W is a set of finite words, we denote by $\Omega(W)$ the subset of all bi-infinite words formed by concatenation of finite words belonging to W. Let S_W denote the shift map on $\Omega(W)$. It is clear $(\Omega(W), S_W)$ is a subshift.

Lemma 6.7.1 *Let W be a set of m finite words of length at least l. Then*

$$h(S_W) \leq \frac{\log m}{l}.$$

Proof Let k be the greatest length of the finite words in W. Let w be a finite word of length n occurring in some word of $\Omega(W)$. Then there exist r finite words $m_1, \ldots, m_r$ of W, a prefix s and suffix p of some finite words in W such that $w = s m_1 \cdots m_r p$. Since $r \leq \frac{n}{l}$, we deduce that there are at most $k^2 m^{2 + \frac{n}{l}}$ finite words of length n in $L_n(\Omega(W))$, which ends the proof. □

In (Boyle and Handelman 1994) the authors use a different lemma. They show that $h(S_W) = \frac{\log m}{l}$, whenever W is a set of m distinct finite words of length l.

An ordering on a Bratteli diagram is a *consecutive ordering* if whenever edges e, f and g have the same range, e and g have the same source and $e \leq f \leq g$, then e and f have the same source (see Figure 6.9).

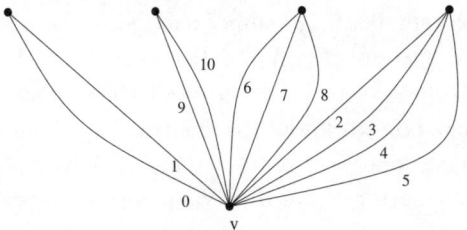

Fig. 6.9 An example of a consecutive ordering viewed from a vertex $v \in V(n+1)$ to $V(n)$.

Proposition 6.7.2 *Let $B = (V, E, \geq)$ be a properly ordered Bratteli diagram where $\geq$ is a consecutive ordering. Suppose that*

$$\lim_{n \to +\infty} \frac{\log(\eta(n+1) \operatorname{Card}(V(n)))}{\eta(n+1)} = 0,$$

where $\eta(n)$, $n \geq 1$, is the minimum number of edges from a vertex at level $n-1$ to a vertex at level n. Then, $h(V_B) = 0$.

Proof Let A_n and X_n be defined as above when we described $h(V_B)$. Let u be a vertex at level n, and let $p_1, \ldots, p_s$ be the paths from level 0 to u, listed in increasing order. We set $W(u) = p_1 \cdots p_s$. We can consider that it belongs to A_n^s. Now, assume that v is a vertex at level $n+1$, that $a_1, \ldots, a_t$ are the edges from u to v, listed in increasing order, and that y is an infinite path in the Bratteli diagram such that $y_1 \cdots y_{n+1} = p_1 a_1$. Then,

$$\left(\pi_n \left(V_B^k(y)\right)\right)_{0 \leq k \leq st-1} = W(u)^t.$$

Note that t is greater than $\eta(n+1)$. Therefore, X_n is included in $\Omega(\mathcal{W})$ where

$$\mathcal{W} = \left\{ W(u)^t \mid u \in V(n),\ \eta(n+1) \leq t < 2\eta(n+1) \right\},$$

and consequently, $h(S_n) \leq h(S_\mathcal{W})$. As $\mathcal{W}$ consists of at most $\operatorname{Card}(V(n))\eta(n+1)$ finite words of length at least $\eta(n+1)$, from Lemma 6.7.1 we obtain

$$h(S_\mathcal{W}) \leq \frac{\log\left(\eta(n+1) \operatorname{Card}(V(n))\right)}{\eta(n+1)}.$$

This achieves the proof. □

Theorem 6.7.3 *Any minimal Cantor dynamical system is strongly orbit equivalent to a minimal Cantor dynamical system of entropy zero.*

Proof From Theorem 6.4.6, it suffices to consider a minimal Bratteli–Vershik dynamical system (X_B, V_B). Let $\eta(n)$, $n \geq 1$, be the minimum number of edges from a vertex at level $n-1$ to a vertex at level n.

From Theorem 6.4.6, we know, by contracting if needed, that we can assume the incidence matrices of $B = (V, E, \geq)$ to have strictly positive entries. Hence, contracting again if needed, we can suppose that

$$\lim_{n \to +\infty} \frac{\log(\eta(n+1)\,\mathrm{Card}(V(n)))}{\eta(n+1)} = 0.$$

Consider $B' = (V, E, \geq_*)$ where $\geq_*$ is a consecutive ordering. Then, from Proposition 6.7.2, $(X_{B'}, V_{B'})$ has zero entropy and, from Theorem 6.6.2, is strongly orbit equivalent to (X_B, V_B). □

In this proof we find all the arguments to prove that all minimal BV-dynamical systems with a consecutive ordering have entropy zero.

This theorem shows that there can be dynamical systems with different entropies in a strong orbit equivalence class. Hence it is natural to ask whether all entropies can be realised inside a given class. M. Boyle and D. Handelman showed in (Boyle and Handelman 1994) that it is true in the class of the odometer $(\mathbb{Z}_2, x \mapsto x+1)$. Later, F. Sugisaki proved in (Sugisaki 2003) that it is true in any strong orbit equivalence class. Finally, he showed that moreover the realisations can be chosen to be subshifts.

Theorem 6.7.4 (Sugisaki 2007) *Let $\alpha \in [1, +\infty[$ and (X, T) a minimal Cantor dynamical system. There exists a minimal subshift of entropy $\log \alpha$ that is strongly orbit equivalent to (X, T).*

6.8 Invariant measures and Bratteli diagrams

In this section we describe the construction of invariant measures through Bratteli diagrams. We know that such measures exists as stated in Section 1.6.2 and proven in Proposition 7.2.4.

6.8.1 How to see invariant measures on Bratteli diagrams

The description we give below is very classical and applies to any measure-theoretic dynamical system defined on a compact space. But we present it in the context we have been working with from the beginning of the present chapter.

Let (X_B, V_B) be the minimal Cantor dynamical system that the properly ordered Bratteli diagram B generates, and let $(\mathcal{P}(n))_n$ be the sequence of KR-partitions that B naturally defines:

$$\mathcal{P}(n) = \left\{ V_B^j B_i(n) \mid 0 \leq j < h_i(n), 1 \leq i \leq \mathrm{Card}(V(n)) \right\}.$$

More precisely:

- $B_i(n)$ is the cylinder set spanned by the unique minimal path from $v_i \in V(n)$ to $V(0)$ where we set $V(n) := \{v_1, v_2, \ldots, v_l\}$,
- $(h_i(n)) = \mathbf{M}(n)\mathbf{M}(n-1)\cdots\mathbf{M}(1)$ where the $\mathbf{M}(k)$ are the incidence matrices of the $E(k)$,
- $\cup_{0 \leq j < h_i(n)} V_B^j B_i(n)$ is the clopen set spanned by all the finite paths from $v_i \in V(n)$ to $V(0)$,
- the unique minimal path u of B satisfies $\{u\} = \cap_{n \in \mathbb{N}} B(n)$ where $B(n) = \cup_{1 \leq i \leq \mathrm{Card}(V(n))} B_i(n)$.

Let μ be a V_B-invariant probability measure. Then $\mu(V_B^j B_i(n))$, $0 \leq j < h_i(n)$, does not depend on j and is equal to $\mu(B_i(n))$. Hence, from standard argument (see for example Theorem 7.1 in (Lang 1993)), the measure μ is completely determined by the values it takes at $B_i(n)$ (for all i and all n). Moreover, a simple computation yields the following fundamental relation:

$$\mu(n) = {}^t\mathbf{M}(n+1)\mu(n+1), \tag{6.7}$$

where $\mu(n)$ is the column vector $(\mu(B_i(n)))_{i \in V(n)}$ for any $n \geq 1$.

Let $\mathcal{M}(X_B, V_B)$ be the of V_B-invariant measures and $\mathcal{E}$ the set of V_B-ergodic measures of (X_B, V_B). Let $\mathcal{C}$ be the set of sequences $(m(n))_{n \geq 0}$ satisfying

$$m(n) \in \mathbb{R}_+^{V(n)}, \; m(n) = {}^t\mathbf{M}(n+1)m(n+1), \; n \geq 0 \text{ and } m(0) = 1.$$

Let us note that it is a convex set. The map F from $\mathcal{M}(X_B, V_B)$ to $\mathcal{C}$ defined by $\mu \mapsto (\mu(n))_n$ is a bijection. Roughly speaking, in order to construct a V_B-invariant measure, it suffices, for all n, to put weights on the minimal paths from every $v \in V(n)$ to $V(0)$ that respect Equality (6.7).

The map F sends extremal points to extremal points. Using this and knowing that the ergodic measures of (X_B, V_B) are the extremal points of the convex set $\mathcal{M}(X_B, V_B)$ (this is proven in Chapter 7, see Proposition 7.2.4), it is not difficult to prove that if (X_B, V_B) has bounded topological rank K, then it has at most K ergodic measures (similar results are proven in Chapter 7).

6.8.2 Invariant measures as homomorphisms of dimension groups

A *state* of the dimension group with order unit (G, G^+, u) is a group homomorphism $w : G \to \mathbb{R}$ such that $w(G^+)$ is included in $[0, +\infty[$ and $w(u) = 1$.

Let (X, T) be a minimal Cantor dynamical system, and denote by $\mathcal{S}(X, T)$ the set of states of its dimension group $DG(X, T)$. We recall that π is the natural projection from $C(X, \mathbb{Z})$ onto the quotient group $C(X, \mathbb{Z})/\beta C(X, \mathbb{Z})$. Let U be a clopen set, and $\mathbb{1}_U$ be its indicator function. Consider the map from $\mathcal{M}(X, T)$ to $\mathcal{S}(X, T)$ defined by

$$\mu \longmapsto \left(\pi(f) \mapsto \int_X f\, d\mu \right).$$

The measure μ being T-invariant, this map is well defined. Moreover it is a bijective affine map between $\mathcal{M}(X, T)$ and $\mathcal{S}(X, T)$.

6.8.3 Linearly recurrent dynamical systems are uniquely ergodic

Below we show how to use the Bratteli diagrams, or more precisely their associated KR-partitions, to control the set of invariant measures. We will prove the following result.

Proposition 6.8.1 *Every linearly recurrent Cantor dynamical system is uniquely ergodic.*

In the sequel (X, T) is a linearly recurrent Cantor dynamical system having a BV-representation (X_B, V_B) with B satisfying (i) and (ii) in Theorem 6.5.10. Using the notation in Section 6.8.1, the Bratteli diagram B induces a KR-partition

$$\mathcal{P}(n) = \left\{ V_B^j B_i(n) \mid 0 \leq j < h_i(n), 1 \leq i \leq \operatorname{Card}(V(n)) \right\}$$

such that $\sup_n \operatorname{Card}(V(n)) \leq K$ and $\sup_{i,j,n} \frac{h_i(n+1)}{h_j(n)} \leq K$ for some K.

Lemma 6.8.2 *Let μ be an invariant measure of (X_B, V_B). Then, for all $n \in \mathbb{N}$ and $k \in V(n)$, we have*

$$h_k(n)\mu(B_k(n)) \geq \frac{1}{K}.$$

Proof Let $k \in V(n)$. By Equation (6.7), since all the entries of $\mathbf{M}(n+1)$ are positive, we get

$$\mu(B_k(n)) \geq \sum_{l \in V(n)} \mu(B_l(n+1)).$$

But, as for every l we have $h_k(n) \geq h_l(n+1)/K$, thus

$$h_k(n)\mu(B_k(n)) \geq \sum_{l \in V(n)} \frac{h_l(n+1)}{K}\mu(B_l(n+1)) = \frac{1}{K}.$$

This completes the proof. $\square$

Proof [Proposition 6.8.1] As in Section 6.8.1, given a V_B-invariant probability measure μ, we define the numbers

$$\mu(n)_k = \mu(B_k(n)), \ n \geq 0, \ k \in V(n).$$

From Lemma 6.8, there exists a constant $\delta > 0$ such that

$$\mu(n)_i \geq \delta \mu(n-1)_k$$

for every $n \geq 1$ and $(i,k) \in V(n) \times V(n-1)$, and every invariant measure μ. Without loss of generality we can assume $\delta < 1/2$.

Let μ, μ' be two invariant measures, and $\mu(n)_k, \mu'(n)_k$ be defined as above. We define

$$S_n = \max_k \frac{\mu'(n)_k}{\mu(n)_k} = \frac{\mu'(n)_i}{\mu(n)_i}, \ s_n = \min_k \frac{\mu'(n)_k}{\mu(n)_k} = \frac{\mu'(n)_j}{\mu(n)_j}, \ \text{and} \ r_n = \frac{S_n}{s_n}$$

for some i,j. Let $m_{l,k}(n)$ be the entries of the incidence matrix $\mathbf{M}(n)$. For every $k \in V(n-1)$, we have:

$$\mu'(n-1)_k = \sum_{l \neq j} \mu'(n)_l m_{l,k}(n) + \mu'(n)_j m_{j,k}(n)$$

$$\leq S_n \sum_{l \neq j} \mu(n)_l m_{l,k}(n) + s_n \mu(n)_j m_{j,k}(n)$$

$$= S_n \mu(n-1)_k - (S_n - s_n)\mu(n)_j m_{j,k}(n)$$

$$\leq S_n \mu(n-1)_k - (S_n - s_n)\mu(n)_j$$

$$\leq \mu(n-1)_k s_n \big(r_n(1-\delta) + \delta\big).$$

And in a similar way, for every $k \in V(n-1)$ we have

$$\mu'(n-1)_k \geq \mu(n-1)_k s_n \big(\delta r_n + (1-\delta)\big).$$

We deduce that

$$r_{n-1} \leq \varphi(r_n) \text{ where } \varphi(x) = \frac{(1-\delta)x + \delta}{\delta x + (1-\delta)}.$$

The function φ is increasing on $[0, +\infty)$ and tends to $(1-\delta)/\delta$ at $+\infty$. Writing $\varphi^m = \varphi \circ \cdots \circ \varphi$ (m times), for every $n, m \in \mathbb{N}$ we have $1 \leq r_n \leq \varphi^m(r_{n+m}) \leq \varphi^{m-1}((1-\delta)/\delta)$. Taking the limit with $m \to +\infty$, we get $r_n = 1$. $\square$

6.9 Eigenvalues of stationary BV-models

6.9.1 Basic knowledge on eigenvalues

Let (X, T) be a topological dynamical system. A complex number λ is a *continuous eigenvalue* of (X, T) if there exists a continuous function $f : X \to \mathbb{C}$, $f \neq 0$, such that $f \circ T = \lambda f$. We say that f is a *continuous eigenfunction* (associated with λ). If (X, T) is minimal, then every continuous eigenvalue is of modulus 1 and every continuous eigenfunction has a constant modulus.

Let μ be a T-invariant probability measure, *i.e.*, $T\mu = \mu$, defined on the Borel σ-algebra $\mathcal{B}_X$ of X. A complex number λ is a $L^2(\mu)$-*eigenvalue* of (X, T) if there exists $f \in L^2(X, \mathcal{B}_X, \mu)$ (shortly denoted $L^2(\mu)$), $f \neq 0$, such that $f \circ T = \lambda f$. We say that f is a $L^2(\mu)$-*eigenfunction* (associated with λ). If the system is ergodic, then every eigenvalue is of modulus 1, and every eigenfunction has a constant modulus. Moreover the eigensubspace of $L^2(\mu)$ associated with λ has dimension 1. To simplify the language we will also say that *an eigenvalue is continuous* when the associated eigenfunction is continuous.

6.9.2 Continuous eigenfunctions versus measurable eigenfunctions

Let (X, T) be a topological dynamical system and μ a T-invariant probability measure. In this section we want to explain how Bratteli diagrams and, or, KR-partitions can be used to study eigenvalues (see Section 6.9.1). We illustrate this by proving a result of B. Host (Host 1986), which firstly gives a characterisation of the eigenvalues having a continuous eigenfunction, and which secondly, deduces from this characterisation that all eigenvalues have one continuous eigenfunction. Since minimal substitution subshifts are uniquely ergodic, let us note that the eigensubspace has dimension 1, and consequently, the result of Host means that in the L^2-equivalence class of each eigenfunction, there is a continuous eigenfunction, *i.e.*, eigenfunctions of minimal substitution subshifts are almost surely equal to continuous eigenfunctions.

Theorem 6.9.1 (Host 1986) *Let $\sigma : A \to A^*$ be a one-to-one primitive substitution. Let (X, S) be the subshift it generates and μ be its unique ergodic measure. Suppose that σ is recognisable. Then:*

(i) *All eigenfunctions of (X, S, μ) are μ-almost everywhere equal to a continuous eigenfunction.*

(ii) *The complex number λ is an eigenvalue of (X, T, μ) if, and only if,* there exists an integer $p > 0$ such that for every letter $a \in A$ the limit

$$h(a) = \lim_{n \to +\infty} \lambda^{|\sigma^{pn}(a)|}$$

exists and h satisfies: there exists $f : A \to \mathbb{C}$ such that for all finite words ab of two letters, we have $f(b) = f(a)h(a)$.

Let us make some comments about this statement. From Theorem 6.5.5 and Proposition 6.8.1 we know that minimal substitution subshifts are uniquely ergodic. In (Mossé 1992), B. Mossé showed that all primitive substitutions with non-periodic fixed points are recognisable. Consequently the recognisability requirement becomes vacuous. The one-to-one assumption is not necessary because the subshift can be generated (when it is not periodic), up to isomorphism, by another primitive substitution which is one-to-one (see Exercise 6.25). Moreover, the use of Bratteli diagrams gives a statement without the requirement on the existence of the integer p and on the function h (see Theorem 6.9.2 and 6.9.3).

We split into two parts the result of B. Host and state it by means of Bratteli diagrams. Before this we need some notations. In the sequel, $\|\cdot\|$ denotes the distance to the nearest integer. For a vector $\mathbf{v} = {}^t(v_1, \ldots, v_m) \in \mathbb{R}^m$, we write

$$\|\mathbf{v}\|_\infty = \max_{1 \leq j \leq m} |v_j| \text{ and } \|\mathbf{v}\| = \max_{1 \leq j \leq m} \|v_j\|.$$

Theorem 6.9.2 *Let B be a properly ordered stationary Bratteli diagram. Let $\mathbf{M}$ be its $C \times C$ incidence matrix and $\mathbf{H}(n) = \mathbf{M}^{n-1}\mathbf{M}(1) = (h_i(n))_{1 \leq i \leq C}$. Then, $\lambda \in \mathbb{C}$ is a continuous eigenvalue of (X_B, V_B) if, and only if,*

$$\lim_{n \to +\infty} \lambda^{h_i(n)} = 1 \text{ for all } i \in \{1, \ldots, C\}.$$

The proof we give later also holds for proper primitive substitutions σ generating a non-periodic subshift, with the necessary and sufficient condition becoming: $\lim_{n \to +\infty} \lambda^{|\sigma^n(a)|} = 1$ for all $a \in A$.

Theorem 6.9.3 *Let (X, S, μ) be a minimal substitution subshift. If $f \in L^2(\mu)$ is an eigenfunction of $\lambda \in \mathbb{C}$, then it is μ-almost surely equal to a continuous eigenfunction (of λ).*

We prove these results in the next section.

In (Ferenczi, Mauduit, and Nogueira 1996) an algebraic characterisation of these eigenvalues is given.

It is in general not true that all eigenvalues of a minimal dynamical system have a continuous eigenfunction as it can be seen for some Toeplitz systems (Iwanik 1996, Downarowicz and Lacroix 1996) and for some interval exchange transformations (Ferenczi, Holton, and Zamboni 2001).

The general question is: When does a measure-theoretic eigenvalue $\lambda \in \mathbb{C}$ of the dynamical system (X, T, μ) have a continuous eigenfunction ?

Such a question also appears in (Nogueira and Rudolph 1997) where the authors show that generically interval exchange transformations are not topologically weakly mixing (*i.e.*, they do not have non-trivial continuous eigenfunctions) and where they 'fully expect' that the same holds for (measure-theoretic) weak mixing (*i.e.*, they do not have non-trivial eigenfunctions). This was proven in (Avila and Forni 2007).

6.9.3 Proofs

We first start by two lemmas whose proofs is left to reader.

Lemma 6.9.4 *Let* $\mathbf{M}$ *be a matrix with integer entries. If* $\mathbf{u}$ *is a real vector such that* $\|\mathbf{M}^n \mathbf{u}\|_\infty \to 0$ *when* $n \to \infty$, *then the convergence is exponential, i.e., there exist* $0 \leq r < 1$ *and a constant* K *such that* $\|\mathbf{M}^n \mathbf{u}\|_\infty \leq K r^n$ *for all* $n \in \mathbb{N}$.

Lemma 6.9.5 *Let* $\mathbf{M}$ *be a matrix with integer entries. Let* u *be a real vector such that* $\|\mathbf{M}^n \mathbf{u}\| \to 0$ *as* $n \to +\infty$. *Then there exist an integer vector* $\mathbf{w}$ *and a real vector* $\mathbf{v}$ *such that*

$$\mathbf{u} = \mathbf{w} + \mathbf{v} \text{ and } \|\mathbf{M}^n \mathbf{v}\|_\infty \to_{n \to +\infty} 0.$$

Proof [Theorem 6.9.2] Let $(\mathcal{P}(n))_n$ be the sequence of KR-partitions that B naturally defines:

$$\mathcal{P}(n) = \left\{ V_B^j B_i(n) \mid 0 \leq j < h_i(n), \ 1 \leq i \leq C \right\}.$$

The unique minimal path u of B verifies $\{u\} = \cap_{n \in \mathbb{N}} B(n)$ where $B(n) = \cup_{1 \leq i \leq C} B_i(n)$. Let g be a continuous eigenfunction of λ of modulus equal to 1. Then, it is uniformly continuous. Let $\varepsilon > 0$. There exists $n_0 \in \mathbb{N}$ such that $|g(y) - g(u)| < \varepsilon/2$ for all $y \in B(n_0)$.

Let $i \in \{1, \ldots, C\}$ and $n \geq n_0$. Let $x \in B_i(n)$ and set $y = V_B^{h_i(n)}(x)$ which belongs to $B(n)$. We thus have

$$\left|\lambda^{h_i(n)} - 1\right| = \left|g\left(V_B^{h_i(n)} x\right) - g(x)\right|$$
$$\leq |g(y) - g(u)| + |g(u) - g(x)| < \varepsilon.$$

This proves the direct implication.

Let $\lambda = \exp(2i\pi\alpha)$. Suppose that $\lim_{n \to +\infty} \max_{1 \leq i \leq C} \left|\lambda^{h_i(n)} - 1\right| = 0$. Recall that $h_i(n)$ is equal to the ith entry of $\mathbf{M}^{n-1}\mathbf{M}(1)$. Then, $\|\alpha \mathbf{M}^{n-1}\mathbf{M}(1)\|$ tends to 0 when n goes to $+\infty$, and from Lemmas 6.9.4 and 6.9.5 (by taking $\mathbf{u} = \alpha \mathbf{M}(1)$), the series $\sum \max_{1 \leq i \leq C} \left|\lambda^{h_i(n)} - 1\right|$ converges.

For every $n \in \mathbb{N}$, let f_n be the function on X_B defined by

$$f_n(x) = \lambda^j \text{ for } x \in V_B^j B_i(n),\ 0 \leq j < h_i(n),\ 1 \leq i \leq C.$$

We compare f_n and f_{n-1}. Let $L = \max_{1 \leq i \leq C} \sum_{1 \leq j \leq C} m_{i,j}$. By construction, for every x, $f_n(x)/f_{n-1}(x) = \lambda^l$ where l is a sum of terms of the form $h_i(n-1)$ and this sum contains at most L terms. We thus obtain

$$\sup_{x \in X_B} |f_n - f_{n-1}| \leq L \max_{1 \leq l \leq C} \left|\lambda^{h_l(n-1)} - 1\right|.$$

Hence $(f_n)_{n \in \mathbb{N}}$ converges uniformly to a continuous function f, which is clearly a continuous eigenfunction for λ. □

Lemma 6.9.6 *Let B be a stationary properly ordered Bratteli diagram, with $C \times C$ incidence matrix $\mathbf{M}$, and μ be the unique ergodic measure of (X_B, V_B). Let $(h_i(n))_i = (\mathbf{M}^{n-1}\mathbf{M}(1))_i$. Suppose that $\mathbf{M}$ has positive entries. Then, there exists K such that for all $n \geq 1$ and $(i,l) \in \{1,\ldots,C\}^2$ we have*

$$h_l(n+1) \leq K h_i(n) \text{ and } h_i(n)\mu(B_i(n)) \geq \frac{1}{K}.$$

Proof It is easy to establish that there exists a constant K such that for every $n \geq 1$ and $(i,l) \in \{1,\ldots,C\}^2$, we have $h_l(n+1) \leq K h_i(n)$. Let $n \geq 1$ and $i \in \{1,\ldots,C\}$. We have the relation $\mu(B_i(n-1)) = \sum_{l=1}^{C} m_{l,i}\, \mu(B_l(n))$. Consequently, the entries of $\mathbf{M}$ being positive, we deduce

$$h_i(n)\mu(B_i(n)) \geq \sum_{l=1}^{C} \frac{h_l(n+1)}{K}\mu(B_l(n+1)) = \frac{1}{K}.$$

This completes the proof. □

Proof [Theorem 6.9.3] Let $(\mathcal{P}(n))_n$ be a sequence of partitions of X that induced a stationary BV-representation of (X,S): $\mathcal{P}(n) = \{S^j B_i(n) \mid 0 \leq j < h_i(n),\ 1 \leq i \leq C\}$. We call $\mathbf{M}$ the incidence matrix from level 2. We can suppose that it has positive entries. Let K be the constant given in Lemma 6.9.6. As there is a unique minimal path and B is stationary, by contracting if needed, we can suppose that $B(n+1)$ is included in $B_1(n)$ for all $n \geq 1$.

Let $g \in L^2(\mu)$ be an eigenfunction for λ, i.e., $g \circ S = \lambda g$ and let g_n be its conditional expectation with respect to $\mathcal{P}(n)$:

$$g_n = \sum_{A \in \mathcal{P}(n)} \frac{1}{\mu(A)} \int_A g d\mu \, \mathbb{1}_A.$$

By ergodicity we can suppose that the modulus of g is equal to 1.

We begin with some remarks. The function g_n is constant in $B_i(n)$ for all $i \in \{1,\ldots,C\}$. We call $d_i(n)$ this constant. For all $y \in S^j B_i(n)$, $0 \le j < h_i(n)$, $g_n(y) = \lambda^j d_i(n)$. From martingale theory we know that the sequence $(g_n)_n$ converges to g in $L^1(\mu)$. We start by proving that $\lim_{n \to +\infty} \max_{1 \le i \le C} |d_i(n)| = 1$. We have

$$\int_X |g - g_n| d\mu = \sum_{i=1}^{C} \sum_{j=0}^{h_i(n)-1} \int_{T^j B_i(n)} |g - g_n| d\mu$$

$$= \sum_{i=1}^{C} \sum_{j=0}^{h_i(n)-1} \int_{B_i(n)} |\lambda^j g - \lambda^j d_i(n)| d\mu$$

$$= \sum_{i=1}^{C} h_i(n) \int_{B_i(n)} |g - d_i(n)| d\mu$$

$$\ge \sum_{i=1}^{C} h_i(n) \mu(B_i(n))(1 - |d_i(n)|) \ge \sum_{i=1}^{C} \frac{1}{K}(1 - |d_i(n)|).$$

Since $|d_i(n)| \le 1$, we obtain that

$$\lim_{n \to +\infty} |d_i(n)| = 1. \tag{6.8}$$

As $B(n+1)$ is included in $B_1(n)$ we have

$$h_i(n+1)\mu(B_i(n+1))|d_i(n+1) - d_1(n)|$$

$$\le h_i(n+1) \int_{B_i(n+1)} (|g - d_i(n+1)| + |g - d_1(n)|) d\mu$$

$$\le K h_1(n) \int_{B_1(n)} |g - d_1(n)| d\mu + h_i(n+1) \int_{B_i(n+1)} |g - d_i(n+1)|) d\mu$$

$$\le (K+1) \int_X |g - g_n| d\mu$$

which tends to 0 when n goes to $+\infty$. Hence $\lim_{n \to +\infty} d_i(n+1)/d_1(n) = 1$ and consequently, for all $k \in \{1,\ldots,C\}$

$$\lim_{n \to +\infty} \frac{d_k(n+1)}{d_i(n+1)} = 1. \tag{6.9}$$

Now remark that we have

$$\sum_{k=1}^{C} h_i(n)\mu(B_i(n) \cap S^{-h_i(n)} B_k(n)) = h_i(n)\mu(B_i(n)) \geq \frac{1}{K}.$$

Hence, there exists k such that

$$h_i(n)\mu(B_i(n) \cap S^{-h_i(n)} B_k(n)) \geq \frac{1}{KC}. \tag{6.10}$$

We set $W(n) = B_i(n) \cap S^{-h_i(n)} B_k(n)$. Thus,

$$h_i(n)\mu(W(n))|d_i(n) - \lambda^{-h_i(n)} d_k(n)|$$

$$= h_i(n) \int_{W(n)} |d_i(n) - \lambda^{-h_i(n)} d_k(n)| d\mu$$

$$\leq h_i(n) \int_{B_i(n)} |d_i(n) - g| d\mu + h_i(n) \int_{S^{-h_i(n)} B_k(n)} |g - \lambda^{-h_i(n)} d_k(n)| d\mu$$

$$\leq h_i(n) \int_{B_i(n)} |d_i(n) - g| d\mu + K h_k(n) \int_{B_k(n)} |g - d_k(n)| d\mu$$

$$\leq (K+1) \int_X |g - g_n| d\mu .$$

From (6.8), (6.9) and (6.10) we deduce that

$$\lim_{n \to +\infty} \max_{1 \leq i \leq C} |\lambda^{h_i(n)} - 1| = 0.$$

We conclude by invoking Theorem 6.9.2. □

6.9.4 In the context of linearly recurrent dynamical systems

The following theorem applies to substitution subshifts and linearly recurrent subshifts and was proven in (Cortez, Durand, Host, et al. 2003) and (Bressaud, Durand, and Maass 2005). It generalises Theorem 6.9.2.

Theorem 6.9.7 *Let (X, T) be a Cantor dynamical system having a BV-representation (X_B, V_B) with B satisfying (i) and (ii) in Theorem 6.5.10. Let $(\mathbf{M}(n))_{n \geq 1}$ be the sequence of incidence matrices of B and μ be the unique invariant measure of (X, T). Let $\lambda = \exp(2i\pi\alpha)$. Then,*

(i) *λ is an $L^2(\mu)$-eigenvalue of (X, T) with respect to μ if, and only if,*

$$\sum_{n \geq 2} \|\alpha \mathbf{M}(n) \mathbf{M}(n-1) \cdots \mathbf{M}(1)\|^2 < \infty.$$

(ii) λ is a continuous eigenvalue of the system (X,T) if, and only if,

$$\sum_{n \geq 2} \|\alpha \mathbf{M}(n)\mathbf{M}(n-1)\cdots \mathbf{M}(1)\| < \infty.$$

Such a result in the context of bounded topological rank remains open.

We remark that for minimal substitution subshifts the two conditions are equivalent. It is no longer true for linearly recurrent subshifts. In (Bressaud, Durand, and Maass 2005), the authors have constructed an example that has a L^2-eigenvalue which is not continuous.

6.9.5 In the context of tilings, Delone sets, $\mathbb{Z}^d$ and $\mathbb{R}^d$-actions

In (Solomyak 1997) B. Solomyak characterised the eigenvalues of $\mathbb{R}^d$-actions on self-affine tiling spaces. As a consequence he also obtained they are all continuous. In (Cortez, Gambaudo, and Maass 2007) the authors characterised the continuous eigenvalues of the free minimal actions on the Cantor set.

6.10 Exercises

Section 6.3

Exercise 6.1 Show that $X_B^{\max}$ and $X_B^{\min}$ are non-empty sets for every ordered Bratteli diagram B.

Exercise 6.2 Show that the Bratteli compactum X_B of a simple Bratteli diagram B is a Cantor space.

Exercise 6.3 Show that Theorem 6.4.9 does not hold for clopen sets instead of cylinders.
Hint. Eigenvalues could help, see Section 6.9.

Section 6.5

Exercise 6.4 Prove that $\mathbb{Z}_{(p_n)}$ is a compact topological ring.

Exercise 6.5 Give a necessary and sufficient condition for an odometer $(\mathbb{Z}_{(p_n)}, R)$ to have the following property: For all clopen U there exists a clopen $V \subseteq U$ such that (V, R_V) is isomorphic to $(\mathbb{Z}_{(p_n)}, R)$. Give an example having this property which is not an odometer.

Exercise 6.6 Prove Lemma 6.4.5.

Exercise 6.7 Prove the map φ in the proof of Theorem 6.4.6 is a homeomorphism.

Exercise 6.8 Prove that the family of stationary BV-dynamical systems is stable under Kakutani equivalence.

Exercise 6.9 Prove that the sequence of partitions given in Section 6.5.1 satisfies (KR1), (KR2) and (KR3).

Exercise 6.10 Suppose that (X, S) is a minimal substitution subshift on the alphabet A. Let $x \in X$, $\varphi : A^* \to B^*$ be a morphism and $y = \varphi(x)$. Consider (Y, S) the subshift generated by y. Prove that it is isomorphic to a substitution subshift (X_σ, S) where σ is primitive.

Exercise 6.11 Prove Proposition 6.5.6.

Exercise 6.12 Show that every sequence of a linearly recurrent subshift is linearly recurrent with the same constant.

Exercise 6.13 Give examples of dynamical systems that satisfy either (i) or (ii) in Theorem 6.5.10, and that are not linearly recurrent.

Exercise 6.14 Prove that odometers are minimal dynamical systems.

Exercise 6.15 Show that Theorem 6.4.9 does not hold for clopen sets instead of cylinder sets.

Exercise 6.16 Characterise the odometers having finitely many induced systems on cylinder sets $[a_0, a_1, \ldots, a_n]$ defined in Section 6.5.1 and prove that they have infinitely many induced systems on clopen sets.

Exercise 6.17 Let (X, S) be a minimal substitutive subshift. Prove that for all clopen sets $U \subset X$, there exists a clopen set $V \subset U$ such that $(U, S_{/U})$ is conjugate to $(V, S_{/V})$.

Section 6.7

Exercise 6.18 In Section 6.8.1 show that the map F sends the extremal points of the set $\mathcal{M}(X_B, V_B)$ to the extremal points of $\mathcal{C}$.

Exercise 6.19 Suppose that W is a set of m distinct finite words of length l. Prove that the entropy $h(\Omega(W))$ is equal to $\frac{\log m}{l}$.

Exercise 6.20 Let $B = (V, E, \geq)$ be a properly ordered Bratteli diagram such that $(V(n))_n$ is bounded. Show that the entropy of (X_B, V_B) is zero.

Exercise 6.21 Let B be a properly ordered Bratteli diagram with a consecutive ordering. Show that $h(V_B) = 0$.

Section 6.8

Exercise 6.22 In Section 6.8.1 show that the map F sends the extremal points of $\mathcal{M}(X, T)$ to the extremal points of $\mathcal{C}$.

Exercise 6.23 Prove that if the topological minimal Cantor dynamical system (X, T) has bounded topological rank K then it has at most K ergodic measures.

Exercise 6.24 Show that the map defined in Section 6.8.2 is a bijection between $\mathcal{M}(X, T)$ and $\mathcal{S}(X, T)$.

Section 6.9

Exercise 6.25 Let $\sigma : A \to A^*$ be a primitive substitution which is not one-to-one and which has a non-periodic fixed point. Let (X, S) be the subshift that it generates. Find a one-to-one primitive substitution $\tau : B \to B^*$ and a morphism $\varphi : B \to A^*$ such that $\sigma = \varphi \circ \tau$. Deduce that (X, S) is isomorphic to a subshift generated by a one-to-one primitive substitution.

Exercise 6.26 Construct a linearly recurrent subshift that has a L^2-eigenvalue which is not continuous.

Exercise 6.27 Let $B = (V, E \prec)$ be a properly ordered Bratteli diagram with $\text{Card}(V(n)) = 2$ for infinitely many n. Suppose that it has a continuous eigenvalue $\exp 2i\pi\alpha$ where α is irrational. Prove that it is uniquely ergodic.

Exercise 6.28 Prove that all eigenvalues of a uniquely ergodic Toeplitz subshift with finite topological rank are continuous.

Exercise 6.29 Show that all eigenvalues of a Toeplitz subshift of finite type are rational.

Exercise 6.30 Construct a Toeplitz subshift that has a L^2-eigenvalue which is not continuous.

7
Infinite words with uniform frequencies, and invariant measures

Sébastien Ferenczi,

Thierry Monteil

A fruitful use of word combinatorics is its contribution to the study of dynamical systems, through the symbolic dynamical systems defined in Section 1.6. Indeed, the study of most dynamical systems in the *topological* and the *measure-theoretic* categories can be reduced, by appropriate coding techniques, to the study of a suitable symbolic system X_x; and the topological properties of a symbolic system X_x (equipped with the product topology on $\Lambda^{\mathbb{N}}$) can be translated into combinatorial properties of the infinite word x.

In this chapter, we shall study the first two combinatorial properties of infinite words which are significant (and indeed, primordial) for symbolic dynamical systems. The first one is the well-known *uniform recurrence* which translates the dynamical property of *minimality*, that is the fact that the topological system cannot be split into smaller systems. The second one is the fact that the topological system has one invariant probability measure; this is called *unique ergodicity*, a somewhat unhappy expression as it suggests a close association with the classical (*i.e.*, measure-theoretic) ergodic theory, though in fact it is a purely topological notion. Thus, for symbolic systems, unique ergodicity translates into the existence of *frequencies* for every finite factor of the infinite word x, and the limit defining these frequencies is a uniform one; thus we propose to say that the infinite word x has *uniform frequencies*. Similarly, the set of *invariant measures* depends only on the topological structure, and combinatorial properties of the word x will give informations on its structure.

Thus we want to explain how the notions of minimality/uniform recurrence and unique ergodicity/uniform frequencies provide an interaction between dynamical systems and word combinatorics, in most cases the dynamics being the main source of questions and the combinatorics the main

Combinatorics, Automata and Number Theory, ed. Valérie Berthé and Michel Rigo. Published by Cambridge University Press. ©Cambridge University Press 2010.

tool for answers. We shall focus more on the second couple of notions, as it has been less studied, and also because it constitutes a very strong property, for the system and for the infinite word: in particular it means that dynamical results, which ergodicians are happy to know *almost everywhere* for a given system, will be valid *everywhere* for a uniquely ergodic system.

In Section 7.1 we study the relationship between symbolic systems and languages, and between minimality and uniform recurrence. In Section 7.2 we describe what is known of the set of invariant measures for a general symbolic system, and detail the notion of unique ergodicity and its consequences. Section 7.3 presents some achievements of word combinatorics, initiated by M. Boshernitzan, which allow us to deduce uniform frequencies (or, more generally, to bound the number of ergodic invariant measures of the system) from simple combinatorial properties of the words. In Section 7.4 we review the known examples of words with uniform frequencies, and in Section 7.5 we give important examples which do not have uniform frequencies. We finish in Section 7.6 by hinting how these basic notions have given birth to very deep problems and high achievements in dynamical systems.

7.1 Basic notions

7.1.1 Languages and subshifts

Let us recall some definitions introduced in Section 1.6.2. Proofs are sketched. Let A be a finite alphabet. The set of infinite words $A^{\mathbb{N}}$ is endowed with the product topology defined in Section 1.2.10. It is a compact metrisable space. Let S denote the *shift map* defined on $A^{\mathbb{N}}$ by

$$S(x_0 x_1 x_2 \cdots) = x_1 x_2 x_3 \cdots.$$

This map is continuous for the product topology.

Definition 7.1.1 A *subshift* (also called *symbolic dynamical system*) is a couple (X, S), where X is a non-empty closed subset of $A^{\mathbb{N}}$, which is stable under S. We still denote by S the restriction $S|_X$. The *orbit* of a point x in X under the shift map S is the set $\mathcal{O}(x) = \{S^n(x) \mid n \in \mathbb{N}\}$.

Given a subshift (X, S), let $L(X)$ denote the *language* of the finite words which occur in some element of X, and let $L_n(X) = L(X) \cap A^n$ (for $n \in \mathbb{N}$). This language is a non-empty subset of A^* with the additional properties of being

(i) *factorial*: any factor of any element of $L(X)$ is also in $L(X)$,
(ii) *extendable*: for any element w of $L(X)$, there exists a letter a in A such that wa is also in $L(X)$.

A subshift is determined by its language, more precisely, we have the following correspondence:

Proposition 7.1.2 *The map $(X, S) \mapsto L(X)$ is a bijection from the set of subshifts to the set of non-empty factorial and extendable languages.*

Proof We can define an inverse map by sending a non-empty factorial and extendable language L to the subshift $(\{y \in A^{\mathbb{N}} \mid L(y) \subseteq L\}, S)$. □

When x is an infinite word, we denote by (X_x, S) the smallest subshift containing x. The set X_x is equal to the closure (in $A^{\mathbb{N}}$) of the orbit of x under S:
$$X_x = \overline{\mathcal{O}(x)} = \overline{\{S^n(x) \mid n \in \mathbb{N}\}}.$$
In terms of languages, if $L(x)$ denotes the set of all finite factors of the word x (see Definition 1.2.8), then (X_x, S) is the subshift whose associated language is $L(x)$: $L(x) = L(X_x)$. Hence, we have,
$$X_x = \{y \in A^{\mathbb{N}} \mid L(y) \subseteq L(x)\}.$$
If $u = u_0 \cdots u_{n-1} \in A^n$ is a finite word, we can define the *cylinder* $[u]_X$ (or simply $[u]$ when the context is clear) as the set of elements of X which admit u as a prefix:
$$[u]_X = \{x \in X \mid \forall i \leq n-1,\ x_i = u_i\}.$$
The cylinders are clopen sets and form a basis of the topology of X. In particular, the characteristic function $\chi_{[u]}$ of a cylinder $[u]$ is continuous. We have,
$$L(X) = \{u \in A^* \mid [u]_X \neq \emptyset\}.$$

7.1.2 Uniform recurrence and minimality

Definition 7.1.3 A symbolic dynamical system (X, S) is *minimal* if it does not contain a smaller subshift.

Definition 7.1.4 The infinite word x is *uniformly recurrent* if every factor of x occurs in an infinite number of places with bounded gaps.

Those two notions are in correspondence, we recall this with some more details:

Proposition 7.1.5 *Let x be an infinite word on the alphabet A. The following assertions are equivalent:*

(i) The word x is uniformly recurrent.
(ii) For any factor $u \in L(x)$, there exists an integer $n \in \mathbb{N}$ such that any element of $L_n(x)$ contains an occurrence of u.
(iii) Any word y of X_x satisfies $L(y) = L(x)$.
(iv) Any word y of X_x satisfies $X_y = X_x$.
(v) The orbit of any element y of X_x under the shift map S is dense in X_x.
(vi) The subshift (X_x, S) is minimal.

Proof (i) $\Leftrightarrow$ (ii) and (iii) $\Leftrightarrow$ (iv) are direct reformulations. For (iv) $\Leftrightarrow$ (v) $\Leftrightarrow$ (vi), it suffices to notice that, for any y in X_x, $X_y = \overline{\mathcal{O}(y)}$ is the smallest subshift containing y and is included in X_x. For (ii) $\Rightarrow$ (iii), let y be an element of X_x, and let u be an element of $L(x)$. Because of the hypothesis, there exists an integer sequence $(\alpha_n)_{n \in \mathbb{N}}$ such that $(S^{\alpha_n}(x))$ converges to y, and there exists an integer $n_0 \in \mathbb{N}$ such that any element of $L_{n_0}(x)$ contains an occurrence of u. Since the sequence $(S^{\alpha_n}(x))$ converges to y, for N big enough, the prefix of length n_0 of $S^{\alpha_N}(x)$ coincides with the prefix of length n_0 of y. Hence y contains an occurrence of u. For $\neg$(ii) $\Rightarrow$ $\neg$(iii), because of the hypothesis, there exists a factor $u \in L(x)$ such that for any integer $n \in \mathbb{N}$, there exists an integer α_n such that u does not occur in $x_{\alpha_n} \cdots x_{\alpha_n + n}$. Let y be an adherent point of the sequence $(S^{\alpha_n}(x))_{n \in \mathbb{N}}$ in X_x (which exists by compactness). The finite word u is not a factor of y, hence $L(y) \subset L(x)$. □

7.1.3 Uniform frequencies

Definition 7.1.6 The infinite word x has *uniform frequencies* if, for every factor w of x, the ratio $\frac{|x_k \cdots x_{k+n}|_w}{n+1}$ (see 1.2.3 for the notation) has a limit $f_w(x)$ when $n \to +\infty$, uniformly in k.

The aim of the next section is to provide a dynamical equivalent of uniform frequencies (see in particular Proposition 7.2.10 below).

7.2 Invariant measures and unique ergodicity
7.2.1 Two frameworks in dynamical systems

We first insist on the difference between topological and measure-theoretic dynamics, unique ergodicity being a topological notion. A good reference for this section is (Denker, Grillenberger, and Sigmund 1976).

Definition 7.2.1 A *measure-theoretic dynamical system* is a quadruple

$(X, \mathcal{B}, \mu, T)$ where $(X, \mathcal{B}, \mu)$ is a probability Lebesgue space and $T : X \to X$ is a measurable function that preserves the measure μ:

$$\forall B \in \mathcal{B}, \ \mu(T^{-1}(B)) = \mu(B).$$

Such a system is *ergodic* if the only T-invariant measurable subsets of X have measure 0 or 1 (a subset B of X is *T-invariant* if $T^{-1}(B) = B$).

A measure-theoretic ergodic dynamical system satisfies the Birkhoff ergodic theorem (see Theorem 1.6.7). We refer the reader to Chapter 5 of (Pytheas Fogg 2002) for a presentation of this fundamental result in the present framework:

$$\forall f \in L^1(X, \mathbb{R}), \ \frac{1}{n} \sum_{k=0}^{n-1} f \circ T^k \xrightarrow[n \to \infty]{\mu-a.e.} \int_X f d\mu.$$

Definition 7.2.2 A *topological dynamical system* is a couple (X, T), where X is a compact metric space and $T : X \to X$ is a continuous function.

Hence, a subshift can be seen as a topological dynamical system.

7.2.2 The set of invariant measures of a subshift

A way to understand a topological dynamical system (X, T) is to study the set $\mathcal{M}(X, T)$ of Borel probability measures on X that are preserved by T: this corresponds to the measure-theoretic dynamical systems that are housed by (X, T). Let $\mathcal{E}(X, T)$ denote the subset of ergodic invariant measures.

We will need some basic material coming from measure theory and functional analysis (the Riesz representation theorem and the Banach–Alaoglu theorem) (Rudin 1987) (Rudin 1991), it is summarised in the following proposition:

Proposition 7.2.3 *The set of Borel probability measures on a compact metrisable space X can be identified with a convex subset of the topological dual of $C^0(X, \mathbb{R})$, endowed with the weak-star topology. This topology is metrisable and compact. A sequence $(\mu_n)_{n \in \mathbb{N}}$ of such measures converges to a measure μ if, and only if, for each continuous function $f \in C^0(X, \mathbb{R})$, $\int_X f d\mu_n$ converges to $\int_X f d\mu$. When X is a closed subset of $A^{\mathbb{N}}$, continuous functions can be replaced with characteristic functions of cylinders, i.e., μ_n converges to μ if, and only if, for each finite word u, $\mu_n([u])$ converges to $\mu([u])$.*

We can now describe some properties and the geometry of the set of invariant measures of a topological dynamical system and its subset of ergodic measures:

Proposition 7.2.4 *Let (X, T) be a topological dynamical system.*

(i) *The set $\mathcal{M}(X, T)$ is a non-empty convex compact subspace of the space of Borel probability measures on X.*

(ii) *Let $\mu \in \mathcal{E}(X, T)$ and $\nu \in \mathcal{M}(X, T)$. If ν is absolutely continuous with respect to μ, then $\mu = \nu$.*

(iii) *The extreme points of $\mathcal{M}(X, T)$ are the ergodic measures.*

(iv) *Two distinct elements of $\mathcal{E}(X, T)$ are mutually singular.*

(v) *A set of n distinct ergodic measures generates an affine space of dimension $n - 1$, i.e., such measures are affinely independent.*

Proof

(i) The set $\mathcal{M}(X, T)$ is clearly convex and closed in the set of Borel probability measures on X. Let us prove that the system (X, T) admits at least one invariant probability measure. Pick a point x in X. Since the set of Borel probability measures on X is metrisable and compact, we can take for μ any accumulation point of the sequence of probability measures defined by $\mu_n = \frac{1}{n}\sum_{k=0}^{n-1} \delta_{T^k(x)}$, where δ stands for the one-point *Dirac measure*. The average ensures that μ is preserved by T.

(ii) Since ν is absolutely continuous with respect to μ, the Radon–Nikodym theorem ensures that there exists a map $f \in L^1(X, \mathbb{R}_+)$ such that for any Borel subset B of X, $\nu(B) = \int_B f d\mu$. Let us show that f is constant μ-almost everywhere. Assume, by contradiction, that the measure of the Borel set $B = \{x \in X \mid f(x) > \int_X f d\mu\}$ belongs to $(0, 1)$. Since μ is ergodic, B is not T-invariant. Therefore, $\mu(T^{-1}B \setminus B) = \mu(B \setminus T^{-1}B) > 0$. We have, $\mu(B \setminus T^{-1}B)\int_X f d\mu < \int_{B \setminus T^{-1}B} f d\mu = \nu(B \setminus T^{-1}B) = \nu(T^{-1}B \setminus B) = \int_{T^{-1}B \setminus B} f d\mu \leq \mu(T^{-1}B \setminus B)\int_X f d\mu$, which is absurd. Hence f is constant with value $\nu(X) = 1$, so $\mu = \nu$.

(iii) Let μ be an element of $\mathcal{E}(X, T)$ that can be written as a convex combination $\mu = \lambda\mu_1 + (1-\lambda)\mu_2$, where μ_1 and μ_2 are two elements of $\mathcal{M}(X, T)$ and $\lambda \in (0, 1)$. Since $\lambda\mu_1 \leq \mu$, the measure μ_1 is absolutely continuous with respect to μ, hence (ii) ensures that $\mu_1 = \mu$, so μ is an extreme point of $\mathcal{M}(X, T)$. Conversely, let μ be an extreme point of $\mathcal{M}(X, T)$. Assume by contradiction that μ is not ergodic: there exists a Borel set B of X which is T-invariant and satisfies $\mu(B) \in$

$(0,1)$. Let us define, for any two Borel sets C and D of X with $\mu(C) > 0$, $\mu_C(D) = \frac{\mu(C \cap D)}{\mu(C)}$. We have $\mu = \mu(B)\mu_B + (1-\mu(B))\mu_{X \setminus B}$, hence μ can be written as a non-trivial convex combination of two elements of $\mathcal{M}(X,T)$, a contradiction.

(iv) Let μ and ν be two distinct ergodic measures for (X,T). Since μ and ν are distinct, there exists a measurable function f such that $\int_X f d\mu \neq \int_X f d\nu$. Let $G_{\mu,f}$ be the set of points of X for which the Birkhoff ergodic sums for f converge to $\int_X f d\mu$ and $G_{\nu,f}$ be the one corresponding to $\int_X f d\nu$. Those two sets are disjoint and satisfy $\mu(G_{\mu,f}) = \nu(G_{\nu,f}) = 1$ and $\mu(G_{\nu,f}) = \nu(G_{\mu,f}) = 0$.

(v) Let $\mu_1, \ldots, \mu_n$ be n distinct elements of $\mathcal{E}(X,T)$. Let $(\beta_1, \ldots, \beta_n)$ be a tuple in $\mathbb{R}^n$ such that $\sum_{k=1}^n \beta_k \mu_k = 0$. Let $1 \leq i \leq n$ be an integer. By (iv), for any $j \neq i$, there exists a Borel set G_j such that $\mu_i(G_j) = 1$ and $\mu_j(G_j) = 0$. We have $\beta_i = \sum_{k=1}^n \beta_k \mu_k(\cap_{j \neq i} G_j) = 0$. Hence, the μ_k are linearly independent, therefore affinely independent. □

Those results remain valid for the particular case of symbolic systems, and can again be interpreted in terms of word combinatorics.

Definition 7.2.5 A *weight function* on a subshift (X,S) is a map $\varphi : L(X) \to \mathbb{R}^+$ such that:

(i) $\varphi(\varepsilon) = 1$,
(ii) $\forall w \in L(X)$, $\varphi(w) = \sum_{a \in A, \, wa \in L(X)} \varphi(wa)$,
(iii) $\forall w \in L(X)$, $\varphi(w) = \sum_{a \in A, \, aw \in L(X)} \varphi(aw)$.

Let us denote by $\mathcal{W}(X,S)$ the set of *weight functions* on (X,S).

Again, we have a correspondence:

Proposition 7.2.6 *The following map is a bijection.*

$$\left(\begin{array}{ccc} \mathcal{M}(X,S) & \longrightarrow & \mathcal{W}(X,S) \\ \mu & \longmapsto & \left(\begin{array}{ccc} L(X) & \longrightarrow & \mathbb{R}^+ \\ w & \longmapsto & \mu([w]) \end{array} \right) \end{array} \right)$$

Proof Despite the symmetry between the second and the third item, they do not play the same role: (ii) follows from the additivity of the measure, whereas (iii) follows from its shift-invariance. The first item tells that the measure is a probability measure, *i.e.*, a measure of total mass 1. Therefore, the map is well defined. Since the set of cylinders generates the Borel σ-algebra on X (it is a basis of the topology of X) and is closed under finite intersection, the map is injective (acccording to Dynkin theorem, see e.g., (Rudin 1987)). The surjectivity is guaranteed by the Carathéodory extension theorem (construction of an outer measure), for example any open set is a disjoint (possibly infinite) union of cylinders, its measure is given by summing the weights of those cylinders. □

The Birkhoff ergodic theorem applied to characteristic functions of cylinders can be restated in terms of frequencies:

Proposition 7.2.7 *Let (X, S) be a subshift, and let μ be an element of $\mathcal{E}(X, S)$. Then for μ-almost any x in X, and for any finite word w in $L(X)$, the frequency $f_w(x)$ exists and is equal to $\mu([w])$.*

Proof Apply the Birkhoff ergodic theorem with the countable family of maps $f = \chi_{[w]}$ and notice that

$$|x_0 \cdots x_{n+|w|-2}|_w = \sum_{k=0}^{n-1} \chi_{[w]} \circ S^k(x).$$

□

The points which satisfy this convergence are said to be *generic* for μ. With the notations of the proof of Proposition 7.2.4(iv), the set of generic points for μ is equal to $\cap_{w \in L(X)} G_{\mu, \chi_{[w]}}$. This proposition lets us imagine a protostrategy to prove that an infinite word x has frequencies, by considering the subshift X_x and looking for the ergodic invariant measures on it. Unfortunately, it is possible that the word x is generic for no element of $\mathcal{E}(X_x, S)$ (see Proposition 7.2.11).

7.2.3 Unique ergodicity

Definition 7.2.8 A topological dynamical system is *uniquely ergodic* if it has only one invariant probability measure.

The unique invariant measure μ is ergodic (it is an extreme point of a singleton by Proposition 7.2.4(iii)). In this extreme case of a uniquely ergodic dynamical system, *all* the orbits are equidistributed:

Proposition 7.2.9 *Let (X, T) be a uniquely ergodic topological dynamical system, whose unique invariant measure is denoted by μ. Let $f : X \to \mathbb{R}$ be a continuous function. Then, the sequence of functions $(\frac{1}{n} \sum_{k=0}^{n-1} f \circ T^k)_{n \in \mathbb{N}}$ converges uniformly to the function with constant value $\int_X f d\mu$.*

Proof Assume by contradiction that there exist $\varepsilon > 0$, a sequence $(x_n)_{n \in \mathbb{N}}$ in X and an *extraction* (*i.e.*, a strictly increasing integer sequence) α such that for any integer n, $\frac{1}{\alpha(n)} \sum_{k=0}^{\alpha(n)-1} f(T^k(x_\alpha(n))) - \int_X f d\mu \geq \varepsilon$. As in the proof of Proposition 7.2.4(i), let ν be an adherent point of the sequence of probability measures $\nu_n = \frac{1}{\alpha(n)} \sum_{k=0}^{\alpha(n)-1} \delta_{T^k(x_\alpha(n))}$. The measure ν is T-invariant and satisfies $\int_X f d\nu - \int_X f d\mu \geq \varepsilon$, this contradicts the uniqueness of μ. □

In the symbolic case, when we apply this result to characteristic functions of cylinder sets as in Proposition 7.2.7, we get the following proposition:

Proposition 7.2.10 *A symbolic system (X_x, S) is uniquely ergodic if, and only if, x has uniform frequencies.*

Given a symbolic dynamical system (X, S), one can imagine the situation where there exists more than one ergodic invariant measure but any x in X has frequencies (that can depend on the point x). A theorem of J. Oxtoby (Oxtoby 1952) ensures that this is not possible in the minimal case, making unique ergodicity a necessary condition for global existence of frequencies:

Proposition 7.2.11 *If (X, S) is a minimal non-uniquely ergodic symbolic system, then there exist an infinite word $x \in X$ and a finite word w such that the frequency of w in x is not defined.*

Proof Assume by contradiction that any word in X has frequencies. Therefore, the function

$$f_w(x) = \lim_{n \to \infty} \frac{1}{n} \sum_{k=0}^{n-1} \chi_{[w]}(S^k(x))$$

is well defined on X, for any finite word w. Since X is not uniquely ergodic, there exist two ergodic measures $\mu \neq \nu$, hence there is a finite word w such that $\mu([w]) \neq \nu([w])$. Let x be a generic point for μ and y be a generic point for ν. We have $f_w(x) = \mu([w]) \neq \nu([w]) = f_w(y)$ and since f_w is constant along the orbits, f_w is nowhere continuous (the minimality implies that both orbits of x and y are dense). But f_w is a simple limit of continuous functions on the complete metric space X (which is even compact), hence the Baire category theorem implies that the points of continuity of f_w must be dense in X. A contradiction. □

It is important to keep in mind that unique ergodicity is *not* a measure-theoretic notion but a topological one. The Jewett, Krieger and Rosenthal theorem (Jewett 1969) (Krieger 1972) (Rosenthal 1988) asserts that unique ergodicity does not imply any restriction on the (unique) associated measure-theoretic ergodic dynamical system:

Proposition 7.2.12 *Every measure-theoretic ergodic dynamical system has a uniquely ergodic topological model, i.e., for any measure-theoretic ergodic dynamical system $(X, \mathcal{A}, \mu, S)$, there exists a uniquely ergodic topological dynamical system (Y, T) such that $(X, \mathcal{A}, \mu, S)$ is isomorphic to $(Y, \mathcal{B}, \mu, T)$, where $\mathcal{B}$ denotes the set of Borel subset of X and μ denotes the unique element of $\mathcal{M}(X, T)$.*

7.2.4 Finitely many ergodic invariant measures

When a subshift admits more than one ergodic invariant measure, we can still get some information from a bound on the number of its ergodic invariant measures. Let us see how the cardinality of $\mathcal{E}(X, S)$ measures the diversity of the behaviours of the orbits in the subshift (X, S).

First, Proposition 7.2.4(iii) tells that $\mathcal{E}(X, S)$ is the set of extreme points of $\mathcal{M}(X, S)$. The Krein–Milman theorem (see *e.g.*, (Rudin 1991)) ensures that, in such a situation, $\mathcal{M}(X, S)$ is the closure of the convex hull of $\mathcal{E}(X, S)$. Moreover, the Choquet theorem (see also (Rudin 1991)) ensures that any invariant measure can be written as an average of ergodic measures. Hence, any invariant measure can be recovered from the ergodic ones. Conversely, Propositions 7.2.4(iv) and 7.2.4(v) roughly tell us that no ergodic measure can be recovered from finitely many other ones (though some of them could be obtained as limit points of infinitely many other ones): all of them are necessary to describe $\mathcal{M}(X, S)$. Concerning the orbits, since any ergodic measure admits some generic points, ergodic measures are also all needed to describe the different behaviours of the typical orbits.

Let us now focus on the case when $\mathcal{E}(X, S)$ is known to be finite. Any invariant measure can be written, in a single manner, as a (finite) convex combination of ergodic invariant measures. However, Proposition 7.2.11 tells us that some orbits may be generic for no ergodic measure (see also Exercises 7.8 and 7.9). Their behaviours are nevertheless not out of control:

Proposition 7.2.13 *Let (X, S) be a subshift and let x be an element of X. Let $(w_j)_{j \in J}$ be a family of elements of $L(X)$. Let α be an extraction such that, for any $j \in J$, the sequence $(|x_0 \cdots x_{\alpha(n)-1}|_{w_j}/\alpha(n))$ converges to a number denoted by $f_{w_j}(x, \alpha)$. Then there exists an invariant measure $\mu \in \mathcal{M}(X, S)$ such that the vector $(f_{w_j}(x, \alpha))_{j \in J}$ is equal to $(\mu[w_j])_{j \in J}$.*

In particular, if (X, S) admits at most K ergodic invariant measures, and if $(x^{(i)})_{i \in I}$ is a family of elements of X, then the set of vectors

$\{(f_{w_j}(x^{(i)}, \alpha^{(i)}))_{j \in J} \mid i \in I\}$ spans an affine space of dimension at most $K - 1$.

Another direct consequence is that, for any $x \in X$ and any $w \in L(X)$,

$$\min_{\mu \in \mathcal{E}(X,S)} \mu([w]) \leq \liminf_{n \to \infty} \frac{|x_0 \cdots x_n|_w}{n+1}$$

$$\leq \limsup_{n \to \infty} \frac{|x_0 \cdots x_n|_w}{n+1} \leq \max_{\mu \in \mathcal{E}(X,S)} \mu([w]).$$

Proof As in the proof of Proposition 7.2.4(i), the sequence of probability measures $\mu_n = \frac{1}{\alpha(n)} \sum_{k=0}^{\alpha(n)-1} \delta_{T^k(x)}$ admits an adherent point $\mu \in \mathcal{M}(X, S)$. We have, $(f_{w_j}(x, \alpha))_{j \in J} = (\mu[w_j])_{j \in J}$. □

7.3 Combinatorial criteria

We describe combinatorial criteria implying unique ergodicity, or a bound on the number of ergodic invariant measures.

7.3.1 Complexity and Boshernitzan's criteria

Definition 7.3.1 The *complexity function* is the function that maps any integer n to the number $p_X(n) = \mathrm{Card}(L_n(X))$.

Theorem 7.3.2 (Boshernitzan 1984) *Let $K \geq 1$ be an integer. A minimal symbolic system (X, S) such that $\left\lfloor \liminf_{n \to \infty} \frac{p_X(n)}{n} \right\rfloor \leq K$ admits at most K ergodic invariant measures.*

Theorem 7.3.2 will follow from Theorem 7.3.7.

Theorem 7.3.3 (Boshernitzan 1984) *A minimal symbolic system (X, S) such that $\limsup_{n \to \infty} \frac{p_X(n)}{n} < 3$ is uniquely ergodic.*

The original proof of Theorem 7.3.3 is too long and technical to be given here. With a careful study of the evolution of the Rauzy graphs (see Section 7.3.2 below), it can be generalised (and the proof simplified) in the following way:

Theorem 7.3.4 (Monteil 2009) *Let $K \geq 3$ be an integer. A minimal symbolic system (X, S) such that $\limsup_{n \to \infty} \frac{p_X(n)}{n} < K$ admits at most $K - 2$ ergodic invariant measures.*

7.3.2 Deconnectability of the Rauzy graphs

Theorem 7.3.2 (and Boshernitzan's proof) can be generalised as follows. Let (X, S) be a symbolic system. Any word in $L_{n+1}(X)$ is naturally linked to two words in $L_n(X)$: its prefix and its suffix of length n. A way to keep track of this factorial structure of the language associated with X is the use of *Rauzy graphs*.

Definition 7.3.5 For any integer n, the nth *Rauzy graph* $G_n(X)$ is the directed graph such that:

(i) the set of vertices of $G_n(X)$ is $L_n(X)$,
(ii) there is an (oriented) edge from u to v in $G_n(X)$ if there exists w in $L_{n+1}(X)$ such that w begins with u and ends with v.

When X is the subshift associated with an infinite word x, the Rauzy graphs can be denoted by $G_n(x)$.

These graphs were first defined by N. G. de Bruijn (de Bruijn 1946) in a particular case: the *de Brujn graphs* are the Rauzy graphs when $L_n(X)$ is made of all the possible words of length n on the alphabet, or equivalently when X is the *full shift* defined in Section 7.5.1 below. In their full generality, they were defined by G. Rauzy in (Rauzy 1983) and independently by M. Boshernitzan, in the first published reference in a refereed journal, (Boshernitzan 1985); they were then named by Rauzy's followers. They should not be confused with the *Rauzy diagrams*, see (Yoccoz 2005) for example, which are graphs describing classes of permutations for interval-exchange transformations.

Definition 7.3.6 If $K \geq 1$, a symbolic system (X, S) is said to be *K-deconnectable* if there exist an extraction α and a constant $K' \geq 1$ such that for all $n \geq 1$ there exists a subset $D_{\alpha(n)} \subseteq L_{\alpha(n)}(X)$ of at most K vertices such that every path in $G_{\alpha(n)}(X) \setminus D_{\alpha(n)}$ is of length less than $K'\alpha(n)$ (in particular it does not contain any cycle).

This means that we can disconnect (in a specific way) infinitely many Rauzy graphs by removing at most K vertices.

Theorem 7.3.7 (Monteil 2005) *A K-deconnectable symbolic system (X, S) has at most K ergodic invariant probability measures.*

Proof We will first build at most K possible candidates and then prove that they are the only ones.

Step 1 *We build the candidates to be the only ergodic invariant probability measures.*

For any integer n, let $d_{1,\alpha(n)}, d_{2,\alpha(n)}, \ldots, d_{K,\alpha(n)}$ be an enumeration of $D_{\alpha(n)}$ (there is no loss of generality to consider that all the $D_{\alpha(n)}$ have exactly K elements). For now, we work in the subshift $(A^\mathbb{N}, S)$. We approximate X from the outside by K sequences of periodic subshifts as follows: for $i \leq K$ and $n \in \mathbb{N}$, we define

$$\mu_{i,\alpha(n)} = \frac{1}{\alpha(n)} \sum_{k=0}^{\alpha(n)-1} \delta_{S^k(d_{i,\alpha(n)}^\omega)},$$

where $d_{i,\alpha(n)}^\omega$ denotes the infinite word $d_{i,\alpha(n)} d_{i,\alpha(n)} d_{i,\alpha(n)} \cdots$ and δ stands for the one-point Dirac measure. The measure $\mu_{i,\alpha(n)}$ is the only element of $\mathcal{M}(A^\mathbb{N}, S)$ that gives full measure to the periodic subshift generated by the periodic word $d_{i,\alpha(n)}^\omega$. By compactness of $\mathcal{M}(A^\mathbb{N}, S)^K$, there exists an extraction β such that for each $i \leq K$,

$$\mu_{i,\alpha\circ\beta(n)} \xrightarrow[n\to\infty]{} \mu_i,$$

for some μ_i in $\mathcal{M}(A^\mathbb{N}, S)$. Note that if X is aperiodic (that is, if it contains no periodic orbit), the measures $\mu_{i,\alpha(n)}$ give measure 0 to X. Anyway,

Step 2 *We show that for $i \leq K$, $\mu_i(X) = 1$.*

Since X is closed in $A^\mathbb{N}$, we have the following approximation by open sets:

$$X = \overline{X} = \bigcap_{n \in \mathbb{N}} \bigcup_{u \in L_n(X)} [u],$$

where $[u]$ is considered as a cylinder in the dynamical system $(A^\mathbb{N}, S)$. Let $k \geq n \geq 1$ be two integers. For $i \in \{0, \ldots, \alpha \circ \beta(k) - n\}$, the finite word $(d_{i,\alpha\circ\beta(k)}^\omega)_i \cdots (d_{i,\alpha\circ\beta(k)}^\omega)_{i+n-1}$ is a factor of the finite word $d_{i,\alpha\circ\beta(k)}$, hence it belongs to $L_n(X)$. Hence, $\mu_{i,\alpha\circ\beta(k)}(\bigcup_{u \in L_n(X)}[u]) \geq (\alpha\circ\beta(k) - n + 1)/\alpha\circ\beta(k)$.

Letting k tend to infinity, since the characteristic function of $\bigcup_{u \in L_n(X)}[u]$ is continuous, we have $\mu_i(\bigcup_{u \in L_n(X)}[u]) = 1$. By countable intersection (n is arbitrary), we have $\mu_i(X) = 1$.

Hence, we can consider μ_i as an element of $\mathcal{M}(X, S)$.

Step 3 *Let μ be an ergodic measure on X. We show that μ is one of the μ_i.*

Since μ is ergodic, we can pick a point x in X that is generic for it, that is, for any u in $L(X)$,

$$\mu([u]) = \lim_{n\to\infty} \frac{|x_0 \cdots x_{n-1}|_u}{n}.$$

Let n be a fixed positive integer. We decompose x into blocks of length $\ell = (K'+1)\alpha \circ \beta(n)$, i.e., $x = b_0.b_1.b_2.b_3.b_4 \cdots$ with $b_j = x_{\ell j} \cdots x_{\ell(j+1)-1}$. Because of the hypothesis, any b_j contains an occurrence of one of the $d_{i,\alpha \circ \beta(n)}$ (b_j can be viewed as a path of length $K'\alpha \circ \beta(n)$ in $G_{\alpha \circ \beta(n)}(X)$). So, there exists some $i_{\alpha \circ \beta(n)}$ such that the upper density of the set

$$\{j \in \mathbb{N} \mid |b_j|_{d_{i_{\alpha \circ \beta(n)},\alpha \circ \beta(n)}} \geq 1\}$$

is at least $1/K$. Let γ be an extraction such that $i_{\alpha \circ \beta \circ \gamma(.)}$ is constant with value denoted by i. We denote by $\tilde{\alpha}$ the extraction $\alpha \circ \beta \circ \gamma$.

Let us show that $\mu = \mu_i$. Let u be a finite word in $L(X)$. Let n be an integer greater than $|u|$. There is an extraction δ such that for any integer m,

$$\operatorname{Card}\{j < \delta(m) \mid |b_j|_{d_{i,\tilde{\alpha}(n)}} \geq 1\} \geq \delta(m)/2K.$$

Therefore,

$$|b_0.b_1 \cdots b_{\delta(m)-1}|_u \geq \frac{\delta(m)}{2K}|d_{i,\tilde{\alpha}(n)}|_u.$$

Hence,

$$\mu([u]) = \lim_{m \to \infty} \frac{|b_0.b_1 \cdots b_{\delta(m)-1}|_u}{(K'+1)\tilde{\alpha}(n)\delta(m)} \geq \frac{|d_{i,\tilde{\alpha}(n)}|_u}{2K(K'+1)\tilde{\alpha}(n)}.$$

Moreover, we can control the frequency of occurrences of u in $d^{\omega}_{i,\tilde{\alpha}(n)}$ by counting separately the occurrences of u that fall in some $d_{i,\tilde{\alpha}(n)}$ and the occurrences of u that appear between two consecutive occurrences of $d_{i,\tilde{\alpha}(n)}$:

$$\mu_{i,\tilde{\alpha}(n)}([u]) = \frac{1}{\tilde{\alpha}(n)} \sum_{k=0}^{\tilde{\alpha}(n)-1} \delta_{S^k(d^{\omega}_{i,\tilde{\alpha}(n)})}([u]) \leq \frac{1}{\tilde{\alpha}(n)}(|d_{i,\tilde{\alpha}(n)}|_u + |u|).$$

Therefore,

$$\mu_{i,\tilde{\alpha}(n)}([u]) \leq \frac{|d_{i,\tilde{\alpha}(n)}|_u}{\tilde{\alpha}(n)} + \frac{|u|}{\tilde{\alpha}(n)} \leq 2K(K'+1)\mu([u]) + \frac{|u|}{\tilde{\alpha}(n)}.$$

Letting n tend to infinity, we have $\mu_i([u]) \leq 2K(K'+1)\mu([u])$, so μ_i is absolutely continuous relatively to μ. Since μ_i is S-invariant and μ is ergodic, Proposition 7.2.4(ii) ensures that $\mu_i = \mu$, hence there are at most K ergodic invariant measures.

$\square$

Proof of theorem 7.3.2 Because of the hypothesis, there is an extraction α such that for every integer n, $p_X(\alpha(n)) \leq (K+1)\alpha(n)$ and $p_X(\alpha(n)+1) - p_X(\alpha(n)) \leq K$. For n in $\mathbb{N}$, let $D_{\alpha(n)}$ be the set of left special factors (see Section 4.5) of length $\alpha(n)$ of X, whose cardinality is not greater than K.

Let $n \in \mathbb{N}$. Any loop O in $G_{\alpha(n)}(X)$ must contain a left special factor. Indeed, since we can assume that X is aperiodic, there exists a finite word u in $L_{\alpha(n)+1}(X) \setminus O$, and since X is minimal, there exists a path from the edge u to any vertex of O, so the first vertex of O that this path meets is a left special factor. Therefore, $G_{\alpha(n)}(X) \setminus D_{\alpha(n)}$ does not contain any loop.

So, a path in $G_{\alpha(n)}(X) \setminus D_{\alpha(n)}$ is necessarily injective and cannot be of length greater than $\mathrm{Card}(L_{\alpha(n)}(X)) \leq (K+1)\alpha(n)$. Therefore, (X,S) is K-deconnectable and Theorem 7.3.7 applies. □

7.3.3 Boshernitzan's ne_n condition

Let (X, S) be a minimal symbolic system. If $\mu \in \mathcal{M}(X, S)$ is a S-invariant probability measure and n is an integer, we denote by $e_n(\mu)$ the minimal measure of the cylinder sets of length n of X.

Theorem 7.3.8 (Boshernitzan 1992) *Let (X, S) be a minimal symbolic system. If there exists μ in $\mathcal{M}(X, S)$ such that $\limsup_{n \to \infty} ne_n(\mu) > 0$, then (X, S) is uniquely ergodic.*

Proof Because of the hypothesis, there exists $c > 0$ such that, for infinitely many integers $n \in \mathbb{N}$, we have $ne_n(\mu) \geq c$. For such n and $w \in L_n(X)$, we have $\mu([w]) > c/n$ and since all those cylinders are disjoint, we have $p_X(n) = \mathrm{Card}\, L_n(X) \leq n/c$. Hence, $\liminf_{n \to \infty} p_X(n)/n \leq 1/c < \infty$. So, Theorem 7.3.2 tells us that (X, S) admits a finite number of ergodic invariant measures.

Assume by contradiction that (X, S) is not uniquely ergodic: it admits at least two distinct ergodic invariant measures. Hence, there exists a finite word w such that the set $E = \{\mu([w]) \mid \mu \in \mathcal{E}(X, S)\}$ has a finite cardinality which is greater than one. We choose two ergodic measures μ_1 and μ_2 which correspond to consecutive elements in E, i.e., such that $\mu_1([w]) < \mu_2([w])$ and $(\mu_1([w]), \mu_2([w])) \cap E = \emptyset$. Let r and s be two real numbers such that $\mu_1([w]) < r < s < \mu_2([w])$.

For $n \geq 1$, let F_n denote the set $\{x \in X \mid \frac{|x_0 \cdots x_{n-1}|_w}{n} \in [r,s]\}$. If ν is an ergodic invariant measure of (X, S), then the Birkhoff ergodic theorem tells us that ν-almost every point $x \in X$ satisfies $\frac{|x_0 \cdots x_{n-1}|_w}{n} \xrightarrow[n \to \infty]{} \nu([w]) \notin [r-s]$, which implies $\nu(\bigcup_{N \in \mathbb{N}} \bigcap_{n \geq N} X \setminus F_n) = 1$. Hence, $\nu(F_N) \leq \nu(\bigcup_{n \geq N} F_n) \xrightarrow[N \to \infty]{} \nu(\bigcap_{N \in \mathbb{N}} \bigcup_{n \geq N} F_n) = 0$. Since μ is a (finite) convex combination of such ergodic invariant measures, we also have $\mu(F_N) \xrightarrow[N \to \infty]{} 0$.

Let y be a generic point for μ_1 and z be a generic point for μ_2: there exists an integer N such that for any $n \geq N$, $\frac{|y_0 \cdots y_{n-1}|_w}{n} < r$ and $\frac{|z_0 \cdots z_{n-1}|_w}{n} > s$. Let n be an integer which is greater than N. Since (X, S) is minimal, the Rauzy graph $G_n(X)$ is strongly connected, so there is a path $y_0 \cdots y_{n-1} = v_1 \to v_2 \to \cdots \to v_{p-1} \to v_p = z_0 \cdots z_{n-1}$ from $y_0 \cdots y_{n-1}$ to $z_0 \cdots z_{n-1}$. We choose a shortest such path (in particular, it has no loop).

For $i \leq p-1$, we have $|v_{i+1}|_w \leq |v_i|_w + 1$, hence $\frac{|v_{i+1}|_w}{n} \leq \frac{|v_i|_w}{n} + \frac{1}{n}$. Since $\frac{|v_1|_w}{n} < r$ and $\frac{|v_p|_w}{n} > s$, the set $I = \{i \leq p \mid \frac{|v_i|_w}{n} \in [r, s]\}$ has cardinality greater than or equal to $n(s-r) - 2$. Hence, F_n contains $\cup_{i \in I}[v_i]$, so its measure is at least $\mu(F_n) \geq (n(s-r) - 2)e_n(\mu) = (s-r)ne_n(\mu) - 2e_n(\mu)$. Hence, $ne_n(\mu) \xrightarrow[n \to \infty]{} 0$, a contradiction. □

7.4 Examples

7.4.1 Classical symbolic systems

Because of Theorem 7.3.3, every *Sturmian* infinite word (see Definition 1.2.13) has uniform frequencies. Exercises 7.12, 7.13 and 7.14 in Section 7.7 deal with some applications of the results of Section 7.3 to some known families of infinite words.

It follows from Theorem 1.6.9 above that a *fixed point of a primitive substitution* has uniform frequencies: this was shown in (Queffélec 1987), and, from the same proof, we can deduce that these words satisfy a strong version of the ne_n condition of Boshernitzan (see also Exercise 7.14). Thus the Thue–Morse word (Example 1.2.21), the Fibonacci word (Example 1.2.22), or the *Rudin–Shapiro word* (defined as the fixed point beginning with a of the substitution $a \mapsto ab, b \mapsto ac, c \mapsto db, d \mapsto dc$) have uniform frequencies.

For the Chacon word defined after Proposition 1.4.6, the substitution is not primitive, but this word has also uniform frequencies; this can be proved directly, see Exercises 1.8 and 7.5; we could also notice that the complexity is $2n - 1$ for $n \geq 2$ (Ferenczi 1995) and apply Theorem 7.3.3, or notice that the dynamical system is topologically isomorphic to the symbolic system associated with the fixed point of a primitive substitution (Ferenczi 1995) and use Theorem 1.6.9. Note that Proposition 6.5.6 generalises this fact, implying that the fixed points of many non-primitive substitutions have uniform frequencies.

7.4.2 Non-uniformly recurrent words

It is a common mistake to believe that uniform frequencies imply uniform recurrence. A simple counter-example is the word $x = baaa\cdots$, which has uniform frequencies ($f_w(x) = 1$ if $w = a^k$ for any k, $f_w(x) = 0$ if w is any other word) but is not uniformly recurrent (b occurs only once).

A more elaborate counter-example was suggested by E. Lesigne: let y be an infinite word which is known to be uniformly recurrent and to have uniform frequencies, for example the Thue–Morse word. Let w be a finite word not occurring in y, for example here $w = aaa$. We build a new infinite word x by inserting the word w into y at places along a sequence of uniform density 0: for example here, for every $k \geq 0$, we put

(i) $x_{2^k+3k+i} = a$, $i = 1, 2, 3$,
(ii) $x_{2^k+3k+3+j} = y_{2^k+j}$, $1 \leq j \leq 2^{k+1} - 2^k$.

Then x is not uniformly recurrent as w does not occur with bounded gaps; but if we compute $\frac{|x_k \cdots x_{k+n}|_{w'}}{n+1}$ for a finite word w' occurring in x, we see that it converges uniformly to $\lim_{n \to +\infty} \frac{|y_k \cdots y_{k+n}|_{w'}}{n+1}$, which is 0 if w' does not occur in y, and is known to exist otherwise. Hence, x has uniform frequencies. See also Exercise 7.16.

This example is fairly typical, and assuming that uniform frequencies imply uniform recurrence is not a very serious mistake, as

Proposition 7.4.1 *Let x be an infinite word with uniform frequencies, with*

$$f_w(x) = \lim_{n \to +\infty} \frac{|x_k \cdots x_{k+n}|_w}{n+1}.$$

Then, there exist infinite words y such that $L(y) = \{w \in L(x) \mid f_w(x) > 0\}$. Any of these y is uniformly recurrent and has uniform frequencies, with

$$f_w(y) = \lim_{n \to +\infty} \frac{|y_k \cdots y_{k+n}|_w}{n+1} = f_w(x)$$

for every $w \in L(y)$.

Moreover, (X_y, S) is the only minimal subshift of (X_x, S), and X_y has full measure in X_x for the invariant measure. For any $z \in X_x$, there exists an infinite sequence n_k such that $\lim_{k \to +\infty} S^{n_k}(z) \in X_y$.

Proof The subshift (X_x, S) is uniquely ergodic, let μ denotes its unique invariant probability measure. The language $L = \{w \in L(x) \mid f_w(x) > 0\}$ is factorial and extendable, hence Proposition 7.1.2 ensures the existence of a subshift $Y \subseteq X_x$ such that $L(Y) = L$.

Let Z be a subshift which is included in X_x. Proposition 7.2.4(i) ensures that (Z,S) admits an invariant measure ν, which can be considered as an invariant measure on X_x, hence $\mu = \nu$. In particular, ν gives positive measure to the cylinders $[w]$ for any $w \in L$, hence $L(Y) \subseteq L(Z)$, hence $Y \subseteq Z$. Therefore, Y is the only minimal subshift which is included in X_x.

Let y be any element of Y: y is uniformly recurrent, and since $y \in X_x$, it has uniform frequencies with $f_w(y) = f_w(x)$ for any $w \in L(y) \subseteq L(x)$. The last assertion comes from the fact that for any $z \in X_x$, X_z contains $Y = X_y$. □

Corollary 7.4.2 *If x is an infinite word with uniform frequencies, x is uniformly recurrent if, and only if, $f_w(x) > 0$ for every factor w of x.*

Proof X_x is minimal if, and only if, $X_x = X_y$, or $L(x) = L(y)$. □

A minimal and uniquely ergodic dynamical system is sometimes called *strictly ergodic*; the symbolic translation of this notion could be *positive uniform frequencies*.

7.4.3 Positive entropy and Grillenberger words

The *topological entropy* of a symbolic dynamical system (X_x, S) can be defined as

$$h(x) = \lim_{n \to +\infty} \frac{\log p_x(n)}{n}.$$

For a word with uniform frequencies, the same limit is also the *metric* (or *measure-theoretic*) *entropy* of the system (X_x, S, μ) equipped with its unique invariant probability measure. All the examples in Sections 7.4.1 and 7.4.2 have entropy zero.

Though the Jewett–Krieger theorem (see Proposition 7.2.12) implies that there exist uniquely ergodic systems of every given topological entropy, it seems difficult to find an infinite word with uniform frequencies and positive entropy, which implies exponential complexity. The standard way to ensure exponential complexity is to concatenate words independently: starting with a family of words $B_1, \ldots, B_r$, we decide that our infinite word will have all the factors $B_{i_1} \cdots B_{i_s}$ for some s and every sequence $(1 \leq i_1 \leq r, \ldots, 1 \leq i_s \leq r)$, and then iterate this process. But the resulting infinite word will not have uniform recurrence or frequencies (if we start from all the 1-letter words, we get the counter-examples of Section 7.5.1).

The first explicit examples with uniform frequencies (and recurrence) and exponential complexity (indeed, with arbitrarily high entropy) were built by

F. Hahn and Y. Katznelson (Hahn and Katznelson 1967); a much simpler construction is due to C. Grillenberger (Grillenberger 1972), who proved

Proposition 7.4.3 (Grillenberger 1972) *For every integer k and every real number $0 \leq h < \log k$, there exists a uniformly recurrent infinite word x on an alphabet of k letters which has uniform frequencies and for which $h(x) = h$.*

The basic construction falls into the general framework of *adic* words studied in Section 7.5.3; it is a clever replacement of the independent concatenation mentioned above by permutations, and of exponentials by factorials; indeed, to build a one-sided infinite word x (as a slight variation from (Grillenberger 1972) which considers two-sided infinite words), we build inductively families of words $B_{n,i}$, $1 \leq i \leq k_n$, where $B_{0,i} = i$, $1 \leq i \leq k$, and, for any permutation π on $\{1, \ldots, k_n\}$

$$B_{n+1,\pi} = B_{n,\pi(1)} \cdots B_{n,\pi(k_n)},$$

the permutations are then ordered lexicographically to number the new words from $B_{n+1,1}$ to $B_{n+1,k_n!}$.

Then the infinite word x beginning with $B_{n,1}$ for every n has uniform frequencies and recurrence and, if $k \geq 3$, exponential complexity. The infinite words in Proposition 7.4.3 are then deduced from one of these x by applying a suitable substitution. Note that on two letters, the above construction yields the Thue–Morse word, which is of entropy zero: if we want exponential complexity on two letters, we can start with the above construction on three letters, and replace them by aa, ab, ba.

7.4.4 Interval exchange maps

Interval exchange maps are one-dimensional geometrical systems introduced by V. I. Oseledec (Oseledec 1966), for the study of which symbolic dynamics proved to be a very efficient tool.

Definition 7.4.4 Let $r \geq 3$. Let Λ_r be the set of vectors $(\lambda_1, \ldots, \lambda_r)$ in $\mathbb{R}^r$ such that $0 \leq \lambda_i \leq 1$ for all i and $\Sigma_{i=1}^{r} \lambda_i = 1$. An *$r$-interval exchange map* is given by a vector $\lambda \in \Lambda_r$ and a permutation π of $\{1, 2, \ldots, r\}$. The map $T_{\lambda,\pi}$ is the piecewise translation defined by partitioning the interval $[0, 1)$ into r sub-intervals of lengths $\lambda_1, \lambda_2, \ldots, \lambda_r$ and rearranging them according to the permutation π or, formally,

$$\Delta_i = \left[\sum_{j<i} \lambda_j, \sum_{j \leq i} \lambda_j \right),$$

$$T_{\lambda,\pi}\xi = \xi + \sum_{\pi^{-1}j < \pi^{-1}i} \lambda_j - \sum_{j<i} \lambda_j$$

if $\xi \in \Delta_i$.

The map $T_{\lambda,\pi}$ satisfies the *i.d.o.c. property* (Keane 1975) if the negative orbits of the discontinuity points $\sum_{j\leq i} \lambda_j$, $1 \leq i \leq r-1$, are infinite and disjoint.

Warning: roughly half the texts on interval exchange maps re-order the subintervals by π^{-1}; as it is not always clear to which half a given text belongs, we insist that the present definition corresponds to the following ordering of $T\Delta_i$: from left to right, $T\Delta_{\pi(1)}, \ldots, T\Delta_{\pi(r)}$. It makes sense to re-order also the Δ_i, thus defining T by two permutations π_0 and π_1 (though of course sometimes π_0^{-1} and π_1^{-1} are used...), see (Yoccoz 2006) for example. Though it might have come useful in the definition of Keane's examples in Section 7.5.2.2 below, we prefer to stick to one permutation.

Definition 7.4.5 A *natural coding* of an r-interval exchange map is any of the words $x(\xi)$ for a point $\xi \in [0,1)$, where $x_i(\xi) = j$ whenever $T^i\xi \in \Delta_j$.

If x is a natural coding of $T_{\lambda,\pi}$, we can consider the symbolic system (X_x, S). Though this system is not topologically conjugate to $([0,1), T_{\lambda,\pi})$, it shares all its properties of minimality and unique ergodicity, and any invariant measure for one of these systems can be carried to the other one. The i.d.o.c. condition ensures that (X_x, S) and $([0,1), T_{\lambda,\pi})$ are minimal, each natural coding x is uniformly recurrent and the language $L(x)$ is the same for all the natural codings. Then the complexity function of any natural coding is $(r-1)n+1$ and thus the number of ergodic invariant probability measures on (X_x, S) or $([0,1), T_{\lambda,\pi})$ is at most $r-2$ by Theorem 7.3.4 (geometrical methods can improve this bound to $\frac{r}{2}$ (Katok 1973)).

We shall call m the normalised Lebesgue measure on Λ_r, and μ the Lebesgue measure on $[0,1)$. It is proved in (Keane 1975) that for an *irreducible* permutation π ($\pi\{1,\ldots k\} \neq \{1,\ldots k\}$ for every $k < r$) the i.d.o.c. property is implied by the *total irrationality* of the λ_i (the λ_i have no rational relation except $\lambda_1 + \cdots + \lambda_r = 1$), and thus for m-almost every $\lambda \in \Lambda_r$, $T_{\lambda,\pi}$ satisfies it, and hence is minimal, or equivalently the $x(\xi)$ are uniformly recurrent.

It follows from Theorem 7.3.3, and it is stated in (Keane 1975), that

Proposition 7.4.6 *For three-interval exchange maps the i.d.o.c. condition implies unique ergodicity (hence uniform frequencies for the natural codings).*

It was conjectured by M. Keane (Keane 1975) that the i.d.o.c. condition implies unique ergodicity for every r. After this conjecture was disproved (see Section 7.5.2.1), a weaker result was considered as a question by the same author (Keane 1977) and proved independently by H. Masur (Masur 1982) and W. Veech (Veech 1982).

Theorem 7.4.7 *For a given irreducible π, $T_{\lambda,\pi}$ is uniquely ergodic (or equivalently the $x(\xi)$ have uniform frequencies) for m-almost every $\lambda \in \Lambda_r$.*

The proofs of W. Veech and H. Masur use deep geometrical methods; but a later proof of M. Boshernitzan uses mainly combinatorial methods; it is published in (Boshernitzan 1985) but can be simplified (and made completely combinatorial) by using the criteria of (Boshernitzan 1984) (Boshernitzan 1992) described in Section 7.3. Thus we give here this simplified proof, with an updated vocabulary.

Proposition 7.4.8 (Boshernitzan 1985) *Let $U_{n,\varepsilon}$ be the set of $\lambda \in \Lambda_r$ such that $e_n(T_{\lambda,\pi}) \leq \frac{\varepsilon}{n}$ (see Section 7.3.3). If $0 < \varepsilon < \frac{1}{r}$, $m(U_{n,\varepsilon}) \leq 3r^3 \varepsilon$.*

Proof Let $G_n(\lambda)$ denote the Rauzy graphs (see Definition 7.3.5) of length n of any infinite word $x(\xi)$ in the natural coding of $T_{\lambda,\pi}$. As the complexity of any $x(\xi)$ is $(r-1)n + 1$, $G_n(\lambda)$ has at most $3r - 3$ branches.

The weight functions of Definition 7.2.5 can be carried over to the Rauzy graphs: ψ is a *weight function on a graph* if it is positive on each vertex, the sum of its values on vertices is 1, and it can be extended to the edges such that for every vertex w

$$\psi(w) = \sum_{\text{incoming edges}} \psi(e) = \sum_{\text{outgoing edges}} \psi(e).$$

And the function ψ_λ, defined on the vertices of $G_n(\lambda)$ by associating with the vertex $w_1 \cdots w_n$ the measure of the cylinder, $\mu[w_1 \cdots w_n]$, is a weight function on the graph $G_n(\lambda)$; the weight of an edge $w_1 \cdots w_{n+1}$ is also $\mu[w_1 \cdots w_{n+1}]$.

We fix now a Rauzy graph G of length n; let $\Lambda(G)$ be the set of $\lambda \in \Lambda_r$ such that $G_n(\lambda) = G$. For a given word $w = w_1 \cdots w_n$, $\psi_\lambda(w_1)$ is just λ_{w_1}; for all $\lambda \in \Lambda(G)$, all the Rauzy graphs $G_i(\lambda)$, $1 \leq i \leq n$, are fixed, and when we look at the defining equalities of the weight function ψ_λ on $G_i(\lambda)$, we see that the measures of cylinders of length $i+1$ are computed by explicit formulas from those of length i; thus the numbers $\psi_\lambda(w_1 \cdots w_i)$, $1 < i \leq n$, can be computed inductively; they depend linearly on λ. Because $T_{\lambda,\pi}$ preserves the measure μ, $\psi_\lambda(w_1 \cdots w_n) = \psi_\lambda(w'_1 \cdots w'_n)$ if $w_1 \cdots w_n$ and

$w_1' \cdots w_n'$ are on the same branch of G; hence, for fixed λ, $\psi_\lambda(w_1 \cdots w_n)$ takes $1 \leq t \leq 3r - 3$ values, which we denote by $\varphi_1(\lambda), \ldots, \varphi_t(\lambda)$; the φ_j are linear functionals, $e_n(T_{\lambda,\pi})$ is just the smallest of the $\varphi_j(\lambda)$, $1 \leq j \leq t$. Furthermore, again through the defining equalities of the successive weight functions on $G_i(\lambda)$, we can retrieve λ from the values $\psi_\lambda(w)$ on all the vertices of G; thus $\Lambda(G)$ is a convex set and every weight function ψ on G yields a $\lambda \in \Lambda(G)$ such that $\psi_\lambda = \psi$.

We want to estimate the measure of $\{\lambda \in \Lambda(G) \mid \varphi_i(\lambda) \leq \frac{\varepsilon}{n}\}$; for this, we use a general result for which we refer the reader to (Boshernitzan 1985), Corollary 7.4: *If φ is the restriction of a linear functional to a convex set K of dimension d, taking values between 0 and A, then, if V denotes the volume,*

$$V(\varphi^{-1}[0, B)) \leq \frac{dB}{A} V(K).$$

We apply it with $K = \Lambda(G)$, restricting ourselves to those with $m(\Lambda(G)) > 0$, $\varphi = \varphi_i$, $B = \frac{\varepsilon}{n}$; the dimension is $r - 1$, the volume is the Lebesgue measure; we need an estimate on A; for this, we claim that *for each vertex s of G, there exists a weight function such that $\psi(s) \geq \frac{1}{rn}$.* To do this, we choose a $\lambda \in \Lambda(G)$ such that $T_{\lambda,\pi}$ is minimal, which is possible as $m(\Lambda(G)) > 0$; this implies that G is strongly connected and thus we can find a loop $s = s_0 \to \cdots \to s_k \to s_0$ in G; by taking it of minimal length, we ensure it has no repetition. Then we define ψ' to be $\frac{1}{k+1}$ on the s_i and 0 on the other vertices; ψ' is not a weight function as it may be 0 on some vertices, but $\psi = (1 - \delta)\psi' + \delta\psi_\lambda$ is a weight function, and as $k \leq (r - 1)n + 1$ we can choose δ such that our claim is proved.

Thus we have $A \geq \frac{1}{rn}$, and thus, for all G with $m(\Lambda(G)) > 0$ and hence for all G,

$$m(\{\lambda \in \Lambda(G) \mid \varphi_i(\lambda) \leq \frac{\varepsilon}{n}\}) \leq (r - 1)r\varepsilon m(\lambda(G)).$$

As $t \leq 3r - 3$,

$$m(\{\lambda \in \Lambda(G) \mid \min_{1 \leq i \leq t} \varphi_i(\lambda) \leq \frac{\varepsilon}{n}\}) \leq 3(r - 1)^2 r\varepsilon m(\lambda(G)),$$

which implies the proposition. □

Proof of Theorem 7.4.7 If $0 < \varepsilon < \frac{1}{r}$ and $n \geq 1$, we put $V_{n,\varepsilon} = \Lambda_r \setminus U_{n,\varepsilon}$, and $V_\varepsilon = \cap_{N \geq 1} \cup_{n > N} V_{n,\varepsilon} \cap \{\lambda \mid T_{\lambda,\pi} \text{ is i.d.o.c.}\}$.

If λ is in V_ε, there are infinitely many n such that $e_n(T_{\lambda,\pi}, \mu) \geq \frac{\varepsilon}{n}$, hence $ne_n(T_{\lambda,\pi}, \mu) \not\to 0$ when $n \to +\infty$, and $T_{\lambda,\pi}$ is uniquely ergodic by Theorem 7.3.8. Thus $m(\{\lambda \mid T_{\lambda,\pi} \text{ is uniquely ergodic}\})$ is at least $m(V_\varepsilon) \geq 1 - 3r^3\varepsilon$), and thus is one as ε is arbitrary. □

The above proof does not use any of the geometrical properties of interval exchange maps.

Explicit examples of infinite words with uniform frequencies coming from coding of four-interval exchanges can be deduced from (Keane 1977) or found in (Ferenczi and Zamboni 2008) or (Cheung and Masur 2006). Examples for higher number of intervals can be deduced from (Sataev 1975), see Proposition 7.5.3 below.

7.5 Counter-examples

7.5.1 The full shift

As we have seen in Section 7.4, *non-uniformly recurrent* words and *words of positive entropy* are generally expected not to have uniform frequencies, unless they have been built for the specific purpose of having them.

Typical examples falling into both these categories are the words with full complexity $p(n) = k^n$ on any finite alphabet A of cardinality k, such as the *Champernowne word* $011011100101110\cdots$, built by concatenating the expansions in base 2 of $0, 1, 2, \ldots, n, \ldots$: they do not have uniform frequencies, and their associated symbolic system is the *full shift* $A^{\mathbb{N}}$.

Proposition 7.5.1 *The full shift has uncountably many ergodic invariant measures.*

Proof Take a probability vector $\pi = (\pi_1, \ldots, \pi_k)$ and assign to cylinders the measure $\mu_\pi([w_1 \cdots w_n]) = \pi_{w_1} \cdots \pi_{w_n}$. These measures are ergodic (see Exercise 7.17). □

The system $(A^{\mathbb{N}}, \mathcal{B}, \mu_\pi, S)$ is then called a (one-sided) *Bernoulli shift*. But there are lots of other ergodic invariant measures for the full shift, for examples the Dirac measure on each periodic orbit (see Exercise 7.11), or the measures arising from the uniquely ergodic examples given in Section 7.4 (that can be considered as invariant measures on the full shift defined on the same alphabet).

The non-trivial *subshifts of finite type* (where we consider all the infinite words in which a prescribed set of finite words does not occur) and the *sofic systems* (which constitute the closure of the subshifts of finite type for a natural notion of homomorphism), defined in (Weiss 1973) behave like the full shift in having positive entropy, with no uniform recurrence or frequencies. See Section 1.6 and 2.3.1 for definitions.

7.5.2 Interval exchange maps again

7.5.2.1 Veech's counter-examples

Systems which are minimal and not uniquely ergodic are not so easy to build; the famous examples of H. Furstenberg (Furstenberg 1961) are defined on multi-dimensional tori, and do not give rise naturally to infinite words.

Then came the examples of W. Veech (Veech 1969), which use ideas of (Furstenberg 1961) together with very involved arithmetic considerations:

Theorem 7.5.2 (Veech 1969) *For two irrationals α and β, let $f = -2\chi_{[0,\beta)} + 1$ and T be the map on $\mathbb{R}/\mathbb{Z} \times \{-1, 1\}$ defined by*

$$T(\xi, e) = (\xi + \alpha, f(\xi)e).$$

If β is not of the form $p\alpha + q$ with p and q integers (independence condition), T is minimal.

If α and β satisfy the coboundary condition

$$ef(\xi)g(\xi) = g(\xi + \alpha),$$

for every $\xi \in [0, 1)$, some measurable function g and number $e = \pm 1$, T is not uniquely ergodic.

For each α with unbounded partial quotients, there exist uncountably many numbers β such that both these conditions are satisfied.

This result does not appear as such in the paper (Veech 1969). Indeed, it is written under a partly symbolic form with T replaced by the map $\overline{T}(y, e) = (S(y), y_0 . e)$ on $X_x \times \{-1, 1\}$, where $x = (x_n)_n$ is defined as follows: we fix $\xi \in [0, 1)$, and $x_n = -1$ if $\xi + n\alpha$ (modulo 1) falls into $[0, \beta)$, $x_n = +1$ otherwise; hence, $X_x \subset \{-1, 1\}^{\mathbb{N}}$. It is then straightforward, though not written in (Veech 1969), that, for α and β satisfying the conditions, the iterates of $(x, 1)$ give an infinite word on the alphabet $\{(-1, -1), (1, -1), (-1, 1), (1, 1)\}$ which is uniformly recurrent and does not have uniform frequencies, and these may be considered as the first infinite words with these properties.

If now we replace $\mathbb{R}/\mathbb{Z} \times \{-1\}$ with $[0, 1)$ and $\mathbb{R}/\mathbb{Z} \times \{1\}$ with $[1, 2)$, and normalise by 2, we see that T is also a *five-interval exchange map* defined (if for example $\beta < 1 - \alpha$) by $\lambda = \frac{1}{2}(\beta, 1 - \alpha - \beta, \alpha + \beta, 1 - \alpha - \beta, \alpha)$ and the permutation $1 \mapsto 3, 2 \mapsto 2, 3 \mapsto 5, 4 \mapsto 1, 5 \mapsto 4$. Under the independence condition, T satisfies the i.d.o.c. condition; and its natural coding on a five-letter alphabet is another uniformly recurrent infinite word, without uniform frequencies if α and β satisfy the coboundary condition.

We can look at another map T', induced (see Definition 7.5.4 below) on $[0, \beta)$ by the rotation of angle α. The map T' is a three-interval exchange map, satisfying the i.d.o.c. condition if α and β satisfy the independence condition. It is an implicit consequence of (Veech 1969), stated in (Veech 1984), that the coboundary condition is equivalent to an unexpected property, namely that T' has -1 as an eigenvalue, meaning that there exists g in $L^2([0,1))$ with $g \circ T' = -g$. A direct proof of this property is given in (Ferenczi, Holton, and Zamboni 2004).

We can also look at the map T'', which is an *exduction* on $[0, \beta)$ of the rotation of angle α: namely $T''\xi = \xi + 1$ for $0 \leq \xi < \beta$, and for $\beta \leq \xi < 1 + \beta$, $T''\xi$ is the representative of $\xi + \alpha$ modulo 1 which falls into $[0, 1)$. It is also a three-interval exchange map, defined after normalisation (if for example $\beta < 1 - \alpha$) by $\lambda = \frac{1}{1+\beta}(\beta, 1 - \alpha - \beta, \alpha + \beta)$ and the permutation $1 \mapsto 3, 2 \mapsto 2, 3 \mapsto 1$. It can be viewed as a dual version of T' (as the rotation is an induced map of T'') and shares the same properties. The map T'' was introduced in (Keynes and Newton 1976) in order to exhibit T''^2 as a non-uniquely ergodic five-interval exchange map.

The coboundary condition is studied further by Y. Cheung (Cheung 2003) where, for fixed β, estimates are given for the Hausdorff dimension of the set of α for which α and β satisfy it; the non-unique ergodicity of T is also seen as the non-ergodicity of some directions for a *billiard flow*.

A nice generalisation of Veech's result appeared a few years later in a paper of E. Sataev. The lack of communication between West and East at that time explains that Sataev apparently did not know the paper (Veech 1969) and that in turn (Sataev 1975) was widely ignored thereafter.

Proposition 7.5.3 (Sataev 1975) *For any integer $r \geq 2$ and any integer $1 \leq k \leq r$, there exist an irrational α, $r - 1$ disjoint intervals $I_j \subseteq [0, 1)$ and $r - 1$ different permutations π_j of $\{1, \ldots, r\}$, $1 \leq j \leq r - 1$, such that U is minimal and has exactly k ergodic invariant probability measures, where U is the map on $\mathbb{T}^1 \times \{1, \ldots, r\}$ defined by $U(\xi, e) = (\xi + \alpha, h(\xi)e)$, and $h(\xi)$ is the permutation π_j when ξ is in I_j and the identity elsewhere.*

This gives exchanges of at least r^2 and at most $2r^2$ intervals which have a prescribed number $1 \leq k \leq r$ of ergodic invariant probability measures.

7.5.2.2 Keane's counter-examples

Then M. Keane (Keane 1977) lowered the number of intervals required for a counter-example to four, which is optimal in view of Proposition 7.4.6. But his paper uses very different techniques, and there appear for the first time two ideas which were to be named and systematically studied later:

one is the *induction*, a different form of which will give the *Rauzy induction* and is the starting point of the geometrical methods mentioned in Section 7.4.4; the other one is the use of *matrices for adic systems*, which will be developed in the next section.

We recall that

Definition 7.5.4 If T is map from a set X to itself, and A a subset of X, the *induced map* of T on A is defined by $T_A(z) = T^{r_A(z)}z$, with $r_A(z) = \min\{n > 0 \mid T^n z \in A\}$, for all $z \in A$ for which $r_A(z)$ is finite.

The induction idea works as follows: M. Keane takes a four-interval exchange for the permutation (in our notation, see Section 7.4.4 above) $1 \to 4, 2 \to 2, 3 \to 1, 4 \to 3$, denoted by π, with a probability vector λ as yet unknown.

Some inequalities on the λ_i ensure that the induced map of $T_{\lambda,\pi}$ on the fourth interval Δ_4 is well defined and is another four-interval exchange map which, after renormalisation and a renumbering of the intervals which reverses their order, can be defined by the permutation π and a vector λ' such that $\lambda = \mathbf{A}_{m,p}\lambda'$, where m and p are integers and $\mathbf{A}_{m,p}$ is the matrix

$$\begin{pmatrix} 0 & 0 & 1 & 1 \\ m-1 & m & 0 & 0 \\ p & p & p-1 & p \\ 1 & 1 & 1 & 1 \end{pmatrix}.$$

This induction process is then iterated.

Proposition 7.5.5 (Keane 1977) *For every infinite sequence of matrices* $\mathbf{A}_{m_k,p_k}$, *if P is the positive cone in $\mathbb{R}^4$, the set $\cap_{k \in \mathbb{N}} \mathbf{A}_{m_1,p_1} \cdots \mathbf{A}_{m_k,p_k} P$ is non-empty.*

Let E be this set normalised by $\lambda_1 + \cdots + \lambda_4 = 1$. For every $\lambda \in E$, the four-interval exchange $T_{\lambda,\pi}$ is such that, if we iterate k times the induction on the fourth interval, after renormalising, and reversing the order if k is even, we get the four-interval exchange $T_{\lambda'_{(k)},\pi}$, with $\lambda = \mathbf{A}_{m_1,p_1} \cdots \mathbf{A}_{m_k,p_k} \lambda'_{(k)}$.

The matrix part of (Keane 1977) will be stated in greater generality as Proposition 7.5.10; it shows that under mild conditions on the m_k and p_k, T is not uniquely ergodic. Minimality for this example can be realised through the i.d.o.c. condition, and also with the stronger requirement of total irrationality, which was not satisfied by Veech's examples.

Further examples of non-uniquely ergodic four-interval exchanges can be found in (Marmi, Moussa, and Yoccoz 2005) and

(Ferenczi and Zamboni 2008), and a generalisation of Keane's examples to r intervals has been done in (Yoccoz 2005) for any $r \geq 4$. In all cases, by natural coding we get uniformly recurrent words without uniform frequencies.

7.5.3 Adic words and languages

7.5.3.1 Adic systems in the symbolic framework

The *Bratteli–Vershik dynamical systems*, initially called *adic systems* by A. Vershik, are defined and studied at length in Chapter 6 of the present book. They do not fit into the framework of symbolic dynamics. Indeed, they are defined on what looks like a space of infinite words, but not as a shift, and some of them are not topologically isomorphic to a symbolic system since they are not *expansive* (see Section 6.5.3). For many of them, however, there is a standard way to code them into a well-defined subshift, see the discussion after Proposition 7.5.8, and we take this coded form as a (somewhat pedestrian) definition of what we call a *symbolic adic system*. More precisely we define an *adic infinite word* as a word whose language is generated by a finite number of families of words, build by recursive concatenation rules:

Definition 7.5.6 An infinite word x is *adic* if there exist finite words $B_{n,1}, \ldots, B_{n,k_n}$, for $n \in N$, such that $L(x)$ is the set of words w for which there exist n and i such that w is a factor of $B_{n,i}$, with the additional conditions

(i) $B_{0,i} = i$, $1 \leq i \leq k_0$,
(ii) for each $1 \leq i \leq k_n$, there exist an integer $t(n,i) > 0$, and $t(n,i)$ integers $1 \leq k_s(n,i) \leq k_{n-1}$ such that
$$B_{n,i} = \Pi_{s=1}^{t(n,i)} B_{n-1,k_s(n,i)},$$
(iii) for every p there exists $N(p)$ such that for every word w in $L(x)$ of length at least $N(p)$ and every decomposition
$$w = U\Pi_{j=1}^{r} B_{p,l_j} V$$
where U is a (proper, possibly empty) suffix of some B_{p,l_0} and V is a (proper, possibly empty) prefix of some $B_{p,l_{r+1}}$, U and the l_i, $1 \leq i \leq r$, depend only on w.

The nth matrix $\mathbf{M}_n(x)$ of the adic word is the matrix which has on its ith line, $1 \leq i \leq k_{n-1}$, and jth column, $1 \leq j \leq k_n$, the number of $1 \leq s \leq t(n,j)$ such that $k_s(n,j) = i$.

The notation B is for *block* which is used more by ergodicians than *word*, and the matrix counts the number of $B_{n-1,i}$ which appear in the defining formula for $B_{n,j}$.

The third condition in the definition is called a condition of *recognisability*; though it is cumbersome to write, it is generally easy to check.

We have already seen examples of adic words in Section 7.4.3, with those of the Grillenberger words which are built explicitly with blocks $B_{n,i}$. As has been stated, the examples of Section 7.5.2.2 fall into this category: it is an easy consequence of Proposition 7.5.5 that (with the notations of Section 7.5.2.2)

Proposition 7.5.7 *For any λ in E, the natural codings of $T_{\lambda,\pi}$ are adic words with $k_n = 4$ for all n and $\mathbf{M}_n = \mathbf{A}_{m_n, p_n}$.*

More generally, natural codings of interval exchange maps are adic words, with explicit constructions being given for some permutations in (Ferenczi, Holton, and Zamboni 2003) (Ferenczi and Zamboni 2009) (Ferenczi and Zamboni 2008); another construction can be deduced from Section 6.5.6.

Recent general results on Bratteli–Vershik systems compute all their invariant probability measures, see for example (Fisher 2009), (Bezuglyi, Kwiatkowski, Medynets, et al. 2009), and Section 6.8 with further references. We give here a simple particular case of these results, adapted to the needs of the present section, together with a sketch of a self-contained proof which does not need the whole Bratteli–Vershik machinery, but still may be skipped by readers who are more interested in word combinatorics than in dynamical systems.

Proposition 7.5.8 *Let x be an adic word with matrices $\mathbf{M}_n$, such that $k_n = k$ and $\det \mathbf{M}_n \neq 0$ for every n, and*

$$\lim_{n \to +\infty} \min\{\Sigma_{a \in C} a \mid C \text{ column of } \mathbf{M}_1 \cdots \mathbf{M}_n\} = +\infty.$$

If P is the positive cone in $\mathbb{R}^k$, each point μ in the set $\cap_{n \in \mathbb{N}} \mathbf{M}_1 \cdots \mathbf{M}_n P$, normalised by $\mu_1 + \cdots + \mu_k = 1$, defines an invariant probability measure on (X_x, S) such that $\mu[i] = \mu_i$; every invariant probability measure on (X_x, S) is of that form, and at most k of them are ergodic.

Proof The recognisability condition ensures that for every n, every infinite word y in X_x admits a unique infinite decomposition $U B_{n,l_1} B_{n,l_2} \cdots$; for $1 \leq i \leq k$ we define the set $F_{n,i}$ to be the set of $y \in X_x$ for which the suffix U is empty and $l_1 = i$. We check that for each given n, the $S^j F_{n,i}$, $1 \leq i \leq k$,

$0 \leq j \leq |B_{n,i}| - 1$ form a partition of X_x (which is indeed a *Kakutani–Rokhlin partition*, see Definition 6.4.1); these partitions are increasing (the atoms of the $(n+1)$th partition are subsets of atoms of the nth partition), and, except possibly for a countable number of points, two points belonging to the same atom of the nth partition for every n are the same (these infinite words coincide on arbitrarily long initial segments, as the condition on the columns of $M_1 \cdots M_n$ ensures that the $|B_{n,i}|$ go to infinity with n); the system (X_x, S) is indeed a Bratteli–Vershik dynamical system and moreover it is of *finite rank*, see (Ferenczi 1997). This is enough to ensure that a measure on X_x is determined by its values on the atoms of these partitions, thus, if it is S-invariant, by its values on the $F_{n,i}$.

It follows from the definitions that $F_{n-1,i}$ is a union of images by S of the $F_{n,j}$, $1 \leq j \leq k$, with an iterate of $F_{n,j}$ appearing in $F_{n-1,i}$ whenever some $k_s(n,i)$ is equal to j in the defining decomposition of $B_{n,i}$ thus, if $\rho_n = (\mu(F_{n,1}), \ldots, \mu(F_{n,k}))$, we get $\rho_{n-1} = M_n \rho_n$. Thus the measure μ is completely determined by the vector ρ_0, and the space of such vectors is of dimension at most k; we can define such a measure if, and only if, all the vectors ρ_n have positive coordinates; thus, after normalising, we get the claimed result. □

Let us mention that if we describe a Bratteli–Vershik dynamical system through a sequence of Kakutani–Rokhlin partitions as in Theorem 6.4.3, this gives immediately a symbolic system as in Definition 7.5.6, by putting the letter l at the jth place of the word $B_{n,i}$ whenever (with the notation of Theorem 6.4.3) $T^j B_i(n)$ falls into $B_i(0)$. Unfortunately, the symbolic system we get is not always isomorphic to the system defined from the KR-partitions, as happens when we start from the dyadic odometer (see Section 6.5.1), which is aperiodic but gives rise to an infinite word defined by $k_n = 1$, $B_0 = 0$, $B_{n+1} = B_n B_n$, which is just the periodic infinite word $0000\cdots$; the isomorphism does however work whenever the words we get satisfy the recognisability condition, which is generally the case – though obviously not for the B_n of the dyadic odometer.

7.5.3.2 Some families of examples

For an adic word, minimality can be ensured by mild properties of *primitivity* of the matrices:

Proposition 7.5.9 *Let x be an adic word with matrices $\mathbf{M}_n$, such that, for every n, there exists $m \geq n$ such that all the entries of $\mathbf{M}_n \cdots \mathbf{M}_m$ are positive; then x is uniformly recurrent.*

Proof Exercise 7.20. □

This gives a cornucopia of uniformly recurrent words without uniform frequencies, by ensuring there is more than one normalised element in $\cap_{n \in \mathbb{N}} M_1 \cdots M_n P$. And first we can state the result which is implicitly proved in the matrix part of (Keane 1977):

Proposition 7.5.10 *An adic word with matrices* $\mathbf{M}_n = \mathbf{A}_{m_n,p_n}$, *with* $p_1 \geq 9$ *and* $3(p_n + 1) \leq m_n \leq \frac{1}{2}(p_{n+1} + 1)$ *for all* n, *does not have uniform frequencies.*

Thus the non-unique ergodicity of Keane's examples comes only from the adic structure of the system and Proposition 7.5.8, which is proved in (Keane 1977) in this particular case (the general case uses basically the same reasoning). And we see from Propositions 7.5.7 and 7.5.8 that there is a duality between the lengths of the intervals and the values of the invariant measures on them: with the notations of Section 7.5.2.2, for any $\lambda \in E$, every invariant probability measure on $([0,1), T_{\lambda,\pi})$ is defined from a vector $\mu \in E$ by giving measure μ_i to the ith interval (it follows from the proofs of both Theorem 7.4.7 and Proposition 7.5.8 that the measures of the four initial intervals determine the measure μ completely). Under the above conditions on the (m_k, p_k), E is not reduced to a point but is a segment, whose two endpoints give the two ergodic invariant measures (note that here the adic structure predicts at most four ergodic invariant measures; but we have seen in Section 7.4.4 that, by Katok's result (Katok 1973), this can be reduced to two for a four-interval exchange). If we choose λ to be in the interior of this segment, these two ergodic measures are absolutely continuous with respect to the Lebesgue measure but different from it; if we choose λ to be an endpoint, one ergodic measure is the Lebesgue measure and the other one is singular; a recent work of J. Chaika (Chaika 2008) has proved that this singular measure can have a support of arbitrarily small Hausdorff dimension.

The non-unique ergodicity for the examples of (Ferenczi and Zamboni 2008) comes also directly from their adic definition.

The paper (Ferenczi, Fisher, and Talet 2009) is devoted to the building of examples with the lowest possible k_n.

Proposition 7.5.11 (Ferenczi, Fisher, and Talet 2009) *Any adic words with the following matrices* $\mathbf{M}_n$ *are uniformly recurrent without uniform frequencies:*

(i) $\begin{pmatrix} q_n & 1 \\ 1 & q_n \end{pmatrix}$ *if* $\Sigma_{n=0}^{+\infty} \frac{1}{q_n} < 1$.

(ii) $\begin{pmatrix} q_n & r_n \\ s_n & t_n \end{pmatrix}$ if $q_0 t_0 > r_0 s_0$, $\Sigma_{n=0}^{+\infty} \frac{q_n r_{n+1}}{r_n t_{n+1}} < +\infty$, $\Sigma_{n=0}^{+\infty} \frac{b_n s_{n+1}}{a_n q_{n+1}} < +\infty$,

where (a_n, b_n) is the first line of the matrix $\mathbf{M}_0 \cdots \mathbf{M}_n$.

(iii) $\begin{pmatrix} q_n & 1 & 0 \\ 1 & q_n & 1 \\ 0 & q_n & 1 \end{pmatrix}$ if $\Sigma_{n=0}^{+\infty} \frac{1}{q_n} < 1$.

(iv) $\begin{pmatrix} q_n & 0 & q_n - 1 \\ r_n & r_n - 1 & r_n \\ 1 & 1 & 1 \end{pmatrix}$ if $r_0 \geq 6$ and $3r_n + 1 \leq 2q_n \leq r_{n+1}$ for all $n > 0$.

The first examples provide probably the simplest words which may be built with uniform recurrence but without uniform frequencies. The last family of examples is an abstract version of Keane's examples of Section 7.5.2.2 with $k = 3$; note that they are not natural codings of three-interval exchanges, because of Proposition 7.4.6.

In the above examples, uniform recurrence and uniform frequencies are ensured by sufficient conditions on the matrix only; however, in general uniform recurrence and uniform frequencies depend on the actual recursion formulas giving the words $B_{n,i}$, see Exercise 7.18.

7.5.3.3 Complexity and the Cassaigne–Kaboré word

The complexity of an adic word depends on the actul recursion formulas; let us mention that if $k_n = k$ for all n the complexity is sub-exponential (the topological entropy is 0), but examples can be built with $\limsup \frac{p_x(n)}{f(n)} = +\infty$ for any given f with subexponential growth: such examples are built in Proposition 3 of (Ferenczi 1996) under a slightly different form; with the notations of that paper, those built for bounded K can be defined as adic words with $k_n = K + 1$ by putting $B_{n,i} = B_n 1^{i-1}$, $1 \leq i \leq K + 1$.

At the other end of the spectrum J. Cassaigne and I. Kaboré (Cassaigne and Kaboré 2009) have investigated the words without uniform frequencies with the lowest possible complexity function.

Proposition 7.5.12 (Cassaigne and Kaboré 2009) *There exists a uniformly recurrent adic word without uniform frequencies such that $p_x(n) \leq 3n + 1$ and $\liminf_{n \to +\infty} \frac{p_x(n)}{n} = 2$.*

This tends to show that the bounds in Theorems 7.3.2 and 7.3.3 are optimal.

The example itself uses sequences of numbers $0 < l_n < m_n < p_n$, with l_n tending to infinity, $\frac{p_n}{m_n}$ growing fast enough, and $\frac{m_n}{l_n}$ faster enough; for example $l_n = 2^{4+2^{n+1}}$, $m_n = 2^{2^{n+3}}$, $p_n = 2^{5 \cdot 2^{n+1}}$. Then, we define $B_{0,1} = 1$, $B_{0,2} = 2$, $B_{n+1,1} = B_{n,1}^{m_n} B_{n,2}^{l_n}$ and $B_{n+1,2} = B_{n,1}^{m_n} B_{n,2}^{p_n}$.

The infinite word x can be defined as the limit of the $B_{n,1}$ when $n \to +\infty$. Its complexity can be computed by using bispecial factors (see Section 4.5): we find that $p_x(n+1) - p_x(n)$ is 2 or 3, the value 2 being taken on intervals $[s_n, t_n]$ with $t_n \geq l_n s_n$, which proves the assertions. The uniform recurrence comes from Proposition 7.5.9, and the absence of uniform frequencies can be proved either by checking we are in the second class of examples in Proposition 7.5.11, or by imitating the simpler proof for the first class of examples in Proposition 7.5.11, or directly, see Exercise 7.19 below.

7.5.3.4 Earlier examples

X. Bressaud (unpublished) used the same formulas as J. Cassaigne and I. Kaboré but with l_n replaced by 1, m_n growing fast and p_n much faster than m_n; his word is also uniformly recurrent and without uniform frequencies.

A. Frid (CANT 2006, unpublished) suggested the following example: let x be the infinite word that is the limit of the finite words B_n defined by $B_0 = 1$ and $B_{n+1} = B_n \sigma(B_n)^{k_n}$, where $\sigma(w)$ denotes the word obtained by exchanging 2 and 1 in w.

If k_n grows sufficiently fast, then A. Frid showed that x is uniformly recurrent and does not have frequencies. More precisely (X_x, S) admits exactly two ergodic invariant measures. Indeed, this is an adic word if we put $B_{n,1} = B_n$, $B_{n,2} = \sigma(B_n)$; it looks different from the previous examples as the higher entries in the matrix are not on the diagonal, but x is also the limit of $B_{2n,1}$, and if we look only at the $B_{2n,1}$ and $B_{2n,2}$ we get an adic word whose matrix is of the same type as those in Proposition 7.5.11.

7.5.3.5 The Pascal-adic language

The reference for this whole section is (Méla and Petersen 2005). As we noticed in the proof of Proposition 7.5.8, all the adic examples of Section 7.5.3, which are built with a constant k_n, could also be described with the older notion of *finite rank* (Ferenczi 1997), and the adic terminology may seem to be just a more fashionable presentation; however, this terminology comes into its own when k_n is unbounded, and the following very interesting example could not have been defined within the framework of earlier notions. Note that here the dynamical system is not defined by one infinite word, but by a language; but this slight generalisation does not change anything to the properties and techniques involved.

Definition 7.5.13 The *Pascal-adic language* is the set L of words w for which there exist n and i such that w is a factor of $B_{n,i}$, where the $B_{n,i}$, $1 \leq i \leq n+2$, are defined by $B_{0,i} = i$, $1 \leq i \leq 2$; for every n, $B_{n,1} = B_{n-1,1}$, $B_{n,n+2} = B_{n-1,n+1}$ and for $2 \leq j \leq n+1$

$$B_{n,j} = B_{n-1,j-1} B_{n-1,j}.$$

The *Pascal-adic system* is the subshift (X_L, S), where X_L is the set of all $x \in \{1,2\}^{\mathbb{N}}$ such that $L(x) \subseteq L$ (see Proposition 7.1.2).

This language is related to the Pascal triangle, as the length of $B_{n,i}$ is the binomial coefficient $\binom{n+1}{i-1}$, and the blocks are built in the same way as these coefficients are built along the Pascal triangle.

The system (X_L, S) is not minimal, as $B_{n,2} = 1^n 2$ for every $n \geq 1$ so $1^\omega \in X_L$. The following result states that it has infinitely many ergodic invariant probability measures.

Theorem 7.5.14 (Méla and Petersen 2005) *For every word w in L, for every real number $0 \leq \alpha \leq 1$, there exists $f_\alpha(w)$ such that for any sequence $k_n \to +\infty$ with $\lim_{n \to +\infty} \frac{k_n}{n} = \alpha$,*

$$\lim_{n \to +\infty} \frac{|B_{n,k_n}|_w}{|B_{n,k_n}|} = f_\alpha(w).$$

Moreover, all the ergodic invariant probability measures on (X_L, S) are the measures μ_α given on the cylinders by $\mu_\alpha([w]) = f_\alpha(w)$.

Thus, this is a very interesting intermediate case where the invariant measures are known (and there are not too many of them), and there are what we could call directional frequencies. As for the complexity, it satisfies $\lim_{n \to +\infty} \frac{p(n)}{n^3} = \frac{1}{6}$.

7.6 Further afield

In spite of the Jewett–Krieger theorem, uniquely ergodic systems may be considered to represent a small and particularly well-behaved class of dynamical systems; however, even more complicated systems can present properties which generalise directly the notion of unique ergodicity: instead of hoping that there will be one invariant measure, we know some explicit invariant measures, and what we want to prove is that these constitute the whole set of invariant measures. For example, a measure-theoretic system (X, T, μ) has *minimal self-joinings* if every ergodic measure on $X \times X$, invariant by $T \times T$, and whose marginals on X are μ, is either the product measure $\mu \times \mu$, or a diagonal measure defined by $\nu(A \times B) = \mu(A \cap T^k B)$ for some k. This notion was defined by

D. J. Rudolph in (Rudolph 1979). The standard example of a system with minimal self-joinings is *Chacon's map*, the symbolic system associated with the Chacon word of Section 1.4. Moreover, the (mostly combinatorial) techniques used in (del Junco, Rahe, and Swanson 1980) to prove this result were generalised in a famous series of papers by M. Ratner, starting from the joinings of *horocycle flows* (Ratner 1983) and culminating in the proof of *Raghunathan's conjecture* on unipotent groups (Ratner 1991).

The results on interval-exchange maps can generally be interpreted into results on more geometrical systems; in particular, Boshernitzan's ne_n criterion is not only a tool in proving Theorem 7.4.7 above, but also a way of knowing that a given interval exchange is uniquely ergodic; thus it is proved by W. Veech (Veech 1999) to imply an earlier criterion of H. Masur (Masur 1992) for the unique ergodicity of some foliations on Riemann surfaces of genus at least two equipped with a holomorphic 1-form. Then he uses this criterion to precise the knowledge of invariant measures for another famous flow, the Teichmüller geodesic flow.

7.7 Exercises

Section 7.1

Exercise 7.1 Prove that an infinite word x is eventually periodic if, and only if, X_x is finite. Prove that x is periodic if, and only if, X_x is finite and $S|_{X_x} : X_x \to X_x$ is onto.

Exercise 7.2 Consider the words $x = 0101101^3 01^4 01^5 \cdots$ and $y = 0100110^3 1^3 0^4 1^4 \cdots$. Describe the subshifts X_x and X_y and the subshifts they contain. Which ones are minimal? Make a picture describing the action of S on X_x and X_y.

Exercise 7.3 Prove that the set of cylinders of a subshift (X, S) is equal to the set of closed balls (any centre, any radius) of X for the distance defined in 1.2.10. Prove that it is also equal to the set of open balls of X for this distance. Prove that any element of a ball is a centre of it.

Exercise 7.4 Let (X, S) be a subshift. Prove that

$$X = \bigcap_{n \in \mathbb{N}} \bigcup_{u \in L_n(X)} [u],$$

where $[u]$ is considered as a cylinder in the dynamical system $A^{\mathbb{N}}$.

Exercise 7.5 Using exercise 1.8, prove directly that the Chacon word has uniform frequencies.
Hint. Show that $f_w(x) = \lim_{n \to +\infty} \frac{|b_n|_w}{|b_n|}$.

Section 7.2

Exercise 7.6 Rephrase the definition of a weight function in terms of Rauzy graphs.

Exercise 7.7 Give an example of a non-minimal non-uniquely ergodic subshift (X, S) such that any infinite word x in X has uniform frequencies.

Exercise 7.8 Let (X, S) be a subshift, and n be a positive integer. Assume that there exist n elements $x_1, \ldots, x_n$ of X and n elements $w_1, \ldots, w_n$ of $L(X)$ such that, for any $i, j \leq n$ satisfying $i \neq j$, we have $f_{w_i}(x_j) > 0$ and $f_{w_i}(x_i) = 0$. Prove that (X, S) admits at least n ergodic invariant measures.

Exercise 7.9 Let (X, S) be a subshift, such that for any letter $a \in A$, there exists a word $x \in X$ such that $\limsup_{n \to \infty} \frac{|x_0 \cdots x_n|_a}{n} > 1/2$. Prove that (X, S) admits at least $\mathrm{Card}(A)$ ergodic invariant measures.

Exercise 7.10 Let $(a_n)_{n \in \mathbb{N}}$ be an integer sequence which grows sufficiently fast. Let x be the word $0^{a_0} 1^{a_1} 0^{a_2} 1^{a_3} 0^{a_4} 1^{a_5} 0^{a_6} 1^{a_7} \cdots$. Show that x does not have frequencies, but all of the minimal subshifts contained in X_x are uniquely ergodic.

Section 7.3

Exercise 7.11 For any element v of A^+, let us define the measure $\mu_v = \frac{1}{|v|} \sum_{k=0}^{|v|-1} \delta_{S^k(v^\omega)}$. Prove that the set $\{\mu_v \mid v \in A^+\}$ is a subset of $\mathcal{E}(A^\mathbb{N}, S)$ which is dense in $\mathcal{M}(A^\mathbb{N}, S)$. In particular, $\overline{\mathcal{E}(A^\mathbb{N}, S)} = \mathcal{M}(A^\mathbb{N}, S)$.
Hint. To approximate an invariant measure μ in $\mathcal{M}(A^\mathbb{N}, S)$, let n be a big integer, let $\{u_1, \ldots, u_k\}$ be an enumeration of $L_n(A^\mathbb{N}) = A^n$, and consider the word $v = u_1^{p_1} \cdots u_k^{p_k}$, where p_i/q is a rational approximation of $\mu([u_i])$.

Section 7.4

Exercise 7.12 An infinite word x is said to be *episturmian* if $L(x)$ is closed under reversal and has at most one right special factor of each length (Droubay, Justin, and Pirillo 2001). Prove that an episturmian infinite word has uniform frequencies.

Exercise 7.13 Let x be an infinite word on A. A finite word $q \in A^*$ is said to be a *quasiperiod* of x if x is covered by the occurrences of q (in particular, q is a prefix of x). An infinite word x is said to be *multi-scale quasiperiodic* if it admits infinitely many quasiperiods (Marcus and Monteil 2006). Prove that a multi-scale quasiperiodic infinite word has uniform frequencies.

Exercise 7.14 An infinite word x is said to be *linearly recurrent* if $\exists K \geq 0$, $\forall u \in L(x)$, $\forall v \in L_{K|u|}(x)$, $|v|_u \geq 1$ (Durand 2000) (see also Section 6.5.3). Prove that a linearly recurrent infinite word has uniform frequencies. Prove that a fixed point of a primitive substitution is linearly recurrent. Prove that an infinite word x is linearly recurrent if, and only if, there exists μ in $\mathcal{M}(X_x, S)$ such that $\liminf_{n \to \infty} n e_n(\mu) > 0$.

Hint. If $v = v_0 \cdots v_n$ is a long element of $L(x)$ having no occurrence of u, assume that $vu \in L(x)$ and compute the measure of $\cup_{i \leq n} [v_i v_{i+1} \cdots v_n u]$ (this trick is due to M. Boshernitzan).

Exercise 7.15 Give an example of a minimal subshift of linear complexity whose invariant measures are not determined by their values on the cylinders of length 1.

Hint. Consider the substitution $0 \mapsto 0011$, $1 \mapsto 0101$.

Exercise 7.16 Give an example of an infinite word x such that, for any k, $f_{0^k}(x) = 1$ and such that X_x has infinitely many ergodic invariant measures. Is it possible to construct a uniformly recurrent example?

Section 7.5

Exercise 7.17 Let $\mu = \mu_\pi$ be the Borel measure defined on the full-shift $(A^\mathbb{N}, S)$ as in the proof of Proposition 7.5.1. Prove that, for any two finite words u and v and for any $n \geq |u|$, we have $\mu([u] \cap S^{-n}([v])) = \mu([u])\mu([v])$. Since μ is outer-regular, we know that for any Borel subset B of $A^\mathbb{N}$ and for any $\varepsilon > 0$, there exist some finite words $w_1, \ldots, w_\ell$ such that B is included in the disjoint union $\sqcup_{i=1}^\ell [w_i]$ and such that $\mu(\sqcup_{i=1}^\ell [w_i]) \leq \mu(B) + \varepsilon$. Prove that, for any two Borel subsets B and C of $A^\mathbb{N}$, we have $\lim_{n \to \infty} \mu(B \cap S^{-n}(C)) = \mu(B)\mu(C)$ (the measure-theoretic dynamical systems satisfying this property are said to be *strongly mixing*). Prove that, for any S-invariant Borel subset B of $A^\mathbb{N}$, we have $\mu(B) = \mu(B)^2$. Conclude that the system $(A^\mathbb{N}, \mathcal{B}, \mu_\pi, S)$ is ergodic.

Exercise 7.18 Let x denote the fixed point of the substitution $a \mapsto aaab$, $b \mapsto b$ which begins with the letter a. Show that X_x is neither minimal nor uniquely ergodic. Show that both x and Chacon's word can be described in the adic framework with the same (constant) sequence of matrices.

Exercise 7.19 Let $B_{n,1}$ and $B_{n,2}$ be as in Section 7.5.3.3. Show that $1 \leq \frac{|B_{n,2}|}{|B_{n,1}|} \leq \frac{p_{n-1}}{l_{n-1}}$. Show that $\frac{|B_{n+1,2}|_2}{|B_{n+1,2}|} \geq \frac{p_n}{m_n + p_n} \geq \frac{|B_{n,2}|_2}{|B_{n,2}|}$. Show that $\frac{|B_{n+1,1}|_1}{|B_{n+1,1}|} \geq \frac{m_n}{m_n + l_n \frac{p_{n-1}}{l_{n-1}}} \frac{|B_{n,1}|_1}{|B_{n,1}|}$. Use infinite products to give lower bounds for $\frac{|B_{n,2}|_2}{|B_{n,2}|}$ and $\frac{|B_{n,1}|_1}{|B_{n,1}|}$. Show that $\frac{|B_{n,2}|_2}{|B_{n,2}|} + \frac{|B_{n,1}|_1}{|B_{n,1}|} \geq \frac{3}{2}$ for good choices of l_n, m_n, p_n. Conclude that this contradicts the existence of uniform frequencies.

Exercise 7.20 Prove Proposition 7.5.9.

7.8 Note: Dictionary between word combinatorics and symbolic dynamics

combinatorics on words	symbolic dynamics
infinite word x	subshift (X_x, S)
factorial and extendable language L	subshift (X, S)
finite word w	cylinder $[w]$
uniform recurrence	minimality
weight function	invariant measure
uniform frequencies	unique ergodicity
positive uniform frequencies	strict ergodicity

8
Transcendence and Diophantine approximation

Boris Adamczewski,

Yann Bugeaud

The aim of this chapter is to present several number-theoretic problems that reveal a fruitful interplay between combinatorics on words and Diophantine approximation. Finite and infinite words occur naturally in Diophantine approximation when we consider the expansion of a real number in an integer base b or its continued fraction expansion. Conversely, with an infinite word a on the finite alphabet $\{0, 1, \ldots, b-1\}$ we associate the real number ξ_a whose base-b expansion is given by a. As well, with an infinite word a on the infinite alphabet $\{1, 2, 3, \ldots\}$, we associate the real number ζ_a whose continued fraction expansion is given by a. It turns out that, if the word a enjoys certain combinatorial properties involving repetitive or symmetric patterns, then this gives interesting information on the arithmetical nature and on the Diophantine properties of the real numbers ξ_a and ζ_a.

We illustrate our results by considering the real numbers associated with two classical infinite words, the *Thue–Morse word* and the *Fibonacci word*, see Example 1.2.21 and 1.2.22. There are several ways to define them. Here, we emphasize the fact that they are fixed points of morphisms.

Consider the morphism σ defined on the set of words on the alphabet $\{0, 1\}$ by $\sigma(0) = 01$ and $\sigma(1) = 0$. Then, we have $\sigma^2(0) = 010, \sigma^3(0) = 01001, \sigma^4(0) = 01001010$, and the sequence $(\sigma^k(0))_{k \geq 0}$ converges to the Fibonacci word

$$f = 010010100100101001010\cdots. \tag{8.1}$$

Consider the morphism τ defined over the same alphabet by $\tau(0) = 01$ and $\tau(1) = 10$. Then, we have $\tau^2(0) = 0110, \tau^3(0) = 01101001$, and the sequence $(\tau^k(0))_{k \geq 0}$ converges to the Thue–Morse word

$$t = 0110100110010110100010\cdots. \tag{8.2}$$

Combinatorics, Automata and Number Theory, ed. Valérie Berthé and Michel Rigo.
Published by Cambridge University Press. ©Cambridge University Press 2010.

For every $n \geq 1$, we denote by f_n the nth letter of f and by t_n the nth letter of t. For an integer $b \geq 2$, we set

$$\xi_f = \sum_{n \geq 1} \frac{f_n}{b^n}$$

and

$$\xi_t = \sum_{n \geq 1} \frac{t_n}{b^n}.$$

We further define Fibonacci and Thue–Morse continued fractions, but, since 0 cannot be a partial quotient, we have to write our words on an alphabet other than $\{0, 1\}$. We take two distinct positive integers a and b, set $f'_n = a$ if $f_n = 0$ and $f'_n = b$ otherwise, and $t'_n = a$ if $t_n = 0$ and $t'_n = b$ otherwise. Then, we define

$$\zeta_{f'} = [a, b, a, a, b, a, b, a, \ldots] = [f'_1, f'_2, f'_3, f'_4, \ldots]$$

and

$$\zeta_{t'} = [a, b, b, a, b, a, a, b, \ldots] = [t'_1, t'_2, t'_3, t'_4, \ldots].$$

Among other results, we will explain how to combine combinatorial properties of the Fibonacci and Thue–Morse words with the Thue–Siegel–Roth–Schmidt method to prove that all these numbers are transcendental. Beyond transcendence, we will show that the Fibonacci continued fractions satisfy spectacular properties regarding a classical problem in Diophantine approximation: the existence of real numbers ξ with the property that ξ and ξ^2 are uniformly simultaneously very well approximable by rational numbers of the same denominator. We will also describe an *ad hoc* construction to obtain explicit examples of pairs of real numbers that satisfy the Littlewood conjecture, which is a major open problem in simultaneous Diophantine approximation.

We use the following convention throughout this chapter. The Greek letter ξ stands for a real number given by its base-b expansion, where b always means an integer at least equal to 2. The Greek letter ζ stands for a real number given by its continued fraction expansion. If its partial quotients take only two different values, these are denoted by a and b, which represent distinct positive integers (here, b is not assumed to be at least 2).

The results presented in this chapter are not the most general statements that can be established by the methods described here. Our goal is not to make an exhaustive review of the state-of-the-art, but rather to emphasize the ideas used. The interested reader is directed to the original papers.

8.1 The expansion of algebraic numbers in an integer base

Throughout the present section, b always denotes an integer at least equal to 2 and ξ is a real number with $0 < \xi < 1$. Recall that there exists a unique infinite word $a = a_1 a_2 \cdots$ defined over the finite set $\{0, 1, \ldots, b-1\}$, called the *base-b expansion* of ξ, such that

$$\xi = \sum_{n \geq 1} \frac{a_n}{b^n} := 0.a_1 a_2 \cdots, \tag{8.3}$$

with the additional condition that a does not terminate in an infinite string of the digit $b - 1$. Obviously, a depends on ξ and b, but we choose not to indicate this dependence.

For instance, in base 10, we have

$$3/7 = 0.(428\,571)^\omega$$

and

$$\pi - 3 = 0.314\,159\,265\,358\,979\,323\,846\,264\,338\,327\cdots.$$

Conversely, if $a = a_1 a_2 \cdots$ is an infinite word defined over the finite alphabet $\{0, 1, \ldots, b-1\}$ such that a does not terminate in an infinite string of the digit $b-1$, there exists a unique real number, denoted by ξ_a, such that

$$\xi_a := 0.a_1 a_2 \cdots.$$

This notation does not indicate in which base ξ_a is written. However, this will be clear from the context and should not cause any difficulty.

In the sequel, we will also sometimes make a slight abuse of notation and, given an infinite word a defined over the finite alphabet $\{0, 1, \ldots, b-1\}$ that could end in an infinite string of $b-1$, we will write

$$0.a_1 a_2 \cdots$$

to denote the infinite sum

$$\sum_{n \geq 1} \frac{a_n}{b^n}.$$

We recall the following fundamental result that can be found in the classical textbook (Hardy and Wright 1985).

Theorem 8.1.1 *A real number is rational if, and only if, its base-b expansion is eventually periodic.*

8.1.1 Normal numbers and algebraic numbers

At the beginning of the twentieth century, Émile Borel (Borel 1909) investigated the following question:

What does the decimal expansion of a randomly chosen real number look like?

This question leads to the notion of normality.

Definition 8.1.2 A real number ξ is called *normal to base b* if, for any positive integer n, each one of the b^n words of length n on the alphabet $\{0, 1, \ldots, b-1\}$ occurs in the base-b expansion of ξ with the same frequency $1/b^n$. A real number is called a *normal number* if it is normal to every integer base.

É. Borel (Borel 1909) proved the following fundamental result regarding normal numbers.

Theorem 8.1.3 *The set of normal numbers has full Lebesgue measure.*

Some explicit examples of real numbers that are normal to a given base have been known for a long time. For instance, the number

$$0.123\,456\,789\,101\,112\,131\,415\cdots, \qquad (8.4)$$

whose sequence of digits is the concatenation of the sequence of all positive integers written in base 10 and ranged in increasing order, was proved to be normal to base 10 in 1933 by D. G. Champernowne (Champernowne 1933).

In contrast, to decide whether a specific number, like e, π or

$$\sqrt{2} = 1.414\,213\,562\,373\,095\,048\,801\,688\,724\,209\cdots,$$

is or is not a normal number remains a challenging open problem. In this direction, the following conjecture is widely believed to be true.

Conjecture 8.1.4 *Every real irrational algebraic number is a normal number.*

8.1.2 Complexity of real numbers

Conjecture 8.1.4 is reputed to be out of reach. We will thus focus our attention to simpler questions. A natural way to measure the *complexity* of the real number ξ (with respect to the base b) is to count the number of distinct blocks of given length in the infinite word a defined in (8.3). For

$n \geq 1$, we set $p(n, \xi, b) = p_a(n)$ with a as above and where p_a denotes the complexity function of a. Then, we have

$$1 \leq p(n, \xi, b) \leq b^n,$$

and both inequalities are sharp (take *e.g.*, the analogue in base b of the number given in (8.4) to show that the right-hand inequality is sharp). A weaker conjecture than Conjecture 8.1.4 reads then as follows.

Conjecture 8.1.5 For every real irrational algebraic number ξ, every positive integer n and every base b, we have $p(n, \xi, b) = b^n$.

Our aim is to prove the following lower bound for the complexity of algebraic irrational numbers, as in (Adamczewski and Bugeaud 2007a), (Adamczewski, Bugeaud, and Luca 2004).

Theorem 8.1.6 *If ξ is an algebraic irrational number, then*

$$\lim_{n \to +\infty} \frac{p(n, \xi, b)}{n} = +\infty.$$

Up to now, Theorem 8.1.6 is the main achievement regarding Conjecture 8.1.5. A notable consequence of this result is the confirmation of a conjecture suggested by A. Cobham in 1968 (Cobham 1968).

Theorem 8.1.7 *The base-b expansion of an algebraic irrational number cannot be generated by a finite automaton.*

Indeed, a classical result of A. Cobham (Cobham 1972) asserts that every infinite word a that can be generated by a finite automaton has a complexity function p_a satisfying $p_a(n) = O(n)$.

Nevertheless, we are still very far away from what is expected, and we are still unable to confirm the following conjecture.

Conjecture 8.1.8 For every algebraic irrational number ξ and every base b with $b \geq 3$, we have $p(1, \xi, b) \geq 3$.

8.1.3 Transcendence and Diophantine approximation: an introduction

In order to prove Theorem 8.1.6, we have to show that irrational real numbers whose base-b expansion has a too low complexity are transcendental.

Throughout this section, we recall several classical results concerning the rational approximation of algebraic irrational real numbers (Liouville's inequality, Roth's theorem, Ridout's theorem) and derive from them several combinatorial transcendence criteria concerning real numbers defined by their base-b expansion. We also apply them to concrete examples.

8.1.3.1 Liouville's inequality

In 1844, J. Liouville (Liouville 1844) was the first to prove that transcendental numbers do exist. Moreover, he constructed explicit examples of such numbers. The numbers $\mathcal{L}_b$ below are usually considered as the first examples of transcendental numbers. This is however not entirely true, since the main part of Liouville's paper is devoted to the construction of transcendental continued fractions.

Theorem 8.1.9 *For every integer $b \geq 2$, the real number*

$$\mathcal{L}_b := \sum_{n=1}^{+\infty} \frac{1}{b^{n!}}$$

is transcendental.

The proof of Theorem 8.1.9 relies on the famous Liouville's inequality recalled below.

Proposition 8.1.10 *Let ξ be an algebraic number of degree $d \geq 2$. Then, there exists a positive real number c_ξ such that*

$$\left| \xi - \frac{p}{q} \right| \geq \frac{c_\xi}{q^d}$$

for every rational number p/q with $q \geq 1$.

Proof Let P denote the minimal defining polynomial of ξ and set

$$c_\xi = 1/(1 + \max_{|\xi - x| < 1} |P'(x)|).$$

If $|\xi - p/q| \geq 1$, then our choice of c_ξ ensures that $|\xi - p/q| \geq c_\xi/q^d$.

Let us now assume that $|\xi - p/q| < 1$. Since P is the minimal polynomial of ξ, it does not vanish at p/q and $q^d P(p/q)$ is a non-zero integer. Consequently,

$$|P(p/q)| \geq \frac{1}{q^d}.$$

Since $|\xi - p/q| < 1$, Rolle's theorem implies the existence of a real number t in $[p/q - 1, p/q + 1]$ such that

$$|P(p/q)| = |P(\xi) - P(p/q)| = \left| \xi - \frac{p}{q} \right| \times |P'(t)|.$$

We thus have
$$\left|\xi - \frac{p}{q}\right| \geq \frac{c_\xi}{q^d},$$
which ends the proof. □

Proof [Proof of Theorem 8.1.9] Let $d \geq 2$ be an integer. Let j be a positive integer with $j \geq d$ and set
$$\frac{p_j}{b^{j!}} := \sum_{n=1}^{j} \frac{1}{b^{n!}}.$$

Observe that
$$\left|\mathcal{L}_b - \frac{p_j}{b^{j!}}\right| = \sum_{n>j} \frac{1}{b^{n!}} < \frac{2}{b^{(j+1)!}} < \frac{1}{(b^{j!})^d}.$$

It then follows from Proposition 8.1.10 that $\mathcal{L}_b$ cannot be algebraic of degree less than d. Since d is arbitrary, $\mathcal{L}_b$ is transcendental. □

8.1.3.2 Roth's theorem

The following famous improvement of Liouville's inequality was established by K. F. Roth (Roth 1955).

Theorem 8.1.11 *Let ξ be a real algebraic number and ε be a positive real number. Then, there are only a finite number of rationals p/q such that $q \geq 1$ and*
$$0 < \left|\xi - \frac{p}{q}\right| < \frac{1}{q^{2+\varepsilon}}.$$

We give a first application of Roth's theorem to transcendence.

Corollary 8.1.12 *For any integer $b \geq 2$, the real number*
$$\sum_{n=1}^{+\infty} \frac{1}{b^{3^n}}$$
is transcendental.

Proof Use the same argument as in the proof of Theorem 8.1.9. □

Actually, the same proof shows that the real number
$$\sum_{n=1}^{+\infty} \frac{1}{b^{\lfloor d^n \rfloor}}$$

Transcendence and Diophantine approximation 417

is transcendental for any real number $d > 2$, but gives no information on the arithmetical nature of the real number

$$\sum_{n=1}^{+\infty} \frac{1}{b^{2^n}},$$

which will be considered in Corollary 8.1.18 below.

Up to now, we have just truncated the infinite series to construct our very good rational approximants, taking advantage of the existence of very long strings of 0 in the base-b expansion of the real numbers involved. We describe below a (slightly) more involved consequence of Roth's theorem. Roughly speaking, instead of just truncating, we truncate and complete by periodicity.

We consider the Fibonacci word f defined at the beginning of this chapter. Unlike the base-b expansions of the number $\mathcal{L}_b$ (defined in Theorem 8.1.9) and of the number defined in Corollary 8.1.12, the word f contains no occurrence of more than two consecutive 0's. However, its combinatorial structure can be used to reveal more hidden good rational approximations by means of which we will derive the following result.

Theorem 8.1.13 *The real number*

$$\xi_f := \sum_{n \geq 1} \frac{f_n}{b^n} = 0.010\,010\,100\,100\,101\,001\,010\cdots.$$

is transcendental.

Before proving Theorem 8.1.13, we need the following result. Let $(F_j)_{j \geq 0}$ denote the Fibonacci sequence, that is, the sequence starting with $F_0 = 0$, $F_1 = 1$, and satisfying the recurrence relation $F_{j+2} = F_{j+1} + F_j$, for every $j \geq 0$.

Lemma 8.1.14 *For every integer $j \geq 4$, the finite word*

$$f_1 f_2 \cdots f_{F_j} f_1 f_2 \cdots f_{F_j} f_1 f_2 \cdots f_{F_{j-1}-2}$$

is a prefix of f.

Proof For every integer $j \geq 2$, set $w_j := f_1 f_2 \cdots f_{F_j}$. Then we recall the following fundamental relation:

$$w_{j+1} = w_j w_{j-1}, \tag{8.5}$$

valid for $j \geq 3$. If a finite word $u := u_1 u_2 \cdots u_r$ has length larger than 2, we set $h(u) := u_1 u_2 \cdots u_{r-2}$.

We now prove by induction that, for every integer $j \geq 3$,

$$h(w_j w_{j-1}) = h(w_{j-1} w_j). \tag{8.6}$$

For $j = 3$ the result follows from the two obvious equalities $w_3 w_2 = aba$ and $w_2 w_3 = aab$. Let us assume now that Equality (8.6) holds for an integer $j \geq 3$. We infer from (8.5) that

$$h(w_{j+1} w_j) = h(w_j w_{j-1} w_j) = w_j h(w_{j-1} w_j).$$

By assumption, we get that

$$h(w_{j-1} w_j) = h(w_j w_{j-1}).$$

Using again Equation (8.5) we obtain that

$$h(w_{j+1} w_j) = w_j h(w_j w_{j-1}) = w_j h(w_{j+1}) = h(w_j w_{j+1}),$$

as claimed.

We now end the proof of the lemma. Let $j \geq 4$ and note that by definition $h(w_{j+2})$ is a prefix of f. On the other hand, we infer from (8.5) and (8.6) that

$$h(w_{j+2}) = h(w_{j+1} w_j) = h(w_j w_{j-1} w_j) = h(w_j w_j w_{j-1}),$$

which ends the proof, since w_{j-1} has at least two letters. □

We are now ready to prove Theorem 8.1.13.

Proof For every integer $j \geq 4$, let us consider the rational number ρ_j defined by

$$\rho_j := 0.(f_1 f_2 \cdots f_{F_j})^\omega.$$

Thus,

$$\rho_j = \frac{f_1}{b} + \frac{f_2}{b^2} + \cdots + \frac{f_{F_j}}{b^{F_j}} + \frac{f_1}{b^{F_j+1}} + \frac{f_2}{b^{F_j+2}} + \cdots + \frac{f_{F_j}}{b^{2F_j}} + \cdots$$

$$= \left(\frac{f_1}{b} + \frac{f_2}{b^2} + \cdots + \frac{f_{F_j}}{b^{F_j}}\right) \frac{b^{F_j}}{b^{F_j} - 1}$$

and there exists an integer p_j such that

$$\rho_j = \frac{p_j}{b^{F_j} - 1}.$$

Now, Lemma 8.1.14 tells us that ρ_j is a very good approximation to ξ_f. Indeed, the first $F_{j+2} - 2$ digits in the base-b expansion of ξ_f and of ρ_j are the same. Furthermore, ρ_j is distinct from ξ_f since the latter is irrational as a consequence of Theorem 8.1.1 and the fact that f is aperiodic. We thus obtain

$$0 < |\xi_f - \rho_j| < \sum_{h \geq -1} b^{-F_{j+2}-h} \leq b^{-F_{j+2}+2}. \tag{8.7}$$

On the other hand, an easy induction shows that $F_{j+2} \geq 1.5\,F_{j+1}$ for every $j \geq 2$. Consequently, we infer from (8.7) that

$$0 < \left|\xi_f - \frac{p_j}{b^{F_j} - 1}\right| < \frac{b^2}{(b^{F_j} - 1)^{2.25}},$$

and it follows from Roth's theorem that ξ_f is transcendental. □

8.1.3.3 Repetitions in infinite words

The key observation in the previous proof is that the infinite word f begins in infinitely many 'more-than-squares'. Let us now formalise what has been done above. For any positive integer ℓ, we write u^ℓ for the word

$$\underbrace{u \cdots u}_{\ell \text{ times}}$$

(ℓ times repeated concatenation of the word u). More generally, for any positive real number x, we denote by u^x the word $u^{\lfloor x \rfloor}u'$, where u' is the prefix of u of length $\lceil (x - \lfloor x \rfloor)|U|\rceil$. We recall that $\lfloor y \rfloor$ and $\lceil y \rceil$ denote the floor and ceiling functions. A *repetition* of the form u^x, with $x > 2$, is called an *overlap*. A repetition of the form u^x, with $x > 1$, is called a *stammering*. For instance the word $ababab = (ab)^3$ is a cube that contains the overlap $ababa = (ab)^{2+1/2}$. The word $1234567891 = (123456789)^{1+1/9}$ is a stammering which is overlap-free. For more on repetitions, see also Section 11.2.2.

Definition 8.1.15 Let $w > 1$ and $c \geq 0$ be real numbers. We say that an infinite word a defined over a finite or an infinite alphabet satisfies Condition $(*)_{w,c}$ if a is not eventually periodic and if there exist two sequences of finite words $(u_j)_{j \geq 1}$, $(v_j)_{j \geq 1}$ such that:

(i) For every $j \geq 1$, the word $u_j v_j^w$ is a prefix of a.

(ii) The sequence $(|u_j|/|v_j|)_{j \geq 1}$ is bounded from above by c.

(iii) The sequence $(|v_j|)_{j \geq 1}$ is strictly increasing.

Let a be an infinite word defined over the alphabet $\{0, 1, \ldots, b-1\}$ and satisfying Condition $(*)_{w,c}$ for some w and c. By definition of Condition

$(*)_{w,c}$, we stress that

$$\xi_a = 0.u_j \underbrace{v_j \cdots v_j}_{\lfloor w \rfloor \text{ times}} v'_j \cdots ,$$

for every $j \geq 1$, where v'_j is the prefix of v_j of length $\lceil (w - \lfloor w \rfloor)|v_j| \rceil$.

We derived above the transcendence of ξ_f by proving that the Fibonacci word f satisfies Condition $(*)_{2.25,0}$. Actually, the proof of Theorem 8.1.13 leads to the following combinatorial transcendence criterion.

Proposition 8.1.16 *If an infinite word a defined over the finite alphabet $\{0, 1, \ldots, b - 1\}$ satisfies Condition $(*)_{w,0}$ for some $w > 2$, then the real number ξ_a is transcendental.*

We leave the proof to the reader.

8.1.3.4 A p-adic Roth theorem

A disadvantage of the use of Roth's theorem in this context is that we need, in order to apply Proposition 8.1.16, that the repetitions occur at the very beginning (otherwise, we would have to assume that w is much larger than 2). We present here an idea of S. Ferenczi and C. Mauduit (Ferenczi and Mauduit 1997) to get a stronger transcendence criterion. It relies on the following non-Archimedean extension of Roth's theorem proved by D. Ridout (Ridout 1957).

In the sequel of this chapter, for every prime number ℓ, the ℓ-adic absolute value $|\cdot|_\ell$ is normalised such that $|\ell|_\ell = \ell^{-1}$.

Theorem 8.1.17 *Let ξ be an algebraic number and ε be a positive real number. Let S be a finite set of distinct prime numbers. Then there are only a finite number of rationals p/q such that $q \geq 1$ and*

$$\left(\prod_{\ell \in S} |p|_\ell \cdot |q|_\ell \right) \cdot \left| \xi - \frac{p}{q} \right| < \frac{1}{q^{2+\varepsilon}}.$$

We point out a first consequence of Ridout's theorem.

Corollary 8.1.18 *For every integer $b \geq 2$, the real number*

$$\mathcal{K}_b := \sum_{n=1}^{+\infty} \frac{1}{b^{2^n}}$$

is transcendental.

Proof Let j be a positive integer and set

$$\rho_j := \sum_{n=1}^{j} \frac{1}{b^{2^n}}.$$

There exists an integer p_j such that $\rho_j = p_j/q_j$ with $q_j = b^{2^j}$. Observe that

$$|\mathcal{K}_b - \rho_j| = \sum_{n>j} \frac{1}{b^{2^n}} < \frac{2}{b^{2^{j+1}}} = \frac{2}{(q_j)^2}$$

and let S be the set of prime divisors of b. Then, an easy computation gives that

$$\left(\prod_{\ell \in S} |q_j|_\ell \cdot |p_j|_\ell\right) \cdot |\mathcal{K}_b - p_j/q_j| < \frac{2}{(q_j)^3}$$

and Theorem 8.1.17 implies that $\mathcal{K}_b$ is transcendental. □

As shown in (Ferenczi and Mauduit 1997), Ridout's theorem yields the following improvement of Proposition 8.1.16.

Proposition 8.1.19 *If an infinite word a defined over $\{0, 1, \ldots, b-1\}$ satisfies Condition $(*)_{w,c}$ for some $w > 2$ and some $c \geq 0$, then the associated real number ξ_a is transcendental.*

An interesting consequence of this combinatorial transcendence criterion, pointed out in (Ferenczi and Mauduit 1997), is that every real number with a *Sturmian* base-b expansion is transcendental. Such a result cannot be proved by using Proposition 8.1.16.

Proof Let a be an infinite word defined over $\{0, 1, \ldots, b-1\}$ and satisfying Condition $(*)_{w,c}$ for some $w > 2$ and some $c \geq 0$. Then, for every $j \geq 1$, the real number

$$\xi_a = 0.u_j \underbrace{v_j \cdots v_j}_{\lfloor w \rfloor \text{ times}} v'_j \cdots,$$

where v'_j is the prefix of v_j of length $\lceil (w - \lfloor w \rfloor)|v_j|\rceil$, is very close to the rational number

$$0.u_j(v_j)^\omega,$$

obtained from ξ_a by truncating its base-b expansion and completing by periodicity. Precisely, as shown by an easy computation, there exist integers p_j, r_j and s_j such that

$$\left|\xi - \frac{p_j}{b^{r_j}(b^{s_j} - 1)}\right| < \frac{2}{b^{r_j + ws_j}}, \qquad (8.8)$$

where $r_j = |u_j|$ and $s_j = |v_j|$.

Note that the rational approximations to ξ_a that we have obtained are very specific: their denominators have a possibly very large part composed of a finite number of fixed prime numbers (namely, the prime divisors of b). More precisely, if S denotes the set of prime divisors of b, we have

$$\prod_{\ell \in S} |b^{r_j}(b^{s_j} - 1)|_\ell = \frac{1}{b^{r_j}}.$$

We thus infer from (8.8) that, setting $q_j := b^{r_j}(b^{s_j} - 1)$, we have

$$\left(\prod_{\ell \in S} |p_j|_\ell \cdot |q_j|_\ell\right) \cdot \left|\xi_a - \frac{p_j}{q_j}\right| < \frac{2}{b^{2r_j + w s_j}}, \qquad (8.9)$$

for every positive integer j.

Set $\varepsilon := (w - 2)/2(c + 1)$. Since $r_j \leq c s_j$, we obtain

$$s_j/(r_j + s_j) \geq 1/(c + 1).$$

Combining

$$\frac{2}{b^{2r_j + w s_j}} \leq \frac{2}{b^{(r_j + s_j)(2 + (w-2)/(c+1))}} < \frac{2}{q_j^{2 + 2\varepsilon}}$$

and (8.9), we deduce that

$$\left(\prod_{\ell \in S} |p_j|_\ell \cdot |q_j|_\ell\right) \cdot \left|\xi_a - \frac{p_j}{q_j}\right| < \frac{1}{q_j^{2+\varepsilon}},$$

for every integer j large enough. Since ε is positive, Theorem 8.1.17 implies that the real number ξ_a is transcendental, concluding the proof. □

8.1.4 The Schmidt subspace theorem

A formidable multidimensional generalisation of the Roth and Ridout theorems is known as the Schmidt subspace theorem (Schmidt 1980b). We state below without proof a simplified p-adic version of this result that will be enough to derive our main result regarding the complexity of algebraic irrational real numbers, namely, Theorem 8.1.6. This result will also play a key role in Section 8.3.

Theorem 8.1.20 *Let $m \geq 2$ be an integer and ε be a positive real number. Let S be a finite set of distinct prime numbers. Let $L_1, \ldots, L_m$ be m linearly* independent linear forms with real algebraic coefficients. Then, the set of solutions $\mathbf{x} = (x_1, \ldots, x_m)$ in $\mathbb{Z}^m$ to the inequality

$$\left(\prod_{i=1}^{m} \prod_{\ell \in S} |x_i|_\ell\right) \cdot \prod_{i=1}^{m} |L_i(\mathbf{x})| \leq (\max\{|x_1|, \ldots, |x_m|\})^{-\varepsilon}$$

lies in finitely many proper subspaces of $\mathbb{Q}^m$.

Before going on, we show how Roth's theorem follows from Theorem 8.1.20. Let ξ be a real algebraic number and ε be a positive real number. Consider the two independent linear forms $\xi X - Y$ and X. Theorem 8.1.20 implies that all the integer solutions (p, q) to

$$|q| \cdot |q\xi - p| < |q|^{-\varepsilon} \tag{8.10}$$

are contained in a finite union of proper subspaces of $\mathbb{Q}^2$. There thus is a finite set of equations $x_1 X + y_1 Y = 0, \ldots, x_t X + y_t Y = 0$ such that, for every solution (p, q) to (8.10), there exists an integer k with $x_k p + y_k q = 0$. If ξ is irrational, this means that there are only finitely many rational solutions to $|\xi - p/q| < |q|^{-2-\varepsilon}$, which is Roth's theorem.

The following combinatorial transcendence criterion was proved by means of Theorem 8.1.20 by B. Adamczewski, Y. Bugeaud and F. Luca (Adamczewski, Bugeaud, and Luca 2004).

Proposition 8.1.21 *If an infinite word a defined over $\{0, 1, \ldots, b-1\}$ satisfies Condition $(*)_{w,c}$ for some $w > 1$ and some $c \geq 0$, then the real number ξ_a is transcendental.*

The strategy to prove this result is the same as for the other criteria, but, in addition, we will take advantage of the specific shape of the factors $b^{s_j} - 1$ in the denominators of the good approximations to ξ_a. In this new criterion, it is not needed any more that squares occur in order to prove the transcendence of our number. Only occurrences of stammerings are enough, provided that they do not occur too far from the beginning. This difference turns out to be the key for applications.

Before proving Proposition 8.1.21, let us quote a first consequence.

Note that the combinatorial structure of the Thue–Morse word t is quite different from that of the Fibonacci word. Indeed, a well-known property of t is that it is overlap-free and thus cannot satisfy Condition $(*)_{w,c}$ for some $w > 2$. The following result, first proved by K. Mahler in 1929 (Mahler 1929) by means of a totally different approach, is also a straightforward consequence of Proposition 8.1.21.

Theorem 8.1.22 *The real number*

$$\xi_t := 0.011\,010\,011\,001\,011\,010\,010\,110\cdots$$

is transcendental.

Proof First, we recall that t is aperiodic. Note that t begins with the word 011. Consequently, for every positive integer j, the word $\tau^j(011)$ is also a prefix of t. Set $u_j := \tau^j(0)$ and $v_j := \tau^j(1)$. Then, for every positive integer j, the word t begins with $u_j v_j^2$. Furthermore, a simple computation shows that

$$|u_j| = |v_j| = 2^j.$$

This proves that t satifies Condition $(*)_{2,1}$. In view of Proposition 8.1.21, the theorem is established. $\square$

Proof [Proof of Proposition 8.1.21] Let a be an infinite word defined over $\{0, 1, \ldots, b-1\}$ and satisfying Condition $(*)_{w,c}$ for some $w > 1$ and some $c \geq 0$. We assume that ξ_a is algebraic, and we will reach a contradiction.

We consider, as in the proof of Proposition 8.1.19, the integers p_j, r_j, s_j and the set S of prime divisors of b.

Consider the three linearly independent linear forms with real algebraic coefficients:

$$\begin{aligned} L_1(X_1, X_2, X_3) &= \xi_a X_1 - \xi_a X_2 - X_3, \\ L_2(X_1, X_2, X_3) &= X_1, \\ L_3(X_1, X_2, X_3) &= X_2. \end{aligned}$$

For $j \geq 1$, evaluating them on the integer triple $\mathbf{x}_j := (b^{r_j + s_j}, b^{r_j}, p_j)$, we obtain

$$\prod_{i=1}^{3} |L_i(\mathbf{x}_j)| \leq 2\, b^{2r_j + s_j - (w-1)s_j}. \tag{8.11}$$

Let us denote by $x_i^{(j)}$ the ith coordonate of the vector $\mathbf{x}_j$. We have

$$\prod_{i=1}^{3} \prod_{\ell \in S} |x_i^{(j)}|_\ell \leq \prod_{\ell \in S} |x_1^{(j)}|_\ell \cdot \prod_{\ell \in S} |x_2^{(j)}|_\ell = b^{-2r_j - s_j}. \tag{8.12}$$

Combining (8.11) and (8.12), we get that

$$\left(\prod_{i=1}^{3} \prod_{\ell \in S} |x_i^{(j)}|_\ell \right) \cdot \prod_{i=1}^{3} |L_i(\mathbf{x})| \leq 2\, (b^{r_j + s_j})^{-(w-1)s_j/(r_j + s_j)}.$$

Set $\varepsilon := (w-1)/2(c+1)$. Since by assumption a satisfies Condition $(*)_{w,c}$, we obtain

$$\left(\prod_{i=1}^{3}\prod_{\ell\in S}|x_i^{(j)}|_\ell\right)\cdot\prod_{i=1}^{3}|L_i(\mathbf{x}_j)|\leq\left(\max\{b^{r_j+s_j},b^{r_j},p_j\}\right)^{-\varepsilon},$$

if the integer j is sufficiently large.

We then infer from Theorem 8.1.20 that all points $\mathbf{x}_j$ lie in a finite number of proper subspaces of $\mathbb{Q}^3$. Thus, there exist a non-zero integer triple (z_1, z_2, z_3) and an infinite set of distinct positive integers $\mathcal{J}$ such that

$$z_1 b^{r_j+s_j} + z_2 b^{r_j} + z_3 p_j = 0, \tag{8.13}$$

for every j in $\mathcal{J}$. Recall that $p_j/b^{r_j+s_j}$ tends to ξ when j tends to infinity. Dividing (8.13) by $b^{r_j+s_j}$ and letting j tend to infinity along $\mathcal{J}$, we get that ξ_a is a rational number. Since by assumption a satisfies Condition $(*)_{w,c}$, it is not eventually periodic. This provides a contradiction, ending the proof. □

8.1.4.1 Proof of Theorem 8.1.6

We are now ready to finish the proof of Theorem 8.1.6.

Proof [Proof of Theorem 8.1.6] Let $a = a_1 a_2 \cdots$ be a non-eventually periodic infinite word defined over the finite alphabet $\{0, 1, \ldots, b-1\}$. We assume that there exists an integer $\kappa \geq 2$ such that its complexity function p_a satisfies

$$p_a(n) \leq \kappa n \quad \text{for infinitely many integers } n \geq 1,$$

and we shall derive that a satisfies Condition $(*)_{w,c}$ for some $w > 1$ and some $c \geq 0$. In view of Proposition 8.1.21, we will obtain that the real number ξ_a is transcendental, concluding the proof.

Let n_k be an integer with $p_a(n_k) \leq \kappa n_k$. Denote by $a(\ell)$ the prefix of a of length ℓ. By the pigeonhole principle, there exists (at least) one word m_k of length n_k which has (at least) two occurrences in $a((\kappa+1)n_k)$. Thus, there are (possibly empty) words b_k, c_k, d_k and e_k, such that

$$a((\kappa+1)n_k) = b_k m_k d_k e_k = b_k c_k m_k e_k \text{ and } |c_k| \geq 1.$$

We observe that $|b_k| \leq \kappa n_k$. We have to distinguish three cases:

(i) $|c_k| > |m_k|$,

(ii) $\lceil |m_k|/3 \rceil \leq |c_k| \leq |m_k|$,

(iii) $1 \leq |c_k| < \lceil |m_k|/3 \rceil$.

(i) Under this assumption, there exists a word f_k such that

$$a((\kappa+1)n_k) = b_k m_k f_k m_k e_k.$$

Since $|e_k| \leq (\kappa-1)|m_k|$, the word $b_k(m_k f_k)^s$ with $s := 1 + 1/\kappa$ is a prefix of a. Furthermore, we observe that

$$|m_k f_k| \geq |m_k| \geq \frac{|b_k|}{\kappa}.$$

(ii) Under this assumption, there exist two words f_k and g_k such that

$$a((\kappa+1)n_k) = b_k m_k^{1/3} f_k m_k^{1/3} f_k g_k.$$

Thus, the word $b_k(m_k^{1/3} f_k)^2$ is a prefix of a. Furthermore, we observe that

$$|m_k^{1/3} f_k| \geq \frac{|m_k|}{3} \geq \frac{|b_k|}{3\kappa}.$$

(iii) In this case, c_k is clearly a prefix of m_k and m_k is a prefix of $c_k m_k$. Consequently, c_k^t is a prefix of m_k, where t is the integer part of $|m_k|/|c_k|$. Observe that $t \geq 3$. Setting $s = \lfloor t/2 \rfloor$, we see that $b_k(c_k^s)^2$ is a prefix of a and

$$|c_k^s| \geq \frac{|m_k|}{4} \geq \frac{|b_k|}{4\kappa}.$$

In each of the three cases above, we have proved that there are finite words u_k, v_k and a positive real number w such that $u_k v_k^w$ is a prefix of a and:

- $|u_k| \leq \kappa n_k$,
- $|v_k| \geq n_k/4$,
- $w \geq 1 + 1/\kappa > 1$.

Consequently, the sequence $(|u_k|/|v_k|)_{k \geq 1}$ is bounded from above by 4κ. Furthermore, it follows from the lower bound $|v_k| \geq n_k/4$ that we can assume without loss of generality that the sequence $(|v_k|)_{k \geq 1}$ is increasing. This implies that the infinite word a satisfies Condition $(*)_{1+1/\kappa, 4\kappa}$, concluding the proof. □

8.2 Basics from continued fractions

We collect in this section several basic results from the theory of continued fractions that will be useful in the rest of this chapter. All these results are stated without proofs. For more details, the reader is referred to classical monographs on continued fractions, such as (Perron 1929), (Khintchine 1963), (Rockett and Szüsz 1992), (Lang 1995), (Schmidt 1980b), (Bugeaud 2004a).

8.2.1 Notations

Every rational number that is not an integer has a unique *continued fraction expansion*

$$a_0 + \cfrac{1}{a_1 + \cfrac{1}{a_2 + \cfrac{1}{\ddots + \cfrac{1}{a_n}}}}$$

where a_0 is an integer and a_i, $i \geq 1$, are positive integers with $a_n \geq 2$. As well, every irrational real number ζ has a unique continued fraction expansion

$$a_0 + \cfrac{1}{a_1 + \cfrac{1}{\ddots + \cfrac{1}{a_n + \cfrac{1}{\ddots}}}}$$

where a_0 is an integer and a_i, $i \geq 1$, are positive integers. For short, we will write $[a_0, a_1, \ldots, a_n]$ to denote a finite continued fraction and $[a_0, a_1, \ldots]$ for an infinite continued fraction.

For instance, we have:

$$\frac{77\,708\,431}{2\,640\,858} = [29, 2, 2, 1, 5, 1, 4, 1, 1, 2, 1, 6, 1, 10, 2, 2, 3],$$

$$\sqrt{2} = [1, 2, 2, 2, 2, 2, 2, 2, \ldots],$$

$$e = [2, 1, 2, 1, 1, 4, 1, 1, 6, 1, 1, 8, 1, 1, 10, 1, 1, 12, \ldots],$$

$$\pi = [3, 7, 15, 1, 292, 1, 1, 1, 12, 1, 3, 1, 14, 2, 1, 1, 2, 2, 2, 2, 1, 84, 2, \ldots].$$

The integers $a_0, a_1, \ldots$ are called the *partial quotients*. For $n \geq 1$, the rational number $p_n/q_n := [a_0, a_1, \ldots, a_n]$ is called the nth *convergent* to ζ. Setting

$$p_{-1} = 1, p_0 = a_0, q_{-1} = 0, q_0 = 1$$

the integers p_n and q_n satisfy, for every non-negative integer n, the fundamental relations

$$p_{n+1} = a_{n+1} p_n + p_{n-1} \quad \text{and} \quad q_{n+1} = a_{n+1} q_n + q_{n-1}. \tag{8.14}$$

The sequence $(p_n/q_n)_{n \geq 0}$ converges to ζ. More precisely, we have

$$\frac{1}{q_n(q_n + q_{n+1})} < \left| \zeta - \frac{p_n}{q_n} \right| < \frac{1}{q_n q_{n+1}} \tag{8.15}$$

for every positive integer n.

Note also that to any sequence $(a_n)_{n \geq 1}$ of positive integers corresponds a unique real irrational number ζ such that

$$\zeta = [0, a_1, a_2, \ldots].$$

8.2.2 Speed of convergence

It follows from inequalities (8.15) that two real numbers having the same first n partial quotients are close to each other. However, they cannot be too close if their $(n+1)$th partial quotients are different.

Lemma 8.2.1 *Let $\zeta = [a_0, a_1, \ldots]$ and $\eta = [b_0, b_1, \ldots]$ be real numbers. Let us assume that there exists a positive integer n such that $a_j = b_j$ for $j = 0, \ldots, n$. Then,*

$$|\zeta - \eta| \leq q_n^{-2},$$

where q_n denotes the nth convergent to ζ. Furthermore, if the partial quotients of ζ and η are bounded by M, and if $a_{n+1} \neq b_{n+1}$, then

$$|\zeta - \eta| \geq \frac{1}{(M+2)^3 q_n^2}.$$

A proof of Lemma 8.2.1 is given in (Adamczewski and Bugeaud 2006a).

8.2.3 Growth of convergents and continuants

The next lemma is an easy consequence of the recurrence relation satisfied by the denominators of the convergents to a real number.

Lemma 8.2.2 *Let $(a_i)_{i \geq 1}$ be a sequence of positive integers and let n be a positive integer. If $p_n/q_n = [0, a_1, \ldots, a_n]$ and $M = \max\{a_1, \ldots, a_n\}$, then*

$$2^{(n-1)/2} \leq q_n \leq (M+1)^n.$$

Given positive integers $a_1, \ldots, a_m$, we denote by $K_m(a_1, \ldots, a_m)$ the denominator of the rational $[0, a_1, \ldots, a_m]$ written in lowest terms. This quantity is called the *continuant* associated with the sequence $a_1, \ldots, a_m$. If $a = a_1 a_2 \cdots a_m$ denotes a finite word defined over the set of positive integers, we also write $K_m(a)$ for the continuant $K_m(a_1, a_2, \ldots, a_m)$, when the context is clear enough to avoid a possible confusion.

Lemma 8.2.3 *Let $a_1, \ldots, a_m$ be positive integers and let k be an integer such that $1 \leq k \leq m-1$. Then,*

$$K_m(a_1, \ldots, a_m) = K_m(a_m, \ldots, a_1)$$

and

$$K_k(a_1, \ldots, a_k) \cdot K_{m-k}(a_{k+1}, \ldots, a_m) \leq K_m(a_1, \ldots, a_m)$$

$$\leq 2\, K_k(a_1, \ldots, a_k) \cdot K_{m-k}(a_{k+1}, \ldots, a_m).$$

This lemma is proved in (Cusick and Flahive 1989). As we will see in the sequel of this chapter, the formalism of continuants is often very convenient to estimate the size of denominators of convergents.

8.2.4 The mirror formula

The *mirror formula*, which can be established by induction on n using the recurrence relations (8.14) giving the sequence $(q_n)_{n \geq 1}$, is sometimes omitted from classical textbooks.

Lemma 8.2.4 *Let $\zeta = [a_0, a_1, \ldots]$ be a real number and $(p_n/q_n)_{n \geq 0}$ be the sequence of convergents to ζ. Then,*

$$\frac{q_n}{q_{n-1}} = [a_n, a_{n-1}, \ldots, a_1]$$

for every positive integer n.

However, this is a very useful auxiliary result, as will become clear in the following sections.

8.2.5 The Euler–Lagrange theorem

It is easily seen that if the continued fraction expansion of ζ is ultimately periodic, then ζ is a quadratic number. The converse is also true.

Theorem 8.2.5 *The continued fraction expansion of a real number is eventually periodic if, and only if, it is a quadratic irrational number.*

In contrast, very little is known on the continued fraction expansion of an algebraic real number of degree at least 3.

8.3 Transcendental continued fractions

All through this section, we use the following notation. If $a = a_1 a_2 \cdots$ is an infinite word defined over the set of positive integers, we denote by ζ_a the associated continued fraction, that is,

$$\zeta_a := [0, a_1, a_2, \ldots].$$

It is widely believed that the continued fraction expansion of any irrational algebraic number ζ either is eventually periodic (and, according to Theorem 8.2.5, this is the case if, and only if, ζ is a quadratic irrational), or it contains arbitrarily large partial quotients. Apparently, this problem was first considered by A. Ya. Khintchine (Khintchine 1963). Some speculations about the randomness of the continued fraction expansion of algebraic numbers of degree at least three have later been made by several authors. However, one shall admit that our knowledge on this topic is very limited.

In this section, we use the Schmidt subspace theorem (Theorem 8.1.20) to prove the transcendence of families of continued fractions involving periodic or symmetric patterns.

8.3.1 The Fibonacci continued fraction

We first prove that continued fractions beginning in arbitrarily large squares are either quadratic or transcendental.

Proposition 8.3.1 *Let a be an infinite word whose letters are positive integers. If a satisfies Condition* $(*)_{w,0}$ *for some $w \geq 2$, then the real number ζ_a is transcendental.*

Let a and b be distinct positive integers and let

$$f' := f'_1 f'_2 f'_3 \cdots = abaababaabaabab \cdots$$

Transcendence and Diophantine approximation 431

be the Fibonacci word defined over the alphabet $\{a, b\}$ as in the introduction of this chapter. Recall that we already proved in Section 8.1.3.2 that f (and thus f') satisfies Condition $(*)_{2.25,0}$. As a direct consequence of Proposition 8.3.1, we obtain the following result.

Theorem 8.3.2 *The real number*

$$\zeta_{f'} := [f'_1, f'_2, \ldots]$$

is transcendental.

Proof [Proof of Proposition 8.3.1] Let $a := a_1 a_2 \cdots$ be an infinite word defined over the infinite alphabet $\{1, 2, \ldots\}$ and satisfying Condition $(*)_{w,0}$ for some $w \geq 2$. Assume that the parameter $w \geq 2$ is fixed, as well as the sequence $(v_j)_{j \geq 1}$ occurring in the definition of Condition $(*)_{w,0}$. Set also $s_j = |v_j|$, for every $j \geq 1$. We want to prove that the real number

$$\zeta_a := [0, a_1, a_2, \ldots]$$

is transcendental.

By definition of Condition $(*)_{w,0}$, the sequence a is aperiodic and it follows from the Euler–Lagrange theorem (Theorem 8.2.5) that ζ_a is not a quadratic number. Furthermore, ζ_a is irrational since the sequence a is infinite.

From now on, we assume that ζ_a is algebraic of degree at least three and we aim at deriving a contradiction. Throughout this proof, the constants implied by $\ll$ are independent of j.

Let $(p_\ell/q_\ell)_{\ell \geq 1}$ denote the sequence of convergents to ζ_a. We infer from (8.15) that

$$|q_{s_j} \zeta_a - p_{s_j}| \leq q_{s_j}^{-1} \tag{8.16}$$

and

$$|q_{s_j-1} \zeta_a - p_{s_j-1}| \leq q_{s_j}^{-1}. \tag{8.17}$$

The key fact for the proof of Proposition 8.3.1 is the observation that ζ_a admits infinitely many good quadratic approximants obtained by truncating its continued fraction expansion and completing by periodicity. Precisely, for every positive integer j, we define the sequence $(b_k^{(j)})_{k \geq 1}$ by

$$b_{h+ks_j}^{(j)} = a_h \text{ for } 1 \leq h \leq s_j \text{ and } k \geq 0.$$

The sequence $(b_k^{(j)})_{k \geq 1}$ is purely periodic with period s_j. Set

$$\alpha_j = [0, b_1^{(j)}, b_2^{(j)}, \ldots]$$

and observe that

$$\alpha_j = [0, a_1, \ldots, a_{s_j}, 1/\alpha_j] = \frac{p_{s_j}/\alpha_j + p_{s_j-1}}{q_{s_j}/\alpha_j + q_{s_j-1}}.$$

Thus, α_j is a root of the quadratic polynomial

$$P_j(X) := q_{s_j-1}X^2 + (q_{s_j} - p_{s_j-1})X - p_{s_j}.$$

By Rolle's theorem and Lemma 8.2.1, for every positive integer j, we have

$$|P_j(\zeta_a)| = |P_j(\zeta_a) - P_j(\alpha_j)| \ll q_{s_j}|\zeta_a - \alpha_j| \ll q_{s_j} q_{2s_j}^{-2}, \qquad (8.18)$$

since the first $2s_j$ partial quotients of ζ_a and α_j are the same. Furthermore, with the notation of Section 8.2.3, we have $q_{s_j} = K_{s_j}(v_j)$ and $q_{2s_j} = K_{2s_j}(v_j v_j)$. Then, we infer from Lemma 8.2.3 that

$$q_{2s_j} \geq q_{s_j}^2.$$

By (8.18), this gives

$$|P_j(\zeta_a)| \ll \frac{1}{q_{s_j}^3}. \qquad (8.19)$$

Consider now the three linearly independent linear forms:

$$L_1(X_1, X_2, X_3) = \zeta_a^2 X_1 + \zeta_a X_2 - X_3,$$
$$L_2(X_1, X_2, X_3) = X_1,$$
$$L_3(X_1, X_2, X_3) = X_3.$$

Evaluating them on the triple $(q_{s_j-1}, q_{s_j} - p_{s_j-1}, p_{s_j})$, we infer from (8.19) that

$$\prod_{1 \leq i \leq 3} |L_i(q_{s_j-1}, q_{s_j} - p_{s_j-1}, p_{s_j})| \ll \frac{1}{\max\{q_{s_j-1}, q_{s_j} - p_{s_j-1}, p_{s_j}\}}. \qquad (8.20)$$

It then follows from Theorem 8.1.20 that the points $(q_{s_j-1}, q_{s_j} - p_{s_j-1}, p_{s_j})$ with $j \geq 1$ lie in a finite number of proper subspaces of $\mathbb{Q}^3$. Thus, there exist a non-zero integer triple (x_1, x_2, x_3) and an infinite set of distinct positive integers $\mathcal{J}_1$ such that

$$x_1 q_{s_j-1} + x_2(q_{s_j} - p_{s_j-1}) + x_3 p_{s_j} = 0, \qquad (8.21)$$

for every j in $\mathcal{J}_1$. Observe that $(x_1, x_2) \neq (0, 0)$, since (x_1, x_2, x_3) is a non-zero triple. Dividing (8.21) by q_{s_j}, we obtain

$$x_1 \frac{q_{s_j-1}}{q_{s_j}} + x_2 \left(1 - \frac{p_{s_j-1}}{q_{s_j-1}} \cdot \frac{q_{s_j-1}}{q_{s_j}}\right) + x_3 \frac{p_{s_j}}{q_{s_j}} = 0. \qquad (8.22)$$

By letting j tend to infinity along $\mathcal{J}_1$ in (8.22), we get that

$$\lim_{\mathcal{J}_1 \ni j \to +\infty} \frac{q_{s_j-1}}{q_{s_j}} = -\frac{x_2 + x_3\zeta_a}{x_1 - x_2\zeta_a} =: \alpha.$$

By definition of α and Equality (8.22), we observe that

$$\left| \alpha - \frac{q_{s_j-1}}{q_{s_j}} \right| = \left| \frac{x_2 + x_3\zeta_a}{x_1 - x_2\zeta_a} - \frac{x_2 + x_3 p_{s_j}/q_{s_j}}{x_1 - x_2 p_{s_j-1}/q_{s_j-1}} \right|,$$

for every j in $\mathcal{J}_1$. As a consequence of (8.16) and (8.17), we get that

$$\left| \alpha - \frac{q_{s_j-1}}{q_{s_j}} \right| \ll \frac{1}{q_{s_j}^2}, \tag{8.23}$$

for every j in $\mathcal{J}_1$. Since q_{s_j-1} and q_{s_j} are coprime and s_j tends to infinity when j tends to infinity along $\mathcal{J}_1$, the real number α is irrational.

Consider now the three linearly independent linear forms:

$$\begin{aligned} L'_1(Y_1, Y_2, Y_3) &= \alpha Y_1 - Y_2, \\ L'_2(Y_1, Y_2, Y_3) &= \zeta_a Y_1 - Y_3, \\ L'_3(Y_1, Y_2, Y_3) &= Y_1. \end{aligned}$$

Evaluating them on the triple $(q_{s_j}, q_{s_j-1}, p_{s_j})$ with $j \in \mathcal{J}_1$, we infer from (8.16) and (8.23) that

$$\prod_{1 \leq j \leq 3} |L'_j(q_{s_j}, q_{s_j-1}, p_{s_j})| \ll q_{s_j}^{-1}.$$

It then follows from Theorem 8.1.20 that the points $(q_{s_j}, q_{s_j-1}, p_{s_j})$ with $j \in \mathcal{J}_1$ lie in a finite number of proper subspaces of $\mathbb{Q}^3$. Thus, there exist a non-zero integer triple (y_1, y_2, y_3) and an infinite set of distinct positive integers $\mathcal{J}_2$, included in $\mathcal{J}_1$, such that

$$y_1 q_{s_j} + y_2 q_{s_j-1} + y_3 p_{s_j} = 0, \tag{8.24}$$

for every j in $\mathcal{J}_2$. Dividing (8.24) by q_{s_j} and letting j tend to infinity along $\mathcal{J}_2$, we get

$$y_1 + y_2 \alpha + y_3 \zeta_a = 0. \tag{8.25}$$

To obtain another equation linking ζ_a and α, we consider the three linearly independent linear forms:

$$\begin{aligned} L''_1(Z_1, Z_2, Z_3) &= \alpha Z_1 - Z_2, \\ L''_2(Z_1, Z_2, Z_3) &= \zeta_a Z_2 - Z_3, \\ L''_3(Z_1, Z_2, Z_3) &= Z_1. \end{aligned}$$

Evaluating them on the triple $(q_{s_j}, q_{s_j-1}, p_{s_j-1})$ with $j \in \mathcal{J}_1$, we infer from (8.17) and (8.23) that

$$\prod_{j=1}^{3} |L_j''(q_{s_j}, q_{s_j-1}, p_{s_j-1})| \ll q_{s_j}^{-1}.$$

It then follows from Theorem 8.1.20 that the points $(q_{s_j}, q_{s_j-1}, p_{s_j-1})$ with $j \in \mathcal{J}_1$ lie in a finite number of proper subspaces of $\mathbb{Q}^3$. Thus, there exist a non-zero integer triple (z_1, z_2, z_3) and an infinite set of distinct positive integers $\mathcal{J}_3$, included in $\mathcal{J}_1$, such that

$$z_1 q_{s_j} + z_2 q_{s_j-1} + z_3 p_{s_j-1} = 0, \qquad (8.26)$$

for every j in $\mathcal{J}_3$. Dividing (8.26) by q_{s_j-1} and letting j tend to infinity along $\mathcal{J}_3$, we get

$$\frac{z_1}{\alpha} + z_2 + z_3 \zeta_a = 0. \qquad (8.27)$$

We infer from (8.25) and (8.27) that

$$(z_3 \zeta_a + z_2)(y_3 \zeta_a + y_1) = y_2 z_1. \qquad (8.28)$$

If $y_3 z_3 = 0$, then (8.25) and (8.27) yield that α is rational, which is a contradiction. Consequently, $y_3 z_3 \neq 0$ and we infer from (8.28) that ζ_a is a quadratic real number, which is again a contradiction. This completes the proof the proposition. □

8.3.2 The Thue–Morse continued fraction

Let a and b be distinct positive integers and let

$$t' := t_1' t_2' t_3' \cdots = abbabaabbaababba \cdots$$

be the Thue–Morse word defined over the alphabet $\{a, b\}$ as in the introduction of this chapter. M. Queffélec (Queffélec 1998) showed that the Thue–Morse continued fraction $\zeta_{t'}$ is transcendental. To prove this result, she used an extension of Roth's theorem to approximation by quadratic numbers, worked out by W. M. Schmidt and which is a consequence of Theorem 8.1.20, combined with the fact that t satisfies Condition $(*)_{1.6,0}$ and that the subshift associated with t is uniquely ergodic.

The purpose of this subsection is to present an alternative and shorter proof of her result, using the fact that t begins in arbitrarily large *palindromes*. Actually, the following combinatorial transcendence criterion obtained in (Adamczewski and Bugeaud 2007c) (see also the paper (Adamczewski and Bugeaud 2007d)) relies, once again, on the Schmidt subspace theorem.

Proposition 8.3.3 *Let a be an infinite word whose letters are positive integers. If a begins with arbitrarily long palindromes, then the real number ζ_a is either quadratic or transcendental.*

As a consequence of Proposition 8.3.3, we obtain the following result.

Theorem 8.3.4 *The real number*
$$\zeta_{t'} := [t_1', t_2', \ldots]$$
is transcendental.

Proof We first recall that the word t (and thus t') is not eventually periodic. The Euler–Lagrange theorem (Theorem 8.2.5) implies that $\zeta_{t'}$ is not a quadratic irrational number. Note that $\tau^2(0) = 0110$ and $\tau^2(1) = 1001$ are palindromes. Now, observe that the Thue–Morse word t begins with the palindrome 0110. Since for every positive integer j, the word $\tau^j(0)$ is a prefix of t, we obtain that, for every positive integer n, the prefix of length 4^n of t (and thus of t') is a palindrome. In view of Proposition 8.3.3, this concludes the proof. □

Proof [Proof of Proposition 8.3.3] Let $a = a_1 a_2 \cdots$ be an infinite word satisfying the assumptions of the proposition and set
$$\zeta_a := [0, a_1, a_2, \ldots].$$

Let us denote by $(n_j)_{j \geq 1}$ the increasing sequence of all lengths of prefixes of a that are palindromes. Let us also denote by p_n/q_n the nth convergent to ζ_a.

In the sequel, we assume that ζ_a is algebraic and our aim is to prove that ζ_a is a quadratic irrational number. Note that ζ_a is irrational since it has an infinite continued fraction expansion.

Let $j \geq 1$ be an integer. Since by assumption the word $a_1 a_2 \cdots a_{n_j}$ is a palindrome, we infer from Lemma 8.2.4 that
$$\frac{q_{n_j - 1}}{q_{n_j}} = \frac{p_{n_j}}{q_{n_j}},$$
that is,
$$q_{n_j - 1} = p_{n_j}.$$

It then follows from (8.15) that
$$|q_{n_j} \zeta_a - q_{n_j - 1}| < \frac{1}{q_{n_j}} \tag{8.29}$$

and
$$|q_{n_j-1}\zeta_a - p_{n_j-1}| < \frac{1}{q_{n_j}}. \tag{8.30}$$

Consider now the three linearly independent linear forms:
$$\begin{aligned} L_1(X_1, X_2, X_3) &= \zeta_a X_1 - X_2, \\ L_2(X_1, X_2, X_3) &= \zeta_a X_2 - X_3, \\ L_3(X_1, X_2, X_3) &= X_1. \end{aligned}$$

Evaluating them on the triple $(q_{n_j}, q_{n_j-1}, p_{n_j-1})$, we infer from (8.29) and (8.30) that
$$\prod_{i=1}^{3} |L_i(q_{n_j}, q_{n_j-1}, p_{n_j-1})| < \frac{1}{\max\{q_{n_j}, q_{n_j-1}, p_{n_j-1}\}}.$$

It then follows from Theorem 8.1.20 that the points $(q_{n_j}, q_{n_j-1}, p_{n_j-1})$, $j \geq 1$, lie in a finite number of proper subspaces of $\mathbb{Q}^3$. Thus, there exist a non-zero integer triple (z_1, z_2, z_3) and an infinite set of distinct positive integers $\mathcal{J}$ such that
$$z_1 q_{n_j} + z_2 q_{n_j-1} + z_3 p_{n_j-1} = 0, \tag{8.31}$$

for every j in $\mathcal{J}$. Dividing (8.31) by q_{n_j}, this gives
$$z_1 + z_2 \frac{q_{n_j-1}}{q_{n_j}} + z_3 \left(\frac{p_{n_j-1}}{q_{n_j-1}} \cdot \frac{q_{n_j-1}}{q_{n_j}} \right) = 0.$$

Letting j tend to infinity along $\mathcal{J}$, we infer from (8.29) and (8.30) that
$$z_1 + z_2 \zeta_a + z_3 \zeta_a^2 = 0.$$

Since (z_1, z_2, z_3) is a non-zero triple, this implies that ζ_a is a quadratic or a rational number. Since we already observed that ζ_a is irrational, this ends the proof. □

8.4 Simultaneous rational approximations to a real number and its square

All along this section, φ will denote the Golden Ratio.

8.4.1 Uniform Diophantine approximation

A fundamental result in Diophantine approximation was obtained by P. G. L. Dirichlet in 1842 (Dirichlet 1842).

Theorem 8.4.1 *For every real number ξ and every real number $X > 1$, the system of inequalities*

$$|x_0\xi - x_1| \leq X^{-1}, \qquad (8.32)$$
$$|x_0| \leq X,$$

has a non-zero solution $(x_0, x_1) \in \mathbb{Z}^2$.

Proof Let t denote the smallest integer greater than or equal to $X - 1$. If ξ is the rational a/b, with a and b integers and $1 \leq b \leq t$, it is sufficient to set $x_1 = a$ and $x_0 = b$. Otherwise, the $t+2$ points $0, \{\xi\}, \ldots, \{t\xi\}$, and 1 are pairwise distinct and they divide the interval $[0,1]$ into $t+1$ subintervals. By the pigeonhole principle, at least one of these has its length at most equal to $1/(t+1)$. This means that there exist integers k, ℓ and m_k, m_ℓ with $0 \leq k < \ell \leq t$ and

$$|(\ell\xi - m_\ell) - (k\xi - m_k)| \leq \frac{1}{t+1} \leq \frac{1}{X}.$$

We get (8.32) by setting $x_1 := m_\ell - m_k$ and $x_0 := \ell - k$, and by noticing that x_0 satisfies $1 \leq x_0 \leq t \leq X$. $\square$

Theorem 8.4.1 implies that every irrational real number is approximable at order at least 2 by rationals, a statement that also follows from (8.15). However, Theorem 8.4.1 gives a stronger result, in the sense that it asserts that the system (8.32) has a solution for every real number $X > 1$, while (8.15) only implies that (8.32) has a solution for arbitrarily large values of X. In Diophantine approximation, a statement like Theorem 8.4.1 is called *uniform*.

Obviously, the quality of approximation strongly depends on whether we are interested in a uniform statement or in a statement valid only for arbitrarily large X. Indeed, for any $w > 1$, there clearly exist real numbers ξ for which, for arbitrarily large values of X, the equation

$$|x_0\xi - x_1| \leq X^{-w}$$

has a solution in integers x_0 and x_1 with $1 \leq x_0 \leq X$. In contrast, it was proved by A. Ya. Khintchine (Khintchine 1926) that there is no irrational real number ξ satisfying a stronger form of Theorem 8.4.1 in which the exponent of X in (8.32) is less than -1.

In the case of rational approximation, these questions are quite well understood, essentially thanks to the theory of continued fractions.

It is a well-known fact that questions of simultaneous Diophantine approximation are in general much more difficult when the quantities we approximate are dependent. A classical example is provided by the simultaneous rational approximation of the first n powers of a transcendental number by rational numbers of the same denominator. In the sequel, we will focus on the 2-dimensional case, that is, on the uniform simultaneous approximation to a real number and its square.

In this framework, Dirichlet's theorem can be extended as follows.

Theorem 8.4.2 *For every real number ξ and every real number $X > 1$, the system of inequalities*

$$\begin{aligned} |x_0 \xi - x_1| &\leq X^{-1/2}, \\ |x_0 \xi^2 - x_2| &\leq X^{-1/2}, \\ |x_0| &\leq X, \end{aligned}$$

has a non-zero solution $(x_0, x_1, x_2) \in \mathbb{Z}^3$.

We omit the proof which follows the same lines as that of Theorem 8.4.1.

It was proved by I. Kubilyus (Kubilius 1949) that, for almost every real number ξ with respect to the Lebesgue measure, the exponent $-1/2$ in the above statement cannot be lowered. It was also expected that this exponent cannot be lowered for a real number that is neither rational nor quadratic. In this direction, a first limitation was obtained by H. Davenport and W. M. Schmidt (Davenport and Schmidt 1968).

Theorem 8.4.3 *Let ξ be a real number that is neither rational nor quadratic. Then, there exists a positive real number c such that the system of inequalities*

$$\begin{aligned} |x_0 \xi - x_1| &\leq cX^{-1/\varphi}, \\ |x_0 \xi^2 - x_2| &\leq cX^{-1/\varphi}, \\ |x_0| &\leq X, \end{aligned}$$

has no solution $(x_0, x_1, x_2) \in \mathbb{Z}^3$ for arbitrarily large real numbers X.

Note that $1/\varphi = 0.618\ldots$ is larger than $1/2$.

8.4.2 Extremal numbers

As we just mentioned, it was expected for a long time that the constant $-1/\varphi$ in Theorem 8.4.3 could be replaced by $-1/2$. This is actually not the case, as was proved by D. Roy (Roy 2004) (see also (Roy 2003a)).

Theorem 8.4.4 *There exist a real number ξ which is neither rational nor quadratic and a positive real number c such that the system of inequalities*

$$\begin{aligned} |x_0 \xi - x_1| &\leq cX^{-1/\varphi}, \\ |x_0 \xi^2 - x_2| &\leq cX^{-1/\varphi}, \\ |x_0| &\leq X, \end{aligned} \quad (8.33)$$

has a non-zero solution $(x_0, x_1, x_2) \in \mathbb{Z}^3$ for every real number $X > 1$.

Such a result is quite surprising, since the volume of the convex body defined by (8.33) tends rapidly to zero as X grows to infinity.

Any real number ξ satisfying an exceptional Diophantine condition as in Theorem 8.4.4 was termed by D. Roy an *extremal number*. He proved that the set of extremal numbers is countable. Furthermore, he also gave some explicit examples of extremal numbers. As we will see in the sequel, if a and b denote two distinct positive integers, the real number $\zeta_{f'}$ defined in the introduction of this chapter is an extremal number.

8.4.3 Simultaneous rational approximations, continued fractions and palindromes

The crucial point for the proof of Theorem 8.4.4 is a surprising connection between simultaneous rational approximation, continued fractions and palindromes. The aim of this section is to describe this connection.

Let $\zeta = [0, a_1, a_2, \ldots]$ be a real number and let p_n/q_n denote the nth convergent to ζ. Let us assume that the word $a_1 \cdots a_n$ is a palindrome. As we already observed in the proof of Proposition 8.3.3, Lemma 8.2.4 then implies that

$$p_n = q_{n-1}.$$

On the other hand, we infer from (8.15) that

$$\left| \zeta - \frac{p_n}{q_n} \right| < \frac{1}{q_n^2}.$$

Since $0 < \zeta < 1$, $a_1 = a_n$ and $q_n \leq (a_n + 1)q_{n-1}$, we obtain

$$\left| \zeta^2 - \frac{p_{n-1}}{q_n} \right| \leq \left| \zeta^2 - \frac{p_{n-1}}{q_{n-1}} \times \frac{p_n}{q_n} \right|$$

$$\leq \left| \zeta + \frac{p_{n-1}}{q_{n-1}} \right| \times \left| \zeta - \frac{p_n}{q_n} \right| + \frac{1}{q_n q_{n-1}}$$

$$\leq 2 \left| \zeta - \frac{p_n}{q_n} \right| + \frac{1}{q_n q_{n-1}} < \frac{a_1 + 3}{q_n^2}.$$

To sum up, if the word $a_1 a_2 \cdots a_n$ is a palindrome, then

$$|q_n \zeta - p_n| < \frac{1}{q_n} \quad \text{and} \quad |q_n \zeta^2 - p_{n-1}| < \frac{a_1 + 3}{q_n}. \tag{8.34}$$

In other words, palindromes provide very good simultaneous rational approximations to ζ and ζ^2.

An essential aspect of the question we consider here is that it is a problem of uniform approximation. Let us assume that the infinite word $a = a_1 a_2 \cdots$ begins with arbitrarily long palindromes, and denote by $(n_j)_{j \geq 1}$ the increasing sequence formed by the lengths of prefixes of a that are palindromes. Set

$$\zeta_a := [0, a_1, a_2, \ldots].$$

If the sequence $(n_j)_{j \geq 1}$ increases sufficiently slowly to ensure the existence of a positive real number c_1 and a real number τ such that $q_{n_{j+1}} \leq c_1 q_{n_j}^\tau$ for every large j, then Inequalities (8.34) ensure that, for every real number X large enough, there exists a positive real number c_2 such that the system of inequalities

$$|x_0| \leq X, \ |x_0 \zeta_a - x_1| \leq c_2 X^{-1/\tau}, \ |x_0 \zeta_a^2 - x_2| \leq c_2 X^{-1/\tau}, \tag{8.35}$$

has a non-zero solution $(x_0, x_1, x_2) \in \mathbb{Z}^3$. Indeed, given a sufficiently large real number X, there exists an integer i such that $q_{n_i} \leq X < q_{n_i+1}$ and the triple $(q_{n_i}, p_{n_i}, p_{n_i-1})$ provides a non-zero solution to the system (8.35).

Consequently, if the continued fraction expansion of a real number ζ_a begins with many palindromes, then ζ_a and ζ_a^2 are uniformly very well simultaneously approximated by rationals. In view of Theorem 8.4.1, this observation is only interesting if there exist infinite words a for which the associated exponent τ is less than 2.

8.4.4 Fibonacci word and palindromes

Let a and b be distinct positive integers. As in Section 8.3.1, we consider the Fibonacci word

$$f' := abaababaabaabab \cdots$$

defined over the alphabet $\{a, b\}$. As in Section 8.1.3.2, $(F_j)_{j \geq 0}$ denotes the Fibonacci sequence.

In this section, we prove that many prefixes of f' are palindromes.

Proposition 8.4.5 *For every integer $j \geq 1$, the prefix of f' of length*

$$n_j := F_{j+3} - 2, \tag{8.36}$$

is a palindrome.

Proof For every integer $j \geq 2$ we set $w_j = f'_1 f'_2 \cdots f'_{F_j}$ and we recall that $w_2 = a$, $w_3 = ab$ and $w_{j+2} = w_{j+1} w_j$. One can show by an easy induction that, for every integer $j \geq 2$, the word w_{2j} ends with ba while w_{2j+1} ends with ab. Furthermore, the length of w_j is equal to F_j for every $j \geq 2$.

Let $j \geq 1$ and φ_j denote the prefix of f' of length n_j. Observe that $\varphi_1 = a$, $\varphi_2 = aba$, $\varphi_3 = abaaba$. We then obtain by induction that

$$\varphi_j = \varphi_{j-1} ba \varphi_{j-2}, \text{ for every even integer } j \geq 4,$$

while

$$\varphi_j = \varphi_{j-1} ab \varphi_{j-2}, \text{ for every odd integer } j \geq 3.$$

Then, we observe that

$$\varphi_j = \varphi_{j-2} ab \varphi_{j-3} ba \varphi_{j-2}, \text{ for every even integer } j \geq 4, \qquad (8.37)$$

while

$$\varphi_j = \varphi_{j-2} ba \varphi_{j-3} ab \varphi_{j-2}, \text{ for every odd integer } j \geq 5. \qquad (8.38)$$

Since φ_1, φ_2 and φ_3 are palindromes, we deduce again by induction that the word φ_j is a palindrome for every positive integer j. This ends the proof. □

8.4.5 Proof of Theorem 8.4.4

We are now ready to complete the proof of Theorem 8.4.4.

Proof [Proof of Theorem 8.4.4] Let a and b be distinct positive integers. We are going to prove that the real number $\zeta_{f'}$ defined in Theorem 8.3.2 is an extremal number.

Note first that, as a consequence of Theorem 8.4.4, the real number $\zeta_{f'}$ is neither rational nor quadratic. Let $(n_j)_{j \geq 1}$ be the sequence of positive integers defined in (8.36). Set also $Q_j = q_{n_j}$, where p_n/q_n denotes the nth convergent to $\zeta_{f'}$.

In view of Proposition 8.4.5 and Inequalities (8.34), there exists a positive real number c_1 such that the system

$$\begin{aligned} |x_0 \zeta_{f'} - x_1| &\leq c_1 Q_j^{-1}, \\ |x_0 \zeta_{f'}^2 - x_2| &\leq c_1 Q_j^{-1}, \\ |x_0| &\leq Q_j, \end{aligned} \qquad (8.39)$$

has a non-zero integer solution $(x_0^{(j)}, x_1^{(j)}, x_2^{(j)})$ for every positive integer

$j \geq 4$. More precisely, we have $(x_0^{(j)}, x_1^{(j)}, x_2^{(j)}) = (q_{n_j}, p_{n_j}, p_{n_j-1})$. In particular, $x_0^{(j)} = Q_j$.

We are now going to prove the existence of a positive real number c such that

$$Q_{j+1} \leq c Q_j^\varphi \tag{8.40}$$

for every positive integer j.

We argue by induction. We infer from (8.37), (8.38) and Lemma 8.2.3 that there exist two positive real numbers c_2 and c_3 such that

$$c_2 < \frac{Q_{j+1}}{Q_j Q_{j-1}} < c_3, \tag{8.41}$$

for every integer $j \geq 2$. Without loss of generality, we can assume that

$$c_3 \geq \left\{ \frac{(c_2 Q_1)^\varphi}{Q_2}, \frac{(c_2 Q_2)^{1/\varphi}}{Q_1} \right\}. \tag{8.42}$$

Set $c_4 := c_2^\varphi/c_3$ and $c_5 := c_3^\varphi/c_2$. We will prove by induction that

$$c_4 \, Q_{j-1}^\varphi \leq Q_j \leq c_5 \, Q_{j-1}^\varphi \tag{8.43}$$

for every integer $j \geq 2$. For $j = 2$, this follows from (8.41) and (8.42). Let us assume that (8.43) is satisfied for an integer $j \geq 2$. By (8.41), we obtain

$$c_2 \, Q_j^\varphi \left(Q_j^{1-\varphi} Q_{j-1} \right) < Q_{j+1} < c_3 \, Q_j^\varphi \left(Q_j^{1-\varphi} Q_{j-1} \right).$$

Since $\varphi(\varphi - 1) = 1$, it follows that

$$c_2 \, Q_j^\varphi \left(Q_j Q_{j-1}^{-\varphi} \right)^{1-\varphi} < Q_{j+1} < c_3 \, Q_j^\varphi \left(Q_j Q_{j-1}^{-\varphi} \right)^{1-\varphi}.$$

We then deduce from (8.43) that

$$\left(c_2 c_5^{1-\varphi} \right) Q_j^\varphi < Q_{j+1} < \left(c_3 c_4^{1-\varphi} \right) Q_j^\varphi.$$

By definition of c_4 and c_5, and since $\varphi(\varphi - 1) = 1$, this gives

$$c_4 \, Q_j^\varphi < Q_{j+1} < c_5 \, Q_j^\varphi,$$

which proves that (8.43) is true for $j + 1$. Consequently, Inequality (8.40) is established.

Let X be a sufficiently large real number. There exists an integer $j \geq 4$ such that $Q_j \leq X < Q_{j+1}$. We infer from (8.39) and (8.40) that

$$\begin{aligned}
\max\left\{ |x_0^{(j)} \zeta_{f'} - x_1^{(j)}|, |x_0^{(j)} \zeta_{f'}^2 - x_2^{(j)}| \right\} &< c_1 Q_j^{-1} \\
&\leq c_1 X^{-\log Q_j / \log X} \\
&\leq c_1 X^{-\log Q_j / \log Q_{j+1}} \\
&\leq c_6 X^{-1/\varphi},
\end{aligned}$$

for a suitable positive real number c_6.

This proves that $\zeta_{f'}$ is an extremal number, concluding the proof. □

8.4.6 Palindromic density of an infinite word

For an infinite word $a = a_1 a_2 \cdots$ let us denote by $n_1 < n_2 < \ldots$ the increasing (finite or infinite) sequence of all the lengths of the prefixes of a that are palindromes. We define the *palindromic density* of a, denoted by $d_p(a)$, by setting $d_p(a) = 0$ if only a finite number of prefixes of a are palindromes and, otherwise, by setting

$$d_p(a) := \left(\limsup_{j \to +\infty} \frac{n_{j+1}}{n_j} \right)^{-1}.$$

Clearly, for every infinite word a we have

$$0 \leq d_p(a) \leq 1.$$

Furthermore, if $a = uu\cdots$ is a periodic word, then either $d_p(a) = 0$ or $d_p(a) = 1$, and the latter holds if, and only if, there exist two (possibly empty) palindromes v and w such that $u = vw$. On the other hand, an eventually periodic word that begins with arbitrarily long palindromes is purely periodic. Thus, the palindromic density of an eventually periodic word is either maximal or minimal.

S. Fischler (Fischler 2006) proved that the Fibonacci word has the highest palindromic density among aperiodic infinite words. We state his result without proof.

Theorem 8.4.6 *Let a be a non-eventually periodic word. Then,*

$$d_p(a) \leq \frac{1}{\varphi},$$

where φ is the Golden Ratio. Furthermore, the bound is sharp and reached by the Fibonacci word.

This explains *a posteriori* why the Fibonacci continued fraction was a good candidate to be an extremal number.

8.5 Explicit examples for the Littlewood conjecture

As we have already seen, the theory of continued fractions ensures that, for every real number ξ, there exist infinitely many positive integers q such that

$$q \cdot \|q\xi\| < 1, \tag{8.44}$$

where $\|\cdot\|$ denotes the distance to the nearest integer. In particular, for all pairs (α, β) of real numbers, there exist infinitely many positive integers q such that

$$q \cdot \|q\alpha\| \cdot \|q\beta\| < 1.$$

In this section we consider the *Littlewood conjecture* (Littlewood 1968), a famous open problem in simultaneous Diophantine approximation. It claims that in fact, for any given pair (α, β) of real numbers, a slightly stronger result holds, namely

$$\inf_{q \geq 1} q \cdot \|q\alpha\| \cdot \|q\beta\| = 0. \qquad (8.45)$$

We will see how the theory of continued fractions and combinatorics on words can be combined to construct a large class of explicit pairs satisfying this conjecture. In the sequel, we denote by **L** the set of pairs of real numbers satisfying Littlewood's conjecture, that is,

$$\mathbf{L} := \left\{ (\alpha, \beta) \in \mathbb{R}^2 \mid \inf_{q \geq 1} q \cdot \|q\alpha\| \cdot \|q\beta\| = 0 \right\}.$$

8.5.1 Two useful remarks

Let us denote by

$$\mathbf{Bad} := \left\{ \xi \in \mathbb{R} \mid \inf_{q \geq 1} q \cdot \|q\xi\| > 0 \right\}$$

the set of *badly approximable* real numbers. A straightforward consequence of Inequalities (8.15) is that a real number belongs to **Bad** if, and only if, the sequence of partial quotients in its continued fraction expansion is bounded.

Our first remark is a trivial observation: If α or β has unbounded partial quotients, then the pair (α, β) satisfies the Littlewood conjecture. Combined with a classical theorem of É. Borel (Borel 1909), a notable consequence of this fact is the following proposition.

Proposition 8.5.1 *The set* **L** *has full Lebesgue measure.*

Our second remark is concerned with pairs of linearly dependent numbers.

Proposition 8.5.2 *Let α and β be two real numbers such that 1, α and β are linearly dependent over $\mathbb{Q}$. Then, the pair (α, β) belongs to* **L**.

Proof Let (α, β) be a pair of real numbers such that $1, \alpha$ and β are linearly dependent over $\mathbb{Q}$. If α or β is rational, then (α, β) belongs to the set **L**. Thus, we assume that α and β are irrational.

Let q be a positive integer such that

$$\|q\alpha\| < \frac{1}{q}.$$

By assumption, there exist integers a, b and c not all zeroes, such that

$$a\alpha + b\beta + c = 0.$$

Since α and β are irrational, a and b are both non-zero. Thus, $\|qa\alpha\| = \|qb\beta\| < a/q$. This gives

$$\|qab\alpha\| < \frac{ab}{q} \quad \text{and} \quad \|qab\beta\| < \frac{a^2}{q}.$$

Setting $Q = qab$, we then obtain

$$Q \cdot \|Q\alpha\| \cdot \|Q\beta\| < \frac{a^5 b^3}{Q}.$$

Since q can be taken arbitrarily large, the pair (α, β) belongs to **L**, concluding the proof. □

From now on, we say that (α, β) is a *non-trivial pair* of real numbers if the following conditions hold:

(i) α and β both belong to **Bad**,
(ii) $1, \alpha$ and β are linearly independent over $\mathbb{Q}$.

In view of the remarks above, it is natural to focus our attention on non-trivial pairs and to ask whether there are examples of non-trivial pairs (α, β) satisfying Littlewood's conjecture.

8.5.2 The problem of explicit examples

Recently, in a important paper, M. Einsiedler, A. Katok and E. Lindenstrauss (Einsiedler, Katok, and Lindenstrauss 2006) used an approach based on the theory of dynamical systems to prove the following outstanding result regarding Littlewood's conjecture.

Theorem 8.5.3 *The complement of* **L** *in* $\mathbb{R}^2$ *has Hausdorff dimension zero.*

Since **Bad** has full Hausdorff dimension, Theorem 8.5.3 implies that non-trivial examples for the Littlewood conjecture do exist. In particular, for every real number α in **Bad**, there are many non-trivial pairs (α, β) in **L**. Unfortunately, this result says nothing about the following simple question:

*Given a real number α in **Bad**, can we construct explicitly a real number β such that (α, β) is a non-trivial pair satisfying the Littlewood conjecture?*

The aim of this section is to answer this question. We will use an elementary construction based on the theory of continued fractions.

Let $\alpha := [0, a_1, a_2, \ldots]$ be a real number whose partial quotients are bounded, say by an integer $M \geq 2$. With any increasing sequence of positive integers $n = (n_i)_{i \geq 1}$ and any sequence $t = (t_i)_{i \geq 1}$ taking its values in $\{M+1, M+2\}$, we associate a real number $\beta_{n,t}$ as follows. For every positive integer j, let us denote by u_j the prefix of length j of the infinite word $a_1 a_2 \cdots$ and let $\widetilde{u}_j$ be the *mirror* of u_j. Then, we set

$$\beta_{n,t} := [0, \widetilde{u_{n_1}}, t_1, \widetilde{u_{n_2}}, t_2, \widetilde{u_{n_3}}, t_3, \ldots].$$

We will show that if the sequence n increases sufficiently rapidly, then the pair $(\alpha, \beta_{n,t})$ provides a non-trivial example for the Littlewood conjecture. More precisely, we will prove the following result established in (Adamczewski and Bugeaud 2006a).

Theorem 8.5.4 *Let ε be a positive real number with $\varepsilon < 1$. Keeping the previous notation and under the additional assumption that*

$$\liminf_{i \to +\infty} \frac{n_{i+1}}{n_i} > \frac{4 \log(M+3)}{\varepsilon \log 2}, \tag{8.46}$$

the pair $(\alpha, \beta_{n,t})$ is a non-trivial pair satisfying

$$q \cdot \|q\alpha\| \cdot \|q\beta_{n,t}\| \leq \frac{1}{q^{1-\varepsilon}}.$$

In particular, the pair $(\alpha, \beta_{n,t})$ belongs to **L**.

Proof We keep the notation of the theorem. Let $(p_j/q_j)_{j \geq 1}$ denote the sequence of convergents to α and let $(r_j/s_j)_{j \geq 1}$ denote the sequence of convergents to $\beta_{n,t}$. Set $m_j = n_1 + n_2 + \ldots + n_j + (j-1)$.

By Lemma 8.2.4, we have

$$\frac{s_{m_j-1}}{s_{m_j}} = [0, a_1, \ldots, a_{n_j}, t_{j-1}, a_1, \ldots, a_{n_{j-1}}, t_{j-2}, a_1, \ldots, t_1, a_1, \ldots, a_{n_1}].$$

Thus, Lemma 8.2.1 implies that

$$\|s_{m_j}\alpha\| \leq s_{m_j} q_{n_j}^{-2}. \tag{8.47}$$

On the other hand, (8.15) ensures that

$$s_{m_j} \cdot \|s_{m_j} \beta_{n,t}\| < 1. \tag{8.48}$$

Combining (8.47) and (8.48), it remains to prove that

$$s_{m_j} q_{n_j}^{-2} < \frac{1}{(s_{m_j})^{1-\varepsilon}},$$

that is,

$$q_{n_j} > (s_{m_j})^{1-\varepsilon/2}. \tag{8.49}$$

In order to prove (8.49), we will use the formalism of continuants introduced in Subsection 8.2.3 combined with the following simple idea: If the sequence n increases very quickly, then the word $a_1 a_2 \cdots a_{n_j}$ is much longer than the word $t_{j-1} a_1 \cdots a_{n_{j-1}} t_{j-2} \cdots t_1 a_1 \cdots a_{n_1}$. Since all these integers are bounded by $M+2$, we obtain that the integer $q_{n_j} = K(a_1, a_2, \ldots, a_{n_j})$ is much larger than

$$K_j := K(t_{j-1}, a_1, \ldots, a_{n_{j-1}}, t_{j-2}, \ldots, t_1, a_1, \ldots, a_{n_1}).$$

Furthermore, since Lemma 8.2.3 implies that

$$q_{n_j} K_j \leq s_{m_j} \leq 2 q_{n_j} K_j, \tag{8.50}$$

we will get the desired result.

Let us now give more details on this computation. Since the partial quotients of $\beta_{n,t}$ are bounded by $M+2$, Lemma 8.2.2 gives that

$$K_j < (M+3)^{m_{j-1}+1}$$

and, also,

$$s_{m_j} \geq 2^{(m_j-1)/2}.$$

Consequently,

$$K_j \leq \frac{1}{2}(s_{m_j})^{\delta_j},$$

where

$$\delta_j := \frac{m_{j-1}+1}{m_j-1} \cdot \frac{2\log(M+3)}{\log 2} + \frac{2}{m_j-1}.$$

On the other hand, an easy computation starting from Inequality (8.46) shows that

$$\liminf_{j \to +\infty} \frac{m_j}{m_{j-1}} > \frac{4\log(M+3)}{\varepsilon \log 2}.$$

This implies that $\delta_j < \varepsilon/2$ for every integer j large enough, and, consequently, that

$$K_j < \frac{1}{2} \cdot (s_{m_j})^{\varepsilon/2},$$

for every integer j large enough. Inequality (8.49) then follows from (8.50).

To end the proof, it now remains to prove that 1, α and $\beta_{n,t}$ are linearly independent over $\mathbb{Q}$. We assume that they are dependent and we aim at deriving a contradiction. In the rest of this proof, the constants implied by the symbols $\gg$ and $\ll$ do not depend on the positive integer j.

By assumption, there exists a non-zero triple of integers (a, b, c) such that

$$a\alpha + b\beta_{n,t} + c = 0 \ .$$

Thus,

$$\|s_{m_j} a\alpha\| = \|s_{m_j} b\beta_{n,t}\| \leq |b| \cdot \|s_{m_j} \beta_{n,t}\| \ll \frac{1}{s_{m_j}} \ll \frac{1}{q_{n_j} K_j}. \qquad (8.51)$$

We infer from Lemma 8.2.2 that

$$|s_{m_j} \alpha - s_{m_j-1}| \gg \frac{s_{m_j}}{q_{n_j}^2} = \frac{K_j}{q_{n_j}}.$$

Note that, for every integer j large enough, we have

$$|s_{m_j} a\alpha - s_{m_j-1} a| < \frac{1}{2},$$

thus,

$$\|s_{m_j} a\alpha\| = |s_{m_j} a\alpha - s_{m_j-1} a| = |a| \cdot |s_{m_j} \alpha - s_{m_j-1}| \gg \frac{K_j}{q_{n_j}}.$$

Since K_j tends to infinity with j, this contradicts (8.51) if j is large enough. This concludes the proof. $\square$

8.6 Exercises and open problems

Exercise 8.1 Prove that there is no irrational real number ξ satisfying a stronger form of Theorem 8.4.1 in which the exponent of X in (8.32) is less than -1. You may use the theory of continued fractions.

Exercise 8.2 Prove that the pair $(\sqrt{2}, e)$ satisfies the Littlewood conjecture. You may use the continued fraction expansion of e.

Exercise 8.3 (Open problem) Prove that the pair $(\sqrt{2}, \sqrt{3})$ satisfies the Littlewood conjecture.

Exercise 8.4 Let α be an irrational real number whose continued fraction expansion begins with arbitrarily large palindromes. Prove that the Littlewood conjecture is true for the pair $(\alpha, 1/\alpha)$ and that, moreover, we have

$$\liminf_{q \to +\infty} q^2 \cdot \|q\alpha\| \cdot \|q/\alpha\| < +\infty.$$

Exercise 8.5 (Open problem) The base-b expansion of an algebraic irrational number cannot be generated by a morphism. To this end, it would be sufficient to establish that, if ξ is an algebraic irrational number, then

$$\limsup_{n \to +\infty} \frac{p(n, \xi, b)}{n^2} = +\infty$$

holds for every integer base $b \geq 2$.

Exercise 8.6 (Open problem) Prove that

$$\lim_{n \to +\infty} p(n, \pi, b) - n = +\infty$$

holds for every integer base $b \geq 2$.

Exercise 8.7 (Open problem) The base-b expansion of an algebraic irrational number cannot begin with arbitrarily large palindromes.

Exercise 8.8 (Open problem) The continued fraction expansion of an algebraic irrational number of degree ≥ 3 cannot be generated by a finite automaton.

8.7 Notes

Section 8.1

Many other examples of normal numbers with respect to a given integer base have been worked out (see for instance (Copeland and Erdős 1946) and (Bailey and Crandall 2002)). In contrast, no natural example of a normal number seems to be known.

Conjecture 8.1.4 is sometimes attributed to É. Borel after he suggested that $\sqrt{2}$ could be normal with respect to the base 10 (Borel 1950).

A consequence of Conjecture 8.1.4 would be that the digits 0 and 1 occur with the same frequency in the binary expansion of any algebraic number. In (Bailey, Borwein, Crandall, et al. 2004) the authors proved that, an algebraic real number α of degree d being given, there exists a positive c such that, for every sufficiently large integer N, there are at least $cN^{1/d}$ non-zero digits among the first N digits of the binary expansion of α.

A famous open problem of K. Mahler (Mahler 1984) asks whether there are irrational algebraic numbers in the triadic Cantor set. This corresponds to a special instance of Conjecture 8.1.8.

As a complement to Theorem 8.1.6, Y. Bugeaud and J.-H. Evertse

(Bugeaud and Evertse 2008) established that, if ξ is an algebraic irrational number, then

$$\limsup_{n\to+\infty} \frac{p(n,\xi,b)}{n(\log n)^{0.09}} = +\infty$$

holds for every integer base $b \geq 2$.

Partial results in the direction of Cobham's conjecture (Theorem 8.1.7) were obtained in (Loxton and van der Poorten 1988). A classical result of G. Christol (Christol 1979) is related to Theorem 8.1.7: given an integer q that is a power of a prime number p, a Laurent power series $\sum_{n=-k}^{\infty} a_n T^n \in \mathbb{F}_q((T))$ is algebraic over the field $\mathbb{F}_q(T)$ if, and only if, the infinite word $a_0 a_1 \cdots$ is p-automatic (see also the paper (Christol, Kamae, Mendès France, et al. 1980)). More references about automatic sequences and automatic real numbers can be found in the monograph (Allouche and Shallit 2003).

A recent application of Proposition 8.1.21 to repetitive patterns that should occur in the binary expansion of algebraic numbers is given in (Adamczewski and Rampersad 2008).

It was also recently observed in (Adamczewski 2009) that the transcendence of real numbers whose base-b expansion is a Sturmian word can be obtained by combining Roth's theorem with some results from (Berthé, Holton, and Zamboni 2006). As a consequence of classical Diophantine results, it follows that the complexity of the number e (and of many other classical transcendental numbers) satisfies $\lim_{n\to+\infty} p(n,e,b) - n = +\infty$. In contrast, the best lower bound for the complexity of π seems to be $p(n,\pi,b) \geq n+1$ for every positive integer n, as follows from (Morse and Hedlund 1938).

The first p-adic version of the Schmidt subspace theorem is due to H. P. Schlickewei (Schlickewei 1976). A survey of recent applications of the Schmidt subspace theorem can be found in (Bilu 2008). See also (Waldschmidt 2006) or (Waldschmidt 2008) for a survey of known results about base-b expansions and continued fraction expansions of algebraic numbers.

Section 8.2

A survey of recent works involving the mirror formula can be found in (Adamczewski and Allouche 2007).

Section 8.3

Proposition 8.3.1 is a special instance of the main result proved in (Adamczewski and Bugeaud 2005). For more general transcendence results regarding continued fractions involving repetitive patterns, see also (Adamczewski and Bugeaud 2007b), (Adamczewski and Bugeaud 2005) as well as (Adamczewski, Bugeaud, and Davison 2006). All these results extend in particular those obtained in (Baker 1962), (Queffélec 1998), (Queffélec 2000) and (Allouche, Davison, Queffélec, et al. 2001).

Some generalisations of Proposition 8.3.3 can be found in (Adamczewski and Bugeaud 2007c). In contrast, there are only a few results about transcendental numbers whose base-b expansion involves some symmetric pattern (see (Adamczewski and Bugeaud 2006b)).

Section 8.4

Theorem 8.4.4 also leads to some results (see (Roy 2003b)) related to a famous conjecture due to E. Wirsing concerning the approximation of real numbers by algebraic numbers of bounded degree (Wirsing 1960).

It was proved in (Bugeaud and Laurent 2005) that ξ and ξ^2 are uniformly simultaneously very well approximated by rational numbers when the real number ξ belongs to a large class of Sturmian continued fractions.

Section 8.5

A classical result regarding the Littlewood conjecture is that any pair of algebraic numbers lying in a same cubic number field satisfies the Littlewood conjecture (Cassels and Swinnerton-Dyer 1955). Note that a weaker result than Theorem 8.5.3 was obtained previously by different techniques in (Pollington and Velani 2000).

9
Analysis of digital functions and applications

Michael Drmota,

Peter J. Grabner

9.1 Introduction: digital functions

Digital functions in a rather informal and general sense are functions defined in a way depending on the digits in some digital representation of the integers. In the simplest case the digital representation is the q-adic representation and the dependence of the function on the digits is additive as for the sum-of-digits function given by

$$s_q\left(\sum_{k=0}^{K} a_k q^k\right) = \sum_{k=0}^{K} a_k,$$

which also serves as the most prominent example for such functions. As a very general reference for results on digital functions, we refer to (Allouche and Shallit 2003). We remark that depending on the point of view such maps $f : \mathbb{N} \to A$ can be seen as (arithmetic) functions or sequences. The aim of this chapter is to study various asymptotic and limiting properties of such functions.

For the convenience of the reader we collect the basic definitions as given in (Allouche and Shallit 2003).

Automatic sequences

Definition 9.1.1 A sequence $(v(n))_{n \in \mathbb{N}}$ is called q-*automatic*, if the collection of sequences

$$K_q(v) = \left\{ \left(v\left(q^k n + \ell\right)\right)_{n \in \mathbb{N}} \mid k \in \mathbb{N}, 0 \leq \ell < q^k \right\}, \quad \text{'the } q\text{-kernel'}, \quad (9.1)$$

is finite.

Combinatorics, Automata and Number Theory, ed. Valérie Berthé and Michel Rigo.
Published by Cambridge University Press. ©Cambridge University Press 2010.

This definition is equivalent to saying that the value $v(n)$ can be determined by a deterministic finite automaton on the q-adic digits of n. Furthermore, this definition is equivalent to saying that the sequence $(v(n))_{n\in\mathbb{N}}$ is the image of a fixed point of a morphism of constant length q on a finite alphabet (see the discussion in (Allouche and Shallit 2003, Chapters 5 and 6), and Sections 1.2.4, 1.3.2 and 3.4). We remark that for this definition the set of values A of v is simply a finite set without any further structure.

Regular sequences

If the set A is a ring (in most of the examples this is $\mathbb{Z}$, $\mathbb{F}_q$, $\mathbb{R}$ or $\mathbb{C}$), then the structure of the ring A can be used for the following definition (cf. (Allouche and Shallit 2003, Chapter 16)).

Definition 9.1.2 Let A be a ring. Then an A-valued sequence $(f(n))_{n\in\mathbb{N}}$ is said to be *q-regular*, if the A-module generated by the q-kernel (9.1) is finitely generated.

This is equivalent to saying that there is a positive integer r, sequences $(f_1(n))_{n\in\mathbb{N}} = (f(n))_{n\in\mathbb{N}}, \ldots, (f_r(n))_{n\in\mathbb{N}}$ and a map $\mathbf{M} : \{0, \ldots, q-1\} \to A^{r\times r}$, such that for all $n \in \mathbb{N}$ and all $b \in \{0, \ldots, q-1\}$

$$\begin{pmatrix} f_1(qn+b) \\ f_2(qn+b) \\ \vdots \\ f_r(qn+b) \end{pmatrix} = \mathbf{M}(b) \begin{pmatrix} f_1(n) \\ f_2(n) \\ \vdots \\ f_r(n) \end{pmatrix}. \tag{9.2}$$

See Section 11.2.2 for an example of a 2-regular sequence.

q-Additive functions

For the following definition it is more convenient to view the sequence of values $(v(n))_{n\in\mathbb{N}}$ as a function on the positive integers taking its values in an abelian group. In the most important examples the groups are $\mathbb{R}$, $\mathbb{T} = \mathbb{R}/\mathbb{Z}$ or $\mathbb{Z}/m\mathbb{Z}$. The group law will be written additively.

Definition 9.1.3 Let A be an abelian group. A function $f : \mathbb{N} \to A$ is called *q-additive*, if for all $n \in \mathbb{N}$, all $k \in \mathbb{N}$, and all $0 \leq \ell < q^k$

$$f\left(q^k n + \ell\right) = f(q^k n) + f(\ell) \tag{9.3}$$

holds. If there is no dependence on k on the right-hand side, i.e.,

$$f\left(q^k n + \ell\right) = f(n) + f(\ell), \tag{9.4}$$

then f is called *completely q-additive*.

From (9.3) it follows by induction that

$$f\left(\sum_{k=0}^{K} a_k q^k\right) = \sum_{k=0}^{K} f(a_k q^k),$$

which shows that a q-additive function f is determined by the values $f(aq^k)$, $a \in \{1, \ldots, q-1\}$, $k \in \mathbb{N}$ and $f(0) = 0$. A completely q-additive function is determined by the values $f(1), \ldots, f(q-1)$.

The values aq^k can be viewed as the (additive) 'building blocks' of the positive integers with respect to q-adic numeration. This is a direct analogy to classical additive functions as studied in analytic number theory (*cf.* (Tenenbaum 1995), (Elliott 1979), (Elliott 1980), (Elliott 1985)). In this case the (multiplicative) 'building blocks' of the positive integers, namely the prime powers are used to define

$$f(p_1^{e_1} p_2^{e_2} \cdots p_k^{e_k}) = f(p_1^{e_1}) + f(p_2^{e_2}) + \cdots + f(p_k^{e_k}).$$

These functions and the properties of their value distribution are usually studied by probabilistic methods ('Kubilius models'). In Section 9.3 analogous models will be presented for q-additive functions.

Remark 9.1.4 A completely q-additive function f taking its values in a ring A is q-regular, since the A-module generated by f and the constant function 1 equals the A-module generated by the q-kernel.

q-Multiplicative functions

Multiplicative functions are defined analogous to additive functions using a multiplicative structure (usually the multiplicative group of the field $\mathbb{R}$ or $\mathbb{C}$ or the multiplicative semigroup of the ring $\mathbb{Z}$).

Definition 9.1.5 Let A be a monoid (written multiplicatively). A function $f : \mathbb{N} \to A$ is called *q-multiplicative*, if for all $n \in \mathbb{N}$, all $k \in \mathbb{N}$, and all $0 \leq \ell < q^k$

$$f\left(q^k n + \ell\right) = f(q^k n) f(\ell) \tag{9.5}$$

holds. If there is no dependence on k on the right-hand side, *i.e.*,

$$f\left(q^k n + \ell\right) = f(n) f(\ell), \tag{9.6}$$

then f is called *completely q-multiplicative*.

From (9.5) it follows by induction that

$$f\left(\sum_{k=0}^{K} a_k q^k\right) = \prod_{k=0}^{K} f(a_k q^k), \qquad (9.7)$$

which shows that a q-additive function f is determined by the values $f(aq^k)$, $a \in \{1, \ldots, q-1\}$, $k \in \mathbb{N}$, and $f(0) = 1$. A completely q-multiplicative function is determined by the values $f(1), \ldots, f(q-1)$.

Remark 9.1.6 A completely q-multiplicative function f taking its values in a ring A is q-regular, since the A-module generated by f equals the A-module generated by the q-kernel.

Remark 9.1.7 If $(f(n))$ is a q-automatic sequence taking its values in the finite set A, the indicator function $\mathbb{1}_{\{a\}}(f(n))$ (for $a \in A$) can be expressed in terms of a matrix product involving the transition matrices of the underlying finite automaton: let $\mathcal{A} = (Q, \{0, \ldots, q-1\}, \delta, \{i_0\}, A, \tau)$ be the DFAO defining f. For $a \in \{0, \ldots, q-1\}$ define the $Q \times Q$-matrix $\mathbf{M}_\delta(a)$ by

$$(m_\delta(a))_{ij} = \begin{cases} 1 & \text{if } \delta(i, a) = j \\ 0 & \text{otherwise} \end{cases}$$

and the vectors $\mathbf{v} = (1, 0, \ldots, 0)$ with the entry 1 in position i_0 and $\mathbf{w}$

$$w_\ell = \begin{cases} 1 & \text{if } \tau(\ell) = a \\ 0 & \text{otherwise.} \end{cases}$$

Then

$$\mathbb{1}_{\{a\}}(f(n)) = \mathbf{v} \prod_{j=0}^{K} \mathbf{M}_\delta(a_j(n)) \mathbf{w} \qquad (9.8)$$

gives this representation.

Since the underlying DFA $(Q, \{0, \ldots, q-1\}, \delta, \{i_0\})$ recognises all q-adic representations, the matrix

$$\mathbf{M}_\delta = \mathbf{M}_\delta(0) + \mathbf{M}_\delta(1) + \cdots + \mathbf{M}_\delta(q-1) \qquad (9.9)$$

has the eigenvalue q and all eigenvalues are of modulus $\leq q$. Furthermore, the eigenvalue q has the same geometric and algebraic multiplicity (*cf.* Section 1.4).

9.2 Asymptotic analysis of digital functions

In this section we study the asymptotic behaviour of summatory functions of various digital functions. We develop several techniques and discuss their strengths and weaknesses. As a general theme, we can say that the values $(f(n))_{n\in\mathbb{N}}$ for digital functions usually vary very irregularly. Nevertheless, the summatory function

$$F(N) = \sum_{n<N} f(n) \qquad (9.10)$$

shows a rather 'smooth' asymptotic behaviour, in many cases there even exists an exact formula for $F(N)$. Usually, this formula involves a periodic continuous function of $\log_q N$.

9.2.1 Completely additive functions

Completely additive functions, such as the sum-of-digits function have been the first and simplest type of digital functions that has been studied. The study of the sum (9.10) is especially easy in this case and will be used to develop several techniques. Furthermore, this will be a preparation for the development in Section 9.3, where we consider f as a random variable on the finite probability space

$$\Omega_N = \{0, \ldots, N-1\}, \quad \mathbb{P}_N(A) = \frac{\text{Card } A}{N}.$$

In this interpretation, $F(N)/N$ is simply the mean of the function f on the space Ω_N.

Theorem 9.2.1 *Let $f : \mathbb{N} \to \mathbb{R}$ be a completely q-additive function given by the values $f(0) = 0, f(1), \ldots, f(q-1)$. Then there exists a continuous periodic function Φ_f of period 1, such that*

$$\sum_{n<N} f(n) = C_f N \log_q N + N\Phi_f(\log_q N) \qquad (9.11)$$

with $C_f = \frac{1}{q}(f(1) + \cdots + f(q-1))$.

Remark 9.2.2 Notice that the fractional parts of $\log_q N$ are dense in the interval $[0, 1]$. This allows us to interpret continuity of Φ in the sense that the discrete set of points

$$\{(\{\log_q N\}, F(N)/N - C_f \log_q N) \mid N \in \mathbb{N}\}$$

can be fit by one (and only one!) graph of a continuous function.

Remark 9.2.3 The function Φ_f is continuous and nowhere differentiable (except for trivial cases). This fact has been observed in (Tenenbaum 1997) for rather general periodic functions occurring in the context of digital functions.

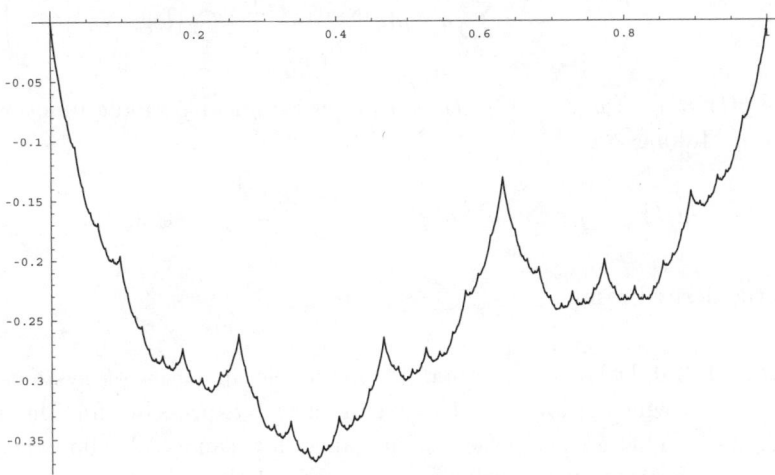

Fig. 9.1 Plot of the function Φ_f for the sum-of-digits function in base 3.

Proof [(Delange 1975)] We will present Delange's method for general completely q-additive functions f. As a first step we rewrite $F(N)$ as

$$\sum_{n<N} f(n) = \sum_{k=0}^{K} \sum_{n<N} f(a_k(n))$$

with $K = \lfloor \log_q N \rfloor = \log_q N - \{\log_q N\}$. The kth digit $a_k(x)$ of a real number x is a periodic function of period q^{k+1}, which is constant on intervals $[nq^k, (n+1)q^k)$. Using $a_k(x) = a_0(xq^{-k})$, the inner sum can be rewritten as an integral

$$\sum_{n<N} f(a_k(x)) = \int_0^N f(a_k(x))\,dx = NC_f + q^k \int_0^{Nq^{-k}} (f(a_0(x)) - C_f)\,dx.$$

Thus we have by inverting the order of summation

$$\sum_{n<N} f(n) = C_f(K+1)N + q^K \sum_{k=0}^{K} q^{-k} \int_0^{q^k Nq^{-K}} (f(a_0(x)) - C_f)\,dx.$$

Since $C_f = \frac{1}{q}\int_0^q f(a_0(x))\,dx$, the last integral vanishes for $k > K$. Thus the sum can be extended to an infinite sum, and we get

$$\sum_{n<N} f(n) = C_f N \log_q N$$

$$+ N\left(q^{-\{\log_q N\}}\sum_{k=0}^{\infty} q^{-k} g(q^{\{\log_q N\}+k}) - C_f\{\log_q N\} + C_f\right),$$

where $g(t) = \int_0^t f(a_0(x)) - C_f)\,dx$ is a piecewise linear periodic function of period q. Taking

$$\Phi_f(x) = q^{-x}\sum_{k=0}^{\infty} q^{-k} g(q^{k+x}) + C_f(1-x) \quad x \in [0,1]$$

gives the desired result. $\square$

Remark 9.2.4 Delange's method is applicable in cases of systems of numeration, which allows us to give a closed expression for the single digits. This is the case for instance for canonical number systems (see Section 2.4), where this method has been applied by (Grabner, Kirschenhofer, and Prodinger 1998) to prove an asymptotic formula for the sum of digits on the Gaussian integers.

We compute the Fourier-coefficients of Φ_f

$$\widehat{\Phi_f}(m) = \int_0^1 \Phi_f(x) e^{-2\pi i m x}\,dx$$

$$= \sum_{k=0}^{\infty} \int_0^1 q^{-k-x} g(q^{k+x}) e^{-2\pi i m x}\,dx - C_f \int_0^1 x e^{-2\pi i m x}\,dx.$$

We substitute $u = q^{k+x}$ in the integral and observe that the ranges of integration for u add then up to $[1,\infty)$

$$\widehat{\Phi_f}(m) = \frac{1}{\log q}\int_1^{\infty} g(u) u^{-\frac{2\pi i m}{\log q}-2}\,du + \frac{C_f}{2\pi i m}.$$

This integral is easily recognised as a Mellin transform and can be computed by partial integration using the fact that the integrand is periodic with period q.

We write

$$\mathcal{M}g(s) = \int_1^{\infty} g(u) u^{s-1}\,du, \tag{9.12}$$

and using the notation $s_m = \frac{2\pi i m}{\log q}$

$$\widehat{\Phi_f}(m) = \frac{1}{\log q} \mathcal{M}g(-1 - s_m) + \frac{C_f}{2\pi i m}.$$

Since g is bounded, the integral (9.12) converges for $\Re s < 0$. By partial integration we get

$$\mathcal{M}g(s) = g(u)\frac{u^s}{s}\bigg|_{u=1}^{\infty} - \frac{1}{s}\int_1^{\infty} (f(a_0(u)) - C_f)u^s \, du$$

$$= \frac{C_f}{s} - \frac{1}{s}\int_1^{\infty} (f(a_0(u)) - C_f)u^s \, du.$$

The function $f(a_0(u))$ is periodic with period q and constant on intervals between consecutive integers, which allows us to compute the last integral

$$\frac{C_f}{s} - \frac{1}{s}\int_1^{\infty} (f(a_0(u)) - C_f)u^s \, du$$

$$= \frac{C_f}{s} - \frac{1}{s(s+1)} \sum_{k=1}^{q-1} f(k) \sum_{n=0}^{\infty} \left((qn+k+1)^{s+1} - (qn+k)^{s+1}\right) - \frac{C_f}{s(s+1)}$$

$$= \frac{C_f}{s+1} - \frac{q^{s+1}}{s(s+1)} \sum_{k=1}^{q-1} f(k) \left(\zeta\left(-s-1, \frac{k+1}{q}\right) - \zeta\left(-s-1, \frac{k}{q}\right)\right),$$
(9.13)

where

$$\zeta(s, \alpha) = \sum_{n=0}^{\infty} \frac{1}{(n+\alpha)^s}$$

denotes the Hurwitz zeta function; the Riemann zeta function is given by $\zeta(s) = \zeta(s, 1)$. The poles of the Hurwitz zeta functions in (9.13) at $s = -2$ cancel; furthermore, the poles at $s = -1$ in (9.13) cancel by the fact that $\zeta(0, \alpha) = \frac{1}{2} - \alpha$.

Putting everything together gives (for $m \neq 0$)

$$\widehat{\Phi_f}(m) = -\frac{1}{s_m(s_m+1)} \sum_{k=1}^{q-1} f(k) \left(\zeta\left(s_m, \frac{k+1}{q}\right) - \zeta\left(s_m, \frac{k}{q}\right)\right)$$

and

$$\widehat{\Phi_f}(0) = \frac{1}{\log q} \int_1^{\infty} g(u) u^{-2} \, du - \frac{C_f}{2}$$

$$= C_f \left(\frac{1}{\log q} - \frac{1}{2}\right) - \frac{1}{\log q} \sum_{k=1}^{q-1} f(k) \log \frac{\Gamma((k+1)/q)}{\Gamma(k/q)}.$$

Here we have used

$$\zeta'(0,\alpha) = -\frac{1}{2}\log(2\pi) + \log\Gamma(\alpha).$$

If $f = s_q$ is the sum of digits function, these formulas for the Fourier-coefficients can be simplified further. Since $s_q(k) = k$ for $k = 0,\ldots,q-1$, Abelian summation yields

$$\widehat{\Phi_{s_q}}(m) = -\frac{(q-1)}{s_m(s_m+1)}\zeta(s_m) \quad \text{for } m \neq 0$$

and

$$\widehat{\Phi_{s_q}}(0) = \frac{q-1}{2}\left(\frac{1}{\log q} - \frac{\log 2\pi}{\log q} - \frac{1}{2}\right) + \frac{1}{2},$$

which is the result of (Delange 1975).

Remark 9.2.5 Finding the maxima and minima of the periodic function Φ_f is a tricky task. For the sum-of-digits function that has been done in (Foster 1987), (Foster 1991), and (Foster 1992). For the counting functions of q-adic digits $\geq d$, the minimum of the corresponding periodic function has been determined in (Grabner 2004).

Remark 9.2.6 There is an elementary method for proving (9.11) which runs along the same lines as the proof of Theorem 9.2.7 below. This method can be generalised to recurrence based systems of numeration and other generalisations of base q numeration systems (cf. (Grabner and Tichy 1990), (Grabner and Tichy 1991) and also (Kirschenhofer and Tichy 1984), (Grabner and Rigo 2003)). Working out this proof of Theorem 9.2.1 is left as an exercise.

9.2.2 Multiplicative functions

Multiplicative functions with respect to numeration have been introduced to study statistical properties of additive functions via exponential sums (cf. (Delange 1972)). This is used in Section 9.3 to derive limit theorems of various kinds for additive functions.

In this section we study the summatory functions of completely q-multiplicative functions by elementary means. The following theorem was proved in (Grabner 1993).

Theorem 9.2.7 Let f be a complex valued completely q-multiplicative function satisfying

$$|1 + f(1) + \ldots + f(q-1)| > \max_{0 \leq a < q} |f(a)|. \tag{9.14}$$

Then there exists a continuous periodic function ψ of period 1, such that

$$F(N) = \sum_{n<N} f(n) = N^\rho e^{i\alpha \log_q N} \psi(\log_q N), \qquad (9.15)$$

where $\rho = \log_q |F(q)|$ and $\alpha = \arg(F(q))$.

Proof We write

$$N = \sum_{k=0}^{K} a_k(N) q^k$$

and set

$$N_\ell = \sum_{k=\ell}^{K} a_k(N) q^k.$$

Then we split the sum for $F(N)$

$$F(N) = \sum_{\ell=0}^{K} \sum_{n<a_\ell(N)q^\ell} f(N_{\ell+1}+n) = \sum_{\ell=0}^{K} f(N_{\ell+1}) F(a_\ell(N) q^\ell). \qquad (9.16)$$

Thus we have reduced the problem to the computation of $F(aq^\ell)$ for $a \in \{0,\ldots,q-1\}$, which can be done using the independence of the digits

$$F(aq^\ell) = \sum_{a_0=0}^{q-1} \cdots \sum_{a_{\ell-1}=0}^{q-1} \sum_{a_\ell=0}^{a-1} f(a_0) \cdots f(a_\ell) = F(q)^\ell F(a).$$

Inserting this into (9.16) gives

$$F(N) = \sum_{\ell=0}^{K} F(q)^\ell F(a_\ell(N)) f(a_{\ell+1}(N)) \cdots f(a_K(N)). \qquad (9.17)$$

We define a function $\varphi : [1,q] \to \mathbb{C}$ by

$$\varphi\left(\sum_{\ell=0}^{\infty} \frac{a_\ell}{q^\ell}\right) = \sum_{\ell=0}^{\infty} F(a_\ell) F(q)^{-\ell} \prod_{k=0}^{\ell-1} f(a_k). \qquad (9.18)$$

Notice that this series is geometrically convergent by the assumption (9.14).

The function φ is well defined and continuous, because

$$\varphi\left(\sum_{\ell=0}^{L}\frac{a_\ell}{q^\ell}\right) = \sum_{\ell=0}^{L} F(a_\ell) F(q)^{-\ell} \prod_{k=0}^{\ell-1} f(a_k)$$

$$= \sum_{\ell=0}^{L-1} F(a_\ell) F(q)^{-\ell} \prod_{k=0}^{\ell-1} f(a_k) + F(a_L - 1) F(q)^{-L} \prod_{k=0}^{L-1} f(a_k)$$

$$+ \prod_{k=0}^{L} f(a_k) \sum_{\ell=L+1}^{\infty} F(q-1) F(q)^{-\ell} f(q-1)^{\ell-L-1}$$

$$= \varphi\left(\sum_{\ell=0}^{L-1}\frac{a_\ell}{q^\ell} + \frac{a_L - 1}{q^L} + \sum_{\ell=L+1}^{\infty}\frac{q-1}{q^\ell}\right).$$

Furthermore, we have

$$\varphi(1) = 1 \quad \text{and} \quad \varphi(q) = F(q).$$

Using the function φ we can rewrite (9.17)

$$F(N) = F(q)^K \varphi\left(N q^{-K}\right) = N^\rho e^{i\alpha \log_q N} F(q)^{-\{\log_q N\}} \varphi\left(q^{\{\log_q N\}}\right).$$

Setting

$$\psi(\log_q N) = F(q)^{-\{\log_q N\}} \varphi\left(q^{\{\log_q N\}}\right)$$

yields the desired result (notice that $\psi(0) = \psi(1)$). $\square$

Remark 9.2.8 Notice that condition (9.14) is trivially satisfied for f taking only positive values. In this case the function ψ is the quotient of a monotonically increasing function and a differentiable function. It is therefore differentiable almost everywhere. Nevertheless, it is not the integral of its derivative (except for trivial cases like $f \equiv 1$). An explanation for this phenomenon is given by (Tenenbaum 1997).

The following corollary describes the behaviour of $F(N)$, if (9.14) is not satisfied.

Corollary 9.2.9 *Let f be a completely q-multiplicative function satisfying*

$$|F(q)| = |1 + f(1) + \ldots + f(q-1)| = \max_{0 \leq a < q} |f(a)| = M, \qquad (9.19)$$

and $Q = \max_{1 \leq a < q} |F(a)|$. Then

$$|F(N)| \leq Q(K+1)M^K \leq Q N^{\log_q M} (\log_q N + 1).$$

If
$$|F(q)| = |1 + f(1) + \cdots + f(q-1)| < \max_{0 \le a < q} |f(a)| = M, \quad (9.20)$$

then
$$|F(N)| \le \frac{QM^{K+1}}{M - |F(q)|} \le \frac{QM}{M - |F(q)|} N^{\log_q M}.$$

Proof Every summand on the right-hand side of (9.17) can be estimated by $Q|F(q)|^\ell M^{K-\ell}$. Considering the two cases $M = |F(q)|$ and $M > |F(q)|$ gives the two assertions. □

Example 9.2.10 (Barbolosi and Grabner 1996) Let p be a prime and write n and k in base p. Then Lucas' congruence asserts that

$$\binom{n}{k} \equiv \binom{a_0(n)}{a_0(k)} \binom{a_1(n)}{a_1(k)} \cdots \binom{a_L(n)}{a_L(k)} \pmod{p}, \quad (9.21)$$

where $n = \sum_{\ell=0}^{L} a_\ell(n) p^\ell$ and $k = \sum_{\ell=0}^{L} a_\ell(k) p^\ell$. From this congruence it follows immediately that (see also (Stein 1989))

$$\text{Card}\left\{0 \le k \le n \mid \binom{n}{k} \not\equiv 0 \pmod{p}\right\} = \prod_{\ell=0}^{L} (1 + a_\ell(n)),$$

which is a completely p-multiplicative function. Furthermore, for any multiplicative character χ modulo p the function

$$f_\chi(n) = \sum_{k=0}^{n} \chi\left(\binom{n}{k}\right) = \prod_{\ell=0}^{L} f_\chi(a_\ell(n)),$$

is completely p-multiplicative.

Using Fourier analysis on the finite group $(\mathbb{Z}/p\mathbb{Z})^*$ we obtain for $a \not\equiv 0 \pmod{p}$

$$\text{Card}\left\{0 \le k \le n < N \mid \binom{n}{k} \equiv a \pmod{p}\right\} = \frac{1}{p-1} \sum_\chi \sum_{n<N} \overline{\chi}(a) f_\chi(n).$$

Since for $\chi \ne \chi_0$ (χ_0 denotes the principal character) $|F_\chi(p)| < F_{\chi_0}(p) = \frac{p(p-1)}{2}$ and $\max_{0 \le a < p} |f_\chi(a)| \le p-1$, the term for $\chi = \chi_0$ is the asymptotic main term. This implies that the binomial coefficients not divisible by p are uniformly distributed in the prime residue classes modulo p.

For $p = 3$ we have the simple formulas (*cf.* also (Wolfram 1984))

$$\text{Card}\left\{0 \le k \le n \mid \binom{n}{k} \equiv 1 \pmod{3}\right\} = \frac{1}{2} 2^{c_1(n)} \left(3^{c_2(n)} + 1\right)$$

and

$$\operatorname{Card}\left\{0 \leq k \leq n \mid \binom{n}{k} \equiv 2 \pmod{3}\right\} = \frac{1}{2} 2^{c_1(n)} \left(3^{c_2(n)} - 1\right),$$

where $c_a(n)$ ($a = 1, 2$) denotes the number of digits a in the base 3 expansion of n.

Example 9.2.11 (Larcher 1996) The number of odd binomial coefficients in the nth row of Pascal's triangle is given by $2^{s_2(n)}$, where $s_2(n)$ denotes the binary sum-of-digits function. This is just the special case $p = 2$ of the last example. In this case the corresponding periodic function ψ has been investigated further by (Harborth 1977). Since the function is singular, the minimum cannot be found by differential calculus (the maximum is easily found to be 1 and to be attained at integer points). (Larcher 1996) gives an algorithm, which finds arbitrarily good approximations to the minimum.

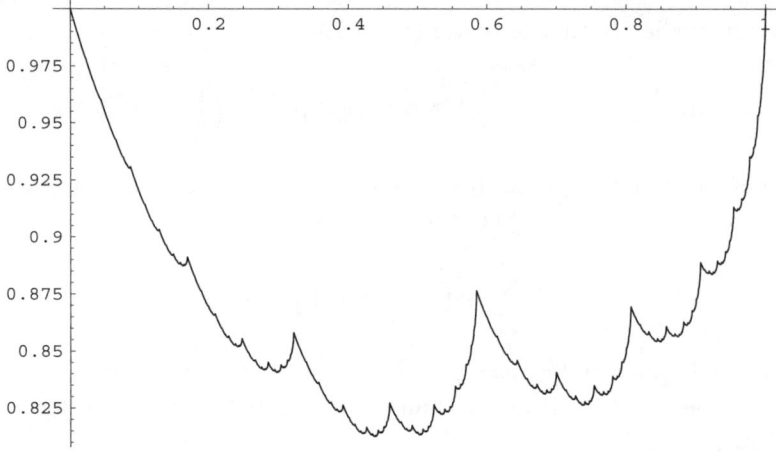

Fig. 9.2 Plot of the function ψ for the multiplicative function $2^{s_2(n)}$.

Remark 9.2.12 The two examples show that the number of binomial coefficients $\binom{n}{k}$ not divisible by p in the region $0 \leq k \leq n < N$ is of order of magnitude N^{ρ_p} with $\rho_p = \log_p \frac{p(p+1)}{2} < 2$, which implies that 'almost all' (in the sense of density) binomial coefficients are divisible by p. This fact and further results on the divisibility of binomial coefficients have been observed in (Singmaster 1974a), (Singmaster 1974b) and (Singmaster 1974c).

9.2.3 Divide-and-conquer recursions and Mellin–Perron techniques

A divide-and-conquer recursion is a relation of the form

$$h_n = \alpha\, h_{\lfloor n/2 \rfloor} + \beta\, h_{\lceil n/2 \rceil} + g_n \qquad (9.22)$$

(with suitable initial conditions). Such kinds of recurrences appear in several applications in the analysis of algorithms, for example in the analysis of the number of comparisons in sorting networks (Bose and Nelson 1962), (Hwang 1998), the Karatsuba multiplication (Knuth 1998), or also the Euclidean matching heuristic (Reingold and Tarjan 1981). The general scheme of all these algorithms is that to perform an operation (for instance, sorting) on a set of data of size n, at first the set is divided into parts of respective sizes $\lceil \frac{n}{2} \rceil$ and $\lfloor \frac{n}{2} \rfloor$. Then the algorithm is applied recursively to these smaller sets of data. The costs for merging the results for the original set of data are measured by the term g_n. A first approach to a unified study of such recurrences was done in (Flajolet and Golin 1993). This was extended further in (Hwang 1998).

Furthermore, several digital functions satisfy a relation of the form (9.22). For example, if we consider the summatory function $S(N) = \sum_{n<N} s(n)$, where $s(n)$ denotes the binary sum-of-digits function, then by using the relations $s(2k) = s(k)$ and $s(2k+1) = s(k) + 1$ we directly get

$$\begin{aligned} S(N) &= \sum_{2k<N} s(2k) + \sum_{2k+1<N} s(2k+1) \\ &= \sum_{k<\lceil N/2 \rceil} s(k) + \sum_{k<\lfloor N/2 \rfloor} (s(k)+1) \\ &= S(\lceil N/2 \rceil) + S(\lfloor N/2 \rfloor) + \lfloor N/2 \rfloor. \end{aligned}$$

It is not difficult to guess the (correct) growth rate of h_n if one knows the growth rate of g_n. However, it is usually a non-trivial problem to get precise asymptotic information on h_n. There is, however, a general method based on Dirichlet series. This method has its origin in classical analytic number theory (*cf.* (Tenenbaum 1995)). Set

$$H(s) = \sum_{n \geq 1} \frac{h_{n+1} - h_n}{n^s} \quad \text{and} \quad G(s) = \sum_{n \geq 1} \frac{g_{n+1} - g_n}{n^s}. \qquad (9.23)$$

Then, by distinguishing between odd and even numbers we get

$$H(s) = \alpha \sum_{n\geq 1} \frac{h_{\lfloor (n+1)/2 \rfloor} - h_{\lfloor n/2 \rfloor}}{n^s} + \beta \sum_{n\geq 1} \frac{h_{\lceil (n+1)/2 \rceil} - h_{\lceil n/2 \rceil}}{n^s} + G(s)$$

$$= \alpha \sum_{k\geq 0} \frac{h_{k+1} - h_k}{(2k+1)^s} + \beta \sum_{k\geq 1} \frac{h_{k+1} - h_k}{(2k)^s} + G(s)$$

$$= \frac{\alpha + \beta}{2^s} H(s) + \alpha(h_1 - h_0) + \alpha \sum_{k\geq 1} (h_{k+1} - h_k) \left(\frac{1}{(2k+1)^s} - \frac{1}{(2k)^s} \right) + G(s)$$

which leads to the representation

$$H(s) = \frac{\alpha(h_1 - h_0) + F(s) + G(s)}{1 - (\alpha + \beta)2^{-s}}, \qquad (9.24)$$

where

$$F(s) = \alpha \sum_{k\geq 1} (h_{k+1} - h_k) \left(\frac{1}{(2k+1)^s} - \frac{1}{(2k)^s} \right)$$

usually has a smaller abscissa of convergence than $H(s)$. Thus the expression (9.24) provides us with the analytic continuation of $H(s)$ to a larger domain together with information about the poles of $H(s)$.

Applying the Mellin–Perron summation formula (*cf.* (Tenenbaum 1995)) we obtain

$$h_n = h_1 + \sum_{k=1}^{n-1} (h_{k+1} - h_k) = h_1 + \frac{1}{2\pi i} \int_{c-i\infty}^{c+i\infty} H(s) \frac{n^s}{s}\, ds. \qquad (9.25)$$

Here c has to be chosen large enough to make the series $H(s)$ absolutely (and therefore uniformly) convergent on the line $\Re(s) = c$. Usually, the integral in (9.25) is only convergent in the sense of a Cauchy principal value, which makes this analysis a bit delicate.

Using the analytic continuation of $H(s)$ makes it possible to deform the contour of integration to the left (in order to make the exponent of n occurring in the integral smaller). This is again a standard technique in analytic number theory. The problem when shifting the line of integration to the left comes from the convergence of the integral in (9.25). As a general fact about Dirichlet series, the growth along vertical lines becomes stronger for smaller values of $\Re(s)$ (*cf.* (Hardy and Riesz 1964)). Thus additional information on the growth of $F(s)$ and $G(s)$ along vertical lines is needed. This is usually the technically most elaborate step of this method.

If the convergence of the integral is proved, then (9.25) can be rewritten as (for $c' < c$)

$$h_n = h_1 + \sum_{\substack{\text{poles of } H(s) \\ \text{with } c' < \Re(s) < c}} \text{Res}\left(\frac{n^s H(s)}{s}\right) + \frac{1}{2\pi i} \int_{c'-i\infty}^{c'+i\infty} H(s) \frac{n^s}{s} \, ds. \quad (9.26)$$

Usually, the poles mainly come from the zeroes of the denominator in (9.24), which are given by

$$s_k = \frac{\log(\alpha + \beta)}{\log 2} + \frac{2k\pi i}{\log 2}, \quad k \in \mathbb{Z}.$$

The residues at these points take the form

$$\text{Res}_{s=s_k}\left(\frac{n^s H(s)}{s}\right) = n^\rho \frac{e^{2k\pi i \log_2 n}}{s_k \log 2} \left(\alpha(h_1 - h_0) + F(s_k) + G(s_k)\right) \quad (9.27)$$

with $\rho = \log_2(\alpha + \beta)$. The sum of these residues is then the main term in the asymptotic expansion of h_n and can be written as

$$h_n = h_1 + n^\rho \sum_{k \in \mathbb{Z}} \frac{e^{2k\pi i \log_2 n}}{s_k \log 2} \left(\alpha(h_1 - h_0) + F(s_k) + G(s_k)\right) + \mathcal{O}\left(n^{c'}\right)$$

$$= n^\rho H(\log_2 n) + \mathcal{O}\left(n^{c'}\right).$$

The series can be interpreted as the Fourier series of a periodic function H in the dyadic logarithm of n. This is the same periodicity phenomenon that we encountered in the study of additive and multiplicative functions before. The Fourier series usually converges only very slowly, which reflects the lack of smoothness of the function H. A collection of arguments, which can be used to prove growth estimates for Dirichlet series in the context of digital functions and divide-and-conquer recurrences as well as arguments for the convergence of the Fourier series of the occurring periodic functions is given in (Grabner and Hwang 2005). The fact that Dirichlet generating functions of q-regular functions have an analytic continuation to the whole complex plane, as well as information on their poles, was derived in (Allouche, Mendès France, and Peyrière 2000).

Example 9.2.13 For the binary sum-of-digits function $s_2(n)$, we have (*cf.* (Flajolet, Grabner, Kirschenhofer, et al. 1994))

$$\sum_{n=1}^{\infty} \frac{s_2(n) - s_2(n-1)}{n^s} = \frac{2^s - 2}{2^s - 1} \zeta(s),$$

where $\zeta(s)$ is the Riemann ζ-function. Using the relevant version of the

Mellin–Perron formula, we get

$$\sum_{n<N} s_2(n) = \frac{1}{2\pi i} \int_{2-i\infty}^{2+i\infty} \frac{2^s-2}{2^s-1} \zeta(s) \frac{N^{s+1}}{s(s+1)} \, ds. \qquad (9.28)$$

Since the growth of $\zeta(\sigma+it)$ for fixed σ and $|t| \to \infty$ is very well understood (cf. (Titchmarsh 1986)), the line of integration can be shifted to $\Re(s) = -\frac{1}{4}$ (it is known that $\zeta(-\frac{1}{4}+it) = \mathcal{O}(|t|^{3/4})$). Collecting residues (there is a double pole at $s=0$, which corresponds to the $N \log N$-term) gives

$$\sum_{n<N} s_2(n)$$
$$= \frac{1}{2} N \log_2 N - \frac{\log_2 \pi}{2} - \frac{1}{2 \log 2} - \frac{1}{4} - N \sum_{k \in \mathbb{Z} \setminus \{0\}} \frac{\zeta(s_k)}{s_k(s_k+1) \log 2} e^{2k\pi i \log_2 N}.$$
$$(9.29)$$

In this case it can be shown that the remainder term vanishes. This is exactly the Fourier series that we got by Delange's method before.

Remark 9.2.14 The vanishing of the remainder term in (9.29) comes from the fact that the integral

$$\frac{1}{2\pi i} \int_{-\frac{1}{4}-i\infty}^{-\frac{1}{4}+i\infty} \frac{2^s-2}{2^s-1} \zeta(s) \frac{N^{s+1}}{s(s+1)} \, ds$$

vanishes for $N \in \mathbb{N}$. There is a rather general theorem, which ensures the vanishing of integrals occurring as remainder terms in this context (cf. (Hwang 1998) and for a slightly more general formulation (Grabner and Hwang 2005)).

In order to overcome the difficulties originating from the slow convergence of the integral in (9.25), in (Grabner and Hwang 2005) double differences and higher-order Mellin–Perron formulas were studied. Instead of the function $H(s)$ in (9.23) the function

$$\tilde{H}(s) = \sum_{n=1}^{\infty} \frac{h_{n+1} - 2h_n + h_{n-1}}{n^s}$$

was used, which allows us to compute h_n by the formula

$$h_n = nh_1 + \frac{1}{2\pi i} \int_{c-i\infty}^{c+i\infty} \tilde{H}(s) \frac{n^{s+1}}{s(s+1)} \, ds.$$

This formula gives a gain in convergence. On the other hand this gain has to be paid for by more complicated expressions and the fact that the poles of $\tilde{H}(s)$ are one unit further to the left of the poles of $H(s)$, which makes the growth of $\tilde{H}(s)$ worse on vertical lines. Nevertheless, in many cases the double differencing technique gives easier proofs for the convergence of the Mellin–Perron integrals.

Even if the integral in (9.25) is only conditionally convergent, in (Drmota, Grabner, and Liardet 2008) rather elaborate estimates could be used to obtain an analogue to Theorem 9.3.17 for the summatory function of a block-multiplicative function on the Gaussian integers. In this case the method was applied to the Gaussian integers, which made alternative techniques still more difficult or even impossible to apply.

9.2.4 Generalisations

9.2.4.1 q-regular functions

The study of the summatory functions of q-regular functions follows the same line of ideas as the study of completely multiplicative functions. The only difference is that the products of scalars used there have to be replaced by matrix products. By the lack of commutativity, this makes the order of factors in all occurring products significant.

More precisely, let $f : \mathbb{N} \to \mathbb{R}$ be a real-valued q-regular function. Then by Definition 9.1.2 and the discussion after the definition there exist functions $f = f_1, f_2, \ldots, f_r$ and a map $\mathbf{M} : \{0, \ldots, q-1\} \to \mathbb{R}^{r \times r}$ such that (9.2) holds. We write $\mathbf{f}(n) = (f_1(n), \ldots, f_r(n))^T$ and use (9.2) to obtain

$$\mathbf{f}\left(\sum_{\ell=0}^{L} a_\ell q^\ell\right) = \prod_{\ell=0}^{L} \mathbf{M}(a_\ell)\mathbf{f}(0), \tag{9.30}$$

which allows us to write

$$f(n) = \mathbf{v}_1 \prod_{\ell=0}^{L} \mathbf{M}(a_\ell(n))\mathbf{v}_2$$

with $\mathbf{v}_1 = (1, 0, \ldots, 0)$ and $\mathbf{v}_2 = \mathbf{f}(0)$.

Then by arguing along the same lines as in the proof of Theorem 9.2.7 we get the following theorem.

Theorem 9.2.15 *Let $f : \mathbb{N} \to \mathbb{R}$ be a q-regular function and let $\mathbf{M}$ be the matrix function related to f by (9.2). Let $\mathbf{F} = \mathbf{M}(0) + \cdots + \mathbf{M}(q-1)$ and assume that $\mathbf{F}$ has a unique eigenvalue $\lambda > 0$ of maximal modulus and that* this eigenvalue has algebraic multiplicity 1. Assume further that

$$\lambda > \max_a \|\mathbf{M}(a)\|$$

for some matrix norm $\|\cdot\|$. Denote by λ_2 the modulus of the second largest eigenvalue. Then there exists a periodic continuous function Φ such that

$$\sum_{n<N} f(n) = N^{\log_q \lambda} \Phi(\log_q N) + \mathcal{O}\left(N^{\log_q \lambda_2}\right) + \mathcal{O}(\log N).$$

Remark 9.2.16 The asymptotic behaviour of $\sum_{n<N} f(n)$ like a pure power of N corresponds to the fact that the part of $\mathbf{F}$ corresponding to λ is diagonalisable. If there are different Jordan-blocks occurring for λ, then powers of the logarithm occur in the asymptotic main terms. This happens, as can be seen from the result for q-additive functions.

Remark 9.2.17 The value $f(n)$ of a q-regular function in terms of the q-adic digits of n is given the matrix product (9.30). Since all possible finite sequences of digits occur as digital expansions of the positive integers, the question of finding the maximal growth order of $f(n)$ is related to extremal matrix products as studied in Chapter 11.

9.2.4.2 q-automatic functions

Let $f(n)$ be an A-valued q-automatic function and $a \in A$. By Remark 9.1.7 the indicator function $\mathbb{1}_{\{a\}}(f(n))$ can be expressed in terms of a matrix valued completely q-multiplicative function by (9.8). This makes the ideas developed before applicable for the computation of

$$F(N) = \sum_{n<N} \mathbb{1}_{\{a\}}(f(n)). \tag{9.31}$$

The question of existence of the limit $\lim_{N\to\infty} F(N)/N$, the density of the set $\{n \in \mathbb{N} \mid f(n) = a\}$, is of special interest in this context (cf. Remark 9.2.20 below).

Applying the same reasoning as above we can prove

Theorem 9.2.18 Let $f(n)$ be an A-valued q-automatic function and $a \in A$. Assume that q is the dominating eigenvalue of the matrix M_δ defined by (9.9) (i.e., all other eigenvalues have modulus $< q$). Then there is a continuous periodic function Ψ of period 1 such that

$$F(N) = N\Psi(\log_q N) + o(N).$$

Corollary 9.2.19 *Under the assumptions of Theorem 9.2.18 let* $\mathbf{A} = \lim_{k\to\infty} q^{-k}\mathbf{M}_\delta^k$. *Then the function* Ψ *is constant, if*

$$\mathbf{AM}_\delta(0) = \mathbf{AM}_\delta(1) = \ldots = \mathbf{AM}_\delta(q-1) = \mathbf{Q} \qquad (9.32)$$

and

$$\mathbf{QM}_\delta(a) = \mathbf{Q} \quad \text{for } a = 0,\ldots,q-1. \qquad (9.33)$$

Proof Using the notation of Remark 9.1.7 we can write for $K = \lfloor \log_q N \rfloor$

$$F(N) = q^K \mathbf{v}\mathbf{G}(Nq^{-K})\mathbf{w} + o(N),$$

where

$$\mathbf{G}\left(\sum_{j=0}^\infty \frac{a_j}{q^j}\right) = \mathbf{A}\left(\mathbf{M}_\delta(1) + \cdots + \mathbf{M}_\delta(q-1)\right)$$

$$+ \sum_{j=1}^\infty q^{-j} \mathbf{A}\left(\mathbf{M}_\delta(0) + \cdots + \mathbf{M}_\delta(a_j-1)\right) \mathbf{M}_\delta(a_{j-1}) \cdots \mathbf{M}_\delta(a_0)$$

$$+ \frac{1}{q-1}\mathbf{A}\left(\mathbf{M}_\delta - \mathbf{M}_\delta(0)\right). \qquad (9.34)$$

Here and in the sequel we omit the subscript δ. This function is continuous on the interval $[1,q]$ and

$$\mathbf{G}(1) = \frac{1}{q-1}\mathbf{A}(\mathbf{M}_\delta - \mathbf{M}_\delta(0)) \quad \text{and} \quad \mathbf{G}(q) = \frac{q}{q-1}\mathbf{A}(\mathbf{M}_\delta - \mathbf{M}_\delta(0)).$$

This proves the theorem.

The function Ψ in the theorem is constant, if $\mathbf{G}(x)$ is proportional to x for $x \in [1,q]$. Inserting the integer values $\{1,\ldots,q\}$ for x gives the conditions (9.32). Inserting $1 + \frac{a}{q}$ gives (9.33). Inserting these two conditions into (9.34) gives that $\mathbf{G}(x) = x\mathbf{Q}$ for $x \in [1,q]$. This proves the corollary. □

Remark 9.2.20 Corollary 9.2.19 gives a condition for the existence of the density of the set

$$S = \{n \in \mathbb{N} \mid f(n) = a\}$$

for a given q-automatic function f. It is known (*cf.* (Allouche and Shallit 2003, Chapter 8)) that the density does not always exist. It is also known that the logarithmic density of S

$$\lim_{N\to\infty} \frac{1}{\log N} \sum_{n<N} \frac{\mathbb{1}_S(n)}{n}$$

always exists (*cf.* (Allouche and Shallit 2003, Theorem 8.4.8)) and that the two densities are equal, if the density exists.

Remark 9.2.21 In the proof of (Allouche and Shallit 2003, Theorem 8.4.8) the logarithmic density is related to the residue of the Dirichlet series

$$\varphi_S(s) = \sum_{n=1}^{\infty} \frac{\mathbb{1}_S(n)}{n^s}$$

at $s = 1$. The logarithmic averaging process has the effect that the (possible) other poles of $\varphi_S(s)$ with $\Re(s) = 1$ (that correspond to the Fourier series of the periodic oscillation occurring in Theorem 9.2.18 as was shown in Section 9.2.3) can be disregarded. This corresponds to the fact that

$$\sum_{n<N} \frac{1}{n^{1+it}} = \begin{cases} \log N + \mathcal{O}(1) & \text{if } t = 0 \\ \mathcal{O}(1) & \text{if } t \neq 0, \end{cases}$$

which means that the logarithmic averaging singles out the Fourier coefficient of index zero (the mean of the periodic function).

9.2.4.3 Block-additive and block-multiplicative functions

Block-additive functions have been introduced in (Cateland 1992) as a more flexible generalisation of q-additive functions. Given a map $f : \{0, \ldots, q-1\}^L \to \mathbb{R}$ with $f(0, \ldots, 0) = 0$ the corresponding block-additive function is given by

$$s_f(n) = \sum_{k=0}^{\infty} f(a_k(n), \ldots, a_{k+L-1}(n)). \tag{9.35}$$

Examples for such functions are block-counting functions. Block-additive functions are q-regular by the observation that the $\mathbb{R}$-module generated by the kernel is generated by the function f and the functions $n \to f(b_1, b_2, b_r, a_0(n), \ldots, a_{L-r-1}(n))$, $1 \leq r \leq L-2$, $b_1, \ldots, b_r \in \{0, \ldots, q-1\}$. As for q-additive functions, we can expect that the dominating eigenvalue occurring in the matrix $\mathbf{F}$ in Theorem 9.2.15 has different algebraic and geometric multiplicity, which makes this theorem inapplicable.

A means to study block-additive functions, and also an object of study in their own right, are block-multiplicative functions given by a map $g : \{0, \ldots, q-1\}^L \to \mathbb{R}$ with $g(0, \ldots, 0) = 1$ and

$$m_g(n) = \prod_{k=0}^{\infty} g(a_k(n), \ldots, a_{k+L-1}(n)). \tag{9.36}$$

Such functions are again q-regular.

A block-multiplicative function defines a $q^L \times q^L$-matrix $\mathbf{U}$ given by

$$u_{B,C} = \begin{cases} \frac{m_g(BC)}{m_g(C)} & \text{for } m_g(C) \neq 0 \\ 0 & \text{otherwise} \end{cases} \quad \text{for } B, C \in \{0, \ldots, q-1\}^L.$$

Here we have used the convention that m_g evaluated at a block of digits is the same as m_g evaluated at the number represented by that block. As usual BC denotes the concatenation of the blocks B and C. As for the (less explicit) matrix $\mathbf{F}$ in Theorem 9.2.15, this matrix U allows us to express $\sum_{n<N} f(n)$ in terms of sums of matrix products involving $\mathbf{U}$. The asymptotic behaviour of $\sum_{n<N} f(n)$ then depends on the dominating eigenvalue λ of $\mathbf{U}$ and its algebraic and geometric multiplicities. For a more detailed discussion we refer to (Barat and Grabner 2001).

Given a block-additive function s_f, $\exp(ts_f(n))$ clearly defines a block-multiplicative function. A simple idea to study the moments of a block-additive function is to use

$$\sum_{n<N} s_f(n)^k = \left(\frac{d}{dt}\right)^k \left(\sum_{n<N} \exp(ts_f(n))\right)\bigg|_{t=0}.$$

The most general theorem that was obtained in (Barat and Grabner 2001) gives an asymptotic formula for the summatory function of a product of several block-additive functions with a multiplicative function.

Proposition 9.2.22 *Let θ be a positive-valued block-multiplicative function and $f_1, \ldots, f_m$ arbitrary real-valued block-additive functions. Then the summatory function F of $\theta(n)f_1(n)\cdots f_m(n)$ satisfies*

$$F(N) = \sum_{n<N} \theta(n)f_1(n)\cdots f_m(n)$$

$$= N^{\log_q \lambda} \sum_{j=0}^{m} (\log_q N)^j \psi_j(\log_q N) + o(N^{\log_q \lambda_2}),$$

where the functions ψ_j are continuous and periodic with period 1, λ and λ_2 are the eigenvalues of the matrix $\mathbf{U}$ corresponding to θ of largest and second largest modulus.

Remark 9.2.23 This result includes, for instance, moments of q-additive functions such as the sum-of-digits function (*cf.* (Coquet 1986)), digital functions occurring in the study of binomial coefficients with given divisibility by a prime power (*cf.* (Carlitz 1967) and Example 9.2.25 below).

Remark 9.2.24 Since θ only attains positive values, the matrix U is fully populated with positive entries, which by the Perron–Frobenius theorem (Theorem 1.4.2) ensures the existence of a dominating eigenvalue λ of multiplicity 1.

Example 9.2.25 (Barat and Grabner 2001) As an application of block-multiplicative functions, further distribution properties of binomial coefficients can be obtained. These use an extension of Lucas' congruence (9.21) to prime powers by (Granville 1997). This congruence involves digital blocks of length L for a congruence (mod p^L). Results of the following kind could then be proved by applying summation formulas for block-additive and block-multiplicative functions. For $(a,p) = 1$ and $v_p(n)$ denoting the p-adic valuation (i.e., the highest power of p dividing n)

$$\operatorname{Card}\left\{(k,n) \mid 0 \leq k \leq n < N,\ v_p\left(\binom{n}{k}\right) = j,\text{ and } p^{-j}\binom{n}{k} \equiv a \bmod p^\ell\right\}$$
$$= \frac{1}{\varphi(p^\ell)} \operatorname{Card}\left\{(k,n) : 0 \leq k \leq n < N \text{ and } v_p\left(\binom{n}{k}\right) = j\right\} + \mathcal{O}\left(N^\beta\right) \tag{9.37}$$
$$= \frac{1}{\varphi(p^\ell)} N^\alpha \sum_{r=0}^{j} \psi_r^{(j)}(\log_p N)(\log_p N)^r + \mathcal{O}\left(N^\beta\right),$$

where $\psi_r^{(j)}$ are continuous periodic functions of period 1 and $\beta < \alpha = \log_p \frac{p(p+1)}{2}$.

9.2.4.4 A measure-theoretic method for the analysis of digital functions

Looking back at the derivation of asymptotic formulas for the summatory functions of regular, multiplicative, or block-multiplicative functions, we observe that the expression for

$$F(q^k) = \sum_{n < q^k} f(n)$$

is usually much simpler and much simpler to obtain than the formula for general N. Note that in (Grabner and Heuberger 2006) and (Grabner, Heuberger, and Prodinger 2005) a rather general technique has been developed, which can even be applied in multidimensional settings. This technique is based on the simple observation that the sequence of measures

$$\mu_k = \frac{1}{F(q^k)} \sum_{n < q^k} f(n) \delta_{nq^{-k}} \tag{9.38}$$

converges weakly to a limiting measure μ. (Here δ_x as usual denotes the Dirac measure, that is, a unit point mass at x.) Then the sum of $f(n)$ can be rewritten in terms of μ_k

$$\sum_{n<N} f(n) = F(q^k)\mu_k\left([0, Nq^{-k})\right),$$

where k has to be chosen such that $q^k > N$. If estimates for the error $|\mu([0,x)) - \mu_k([0,x))|$ are known, the right-hand side can be rewritten as

$$\sum_{n<N} f(n) = F(q^k)\mu\left([0, Nq^{-k})\right) + o(F(q^k)).$$

Estimates for the difference of measures of intervals can be obtained by (versions of) the Berry–Esseen inequality, which estimates $|\mu([0,x)) - \mu_k([0,x))|$ in terms of the difference of the Fourier transforms of the measures. By the product structure of the functions f and $F(q^k)$, the Fourier transform of μ_k can be expressed as a product (of scalar or matrix functions).

We explain the technique by an example that is motivated by applications of digital expansions in cryptography. For a more detailed explanation of the background and for all the details left out in the exposition we refer to (Grabner and Heuberger 2006).

Every positive integer n can be represented in the form

$$n = \sum_{k=0}^{K} a_k 2^k \text{ with } a_k \in \{-1, 0, 1\}.$$

Adding the extra digit -1 introduces some freedom, which is used to minimise the number of non-zero digits (the 'weight' of the representation). See also Section 2.2.2.2 and 2.2.2.3. The weight corresponds to the number of additions needed for multiplication by n in an abelian group using Horner's scheme. In cryptographic applications, especially in elliptic curve cryptography, the number of operations needed for the computation of multiples is an important parameter (see for instance the discussion of optimal multiplication algorithms in (Cohen, Frey, Avanzi, et al. 2006)).

In (Heuberger and Prodinger 2006) it was shown that the automaton in Figure 9.3 recognises all representations of minimal weight. We define $f(n)$ as the number of representations of n recognised by this automaton. By a careful investigation of the transitions in the automaton it can be proved that

$$f(n) = \mathcal{O}(n^\rho) \text{ with } \rho = \log_4\left(\frac{1+\sqrt{5}}{2}\right). \tag{9.39}$$

Remark 9.2.26 As was pointed out earlier, the question of determining

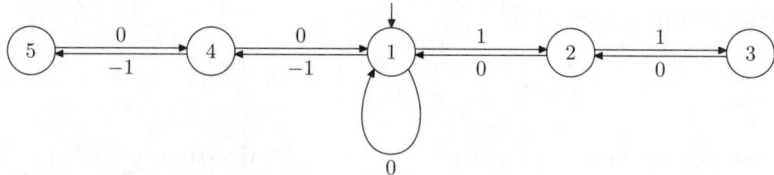

Fig. 9.3 Automaton recognising signed binary expansions of minimal weight from right to left. All states are terminal.

the maximal growth order of such functions is related to extremal matrix products as studied in Chapter 11 and therefore is rather hard in general.

Let $f_n(k)$ denote the number of representations of an integer k of minimal weight and length at most n. Since any representation of minimal weight is at most 1 digit longer than the usual binary expansion, $f_n(k) = f_{\lfloor \log_2 |k| \rfloor + 2}(k) = f(k)$ for $n \geq \lfloor \log_2 |k| \rfloor + 2$. We define a sequence of measures by

$$\mu_n = \frac{1}{M_n} \sum_{k \in \mathbb{Z}} f_n(k) \delta_{k 2^{-n}}, \qquad (9.40)$$

where δ_x denotes the unit point mass concentrated in x and

$$M_n = \sum_{k \in \mathbb{Z}} f_n(k).$$

We notice that all points $k 2^{-n}$ with $f_n(k) > 0$ lie in the interval $[-1, 1]$.

In order to compute the characteristic function of μ_n we consider the weighted adjacency matrix of the automaton in Figure 9.3 (using the notation $e(t) = e^{2\pi i t}$):

$$\mathbf{A}(t) = \begin{pmatrix} 1 & e(t) & 0 & e(-t) & 0 \\ 1 & 0 & e(t) & 0 & 0 \\ 0 & 1 & 0 & 0 & 0 \\ 1 & 0 & 0 & 0 & e(-t) \\ 0 & 0 & 0 & 1 & 0 \end{pmatrix}.$$

In the matrix $\mathbf{A}(t)$ a transition with label ℓ is represented by an entry $e(\ell t)$. Then we have

$$\widehat{\mu}_n(t) = \frac{1}{M_n} \sum_{k \in \mathbb{Z}} f_n(k) e\left(k 2^{-n} t\right)$$

$$= \frac{1}{M_n} v_1 \mathbf{A}\left(t 2^{-n}\right) \mathbf{A}\left(t 2^{-n+1}\right) \cdots \mathbf{A}(t/2) v_2 \quad (9.41)$$

with $v_1 = (1, 0, 0, 0, 0)$ and $v_2 = (1, 1, 1, 1, 1)^T$.

We notice that

$$M_n = (1,0,0,0,0)\mathbf{A}(0)^n(1,1,1,1,1)^T = C\alpha^n + \mathcal{O}(|\alpha_2|^n), \qquad (9.42)$$

where α and α_2 are the largest and second largest roots of the characteristic polynomial of $\mathbf{A}(0)$ given by

$$(x-1)(x+1)(x^3 - x^2 - 3x + 1),$$

and $C = \frac{1}{37}(14\alpha^2 + 5\alpha - 22)$, numerically

$$\alpha = 2.17009\ldots, \qquad \alpha_2 = -1.48119\ldots, \qquad C = 1.48055\ldots.$$

We will prove that (μ_n) weakly tends to a limit measure by showing that $\widehat{\mu}_n(0) = 1$ for all n and that $\widehat{\mu}_n(t)$ tends to a limit $\widehat{\mu}(t)$.

Lemma 9.2.27 *The sequence of measures μ_n defined by (9.40) converges weakly to a probability measure μ. The characteristic functions satisfy the inequality*

$$|\widehat{\mu}_n(t) - \widehat{\mu}(t)| = \begin{cases} \mathcal{O}(|t|2^{-\eta n}) & \text{for } |t| \leq 1 \\ \mathcal{O}(|t|^\eta 2^{-\eta n}) & \text{for } |t| \geq 1 \end{cases} \qquad (9.43)$$

with

$$\eta = \frac{\log \alpha - \log |\alpha_2|}{\log 2 + \log \alpha - \log |\alpha_2|} = 0.355251\ldots.$$

The constants implied by the $\mathcal{O}$-symbol are absolute.

Proof Equation (9.41) allows us to express $\widehat{\mu}_n(t)$ in terms of matrix products. Standard analysis of such products allows us to give estimates for $|\widehat{\mu}_m(t) - \widehat{\mu}_n(t)|$ for $n > m$, which give the desired estimates by letting n tend to infinity. The exponent η comes from a balancing argument used in an intermediate estimate. $\square$

In the next lemma we prove continuity of the measure μ. Our study of the Fourier expansion of the periodic main term as well as the remainder term estimate in (9.10) will depend on the modulus of continuity given here.

Lemma 9.2.28 *The measure μ satisfies*

$$\mu([x,y]) = \mathcal{O}\left((y-x)^\beta\right) \qquad (9.44)$$

for $\beta = \log_2 \alpha - \log_4\left(\frac{1+\sqrt{5}}{2}\right) = 0.770632\ldots > \frac{1}{2}$.

Proof Every interval $[x,y]$ can be approximated by dyadic intervals $[2^{-n}\lfloor x2^n\rfloor, 2^{-n}\lceil y2^n\rceil]$. The measure μ of such intervals can be computed as the limit of the measures μ_k of these intervals. The estimate (9.39) is used as a trivial bound for the measure. □

In order to give an error bound for the rate of convergence of the measures μ_n to the measure μ, we will use the following version of the Berry–Esseen inequality, which was proved in (Grabner 1997).

Proposition 9.2.29 *Let μ_1 and μ_2 be two probability measures with their Fourier transforms defined by*

$$\widehat{\mu}_k(t) = \int_{-\infty}^{\infty} e^{2\pi i t x}\, d\mu_k(x), \quad k=1,2.$$

Suppose that $(\widehat{\mu}_1(t)-\widehat{\mu}_2(t))t^{-1}$ is integrable on a neighbourhood of zero and μ_2 satisfies

$$\mu((x,y)) \le c|x-y|^\beta$$

for some $0 < \beta < 1$. Then the following inequality holds for all real x and all $T > 0$

$$|\mu_1((-\infty,x)) - \mu_2((-\infty,x))|$$

$$\le \left|\int_{-T}^{T} \widehat{J}(T^{-1}t)(2\pi i t)^{-1}\left(\widehat{\mu}_1(t)-\widehat{\mu}_2(t)\right)e^{-2\pi i x t}\,dt\right|$$

$$+ \left(c+\frac{1}{\pi^2}\right)T^{-\frac{2\beta}{2+\beta}} + \left|\frac{1}{2T}\int_{-T}^{T}\left(1-\frac{|t|}{T}\right)\left(\widehat{\mu}_1(t)-\widehat{\mu}_2(t)\right)e^{-2\pi i x t}\,dt\right|,$$

where

$$\widehat{J}(t) = \pi t(1-|t|)\cot \pi t + |t|.$$

Lemma 9.2.30 *The measures μ_n satisfy*

$$|\mu_n((x,y)) - \mu((x,y))| = \mathcal{O}\left(2^{-\theta n}\right) \tag{9.45}$$

uniformly for all $x,y \in \mathbb{R}$ with $\theta = \frac{2\beta\eta}{\eta(\beta+2)+2\beta} = 0.2168\ldots$.

Proof We apply Proposition 9.2.29 to the measures μ_n and μ. For this purpose we use the inequalities (9.43) to obtain

$$|\mu_n((-\infty, x)) - \mu((-\infty, x))| \ll 2^{-\eta n} \int_{-1}^{1} dt + 2^{-\eta n} \int_{1 \leq |t| \leq T} |t|^{\eta-1} dt$$

$$+ T^{-\frac{2\beta}{2+\beta}} + 2^{-\eta n} \frac{1}{T} \int_{-1}^{1} |t| \, dt + 2^{-\eta n} \frac{1}{T} \int_{1 \leq |t| \leq T} |t|^{\eta} \, dt \ll 2^{-\theta n}$$

by choosing $T = 2^{\theta \frac{2+\beta}{2\beta} n}$. □

Putting everything together, we have found

$$\sum_{n<N} f(n) = M_k \mu_k([0, n2^{-k})) = C\alpha^k \mu([0, n2^{-k})) + \mathcal{O}(|\alpha_2|^k) + \mathcal{O}(\alpha^k 2^{-\theta k})$$
(9.46)

with $k = \lfloor \log_2 N \rfloor + 2$. We set $\Phi(x) = C\alpha^{-x+2} \mu([0, 2^{x-2}])$ for $0 \leq x \leq 1$ and observe that $\Phi(0) = \Phi(1)$ by the fact that we can also choose $= \lfloor \log_2 N \rfloor + 3$ by the discussion in the beginning. Thus we can write

$$\sum_{n<N} f(n) = N^{\log_2 \alpha} \Phi(\{\log_2 N\}) + \mathcal{O}(N^{\log_2 \alpha - \theta}).$$
(9.47)

Remark 9.2.31 The main ingredients for the method to work are the following

(i) a good understanding of the Fourier-transforms of $\widehat{\mu}_k$ and $\widehat{\mu}$, for instance in terms of products of scalar or matrix functions, which come from the underlying q-regular or block-multiplicative structure of the function f
(ii) an estimate for the difference $|\widehat{\mu}_k(t) - \widehat{\mu}(t)|$ (Lemma 9.2.27)
(iii) an estimate for the measure-dimension of μ (Lemma 9.2.28) is needed in the Berry–Esseen-type inequality (Proposition 9.2.29). This estimate can be obtained by *a priori* estimates for $f(n)$ (like (9.39)), which can be tricky (*cf.* Chapter 11).
(iv) In higher-dimensional applications different versions of the Berry–Esseen inequality, for instance for balls in Euclidean space, are needed (Grabner, Heuberger, and Prodinger 2005, Proposition 1).

The method usually produces rather weak error terms, since the estimate for $|\widehat{\mu}_k(t) - \widehat{\mu}(t)|$ may not be the best possible, and this estimate is pulled through the Berry–Esseen inequality, which needs one further balancing. Nevertheless, the method avoids the somewhat intricate computations with the complicated explicit expressions for f needed for other approaches.

Remark 9.2.32 In (Okada, Sekiguchi, and Shiota 1995) suitably defined measures on $[0,1]$ were used to give exact formulas for the moments of the binary sum-of-digits function. Their construction uses the measure μ related to the function $e^{ts_2(n)}$. Since in this case $\mu_k([0, a2^{-k}]) = \mu([0, a2^{-k}])$ the approximating measures μ_k are not needed, as well as the application of the Berry–Esseen argument.

Remark 9.2.33 The measures μ occurring in this context in many cases can be interpreted as distributions of infinite series of dependent random variables. The dependence is coded by the underlying matrix structure and is therefore of a Markov type. This relates the measures to Bernoulli convolutions (*i.e.*, infinite series of independent random variables) such as studied by (Erdős 1939) and (Erdős 1940). For a survey on this subject we refer to (Peres, Schlag, and Solomyak 2000). Furthermore, we remark that all these measures are of pure type (either purely absolutely continuous, purely singular continuous, or consists only of point masses) by the Jessen–Wintner theorem (*cf.* (Elliott 1979, Lemma 1.22)).

9.3 Statistics on digital functions

Let f be a q-additive function and define the shorthand notation $f_j(n) = f(a_j(n)q^j)$. Then $f(n) = \sum_{k=0}^{K} f_j(n)$ with $K = \lfloor \log_q n \rfloor$.

We will make now extensive use of the probabilistic interpretation of f as a random variable. As above the underlying probability space is $\Omega_N = \{0, 1, \ldots, N-1\}$ with the uniform distribution, that is, every $n \in \Omega_N$ is equally likely.

The digits $a_j(n)$ and also $f_j(n)$ are then random variables, too. However, the essential observation is that the digits $a_j(n)$, $j \leq K$, are almost independent (in what follows, we will make this more precise). Thus, we can consider $f(n)$ as a sum of K almost independent random variables. It is therefore not unexpected that several results from sums of independent random variables transfer to asymptotic and distributional properties of q-additive functions. Note that for $N = q^L$ for some positive integer L the digits $(a_0(n), \ldots, a_{L-1}(n))$ are actually independent.

We will first survey general distributional results on (general) q-additive functions. In Section 9.3.3 we focus on completely q-additive functions where we can get much more precise results by using a generating function approach. Note that completely q-additive functions correspond to sums of almost independent and identically distributed random variables. More precise results are thus not unexpected.

There are several different types of distribution results known for q-additive functions that can be unified to some extent:

(i) existence of an asymptotic distribution of the values of f on $\mathbb{R}$ (Erdős–Wintner-type theorems): Section 9.3.1.3,
(ii) existence of a normal limit distribution for suitably renormalised values of f on $\mathbb{R}$ (central limit theorems): Section 9.3.1.4,
(iii) some results of these kinds are also known for n ranging through subsequences of the integers, such as the values of a polynomial or the primes: Section 9.3.2,
(iv) precise estimates for the number of n, where $f(n)$ attains a fixed value, for integer valued f (local limit theorems): Section 9.3.3,
(v) uniform distribution of the values of f in a compact abelian group (usually $\mathbb{Z}/m\mathbb{Z}$ and $\mathbb{T}$): Section 9.3.4.

9.3.1 General distributional results for additive functions

9.3.1.1 Approximation of digits by independent random variables

Our first goal is to make the statement that a q-additive function is a sum of almost independent random variables more precise. For this purpose we introduce an analogue to the number-theoretic Kubilius model (see (Elliott 1979), (Elliott 1980)) to the digital situation which was formulated by (Manstavičius 1997).

We start by considering infinite subsets of the non-negative integers $\mathbb{N} = \{0, 1, 2, \ldots\}$ that are defined by *digital restrictions*. For $0 \leq d < q$ and $j \geq 0$ set

$$E_j(d) = \{n \in \mathbb{N} \mid a_j(n) = d\}.$$

Furthermore, for every non-negative integer $k < q^{r+1}$ we consider the sets

$$K_r(k) = \bigcap_{j \leq r} E_j(a_j(k)) = \{n \in \mathbb{N} \mid n \equiv k \bmod q^{r+1}\}$$

that consist exactly of those $n \in \mathbb{N}$ with $a_j(n) = a_j(k)$ for all $j \leq r$. Note that the sets $K_r(k)$, $0 \leq k < q^{r+1}$, are disjoint arithmetic progressions. It is clear that the algebra $\mathcal{F}_r$ of subsets of $\mathbb{N}$ generated by the sets $E_j(d)$ for $0 \leq d < q$ and $j \leq r$ are precisely sets of the form

$$A = \bigcup_{k \in I} K_r(k), \qquad (9.48)$$

where I is any subset of $\{k \in \mathbb{N} \mid k < q^{r+1}\}$. Furthermore the asymptotic density of the sets $K_r(k)$ in $\mathbb{N}$ equals q^{-r-1}. It is therefore natural to define the probability of $A \subset \mathcal{F}_r$ by

$$\mathbb{P}(A) = \mathbb{P}_r(A) = \frac{\operatorname{Card} I}{q^{r+1}}.$$

By this definition $(\mathbb{N}, \mathcal{F}_r, \mathbb{P}_r)$ is a finite probability space, where the events $E_0(d_0), E_1(d_1), \ldots, E_r(d_r)$ are independent for any choice of numbers $0 \leq d_j < q$, $j \leq r$. Namely if we set $k_0 = d_0 + d_1 q + \cdots + d_r q^r$ then

$$\mathbb{P}_r \left(\bigcap_{0 \leq j \leq r} E_j(d_j) \right) = \mathbb{P}_r(K_k(k_0)) = q^{-r-1} = \prod_{0 \leq j \leq r} \mathbb{P}_r(E_j(d_j)).$$

Next observe that the function $f_j(n) = f(a_j(n) q^j)$ just depends on the j-digit $a_j(n)$ and is thus a $\mathcal{F}_r$-measurable function $f_j : \mathbb{N} \to \mathbb{R}$ (for $j \leq r$). Hence, it can be considered as a random variable Y_j. Due to the independence property of the sets $E_j(d_j)$ the random variables $Y_0, Y_1, \ldots, Y_r$ are independent, too.

The following *Fundamental Lemma* that is due to (Manstavičius 1997) quantifies the difference between $\mathbb{P}_r$ and the counting measure.

Lemma 9.3.1 *Let $N \geq 1$ be given and set $K = \lfloor \log_q N \rfloor$. Then we have uniformly for all $r < K$ and all sets $A \in \mathcal{F}_r$*

$$\frac{1}{N} \operatorname{Card} \{ n < N \mid n \in A \} = \mathbb{P}_r(A) + \mathcal{O} \left(\frac{q^r}{N} \right), \tag{9.49}$$

where the constant implied by the error term is universal.

Proof Let A be a set of the form (9.48) for some $r < K$. Since

$$\operatorname{Card} \{ n < N \mid n \in K_r(k) \} = \left\lfloor \frac{N-k}{q^{r+1}} \right\rfloor + \theta_{r,k,N}$$

with $\theta_{r,k,N} \in \{0, 1\}$ we hence obtain

$$\frac{1}{N} \operatorname{Card} \{ n < N \mid n \in A \} = \frac{1}{N} \sum_{k \in I} \left(\left\lfloor \frac{N-k}{q^{r+1}} \right\rfloor + \theta_{r,k,N} \right)$$

$$= \frac{\operatorname{Card} I}{q^{r+1}} + \mathcal{O} \left(\frac{q^{r+1}}{N} \right)$$

and $\frac{1}{N} \operatorname{Card} \{ n < N \mid n \in A \} = \mathbb{P}_r(A) + \mathcal{O} \left(\frac{q^r}{N} \right)$. $\square$

In particular, if $A = \left\{ n \in \mathbb{N} \mid \sum_{j \leq r} f_j(n) \in B \right\}$ (for some Borel set B)

then we obtain

$$\frac{1}{N}\operatorname{Card}\left\{n<N\mid \sum_{j\leq r}f_j(n)\in B\right\}$$
$$=\mathbb{P}_r\left\{n\in\mathbb{N}\mid \sum_{j\leq r}f_j(n)\in B\right\}+\mathcal{O}\left(\frac{q^r}{N}\right)=\mathbb{P}_r\left\{\sum_{j\leq r}Y_j\in B\right\}+\mathcal{O}\left(\frac{q^r}{N}\right).$$
(9.50)

Note that (9.50) gives very precise bounds for partial sums $\sum_{j\leq r}f_j(n)$ but not for $f(n)$. However, it is easy to extend the above model.

Let $a_K(N)\geq 1$ denote the leading digit of N and let $\mathcal{F}$ be the algebra generated by $E_j(d)$ ($0\leq d<q$, $j<K$) and $E_K(d)$ ($0\leq d\leq a_K(N)$), where we also set

$$\mathbb{P}(E_K(d))=\frac{1}{a_K(N)},\qquad 0\leq d\leq a_K(N).$$

In this new probability space the Kth term $f_K(n)=f(a_K(n)q^K)$ is also a random variable and $f(n)$ can be considered, too. Note also that $\mathbb{P}_r(A)=\mathbb{P}(A)$ for all $A\in\mathcal{F}_r$ and $r<K$. In particular if follows easily that for all $A\in\mathcal{F}$ (see (Manstavičius 1997))

$$\frac{1}{N}\operatorname{Card}\{n<N\mid n\in A\}\leq 2\,\mathbb{P}(A).$$

Consequently

$$\frac{1}{N}\operatorname{Card}\{n<N\mid f(n)\in B\}\leq 2\,\mathbb{P}\{n\in\mathbb{N}\mid f(n)\in B\}.$$

9.3.1.2 A Turán–Kubilius inequality for additive functions

Let $f(n)$ denote a q-additive function and set

$$m_{j,q}:=\mathbb{E}\,Y_j=\frac{1}{q}\sum_{d=1}^{q-1}f(dq^j),$$

$$m_{2;j,q}^2:=\mathbb{E}\,Y_j^2=\frac{1}{q}\sum_{d=1}^{q-1}f(dq^j)^2,$$

where $Y_j=f_j$ is the random variable related to the probability space $(\mathbb{N},\mathcal{F}_r,\mathbb{P}_r)$ for some $j\leq r$ and

$$M_q(N):=\sum_{j=0}^{\lfloor\log_q N\rfloor}m_{j,q},\qquad D_q^2(N)=\sum_{j=0}^{\lfloor\log_q N\rfloor}\left(m_{2;j,q}^2-m_{j,q}^2\right).$$

Then the following general property holds which can be seen as an analogue of the celebrated Turán–Kubilius inequality (see (Kubilius 1964)) which has many applications in number theory.

Theorem 9.3.2 *Let f be a q-additive function. Then we have*

$$\frac{1}{N} \sum_{n<N} (f(n) - M_q(N))^2 \le 2D_q^2(N). \tag{9.51}$$

Proof We use the relation

$$\mathbb{E} Y^2 = \int_0^\infty \mathbb{P}\{|Y| \ge u\} \, 2u \, du.$$

Hence, if we integrate the inequality

$$\frac{1}{N} |\{n < N \mid |f(n) - M_q(N)| \ge u\}|$$
$$\le 2 \mathbb{P}\{n \in \mathbb{N} \mid |f(n) - M_q(N)| \ge u\}$$

with respect to $2u \, du$ we obtain the proposed result (compare also with (Ruzsa 1984)). □

A direct application of Theorem 9.3.2 is a very general property for the mean value of q-additive functions.

Corollary 9.3.3 *Let f be a q-additive function. Then we have*

$$\frac{1}{N} \sum_{n<N} f(n) = M_q(N) + \mathcal{O}(D_q(N)).$$

Note that this corollary is in accordance with Theorem 9.2.1. If f is completely q-additive then

$$M_q(N) = \left(\lfloor \log_q N \rfloor + 1\right) C_f \sim C_f \log_q N$$

and

$$D_q(N)^2 = \left(\frac{1}{q} \sum_{d=1}^{q-1} f(d)^2 - C_f^2\right) \log_q N.$$

9.3.1.3 An Erdős–Wintner theorem for additive functions

The above inequalities provide only a very rough idea of the overall behaviour of q-additive functions. We are now interested in conditions which ensure that the values of f have an asymptotic limiting distribution. In the context of classical additive functions Erdős and Wintner proved a necessary and sufficient condition for the existence of a limiting distribution (*cf.* (Elliott 1979)).

For additive functions the situation is very similar. (Delange 1972) could prove the following theorem, which is the analogue of Erdős' and Wintner's theorem for q-additive functions.

Theorem 9.3.4 *Let $f(n)$ be a q-additive function. Then $f(n)$ has a distribution function $G(y)$, that is*

$$\lim_{N\to\infty} \frac{1}{N} \text{Card}\{n < N \mid f(n) < y\} = G(y), \qquad (9.52)$$

if and only if the two series

$$\sum_{j\geq 0}\sum_{d=1}^{q-1} f(dq^j) \quad \text{and} \quad \sum_{j\geq 0}\sum_{d=1}^{q-1} f(dq^j)^2 \qquad (9.53)$$

converge.

Proof The idea of the original proof of (Delange 1972) is to discuss convergence properties of q-multiplicative functions $F(n) = e(tf(n))$, in particular by using the identity

$$\sum_{n<q^L} F(n) = \prod_{j<L}\left(1 + \sum_{d=1}^{q-1} F(dq^j)\right).$$

Since $e(u) = 1 + 2\pi i u + \mathcal{O}(u^2)$ for real u we have

$$\log\left(\frac{1 + \sum_{d=1}^{q-1} F(dq^j)}{q}\right) = \log\left(1 + 2\pi i t m_{j,q} + \mathcal{O}\left(t^2 m_{2;j,q}^2\right)\right)$$

$$= 2\pi i t m_{j,q} + \mathcal{O}\left(t^2 m_{2;j,q}^2\right).$$

Hence the limit

$$\lim_{L\to\infty} \frac{1}{q^L} \prod_{j<L}\left(1 + \sum_{d=1}^{q-1} F(dq^j)\right) = \prod_{j=0}^{\infty} \frac{1}{q}\left(1 + \sum_{d=1}^{q-1} e(tf(dq^j))\right) \qquad (9.54)$$

exists if the two series (9.53) converge. The converse statement is also true. Finally this easily extends to the convergence of

$$\frac{1}{N}\sum_{n<N} F(n) = \frac{1}{N}\sum_{n<N} e(tf(n)), \qquad (9.55)$$

by comparing the sums with partial products of (9.54). By Lévy's criterion this is equivalent to the existence of a distribution function. □

Remark 9.3.5 The expression (9.54) for the characteristic function of the limiting distribution $G(y)$ shows that this distribution can be interpreted as an infinite Bernoulli convolution. The Theorem of Jessen and Wintner asserts that such distribution measure given by $G(y)$ is either purely absolutely continuous, purely singular continuous, or consists only of point masses. A theorem of Lévy applied to the present setting asserts that the last alternative can only occur, if there exists a J such that $f(dq^j) = 0$ for $j > J$. In this case the distribution consists only of finitely many point masses. The two theorems cited are the contents of Lemma 1.22 in (Elliott 1979).

Remark 9.3.6 A totally different proof of Theorem 9.3.4 was given in (Barat and Grabner 2008). There the addition-by-one map τ (also called *odometer*) is studied on the compact space $\mathbb{Z}_q = \text{proj lim}_j \mathbb{Z}/q^j\mathbb{Z}$, which can be viewed as the compactification of $\mathbb{N}$ associated to the q-adic expansion. See also Section 6.5.1. By Kolmogorov's three series theorem the conditions (9.53) are necessary and sufficient that the random series

$$\sum_{j=0}^{\infty} f(X_j q^j)$$

converges almost surely for $X_j \in \{0, \ldots, q-1\}$ independent and identically uniformly distributed random variables (the convergence of the third series in the three series theorem is trivial in this case). In the setting of the dynamical system $(\mathbb{Z}_q, \tau)$ this simply means that f can be extended to an almost everywhere defined measurable function on $\mathbb{Z}_q$. Since $(\mathbb{Z}_q, \tau)$ is ergodic with respect to the Haar measure μ on $\mathbb{Z}_q$ Birkhoff's ergodic theorem (Theorem 1.6.7) asserts that

$$\lim_{N \to \infty} \frac{1}{N} \text{Card}\{n < N \mid f(\tau^n(x)) < t\} = \mu\left(\{y \in \mathbb{Z}_q \mid f(y) < t\}\right) = G(t),$$

for μ-almost all $x \in \mathbb{Z}_q$. It remains to prove that 0 is one of the points for which this is valid (*i.e.*, 0 is a *generic point*, see Section 7.2.2).

This point of view allows us to generalise Delange's theorem to other types of digital expansions, such as expansions with linear recurrent base sequences, which involve dependent digits.

Finally we want to remark that there is also an alternative proof by (Manstavičius 1997) that uses the approximation properties of the form stated in Lemma 9.3.1 and applies for Cantor expansions.

Remark 9.3.7 Theorem 9.3.4 was generalised by (Kátai 1992) who proved that there exists a distribution function $G(y)$ such that,

$$\lim_{N \to \infty} \frac{1}{N} \text{Card}\left\{n < N \mid f(n) - M_q(N) < y\right\} = G(y)$$

if and only if the series $\sum_{j\geq 0}\sum_{d=1}^{q-1} f(dq^j)^2$ converges.

Example 9.3.8 The q-additive function

$$v(n) = \sum_{j=0}^{\infty} \frac{a_j(n)}{q^{j+1}}$$

defines the van der Corput sequence (*cf.* (Kuipers and Niederreiter 1974) and (Drmota and Tichy 1997)). It is easy to see that the distribution of this sequence is the uniform distribution on $[0,1]$. This sequence and related constructions are used in numerical integration to define sequences of low discrepancy, which give a small error in integration (*cf.* (Niederreiter 1992)).

9.3.1.4 A general central limit theorem for additive functions

Theorem 9.3.4 and its variant by Kátai do not apply for completely q-additive functions $f(n)$ or for functions where f_j does not converge to 0. In these cases we expect a central limit theorem which is ubiquitous in the context of sums of independent random variables.

The most general central limit theorem for q-additive functions is due to (Manstavičius 1997).

Theorem 9.3.9 *Suppose that, as $N \to \infty$,*

$$\max_{j \leq \log_q N} \max_{0 \leq d < q} |f(dq^j)| = o(D_q(N)) \tag{9.56}$$

and $D_q(N) \to \infty$. Then,

$$\lim_{N \to \infty} \frac{1}{N} \operatorname{Card}\left\{ n < N \mid \frac{f(n) - M_q(N)}{D_q(N)} < y \right\} = \Phi(y),$$

where Φ is the normal distribution function.

Proof For $N \geq 1$ let $(\mathbb{N}, \mathcal{F}, \mathcal{P})$ be the probability space constructed after the proof of Lemma 9.3.1 for which the random variables $Y_j = f_j$, $0 \leq j \leq K = \lfloor \log_q N \rfloor$, are independent. For $r \leq K$ let

$$F_{N,r}(y) = \frac{1}{N}\left\{ n < N \mid \frac{\sum_{j \leq r} f_j(n) - M_q(N)}{D_q(N)} \leq y \right\},$$

$$V_{N,r}(y) = \mathbb{P}\left\{ n \in \mathbb{N} \mid \frac{\sum_{j \leq r} f_j(n) - M_q(N)}{D_q(N)} \leq y \right\}$$

denote the distribution functions of the normalised and truncated functions and

$$F_N(y) = F_{N,K}(y) \quad \text{and} \quad V_N(y) = V_{N,K}(y)$$

the distribution function of (normalised) $f(n)$ according to the counting measure and to the measure $\mathbb{P}$, respectively.

By the central limit theorem for sums of independent random variables (Billingsley 1968, Theorem 7.2) it is obvious that $V_N(y) \to \Phi(y)$, where $\Phi(y)$ denotes the distribution of the standard normal distribution, since the assumption (9.56) implies the Lindeberg condition.

Thus, it remains to show that $V_N(y)$ and $F_N(y)$ are close. For this purpose one can use the Lévy metric

$$L(F, G) = \inf\{\varepsilon > 0 \mid \forall y \in \mathbb{R}: F(y - \varepsilon) - \varepsilon \leq G(y) \leq F(y + \varepsilon) + \varepsilon\}$$

between two distribution functions F and G which quantifies and characterises weak convergence.

By the triangle inequality we have

$$L(F_N, V_N) \leq L(F_N, F_{N,K-r}) + L(F_{N,K-r}, V_{N,K-r}) + L(V_{n,K-r}, V_N).$$

First by Lemma 9.3.1 it follows for all $r > 0$

$$L(F_{N,K-r}, V_{N,K-r}) = \mathcal{O}(q^{-r}).$$

Furthermore, for every $r > 0$ we obtain for every $\varepsilon > 0$ by another application of (9.56), as $N \to \infty$,

$$L(V_{n,K-r}, V_N) \leq \varepsilon + \mathbb{P}\left\{n \in \mathbb{N} \;\middle|\; \left|\sum_{K-r < j \leq K} f_j(n)\right| \geq \varepsilon D_q(N)\right\}$$
$$= \varepsilon + o_\varepsilon(1).$$

A similar estimate holds for the distance $L(F_N, F_{N,K-r})$. Hence, we obtain $L(F_N, V_N) \to 0$ as $N \to \infty$ and consequently $L(F_N, \Phi) \to 0$. □

9.3.2 A central limit theorem for subsequences

The advantage of Theorem 9.3.9 is its generality. However, it cannot be applied if we are dealing with certain subsequences of the integers, that is, the underlying probability space $\Omega_N = \{0, 1, \ldots, N-1\}$ is replaced by a certain subset of integers, for example by $\Omega_N = \{2, 3, 5, \ldots, p_N\}$, the first N primes, or by $\Omega_N = \{0^2, 1^2, 2^2, \ldots, (N-1)^2\}$, the first N squares.

In this section we describe a general method that is due to (Bassily and Kátai 1995). In particular they could cover polynomial sequences and polynomial sequences of primes.

Theorem 9.3.10 *Let f be a q-additive function such that*

$$\sup_{j\geq 0} \max_{0\leq d<q} f(dq^j) = \mathcal{O}(1)$$

Assume that $\frac{D_q(N)}{(\log N)^\eta} \to \infty$ as $N \to \infty$ for some $\eta > 0$ and let $P(x)$ be a polynomial with integer coefficients, degree r, and positive leading term. Then,

$$\lim_{N\to\infty} \frac{1}{N} \text{Card}\left\{ n < N \mid \frac{f(P(n)) - M_q(N^r)}{D_q(N^r)} < y \right\} = \Phi(y)$$

and

$$\lim_{N\to\infty} \frac{1}{\pi(N)} \text{Card}\left\{ p < N \mid p \text{ prime}, \frac{f(P(p)) - M_q(N^r)}{D_q(N^r)} < y \right\} = \Phi(y).$$

In what follows we present the framework of their method in a slightly more general setting. We consider general subsets Ω_N of the non-negative integers of size N that satisfy the following property. It has a similar flavour to Lemma 9.3.1, it quantifies the difference between the counting measure and and a measure with independent digits. However, it only needs properties of finitely many different digits.

Property 9.3.11 (BK-Property) *Let Ω_N be subsets of the non-negative integers of size N. We assume that $M_N = \max \Omega_N = \mathcal{O}(N^k)$ for some $k \geq 1$ and that for every $\eta > 0$, $\lambda > 0$ and for every integer $L \geq 1$ we have*

$$\frac{1}{N} \text{Card}\{n \in \Omega_N \mid a_{j_1}(n) = \ell_1, \ldots, a_{j_L}(n) = \ell_L\} = q^{-L} + \mathcal{O}\left((\log N)^{-\lambda}\right),$$

uniformly for all $j_1, \ldots, j_L$ with

$$(\log N)^\eta \leq j_1 < j_2 < \cdots < j_L \leq \log_q M_N - (\log N)^\eta$$

and for all $\ell_1, \ldots, \ell_L \in \{0, 1, \ldots, q-1\}$.

Note that the constant implied by the error term might depend on η, λ, and L.

Example 9.3.12 *Let $\Omega_N = \{0, 1, \ldots, N-1\}$.* Then the BK-Property is trivially satisfied. We actually have an error bound of the form

$$\mathcal{O}\left(\frac{q^{j_L}}{N}\right) = \mathcal{O}\left(q^{-(\log N)^\eta}\right) = \mathcal{O}\left((\log N)^{-\lambda}\right),$$

since $j_L \leq \log_q N - (\log N)^\eta$ and $\eta > 0$.

The essential observation is that the BK-Property implies a central limit theorem.

Theorem 9.3.13 *Suppose that f satisfies the same assumptions as in Theorem 9.3.10 and for every $N \geq 1$ let Ω_N be a subset of the non-negative integers of size N.*

If the BK-Property holds, then we have

$$\lim_{N \to \infty} \frac{1}{N} \operatorname{Card} \left\{ n \in \Omega_N \mid \frac{f(n) - M_q(M_N)}{D_q(M_N)} < y \right\} = \Phi(y).$$

Proof The idea of the proof is to compare moments, however, one has to be careful. We choose $0 < \eta' < \eta/(2k)$ (where $\eta > 0$ satisfies $D_q(N)/(\log N)^\eta \to \infty$) and replace f by $\tilde{f}$ that is defined by

$$\tilde{f}(n) = \sum_{(\log N)^{\eta'} \leq j \leq \log M_N - (\log N)^{\eta'}} f(a_j(n)q^j),$$

that is, we cut off some of the first and of the last digits. Let $\tilde{T}_N$ denote the random variable associated to $\tilde{f}$ and the counting measure on Ω_N, that is, the distribution function $F_{\tilde{T}_N}$ is given by

$$F_{\tilde{T}_N}(u) = \frac{1}{N} \operatorname{Card} \left\{ n \in \Omega_N \mid \tilde{f}(n) \leq u \right\}.$$

Furthermore set

$$\tilde{S}_{M_N} = \sum_{(\log N)^{\eta'} \leq j \leq \log M_N - (\log N)^{\eta'}} Y_j,$$

where $Y_j = f_j$ are the independent random variables from above. We also set

$$\tilde{M}_q(M_N) = \mathbb{E}\, \tilde{S}_{M_N} = \sum_{(\log N)^\eta \leq j \leq \log M_N - (\log N)^\eta} m_{j;q}$$

$$\tilde{D}_q(M_N)^2 = \mathbb{V}\, \tilde{S}_{M_N} = \sum_{(\log N)^\eta \leq j \leq \log M_N - (\log N)^\eta} \left(m_{2,j;q}^2 - m_{j;q}^2 \right).$$

Note that by assumption $\tilde{D}_q(M_N) \sim D_q(M_N)$ as $N \to \infty$.

Next we expand the difference

$$\delta_L = \mathbb{E} \left(\tilde{T}_N - \tilde{M}_q(M_N) \right)^L - \mathbb{E} \left(\tilde{S}_N - \tilde{M}_q(M_N) \right)^L$$

in terms of the probabilities

$$\frac{1}{N} \operatorname{Card}\{ n \in \Omega_N \mid a_{j_1}(n) = \ell_1, \ldots, a_{j_L}(n) = \ell_L \}$$

and compare them with help of the BK-Property. In fact, we have to take into account $\leq (q \log_q M_N)^L$ terms and, thus, we get

$$|\delta_L| = \mathcal{O}\left((q \log_q M_N)^L (\log N)^{-\lambda} \right) = \mathcal{O}\left((\log N)^{kL - \lambda} \right).$$

By assumption it follows that

$$\mathbb{E}\left(\frac{\tilde{T}_N - \tilde{M}_q(M_N)}{\tilde{D}_q(M_N)}\right)^L - \mathbb{E}\left(\frac{\tilde{S}_{M_N} - \tilde{M}_q(M_N)}{\tilde{D}_q(M_N)}\right)^L \to 0.$$

By standard tools in probability (see (Billingsley 1968)) we know that $(\tilde{S}_{M_N} - \tilde{M}_q(M_N))/\tilde{D}_q(M_N)$ converges to the Gaussian distribution $N(0,1)$ and we have convergence of all moments. Hence, the same is true for $(\tilde{T}_N - \tilde{M}_q(M_N))/\tilde{D}_q(M_N)$. Finally, since $f(n) - \tilde{f}(n) = \mathcal{O}((\log N)^{\eta'})$, $M_q(N) - \tilde{M}_q(N) = \mathcal{O}((\log N)^{\eta'})$ and $\tilde{D}_q(M_N) \sim D_q(M_N) \geq (\log N)^{2\eta'}$ we also deduce a central limit theorem for f. □

It remains to verify the BK-Property in several examples. Interestingly enough, proper exponential sum estimates are sufficient to derive this property.

Lemma 9.3.14 *Suppose that for all $\eta > 0$ and $\lambda > 0$ the exponential sum estimate*

$$\frac{1}{N}\sum_{n\in\Omega_N} e\left(\frac{a}{q^r}n\right) = \mathcal{O}\left((\log N)^{-\lambda}\right) \tag{9.57}$$

holds uniformly for $(\log N)^\eta \leq r \leq \log_q M_N - (\log N)^\eta$, for all integers a with $1 \leq a < (\log N)^\lambda$ and for all integers a with $1 \leq a < q^r$ which are not divisible by q.

Then the BK-Property holds.

Proof We just sketch the idea of the proof. For a detailed analysis we refer to (Bassily and Kátai 1995).

Since $a_j(n) = \ell$ if and only if $\{nq^{-j-1}\} = \ell/q$ we have

$$\frac{1}{N}\operatorname{Card}\{n \in \Omega_N \mid a_j(n) = \ell\} = \frac{1}{N}\sum_{n\in\Omega_N} \mathbb{1}_{[\ell/q,(\ell+1)/q)}(\{nq^{-j-1}\}),$$

where $\mathbb{1}_S$ denotes the indicator function of the set S. Let

$$\mathbb{1}_{[\ell/q,(\ell+1)/q)}(x) = \sum_{k\in\mathbb{Z}} c_k e(kx)$$

denote the Fourier series; note that $c_0 = 1/q$. Then it also follows that

$$\frac{1}{N}\operatorname{Card}\{n \in \Omega_N \mid a_j(n) = \ell\} = \sum_{k\in\mathbb{Z}} c_k \frac{1}{N}\sum_{n\in\Omega_N} e\left(\frac{kn}{q^{j+1}}\right)$$

$$= \frac{1}{q} + \sum_{k\neq 0} c_k \frac{1}{N}\sum_{n\in\Omega_N} e\left(\frac{kn}{q^{j+1}}\right).$$

Hence, exponential sum estimates provide asymptotic information for the probabilities $\frac{1}{N}\,\text{Card}\,\{n \in \Omega_N \mid a_j(n) = \ell\}$.

However, one has to be more precise since the Fourier series of the characteristic function is not absolutely convergent. In fact one can use smoothing arguments so that only a few exponential sums are sufficient. Furthermore, this method extends to several digits (see (Bassily and Kátai 1995)). □

Example 9.3.15 Let P be a polynomial. By using standard estimates for exponential sums for polynomials and polynomials of primes (see (Iwaniec and Kowalski 2004)) it is clear that (9.57) is satisfied for the sets

$$\Omega_N = \{P(n) \mid n < N\}$$

and

$$\Omega_N = \{P(p_1), \ldots, P(p_N)\},$$

where $p_1, \ldots, p_N$ are the first N primes.

Example 9.3.16 Let $c > 1$ be a real non-integral number and set

$$\Omega_N = \{\lfloor n^c \rfloor \mid n < N\}.$$

In this case we first observe that the digits $a_0, a_1, \ldots$ coincide for $\lfloor n^c \rfloor$ and n^c in the q-ary digital expansion. Furthermore we have $a_j(n^c) = \ell$ if and only if $\{n^c q^{-j-1}\} = \ell/q$. Thus, we can replace the exponential sums from (9.57) by the sums

$$\frac{1}{N} \sum_{n<N} e\left(\frac{a}{q^r} n^c\right).$$

For non-integral c these kinds of exponential sums can be easily estimated by Van der Corput's theorem (Iwaniec and Kowalski 2004, Theorem 8.20) and provide upper bounds which are of the same kind as those from (9.57). Hence, the BK-Property holds, too.

An important feature of the method of Bassily and Kátai is its flexibility. It also applies for so-called block-additive functions as well as for other digital expansions. We will comment on this in Section 9.4.

9.3.3 A generating function approach to completely q-additive functions

We have observed that due to the (almost) independence properties of the digits, a central limit theorem appears in very general situations.

We now concentrate on a very special situation, where we can obtain much more precise results. We discuss properties of the *generating function*

$$S(N, x) = \sum_{n<N} x^{f(n)},$$

where $x \ne 0$ is a complex variable. If we assume that f is integer valued then $S(N, x)$ can be rewritten as

$$S(N, x) = \sum_{k \in \mathbb{Z}} \operatorname{Card}\{n < N \mid f(n) = k\} x^k$$

which explains the notion generating function. In fact, we will use this interpretation in the proof of Corollary 9.3.21.

Obviously, the function $n \mapsto x^{f(n)}$ is a completely q-multiplicative function. Thus, we can apply the method of Theorem 9.2.7 and obtain the following representation.

Theorem 9.3.17 *Suppose that f is a completely q-additive function and let $G \subseteq \mathbb{C}$ be defined by*

$$G = \left\{ x \in \mathbb{C} \,\Big|\, \left|1 + x^{f(1)} + \cdots + x^{f(q-1)}\right| > \max_{0 \le d < q} |x^{f(d)}| \right\}.$$

Then there exists a function $\Psi(x, t)$ $(x \in G, t \in \mathbb{R})$ that is analytic for $x \in G$ and Hölder continuous and periodic in t with period 1 such that

$$S(N, x) = \Psi(x, \log_q N) \left(1 + x^{f(1)} + \cdots + x^{f(q-1)}\right)^{\log_q N}. \tag{9.58}$$

Furthermore there exists a continuous function $C(x)$ $(x \ne 0)$ such that

$$|S(N, x)| \le C(x) \sum_{k \le \log_q N} \left|1 + x^{f(1)} + \cdots + x^{f(q-1)}\right|^k. \tag{9.59}$$

Proof The proof of (9.58) is just a refinement of the proof of Theorem 9.2.7.

The proof of (9.59) is similar but even easier, since we are only interested in upper bounds, compare with Corollary 9.2.9. □

It is an important feature of this lemma that the function $\Psi(x, t)$ represents an analytic function in x if x is sufficiently close to the real axis. It is interesting that Theorem 9.3.17 has several corollaries (compare with (Drmota and Rivat 2005), (Drmota, Grabner, and Liardet 2008)). We start with a representation for moments.

Corollary 9.3.18 *Suppose that f is a completely q-additive function. Then* for every integer $r \geq 1$ we have

$$\frac{1}{N} \sum_{n<N} f(n)^r = C_f (\log_q N)^r + \sum_{\ell=0}^{r-1} \Psi_{r,\ell}(\log_q N) \cdot (\log_q N)^\ell, \qquad (9.60)$$

where the functions $\Psi_{r,\ell}(t)$ $(0 \leq \ell < r)$ are continuous and periodic (with period one).

Proof We just set $x = e^t$ in (9.58) and evaluate the rth derivative (compare also with Proposition 9.2.22). Furthermore, note that $\Psi(1,t) = C_f$. Hence, the asymptotic leading term is given by $C_f (\log_q N)^r$ and has no periodic fluctuations. $\square$

Remark 9.3.19 The idea of taking the derivative also applies if a formula of the kind (9.58) is not exact but has an error term that is uniform in a neighbourhood of $x = 1$. Due to analyticity in x one can take derivatives at $x = 1$ at arbitrary order by using the formula

$$G^{(r)}(1) = \frac{r!}{2\pi i} \int_{|x-1|=\delta} \frac{G(x)}{(x-1)^{r+1}} dx.$$

Next we derive a global and a local central limit theorem.

Corollary 9.3.20 *Suppose that f is a completely q-additive function and suppose that*

$$D_f^2 = \frac{1}{q} \sum_{d=1}^{q-1} f(d)^2 - C_f^2 > 0.$$

Then,

$$\lim_{N \to \infty} \frac{1}{N} \operatorname{Card} \left\{ n < N \mid \frac{f(n) - C_f \log_q N}{\sqrt{D_f^2 \log_q N}} < y \right\} = \Phi(y), \qquad (9.61)$$

and for all $r \geq 1$

$$\frac{1}{N} \sum_{n<N} \left(\frac{f(n) - C_f \log_q N}{\sqrt{D_f^2 \log_q N}} \right)^r = \frac{1}{\sqrt{2\pi}} \int_{-\infty}^{\infty} u^r e^{-\frac{1}{2}u^2} \, du + o(1). \qquad (9.62)$$

Furthermore, we have exponential tail estimates of the form

$$\frac{1}{N} \operatorname{Card} \left\{ n < N \mid |f(n) - C_f \log_q N| \geq y \sqrt{\log_q N} \right\} \qquad (9.63)$$
$$\ll \min \left(e^{-cy}, e^{-cy^2 + \mathcal{O}(y^3/\sqrt{\log N})} \right)$$

for some constant $c > 0$.

Proof Let T_N denote the random variable that is induced by the distribution of $f(n)$ on $\Omega_N = \{0, 1, \ldots, N-1\}$. Then the moment generating function $\mathbb{E}\,e^{tT_N}$ is given by

$$\mathbb{E}\,e^{tT_N} = \frac{1}{N}\sum_{n<N} e^{tf(n)} = \frac{1}{N}S(N, e^t)$$

$$= \Phi(e^t, \log_q N)\left(1 + e^{tf(1)} + \cdots + e^{tf(q-1)}\right)^{\log_q N}.$$

Hence, by using the local expansion

$$\log\left(1 + e^{tf(1)} + \cdots + e^{tf(q-1)}\right) = \log q + C_f t + \frac{D_f^2}{2}t^2 + \mathcal{O}(t^3)$$

we obtain that the moment generating function of the normalised random variable

$$Z_N = \frac{T_N - C_f \log_q N}{\sqrt{D_f^2 \log_q N}}$$

is given by

$$\mathbb{E}\,e^{tZ_N} = e^{t(C_f/D_f)\sqrt{\log_q N}}\,\mathbb{E}\,e^{(t/\sqrt{D_f^2 \log_q N})Y_N}$$
$$= e^{\frac{1}{2}t^2 + \mathcal{O}(t^3/\sqrt{\log N})}.$$

Of course, this translates to (9.61).

Further, convergence of the moment generating function also implies convergence of all moments, that is, we get (9.62). Finally, the tail estimates (9.63) are a direct consequence of Chernov type inequalities. □

Corollary 9.3.21 *Suppose that f is an integer valued completely q-additive function such that*

$$d = \gcd\{f(c) \mid 0 \le c < q\} = 1. \tag{9.64}$$

Set

$$\mu(x) = \frac{x\lambda'(x)}{\lambda(x)} \quad \text{and} \quad \sigma^2(x) = \frac{x^2\lambda''(x)}{\lambda(x)} + \mu(x) - \mu(x)^2,$$

where $\lambda(x)$ abbreviates $\lambda(x) = 1 + x^{f(1)} + \cdots + x^{f(q-1)}$. Furthermore, for $k \in K(N) = \mathbb{Z} \cap [\delta \log_q N, (1-\delta)\log_q N]$ we define $x_{k,N}$ by $\mu(x_{k,N}) = k/\log_q N$, where $\delta > 0$ is arbitrary. Then we have uniformly for $k \in K(N)$

$$\operatorname{Card}\{n < N, \ f(n) = k\}$$
$$= \frac{\Phi(x_{k,N}, \log_q N)}{\sqrt{2\pi\sigma^2(x_{k,N})\log_q N}}\, N^{\log_q \lambda(x_{k,N})}\, x_{k,N}^{-k}\left(1 + \mathcal{O}\left(\frac{1}{\log N}\right)\right).$$

$$\tag{9.65}$$

Furthermore, if $|k - C_f \log_q N| \leq C\sqrt{\log_q N}$ (for some $C > 0$) we also have

$$\mathrm{Card}\{n < N, \ f(n) = k\}$$
$$= \frac{\pi N}{\sqrt{2\pi\sigma^2 \log_q N}} \exp\left(-\frac{(k - C_f \log_q N)^2}{2 D_f^2 \log_q N}\right)\left(1 + \mathcal{O}\left(\frac{1}{\sqrt{\log N}}\right)\right).$$
(9.66)

Note that $C_f = \mu(1)$ and $D_f^2 = \sigma^2(1)$.

Proof We apply (9.58) and (9.59) and use Cauchy's formula:

$$\mathrm{Card}\{n < N, \ f(n) = k\} = \frac{1}{2\pi i} \int_{|x| = x_{k,N}} \left(\sum_{n < N} x^{f(n)}\right) x^{-k-1}\, dx,$$

where $x_{k,N}$ is the saddle point of the asymptotic leading term of the integrand:

$$\lambda(x)^{\log_q N} x^{-k} = e^{\log \lambda(x) \cdot \log_q N - k \log x}.$$

We do not work out the details of standard saddle point techniques. We just refer to (Mauduit and Sárközy 1997) and (Drmota and Rivat 2005), where problems of almost the same kind have been discussed. □

9.3.4 Uniform distribution of q-additive functions

A last type of distribution results for additive functions is the distribution of values in a compact abelian group. Usually, the group under consideration is $\mathbb{T} = \mathbb{R}/\mathbb{Z}$ or $\mathbb{Z}/m\mathbb{Z}$.

Definition 9.3.22 Let A be a compact abelian group equipped with its Haar measure λ and $f : \mathbb{N} \to A$ an A-valued arithmetic function. The sequence $(f(n))_{n \in \mathbb{N}}$ is called uniformly distributed, if for all measurable subsets $B \subseteq A$ with $\lambda(\partial B) = 0$

$$\lim_{N \to \infty} \frac{1}{N} \sum_{n < N} \mathbb{1}_B(f(n)) = \lambda(B).$$

By harmonic analysis on the group A this is equivalent to saying that for all characters $\chi \in \widehat{A} \setminus \{\chi_0\}$ ($\chi_0 \equiv 1$ denotes the trivial character) one has

$$\lim_{N \to \infty} \frac{1}{N} \sum_{n < N} \chi(f(n)) = 0 \qquad (9.67)$$

(Weyl's criterion, *cf.* (Kuipers and Niederreiter 1974)).

For a q-additive function f the function $\chi \circ f$ is a $\mathbb{C}$-valued q-multiplicative function, for which we can apply Theorems 9.2.7 and 9.3.17 or Corollary 9.2.9 to obtain the following corollaries.

Corollary 9.3.23 *Suppose that f is an integer-valued completely q-additive function and that (9.64) holds. Then for every integer $M \geq 1$ and all $m \in \{0, 1, \ldots, M-1\}$ we have*

$$\frac{1}{N} \operatorname{Card}\{n < N \mid f(n) \equiv m \bmod M\} = \frac{1}{M} + \mathcal{O}(N^{-\eta})$$

for some $\eta > 0$.

Remark 9.3.24 Alternatively to condition (9.64) we can assume that f attains a value that is relatively prime to M. Then the same assertion holds.

Proof We use (9.59) for all Mth roots of unity $x = e^{2\pi i m/M}$ and apply simple discrete Fourier techniques. The exponent η comes from Corollary 9.2.9 or Theorem 9.2.7 as

$$\eta = 1 - \log_q \left(\max_{1 \leq h < M} \left| \sum_{\ell=0}^{q-1} e\left(\frac{h}{M} f(\ell)\right) \right| \right) > 0.$$

$\square$

Corollary 9.3.25 *Let f be a real-valued completely q-additive function which attains one irrational value. Then the sequence $(f(n))_{n \geq 0}$ is uniformly distributed modulo 1.*

Remark 9.3.26 Note that Corollary 9.3.25 particularly applies to sequences of the kind $(\alpha f(n))_{n \geq 0}$ if f is integer valued and if α is irrational.

Proof We just set $x = e^{2\pi i h}$ for a non-zero integer h and use (9.59) to show that

$$\lim_{N \to \infty} \frac{1}{N} \sum_{n < N} e(hf(n)) = 0.$$

This is just Weyl's criterion (9.67) for the group $\mathbb{T}$. $\square$

Corollary 9.3.27 *Let f be a real-valued completely additive function, which attains one irrational value and $\beta \in \mathbb{R} \setminus \mathbb{Q}$. Then the sequence $(n\beta \pmod 1, f(n) \pmod 1)$ is uniformly distributed in $\mathbb{T}^2$.*

Proof Simply realise that $n \mapsto n$ is q-additive and study the exponential sum

$$\sum_{n<N} e(h_1 \beta n + h_2 f(n))$$

for $(h_1, h_2) \in \mathbb{Z}^2 \setminus \{(0,0)\}$ using the same ideas as in the proof of Theorem 9.3.4. □

Remark 9.3.28 After the explanation of the probabilistic point of view in Section 9.3.1.1 such results can also be seen as limiting distribution results for sums of independent random variables taking values in a compact group. For this point of view we refer to (Bártfai 1966).

Remark 9.3.29 As it was explained in Remark 9.3.6 distribution results can also be seen from an ergodic point of view. Contrary to the situation explained there, the function f cannot be extended in a consistent way to $\mathbb{Z}_q$. The following idea originating from (Kamae 1977), (Kamae 1978), (Kamae 1987) defines a *cocycle* by

$$a_f(x, n) = \lim_{\substack{m \to x \\ m \in \mathbb{N}}} f(m+n) - f(n),$$

which is easily seen to exist almost everywhere in $\mathbb{Z}_q$. Then the dynamical system $(\mathbb{Z}_q \times A, T_a, \mu_q \otimes \lambda)$ (*cf.* Section 1.6.2) with

$$T_a(x, \alpha) = (x+1, \alpha + a_f(x, 1))$$

is used to study the distribution properties of f. Such systems are called *skew-products*. For further examples of such dynamical systems we refer to Section 7.5.2.1. Then we have $T_a^n(0,0) = (n, f(n))$, which motivates the definition of T_a. By arguments explained in (Grabner and Liardet 1999) and (Drmota, Grabner, and Liardet 2008) a special property of the nullset, where a_f is not defined ('uniform negligibility', *cf.* (Liardet 1978)), is used to prove that all points are generic for the dynamical system, *i.e.*, the ergodic theorem holds for all points, especially $(0,0)$.

This point of view has the advantage that it generalises to other situations, such as multidimensional settings. Furthermore, (Kamae 1987) used this idea to prove spectral disjointness for such dynamical systems with respect to multiplicatively independent bases. This was generalised to the Gaussian integers in (Grabner, Liardet, and Tichy 2005).

9.4 Further results

9.4.1 Gelfond Problems

In 1968 (Gelfond 1968) proved for the q-ary sum-of-digits function $s_q(n)$

$$\text{Card}\{n < N \mid n \equiv \ell \bmod r, \; s_q(n) \equiv a \bmod m\} = \frac{N}{mr} + \mathcal{O}\left(N^\lambda\right)$$

provided that $\gcd(m, q-1) = 1$, where

$$\lambda = \frac{1}{2 \log q} \log \frac{q \sin(\pi/2m)}{\sin(\pi/2mq)}.$$

This means that the sum-of-digits $s_q(n)$ is asymptotically uniformly distributed modulo m if we restrict n to arithmetic subsequences $n = \ell + kr, k \geq 0$. The proof is based on a subtle but elementary analysis of the expression

$$\sum_{n < q^K} e\left(\alpha n + \beta s_q(n)\right) = \prod_{r < K} \left(1 + e(\alpha 2^k + \beta) + \cdots + e((q-1)(\alpha 2^k + \beta))\right). \tag{9.68}$$

In this paper Gelfond formulated three problems. He first conjectured that a corresponding property is true if one uses two coprime bases q_1, q_2 at once, namely

$$\text{Card}\{n < N \mid s_{q_1}(n) \equiv h_1 \bmod m_1, \; s_{q_2}(n) \equiv h_2 \bmod m_2\}$$
$$= \frac{N}{m_1 m_2} + \mathcal{O}\left(N^\lambda\right) \tag{9.69}$$

provided that $\gcd(m_1, q_1 - 1) = 1$ and $\gcd(m_2, q_2 - 1) = 1$. A few years later (Bésineau 1972) proved this property, however, without an error term. Eventually (Kim 1999) provided also the proposed error term (even for a system of completely q-additive functions). It is interesting that these methods can be extended to non-trivial exponential sum estimates for

$$\sum_{n < N} e\left(\alpha n^k + \beta s_q(n)\right)$$

which were used in (Thuswaldner and Tichy 2005) to discuss Waring's problem under digital congruence conditions.

Gelfond also asked what can be said about the number of primes $p < N$ for which $s_q(p) \equiv a \bmod m$ and about the number $n < N$ for which $s_q(P(n)) \equiv a \bmod m$, where $P(x)$ is an integer polynomial. These challenging problems were unsolved for almost 40 years and only partial results have been obtained (see (Fouvry and Mauduit 1996a), (Fouvry and Mauduit 1996b), (Drmota and Rivat 2005) and also (Dartyge and Tenenbaum 2005), (Dartyge and Tenenbaum 2006)). Finally the problem for the subsequence of primes was completely solved by

(Mauduit and Rivat 2009b). The problem for the subsequence of squares (more precisely on $s_q(n^2)$) was solved by (Mauduit and Rivat 2009a). Interestingly the approach of Mauduit and Rivat again uses a subtle analysis of the product representation (9.68) combined with Fourier analytic tools and tricky but classical exponential sum techniques. Basically they showed that

$$\sum_{p<N} e(\alpha s_q(p)) \ll N^{1-\eta} \qquad (9.70)$$

and

$$\sum_{n<N} e(\alpha s_q(n^2)) \ll N^{1-\eta}, \qquad (9.71)$$

where $\eta = \eta(\alpha) > 0$ if $\alpha(q-1) \notin \mathbb{Z}$. With the help of these estimates one obtains asymptotic distributions in residue classes (as asked by Gelfond) and also uniform distribution modulo 1 for the sequences $(\alpha s_q(p))$ and $(\alpha s_q(n^2))$ for irrational α.

In (Drmota, Rivat, and Stoll 2008), (Drmota, Mauduit, and Rivat 2009) there are already some extensions of these results. The main open problem in this context is to generalise (9.71) to polynomials $P(x)$ of degree ≥ 3.

9.4.2 Odometers and systems of numeration

The q-adic digital representation presented here is one special case of a rather general definition of numeration system. As recalled in Section 2.3.3 and Section 3.1, every strictly increasing sequence of positive integers $(G_k)_{k \in \mathbb{N}}$ with $G_0 = 1$ gives rise to a representation for all $n \in \mathbb{N}$. Every $n \in \mathbb{N}$ can be written in the form

$$n = \sum_{k=0}^{K} a_k(n) G_k \quad \text{with } \forall k : 0 \leq a_k(n) < \frac{G_{k+1}}{G_k}.$$

The additional requirement

$$\forall k, \ a_0(n)G_0 + \cdots a_k(n)G_k < G_{k+1} \qquad (9.72)$$

makes this representation unique. The representation satisfying (9.72) can be determined by the *greedy algorithm*. The main difference between this general notion of digital expansion and the q-adic case presented here is the dependence between the digits given by (9.72).

One possible approach to extend distribution results of various kinds to this more general setting is to define an according compactification of $\mathbb{N}$ by

$$\mathcal{K}_G = \{(a_0, a_1, \ldots) \mid \forall k : a_0 G_0 + \cdots a_k G_k < G_{k+1}\}$$

$$\subseteq \prod_{k=0}^{\infty} \left\{0, 1, \ldots, \left\lceil \frac{G_{k+1}}{G_k} \right\rceil - 1\right\}.$$

This space is equipped with the product topology of the discrete spaces and therefore compact. All representations of positive integers are then in $\mathcal{K}_G$ by (9.72). The addition-by-one map τ on $\mathbb{N}$ can then be extended to $\mathcal{K}_G$ by

$$\tau(x) = \lim_{\substack{n \to x \\ n \in \mathbb{N}}} \tau(n).$$

See also Section 2.3.3.3. Under additional conditions on the sequence $(G_k)_{k \in \mathbb{N}}$ there exists a unique τ-invariant measure μ_G on $\mathcal{K}_G$ (cf. (Barat, Downarowicz, and Liardet 2002)). The properties of the dynamical system $(\mathcal{K}_G, \tau, \mu_G)$ (the *odometer*) have been studied from combinatorial, topological, and dynamical points of view by (Grabner, Liardet, and Tichy 1995), (Barat, Downarowicz, Iwanik, et al. 2000), and (Barat, Downarowicz, and Liardet 2002). For a discussion of odometers as special cases of dynamical systems defined by Bratteli diagrams we refer to Section 6.5.1.

A different point of view was taken in (Lecomte and Rigo 2001), where a regular language $\mathcal{L}$ on an ordered alphabet was used to define numeration: the ordering on the alphabet induces the genealogical ordering (see Definition 1.2.15) on the language $\mathcal{L}$, and the positive integer n is then represented by the nth word in the language $\mathcal{L}$. For a detailed description of this numeration we refer to Chapter 3.

This generalises the numeration systems with linear recurrent base sequence. Again additive functions on such numeration systems can be defined. In (Grabner and Rigo 2003) it was shown that theorems analogous to Theorem 9.2.1 do not hold in this very general setting, but only under additional combinatorial assumptions on the language $\mathcal{L}$. Furthermore, in (Grabner and Rigo 2007) limiting distributions of additive functions on regular languages were studied. Again these distributions exist only under additional assumptions on the language $\mathcal{L}$; there are cases, where the limiting distribution is not Gaussian. In (Berthé and Rigo 2007b) a compactification of $\mathbb{N}$ is constructed from this type of number representations, and the according odometer is studied.

Another different approach to numeration with respect to (certain) linear recurring sequences uses substitutions. Let σ be a primitive substitution on the alphabet A such that for some $a \in A$, a is a prefix of $\sigma(a)$. A sequence of words $m_1, \ldots, m_k$ is called a-admissible, if there exist letters

$a = a_0, a_1, \ldots, a_k$ such that $m_i a_i$ is a prefix of $\sigma(a_{i-1})$ for $i = 1, \ldots, k$. Then every positive integer n can be represented by an a-admissible sequence of words $m_1, \ldots, m_k$ satisfying

$$n = |\sigma^{k-1}(m_k)| + \cdots + |\sigma^0(m_1)|. \tag{9.73}$$

Notice that by definition the length of the occurring words m_i is bounded. The words $m_1, \ldots, m_k$ are considered as the digits in this representation. We thus get the so-called Dumont–Thomas numeration. Additive functions are then defined as

$$s_f(n) = \sum_{\ell=1}^{k} f(m_\ell).$$

In a series of papers J.-M. Dumont and A. Thomas (Dumont 1990), (Dumont and Thomas 1991), (Dumont and Thomas 1993), and (Dumont and Thomas 1997) derived analogues of the theorems in Sections 9.2 and 9.3.1 for this notion of additive functions.

The study of digital functions in the context of harmonic analysis dates back to (Mahler 1927) and (Wiener 1927). They computed what one would call today the Fourier coefficients of the spectral measure associated to a dynamical system given by the sum-of-digits function. Their work was then continued by M. Mendès France, J. Coquet and P. Liardet who worked out the aspect of dynamical systems and uniform distribution in a series of papers (Mendès France 1967), (Mendès France 1970), (Mendès France 1973b), (Mendès France 1973a), (Coquet and Mendès France 1977), (Coquet, Kamae, and Mendès France 1977), (Liardet 1978), and (Coquet 1979). As overviews of this aspect we refer to (Queffélec 1987) and (Barat, Berthé, Liardet, et al. 2006).

9.4.3 Distributional results for general numeration systems

Following A. O. Gelfond's question (9.69) on the joint distribution of the sum-of-digits functions $s_{q_1}(n)$ and $s_{q_2}(n)$ (for coprime bases q_1, q_2) it is natural to consider the joint distribution of a q_1-additive function $f(n)$ and a q_2-additive function $g(n)$.

It turns out that Theorem 9.3.4 directly extends to several pairwise coprime bases. For example one has

$$\lim_{N \to \infty} \frac{1}{N} \operatorname{Card} \{n < N \mid f(n) < y_1,\, g(n) < y_2\} = G(y_1)G(y_2)$$

for certain distribution function $G_1(y)$, $G_2(y)$ if and only if the corresponding series (9.53) for $f(n)$ and $g(n)$ converge. This was observed by Hildebrandt (personal communication), the only additional ingredient for the proof is the Chinese remainder theorem.

Interestingly the general central limit theorem (Theorem 9.3.9) has no direct analogue. The reason is that the Fundamental Lemma 9.3.1 only generalises (with the help of the Chinese remainder theorem) to a property of the kind

$$\frac{1}{N} \operatorname{Card} \{n < N \mid n \in A\} = \mathbb{P}_{r_1, r_2}(A) + \mathcal{O}\left(\frac{q_1^{r_1} q_2^{r_2}}{N}\right), \qquad (9.74)$$

where A is a set depending on the first r_1 q_1-ary digits and on the first r_2 q_2-ary digits of n and $\mathbb{P}_{r_1, r_2}$ is the *natural measure* for these sets. Thus (9.74) just provides a proper approximation for half the range. Nevertheless, the BK-Property 9.3.11 generalises to two (coprime) expansions, see (Drmota 2001), (Drmota, Fuchs, and Manstavičius 2003). In particular, Theorem 9.3.10 has a bivariate extension. In contrast to the Fundamental Lemma this method is based on exponential sum estimates of the form of Lemma 9.3.14, where one can apply Baker's theory on linear forms of logarithms of algebraic numbers.

Another problem is to generalise distributional results (Theorems 9.3.4, 9.3.9, 9.3.10) to numeration systems $(G_k)_{k \in \mathbb{N}}$ that have been described in Section 9.4.2. One of the easiest extensions of the q-ary system is the Cantor system, where $G_k = q_1 q_2 \cdots q_k$ with integers $q_j \geq 2$. Here the digits $a_j(n)$ can be independently chosen from the sets $a_j(n) \in \{0, 1, \ldots, q_j - 1\}$. This independence property gives also rise to a corresponding Kubilius model (compare with Section 9.3.1.1) so that Theorems 9.3.4 and 9.3.9 directly extend to the Cantor case provided that the q_j are uniformly bounded.

Numeration systems $(G_k)_{k \in \mathbb{N}}$, where the sequence G_k satisfies a linear recurrence

$$G_k = c_1 G_{k-1} + c_2 G_{k-2} + \cdots + c_d G_{k-d}, \quad k \geq d, \qquad (9.75)$$

with constant coefficients $c_1, \ldots, c_d$ are also very well studied. The most prominent one is the Zeckendorf system ($d = 2$, $c_1 = c_2 = 1$) that is based on the Fibonacci numbers. Assume that the coefficients satisfy the relations

$$(c_j, c_{j+1}, \ldots, c_d) \leq (c_1, c_2, \ldots, c_{d-j+1}), \quad 2 \leq j \leq d,$$

where $\leq$ denotes the lexicographic order. Then every non-negative integer n has the unique (greedy) expansion $n = \sum_{j \geq 0} a_j(n) G_j$ with digits $a_j(n)$ if and only if

$$(a_k(n), a_{k-1}(n), \ldots) < (c_1, c_2, \ldots, c_k), \quad k \geq 0.$$

This already shows that the digits are not independent. However, this system is closely related to a digital representation associated to substitutions and is thus related to a Markov process (see (Dumont and Thomas 1997)). In particular, this implies that a completely additive function (related to such a numeration system) satisfies a central limit theorem. Theorems similar to Theorem 9.3.4 were proved in (Barat and Grabner 1996) and (Barat and Grabner 2008).

However, if one additionally assumes that the characteristic polynomial of the recurrence (9.75) is irreducible and that its dominant root is a Pisot unit then there is an analogue of Theorem 9.3.10 (see (Drmota and Steiner 2002) (Steiner 2002)). For example, if $c_1 \geq c_2 \geq \cdots \geq c_d = 1$ then these assumptions are satisfied (cf. Theorem 2.3.25). The essential point is again a proper variant of the BK-Property 9.3.11 which can be proved with the help of exponential sum estimates. However, at this stage one has to use an interesting relation to Rauzy fractals (cf. Chapter 5). In the q-ary case the q-ary digit $a_j(n)$ can be determined by considering the fractional part of n/q^{j+1}, that is, $a_j(n) = \ell$ if and only if $\{n/q^{j+1}\} \in [\ell/q, (\ell+1)/q)$. In the Pisot case there is a tiling $(T_\ell)_{0 \leq \ell \leq c_1}$ of $\mathbb{R}^d$ that is deduced from the Rauzy fractal related to the α-shift with the property that

$$\text{dist}\left(\mathbf{v}(n,k), T_{a_k(n)}\right) = \mathcal{O}(\alpha^{-k}),$$

where

$$\mathbf{v}(n,k) = \frac{n}{\alpha^k} \frac{\alpha - 1}{\alpha^d - 1} \left(\alpha^{d-1}, \ldots, \alpha, 1\right).$$

This means that the digits $a_k(n)$ can be almost determined by looking at n/α^k modulo the tiling. Thus a Fourier series approach similarly to that of Lemma 9.3.14 applies.

10
The equality problem for purely substitutive words

Juha Honkala

The aim of this chapter is to prove the decidability of the equality problem for purely substitutive words. This problem, also known as the D0L ω-equivalence problem, was first solved in (Culik and Harju 1984). Our presentation follows the simpler approach given in (Honkala 2007a).

10.1 Purely substitutive words and D0L systems

We start by recalling from Chapter 1 the definition of purely substitutive words. Let A be an alphabet, let $\sigma : A^* \to A^*$ be a morphism and let $a \in A$ be a letter. Assume that $\sigma(a) = au$, where $u \in A^+$. Finally, assume that $\lim_{n \to +\infty} |\sigma^n(a)| = +\infty$. Then σ is said to be *prolongable* on a and the infinite word $\sigma^\omega(a)$ is given by

$$\sigma^\omega(a) := \lim_{n \to +\infty} \sigma^n(a) = a\, u\, \sigma(u)\, \sigma^2(u)\, \sigma^3(u) \cdots.$$

An infinite word obtained in this way is called *purely substitutive*. Various examples of such words are given in Chapter 1. We furthermore stress the fact that morphisms might be erasing.

In this chapter we discuss the equality problem for purely substitutive words. In other words, we give an algorithm to decide whether or not $\sigma_1^\omega(a) = \sigma_2^\omega(a)$ if $\sigma_i : A^* \to A^*$ ($i = 1, 2$) are morphisms prolongable on $a \in A$.

The equality problem for purely substitutive words was first studied in connection with D0L systems, which were introduced by A. Lindenmayer as models of biological growth.

Definition 10.1.1 A *D0L system* is a triple $G = (A, \sigma, w)$, where A is an alphabet, $\sigma : A^* \to A^*$ is a morphism and $w \in A^*$ is a word. Then G is called *prolongable* if w is a prefix of $\sigma(w)$ and $\lim_{n \to +\infty} |\sigma^n(w)| = +\infty$. If G is prolongable, then G generates the infinite word

$$\omega(G) = \sigma^\omega(w) := \lim_{n \to +\infty} \sigma^n(w).$$

An infinite word obtained in this way is called an *infinite D0L word*. The equality problem for infinite D0L words is called the *D0L ω-equivalence problem*. If G_1 and G_2 are prolongable D0L systems and $\omega(G_1) = \omega(G_2)$, then G_1 and G_2 are called *ω-equivalent*.

Purely substitutive words are infinite D0L words but the converse is not true.

Example 10.1.2 Let $A = \{a, b\}$ and define the morphism $\sigma : A^* \to A^*$ by $\sigma(a) = a$, $\sigma(b) = b^2$. Then the D0L system $G = (A, \sigma, aab)$ is prolongable and $\sigma^\omega(aab) = aab^\omega$. However, there is no morphism $\tau : A^* \to A^*$ such that $\tau^\omega(a) = aab^\omega$. Therefore aab^ω is an infinite D0L word which is not a purely substitutive word.

The following lemma shows how to modify a given infinite D0L word to get a purely substitutive word. The lemma also shows that the decidability of the equality problem for purely substitutive words implies the decidability of the D0L ω-equivalence problem.

Lemma 10.1.3 Let $G = (B, \sigma, w)$ be a prolongable D0L system and let $\sigma^\omega(w) = uv$, where $|u| \geq |w|$. Choose a new letter $a \notin B$. Then av is a purely substitutive word.

Proof Because $|u| \geq |w|$, u is a proper prefix of $\sigma(u)$ (see Exercise 10.3). Let $\sigma(u) = uy$, where $y \in B^+$. Then

$$\sigma^\omega(w) = \sigma^\omega(u) = u\, y\, \sigma(y)\, \sigma^2(y) \cdots.$$

Let $A = \{a\} \cup B$ and define the morphism $\tau : A^* \to A^*$ by $\tau(a) = ay$ and $\tau(b) = \sigma(b)$ if $b \in B$. Then

$$\tau^\omega(a) = a\, y\, \tau(y)\, \tau^2(y) \cdots = a\, y\, \sigma(y)\, \sigma^2(y) \cdots = a\, v.$$

Hence av is a purely substitutive word. □

We next define two important classes of D0L systems.

Definition 10.1.4 An infinite word u is called *quasi-recurrent* if the first letter of u occurs only once in u but all other letters of u occur infinitely many times. Let $G = (A, \sigma, a)$ be a prolongable D0L system, where $a \in A$.

Then G is a *1-system* if $\text{alph}(\sigma^\omega(a)) = A$ and $\sigma^\omega(a)$ is quasi-recurrent. If $G = (A, \sigma, a)$ is a 1-system we always denote $A_1 = A \setminus \{a\}$. A 1-system $G = (A, \sigma, a)$ is *nearly primitive* if the restriction of σ on A_1 is primitive.

Nearly primitive D0L systems are called 1-simple systems in (Culik and Harju 1984).

Lemma 10.1.5 *The equality problem is decidable for purely substitutive words if it is decidable for quasi-recurrent purely substitutive words.*

Proof Let $\sigma_i : B^* \to B^*$, $i = 1, 2$, be morphisms prolongable on $b \in B$. Let $\sigma_i^\omega(b) = u_i v_i$, where $u_i \in B^+$ and each letter of v_i occurs infinitely many times in v_i, $i = 1, 2$. The words u_1 and u_2 can be computed effectively (see Exercise 10.1). Without loss of generality assume that $|u_1| = |u_2|$. If $u_1 \neq u_2$, then the words $\sigma_1^\omega(b)$ and $\sigma_2^\omega(b)$ are not equal. Assume that $u_1 = u_2$. Choose a new letter $a \notin B$. By Lemma 10.1.3 the words av_1 and av_2 are purely substitutive words. Clearly, av_1 and av_2 are quasi-recurrent. This proves the lemma. □

We recall that the width $\|\sigma\|$ of a morphism σ is defined as

$$\|\sigma\| = \max_{a \in A} |\sigma(a)|.$$

(see Definition 1.2.20).

10.2 Substitutive words and HD0L sequences

As explained in Chapter 1 substitutive words are obtained by applying codings to purely substitutive words. Hence, if w is a substitutive word there are a morphism $\sigma : A^* \to A^*$ prolongable on a letter $a \in A$ and a coding $\tau : A^* \to B^*$ such that $w = \tau(\sigma^\omega(a))$. Chapter 1 should again be consulted for examples of substitutive words.

If $\sigma : A^* \to A^*$ is prolongable on $a \in A$ and $\tau : A^* \to B^*$ is a coding, then we call the sequence $(\tau\sigma^n(a))_{n \geq 0}$ a *substitutive word sequence*. Hence each substitutive word is a limit of a substitutive word sequence. Substitutive word sequences are closely related to HD0L systems and sequences.

Definition 10.2.1 An *HD0L system* is a 5-tuple $H = (A, B, \sigma, \tau, w)$ where (A, σ, w) is a D0L system, B is a finite alphabet and $\tau : A^* \to B^*$ is a morphism. The *sequence* $S(H)$ generated by H consists of the words

$$\tau(w), \tau\sigma(w), \tau\sigma^2(w), \tau\sigma^3(w), \ldots.$$

An HD0L system $H = (A, B, \sigma, \tau, w)$ is called *prolongable* if w is a prefix of

$\sigma(w)$ and $\lim_{n \to +\infty} |\tau\sigma^n(w)| = +\infty$. If H is prolongable, the infinite word $\omega(H)$ generated by H is given by

$$\omega(H) := \tau(\sigma^\omega(w)).$$

An infinite word obtained in this way is called an *infinite HD0L word*.

For the proof of the following result see (Cobham 1968), (Pansiot 1983), (Allouche and Shallit 2003) and (Honkala 2009b).

Proposition 10.2.2 *Let u be an infinite word. Then u is an infinite HD0L word if, and only if, u is a substitutive word.*

(Dekking 1994) has generalised Proposition 10.2.2 by showing that a sequential transducer maps a substitutive word to a finite or substitutive word. A similar result holds for sequences of finite words.

Proposition 10.2.3 *Let $H = (A, B, \sigma, \tau, w)$ be an HD0L system and let $S(H) = (v_n)_{n \geq 0}$. Let $f : B^* \to C^*$ be a total rational function. Then one can construct an HD0L system H_1 such that $S(H_1) = (f(v_n))_{n \geq 0}$.*

We recall that a (partial) function from B^* into C^* is called a *rational function* if its graph is a rational subset of the product monoid $B^* \times C^*$. The word 'total' is used here to emphasise the fact that the function is assumed to be defined for all words of B^*. For the proof of Proposition 10.2.3 see (Honkala 2005).

The methods used in the following sections to solve the equality problem for purely substitutive words do not give a method to decide the equality of two given substitutive words. In fact, the equality problem for substitutive words is an open problem. On the other hand, equality is decidable for substitutive word sequences. This result will also be needed for the solution of the equality problem for purely substitutive words.

Theorem 10.2.4 *It is decidable, given two HD0L systems H_1 and H_2, whether or not $S(H_1) = S(H_2)$.*

Theorem 10.2.4 is due to (Ruohonen 1986b). For simpler proofs see (Ruohonen 1986a) and (Honkala 2000). The notation $|w|_C$, for C alphabet and w a finite word, stands for the number of occurrences of letters of C in u.

Example 10.2.5 Let $H_i = (A_i, B, \sigma_i, \tau_i, a_i)$, $i = 1, 2$, be prolongable HD0L systems. Assume that $C \subseteq B$ such that

$$|\tau_1 \sigma_1^n(a_1)|_C = |\tau_2 \sigma_2^n(a_2)|_C \geq 1$$

for all $n \geq 0$. Assume also that

$$\lim_{n \to +\infty} |\tau_1 \sigma_1^n(a_1)|_C = +\infty.$$

Then Proposition 10.2.3 and Theorem 10.2.4 can be used to decide whether or not $\tau_1(\sigma_1^\omega(a_1)) = \tau_2(\sigma_2^\omega(a_2))$.

First, define the rational function $f : B^* \to B^*$ as follows. If $w \in B^*$ and $|w|_C = 0$, then $f(w) = w$. If $w = w_1 c w_2$, where $w_1, w_2 \in B^*$, $c \in C$ and $|w_2|_C = 0$, then $f(w) = w_1 c$.

Now $\tau_1(\sigma_1^\omega(a_1)) = \tau_2(\sigma_2^\omega(a_2))$ if, and only if, we have

$$f(\tau_1 \sigma_1^n(a_1)) = f(\tau_2 \sigma_2^n(a_2)) \quad \text{for all } n \geq 0. \tag{10.1}$$

The validity of (10.1) can be decided by Proposition 10.2.3 and Theorem 10.2.4.

We will use the idea described in Example 10.2.5 to solve the equality problem for purely substitutive words.

10.3 Elementary morphisms

Elementary morphisms were introduced in (Ehrenfeucht and Rozenberg 1978b). They are a powerful tool in the study of iterated morphisms. (Ehrenfeucht and Rozenberg 1978a) used elementary morphisms to solve the D0L sequence equivalence problem. These morphisms have also been used, e.g., in (Linna 1977) to solve the D0L prefix problem and in (Pansiot 1986) to prove that it is decidable whether or not a purely substitutive word is eventually periodic. Elementary morphisms can also be used to prove Proposition 10.2.2 above.

Definition 10.3.1 A morphism $\sigma : A^* \to B^*$ is *simplifiable* if there exist an alphabet C with $\text{Card}(C) < \text{Card}(A)$ and morphisms $\sigma_1 : A^* \to C^*$ and $\sigma_2 : C^* \to B^*$ such that $\sigma = \sigma_2 \sigma_1$. By convention, if $A = \{a\}$ is a one-letter alphabet, then the trivial morphism $\sigma : A^* \to B^*$ defined by $\sigma(a) = \varepsilon$ is simplifiable. If a morphism σ is not simplifiable, it is called *elementary*.

Example 10.3.2 The Thue–Morse morphism $\sigma : a \mapsto ab, b \mapsto ba$ and the Fibonacci morphism $\sigma : a \mapsto ab, b \mapsto a$ are elementary. The morphism $\tau : 0 \mapsto 0012, 1 \mapsto 12, 2 \mapsto 012$ discussed in (Pytheas Fogg 2002) is simplifiable. Indeed, we have $\tau = \tau_2 \tau_1$, where $\tau_1 : 0 \mapsto ccd, 1 \mapsto d, 2 \mapsto cd$ and $\tau_2 : c \mapsto 0, d \mapsto 12$.

Elementary morphisms are injective on finite and on infinite words, see (Rozenberg and Salomaa 1980) and (Lothaire 2002).

Definition 10.3.3 Let $\sigma : A^* \to A^*$ be a morphism. The set of *cyclic letters* is defined by

$$\text{CYCLIC}(\sigma) = \{a \in A \mid |\sigma(a)|_a \geq 1\}.$$

If $\text{CYCLIC}(\sigma) = A$, then σ is called *cyclic*.

Lemma 10.3.4 *Let $\sigma : A^* \to A^*$ be an elementary morphism. Then there is a positive integer s such that σ^s is cyclic.*

Proof To prove the lemma it suffices to prove that if $a \in A$, then there is an integer $k \geq 1$ such that

$$|\sigma^k(a)|_a \geq 1.$$

Suppose on the contrary that there is a letter $a \in A$ such that $|\sigma^n(a)|_a = 0$ for all $n \geq 1$. Define

$$B = \bigcup_{n \geq 1} \text{alph}(\sigma^n(a))$$

and

$$C = B \cup \{\sigma(b) \mid b \in A \setminus B,\ b \neq a\}.$$

Because $\sigma(B) \subseteq B^*$ we have $\sigma(A) \subseteq C^*$. This contradicts the elementariness of σ because $\text{Card}(C) < \text{Card}(A)$. $\square$

Next we discuss how elementary morphisms can be used to simplify a given pair of morphisms.

Definition 10.3.5 Let $\sigma_i : A^* \to A^*$, $i = 1, 2$, be morphisms. Then the triple $(\alpha, \beta_1, \beta_2)$ *simplifies* the pair (σ_1, σ_2) if the following conditions hold:

(i) There is an alphabet B such that $\alpha : A^* \to B^*$ and $\beta_i : B^* \to A^*$, $i = 1, 2$, are morphisms.
(ii) There exist sequences $i_{11}, \ldots, i_{1k}$ and $i_{21}, \ldots, i_{2k}$ of elements from $\{1, 2\}$ such that

$$\sigma_1 \sigma_{i_{11}} \cdots \sigma_{i_{1k}} = \beta_1 \alpha, \quad \sigma_2 \sigma_{i_{21}} \cdots \sigma_{i_{2k}} = \beta_2 \alpha.$$

(iii) The morphisms β_i and $\alpha \beta_i$, $i = 1, 2$, are elementary.
(iv) The morphisms $\alpha \beta_i$, $i = 1, 2$, are cyclic.

Whenever we consider a triple $(\alpha, \beta_1, \beta_2)$ simplifying a pair (σ_1, σ_2) it is tacitly assumed that B, k and $i_{11}, \ldots, i_{1k}, i_{21}, \ldots, i_{2k}$ are as above.

The equality problem for purely substitutive words   511

A morphism $\sigma : A^* \to A^*$ is called *non-trivial* if $\sigma(A) \neq \{\varepsilon\}$. If $\sigma_i : A^* \to A^*$, $i = 1, 2$, are morphisms, the pair (σ_1, σ_2) is called *non-trivial* if all products of σ_1 and σ_2 are non-trivial.

The following lemma is essentially due to (Ehrenfeucht and Rozenberg 1978a).

Lemma 10.3.6 *Let $\sigma_i : A^* \to A^*$, $i = 1, 2$, be morphisms such that the pair (σ_1, σ_2) is non-trivial. Then there exists a triple $(\alpha, \beta_1, \beta_2)$ which simplifies the pair (σ_1, σ_2).*

Proof Assume $\tau : A^* \to A^*$ is a non-trivial morphism. Then the *defect* of τ is the greatest integer d such that there exist an alphabet B with $\mathrm{Card}(B) = \mathrm{Card}(A) - d$ and two morphisms $\tau_1 : A^* \to B^*$ and $\tau_2 : B^* \to A^*$ such that $\tau = \tau_2 \tau_1$.

Next, let $\sigma_{i_1} \cdots \sigma_{i_k}$, where $i_1, \ldots, i_k \in \{1, 2\}$, be a product of σ_1 and σ_2 such that no product of σ_1 and σ_2 has a defect greater than $\sigma_{i_1} \cdots \sigma_{i_k}$. Let the defect of $\sigma_{i_1} \cdots \sigma_{i_k}$ be d. Then there exist an alphabet B and morphisms $\alpha : A^* \to B^*$, $\beta : B^* \to A^*$ such that

$$\sigma_{i_1} \cdots \sigma_{i_k} = \beta \alpha$$

and $\mathrm{Card}(B) = \mathrm{Card}(A) - d$.

For any $\ell \geq 1$ and $i = 1, 2$, the morphism $(\sigma_i \beta \alpha)^\ell$ is a product of σ_1 and σ_2 and its defect is d. It follows that $\alpha \sigma_i \beta$ is elementary. By Lemma 10.3.4 there is a positive integer s such that $(\alpha \sigma_i \beta)^s$ is cyclic, $i = 1, 2$.

Define $\beta_i = (\sigma_i \beta \alpha)^{s-1} \sigma_i \beta$, $i = 1, 2$. Then the triple $(\alpha, \beta_1, \beta_2)$ simplifies the pair (σ_1, σ_2). The elementariness of β_i and $\alpha \beta_i$, $i = 1, 2$, follows because otherwise the defect of $(\sigma_i \beta \alpha)^{s+1}$ would exceed d. □

The final lemma of this section shows that if we simplify a pair of ω-equivalent 1-systems, then also the resulting systems are ω-equivalent 1-systems.

Lemma 10.3.7 *Let $G_i = (A, \sigma_i, a)$, $i = 1, 2$, be ω-equivalent 1-systems and let $(\alpha, \beta_1, \beta_2)$ simplify the pair (σ_1, σ_2). Then $\alpha(a)$ is a non-empty word. Let the first letter of $\alpha(a)$ be b. Then the D0L systems $F_i = (B, \alpha \beta_i, b)$, $i = 1, 2$, are ω-equivalent 1-systems.*

Proof Let $i = 1$ or $i = 2$. Because $\beta_i \alpha(a) \in a A^*$, the word $\alpha(a)$ is not empty. Because the first letter of $\alpha(a)$ is b, the first letter of $\beta_i(b)$ is a. If $c \in B$, then $|\alpha \beta_i(c)|_c \geq 1$. Therefore $\mathrm{alph}(\alpha(A)) = B$.

Next, if $\{(\alpha \beta_i)^n(b) \mid n \geq 0\}$ were finite so would be $\{(\beta_i \alpha)^n(a) \mid n \geq 0\}$

which is not true. Hence F_i is prolongable. Because $(\alpha\beta_i)^n(b)$ is a prefix of $(\alpha\beta_i)^n\alpha(a)$ for all $n \geq 0$, it follows also that

$$\omega(F_i) = \alpha(\omega(G_1)) \ . \tag{10.2}$$

In particular, every letter of B occurs in $\omega(F_i)$. Denote $\alpha\beta_i(b) = bu$, where $u \in B^+$. Because

$$\omega(F_i) = b\,u\,(\alpha\beta_i)(u)\,(\alpha\beta_i)^2(u)\cdots$$

and $\alpha\beta_i$ is cyclic, every letter of $B \setminus \{b\}$ which occurs in $\omega(F_i)$ occurs there infinitely many times.

This concludes the proof that F_1 and F_2 are 1-systems. They are ω-equivalent because (10.2) holds for $i = 1$ and $i = 2$. □

10.4 Nearly primitive D0L systems

By Theorem 1.4.5 the frequencies of the letters exist for the fixed points of primitive prolongable substitutions. More precisely, if σ is a primitive prolongable substitution and $\mathbf{M}_\sigma$ is the matrix associated with σ, then the frequencies are given by the coordinates of the positive normalised eigenvector associated with the Perron–Frobenius eigenvalue of $\mathbf{M}_\sigma$. In this section we generalise this result for the words generated by nearly primitive D0L systems.

In the sequel $\mathbf{e}$ is the row vector whose all coordinates equal 1. A vector $\mathbf{v}$ is called *normalised* if the sum of its coordinates equals 1. If $\mathbf{M}$ is a real square matrix, we recall that the *spectral radius* of $\mathbf{M}$ is equal to

$$\rho(\mathbf{M}) = \max\{|\lambda| \mid \lambda \text{ is an eigenvalue of } \mathbf{M}\}.$$

If $\mathbf{M}$ and $\mathbf{N}$ are square matrices of the same size, then $\rho(\mathbf{MN}) = \rho(\mathbf{NM})$.

The following result is a corollary of Theorem 1.4.2, see, *e.g.*, (Allouche and Shallit 2003).

Theorem 10.4.1 *Let $\mathbf{M}$ be a primitive $k \times k$ matrix. Let $\mathbf{v}$ be the nonnegative normalised eigenvector associated with the Perron–Frobenius eigenvalue β of $\mathbf{M}$. If $\mathbf{u} \in \mathbb{R}^k$ is a non-zero non-negative vector, then there is a positive real number d such that*

$$\lim_{n\to+\infty} \frac{\mathbf{M}^n \mathbf{u}}{\beta^n} = d\mathbf{v}.$$

As a corollary of Theorem 10.4.1 we obtain a result closely related to Theorem 1.4.5.

Proposition 10.4.2 *Let $\sigma : A^* \to A^*$ be a primitive morphism and let $w \in A^*$ be a non-empty word. Let $\mathbf{v}_\sigma$ be the non-negative normalised eigenvector associated with the Perron–Frobenius eigenvalue β of $\mathbf{M}_\sigma$. Then*

$$\lim_{n \to +\infty} \frac{\mathbf{P}(\sigma^n(w))}{|\sigma^n(w)|} = \mathbf{v}_\sigma.$$

Proof We have

$$\lim_{n \to +\infty} \frac{\mathbf{P}(\sigma^n(w))}{|\sigma^n(w)|} = \left(\lim_{n \to +\infty} \frac{\mathbf{e}\mathbf{M}_\sigma^n \mathbf{P}(w)}{\beta^n} \right)^{-1} \cdot \lim_{n \to +\infty} \frac{\mathbf{M}_\sigma^n \mathbf{P}(w)}{\beta^n} = \mathbf{v}_\sigma.$$

Here the second equality follows by Theorem 10.4.1. □

In the rest of this section we assume that $G = (A, \sigma, a)$ is a nearly primitive D0L system. We order the letters of A such that the special letter a is the first letter of A. Then we can write

$$\mathbf{M}_\sigma = \begin{pmatrix} 1 & \mathbf{0} \\ \mathbf{C} & \mathbf{N} \end{pmatrix},$$

where $\mathbf{N}$ is a primitive matrix. Let β be the Perron–Frobenius eigenvalue of $\mathbf{M}_\sigma$. Then β is also the Perron–Frobenius eigenvalue of $\mathbf{N}$. Let $\mathbf{v}_1$ be the non-negative normalised eigenvector of $\mathbf{N}$ associated with β. Then

$$\mathbf{v}_\sigma = \begin{pmatrix} 0 \\ \mathbf{v}_1 \end{pmatrix}$$

is a non-negative normalised eigenvector of $\mathbf{M}_\sigma$ associated with β. By Proposition 10.4.2 we have

$$\lim_{n \to +\infty} \frac{\mathbf{P}(\sigma^n(w))}{|\sigma^n(w)|} = \mathbf{v}_\sigma \qquad (10.3)$$

for a non-empty word $w \in A_1^*$.

If the finite word u is a prefix of the (possibly infinite) word v, we denote $u \leq_p v$.

Lemma 10.4.3 *Let $G = (A, \sigma, a)$ be a 1-system. If $w \leq_p \omega(G)$, then there exist a non-negative integer n and words $v_0 \in A^*$, $v_1, \ldots, v_n \in A_1^*$ of lengths less than $2\|\sigma\|$ such that*

$$\mathbf{P}(w) = \mathbf{P}(v_0 \sigma(v_1) \sigma^2(v_2) \cdots \sigma^n(v_n)).$$

Note that Lemma 10.4.3 can be compared with Equation (9.73) for Dumont–Thomas numeration (see Section 9.4.2 and 5.11 for more details).

Proof It suffices to prove that if v is a proper prefix of $\sigma^n(b)$, where $b \in A_1$, then there exist words $v_0, \ldots, v_{n-1} \in A_1^*$ of lengths less than $\|\sigma\|$ such that

$$\mathbf{P}(v) = \mathbf{P}(v_0 \sigma(v_1) \cdots \sigma^{n-1}(v_{n-1})).$$

We use induction on n. If $n = 0$, there is nothing to prove. Assume the claim is true for $n \geq 0$ and assume that v is a proper prefix of $\sigma^{n+1}(b)$. Let $\sigma(b) = b_1 b_2 \cdots b_t$ where $b_1, \ldots, b_t \in A_1$ and $t \leq \|\sigma\|$. Because v is a proper prefix of $\sigma^n(b_1 b_2 \cdots b_t)$, we have

$$v = \sigma^n(b_1) \cdots \sigma^n(b_j) w_1,$$

where $0 \leq j < t$ and w_1 is a proper prefix of $\sigma^n(b_{j+1})$. Now the claim follows inductively. □

Proposition 10.4.4 *Assume that $G = (A, \sigma, a)$ is a nearly primitive D0L system. Let α_n be the prefix of $\omega(G)$ having length n. Then*

$$\lim_{n \to +\infty} \frac{\mathbf{P}(\alpha_n)}{n} = \mathbf{v}_\sigma.$$

Proof If $\mathbf{y} = {}^t(y_1, \ldots, y_k) \in \mathbb{R}^k$, we denote $|\mathbf{y}| = {}^t(|y_1|, \ldots, |y_k|)$.

Fix $\delta > 0$. By (10.3) there exists an integer p such that

$$\left| \frac{\mathbf{P}(\sigma^i(v))}{|\sigma^i(v)|} - \mathbf{v}_\sigma \right| < \frac{1}{2} \delta^t \mathbf{e}$$

if $v \in A_1^*$ is a non-empty word of length less than $2\|\sigma\|$ and $i \geq p$. Also, there is an integer $q > p$ such that

$$\left| \frac{\mathbf{P}(v_0 \sigma(v_1) \cdots \sigma^p(v_p)) - |v_0 \sigma(v_1) \cdots \sigma^p(v_p)| \mathbf{v}_\sigma}{q} \right| < \frac{1}{2} \delta^t \mathbf{e}$$

if $v_0, \ldots, v_p \in A^*$ are words of lengths less than $2\|\sigma\|$. Let then w be a long prefix of $\omega(G)$ of length at least q. By Lemma 10.4.3 there exist words $v_0 \in A^*$, $v_1, \ldots, v_n \in A_1^*$ of lengths less than $2\|\sigma\|$ such that

$$\mathbf{P}(w) = \mathbf{P}(v_0 \sigma(v_1) \cdots \sigma^n(v_n)).$$

Denote $\mathbf{a}_i = \mathbf{P}(\sigma^i(v_i))$, $b_i = |\sigma^i(v_i)|$ for $i = 0, 1, \ldots, n$. Then

$$\begin{aligned}
\left| \frac{\mathbf{P}(w)}{|w|} - \mathbf{v}_\sigma \right| &= \left| \frac{\mathbf{a}_0 + \mathbf{a}_1 + \cdots + \mathbf{a}_n}{b_0 + b_1 + \cdots + b_n} - \mathbf{v}_\sigma \right| \\
&\leq \frac{|\mathbf{a}_0 + \mathbf{a}_1 + \cdots + \mathbf{a}_p - (b_0 + b_1 + \cdots + b_p)\mathbf{v}_\sigma|}{b_0 + b_1 + \cdots + b_n} \\
&+ \sum_{i=p+1}^n \frac{|\mathbf{a}_i - b_i \mathbf{v}_\sigma|}{b_0 + b_1 + \cdots + b_n} < \delta^t \mathbf{e}.
\end{aligned}$$

This proves the claim. □

Proposition 10.4.4 is from (Culik and Harju 1984).

As a corollary of Proposition 10.4.4 we get a result which is the first step in proving that from two ω-equivalent D0L systems we can construct two systems which have a similar growth in a sense to be explained.

Lemma 10.4.5 *Let $G = (A, \sigma, a)$ and $H = (A, \tau, a)$ be nearly primitive D0L systems such that $\omega(G) = \omega(H)$. If $\rho(\mathbf{M}_\sigma) = \rho(\mathbf{M}_\tau)$, then*

$$\lim_{n \to +\infty} \frac{|\tau\sigma^n(b)| - |\sigma\sigma^n(b)|}{|\sigma^n(b)|} = 0$$

for all $b \in A$.

Proof Let α_n be the prefix of $\omega(G) = \omega(H)$ having length n. By Proposition 10.4.4 we have

$$\mathbf{v}_\sigma = \lim_{n \to +\infty} \frac{\mathbf{P}(\alpha_n)}{n} = \mathbf{v}_\tau.$$

(Here $\mathbf{v}_\tau$ is a non-negative normalised eigenvector of $\mathbf{M}_\tau$ associated with its Perron–Frobenius eigenvalue.) Denote $\mathbf{v} = \mathbf{v}_\sigma = \mathbf{v}_\tau$ and $\rho = \rho(\mathbf{M}_\sigma) = \rho(\mathbf{M}_\tau)$. Then

$$\mathbf{M}_\sigma \mathbf{v} = \rho \mathbf{v} \quad \text{and} \quad \mathbf{M}_\tau \mathbf{v} = \rho \mathbf{v}.$$

If $b \in A$, then

$$\lim_{n \to +\infty} \frac{\mathbf{P}(\sigma^n(b))}{|\sigma^n(b)|} = \mathbf{v}.$$

This follows by Proposition 10.4.4 if $b = a$ and by (10.3) if $b \in A_1$. Hence, if $b \in A$, then

$$\frac{|\tau\sigma^n(b)| - |\sigma\sigma^n(b)|}{|\sigma^n(b)|} = \mathbf{e}(\mathbf{M}_\tau - \mathbf{M}_\sigma) \cdot \frac{\mathbf{P}(\sigma^n(b))}{|\sigma^n(b)|} \to 0$$

when $n \to +\infty$. □

10.5 Periodic and nearly periodic words

Recall from Chapter 1 that an infinite word u is called *eventually periodic* if there exist words v and w such that $u = vw^\omega$. We next define a more general notion.

Definition 10.5.1 An infinite word u is *nearly periodic* if there exist a primitive word w, words w_i and positive integers γ_i, $i = 1, 2, \ldots$, such that

$$u = w_1 w^{\gamma_1} w_2 w^{\gamma_2} \cdots w_n w^{\gamma_n} \cdots,$$

where $\lim_{n \to +\infty} \gamma_n = +\infty$ and $|w_i| < 2|w|$ for $i \geq 2$.

Suppose u is a nearly periodic word as in Definition 10.5.1. If there is a positive integer j such that $w_i \in w^*$ for all $i \geq j$, then u is eventually periodic. If there are infinitely many w_i such that $w_i \notin w^*$, then u is not eventually periodic.

Example 10.5.2 Let $A = \{a, b, c\}$ and define the morphism $\sigma : A^* \to A^*$ by $\sigma(a) = a^2$, $\sigma(b) = ba^2c$, $\sigma(c) = c$. Then
$$\sigma^\omega(b) = ba^2ca^4ca^8c\cdots$$
is nearly periodic.

The following proposition shows that the infinite word generated by a nearly primitive D0L system $G = (A, \sigma, a)$ is nearly periodic if there is a letter $b \in A$ and a positive integer d such that the period of $\sigma^d(b)$ is small with respect to the length of $\sigma^d(b)$.

The *period* of a word w is denoted by $\mathrm{PER}(w)$. The words w_1 and w_2 are called *conjugates* if there are words u and v such that $w_1 = uv$ and $w_2 = vu$.

Proposition 10.5.3 *Let $G = (A, \sigma, a)$ be a nearly primitive D0L system such that $CYCLIC(\sigma) = A$. Assume that there is a letter $b \in A$ and a positive integer d such that*
$$\mathrm{PER}(\sigma^d(b)) < \frac{1}{2\|\sigma\|}|\sigma^d(b)|. \tag{10.4}$$
Then $\omega(G)$ is nearly periodic.

Proof Let w be a primitive word such that
$$\sigma^d(b) \leq_p w^\omega \quad \text{and} \quad |w| < \frac{1}{2\|\sigma\|}|\sigma^d(b)|.$$

Because b is a factor of $\sigma(b)$, the word $\sigma^{d-1}(b)$ is a factor of $\sigma^d(b)$. Furthermore,
$$|\sigma^{d-1}(b)| \geq \frac{1}{\|\sigma\|}|\sigma^d(b)| > 2|w|.$$

Hence there is a conjugate w_1 of w such that
$$w_1^2 \leq_p \sigma^{d-1}(b).$$

Therefore
$$\sigma(w_1)^2 \leq_p w^\omega,$$

which implies that there is an integer $p \geq 2$ such that

$$\sigma(w_1) = w^p.$$

It follows that for all $n \geq 0$ and all $c \in A_1$, the word $\sigma^n(c)$ is a factor of w^ω. This implies that the word $\omega(G)$, which looks like

$$a\, u\, \sigma(u)\, \sigma^2(u) \cdots \sigma^n(u) \cdots$$

for some $u \in A_1^+$, is nearly periodic. $\square$

The validity of (10.4) for a morphism σ depends on whether or not some power of σ generates a periodic word. We discuss this fact more generally.

Definition 10.5.4 Let $\sigma : A^* \to A^*$ be a morphism. Then σ is *looping* if there exist a positive integer k, a letter $a \in A$ and a non-empty word u such that $(\sigma^k)^\omega(a)$ exists and

$$(\sigma^k)^\omega(a) = u^\omega.$$

If σ is not looping, then σ is called *loop-free*.

In Definition 10.5.4, $(\sigma^k)^\omega(a)$ is the word obtained by iterating σ^k on a. For example, the Thue–Morse morphism is loop-free.

Definition 10.5.5 Let $\sigma : A^* \to A^*$ be a morphism. Then σ is *growing* if

$$\lim_{n \to +\infty} |\sigma^n(a)| = +\infty$$

for all $a \in A$.

Proposition 10.5.6 *Let A be an alphabet having n letters and let $\sigma : A^* \to A^*$ be a growing morphism. Then σ is loop-free if, and only if,*

$$PER(\sigma^d(b)) \geq \frac{1}{2\|\sigma\|^n} |\sigma^d(b)|$$

for all $b \in A$ and $d \geq 1$.

For the proof of Proposition 10.5.6 see (Honkala 2007b).

The following proposition gives a property of nearly periodic words.

Proposition 10.5.7 *Let u be a nearly periodic word and let σ and τ be non-erasing morphisms such that $\sigma(u) = \tau(u) = u$. Then there exists a positive integer C such that*

$$||\sigma\tau(u_1)| - |\tau\sigma(u_1)|| \leq C$$

whenever u_1 is a prefix of u.

Proof Suppose
$$u = w_1 w^{\gamma_1} w_2 w^{\gamma_2} \cdots w_n w^{\gamma_n} \cdots$$
where w is a primitive word, w_i are words and γ_i are positive integers for $i = 1, 2, \ldots$. Furthermore, $\lim_{n \to +\infty} \gamma_n = +\infty$ and $|w_i| < 2|w|$ for $i \geq 2$. Without loss of generality we assume that no w_i, $i \geq 2$, begins or ends with w. Let α be any non-erasing morphism such that $\alpha(u) = u$. Then there exist words v_1, v_2 and a positive integer t such that $\alpha(w) = (v_2 v_1)^t$ and $v_1 v_2 = w$. In particular, $|\sigma\tau(w)| = |\tau\sigma(w)|$. If u is eventually periodic this already implies the claim. Assume that u is not eventually periodic. Without loss of generality assume that $w_i \notin w^*$ for $i \geq 2$.

Now, for all large n, let $\Lambda_\alpha(n)$ be the unique integer k such that
$$\alpha(w_1 w^{\gamma_1} \cdots w_n w^{\gamma_n}) \in w_1 w^{\gamma_1} \cdots w_k w^* v_1. \qquad (10.5)$$
Then for large n,
$$\alpha(w_1 w^{\gamma_1} \cdots w_{n+1} w^{\gamma_{n+1}}) \in w_1 w^{\gamma_1} \cdots w_k w^* v_1 \alpha(w_{n+1}) v_2 w^* v_1.$$
Hence
$$\Lambda_\alpha(n) \leq \Lambda_\alpha(n+1) \leq \Lambda_\alpha(n) + 1.$$
Because α is non-erasing, $\Lambda_\alpha(n) \geq n$ for all n. Consequently, there are only finitely many n such that $\Lambda_\alpha(n) = \Lambda_\alpha(n+1)$. Hence, for (sufficiently) large n we have
$$\Lambda_\alpha(n+1) = \Lambda_\alpha(n) + 1.$$
This implies that there is an integer r such that for all large n
$$\Lambda_\alpha(n) = n + r.$$

Applying these observations to σ and τ we see that there are integers p and q such that
$$\Lambda_\sigma(n) = n + p, \quad \Lambda_\tau(n) = n + q$$
for all large n. This together with (10.5) implies that there is an integer B_1 such that for all large n,
$$||\sigma(w_1 w^{\gamma_1} \cdots w_n)| - |w_1 w^{\gamma_1} \cdots w_{n+p}|| \leq B_1,$$
and
$$||\tau(w_1 w^{\gamma_1} \cdots w_n)| - |w_1 w^{\gamma_1} \cdots w_{n+q}|| \leq B_1.$$
Hence there is an integer B_2 such that for all large n,
$$||\tau\sigma(w_1 w^{\gamma_1} \cdots w_n)| - |\sigma\tau(w_1 w^{\gamma_1} \cdots w_n)|| \leq B_2.$$
This implies the claim because $|\tau\sigma(w)| = |\sigma\tau(w)|$. $\square$

10.6 A balance property for ω-equivalent 1-systems

In this section we prove a fundamental balance property for ω-equivalent 1-systems. In the next section we use this balance property together with Theorem 10.2.4 and Proposition 10.2.3 to solve the equality problem for purely substitutive words.

If $\sigma_i : A^* \to A^*$, $i = 1, 2$, are morphisms, then $\langle \sigma_1, \sigma_2 \rangle$ is the monoid consisting of all products of σ_1 and σ_2. Further, $\sigma_i \langle \sigma_1, \sigma_2 \rangle$, $i = 1, 2$, is the set consisting of all non-trivial products of σ_1 and σ_2 starting with σ_i.

Definition 10.6.1 Let $G_i = (A, \sigma_i, a)$, $i = 1, 2$, be 1-systems and let $Z \subseteq A$. We say that G_1 and G_2 are Z-balanced if there exist $\tau_i \in \sigma_i \langle \sigma_1, \sigma_2 \rangle$, $i = 1, 2$, and a word $u \in A^*$ such that the following conditions hold:

(1) Z contains a letter of $A_1 = A \setminus \{a\}$,
(2) u is a non-empty prefix of $\omega(G_1)$,
(3) there is a positive integer M such that

$$||\tau_1 \tau_2^n(u)|_Z - |\tau_2 \tau_2^n(u)|_Z| \leq M \qquad (10.6)$$

for all $n \geq 0$.

We will prove the following result.

Theorem 10.6.2 Let $G_i = (A, \sigma_i, a)$, $i = 1, 2$, be ω-equivalent 1-systems. Then there exists a set $Z \subseteq A$ such that G_1 and G_2 are Z-balanced.

We will prove Theorem 10.6.2 in three steps.

Lemma 10.6.3 Let $G = (A, \sigma, a)$ and $H = (A, \tau, a)$ be ω-equivalent nearly primitive D0L systems such that $CYCLIC(\sigma) = CYCLIC(\tau) = A$ and $\rho(\mathbf{M}_\sigma) = \rho(\mathbf{M}_\tau)$. Then G and H are A-balanced.

Proof Consider the inequality

$$\text{PER}(\sigma^d(b)) \geq \frac{1}{2\|\sigma\|} |\sigma^d(b)|. \qquad (10.7)$$

If there exist a letter $b \in A$ and a positive integer d such that (10.7) does not hold, Proposition 10.5.3 implies that $\omega(G)$ is nearly periodic and the claim follows by Proposition 10.5.7. Assume then that (10.7) holds for all $b \in A$ and $d \geq 1$.

Let $k = \text{Card}(A)$. Without loss of generality assume that $k > 2$. Let c_1, c_2, ρ be positive real numbers such that

$$c_1 \rho^i < |\sigma^i(b)| < c_2 \rho^i$$

for all $b \in A$ and for all $i \geq 0$. Choose a positive integer q such that all factors of $\omega(G)$ having length two occur in the prefix of $\omega(G)$ of length q. Let $\delta = (5\|\sigma\|qc_2\rho^{2k-1})^{-1}c_1$. Define the mapping $\beta : A^* \to \mathbb{Z}$ by

$$\beta(w) = |\sigma(w)| - |\tau(w)|, \quad w \in A^*.$$

By Lemma 10.4.5 there is a positive integer d such that

$$|\beta(\sigma^i(b))| < \delta|\sigma^i(b)|$$

for all $b \in A$ and for all $i \geq d$. Without loss of generality we assume that $c_1\rho^d > 20\|\sigma\|$. Let

$$m = \lfloor \frac{c_1\rho^d}{4\|\sigma\|} \rfloor$$

and let $e = d + 2k - 1$. Observe that we have $|\sigma^d(b)| > 4m\|\sigma\|$ for all $b \in A$.

To prove the lemma it suffices to show that

$$|\beta(\sigma^e(v))| \leq m \quad \text{if} \quad v \leq_p \omega(G). \tag{10.8}$$

We prove (10.8) by induction on the length of v. Assume first that $|v| \leq q$. Then

$$|\beta(\sigma^e(v))| < \delta|\sigma^e(v)| < \delta qc_2\rho^e < m.$$

Next, assume that $v = w_1xy$ where $w_1 \in A^*$, $x,y \in A_1$, has length greater than q and that (10.8) is true for all proper prefixes of v. By the choice of q there exist words $w_2, w_3 \in A^*$ such that $v = w_2xyw_3$ and $w_3 \neq \varepsilon$. By the inductive assumption we get

$$|\beta(\sigma^e(w_ix))| \leq m \quad \text{for} \quad i = 1,2.$$

Let u be the prefix of $\tau\sigma^e(y)$ of length m and let u_i, $i = 1,2$, be words such that

$$\tau\sigma^e(w_ix)u = \sigma\sigma^e(w_ix)u_i \quad \text{for} \quad i = 1,2.$$

Then $|u_i| \leq 2m$.

Let now z be the last letter of $\sigma^{2k}(x)$. Then there is an integer s, $k \leq s < 2k$, such that $\sigma^s(x) = wz$ where w is a non-empty word. Consequently

$$|\tau\sigma^e(x)u| > |\sigma^e(x)| = |\sigma^{e-s}(wz)| \geq |\sigma^d(wz)|$$
$$> 4m\|\sigma\| + |\sigma^d(z)| > |\sigma^d(z)u_i|$$

for $i = 1,2$. Hence one of the words $\sigma^d(z)u_1$ and $\sigma^d(z)u_2$ is a suffix of the other. Without loss of generality assume that $\sigma^d(z)u_2$ is a suffix of $\sigma^d(z)u_1$

and let u_3 be a word such that $u_1 = u_3 u_2$. Then $\sigma^d(z)$ is a suffix of $\sigma^d(z)u_3$. Now, if $u_3 \neq \varepsilon$, we get

$$\mathrm{PER}(\sigma^d(z)) \leq |u_3| \leq 2m < \frac{1}{2\|\sigma\|}|\sigma^d(z)|$$

which contradicts (10.7). Hence $u_3 = \varepsilon$ and $u_1 = u_2$. This implies that

$$\beta(\sigma^e(w_1 xy)) = \beta(\sigma^e(w_2 xy)).$$

Because $w_2 xy$ is a proper prefix of v, the inductive assumption yields

$$|\beta(\sigma^e(v))| \leq m,$$

which concludes the proof of (10.8). $\qquad\square$

If A is an alphabet and Z is a subset of A then the *projection* from A^* onto Z^* is denoted by π_Z.

Lemma 10.6.4 *Let $G_i = (A, \sigma_i, a)$, $i = 1, 2$, be ω-equivalent 1-systems such that $CYCLIC(\sigma_i) = A$ for $i = 1, 2$. Then there exists a set $Z \subseteq A$ such that G_1 and G_2 are Z-balanced.*

Proof If $\mathrm{Card}(A) = 2$ the claim holds. Indeed, if $A = \{a, b\}$ we have $\sigma_1 \sigma_2(b) = \sigma_2 \sigma_1(b)$. Let then $k \geq 2$ and suppose inductively that the claim holds if A contains at most k letters and consider an alphabet A having $k+1$ letters. Denote $G_{12} = (A, \sigma_1 \sigma_2, a)$ and $G_{21} = (A, \sigma_2 \sigma_1, a)$. If G_{12} and G_{21} are nearly primitive the claim follows by Lemma 10.6.3. We continue with the assumption that G_{12} is not nearly primitive.

Let the matrix of the restriction of $\sigma_1 \sigma_2$ to $A_1 = A \setminus \{a\}$ be $\mathbf{M}$. Because $CYCLIC(\sigma_1 \sigma_2) = A$ we have

$$\mathbf{I} \leq \mathbf{M} \leq \mathbf{M}^2 \leq \cdots .$$

In other words, $(\mathbf{M}^n)_{ij} \leq (\mathbf{M}^{n+1})_{ij}$ for all $n \geq 0$, $1 \leq i, j \leq \mathrm{Card}(A_1)$. Because G_{12} is not nearly primitive there exist integers $1 \leq i, j \leq \mathrm{Card}(A_1)$ such that

$$(\mathbf{M}^n)_{ij} = 0$$

for all $n \geq 0$. Hence there is a letter $b \in A_1$ such that the set

$$A_2 = \mathrm{alph}\{(\sigma_1 \sigma_2)^n(b) \mid n \geq 0\}$$

is a proper subset of A_1. Clearly $\sigma_1 \sigma_2(A_2^*) \subseteq A_2^*$. Because we have $CYCLIC(\sigma_i) = A$ for $i = 1, 2$, we get

$$\sigma_1(A_2^*) \subseteq A_2^* \quad \text{and} \quad \sigma_2(A_2^*) \subseteq A_2^*.$$

Denote $A_3 = A \setminus A_2$. Let $\alpha_i : A_3^* \to A_3^*$, $i = 1, 2$, be morphisms such that

$$\alpha_i(b) = \pi_{A_3}(\sigma_i(b))$$

for $b \in A_3$. Then we have $\alpha_i(\pi_{A_3}(b)) = \pi_{A_3}(\sigma_i(b))$ for all $b \in A$. Denote $F_i = (A_3, \alpha_i, a)$ for $i = 1, 2$. Because

$$\alpha_i^n(a) = \pi_{A_3}(\sigma_i^n(a))$$

for all $n \geq 0$, $i = 1, 2$, we see that F_1 and F_2 are ω-equivalent 1-systems. Further, CYCLIC$(\alpha_i) = A_3$ for $i = 1, 2$ and Card$(A_3) \leq k$. By the inductive hypothesis, there is a set $Z \subseteq A_3$ such that F_1 and F_2 are Z-balanced. Hence there exist a positive integer M, morphisms $\beta_i \in \alpha_i \langle \alpha_1, \alpha_2 \rangle$, $i = 1, 2$, and a non-empty prefix u_1 of $\omega(F_1)$ such that

$$||\beta_1 \beta_2^n(u_1)|_Z - |\beta_2 \beta_2^n(u_1)|_Z| \leq M$$

for all $n \geq 0$. Suppose

$$\beta_1 = \alpha_{i_1} \cdots \alpha_{i_r} \quad \text{and} \quad \beta_2 = \alpha_{j_1} \cdots \alpha_{j_s}$$

where $i_1, \ldots, i_r, j_1, \ldots, j_s \in \{1, 2\}$, $i_1 = 1$ and $j_1 = 2$. Denote

$$\tau_1 = \sigma_{i_1} \cdots \sigma_{i_r} \quad \text{and} \quad \tau_2 = \sigma_{j_1} \cdots \sigma_{j_s}$$

and let u be the shortest prefix of $\omega(G_1)$ such that $\pi_{A_3}(u) = u_1$. Then we have

$$||\tau_1 \tau_2^n(u)|_Z - |\tau_2 \tau_2^n(u)|_Z| = ||\beta_1 \beta_2^n(u_1)|_Z - |\beta_2 \beta_2^n(u_1)|_Z| \leq M$$

for all $n \geq 0$, which implies that G_1 and G_2 are Z-balanced. □

Now we are ready to prove Theorem 10.6.2.

Two words are called *comparable* if one of them is a prefix of the other.

Proof [Proof of Theorem 10.6.2] Assume that $G_i = (A, \sigma_i, a)$, $i = 1, 2$, are ω-equivalent 1-systems. By Lemma 10.3.6 there exists a triple $(\alpha, \beta_1, \beta_2)$ which simplifies the pair (σ_1, σ_2). Let the first letter of $\alpha(a)$ be b. Denote $B_1 = B \setminus \{b\}$ and $F_i = (B, \alpha\beta_i, b)$, $i = 1, 2$. Then F_1 and F_2 are ω-equivalent 1-systems by Lemma 10.3.7. Hence Lemma 10.6.4 implies that there exists a set $Z_1 \subseteq B$ such that F_1 and F_2 are Z_1-balanced. In other words, there exist a positive integer M, morphisms $\alpha_i \in \alpha\beta_i \langle \alpha\beta_1, \alpha\beta_2 \rangle$, $i = 1, 2$, and a non-empty prefix u_1 of $\omega(F_1)$ such that

$$||\alpha_1 \alpha_2^n(u_1)|_{Z_1} - |\alpha_2 \alpha_2^n(u_1)|_{Z_1}| \leq M$$

for all $n \geq 0$. Furthermore, Z_1 contains a letter of B_1. Suppose

$$\alpha_1 = \alpha\beta_{i_1} \cdots \alpha\beta_{i_r} \quad \text{and} \quad \alpha_2 = \alpha\beta_{j_1} \cdots \alpha\beta_{j_s}$$

where $i_1, \ldots, i_r, j_1, \ldots, j_s \in \{1, 2\}$, $i_1 = 1$ and $j_1 = 2$. Denote

$$\tau_1 = \beta_{i_1}\alpha \cdots \beta_{i_r}\alpha \quad \text{and} \quad \tau_2 = \beta_{j_1}\alpha \cdots \beta_{j_s}\alpha$$

and

$$u = \beta_{j_1}\alpha \cdots \beta_{j_s}(u_1).$$

Further, denote

$$Z = \{c \in A \mid \alpha(c) \text{ contains a letter of } Z_1\}.$$

Let $n \geq 0$. Because $\omega(G_1) = \omega(G_2)$ the words $\tau_1\tau_2^n(u)$ and $\tau_2\tau_2^n(u)$ are comparable. Suppose $\tau_1\tau_2^n(u) = \tau_2\tau_2^n(u)v$ where $v \in A^*$. Then we have $\alpha\tau_1\tau_2^n(u) = \alpha\tau_2\tau_2^n(u)\alpha(v)$. Therefore $\alpha_1\alpha_2^{n+1}(u_1) = \alpha_2\alpha_2^{n+1}(u_1)\alpha(v)$. Hence $|\alpha(v)|_{Z_1} \leq M$ implying that

$$||\tau_1\tau_2^n(u)|_Z - |\tau_2\tau_2^n(u)|_Z| = |v|_Z \leq M.$$

Therefore G_1 and G_2 are Z-balanced. This concludes the proof of Theorem 10.6.2. □

10.7 The equality problem for purely substitutive words

We show first that ω-equivalence is decidable for Z-balanced 1-systems. For this we need the following simple lemma.

Lemma 10.7.1 *Let $G_i = (A, \sigma_i, a)$, $i = 1, 2$, be 1-systems. Suppose*

$$\omega(G_i) = wb_i \cdots$$

for $i = 1, 2$, where $w \in A^*$, $b_i \in A$ and $b_1 \neq b_2$. If $i_1, \ldots, i_k \in \{1, 2\}$, then either $\sigma_{i_k} \cdots \sigma_{i_1}(a)$ is a prefix of w or has prefix wb_{i_k}.

Proof Consider the word $\sigma_{i_k} \cdots \sigma_{i_1}(a)$ and assume that the claim holds for $\sigma_{i_{k-1}} \cdots \sigma_{i_1}(a)$. If $\sigma_{i_{k-1}} \cdots \sigma_{i_1}(a)$ is a prefix of w, then $\sigma_{i_k} \cdots \sigma_{i_1}(a)$ is a prefix of $\sigma_{i_k}(w)$ and the claim holds. Assume that w is a prefix of $\sigma_{i_{k-1}} \cdots \sigma_{i_1}(a)$. Then $\sigma_{i_k}(w)$ is a prefix of $\sigma_{i_k} \cdots \sigma_{i_1}(a)$. The claim follows because $\sigma_{i_k}(w)$ is a prefix of $\omega(G_{i_k})$ which is longer than w. □

Lemma 10.7.2 *Let $G_i = (A, \sigma_i, a)$, $i = 1, 2$, be 1-systems and let $Z \subseteq A$. Assume that G_1 and G_2 are Z-balanced. Then we can decide whether or not $\omega(G_1) = \omega(G_2)$.*

Proof Because G_1 and G_2 are Z-balanced we can effectively find $\tau_i \in \sigma_i \langle \sigma_1, \sigma_2 \rangle$, $i = 1, 2$, a non-empty prefix u of $\omega(G_1)$ and a positive integer M such that (10.6) holds for all $n \geq 0$. Then the sequence $(\ell_n)_{n \geq 0}$ defined by

$$\ell_n = |\tau_1 \tau_2^n(u)|_Z - |\tau_2 \tau_2^n(u)|_Z$$

for $n \geq 0$, is $\mathbb{Z}$-rational and its image $\{\ell_n \mid n \geq 0\}$ is a finite set. Hence there exist integers $A \geq 1$, $B \geq 0$ and N such that for $n \geq 0$ we have

$$|\tau_1 \tau_2^{An+B}(u)|_Z = |\tau_2 \tau_2^{An+B}(u)|_Z + N.$$

If $(|\tau_1 \tau_2^{An+B}(u)|_Z)_{n \geq 0}$ or $(|\tau_2 \tau_2^{An+B}(u)|_Z)_{n \geq 0}$ is bounded above or contains infinitely many zero terms, we have $\omega(G_1) \neq \omega(G_2)$. (Recall that each letter of $A \setminus \{a\}$ occurs infinitely many times in $\omega(G_1)$ and $\omega(G_2)$.) We continue with the assumption that each of these sequences is not bounded above and contains no zero term. Suppose that $N \geq 0$.

Next, if j is a positive integer, define the rational function $f_j : A^* \to A^*$ as follows. If $|w|_Z < j$, then $f_j(w) = w$. If $|w|_Z \geq j$, then $f_j(w)$ is the longest prefix of w containing $|w|_Z - j$ letters of Z.

We claim that $\omega(G_1) = \omega(G_2)$ if, and only if,

$$f_{N+1}(\tau_1 \tau_2^{An+B}(u)) = f_1(\tau_2 \tau_2^{An+B}(u)) \text{ for all } n \geq 0. \tag{10.9}$$

Indeed, if $\omega(G_1) = \omega(G_2)$, then (10.9) holds. Conversely, if (10.9) holds, the words $\tau_1 \tau_2^{An+B}(u)$ and $\tau_2 \tau_2^{An+B}(u)$ have arbitrarily long common prefixes for large values of n which is not possible by Lemma 10.7.1 if $\omega(G_1) \neq \omega(G_2)$.

Now the lemma follows because we can decide the validity of (10.9) by Proposition 10.2.3 and Theorem 10.2.4. $\square$

Now we are in a position to solve the equality problem for purely substitutive words. The proof is done by constructing two semi-algorithms, one for equality and one for non-equality. By a semi-algorithm for equality we understand an effective procedure that terminates if the two given purely substitutive words are equal but which may run forever if they are not. A semi-algorithm for non-equality is defined analogously.

Theorem 10.7.3 *It is decidable whether or not two given purely substitutive words are equal.*

Proof By Lemma 10.1.5 it suffices to show that we can decide whether or not two given 1-systems are ω-equivalent.

Let $G_i = (A, \sigma_i, a)$, $i = 1, 2$, be 1-systems. To decide whether or not $\omega(G_1) = \omega(G_2)$ we run concurrently two semi-algorithms. The first semi-algorithm tries to show that $\omega(G_1) \neq \omega(G_2)$ by computing finite prefixes of $\omega(G_1)$ and $\omega(G_2)$. The second semi-algorithm first tries to find $Z \subseteq A$, $\tau_i \in \sigma_i\langle\sigma_1,\sigma_2\rangle$ for $i = 1,2$ and $u \in A^*$ such that conditions (1)–(3) of Definition 10.6.1 hold. When such Z, τ_1, τ_2 and u are found, the second semi-algorithm uses Lemma 10.7.2 to decide whether or not G_1 and G_2 are ω-equivalent. By Theorem 10.6.2 the second semi-algorithm terminates if $\omega(G_1) = \omega(G_2)$. □

10.8 Automatic words

In this section we consider sets of integers which are recognised by finite automata in various ways, the infinite words associated with these sets and their equality problem.

We start with two problems concerning abstract numeration systems. For the notation concerning abstract numeration systems we refer to Chapter 3.

Problem 10.8.1 Let $S_i = (L_i, X_i, <_i)$, $i = 1,2$, be two abstract numeration systems. Is it decidable, given an S_1-automatic word u_1 and an S_2-automatic word u_2, whether or not $u_1 = u_2$?

Problem 10.8.2 Let $S_i = (L_i, X_i, <_i)$, $i = 1,2$, be two abstract numeration systems. Is it decidable, given regular languages $K_i \subseteq L_i$, $i = 1,2$, whether or not

$$\mathrm{val}_{S_1}(K_1) = \mathrm{val}_{S_2}(K_2)?$$

Both of these problems are open. Because an infinite word u is substitutive if, and only if, there is an abstract numeration system S such that u is S-automatic (see Theorem 3.4.1) Problem 10.8.1 is equivalent to the equality problem of substitutive words (see also Remark 3.4.25). Furthermore, Problems 10.8.1 and 10.8.2 are equivalent by results concerning abstract numeration systems. In fact, if $S = (L, X, <)$ is an abstract numeration system and u is an S-automatic word, then for every letter $a \in \mathrm{alph}(u)$, there is a regular language $K_a \subseteq L$ such that

$$\mathrm{val}_S(K_a) = \{i \mid u_i = a\}.$$

(Here u_i is the ith letter of u.) Conversely, if $K \subseteq L$ is a regular language, then the characteristic word of $\mathrm{val}_S(K)$ is S-automatic.

Problem 10.8.2 is decidable if the languages K_1 and K_2 are assumed to be slender. By definition, a language K is *slender* if there exists an integer n_0 such that for all n, K contains at most n_0 words having length n.

Theorem 10.8.3 *Let $S_i = (L_i, X_i, <_i)$, $i = 1, 2$, be two abstract numeration systems. It is decidable, given slender regular languages $K_i \subseteq L_i$, whether or not*

$$\mathrm{val}_{S_1}(K_1) = \mathrm{val}_{S_2}(K_2).$$

For the proof of Theorem 10.8.3 see (Honkala and Rigo 2004).

If instead of abstract numeration systems we consider the classical base k representation of integers, we get the following result.

Theorem 10.8.4 *Let $k_1 \geq 2$ and $k_2 \geq 2$ be positive integers. It is decidable, given a k_1-automatic word u_1 and a k_2-automatic word u_2, whether or not $u_1 = u_2$.*

Proof There are two cases to consider. Suppose first that k_1 and k_2 are multiplicatively dependent. Then both u_1 and u_2 are effectively k_1-automatic and the claim follows because we can decide whether or not two given finite automata with outputs are equivalent.

Suppose then that k_1 and k_2 are multiplicatively independent. Then Cobham's theorem implies that $u_1 \neq u_2$ if u_1 is not eventually periodic. Hence the equality of u_1 and u_2 can be decided by running concurrently two semi-algorithms. The first semi-algorithm tries to show that $u_1 \neq u_2$ by computing finite prefixes of u_1 and u_2. The second semi-algorithm tries to find an eventually periodic word v such that $u_1 = v$ and $u_2 = v$. The equality $u_1 = v$ can be decided by the first part of the proof since every eventually periodic word is k_1-automatic. □

10.9 Complexity questions

Very little is known about the complexity of the equality problem for purely substitutive words. In fact, we are not able to offer any examples which would show that the problem is difficult.

Let $\sigma_i : A^* \to A^*$, $i = 1, 2$, be morphisms prolongable on $a \in A$. If $\sigma_1^\omega(a) \neq \sigma_2^\omega(a)$, then we can usually show this by comparing short prefixes of $\sigma_1^\omega(a)$ and $\sigma_2^\omega(a)$. We do not know whether this always holds.

Somewhat more is known if we consider purely substitutive words obtained by iterating primitive morphisms.

If $\sigma : A^* \to A^*$ and $\tau : A^* \to A^*$ are morphisms, define

$$\mathrm{BAL}(\sigma, \tau) = \max\{||\sigma\sigma^k(b)| - |\tau\sigma^k(b)|| \mid b \in A, k \geq 0\},$$

where the right-hand side is a non-negative integer or ∞.

For the proof of the following theorem see (Honkala 2009a).

Theorem 10.9.1 Let A be an alphabet having $n \geq 2$ letters. Define $K(n) = \lfloor 9n^3 \sqrt{n \log n} \rfloor$. Let $\sigma : A^* \to A^*$ and $\tau : A^* \to A^*$ be primitive morphisms prolongable on $a \in A$. Define $\alpha_1 = \sigma^{2n-2} \tau^{2n-2}$ and $\alpha_2 = \tau^{2n-2} \sigma^{2n-2}$. Then

$$\sigma^\omega(a) = \tau^\omega(a)$$

if, and only if,

$$BAL(\alpha_1, \alpha_2) < \infty$$

and

$$\alpha_1 \alpha_1^{K(n)}(a) \text{ and } \alpha_2 \alpha_1^{K(n)}(a) \text{ are comparable.}$$

The next result concerns primitive invertible substitutions. Let $A = \{a, b\}$ and let $\sigma : A^* \to A^*$ be a morphism. Then σ is called *invertible* if σ is an automorphism of the free group generated by A.

Theorem 10.9.2 Let $A = \{a, b\}$ and let $\sigma_1 : A^* \to A^*$ and $\sigma_2 : A^* \to A^*$ be primitive invertible morphisms prolongable on $a \in A$. Then $\sigma_1^\omega(a) = \sigma_2^\omega(a)$ if, and only if, there exists a primitive invertible morphism $\sigma : A^* \to A^*$ such that $\sigma_1 = \sigma^m$ and $\sigma_2 = \sigma^n$ for some $m, n \geq 1$.

Theorem 10.9.2 is proved in (Rao and Wen 2009). The proof uses results from (Berthé, Ei, Ito, et al. 2007). In particular, it uses Rauzy fractals discussed in Chapter 5.

10.10 Exercises

Section 10.1

Exercise 10.1 Let $\sigma : A^* \to A^*$ be a morphism and let $w \in A^*$ be a word. Define

$$B_n = \text{alph}(\sigma^n(w))$$

for $n \geq 0$. Show that the sequence $(B_n)_{n \geq 0}$ is eventually periodic and explain how one can compute its period and preperiod.

Exercise 10.2 Let $A = \{a, b, c, d\}$ and define $\sigma : A^* \to A^*$ by $\sigma(a) = ab^2 dc^2 d$, $\sigma(b) = b^2 d$, $\sigma(c) = c^2 d$, $\sigma(d) = \varepsilon$. Prove that there does not exist a non-erasing morphism $\tau : A^* \to A^*$ such that $\tau^\omega(a) = \sigma^\omega(a)$.

Exercise 10.3 Let $G = (A, \sigma, w)$ be a prolongable D0L system. Let u be a non-empty prefix of $\omega(G)$ with $|u| \geq |w|$. Show that u is a proper prefix of $\sigma(u)$.

Exercise 10.4 Let $G_1 = (A, \sigma_1, a)$ and $G_2 = (A, \sigma_2, a)$, where $a \in A$, be prolongable D0L systems such that $\omega(G_1) = \omega(G_2)$. Let $k \geq 1$ and let $i_1, \ldots, i_k \in \{1, 2\}$. Define

$$G = (A, \sigma_{i_1}\sigma_{i_2}\ldots\sigma_{i_k}, a).$$

Show that G is a prolongable D0L system and $\omega(G) = \omega(G_1)$.

Section 10.3

Exercise 10.5 Let $\sigma : A^* \to A^*$ be a morphism and let $u, v \in A^\mathbb{N}$. Suppose that there is a positive integer n such that

$$\sigma^n(u) = \sigma^n(v).$$

Prove that

$$\sigma^{\text{Card}(A)}(u) = \sigma^{\text{Card}(A)}(v).$$

(See (Honkala 2008).)

Exercise 10.6 Let $\sigma : A^* \to A^*$ be a morphism and let $u \in A^\mathbb{N}$. Suppose that there is a positive integer n such that

$$u = \sigma^n(u).$$

Prove that

$$u = \sigma^{\text{Card}(A)!}(u).$$

(See (Honkala 2008).)

Section 10.5

Exercise 10.7 Let A be an alphabet having k letters. Let $\sigma : A^* \to A^*$ be a morphism prolongable on $a \in A$. Let $\sigma(a) = au$ where $u \in A^*$. Let

$$w_1 = au\sigma(u) \cdots \sigma^{k-1}(u)$$

and let w_2 be the primitive root of

$$\sigma^k(u)\sigma^{k+1}(u) \cdots \sigma^{k+k!-1}(u).$$

Prove the following results.

a) If $\sigma^\omega(a)$ is eventually periodic, then $\sigma^\omega(a) = w_1 w_2^\omega$.

b) The word $\sigma^\omega(a)$ is eventually periodic if, and only if, there exist integers $i \geq 0$, $j \geq 1$ and words $w_3, w_4 \in A^*$ such that

$$\sigma(w_1) = w_1 w_2^i w_3, \quad \sigma(w_2) = (w_4 w_3)^j, \quad w_2 = w_3 w_4.$$

(See (Honkala 2008). This exercise implies the decidability of periodicity for infinite D0L words due to (Harju and Linna 1986) and (Pansiot 1986).)

Exercise 10.8 Prove that it is decidable whether or not a given morphism is loop-free.

Exercise 10.9 Let $\sigma : A^* \to A^*$ be a primitive morphism prolongable on $a \in A$. Prove that $\sigma^\omega(a)$ is periodic if, and only if, $\sigma^\omega(a)$ is nearly periodic.

Section 10.8

Exercise 10.10 Let v be a fixed eventually periodic word. Show that it is decidable whether or not a given substitutive word equals v.

Exercise 10.11 Let $\sigma : A^* \to A^*$ be a morphism prolongable on $a \in A$ and let $\tau : A^* \to B^*$ be a coding. We say that the substitutive word $u = \tau(\sigma^\omega(a))$ is locally polynomial if there is a letter $b \in \mathrm{alph}(\sigma^\omega(a))$ and polynomial $P(n)$ such that $|\sigma^n(b)| \leq P(n)$ for all $n \geq 0$ and $\lim_{n \to +\infty} |\sigma^n(b)| = +\infty$. Prove that it is decidable whether or not a locally polynomial substitutive word is eventually periodic.

Section 10.9

Exercise 10.12 Let $\alpha_1, \ldots, \alpha_n : A_1^* \to A_1^*$ and $\beta_1, \ldots, \beta_n : A_2^* \to A_2^*$ be uniform morphisms. Let also $\alpha : A_1^* \to B^*$ and $\beta : A_2^* \to B^*$ be uniform morphisms. Let $w_1 \in A_1^*$ and $w_2 \in A_2^*$ be words of equal length. Show that

$$\alpha \alpha_{i_k} \ldots \alpha_{i_1}(w_1) = \beta \beta_{i_k} \ldots \beta_{i_1}(w_2) \qquad (10.10)$$

holds for all $k \geq 0$ and $i_k, \ldots, i_1 \in \{1, \ldots, n\}$ if, and only if, (10.10) holds whenever $0 \leq k \leq \max\{1, \mathrm{card}(A_1) + \mathrm{card}(A_2) - 2\}$ and $i_k, \ldots, i_1 \in \{1, \ldots, n\}$ (see (Honkala 2001a)).

11
Extremal matrix products and the finiteness property

Vincent D. Blondel,

Raphaël M. Jungers

The *joint spectral radius* of a set of matrices is the maximal growth rate that can be obtained by forming long products of matrices taken in the set. This quantity and its minimal growth counterpart, the *joint spectral subradius*, have proved useful for studying several problems from combinatorics and number theory. For instance, they characterise the growth of certain classes of languages, the capacity of forbidden difference constraints on languages, and the trackability of sensor networks. In Section 11.2 we describe some of these applications.

While the joint spectral radius and related notions have applications in combinatorics and number theory, these disciplines have in turn been helpful to improve our understanding of problems related to the joint spectral radius. As an example, we present in Section 11.3 a central result that has been proved with the help of techniques from combinatorics on words: the falseness of the finiteness conjecture.

In practice, computing a joint spectral radius is not an easy task. As we will see, this quantity is NP-hard to approximate in general, and the simple question of knowing, given a set of matrices, if its joint spectral radius is larger than one is even algorithmically undecidable. However, in recent years, approximation algorithms have been proposed that perform well in practice. Some of these algorithms run in exponential time while others provide no accuracy guarantee. In practice, by combining the advantages of the different algorithms, it is often possible to obtain satisfactory estimates. Some approximation methods are presented in Section 11.4.

We finally conclude in Section 11.5, and we describe some open questions.

Combinatorics, Automata and Number Theory, ed. Valérie Berthé and Michel Rigo.
Published by Cambridge University Press. ©Cambridge University Press 2010.

11.1 The joint spectral characteristics
11.1.1 Definitions

The joint spectral radius characterises the maximal asymptotic growth rate of the norms of long products of matrices taken in a set Σ. Let $\|\cdot\|$ be a matrix norm, and $\mathbf{A} \in \mathbb{R}^{n \times n}$ be a real matrix[†]. We recall that the *spectral radius* of $\mathbf{A}$, that is, the largest magnitude of its eigenvalues, represents the asymptotic growth rate of the norm of the successive powers of A:

$$\rho(\mathbf{A}) = \lim_{t \to \infty} \|\mathbf{A}^t\|^{1/t}.$$

This quantity does provably not depend on the norm used and characterises the asymptotic rate of growth for the norm of a state $\mathbf{x}_t$ subject to the linear update $\mathbf{x}_{t+1} = \mathbf{A}\mathbf{x}_t$. In order to generalise this notion to a set of matrices Σ, let us introduce the following notation:

$$\Sigma^0 := I \qquad \Sigma^t := \{\mathbf{A}_1 \cdots \mathbf{A}_t \mid \mathbf{A}_i \in \Sigma\}.$$

Also, we denote by Σ^* the *monoid generated by* Σ:

$$\Sigma^* := \bigcup_{t=0}^{\infty} \Sigma^t.$$

If we exclude Σ^0 from the above definition, we obtain Σ^+, the *semigroup generated by* Σ:

$$\Sigma^+ := \bigcup_{t=1}^{\infty} \Sigma^t.$$

We define the two following quantities that are good candidates to quantify the 'maximal size' of products of length t:

$$\hat{\rho}_t(\Sigma) := \sup\{\|\mathbf{A}\|^{1/t} \mid \mathbf{A} \in \Sigma^t\},$$
$$\rho_t(\Sigma) := \sup\{\rho(\mathbf{A})^{1/t} \mid \mathbf{A} \in \Sigma^t\}.$$

For a matrix $\mathbf{A} \in \Sigma^t$, we call $\|\mathbf{A}\|^{1/t}$ and $\rho(\mathbf{A})^{1/t}$ respectively the *averaged norm* and the *averaged spectral radius*, in the sense that it is averaged with respect to the length of the product.

The *joint spectral radius* has been introduced in (Rota and Strang 1960) as the limit:

$$\hat{\rho}(\Sigma) := \lim_{t \to \infty} \hat{\rho}_t(\Sigma). \qquad (11.1)$$

[†] In this chapter all matrix norms are supposed to be submultiplicative, that is, $\forall \mathbf{A}, \mathbf{B}, \|\mathbf{AB}\| \leq \|\mathbf{A}\| \cdot \|\mathbf{B}\|$. Also, we restrict ourself to real matrices.

In order to prove that this limit indeed exists, let us first recall a classical result, known as *Fekete's Lemma*.

Lemma 11.1.1 (Fekete 1923) *Let $(a_n)_{n \geq 1}$ be a sequence of real numbers such that*

$$a_{m+n} \leq a_m + a_n.$$

Then the limit

$$\lim_{n \to \infty} \frac{a_n}{n}$$

exists and is equal to $\inf\{\frac{a_n}{n}\}$.

Lemma 11.1.2 *For any bounded set $\Sigma \subset \mathbb{R}^{n \times n}$, the sequence $(\hat{\rho}_t(\Sigma))_t$ converges and the limit does not depend on the norm used in the definition of $\hat{\rho}_t(\Sigma)$. Moreover,*

$$\lim_{t \to \infty} \hat{\rho}_t(\Sigma) = \inf\{\hat{\rho}_t(\Sigma)\}.$$

Proof Since the norms considered are submultiplicative, the sequence with general term

$$\log \sup \{\|\mathbf{A}\| \mid \mathbf{A} \in \Sigma^t\} = \log \hat{\rho}_t^t$$

is subadditive, that is,

$$\log \hat{\rho}_{t+t'}^{t+t'} \leq \log \hat{\rho}_t^t + \log \hat{\rho}_{t'}^{t'}.$$

If for all t, $\hat{\rho}_t \neq 0$, then by Fekete's lemma, the sequence with general term $\frac{1}{t} \log \hat{\rho}_t^t = \log \hat{\rho}_t$ converges, and is equal to $\inf\{\log \hat{\rho}_t\}$. If there is an integer t such that $\hat{\rho}_t = 0$, then clearly, for all $t' \geq t, \hat{\rho}_{t'} = 0$, and the proof is done. Finally, the limit is independent of the norm used by the equivalence of the norms in $\mathbb{R}^n$. □

Unlike $\hat{\rho}_t$, the behaviour of ρ_t is not that simple. Daubechies and Lagarias have defined in 1992 the *generalised spectral radius* as (Daubechies and Lagarias 1992):

$$\rho(\Sigma) := \limsup_{t \to \infty} \rho_t.$$

In general, the lim sup in this definition of $\rho(\Sigma)$ cannot be replaced by a limit. However, a fundamental theorem says that the joint spectral radius and the generalised spectral radius coincide.

Theorem 11.1.3 (Joint Spectral Radius Theorem) *For any bounded set of matrices Σ,*

$$\limsup_{t\to\infty} \sup \{\rho(\mathbf{A})^{1/t} \mid \mathbf{A} \in \Sigma^t\} = \lim_{t\to\infty} \sup \{||\mathbf{A}||^{1/t} \mid \mathbf{A} \in \Sigma^t\}.$$

No elementary proof is known for this theorem. Elsner (Elsner 1995) provides a self-contained proof that is somewhat simpler than (though inspired by) the original proof of (Berger and Wang 1992). In the exercises we sketch some results that are useful for the proof. See (Jungers 2009) for a complete proof. From the above theorem, we can now define:

Definition 11.1.4 The *joint spectral radius* of a bounded set of matrices Σ is defined by:

$$\rho(\Sigma) := \limsup_{t\to\infty} \rho_t(\Sigma) = \lim_{t\to\infty} \hat{\rho}_t(\Sigma).$$

Example 11.1.5 Let us consider the following set of matrices:

$$\Sigma = \left\{ \begin{pmatrix} 1 & 1 \\ 0 & 0 \end{pmatrix}, \begin{pmatrix} 1 & 0 \\ 1 & 0 \end{pmatrix} \right\}.$$

Both matrices have a spectral radius equal to one. However, by simple multiplication of the two matrices, one obtains the matrix

$$\mathbf{A} = \begin{pmatrix} 2 & 0 \\ 0 & 0 \end{pmatrix},$$

whose spectral radius is equal to two. Since

$$\lim_{t\to\infty} \hat{\rho}_t(\Sigma) \geq \lim ||\mathbf{A}^{t/2}||^{1/t} = \sqrt{2},$$

it follows that $\rho(\Sigma) \geq \sqrt{2}$. Now, Lemma 11.1.2 implies that $\rho \leq \hat{\rho}_2 = \sqrt{2}$ (where we have chosen the maximum column-sum for the norm), and so $\rho(\Sigma) = \sqrt{2}$.

We now consider the minimal rate of growth. Define

$$\check{\rho}_t(\Sigma) := \inf \{||\mathbf{A}||^{1/t} \mid \mathbf{A} \in \Sigma^t\},$$
$$\underline{\rho}_t(\Sigma) := \inf \{\rho(\mathbf{A})^{1/t} \mid \mathbf{A} \in \Sigma^t\}.$$

Then, the *joint spectral subradius* is defined by the limit:

$$\check{\rho}(\Sigma) := \lim_{t\to\infty} \check{\rho}_t(\Sigma). \tag{11.2}$$

This quantity is again independent of the norm used. We similarly define the *generalised spectral subradius* as

$$\underline{\rho}(\Sigma) := \lim_{t\to\infty} \underline{\rho}_t.$$

These notions were introduced later than the joint spectral radius ((Gurvits 1995), see also (Tsitsiklis and Blondel 1997)). Both are well defined (*i.e.*, the limits exist).

Proposition 11.1.6 *For any set $\Sigma \subset \mathbb{R}^{n \times n}$, the sequence $(\check{\rho}_t(\Sigma))_t$ is convergent, and*

$$\lim_{t\to\infty} \check{\rho}_t(\Sigma) = \inf \check{\rho}_t(\Sigma).$$

Moreover, the sequence $(\underline{\rho}_t(\Sigma))_t$ also converges, and

$$\lim_{t\to\infty} \underline{\rho}_t(\Sigma) = \inf \underline{\rho}_t(\Sigma).$$

We also have the equality between $\check{\rho}$ and $\underline{\rho}$.

Theorem 11.1.7 (Theys 2005) *For any set of matrices Σ,*

$$\check{\rho}(\Sigma) = \lim_{t\to\infty} \inf \{\rho(\mathbf{A})^{1/t} \mid \mathbf{A} \in \Sigma^t\} = \lim_{t\to\infty} \inf \{||\mathbf{A}||^{1/t} \mid \mathbf{A} \in \Sigma^t\}.$$

Proof Clearly,

$$\lim_{t\to\infty} \inf \{\rho(\mathbf{A})^{1/t} \mid \mathbf{A} \in \Sigma^t\} \leq \lim_{t\to\infty} \inf \{||\mathbf{A}||^{1/t} \mid \mathbf{A} \in \Sigma^t\}$$

because for any matrix $\mathbf{A}$, $\rho(\mathbf{A}) \leq ||\mathbf{A}||$.
Now, for any matrix $\mathbf{A} \in \Sigma^t$ with averaged spectral radius r close to $\underline{\rho}(\Sigma)$, the product $\mathbf{A}^k \in \Sigma^{kt}$ is such that $||\mathbf{A}^k||^{1/kt} \to r$ so that

$$\lim_{k\to\infty} \inf\{||\mathbf{A}||^{1/kt} \mid \mathbf{A} \in \Sigma^{kt}\} \leq r.$$

□

We thus introduce the following definition.

Definition 11.1.8 The *joint spectral subradius* of a set of matrices Σ is defined by:

$$\check{\rho}(\Sigma) := \lim_{t\to\infty} \check{\rho}_t(\Sigma) = \lim_{t\to\infty} \underline{\rho}_t(\Sigma).$$

Example 11.1.9 Let us consider the following set of matrices:

$$\Sigma = \left\{ \begin{pmatrix} 2 & 1 \\ 0 & 0 \end{pmatrix}, \begin{pmatrix} 0 & 1 \\ 0 & 3 \end{pmatrix} \right\}.$$

The spectral radius of both matrices is greater than one. However, by multiplying them, one obtains the zero matrix, and thus the joint spectral subradius is equal to zero.

The above example is simple but in general the situation may be much more complex to analyse. More involved examples can be found in the following sections and in the exercises. We add that from now on and unless explicitly stated we restrict our attention to finite sets of matrices.

11.1.2 Basic properties

In this section, we present basic results on the joint spectral characteristics, that allow us to understand what they really represent, and how they can be used in practice.

Proposition 11.1.10 (Scaling property) *For any set* $\Sigma \in \mathbb{R}^{n \times n}$ *and for any real number* α,

$$\hat{\rho}(\alpha \Sigma) = |\alpha| \hat{\rho}(\Sigma),$$

$$\check{\rho}(\alpha \Sigma) = |\alpha| \check{\rho}(\Sigma).$$

Proof This is a simple consequence of the relation $||\alpha \mathbf{A}|| = |\alpha| \, ||\mathbf{A}||$. □

Proposition 11.1.11 (Joint spectral characteristics of the powers) *For any set* $\Sigma \in \mathbb{R}^{n \times n}$

$$\hat{\rho}(\Sigma^t) = \hat{\rho}(\Sigma)^t,$$

$$\check{\rho}(\Sigma^t) = \check{\rho}(\Sigma)^t.$$

Proof This is a consequence of the definitions, and the fact that the sequences $(\hat{\rho}_t)$ and $(\check{\rho}_t)$ are convergent. □

Proposition 11.1.12 (Invariance under similarity) *For any bounded set of matrices* Σ, *and any invertible matrix* $\mathbf{T}$,

$$\hat{\rho}(\Sigma) = \hat{\rho}(\mathbf{T}\Sigma\mathbf{T}^{-1}),$$

$$\check{\rho}(\Sigma) = \check{\rho}(\mathbf{T}\Sigma\mathbf{T}^{-1}).$$

Proof This is due to the fact that for any product $\mathbf{A}_1 \cdots \mathbf{A}_t \in \Sigma^t$, the corresponding product in $\mathbf{T}\Sigma\mathbf{T}^{-1}$ is $\mathbf{T}\mathbf{A}_1 \cdots \mathbf{A}_t\mathbf{T}^{-1}$, and has equal spectral radius. □

The following result has been known for a long time. Indeed it was already present in the seminal paper of Rota and Strang (Rota and Strang 1960). Nevertheless, it is most useful, as it characterises the joint spectral radius in terms of the matrices in Σ, without considering any product of the matrices.

Proposition 11.1.13 (Infimum over all possible norms) *Let Σ be a bounded set such that $\hat{\rho}(\Sigma) \neq 0$. The joint spectral radius is given by*

$$\hat{\rho}(\Sigma) = \inf_{||\cdot||} \sup_{\mathbf{A} \in \Sigma} \{||\mathbf{A}||\}.$$

Proof Let us fix $\varepsilon > 0$, and consider the set $\tilde{\Sigma} = (1/(\hat{\rho} + \varepsilon))\Sigma$. Then, all products of matrices in $\tilde{\Sigma}^*$ are uniformly bounded, and one can define a norm $||\cdot||$ on $\mathbb{R}^n$ in the following way: $||\mathbf{x}|| := \max\{||\mathbf{A}\mathbf{x}||_2 \mid \mathbf{A} \in \tilde{\Sigma}^*\}$, where $||\cdot||_2$ is the Euclidean vector norm. Remark that in the above definition, the maximum can be used instead of the supremum, because $\rho(\tilde{\Sigma}) < 1$. The matrix norm induced by this latter vector norm, that is, the norm defined by

$$||\mathbf{A}|| = \max_{||\mathbf{x}||=1} \{||\mathbf{A}\mathbf{x}||\},$$

clearly satisfies $\sup_{\mathbf{A} \in \tilde{\Sigma}} \{||\mathbf{A}||\} \leq 1$, and so $\sup_{\mathbf{A} \in \Sigma} \{||\mathbf{A}||\} \leq \hat{\rho} + \varepsilon$. □

Common reducibility. We will say that a set of matrices is *commonly reducible*, or simply *reducible* if there is a non-trivial linear subspace (*i.e.*, different from $\{0\}$ and $\mathbb{R}^n$) that is invariant under all matrices in Σ. This property is equivalent to the existence of an invertible matrix T that 'block-triangularises simultaneously' all matrices in Σ :

$$\Sigma \text{ reducible} \quad \Leftrightarrow$$

$$\exists \mathbf{T}, 1 \leq n' \leq n-1, \forall \mathbf{A}_i \in \Sigma, \mathbf{T}\mathbf{A}_i\mathbf{T}^{-1} = \begin{pmatrix} \mathbf{B}_i & \mathbf{C}_i \\ 0 & \mathbf{D}_i \end{pmatrix}, \mathbf{D}_i \in \mathbb{R}^{n' \times n'}.$$

We will say that a set of matrices is *commonly irreducible*, or simply *irreducible* if it is not commonly reducible. Note that if Σ is reduced to a single matrix, one recovers the notion of irreducibility introduced for non-negative matrices in Section 1.4.

Proposition 11.1.14 *With the notation defined above, if Σ is bounded and reducible,*

$$\rho(\Sigma) = \max\{\rho(\{\mathbf{B}_i\}), \rho(\{\mathbf{D}_i\})\},$$

$$\check{\rho}(\Sigma) \geq \max\{\check{\rho}(\{\mathbf{B}_i\}), \check{\rho}(\{\mathbf{D}_i\})\}.$$

Proof The result follows from the invariance under similarity (Proposition 11.1.12), together with the following elementary facts:

$$\begin{pmatrix} \mathbf{B}_1 & \mathbf{C}_1 \\ 0 & \mathbf{D}_1 \end{pmatrix} \cdot \begin{pmatrix} \mathbf{B}_2 & \mathbf{C}_2 \\ 0 & \mathbf{D}_2 \end{pmatrix} = \begin{pmatrix} \mathbf{B}_1\mathbf{B}_2 & \mathbf{B}_1\mathbf{C}_2 + \mathbf{C}_1\mathbf{D}_2 \\ 0 & \mathbf{D}_1\mathbf{D}_2 \end{pmatrix},$$

$$\rho\left(\begin{pmatrix} \mathbf{B} & \mathbf{C} \\ 0 & \mathbf{D} \end{pmatrix}\right) = \max\{\rho(\mathbf{B}), \rho(\mathbf{D})\}.$$

$\square$

It is straightforward that the above proposition generalises inductively to the case where there are more than two blocks on the diagonal.

Proposition 11.1.15 (Three members inequality) *For any bounded set $\Sigma \in \mathbb{R}^{n \times n}$ and $t \geq 1$,*

$$\rho_t(\Sigma) \leq \rho(\Sigma) \leq \hat{\rho}_t(\Sigma). \tag{11.3}$$

Proof The left-hand side inequality is due to the fact that $\rho(\mathbf{A}^k) = \rho(\mathbf{A})^k$. The right-hand side is from Fekete's lemma (Lemma 11.1.1). $\square$

For the joint spectral subradius, both quantities $\underline{\rho}_t$ and $\check{\rho}_t$ are in fact upper bounds.

Proposition 11.1.16 *For any bounded set $\Sigma \subset \mathbb{R}^{n \times n}$ and $t \geq 1$,*

$$\check{\rho}(\Sigma) \leq \underline{\rho}_t(\Sigma) \leq \check{\rho}_t(\Sigma). \tag{11.4}$$

Proof The left-hand side inequality is due to the fact that $\rho(\mathbf{A}^k) = \rho(\mathbf{A})^k$, implying that $\bar{\rho}_{kt} \leq \bar{\rho}_t$. The right-hand side is a straightforward consequence of the property $\rho(\mathbf{A}) \leq \|\mathbf{A}\|$. $\square$

Existence of an optimal long product. A natural (and important) question is to ask whether the successive products that maximise the norm in the definition of the joint spectral radius are different from each other, or if they are in some sense approaching an optimal product. In other words, does there exist a left-infinite word w on Σ such that the joint spectral radius is the average growth of the norm of the successive prefixes of w? The next theorem shows that the answer to this question is affirmative.

Theorem 11.1.17 (Berger and Wang 1992) *For any bounded set of matrices Σ, there exists a left-infinite product $\cdots \mathbf{A}_2 \mathbf{A}_1$ such that $\rho(\Sigma) = \lim_{t \to \infty} ||\mathbf{A}_t \cdots \mathbf{A}_1||^{1/t}$.*

The proof of this theorem is not trivial, see (Jungers 2009).

Remark 11.1.18 Theorem 11.1.17 shows that the joint spectral radius rules the stability of switched dynamical systems. A switched linear system in discrete time is characterised by the equation

$$\mathbf{x}_{t+1} = \mathbf{A}_t \mathbf{x}_t, \quad \mathbf{A}_t \in \Sigma$$

where Σ is a set of real $n \times n$ matrices. We are interested in the following question: 'Does the state $\mathbf{x}_t$ converge to zero for all possible choices of initial state $\mathbf{x}_0$ and sequences of matrices $\mathbf{A}_t$?'. If this is the case, we say that the system is *stable*. Theorem 11.1.17 allows for the following corollary:

Corollary 11.1.19 *For a bounded set of matrices Σ, the corresponding switched dynamical system is stable if, and only if, $\rho(\Sigma) < 1$.*

The same property holds for the joint spectral subradius.

Theorem 11.1.20 (Jungers 2009) *For any (possibly unbounded) set of matrices Σ, there exists a left-infinite product $\cdots \mathbf{A}_2 \mathbf{A}_1$ such that $\check{\rho}(\Sigma) = \lim_{t \to \infty} ||\mathbf{A}_t \cdots \mathbf{A}_1||^{1/t}$.*

11.1.3 Complexity

The joint spectral radius and subradius are difficult to compute, at least theoretically. In this subsection we briefly describe negative complexity results.

NP-hardness. The first theorem we present is on the NP-hardness of the joint spectral radius approximation, and is valid even for binary matrices. This is bad news, because many examples in combinatorics or number theory make use of binary (or non-negative integer) matrices. However, we will see in Subsection 11.1.4 that for these matrices the situation is not that catastrophic. In the following we call the *size* of a number ε its bit size, that is, for instance, if $\varepsilon = p/q$ (p, q relatively prime), its size is equal to $\log(pq)$.

Theorem 11.1.21 (Tsitsiklis and Blondel 1997) *Unless $P = NP$, there is no algorithm that, given a set of matrices Σ and a relative accuracy ε, returns an estimate $\tilde{\rho}$ of $\rho(\Sigma)$ such that $|\tilde{\rho} - \rho| \leq \varepsilon \rho$ in a number of steps that is polynomial in the size of Σ and ε. This is true even if the matrices in Σ have binary entries.*

Proof The proof proceeds by reduction from the NP-complete problem SAT. □

Non-algebraicity. The next theorem, due to Kozyakin (Kozyakin 1990), states that there is no algebraic criterion that allows the stability of a switched linear system to be decided. To state this theorem properly, we consider a finite set of m $n \times n$ matrices as a point $\mathbf{x} \in \mathbb{R}^{mn^2}$. So we can talk about the *joint spectral radius of the point* x as the joint spectral radius of the associated set of matrices. We are interested in the set of all such points corresponding to $\rho(\mathbf{x}) < 1$. For these sets to be easily recognisable, one would like them to be expressed in terms of simple constraints such as, for example, polynomial constraints. This is the notion of *semi-algebraic sets*.

Definition 11.1.22 A subset of $\mathbb{R}^n$ is *semi-algebraic* if it is a finite union of sets that can be expressed by a finite number of polynomial equalities and inequalities.

Theorem 11.1.23 (Kozyakin 1990), (Theys 2005) *For all $m, n \geq 2$, the set of points $\mathbf{x} \in \mathbb{R}^{mn^2}$ for which $\rho(\mathbf{x}) < 1$ and the set of points for which the semigroup of matrices is bounded are not semi-algebraic.*

Proof The proof makes use of a parametric set of matrices $\Sigma(t) = \{\mathbf{G}(t), \mathbf{H}(t)\} \mid t \in]0, 1[$ and shows that the set of admissible values for t (that involve stability of the corresponding set of matrices) is not a finite number of intervals:

$$\mathbf{G}(t) = (1-t^4)\begin{pmatrix} 1 & -\frac{t}{\sqrt{1-t^2}} \\ 0 & 0 \end{pmatrix}, \quad \mathbf{H}(t) = (1-t^4)\begin{pmatrix} \sqrt{1-t^2} & -t \\ t & \sqrt{1-t^2} \end{pmatrix}.$$
(11.5)

The set $\Sigma(t)$ is unstable for all $t = \sin(2\pi/(2k))$ (*i.e.*, $\rho(\Sigma(t)) > 1$), while it is stable for all $t = \sin(2\pi/(2k+1))$ (see (Theys 2005) for a proof). Since $W = \{\Sigma(t)\}$ is an algebraic set in $\mathbb{R}^8$, the intersection $W \cap E$ should be made of a finite number of connected components if E was semi-algebraic. Taking E the set of points corresponding to stable sets, or to sets generating a bounded semigroup of matrices, we have a contradiction. □

Undecidability. The results that we now present are in a sense even worse than the previous ones, since they establish that there does not exist in general any algorithm that allows a joint spectral radius to be computed in finite time.

Theorem 11.1.24 (Blondel and Canterini 2003), (Blondel and Tsitsiklis 2000a) *The problem of determining, given a set of matrices Σ, if the semigroup generated by Σ is bounded is not algorithmically decidable. The same is true for the problem $\rho(\Sigma) \leq 1$ and undecidability remains even if the matrices in Σ contain only non-negative rational entries.*

For the joint spectral subradius, it is even not possible to obtain *approximation* algorithms. Let us define a wide class of approximation algorithms, and show that none is possible for approximating the joint spectral subradius. An algorithm providing the value $\tilde{\rho}$ as an approximation of the joint spectral subradius $\check{\rho}$ of a given set is said to be a (K, L)-approximation algorithm if $|\tilde{\rho} - \check{\rho}| < K + L\check{\rho}$. The following theorem makes use of the undecidability of Post's correspondence problem (see (Papadimitriou 1994) for an introduction).

Theorem 11.1.25 (Tsitsiklis and Blondel 1997, Theorem 2) *Let n_p be a number of pairs of words for which Post's correspondence problem is undecidable (one may take $n_p = 7$). Fix any $K > 0$ and $0 < L < 1$. Unless P=NP, there exists no (K, L)-approximation algorithm for computing the joint spectral subradius. This is true even for the special case of sets of matrices consisting of one $(6n_p + 7) \times (6n_p + 7)$ integer matrix and one $(6n_p + 7) \times (6n_p + 7)$ integer diagonal matrix.*

We close this section with an open problem of great practical interest.

Problem 11.1.26 Is there an algorithm that, given a finite set of matrices Σ, decides whether $\rho < 1$?

This problem is important in practice, since it is equivalent to the stability problem of switched linear systems. This question is closely related to the finiteness property that will be the main topic of a subsequent section.

11.1.4 Non-negative integer matrices

As stated above, the case of non-negative integer matrices (*i.e.*, sets of matrices in $\mathbb{N}^{n \times n}$) is of particular importance for several reasons. First, a number of applications, and especially applications in number theory and combinatorics involve matrices that take their values in $\mathbb{N}$, or in $\{0,1\}$. This is for instance the case for the applications presented in the next section. Second, these sets of matrices appear to have good properties, and they are much easier to handle as is shown in the next theorem. Finally, interesting open problems exist about these particular sets (see for instance the Conjecture 11.3.4 below). The main advantage one has when coping with non-negative integer matrices is the following theorem.

Theorem 11.1.27 *For finite sets of non-negative integer $n \times n$ matrices Σ there is a polynomial-time algorithm that decides between the following four cases:*

- *$\rho = 0$,*
- *$\rho = 1$ and $\hat{\rho}_t(\Sigma)$ is bounded,*
- *the function $\hat{\rho}_t(\Sigma)$ grows polynomially,*
- *the function $\hat{\rho}_t(\Sigma)$ grows exponentially.*

Moreover, in the third case, there exist constants C_1, C_2, such that $C_1 t^k \leq \hat{\rho}_t(\Sigma) \leq C_2 t^k$ for all t. The rate of growth k is an integer satisfying $1 \leq k \leq n-1$, and there is a polynomial time algorithm for computing k. This algorithm performs at most $O(n^5)$ operations, where n is the dimension of the matrices.

Proof See (Jungers, Protasov, and Blondel 2008b) for a proof. □

11.1.5 Conclusion

The goal of this section was to introduce the notions of joint spectral radius and subradius. Some basic facts, such as the equivalence between the joint and generalised spectral radii, require advanced results and we have decided not to present their proof here. Further elementary properties of the joint spectral radius of sets of matrices can be found in (Wirth 2000), (Wirth 2005), (Protasov 1996), (Blondel and Tsitsiklis 2000b).

11.2 Applications

11.2.1 Euler's binary partition function, and generalisations

Binary partition functions have attracted the attention of number theorists since Leonhard Euler (Euler 1922). See (Reznick 1990, de Bruijn 1948) for recent references. The binary partition function with parameters m, d, denoted $b_{m,d}(n)$, is defined as:

$$b_{m,d}(n) = \left|\left\{(a_0, a_1, \ldots) \mid n = \sum a_i m^i, \ 0 \le a_i \le d\right\}\right|.$$

Thus, $b_{m,d}$ is the number of ways one can write n as a sum of powers of m with coefficients that are less than d. It has been proved that the asymptotic behaviour of these functions is ruled by the joint spectral characteristics of certain sets of matrices $\Sigma_{m,d}$, with binary entries.

Theorem 11.2.1 (Protasov 2000) *For all pairs $(m, d) \in \mathbb{N}^2$, there exist constants $C_1, C_2, \lambda_1, \lambda_2$ such that the following holds:*

$$C_1 n^{\lambda_1} \le b_{m,d}(n) \le C_2 n^{\lambda_2}. \tag{11.6}$$

Moreover, there exists a set of binary matrices $\Sigma_{m,d}$, such that the maximal λ_1 and the minimal λ_2 for which Equation (11.6) holds are respectively the joint spectral subradius and the joint spectral radius of $\Sigma_{m,d}$.

In other words, the function $b_{m,d}(n)$ does not behave like a polynomial, but oscillates between two polynomial curves whose rates of growth are given by $\log_m \check{\rho}_{m,d}$ and $\log_m \hat{\rho}_{m,d}$. In (Protasov 2000), these joint spectral characteristics are analysed for $m = 2$. Some sets for $m \ge 3$ have been studied in (Protasov, Jungers, and Blondel 2009). Recent techniques for approximating the joint spectral radius and subradius allow us to approximate λ_1 and λ_2 for larger values of m, d, as in the example below.

Example 11.2.2 (Protasov, Jungers, and Blondel 2009) Let us consider the case $m = 3, d = 14$. For this case it can be shown that

$$\Sigma_{3,14} = \left\{ \begin{pmatrix} 1 & 1 & 1 & 1 & 1 & 0 & 0 \\ 0 & 1 & 1 & 1 & 1 & 0 & 0 \\ 0 & 1 & 1 & 1 & 1 & 1 & 0 \\ 0 & 1 & 1 & 1 & 1 & 1 & 0 \\ 0 & 0 & 1 & 1 & 1 & 1 & 0 \\ 0 & 0 & 1 & 1 & 1 & 1 & 1 \\ 0 & 0 & 1 & 1 & 1 & 1 & 1 \end{pmatrix}, \begin{pmatrix} 1 & 1 & 1 & 1 & 1 & 0 & 0 \\ 1 & 1 & 1 & 1 & 1 & 0 & 0 \\ 0 & 1 & 1 & 1 & 1 & 0 & 0 \\ 0 & 1 & 1 & 1 & 1 & 1 & 0 \\ 0 & 1 & 1 & 1 & 1 & 1 & 0 \\ 0 & 0 & 1 & 1 & 1 & 1 & 0 \\ 0 & 0 & 1 & 1 & 1 & 1 & 1 \end{pmatrix}, \begin{pmatrix} 1 & 1 & 1 & 1 & 0 & 0 & 0 \\ 1 & 1 & 1 & 1 & 1 & 0 & 0 \\ 1 & 1 & 1 & 1 & 1 & 0 & 0 \\ 0 & 1 & 1 & 1 & 1 & 0 & 0 \\ 0 & 1 & 1 & 1 & 1 & 1 & 0 \\ 0 & 1 & 1 & 1 & 1 & 1 & 0 \\ 0 & 0 & 1 & 1 & 1 & 1 & 0 \end{pmatrix} \right\}. \tag{11.7}$$

In (Protasov, Jungers, and Blondel 2009), the authors obtain the following inequalities with methods similar to those presented in Section 11.4:

$$\begin{aligned} 4.525 &\leq \check{\rho}(\Sigma) \leq 4.6105, \\ 4.72 &\leq \rho(\Sigma) \leq 4.8. \end{aligned} \tag{11.8}$$

Thus, one obtains the corresponding bounds on λ_i

$$\begin{aligned} 1.3741 &\leq \lambda_1 \leq 1.3912, \\ 1.4125 &\leq \lambda_2 \leq 1.4278. \end{aligned} \tag{11.9}$$

11.2.2 Repetition-free words

In this subsection we study the asymptotic behaviour of repetition-free languages, with an emphasis on overlap-free words. A γ *repetition*, with $\gamma \in \mathbb{Q}$, is a word z that can be written $x^p y$ for some integer p, where y is a prefix of x, and such that $\gamma = |z|/|x|$ (see also Section 8.1.3.3). So, a two repetition is a square, a three repetition is a cube, *etc.* We study here γ *repetition-free words* (respectively γ^+ *repetition-free words*), that is, words that do not contain as a factor a δ repetition with $\delta \geq \gamma$ (respectively $\delta > \gamma$). See (Berstel 2005) for a recent survey.

An *overlap* is a 2^+ repetition, that is, a word on the alphabet $\{a,b\}$ of the form $xuxux$, where x is a or b, and u is a word that can be empty. For instance, the word *baabaab* is an overlap. Thue (Thue 1906), (Thue 1912) proved in 1906 that there are infinitely many overlap-free words. Indeed, the well-known Thue–Morse sequence (see Example 1.2.21) is overlap-free, and so the set of its factors provides an infinite number of different overlap-free words. The study of the asymptotics of the number u_n of such words has a long history[†](Brlek 1989), (Restivo and Salemi 1985), (Kfoury 1988), (Kobayashi 1988), (Lepistö 1995). In Figure 11.1 (a) we show the values of the sequence u_n for $1 \leq n \leq 200$ and in Figure 11.1 (b) we show the behaviour of $\log u_n / \log n$ for larger values of n. One can observe that the sequence (u_n) is not monotonic, but is globally increasing with n. Moreover the sequence does not seem to have a polynomial growth since the sequence $(\log u_n / \log n)$ does not seem to converge. It has been known for a long time as mentioned in (Berstel 2005) that there exist constants C_1, C_2 such that

$$C_1 \, n^{1.155} \leq u_n \leq C_2 \, n^{1.37}.$$

But actually a lot more can be said on the sequence (u_n). The following result appears in (Cassaigne 1993).

† The number of overlap-free words of length n is referenced in the On-Line Encyclopedia of Integer Sequences under the code A007777. See (Sloane). The sequence starts 1, 2, 4, 6, 10, 14, 20, 24, 30, 36, 44, 48, 60, 60, 62, 72, ...

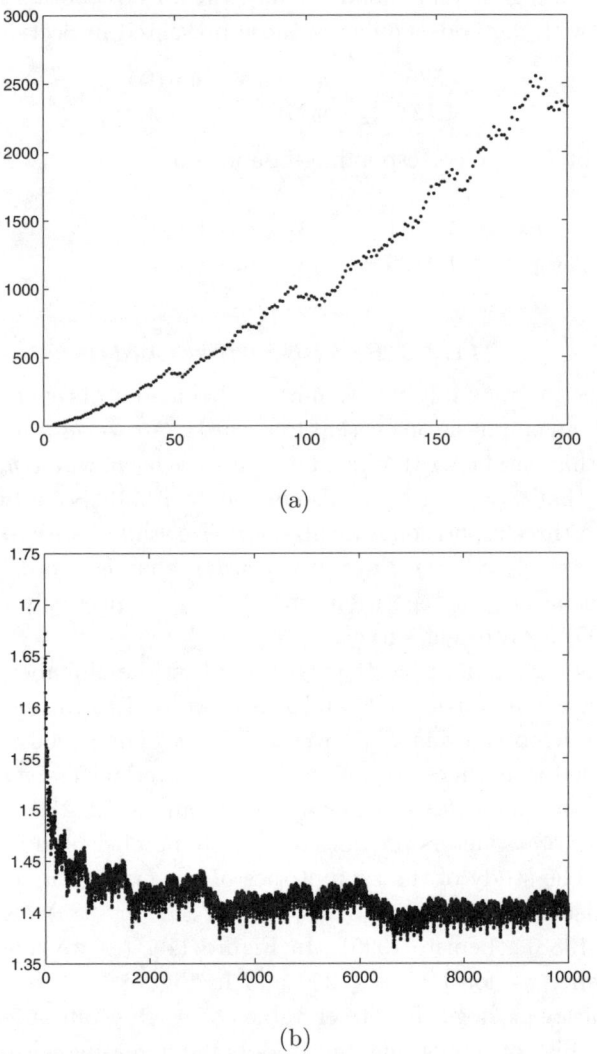

Fig. 11.1 The first values of u_n (a) and the first values of $\log u_n / \log n$ (b).

Theorem 11.2.3 *There exist matrices* $\mathbf{F}_0, \mathbf{F}_1 \in \mathbb{R}^{30 \times 30}$, *and vectors* $\mathbf{w}, \mathbf{y}_8, \ldots, \mathbf{y}_{15} \in \mathbb{R}^{30}_+$ *such that for* $n \geq 16$, *defining* $\mathbf{y}_n$ *as the solution of the following recurrence equations*

$$\begin{aligned} \mathbf{y}_{2n} &= \mathbf{F}_0 \mathbf{y}_n, \\ \mathbf{y}_{2n+1} &= \mathbf{F}_1 \mathbf{y}_n, \end{aligned}$$

then, for any $n \geq 9$, *the number of overlap-free words of length* n *is given*

by:
$$u_n = {}^t\mathbf{w}\mathbf{y}_{n-1}.$$

It follows from this result that the number u_n of overlap-free words of length $n \geq 16$ can be obtained by first computing the binary expansion $d_k \cdots d_1$ of $n-1$, i.e., $n-1 = \sum_{j=0}^{k-1} d_{j+1} 2^j$, and then

$$u_n = {}^t\mathbf{w}\mathbf{F}_{d_1} \cdots \mathbf{F}_{d_{k-4}} \mathbf{y}_m$$

where $m = d_{k-3} + d_{k-2} 2 + d_{k-1} 2^2 + d_k 2^3$ (and $d_k = 1$). Theorem 11.2.3 actually translates in matrix terms the fact that the sequence u_n is 2-regular (see (Allouche and Shallit 2003) for a definition of 2-regular sequences, see also Definition 9.1.2). Let us add that this latter fact had already been proved by Carpi (Carpi 1993), but it was not trivial from his work how to actually construct the matrices.

In view of this, let us introduce, as in (Cassaigne 1993), the lower and the upper exponents of growth:

$$\begin{aligned} \alpha &= \sup\{r \mid \exists C > 0, u_n \geq Cn^r\}, \\ \beta &= \inf\{r \mid \exists C > 0, u_n \leq Cn^r\}. \end{aligned} \quad (11.10)$$

The following theorem appears in (Jungers, Protasov, and Blondel 2009).

Theorem 11.2.4 *There exist matrices $\mathbf{A}_0, \mathbf{A}_1 \in \mathbb{R}_+^{20 \times 20}$ such that*

$$\alpha = \log_2 \check{\rho}(\mathbf{A}_0, \mathbf{A}_1) \quad \text{and} \quad \beta = \log_2 \hat{\rho}(\mathbf{A}_0, \mathbf{A}_1). \quad (11.11)$$

Thanks to the above results, the following estimates have been obtained for α and β

$$\begin{aligned} 1.2690 &< \alpha < 1.2736, \\ 1.3322 &< \beta < 1.3326. \end{aligned} \quad (11.12)$$

The inequality $1.2690 < \alpha$ was obtained with recent techniques for computing the joint spectral subradius, that are similar to the ones presented in Section 11.4. For the upper bound on α, it comes from the fact that the product $\mathbf{F}_1^{10} \mathbf{F}_0$ satisfies:

$$\alpha \leq \log_2 \left[\rho(\mathbf{F}_1^{10} \mathbf{F}_0)^{1/11}\right] = 1.2735\ldots$$

The lower bound on β comes from the equality

$$\rho(\mathbf{F}_0 \mathbf{F}_1)^{1/2} = 1.3322\ldots,$$

and the upper bound can be obtained with the technique presented in Theorem 11.4.9.

These results have recently been generalised to repetition-free words, for any value of the forbidden exponent γ between 2^+ and $7/3$. The fact that the sequence $u_\gamma(n)$ is 2-regular still holds. It is obviously still possible to define r_γ^+, (respectively r_γ^-) as the supremum (respectively the infimum) such that

$$C_1 n^{r_\gamma^-} \leq u_\gamma(l) \leq C_2 n^{r_\gamma^+}.$$

The following results appear in (Blondel, Cassaigne, and Jungers 2009).

Theorem 11.2.5 *Let $2 < \gamma \leq 7/3$ or $\gamma = \delta^+$, $2 \leq \delta < 7/3$. There exist matrices $\mathbf{F}_0, \mathbf{F}_1$ such that*

$$r_\gamma^+ = \log_2 \hat{\rho}(\{\mathbf{F}_0, \mathbf{F}_1\}),$$

$$r_\gamma^- \leq \log_2 \check{\rho}(\{\mathbf{F}_0, \mathbf{F}_1\}).$$

As an application, we have the following estimates for the case $\gamma = 7/3$ (Blondel, Cassaigne, and Jungers 2009):

$$r_{7/3}^- < 2.0035$$

$$2.0121 < r_{7/3}^+ < 2.1050.$$

11.2.3 The capacity of forbidden differences

In certain coding applications one is interested in binary codes whose elements avoid a set of forbidden patterns. This problem is classical and has been widely studied in the past century, see (Lind and Marcus 1995). A more complicated problem arises when it is desirable to find code words whose *differences* avoid forbidden patterns. This is necessary for minimising the error probability of some particular magnetic-recording systems (see for instance (Moision, Orlitsky, and Siegel 1999)). Let $\{0,1\}^t$ denote the set of words of length t over $\{0,1\}$ and let $u, v \in \{0,1\}^t$. The difference $u - v$ is a word of length t over $\{-1, 0, +1\}$ (as a shorthand we shall use $\{-, 0, +\}$ instead of $\{-1, 0, +1\}$). The difference $u - v$ is obtained from u and v by symbol-by-symbol subtraction so that, for example, $0110 - 1011 = -+0-$. Consider now a finite set D of words over $\{-, 0, +\}$. We think of D as a set of *forbidden difference patterns*. A set (or *code*) $C \subseteq \{0,1\}^t$ is said to *avoid* the set D if none of the differences of words in C contains a word from D as subword, that is, none of the differences $u - v$ with $u, v \in C$ can be written as $u - v = xdy$ for $d \in D$ and some (possibly empty) words x and y over $\{-, 0, +\}$.

We are interested in the largest cardinality, which we denote by $\delta_t(D)$, of sets of words of length t whose differences avoid the forbidden patterns in D:

$$\delta_t(D) = \max_{W \subset \{0,1\}^t,\ W \text{ avoids } D} |W|.$$

If the set D is empty, then there are no forbidden patterns and $\delta_t(D) = 2^t$. When D is non-empty, $\delta_t(D)$ grows exponentially with the word length t and is asymptotically equal to $2^{\text{cap}(D)t}$ where the scalar $0 \leq \text{cap}(D) \leq 1$ is the *capacity* of the set D and is defined by

$$\text{cap}(D) = \lim_{t \to \infty} \frac{\log_2 \delta_t(D)}{t}. \tag{11.13}$$

The existence of this limit is a simple consequence of the formulation of the problem with a joint spectral radius, which we now present (for the sake of clarity we suppose in the following that the forbidden patterns have the same length).

Moision et al. show in (Moision, Orlitsky, and Siegel 2001) how to represent codes submitted to a set of difference constraints D as products of matrices taken in a finite set $\Sigma(D)$. The idea is to make use of De Bruijn graphs†. De Bruijn graphs were introduced in (de Bruijn 1946). For an introduction, see for instance (Lind and Marcus 1995). In these graphs, the vertices represent all words of length $M - 1$ (for a fixed M) and the edges represent all words of length M. The edge associated with the word xyz, with x, z letters of the alphabet, and y a word of length $M - 2$ connects the vertex corresponding to the word xy to the one corresponding to the word yz. Paths of length l represent words of length $l + M - 1$.

Let us construct the De Bruijn graph of binary words of length m equal to the lengths of the forbidden patterns. Edges in these graphs represent words of length m, and since some pairs of words cannot appear together, a subgraph of the De Bruijn graph is said to be *admissible* if it does not contain two edges that represent words of length m whose difference is forbidden. Figure 11.2 (a) represents a De Bruijn graph that is admissible for the forbidden pattern $D = \{++-\}$. Now, if one defines a sequence of admissible graphs $\mathcal{G}_t$, for $1 \leq t \leq T$, the set of paths whose tth edge belongs to $\mathcal{G}_t$ represents an admissible code of length $m + T - 1$. For convenience we represent these graphs as bipartite graphs (by duplicating every node), so that we can concatenate them to represent longer codes (see Figure 11.2). Since graphs can be represented by binary matrices, we have the following theorem.

† These graphs are also called Rauzy graphs. See Definition 7.3.5 and after for more information.

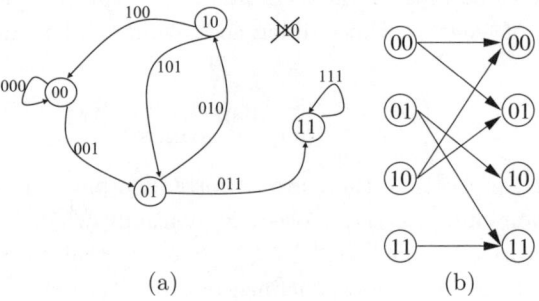

Fig. 11.2 An admissible De Bruijn graph for $D = \{++-\}$ (a), and the same graph under its bipartite form (b).

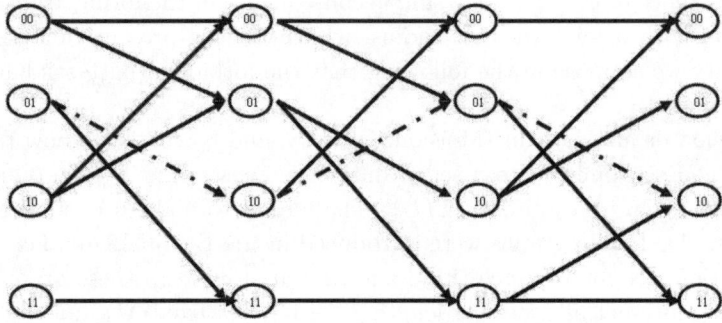

Fig. 11.3 An admissible graph that represents a maximal set of admissible words of length 5 for $D = \{++-\}$. For example, the path on the top represents the word 00000 and the dashed path represents the word 01010. The graph is maximal in the sense that no word can be added to this code, but perhaps another choice of elementary graphs would generate more paths.

Theorem 11.2.6 *Associated with any set D of forbidden patterns of length at most T, there exists a finite set $\Sigma(D)$ of binary matrices for which*

$$\delta_{T-1+t} = \hat{\rho}_t^t(\Sigma(D)) = \max\{\|\mathbf{A}_1 \cdots \mathbf{A}_t\| \mid \mathbf{A}_i \in \Sigma(D)\},$$

where $\|\mathbf{A}\|$ is the sum of the entries of $\mathbf{A}$.

Corollary 11.2.7 *Let D be a set of forbidden patterns and $\Sigma(D)$ be the set of binary matrices constructed as described above, then*

$$cap(D) = \log_2 \rho(\Sigma(D)).$$

Example 11.2.8 Let $D = \{++ +-\}$. The set $\Sigma(D)$ contains two matrices:

$$\mathbf{A}_0 = \begin{pmatrix} 1 & 1 & 0 & 0 & 0 & 0 & 0 & 0 \\ 0 & 0 & 1 & 1 & 0 & 0 & 0 & 0 \\ 0 & 0 & 0 & 0 & 1 & 1 & 0 & 0 \\ 0 & 0 & 0 & 0 & 0 & 0 & 1 & 1 \\ 1 & 1 & 0 & 0 & 0 & 0 & 0 & 0 \\ 0 & 0 & 1 & 1 & 0 & 0 & 0 & 0 \\ 0 & 0 & 0 & 0 & 1 & 1 & 0 & 0 \\ 0 & 0 & 0 & 0 & 0 & 0 & 0 & 1 \end{pmatrix},$$

$$\mathbf{A}_1 = \begin{pmatrix} 1 & 0 & 0 & 0 & 0 & 0 & 0 & 0 \\ 0 & 0 & 1 & 1 & 0 & 0 & 0 & 0 \\ 0 & 0 & 0 & 0 & 1 & 1 & 0 & 0 \\ 0 & 0 & 0 & 0 & 0 & 0 & 1 & 1 \\ 1 & 1 & 0 & 0 & 0 & 0 & 0 & 0 \\ 0 & 0 & 1 & 1 & 0 & 0 & 0 & 0 \\ 0 & 0 & 0 & 0 & 1 & 1 & 0 & 0 \\ 0 & 0 & 0 & 0 & 0 & 0 & 1 & 1 \end{pmatrix}.$$

It is possible to show that $\mathrm{cap}(D) = 0.9005\ldots$. The product that provides this value is simply $\mathbf{A}_0 \mathbf{A}_1$ (Blondel, Jungers, and Protasov 2006).

Little is known on the capacity of codes that avoid forbidden difference patterns, except for its formulation in terms of a joint spectral radius. It is clear that capacity approximation algorithms based on the direct computation of the set $\Sigma(D)$ will not run in polynomial time. Indeed, the number of matrices in $\Sigma(D)$ can be doubly exponential in m and the matrices in $\Sigma(D)$ have dimension $2^{m-1} \times 2^{m-1}$. The situation is even worse if all patterns in D do not have the same size (see (Jungers 2009) for more details). However, this does not preclude the existence of approximation/computation algorithms for the capacity that would be based on other techniques.

Problem 11.2.9 Is the capacity computation/approximation NP-hard?

It has recently been shown that it is possible to decide in polynomial time if the capacity is exactly equal to zero. We present the proof of this result, which makes use of the Aho–Corasick automaton of the set of forbidden patterns (Blondel, Jungers, and Protasov 2006).

Theorem 11.2.10 *Let D be a set of forbidden patterns of lengths at most m. Then $\mathrm{cap}(D) > 0$ if, and only if, there exists a word on the alphabet $\{+, -, 0\}$ that does not contain any word of $D \cup -D$ as a factor and that has a prefix 0^m and a suffix $+0^{m-1}$.*

Proof Let us first suppose $0^m \notin D$. The capacity is positive if, and only if, $\rho(\Sigma(D)) > 1$. For binary matrices this is equivalent to the fact that there is a product in Σ^* that has a diagonal entry larger than one (Jungers, Protasov, and Blondel 2008b). In turn, by construction of the set $\Sigma(D)$, this is equivalent to the existence of two words with the same $m-1$ first characters, and the same $m-1$ last characters, whose difference avoids the forbidden patterns. Now, this latter fact is possible if, and only if, there is a non-trivial sequence on $\{0, +, -\}$ of the shape $0^{m-1}d0^{m-1}$ that avoids $D \cup -D$.

If $0^m \in D$, then $\text{cap}(D) = 0$, and obviously all words of the form $0^m d 0^{m-1}$ contains a factor in D. □

Now, the criterion provided by Theorem 11.2.10 can be checked by a simple application of the Aho–Corasick algorithm (see (Lothaire 1983)), that checks for the unavoidability of a set of words on an arbitrary alphabet.

Corollary 11.2.11 *The problem of determining if the capacity of a given set of forbidden patterns is positive can be solved in polynomial time.*

11.3 The finiteness property

In this section we review the finiteness property for the joint spectral radius. This subject has led to intense research activity in the last decade. The basic question is rather intuitive. As we have seen, the three members inequality (11.3) provides a straightforward way to approximate the joint spectral radius up to any desired accuracy: evaluate the maximal spectral radius, and the maximal norm for products of increasing length t, which are upper and lower bounds on ρ, until the joint spectral radius is squeezed in a sufficiently small interval and the desired accuracy is reached.

Since $\rho(\mathbf{A}^k) = \rho(\mathbf{A})^k$ the left-hand side always provides the exact value when the set Σ consists of only one matrix and one can thus hope to reach the exact value of the joint spectral radius by evaluating the maximal spectral radii of products of increasing length. If for some t and $\mathbf{A} \in \Sigma^t$ we have $\rho(\mathbf{A})^{1/t} = \rho(\Sigma)$, then the value of the joint spectral radius is reached. We say in that case that Σ has the finiteness property.

Definition 11.3.1 A set Σ of matrices is said to have the *finiteness property* if there exists some product $\mathbf{A} = \mathbf{A}_1 \cdots \mathbf{A}_t$ with $\mathbf{A}_i \in \Sigma$ for which $\rho(\Sigma) = \rho(\mathbf{A})^{1/t}$.

One of the interests of the finiteness property arises from its connection with the stability question for a set of matrices. A set of matrices Σ is *stable* if all long products of matrices taken from Σ converge to zero. As mentioned in Section 11.1.3, there are no known algorithms for deciding stability of a set of matrices and it is unknown if this problem is algorithmically decidable. We have also seen that stability of the set Σ is equivalent to the condition $\rho(\Sigma) < 1$ and we may therefore hope to decide stability as follows: for increasing values of t evaluate $\rho_t = \max\{\rho(\mathbf{A})^{1/t} \mid \mathbf{A} \in \Sigma^t\}$ and $\hat{\rho}_t = \max\{||\mathbf{A}||^{1/t} \mid \mathbf{A} \in \Sigma^t\}$. Since we know that $\rho_t \leq \rho \leq \hat{\rho}_t$, as soon as a t is reached for which $\hat{\rho}_t < 1$ we stop and declare the set stable, and if a t is reached for which $\rho_t \geq 1$ we stop and declare the set unstable. This procedure will always stop unless $\rho = 1$ and $\rho_t < 1$ for all t. But this last situation never occurs for sets of matrices that satisfy the finiteness property and so we conclude with the following result.

Proposition 11.3.2 *Stability is algorithmically decidable for sets of matrices that have the finiteness property.*

The finiteness property is known to hold in a number of particular cases including the case where the matrices are symmetric, or if the Lie algebra associated with the set of matrices is solvable, since in these cases the joint spectral radius is simply equal to the maximum of the spectral radii of the matrices (see (Liberzon, Hespanha, and Morse 1999, Gurvits 1995) for more information). The finiteness property also holds if the set of matrices admits a complex polytope extremal norm (Guglielmi, Wirth, and Zennaro 2005). It was first conjectured in 1995 by Lagarias and Wang that all sets of real matrices have the finiteness property (Lagarias and Wang 1995). This conjecture can be restated as saying that the convergence to zero of all periodic products implies the same for all infinite products. This conjecture, known as the *finiteness conjecture*, has attracted intense attention and several counterproofs have been provided in recent years (Kozyakin 2005, Bousch and Mairesse 2002, Blondel, Theys, and Vladimirov 2003).

Theorem 11.3.3 (Bousch and Mairesse 2002) *There exist values for the real parameters κ, h_0, h_1, ν for which the pair of matrices*

$$\Sigma = \left\{ \begin{pmatrix} 1+e^{\kappa h_0} & 0 \\ e^{\kappa} & 1 \end{pmatrix}, e^{\kappa \nu} \begin{pmatrix} 1 & e^{\kappa} \\ 0 & 1+e^{\kappa h_1} \end{pmatrix} \right\}$$

does not satisfy the finiteness conjecture.

The proof makes use of results on *iterated function systems*, *topical maps* and *Sturmian sequences*.

So far all proofs are non-constructive, and all sets of matrices whose joint spectral radius is known exactly satisfy the finiteness property. In fact, all counterproofs describe families of sets of matrices in which there are counter-examples, but no explicit counter-example has yet been exhibited. Thus, many questions remain open about this property: When does the finiteness property hold? Is it decidable to determine if a given set of matrices satisfies the finiteness property? In many applications one has to deal with non-negative matrices, or with binary matrices. Among the applications presented in Section 11.2, two deal with binary matrices and the other with non-negative integer matrices. Based on extensive numerical evidence, the following conjecture appears in (Blondel, Jungers, and Protasov 2006).

Conjecture 11.3.4 Pairs of binary matrices satisfy the finiteness property.

Remarkably, the conjecture is equivalent for binary and for non-negative integer matrices.

Theorem 11.3.5 (Jungers and Blondel 2008) *The finiteness property holds for all sets of non-negative rational matrices if, and only if, it holds for all pairs of binary matrices.*

So, proving that pairs of binary matrices have the finiteness property would imply that stability is actually decidable for sets of non-negative rational matrices.

11.4 Approximation algorithms

Despite the many negative theoretical results on the computability and the approximability of joint spectral characteristics, many approximation algorithms have been proposed. Some of them are simple heuristics, others run in exponential time but have guaranteed accuracy, and some others only work for particular sets of matrices. In practice, one is often able to reach a satisfactory precision by combining the advantages of the different methods. In the first subsection we cite some results that allow the exact value in polynomial time to be reached if the matrices satisfy particular properties. We then present intuitive 'brute force' methods. In the last two subsections, we present two methods that allow to reach any required accuracy.

The goal of this section is not to provide an exhaustive survey on algorithms allowing for the joint spectral radius computation, but rather to provide the practitioner with a selection of methods that often allow one to compute joint spectral characteristics up to a satisfactory accuracy.

11.4.1 Particular cases

There is a polynomial time algorithm to decide whether the joint spectral radius of a set of matrices is zero. This is a consequence of the following proposition.

Proposition 11.4.1 *Let* $\Sigma = \{\mathbf{A}_1, \ldots, \mathbf{A}_m\} \subset \mathbb{R}^{n \times n}$, *Then* $\rho(\Sigma) = 0$ *if, and only if,*

$$\Sigma^n = \{0\}.$$

The proof is based on the following lemma.

Lemma 11.4.2 *If* Σ *is irreducible, then* $\rho(\Sigma) > 0$.

A proof of the above lemma is sketched in the exercises (see Exercise 11.12). Now the proof of Proposition 11.4.1 becomes easy.

Proof The if part is trivial.
The proof of the only if part is by induction on the dimension n: it is true for scalars. Now suppose it is true for sets of matrices of dimension less than n. Let $\Sigma \in \mathbb{R}^{n \times n}, \rho(\Sigma) = 0$. By the previous lemma, we can suppose Σ reducible, and for all $\mathbf{A}_i \in \Sigma$, we can write without loss of generality:

$$\mathbf{A}_i = \begin{pmatrix} \mathbf{B}_i & \mathbf{C}_i \\ \mathbf{0} & \mathbf{D}_i \end{pmatrix}, \quad \mathbf{D}_i \in \mathbb{R}^{n' \times n'},$$

where $\rho(\{\mathbf{B}_i\}) = \rho(\{\mathbf{D}_i\}) = 0$. Now, consider a product of length n. By applying twice the induction hypothesis on n' and $n - n'$, we have:

$$\mathbf{A}_n \cdots \mathbf{A}_{n'+1} \mathbf{A}_{n'} \cdots \mathbf{A}_1 = \begin{pmatrix} \mathbf{0} & \mathbf{C} \\ \mathbf{0} & \mathbf{D} \end{pmatrix} \begin{pmatrix} \mathbf{B}' & \mathbf{C}' \\ \mathbf{0} & \mathbf{0} \end{pmatrix},$$

for some (potentially zero) matrices $\mathbf{C}, \mathbf{C}', \mathbf{B}', \mathbf{D}$ and this latter product vanishes. □

The following theorem gives then a decision procedure that runs in polynomial time.

Theorem 11.4.3 (Gurvits 1996) *Let* $\Sigma = \{\mathbf{A}_1, \ldots, \mathbf{A}_m\}$. *The following procedure allows us to decide whether the joint spectral radius is zero:*

$$\mathbf{X}_0 = \mathbf{I}, \qquad (11.14)$$

$$\mathbf{X}_k = \sum_1^m {}^t\mathbf{A}_i \mathbf{X}_{k-1} \mathbf{A}_i, \qquad (11.15)$$

and $\mathbf{X}_n = 0$ *if, and only if,* $\rho(\Sigma) = 0$.

Proof One has $\mathbf{X}_n = \sum_{\mathbf{A} \in \Sigma^n} {}^t\mathbf{A}\mathbf{A}$, and this matrix is computable in polynomial time. Moreover $\mathbf{X}_n$ is equal to zero if, and only if, $\Sigma^n = \{0\}$. □

Interestingly, the situation is far worse for the joint spectral subradius (Tsitsiklis and Blondel 1997).

Proposition 11.4.4 *The problem of deciding whether or not the joint spectral subradius of a set of matrices is equal to zero is not algorithmically decidable.*

Another case where it is sometimes easy to compute the joint spectral radius is when one of the matrices in the set Σ is optimal. Sufficient conditions for this to be the case are given in the following theorem (see (Jungers 2009) for more information).

Theorem 11.4.5 *The joint spectral radius of a set Σ is equal to the largest spectral radius of the matrices in Σ in the following cases:*

- *Σ is a set of normal matrices (in particular, Σ is a set of symmetric matrices),*
- *Σ is a set of upper (or lower) triangular matrices,*
- *the matrices in Σ are commonly upper triangularisable, that is, if there exists an invertible matrix $\mathbf{T}$ such that for all $\mathbf{A} \in \Sigma$, $\mathbf{T}\mathbf{A}\mathbf{T}^{-1}$ is upper triangular.*

11.4.2 Brute force methods

The first methods that appeared in the literature for computing the joint spectral radius are very natural: they consist in applying the three members inequality (11.3)

$$\rho_t(\Sigma) \leq \rho(\Sigma) \leq \hat{\rho}_t(\Sigma).$$

The corresponding inequalities of Equation (11.4) can also be used for the joint spectral subradius

$$\check{\rho}(\Sigma) \leq \underline{\rho}_t(\Sigma) \leq \check{\rho}_t(\Sigma).$$

Remember that the successive bounds tend to the required value when $t \to \infty$. So an immediate algorithm is as follows: fix a t, compute all products of length t, take the maximal spectral radius as a lower bound, and the maximal norm (for a fixed norm) as an upper bound. This algorithm will converge to the desired value as t increases.

The main problem with the previous algorithm is the combinatorial explosion of the number of products of length t: there are m^t of them (with m the number of matrices in Σ).

For the joint spectral radius, this method has been known for a long time and some ideas have been proposed to attenuate its exponential growth. In (Gripenberg 1996), a branch and bound algorithm is proposed, which allows us to approximate asymptotically the joint spectral radius up to a given fixed absolute error. More precisely, for a fixed precision δ, the algorithm computes iteratively successive bounds α_t and β_t such that $\alpha_t \leq \rho \leq \beta_t$ and $\lim \beta_t - \alpha_t < \delta$. The algorithm builds longer and longer products, based on the ones previously constructed, but removing at each step products that are provably not necessary to reach the required accuracy.

Also, if the matrices in Σ have non-negative entries, there is an obvious way of disregarding some products: if $\mathbf{A}, \mathbf{B}$ are products of length t and $\mathbf{A} \leq \mathbf{B}$ (where the inequality has to be understood entry-wise), then one does not have to keep $\mathbf{A}$ in order to have better and better approximations of the joint spectral radius. Indeed, in any product of length $T > t$, one can always replace the subproduct $\mathbf{A}$ with $\mathbf{B}$, and by doing this the approximation of the joint spectral radius will be at least as good as with the other product.

Even though these algorithms perform quite well for certain applications, it is clear that none of them provides approximations of the joint spectral radius in polynomial time.

11.4.3 Convex combination method for the joint spectral radius

The following result provides a quick lower bound on the joint spectral radius. Recall that a cone is said to be *proper* if it is closed, solid, convex and pointed.

Proposition 11.4.6 (Blondel and Nesterov 2005) *Let Σ be an arbitrary set $\{\mathbf{A}_1, \ldots, \mathbf{A}_m\}$ of matrices in $\mathbb{R}^{n \times n}$. The following simple lower bound on the joint spectral radius holds:*

$$\rho(\mathbf{A}_1 + \cdots + \mathbf{A}_m)/m \leq \rho(\Sigma).$$

If moreover the matrices in Σ leave a proper cone invariant, then

$$\rho(\Sigma) \leq \rho(\mathbf{A}_1 + \cdots + \mathbf{A}_m).$$

Proof The first inequality comes from the fact that $(\mathbf{A}_1 + \cdots + \mathbf{A}_m)/m \in \operatorname{conv} \Sigma$, and it is known that $\rho(\Sigma) = \rho(\operatorname{conv} \Sigma)$ (see Exercise 11.13 for a sketch of the proof). The second inequality comes from the fact that associated with an invariant cone K, there exists a norm $||\cdot||_K$, depending on K, such that for all $\mathbf{A}, \mathbf{B} \in \Sigma$, $||\mathbf{A}||_K \le ||\mathbf{A}+\mathbf{B}||_K$. This norm is given by

$$||\mathbf{A}||_K = \max_{\mathbf{v}\in K,\ \mathbf{w}\in K^*,\ ||\mathbf{v}||, ||\mathbf{w}||=1} {}^t\mathbf{w}\mathbf{A}\mathbf{v},$$

where K^* denotes the dual of the cone K (see (Blondel and Nesterov 2005) for details). □

So, if Σ leaves a cone invariant, it is possible to bound ρ between two bounds. This result might look quite weak, due to the necessity for Σ to leave a cone invariant. In fact, this is not a problem, as pointed out in (Blondel, Nesterov, and Theys 2005). Indeed, if no cone is left invariant, a simple procedure allows one to obtain an associated set of matrices that admits an invariant cone, while keeping essentially the same joint spectral characteristics.

Proposition 11.4.7 *Let Σ be a set of matrices. The semidefinite lifting $\tilde{\Sigma}$ of Σ:*

$$\tilde{\Sigma} = \{\tilde{\mathbf{A}} : \mathbb{R}^{n^2} \to \mathbb{R}^{n^2},\ \mathbf{X} \to {}^t\mathbf{A}\mathbf{X}\mathbf{A}\} \tag{11.16}$$

is a set of linear operators that leave the cone of symmetric positive semidefinite matrices invariant. Moreover it satisfies $\rho(\tilde{\Sigma}) = \rho(\Sigma)^2, \check{\rho}(\tilde{\Sigma}) = \check{\rho}(\Sigma)^2$.

This useful result, together with the simple Proposition 11.1.11 allows for an approximation scheme for the joint spectral radius, see for instance (Parrilo and Jadbabaie 2008).

Theorem 11.4.8 *We consider a set of matrices $\Sigma = \{\mathbf{A}_1, \ldots, \mathbf{A}_m\} \in \mathbb{R}^{n \times n}$. For any $d \in \mathbb{N}_{>0}$, let us denote*

$$\rho_{conv,d} = \rho\Big(\sum_{\mathbf{A}_i \in \tilde{\Sigma}^d} \mathbf{A}_i\Big)^{1/2d}.$$

We have:

$$m^{-\frac{1}{2d}} \rho_{conv,d} \le \rho(\Sigma) \le \rho_{conv,d}.$$

Thus, for any required accuracy, one can compute a d such that Theorem 11.4.8 provides such an accuracy.

11.4.4 Advanced conic programming method

A new class of approximation methods has recently been proposed. These methods seem to be very efficient, at least on several particular applications but they are still an important research topic in themselves, and many questions related to them are not yet settled. Nevertheless, their great practical efficiency, together with their simplicity of implementation, make their presentation interesting. Different methods have been presented in different contexts, with different results in terms of approximation guarantees. Some involve so-called sum-of-square programming (Parrilo and Jadbabaie 2008), others make use of semidefinite programming (Blondel, Nesterov, and Theys 2005, Ando and Shih 1998). Finally (Jungers, Protasov, and Blondel 2009) puts these methods in the more general framework of conic programming. We briefly cite two theorems in this family, that are directly applicable in practice. See (Jungers 2008) for more information.

Theorem 11.4.9 (Parrilo and Jadbabaie 2008) *Let $\Sigma \subset \mathbb{R}^{n \times n}$ be a set of matrices and let $p(x)$ be a homogeneous multivariate polynomial in the n variables $x_1, \ldots, x_n$, of degree $2d$. Suppose moreover that $p(x)$ is a strictly positive sum of squares:*

$$p(x) = q_1(x)^2 + \cdots + q_p(x)^2 > 0$$

for some polynomials $q_i(x)$. If for all $\mathbf{A}_i \in \Sigma$,

$$p(\mathbf{A}_i x) \leq \gamma^{2d} p(x), \qquad (11.17)$$

then $\rho(\Sigma) \leq \gamma$.

For any $d \geq 0$, let us denote by $\rho_{SOS,d}^d$ the minimal γ such that there exists a polynomial p that satisfies Equation (11.17). Then,

$$m^{-\frac{1}{2d}} \rho_{SOS,d} \leq \rho(\Sigma) \leq \rho_{SOS,d},$$

$$\binom{n+d-1}{d}^{-\frac{1}{2d}} \rho_{SOS,d} \leq \rho(\Sigma) \leq \rho_{SOS,d}.$$

For efficient methods to find the minimal γ in Equation (11.17), see (Boyd and Vandenberghe 2004, Parrilo 2000). Once more, these methods do not run in a time which is polynomial in the size of Σ and in the accuracy, but in practice it is often possible to reach a satisfactory precision in a reasonable time.

We now present an algorithm that allows us in some cases to approximate the joint spectral subradius. As we have seen no algorithm exists that approximates the joint spectral subradius of an arbitrary set up to a required accuracy. However, the following theorem has provided good approximations in several practical cases. Moreover it has a guaranteed approximation ratio for a large class of sets of matrices (see (Protasov, Jungers, and Blondel 2009)).

Theorem 11.4.10 (Protasov, Jungers, and Blondel 2009) *Let Σ be a set of matrices that leaves a cone K invariant. We denote by $\check{\sigma}_K(\Sigma)$ the maximum $\lambda \geq 0$ such that*

$$\exists\, v \in K, v \neq 0, \quad \mathbf{A}v - \lambda v \in K \quad \forall \mathbf{A} \in \Sigma.$$

Then,

$$\check{\sigma}_K(\Sigma) \leq \check{\rho}.$$

Again, for classical cones (such as the positive orthant or the set of positive semidefinite matrices for example), efficient methods exist that allow us to find $\check{\sigma}_K(\Sigma)$ efficiently (Boyd and Vandenberghe 2004, Parrilo 2000). The assumptions of Theorem 11.4.10 require that the set Σ leaves a cone invariant. If it is not the case, one can first apply the semidefinite lifting (11.16) before making use of Theorem 11.4.10.

This closes the section on approximation algorithms. This section was not intended to be exhaustive, but gives some tools that often appear sufficient to cope with practical problems. For more on the joint spectral characteristics computation, see (Jungers 2009).

11.5 Conclusions

Although they have first been used in the context of dynamical systems, joint spectral characteristics appear to have intricate connections with a number of subjects in combinatorics and number theory. In some cases they are the right tool to solve problems that arise from these disciplines, like for instance the case of overlap-free words, or the partition function (see Section 11.2). In other cases, these disciplines have proved useful to solve problems related to joint spectral characteristics, like for instance in the counterproof of the finiteness conjecture.

Many problems are still open concerning the joint spectral characteristics. Does the finiteness property hold for binary or for non-negative rational matrices? Or perhaps it holds for matrices that arise in particular applications, such as that of the partition functions or the capacity of codes? Given a set of matrices, is it decidable whether it satisfies the finiteness property? Also, the connections between joint spectral characteristics and repetition-free words are not totally clear and need further investigations.

11.6 Exercises

Exercise 11.1 Prove that the joint spectral radius of the following set:
$$\Sigma = \left\{ \begin{pmatrix} 1 & 1 \\ 1 & 0 \end{pmatrix}, \begin{pmatrix} 0 & 1 \\ 0 & 1 \end{pmatrix} \right\}$$
is equal to the Golden Ratio $(1+\sqrt{5})/2$.
Hint. Prove that the second matrix does not appear in the optimal product.

Exercise 11.2 Prove that the joint spectral radius of the following set:
$$\Sigma = \left\{ \begin{pmatrix} 0 & 1 \\ 0 & 0 \end{pmatrix}, \begin{pmatrix} 1 & 0 \\ 1 & 1 \end{pmatrix} \right\}$$
is equal to $\sqrt[5]{4}$.
Hint. First prove that the optimal infinite product is of the form $\mathbf{A}_1^{t_1} \mathbf{A}_0 \mathbf{A}_1^{t_2} \mathbf{A}_0 \ldots$.

Exercise 11.3 Compute the joint spectral subradius of the following set:
$$\Sigma = \left\{ \begin{pmatrix} 2 & 1 \\ 0 & 4 \end{pmatrix}, \begin{pmatrix} 4 & 0 \\ 2 & 0 \end{pmatrix} \right\}.$$

Exercise 11.4 We have seen that the definition of the joint spectral radius is well posed, that is, the quantity
$$\lim_{t \to \infty} \hat{\rho}_t(\Sigma, \|\cdot\|),$$
does not depend on the norm chosen. Among all the norms that one can choose, some play a crucial role. Consider the following definition:

Definition 11.6.1 A norm $\|\cdot\|$ on $\mathbb{R}^{n \times n}$ is *extremal* for a set of matrices Σ if for all $\mathbf{A} \in \Sigma$,
$$\|\mathbf{A}\| \leq \rho(\Sigma).$$

Let us note that the above definition, together with the three members inequality (11.3) implies that for an extremal norm we have
$$\sup_{\mathbf{A} \in \Sigma} \|\mathbf{A}\| = \rho.$$

Prove that the set

$$\Sigma = \left\{ \begin{pmatrix} 1 & 1 \\ 0 & 1 \end{pmatrix} \right\}$$

does not admit an extremal norm.

Exercise 11.5 Prove that for sets of normal matrices the Euclidean norm is always an extremal norm (see Exercise 11.4 for definitions).

Exercise 11.6 We say that a set of matrices Σ is *non-defective* if the normalised semigroup $(\Sigma/\rho)^*$ is bounded. Prove the following theorem.

Theorem 11.6.2 (Kozyakin 1990, Barabanov 1988) *A bounded set $\Sigma \in \mathbb{R}^{n \times n}$ admits an extremal norm if, and only if, it is non-defective.*

Hint. Consider the following application:

$$n\colon \mathbb{R}^n \to \mathbb{R}_+, \quad \mathbf{x} \mapsto \sup\{\|\mathbf{A}\mathbf{x}\| \mid \mathbf{A} \in \Sigma^*\}.$$

Exercise 11.7 Prove the following proposition (Protasov 2000).

Proposition 11.6.3

$$\begin{aligned} b_{2,2}(n) &= 1, \\ b_{2,3}(n) &= s(n+1), \\ b_{2,4}(n) &= \lfloor n/2 \rfloor + 1, \end{aligned}$$

where $b_{2,d}$ is the binary partition function with coefficients between zero and $d-1$, and $s(x)$ is the Stern sequence which is defined recursively as follows: $s(0) = 0, s(1) = 1, s(2x) = s(x), s(2x+1) = s(x) + s(x+1)$.

Exercise 11.8 Recall that the Thue–Morse sequence is the infinite word obtained as the limit of $\theta^n(a)$ as $n \to \infty$ with $\theta(a) = ab$, $\theta(b) = ba$. Prove that the Thue–Morse sequence is overlap-free.

Exercise 11.9 By using the definition of the capacity of forbidden sets, prove the following result (Blondel, Jungers, and Protasov 2006)

$$\mathrm{cap}(\{+,-\}^m) = (m-1)/m.$$

Exercise 11.10 Prove that the finiteness property holds for rational matrices if, and only if, it holds for integer matrices.

Exercise 11.11 Prove that the finiteness property holds for sets of symmetric matrices.

Exercise 11.12 Prove that if Σ is irreducible, then $\rho(\Sigma) > 0$.
Hint. Use the compactness of the unit ball in $\mathbb{R}^n$ to provide a lower bound on the growth rate $\|\mathbf{A}\mathbf{x}\|/\|\mathbf{x}\|$.

Exercise 11.13 Prove the following proposition.

Proposition 11.6.4 *For any set of matrices Σ,*
$$\rho(\Sigma) = \rho(conv\,\Sigma),$$
where $conv\,\Sigma$ is the convex hull of Σ.

Hint. Express the norm of a product in $conv\,\Sigma$ as a function of products in Σ.

Exercise 11.14 Compute the semidefinite lifting of the following set
$$\Sigma = \left\{ \begin{pmatrix} -2 & 1 \\ 1 & -1 \end{pmatrix}, \begin{pmatrix} 2 & -2 \\ -1 & -1 \end{pmatrix} \right\}.$$

Exercise 11.15 Prove Theorem 11.4.9.
Hint. Use the compactness of the unit ball in $\mathbb{R}^n$ and the continuity of polynomials.

11.7 Notes

The notion of joint spectral radius was introduced in the early sixties (Rota and Strang 1960). It had not received much attention until it was shown to be the appropriate quantity to quantify the continuity of wavelets (Daubechies and Lagarias 1992). Since then, researchers from different mathematical communities have studied this quantity from various perspectives. A list of over hundred related contributions is given in (Strang 2001), a reference that is almost a decade old. Strangely enough, the minimal growth counterpart, the joint spectral subradius, appeared much later in the literature. The first reference seems to be (Gurvits 1995).

Theorem 11.1.23 first appeared in (Kozyakin 1990). However, there is a flaw in the proof, which is corrected in (Theys 2005).

In (Jungers, Protasov, and Blondel 2008a) it is conjectured that the exponent β of maximal rate of growth of the number of overlap-free words is equal to $\rho(\mathbf{F}_0 \mathbf{F}_1)^{1/2} = 1.3322\ldots$.

An important corollary of Theorem 11.2.4 is that the sequence u_n does not have a constant rate of growth, and the value $\frac{\log u_n}{\log n}$ does not converge as $n \to \infty$ (this was already noted in (Cassaigne 1993)). Nevertheless, the value $\frac{\log u_n}{\log n}$ actually has a limit as $n \to \infty$, not along all the natural numbers $n \in \mathbb{N}$, but along a subsequence of $\mathbb{N}$ of density 1. In other terms, the sequence converges with probability 1. The limit, which differs from both α and β can be expressed by the so-called Lyapunov exponent $\bar{\rho}$ of the matrices $\mathbf{A}_0, \mathbf{A}_1$. This is a consequence of Oseledec's theorem (Oseledec 1968). See (Jungers, Protasov, and Blondel 2009) for more information. For the sake of completeness, we give the following estimate for the constant σ:

$$1.3005 < \sigma < 1.3098.$$

Recent results on the capacity of codes allow its computation time to be improved a bit, by exhibiting particular algebraic properties of the matrices (Blondel, Jungers, and Protasov 2006). Also, (Blondel, Jungers, and Protasov 2006) introduces a closely related problem and proves that this result is NP-hard, which is another argument for the difficulty of the problem (see also (Blanchet-Sadri, Jungers, and Palumbo 2009) for a generalised version of the problem, which is also NP-hard). It has been conjectured (Blondel, Jungers, and Protasov 2006) that the sets of matrices with binary entries, and, in particular, those constructed in order to compute a capacity do always possess the finiteness property.

Problem 11.7.1 Do matrices that arise in the context of capacity computation satisfy the finiteness property?

Numerical results in (Moision, Orlitsky, and Siegel 2001), (Jungers 2005), and in this chapter seem to support this conjecture, and moreover the length of the period seems to be very short: it seems to be of the order of the size of the forbidden patterns, which would be surprising, because this length would be logarithmic in the size of the matrices. Finally, the capacity problem has been generalised in (Moision, Orlitsky, and Siegel 2007).

The finiteness conjecture appears under a different guise in (Gurvits 1991) where it is attributed to E. S. Pyatnicky. The algorithm in Theorem 11.4.3 is mentioned in (Gurvits 1996) without proof. The proof presented here is from (Jungers 2008).

As pointed out in (Parrilo and Jadbabaie 2008), there are interesting connections between semidefinite methods to compute the joint spectral radius and formerly introduced geometric methods (Protasov 1997).

Recently, a new class of approximation algorithms have been proposed, which have been called *conic programming methods*. They encapsulate some methods that are described in Section 11.4. See the recent preprints (Blondel and Nesterov 2009, Protasov, Jungers, and Blondel 2009) for more information.

References

[Aberkane 2001] Aberkane, A. Exemples de suites de complexité inférieure à $2n$. *Bull. Belg. Math. Soc.* **8**, (2001) 161–180.

[Aberkane 2003] Aberkane, A. Words whose complexity satisfies $\lim \frac{p(n)}{n} = 1$. *Theoret. Comput. Sci.* **307**, (2003) 31–46.

[Aberkane and Brlek 2002] Aberkane, A. and Brlek, S. Suites de même complexité que celle de Thue-Morse. In *Actes des Journées Montoises d'informatique théorique*. LIRMM, Montpellier, 2002.

[Adamczewski 2009] Adamczewski, B. On the expansion of some exponential periods in an integer base. *Math. Annalen*, **346**, (2010), 107–116.

[Adamczewski and Allouche 2007] Adamczewski, B. and Allouche, J.-P. Reversals and palindromes in continued fractions. *Theoret. Comput. Sci.* **380**, (2007) 220–237.

[Adamczewski and Bugeaud 2005] Adamczewski, B. and Bugeaud, Y. On the complexity of algebraic numbers II. Continued fractions. *Acta Math.* **195**, (2005) 1–20.

[Adamczewski and Bugeaud 2006a] Adamczewski, B. and Bugeaud, Y. On the Littlewood conjecture in simultaneous Diophantine approximation. *J. London Math. Soc.* **73**, (2006) 355–366.

[Adamczewski and Bugeaud 2006b] Adamczewski, B. and Bugeaud, Y. Real and p-adic expansions involving symmetric patterns. *Int. Math. Res. Not.* **2006**, (2006) Art. ID 75968.

[Adamczewski and Bugeaud 2007a] Adamczewski, B. and Bugeaud, Y. On the complexity of algebraic numbers. I. Expansions in integer bases. *Ann. of Math. (2)* **165**, (2007) 547–565.

[Adamczewski and Bugeaud 2007b] Adamczewski, B. and Bugeaud, Y. On the Maillet-Baker continued fractions. *J. Reine Angew. Math.* **606**, (2007) 105–121.

[Adamczewski and Bugeaud 2007c] Adamczewski, B. and Bugeaud, Y. Palindromic continued fractions. *Ann. Inst. Fourier (Grenoble)* **57**, (2007) 1557–1574.

[Adamczewski and Bugeaud 2007d] Adamczewski, B. and Bugeaud, Y. A short proof of the transcendence of Thue-Morse continued fractions. *Amer. Math. Monthly* **114**, (2007) 536–540.

[Adamczewski, Bugeaud, and Davison 2006] Adamczewski, B., Bugeaud, Y., and Davison, L. Continued fractions and transcendental numbers. *Ann. Inst. Fourier (Grenoble)* **56**, (2006) 2093–2113.

[Adamczewski, Bugeaud, and Luca 2004] Adamczewski, B., Bugeaud, Y., and

Luca, F. Sur la complexité des nombres algébriques. *C. R. Acad. Sci. Paris* **339**, (2004) 11–14.

[Adamczewski, Frougny, Siegel, et al. 2010] Adamczewski, B., Frougny, Ch., Siegel, A., and Steiner, W. Rational numbers with purely periodic beta-expansion. *Bull. London Math. Soc.*, doi:10.1112/blms/bdq019.

[Adamczewski and Rampersad 2008] Adamczewski, B. and Rampersad, N. On patterns occurring in binary algebraic numbers. *Proc. Amer. Math. Soc.* **136**, (2008) 3105–3109.

[Akiyama 1999] Akiyama, S. Self affine tiling and Pisot numeration system. In S. Kanemitsu and K. Györy, eds., *Number Theory and Its Applications*, pp. 7–17. Kluwer, 1999.

[Akiyama 2000] Akiyama, S. Cubic Pisot units with finite beta expansions. In F. Halter-Koch and R. F. Tichy, eds., *Algebraic Number Theory and Diophantine Analysis*, pp. 11–26. Walter de Gruyter, 2000.

[Akiyama 2002] Akiyama, S. On the boundary of self affine tilings generated by Pisot numbers. *J. Math. Soc. Japan* **54**(2), (2002) 283–308.

[Akiyama, Barat, Berthé, et al. 2008] Akiyama, S., Barat, G., Berthé, and Siegel, A. Boundary of central tiles associated with Pisot beta-numeration and purely periodic expansions. *Monatsh. Math.* **155**, (2008) 377–419.

[Akiyama, Borbély, Brunotte, et al. 2005] Akiyama, S., Borbély, T., Brunotte, H., Pethő, A., and Thuswaldner, J. M. Generalized radix representations and dynamical systems I. *Acta Math. Hungarica* **108**, (2005) 207–238.

[Akiyama, Frougny, and Sakarovitch 2008] Akiyama, S., Frougny, Ch., and Sakarovitch, J. Powers of rationals modulo 1 and rational base number systems. *Israel J. Math.* **168**, (2008) 53–91.

[Akiyama and Pethő 2002] Akiyama, S. and Pethő, A. On canonical number systems. *Theoret. Comput. Sci.* **270**, (2002) 921–933.

[Akiyama and Rao 2005] Akiyama, S. and Rao, H. New criteria for canonical number systems. *Acta Arith.* **111**, (2005) 5–25.

[Akiyama, Rao, and Steiner 2004] Akiyama, S., Rao, H., and Steiner, W. A certain finiteness property of Pisot number systems. *J. Number Theory* **107**, (2004) 135–160.

[Akiyama and Scheicher 2005] Akiyama, S. and Scheicher, K. From number systems to shift radix systems. *Nihonkai Math. J.* **16**, (2005) 95–106.

[Akiyama and Thuswaldner 2004] Akiyama, S. and Thuswaldner, J. M. A survey on topological properties of tiles related to number systems. *Geometriae Dedicata* **109**, (2004) 89–105.

[Akiyama and Thuswaldner 2005] Akiyama, S. and Thuswaldner, J. M. On the topological structure of fractal tilings generated by quadratic number systems. *Comput. Math. Appl.* **49**, (2005) 1439–1485.

[Allouche 1987] Allouche, J.-P. Automates finis en théorie des nombres. *Exposition. Math.* **5**, (1987) 239–266.

[Allouche 1992] Allouche, J.-P. The number of factors in a paperfolding sequence. *Bull. Austral. Math. Soc.* **46**, (1992) 23–32.

[Allouche 1994] Allouche, J.-P. Sur la complexité des suites infinies. *Bull. Belg. Math. Soc.* **1**, (1994) 133–143.

[Allouche, Davison, Queffélec, et al. 2001] Allouche, J.-P., Davison, L. J., Queffélec, M., and Zamboni, L. Q. Transcendence of Sturmian or morphic continued fractions. *J. Number Theory* **91**, (2001) 39–66.

[Allouche and Mendès France 1995] Allouche, J.-P. and Mendès France, M. Automata and automatic sequences. In F. Axel and D. Gratias, eds., *Beyond Quasicrystals*, pp. 293–367. Les Éditions de Physique; Springer, 1995.

[Allouche, Mendès France, and Peyrière 2000] Allouche, J.-P., Mendès France, M., and Peyrière, J. Automatic Dirichlet series. *J. Number Theory* **81**, (2000) 359–373.

[Allouche, Rampersad, and Shallit 2009] Allouche, J.-P., Rampersad, N., and Shallit, J. Periodicity, repetitions, and orbits of an automatic sequence. *Theoret. Comput. Sci.* **410**(30-32), (2009) 2795–2803.

[Allouche and Shallit 1999] Allouche, J.-P. and Shallit, J. O. The ubiquitous Prouhet-Thue-Morse sequence. In C. Ding, T. Helleseth, and H. Niederreiter, eds., *Sequences and Their Applications, Proceedings of SETA '98*, pp. 1–16. Springer-Verlag, 1999.

[Allouche and Shallit 2003] Allouche, J.-P. and Shallit, J. O. *Automatic Sequences, Theory, Applications, Generalizations*. Cambridge University Press, 2003.

[Ando and Shih 1998] Ando, T. and Shih, M.-H. Simultaneous contractibility. *SIAM J. Matrix Anal. A.* **19**(2), (1998) 487–498.

[Angrand and Sakarovitch] Angrand, P.-Y. and Sakarovitch, J. Radix enumeration of rational languages. *RAIRO Inform. Théor. App.* **44**(1), (2010) 19–36.

[Arno and Wheeler 1993] Arno, S. and Wheeler, F. S. Signed digit representations of minimal Hamming weight. *IEEE Trans. Comput.* **42**, (1993) 1007–1010.

[Arnoux, Berthé, and Ito 2002] Arnoux, P., Berthé, V., and Ito, S. Discrete planes, $\mathbb{Z}^2$-actions, Jacobi-Perron algorithm and substitutions. *Ann. Inst. Fourier (Grenoble)* **52**(2), (2002) 305–349.

[Arnoux, Berthé, and Siegel 2004] Arnoux, P., Berthé, V., and Siegel, A. Two-dimensional iterated morphisms and discrete planes. *Theoret. Comput. Sci.* **319**, (2004) 145–176.

[Arnoux and Ito 2001] Arnoux, P. and Ito, S. Pisot substitutions and Rauzy fractals. *Bull. Belg. Math. Soc.* **8**, (2001) 181–207.

[Avila and Forni 2007] Avila, A. and Forni, G. Weak mixing for interval exchange transformations and translation flows. *Ann. Math.* **165**, (2007) 637–664.

[Avizienis 1961] Avizienis, A. Signed-digit number representations for fast parallel arithmetic. *IRE Trans. Electron. Comput.* **10**, (1961) 389–400.

[Bailey, Borwein, Crandall, et al. 2004] Bailey, D. H., Borwein, J. M., Crandall, R. E., and Pomerance, C. On the binary expansions of algebraic numbers. *J. Théorie Nombres Bordeaux* **16**, (2004) 487–518.

[Bailey and Crandall 2002] Bailey, D. H. and Crandall, R. E. Random generators and normal numbers. *Experimental Math.* **11**, (2002) 527–546.

[Baker 1962] Baker, A. Continued fractions of transcendental numbers. *Mathematika* **9**, (1962) 1–8.

[Baker, Barge, and Kwapisz 2006] Baker, V., Barge, M., and Kwapisz, J. Geometric realization and coincidence for reducible non-unimodular Pisot tiling spaces with an application to beta-shifts. *Ann. Inst. Fourier (Grenoble)* **56**(7), (2006) 2213–2248.

[Balogh and Bollobás 2005] Balogh, J. and Bollobás, B. Hereditary properties of words. *RAIRO Inform. Théor. App.* **39**(1), (2005) 49–65.

[Bar-Hillel, Perles, and Shamir 1961] Bar-Hillel, Y., Perles, M., and Shamir, E. On formal properties of simple phrase structure grammars. *Z. Phonetik. Sprachwiss. Kommuniationsforsch.* **14**, (1961) 143–172.

[Barabanov 1988] Barabanov, N. Lyapunov indicators of discrete inclusions I-III. *Automat. Rem. Contr.* **49**, (1988) 152–157, 283–287, 558–565.

[Barat, Berthé, Liardet, et al. 2006] Barat, G., Berthé, V., Liardet, P., and Thuswaldner, J. M. Dynamical directions in numeration. *Ann. Inst. Fourier (Grenoble)* **56**, (2006) 1987–2092.

[Barat, Downarowicz, Iwanik, et al. 2000] Barat, G., Downarowicz, T., Iwanik, A., and Liardet, P. Propriétés topologiques et combinatoires des échelles de

numération. *Colloq. Math.* **84/85**, (2000) 285–306. Dedicated to the memory of Anzelm Iwanik.
[Barat, Downarowicz, and Liardet 2002] Barat, G., Downarowicz, T., and Liardet, P. Dynamiques associées à une échelle de numération. *Acta Arith.* **103**, (2002) 41–78.
[Barat and Grabner 1996] Barat, G. and Grabner, P. J. Distribution properties of G-additive functions. *J. Number Theory* **60**, (1996) 103–123.
[Barat and Grabner 2001] Barat, G. and Grabner, P. J. Distribution of binomial coefficients and digital functions. *J. London Math. Soc. (2)* **64**, (2001) 523–547.
[Barat and Grabner 2008] Barat, G. and Grabner, P. J. Limit distribution of Q-additive functions from an ergodic point of view. *Ann. Univ. Sci. Budapest. Sect. Comput.* **28**, (2008) 55–78.
[Barbier 1887] Barbier, E. On suppose écrite la suite naturelle des nombres; quel est le $(10^{1000})^{\text{ième}}$ chiffre écrit? *C. R. Acad. Sci. Paris* **105**, (1887) 795–798.
[Barbolosi and Grabner 1996] Barbolosi, D. and Grabner, P. J. Distribution des coefficients multinomiaux et q-binomiaux modulo p. *Indag. Math.* **7**, (1996) 129–135.
[Barge and Diamond 2002] Barge, M. and Diamond, B. Coincidence for substitutions of Pisot type. *Bull. Soc. Math. France* **130**, (2002) 619–626.
[Barge and Diamond 2007] Barge, M. and Diamond, B. Proximality in Pisot tiling spaces. *Fundamenta Math.* **194**(3), (2007) 191–238.
[Barge, Diamond, and Swanson 2009] Barge, M., Diamond, B., and Swanson, R. The branch locus for one-dimensional Pisot tiling spaces. *Fundamenta Math.* **204**(3), (2009) 215–240.
[Barge and Kwapisz 2005] Barge, M. and Kwapisz, J. Elements of the theory of unimodular Pisot substitutions with an application to β-shifts. In *Algebraic and topological dynamics*, vol. 385 of Contemporary Mathematics, pp. 89–99. Amer. Math. Soc., 2005.
[Barge and Kwapisz 2006] Barge, M. and Kwapisz, J. Geometric theory of unimodular Pisot substitution. *Amer. J. Math.* **128**, (2006) 1219–1282.
[Bártfai 1966] Bártfai, P. Grenzverteilungssätze auf der Kreisperipherie und auf kompakten Abelschen Gruppen. *Studia Sci. Math. Hungar.* **1**, (1966) 71–85.
[Bassily and Kátai 1995] Bassily, N. L. and Kátai, I. Distribution of the values of q-additive functions on polynomial sequences. *Acta Math. Hung.* **68**, (1995) 353–361.
[Béal 1993] Béal, M.-P. *Codage symbolique*. Masson, 1993.
[Bell, Charlier, Fraenkel, et al. 2009] Bell, J., Charlier, E., Fraenkel, A. S., and Rigo, M. A decision problem for ultimately periodic sets in non-standard numeration systems. *Internat. J. Algebra Comput.* **19**, (2009) 809–839.
[Berend and Frougny 1994] Berend, D. and Frougny, Ch. Computability by finite automata and Pisot bases. *Math. Systems Theory* **27**, (1994) 275–282.
[Berger and Wang 1992] Berger, M. A. and Wang, Y. Bounded Semigroups of Matrices. *Linear Algebra Appl.* **166**, (1992) 21–27.
[Berstel 1995] Berstel, J. *Axel Thue's Papers on Repetitions in Words: a Translation*. No. 20 in Publications du Laboratoire de Combinatoire et d'Informatique Mathématique. Université du Québec à Montréal, 1995.
[Berstel 2005] Berstel, J. Growth of repetition-free words–a review. *Theoret. Comput. Sci.* **340**(2), (2005) 280–290.
[Berstel, Aaron, Reutenauer, et al. 2008] Berstel, J., Aaron, A., Reutenauer, C., and Saliola, F. V. *Combinatorics on words*, vol. 27 of CRM Monographs Series. American Mathematical Society, 2008.

[Berstel and Boasson 1997] Berstel, J. and Boasson, L. The set of minimal words of a context-free language is context-free. *J. Comput. System Sci.* **55**, (1997) 477–488.
[Berstel, Boasson, Carton, et al. 2006] Berstel, J., Boasson, L., Carton, O., Petazzoni, B., and Pin, J.-E. Operation preserving regular languages. *Theoret. Comput. Sci.* **354**, (2006) 405–420.
[Berstel and Karhumäki 2003] Berstel, J. and Karhumäki, J. Combinatorics on words: a tutorial. *Bull. European Assoc. Theor. Comput. Sci.*, No. 79, (2003), 178–228.
[Berstel and Perrin 2007] Berstel, J. and Perrin, D. The origins of combinatorics on words. *European J. Combinatorics* **28**, (2007) 996–1022.
[Berstel and Reutenauer 1988] Berstel, J. and Reutenauer, C. *Rational Series and Their Languages*, vol. 12 of EATCS Monographs on Theoretical Computer Science. Springer-Verlag, 1988.
[Berthé, Ei, Ito, et al. 2007] Berthé, V., Ei, H., Ito, S., and Rao, H. On substitution invariant Sturmian words: an application of Rauzy fractals. *Theoret. Informatics Appl.* **41**, (2007) 329–349.
[Berthé, Holton, and Zamboni 2006] Berthé, V., Holton, C., and Zamboni, L. Q. Initial powers of Sturmian sequences. *Acta Arith.* **122**, (2006) 315–347.
[Berthé and Rigo 2007a] Berthé, V. and Rigo, M. Abstract numeration systems and tilings. In *Proc. 32nd Symposium, Mathematical Foundations of Computer Science 2007*, vol. 3618 of Lecture Notes in Computer Science, pp. 131–143. Springer-Verlag, 2007.
[Berthé and Rigo 2007b] Berthé, V. and Rigo, M. Odometers on regular languages. *Theory Comput. Syst.* **40**, (2007) 1–31.
[Berthé and Siegel 2005] Berthé, V. and Siegel, A. Tilings associated with beta-numeration and substitutions. *Integers* **5** (2005), #A02 (electronic), www.integers-ejcnt.org/vol5.html
[Berthé, Siegel, Steiner, et al. 2009] Berthé, V., Siegel, A., Steiner, W., Surer, P., and Thuswaldner, J. M. Fractal tiles associated with shift radix systems, 2009. Preprint.
[Berthé and Vuillon 2000] Berthé, V. and Vuillon, L. Tilings and rotations on the torus: a two-dimensional generalization of Sturmian sequences. *Discrete Math.* **223**, (2000) 27–53.
[Bertrand 1977] Bertrand, A. Développements en base de Pisot et répartition modulo 1. *C. R. Acad. Sci. Paris Sér. A-B* **285**(6), (1977) A419–A421.
[Bertrand-Mathis 1986] Bertrand-Mathis, A. Développements en base θ, répartition modulo un de la suite $(x\theta^n)_{n \geq 0}$, langages codés et θ-shift. *Bull. Soc. Math. France* **114**(3), (1986) 271–323.
[Bertrand-Mathis 1989] Bertrand-Mathis, A. Comment écrire les nombres entiers dans une base que n'est pas entière. *Acta Math. Hung.* **54**, (1989) 237–241.
[Bès 2000] Bès, A. An extension of the Cobham-Semenov Theorem. *J. Symbolic Logic* **65**, (2000) 201–211.
[Bésineau 1972] Bésineau, J. Indépendance statistique d'ensembles liés à la fonction 'somme des chiffres'. *Acta Arith.* **20**, (1972) 401–416.
[Bezuglyi, Kwiatkowski, and Medynets 2008] Bezuglyi, S., Kwiatkowski, J., and Medynets, K. Aperiodic substitutional systems and their Bratteli diagrams. *Ergod. Th. & Dynam. Sys.* **29**, (2008) 37–72.
[Bezuglyi, Kwiatkowski, Medynets, et al. 2009] Bezuglyi, S., Kwiatkowski, J., Medynets, K., and Solomyak, B. Invariant measures on stationary Bratteli diagrams. *Ergod. Th. & Dynam. Sys.*, to appear.
[Billingsley 1968] Billingsley, P. *Convergence of Probability Measures*. John Wiley & Sons Inc., 1968.

[Bilu 2008] Bilu, Y. F. The many faces of the subspace theorem (After Adamczewski, Bugeaud, Corvaja, Zannier...). *Astérisque* **317**, (2008) 1–38. Séminaire Bourbaki. Vol. 2006/2007, Exp. No. 967.
[Blanchard 1989] Blanchard, F. β-expansions and symbolic dynamics. *Theoret. Comput. Sci.* **65**, (1989) 131–141.
[Blanchard and Hansel 1986] Blanchard, F. and Hansel, G. Systèmes codés. *Theoret. Comput. Sci.* **44**, (1986) 17–49.
[Blanchet-Sadri, Jungers, and Palumbo 2009] Blanchet-Sadri, F., Jungers, R. M., and Palumbo, J. Testing avoidability of sets of partial words is hard. *Theoret. Comput. Sci.* **410**, (2009) 968–972.
[Blondel and Canterini 2003] Blondel, V. D. and Canterini, V. Undecidable problems for probabilistic automata of fixed dimension. *Theory Comput. Systems* **36**(3), (2003) 231–245.
[Blondel, Cassaigne, and Jungers 2009] Blondel, V. D., Cassaigne, J., and Jungers, R. M. On the number of α-power-free words for $2 < \alpha \leq 7/3$. *Theoret. Comput. Sci.* **410**, (2009) 2823–2833.
[Blondel, Jungers, and Protasov 2006] Blondel, V. D., Jungers, R., and Protasov, V. On the complexity of computing the capacity of codes that avoid forbidden difference patterns. *IEEE Trans. Inform. Theory* **52**(11), (2006) 5122–5127.
[Blondel and Nesterov 2005] Blondel, V. D. and Nesterov, Y. Computationally efficient approximations of the joint spectral radius. *SIAM J. Matrix Anal.* **27**(1), (2005) 256–272.
[Blondel and Nesterov 2009] Blondel, V. D. and Nesterov, Y. Polynomial-time computation of the joint spectral radius for some sets of nonnegative matrices. *SIAM J. Matrix Anal.* **31**, (2009) 865–876.
[Blondel, Nesterov, and Theys 2005] Blondel, V. D., Nesterov, Y., and Theys, J. On the accuracy of the ellipsoid norm approximation of the joint spectral radius. *Linear Algebra Appl.* **394**(1), (2005) 91–107.
[Blondel, Theys, and Vladimirov 2003] Blondel, V. D., Theys, J., and Vladimirov, A. An elementary counterexample to the finiteness conjecture. *SIAM J. Matrix Anal.* **24**(4), (2003) 963–970.
[Blondel and Tsitsiklis 2000a] Blondel, V. D. and Tsitsiklis, J. N. The boundedness of all products of a pair of matrices is undecidable. *Syst. Control Lett.* **41**(2), (2000) 135–140.
[Blondel and Tsitsiklis 2000b] Blondel, V. D. and Tsitsiklis, J. N. A survey of computational complexity results in systems and control. *Automatica* **36**(9), (2000) 1249–1274.
[Boasson and Nivat 1980] Boasson, L. and Nivat, M. Adherences of languages. *J. Comput. System Sci.* **20**, (1980) 285–309.
[Boigelot and Brusten 2009] Boigelot, B. and Brusten, J. A Generalization of Cobham's theorem to automata over real numbers. *Theoret. Comput. Sci.* **410**, (2009) 1694–1703.
[Boigelot, Brusten, and Bruyère 2008] Boigelot, B., Brusten, J., and Bruyère, V. On the sets of real numbers recognized by finite automata in multiple bases. In *ICALP 2008*, vol. 5126 of Lecture Notes in Computer Science, pp. 112–123. Springer, 2008.
[Booth 1951] Booth, A. D. A signed binary multiplication technique. *Quart. J. Mech. Appl. Math.* **4**, (1951) 236–240.
[Borel 1909] Borel, É. Les probabilités dénombrables et leurs applications arithmétiques. *Rendiconti Circ. Mat. Palermo* **27**, (1909) 247–271.
[Borel 1950] Borel, É. Sur les chiffres décimaux de $\sqrt{2}$ et divers problèmes de probabilités en chaîne. *C. R. Acad. Sci. Paris* **230**, (1950) 591–593.

[Bose and Nelson 1962] Bose, R. C. and Nelson, R. J. A sorting problem. *J. Assoc. Comput. Mach.* **9**, (1962) 282–296.
[Boshernitzan 1984] Boshernitzan, M. A unique ergodicity of minimal symbolic flows with linear block growth. *J. Analyse Math.* **44**, (1984) 77–96.
[Boshernitzan 1985] Boshernitzan, M. A condition for minimal interval exchange maps to be uniquely ergodic. *Duke Math. J.* **52**(3), (1985) 723–752.
[Boshernitzan 1992] Boshernitzan, M. A condition for unique ergodicity of minimal symbolic flows. *Ergod. Th. & Dynam. Sys.* **12**(3), (1992) 425–428.
[Boshernitzan and Carroll 1997] Boshernitzan, M. and Carroll, C. R. An extension of Lagrange's theorem to interval exchange transformations over quadratic fields. *J. Analyse Math.* **72**, (1997) 21–44.
[Bosma 2001] Bosma, W. Signed bits and fast exponentiation. *J. Théorie Nombres Bordeaux* **13**, (2001) 27–41.
[Bousch and Mairesse 2002] Bousch, T. and Mairesse, J. Asymptotic height optimization for topical IFS, Tetris heaps, and the finiteness conjecture. *J. Amer. Math. Soc.* **15**(1), (2002) 77–111.
[Boyd 1989] Boyd, D. W. Salem numbers of degree four have periodic expansions. In J. H. de Coninck and C. Levesque, eds., *Number Theory*, pp. 57–64. Walter de Gruyter, 1989.
[Boyd and Vandenberghe 2004] Boyd, S. and Vandenberghe, L. *Convex Optimization*. Cambridge University Press, 2004.
[Boyle 1983] Boyle, M. Topological orbit equivalence and factor maps in symbolic dynamics. Ph.D. thesis, University of Washington, Seattle, 1983.
[Boyle and Handelman 1994] Boyle, M. and Handelman, D. Entropy versus orbit equivalence for minimal homeomorphisms. *Pacific J. Math.* **164**, (1994) 1–13.
[Bratteli 1972] Bratteli, O. Inductive limits of finite-dimensional C^*–algebras. *Trans. Amer. Math. Soc.* **171**, (1972) 195–234.
[Bressaud, Durand, and Maass 2005] Bressaud, X., Durand, F., and Maass, A. Necessary and sufficient conditions to be an eigenvalue for linearly recurrent dynamical Cantor systems. *J. London Math. Soc.* **72**, (2005) 799–816.
[Brlek 1989] Brlek, S. Enumeration of factors in the Thue-Morse word. *Discrete Appl. Math.* **24**, (1989) 83–96.
[de Bruijn 1946] de Bruijn, N. G. A combinatorial problem. *Proc. Konin. Neder. Akad. Wet.* **49**, (1946) 758–764.
[de Bruijn 1948] de Bruijn, N. On Mahler's partition problem. *Indag. Math.* **10**, (1948) 210–220.
[de Bruijn 1981] de Bruijn, N. G. *Asymptotic Methods in Analysis*. Dover, 1981.
[Bruin and Troubetzkoy 2003] Bruin, H. and Troubetzkoy, S. The Gauss map on a class of interval translation mappings. *Israel J. Math.* **137**, (2003) 125–148.
[Brunotte 2001] Brunotte, H. On trinomial bases of radix representation of algebraic integers. *Acta Sci. Math. (Szeged)* **67**, (2001) 553–559.
[Brunotte 2002] Brunotte, H. Characterisation of CNS trinomials. *Acta Sci. Math. (Szeged)* **68**, (2002) 673–679.
[Brunotte, Huszti, and Pethő 2006] Brunotte, H., Huszti, A., and Pethő, A. Bases of canonical number systems in quartic algebraic number fields. *J. Théorie Nombres Bordeaux* **18**, (2006) 537–557.
[Bruyère and Hansel 1997] Bruyère, V. and Hansel, G. Bertrand numeration systems and recognizability. *Theoret. Comput. Sci.* **181**, (1997) 17–43.
[Bruyère, Hansel, Michaux, et al. 1994] Bruyère, V., Hansel, G., Michaux, C., and Villemaire, R. Logic and p-recognizable sets of integers. *Bull. Belg. Math. Soc.* **1**, (1994) 191–238. Corrigendum, *Bull. Belg. Math. Soc.* **1** (1994), 577.
[Bugeaud 2004a] Bugeaud, Y. *Approximation by algebraic numbers*, vol. 160 of *Cambridge Tracts in Mathematics*. Cambridge University Press, 2004.

[Bugeaud 2004b] Bugeaud, Y. Linear mod one transformations and the distribution of fractional parts $\{\xi(p/q)^n\}$. *Acta Arith.* **114**, (2004) 301–311.

[Bugeaud and Evertse 2008] Bugeaud, Y. and Evertse, J.-H. On two notions of complexity of algebraic numbers. *Acta Arith.* **133**, (2008) 221–250.

[Bugeaud and Laurent 2005] Bugeaud, Y. and Laurent, M. Exponents of Diophantine approximation and Sturmian continued fractions. *Ann. Inst. Fourier (Grenoble)* **55**, (2005) 773–804.

[Canterini and Siegel 2001a] Canterini, V. and Siegel, A. Automate des préfixes-suffixes associé à une substitution primitive. *J. Théorie Nombres Bordeaux* **13**, (2001) 353–369.

[Canterini and Siegel 2001b] Canterini, V. and Siegel, A. Geometric representation of substitutions of Pisot type. *Trans. Amer. Math. Soc.* **353**, (2001) 5121–5144.

[Carlitz 1967] Carlitz, L. The number of binomial coefficients divisible by a fixed power of a prime. *Rend. Circ. Matem. Palermo* **16**, (1967) 299–320.

[Carpi 1993] Carpi, A. Overlap-free words and finite automata. *Theoret. Comput. Sci.* **115**(2), (1993) 243–260.

[Carton 2009] Carton, O. Left and right synchronous relations. In *Developments in Language Theory 2009*, vol. 5583 of Lecture Notes in Computer Science, pp. 170–182. Springer-Verlag, 2009.

[Carton and Thomas 2002] Carton, O. and Thomas, W. The monadic theory of morphic infinite words and generalizations. *Inform. Comput.* **176**, (2002) 51–65.

[Cassaigne 1993] Cassaigne, J. Counting overlap-free binary words. In P. Enjalbert, A. Finkel, and K. W. Wagner, eds., *STACS 93, Proc. 10th Symp. Theoretical Aspects of Comp. Sci.*, vol. 665 of Lecture Notes in Computer Science, pp. 216–225. Springer-Verlag, 1993.

[Cassaigne 1996] Cassaigne, J. Special factors of sequences with linear subword complexity. In J. Dassow, G. Rozenberg, and A. Salomaa, eds., *Developments in Language Theory II*, pp. 25–34. World Scientific, 1996.

[Cassaigne 1997] Cassaigne, J. Complexité et facteurs spéciaux. *Bull. Belg. Math. Soc.* **4**, (1997) 67–88.

[Cassaigne and Chekhova 2006] Cassaigne, J. and Chekhova, N. Fonctions de récurrence des suites d'Arnoux-Rauzy et réponse à une question de Morse et Hedlund. *Ann. Inst. Fourier (Grenoble)* **56**.

[Cassaigne and Kaboré 2009] Cassaigne, J. and Kaboré, I. Un mot sans fréquences de faible complexité (provisional title), 2009. In preparation.

[Cassaigne, Mauduit, and Nicolas 2008] Cassaigne, J., Mauduit, C., and Nicolas, F. Asymptotic behavior of growth functions of D0L-systems. Tech. rep., The Computing Research Repository, 2008.

[Cassaigne and Nicolas 2003] Cassaigne, J. and Nicolas, F. Quelques propriétés des mots substitutifs. *Bull. Belg. Math. Soc. Simon Stevin* **10**, (2003) 661–676.

[Cassaigne and Nicolas 2009] Cassaigne, J. and Nicolas, F. On the Morse-Hedlund complexity gap. Tech. rep., The Computing Research Repository, 2009.

[Cassels and Swinnerton-Dyer 1955] Cassels, J. W. S. and Swinnerton-Dyer, H. P. F. On the product of three homogeneous linear forms and the indefinite ternary quadratic forms. *Philos. Trans. Roy. Soc. London. Ser. A.* **248**, (1955) 73–96.

[Cateland 1992] Cateland, E. Suites digitales et suites k-régulières. Ph.D. thesis, Université Bordeaux I, 1992.

[Cauchy 1840] Cauchy, A. Sur les moyens d'éviter les erreurs dans les calculs numériques. *C. R. Acad. Sci. Paris* **11**, (1840) 789–798. Reprinted in A.

Cauchy, *Œuvres complètes*, 1ère série, Tome V, Gauthier-Villars, 1885, pp. 431-442.
[Chaika 2008] Chaika, J. Hausdorff dimension for ergodic measures of interval exchange transformations. *J. Mod. Dyn.* **2**(3), (2008) 457–464.
[Champernowne 1933] Champernowne, D. G. The construction of decimals normal in the scale of ten. *J. London Math. Soc.* **8**, (1933) 254–260.
[Charlier 2009] Charlier, E. Abstract numeration systems: recognizability, decidability and multi-dimensional S-automatic words, and real numbers. Ph.D. thesis, University of Liège, 2009.
[Charlier, Kärki, and Rigo 2010] Charlier, E., Kärki, T., and Rigo, M. Multidimensional generalized automatic sequences and shape-symmetric morphic words. *Discrete Math.* **310**, (2010) 1238–1252.
[Charlier, Le Gonidec, and Rigo] Charlier, E., Le Gonidec, M., and Rigo, M. Representation of real numbers in generalized numeration system. Submitted.
[Charlier, Rigo, and Steiner 2008] Charlier, E., Rigo, M., and Steiner, W. Abstract numeration systems on bounded languages and multiplication by a constant. *Integers* **8**, (2008) A35, 19.
[Chekhova, Hubert, and Messaoudi 2001] Chekhova, N., Hubert, P., and Messaoudi, A. Propriétés combinatoires, ergodiques et arithmétiques de la substitution de Tribonacci. *J. Théorie Nombres Bordeaux* **13**, (2001) 371–394.
[Cheung 2003] Cheung, Y. Hausdorff dimension of the set of nonergodic directions. *Ann. of Math. (2)* **158**(2), (2003) 661–678. With an appendix by M. Boshernitzan.
[Cheung and Masur 2006] Cheung, Y. and Masur, H. A divergent Teichmüller geodesic with uniquely ergodic vertical foliation. *Israel J. Math.* **152**, (2006) 1–15.
[Choffrut 1977] Choffrut, C. Une caractérisation des fonctions séquentielles et des fonctions sous-séquentielles en tant que relations rationnelles. *Theoret. Comput. Sci.* **5**, (1977) 325–337.
[Choffrut and Goldwurm 1995] Choffrut, C. and Goldwurm, M. Rational transductions and complexity of counting problems. *Math. Systems Theory* **28**, (1995) 437–450.
[Choffrut and Karhumäki 1997] Choffrut, C. and Karhumäki, J. Combinatorics of words. In G. Rozenberg and A. Salomaa, eds., *Handbook of Formal Languages*, vol. 1, pp. 329–438. Springer-Verlag, 1997.
[Chomsky and Miller 1958] Chomsky, N. and Miller, G. A. Finite state languages. *Inform. Control* **1**, (1958) 91–112.
[Chow and Robertson 1978] Chow, C. and Robertson, J. Logical design of a redundant binary adder. In *Proceedings of the 4th Symposium on Computer Arithmetic*, pp. 109–115. IEEE Computer Society Press, 1978.
[Christol 1979] Christol, G. Ensembles presque périodiques k-reconnaissables. *Theoret. Comput. Sci.* **9**, (1979) 141–145.
[Christol, Kamae, Mendès France, et al. 1980] Christol, G., Kamae, T., Mendès France, M., and Rauzy, G. Suites algébriques, automates et substitutions. *Bull. Soc. Math. France* **108**, (1980) 401–419.
[Clark and Sadun 2006] Clark, A. and Sadun, L. When shape matters: deformations of tiling spaces. *Ergod. Th. & Dynam. Sys.* **26**, (2006) 69–86.
[Cobham 1968] Cobham, A. On the Hartmanis-Stearns problem for a class of tag machines. In *IEEE Conference Record of 1968 Ninth Annual Symposium on Switching and Automata Theory*, pp. 51–60. 1968. Also appeared as IBM Research Technical Report RC-2178, August 23 1968.
[Cobham 1969] Cobham, A. On the base-dependence of sets of numbers recognizable by finite automata. *Math. Systems Theory* **3**, (1969) 186–192.

[Cobham 1972] Cobham, A. Uniform tag sequences. *Math. Systems Theory* **6**, (1972) 164–192.

[Cohen 1981] Cohen, D. On holy wars and a plea for peace. *Computer* **14**, (1981) 48–54.

[Cohen, Frey, Avanzi, et al. 2006] Cohen, H., Frey, G., Avanzi, R., Doche, C., Lange, T., Nguyen, K., and Vercauteren, F. *Handbook of Elliptic and Hyperelliptic Curve Cryptography*. Discrete Mathematics and its Applications. Chapman & Hall/CRC, 2006.

[Copeland and Erdős 1946] Copeland, A. H. and Erdős, P. Note on normal numbers. *Bull. Amer. Math. Soc.* **52**, (1946) 857–860.

[Coquet 1979] Coquet, J. Sur la mesure spectrale des suites q-multiplicatives. *Ann. Inst. Fourier (Grenoble)* **29**, (1979) 163–170.

[Coquet 1986] Coquet, J. Power sums of digital sums. *J. Number Theory* **22**, (1986) 161–176.

[Coquet, Kamae, and Mendès France 1977] Coquet, J., Kamae, T., and Mendès France, M. Sur la mesure spectrale de certaines suites arithmétiques. *Bull. Soc. Math. France* **105**, (1977) 369–384.

[Coquet and Mendès France 1977] Coquet, J. and Mendès France, M. Suites à spectre vide et suites pseudo-aléatoires. *Acta Arith.* **32**, (1977) 99–106.

[Cornfeld, Fomin, and Sinaĭ 1982] Cornfeld, I. P., Fomin, S. V., and Sinaĭ, Y. G. *Ergodic theory*. Springer-Verlag, 1982. Translated from the Russian by A. B. Sosinskiĭ.

[Cortez, Durand, Host, et al. 2003] Cortez, M. I., Durand, F., Host, B., and Maass, A. Continuous and measurable eigenfunctions of linearly recurrent dynamical Cantor systems. *J. London Math. Soc.* **67**, (2003) 790–804.

[Cortez, Gambaudo, and Maass 2007] Cortez, M. I., Gambaudo, J.-M., and Maass, A. Rotation topological factors of minimal $\mathbb{Z}^d$ actions of the Cantor set. *Trans. Amer. Math. Soc.* **359**, (2007) 2305–2315.

[Culik and Harju 1984] Culik, K., II and Harju, T. The ω-sequence equivalence problem for D0L systems is decidable. *J. Assoc. Comput. Mach.* **31**, (1984) 282–298.

[Cusick and Flahive 1989] Cusick, T. W. and Flahive, M. E. *The Markoff and Lagrange spectra*, vol. 30 of Mathematical Surveys and Monographs. American Mathematical Society, 1989.

[Damanik and Lenz 2006] Damanik, D. and Lenz, D. Substitutional dynamical systems: Characterization of linear repetitivity and applications. *J. Math. Anal. Appl.* **321**, (2006) 766–780.

[Dartnell, Durand, and Maass 2000] Dartnell, P., Durand, F., and Maass, A. Orbit equivalence and Kakutani equivalence with Sturmian subshifts. *Studia Math.* **142**, (2000) 25–45.

[Dartyge and Tenenbaum 2005] Dartyge, C. and Tenenbaum, G. Sommes des chiffres de multiples d'entiers. *Ann. Inst. Fourier (Grenoble)* **55**(7), (2005) 2423–2474.

[Dartyge and Tenenbaum 2006] Dartyge, C. and Tenenbaum, G. Congruences de sommes de chiffres de valeurs polynomiales. *Bull. London Math. Soc.* **38**(1), (2006) 61–69.

[Daubechies and Lagarias 1992] Daubechies, I. and Lagarias, J. C. Sets of matrices all infinite products of which converge. *Linear Algebra Appl.* **161**, (1992) 227–263.

[Davenport and Schmidt 1968] Davenport, H. and Schmidt, W. M. Approximation to real numbers by algebraic integers. *Acta Arith.* **15**, (1968/1969) 393–416.

[Dekking 1978] Dekking, F. M. The spectrum of dynamical systems arising from substitutions of constant length. *Z. Wahrscheinlichkeitstheorie und verw. Gebiete* **41**, (1978) 221–239.

[Dekking 1994] Dekking, F. M. Iteration of maps by an automaton. *Discrete Math.* **126**, (1994) 81–86.

[Delange 1972] Delange, H. Sur les fonctions q-additives ou q-multiplicatives. *Acta Arith.* **21**, (1972) 285–298.

[Delange 1975] Delange, H. Sur la fonction sommatoire de la fonction 'somme des chiffres'. *Enseign. Math.* **21**, (1975) 31–47.

[Denker, Grillenberger, and Sigmund 1976] Denker, M., Grillenberger, C., and Sigmund, K. *Ergodic theory on compact spaces*. Lecture Notes in Mathematics, vol. 527. Springer-Verlag, 1976.

[Devyatov 2008] Devyatov, R. On subword complexity of morphic sequences. In *Proc. 3rd International Computer Science Symposium in Russia (CSR 2008)*, no. 5010 in Lecture Notes in Computer Science, pp. 146–157. Springer-Verlag, 2008.

[Dirichlet 1842] Dirichlet, L. G. P. Verallgemeinerung eines Satzes aus der Lehre von den Kettenbruechen nebst einige Anwendungen auf die Theorie der Zahlen. *S.-B. Preuss. Akad. Wiss.* , (1842) 93–95.

[Downarowicz 2005] Downarowicz, T. Survey of odometers and Toeplitz flows. In S. Kolyada, Y. Manin, and T. Ward, eds., *Algebraic and topological dynamics*, vol. 385 of *Contemporary Math.*, pp. 7–37. Amer. Math. Soc., 2005.

[Downarowicz and Lacroix 1996] Downarowicz, T. and Lacroix, Y. A non-regular Toeplitz flow with preset pure point spectrum. *Studia Math.* **120**, (1996) 235–246.

[Downarowicz and Maass 2008] Downarowicz, T. and Maass, A. Finite rank Bratteli-Vershik diagrams are expansive. *Ergod. Th. & Dynam. Sys.* **28**, (2008) 739–747.

[Drmota 2001] Drmota, M. The joint distribution of q-additive functions. *Acta Arith.* **100**(1), (2001) 17–39.

[Drmota, Fuchs, and Manstavičius 2003] Drmota, M., Fuchs, M., and Manstavičius, E. Functional limit theorems for digital expansions. *Acta Math. Hungar.* **98**(3), (2003) 175–201.

[Drmota, Grabner, and Liardet 2008] Drmota, M., Grabner, P. J., and Liardet, P. Block additive functions on the Gaussian integers. *Acta Arith.* **135**, (2008) 299–332.

[Drmota, Mauduit, and Rivat 2009] Drmota, M., Mauduit, C., and Rivat, J. Primes with an average sum of digits. *Compositio Math.* **145**, (2009) 271–292.

[Drmota and Rivat 2005] Drmota, M. and Rivat, J. The sum-of-digits function of squares. *J. London Math. Soc.* **72**(2), (2005) 273–292.

[Drmota, Rivat, and Stoll 2008] Drmota, M., Rivat, J., and Stoll, T. The sum of digits of primes in $\mathbb{Z}[i]$. *Monatshefte Math.* **155**(3–4), (2008) 317–347. Special volume dedicated to the conference *Journées de Numération*, Graz, April 2007.

[Drmota and Steiner 2002] Drmota, M. and Steiner, W. The Zeckendorf expansion of polynomial sequences. *J. Théor. Nombres Bordeaux* **14**, (2002) 439–475.

[Drmota and Tichy 1997] Drmota, M. and Tichy, R. F. *Sequences, Discrepancies, and Applications*, vol. 1651 of *Lecture Notes in Mathematics*. Springer-Verlag, 1997.

[Droubay, Justin, and Pirillo 2001] Droubay, X., Justin, J., and Pirillo, G. Episturmian words and some constructions of de Luca and Rauzy. *Theoret. Comput. Sci.* **255**(1-2), (2001) 539–553.

[Duchêne, Fraenkel, Nowakowski, et al. 2009] Duchêne, E., Fraenkel, A., Nowakowski, R., and Rigo, M. Extensions and restrictions of Wythoff's game preserving Wythoff's sequence as set of P-positions. *J. Combin. Theory. Ser. A* **117**, (2010) 545–567.

[Duchêne and Rigo 2008a] Duchêne, E. and Rigo, M. A morphic approach to combinatorial games : the Tribonacci case. *Theor. Inform. Appl.* **42**, (2008) 375–393.

[Duchêne and Rigo 2008b] Duchêne, E. and Rigo, M. Cubic Pisot unit combinatorial games. *Monatsh. Math.* **155**, (2008) 217–249.

[Dumont 1990] Dumont, J. M. Summation formulae for substitutions on a finite alphabet. In J.-M. Luck, P. Moussa, and M. Waldschmidt, eds., *Number Theory and Physics*, vol. 47 of Springer Proceedings in Physics, pp. 185–194. Springer-Verlag, 1990.

[Dumont and Thomas 1989] Dumont, J.-M. and Thomas, A. Systèmes de numération et fonctions fractales relatifs aux substitutions. *Theoret. Comput. Sci.* **65**, (1989) 153–169.

[Dumont and Thomas 1991] Dumont, J.-M. and Thomas, A. Digital sum problems and substitutions on a finite alphabet. *J. Number Theory* **39**, (1991) 351–366.

[Dumont and Thomas 1993] Dumont, J.-M. and Thomas, A. Digital sum moments and substitutions. *Acta Arith.* **64**, (1993) 205–225.

[Dumont and Thomas 1997] Dumont, J.-M. and Thomas, A. Gaussian asymptotic properties of the sum-of-digits function. *J. Number Theory* **62**, (1997) 19–38.

[Durand 1998a] Durand, F. A generalization of Cobham's theorem. *Theory Comput. Systems* **31**, (1998) 169–185.

[Durand 1998b] Durand, F. A characterization of substitutive sequences using return words. *Discrete Math.* **179**, (1998) 89–101.

[Durand 1998c] Durand, F. Sur les ensembles d'entiers reconnaissables. *J. Théorie Nombres Bordeaux* **10**, (1998) 65–84.

[Durand 2000] Durand, F. Linearly recurrent subshifts have a finite number of non-periodic subshift factors. *Ergod. Th. & Dynam. Sys.* **20**, (2000) 1061–1078.

[Durand 2002] Durand, F. A theorem of Cobham for non primitive substitutions. *Acta Arith.* **104**, (2002) 225–241.

[Durand 2003] Durand, F. Corrigendum and addendum to 'Linearly recurrent subshifts have a finite number of non-periodic subshift factors'. *Ergod. Th. & Dynam. Sys.* **23**, (2003) 663–669.

[Durand, Host, and Skau 1999] Durand, F., Host, B., and Skau, C. Substitutive dynamical systems, Bratteli diagrams and dimension groups. *Ergod. Th. & Dynam. Sys.* **19**, (1999) 953–993.

[Durand and Messaoudi 2009] Durand, F. and Messaoudi, A. Boundary of the Rauzy fractal sets in $\mathbb{R} \times \mathbb{C}$ generated by $P(x) = x^4 - x^3 - x^2 - x - 1$. *Osaka J. Math.*, to appear.

[Durand and Rigo 2009] Durand, F. and Rigo, M. Syndeticity and independent substitutions. *Adv. in Appl. Math.* **42**, (2009) 1–22.

[Ehrenfeucht, Lee, and Rozenberg 1975] Ehrenfeucht, A., Lee, K. P., and Rozenberg, G. Subword complexities of various classes of deterministic developmental languages without interaction. *Theoret. Comput. Sci.* **1**, (1975) 59–75.

[Ehrenfeucht and Rozenberg 1978a] Ehrenfeucht, A. and Rozenberg, G. Elementary homomorphisms and a solution of the D0L sequence equivalence problem. *Theoret. Comput. Sci.* **7**, (1978) 169–183.

[Ehrenfeucht and Rozenberg 1978b] Ehrenfeucht, A. and Rozenberg, G. Simplifications of homomorphisms. *Inform. Control* **38**, (1978) 298–309.

[Ehrenfeucht and Rozenberg 1982] Ehrenfeucht, A. and Rozenberg, G. On the subword complexities of homomorphic images of languages. *RAIRO Inform. Théor. App.* **16**, (1982) 303–316.

[Ei and Ito 2005] Ei, H. and Ito, S. Tilings from some non-irreducible Pisot substitutions. *Discrete Math. & Theoret. Comput. Sci.* **7**, (2005) 81–122.
[Ei, Ito, and Rao 2006] Ei, H., Ito, S., and Rao, H. Atomic surfaces, tilings and coincidences II. Reducible case. *Ann. Inst. Fourier (Grenoble)* **56**, (2006) 2285–2313.
[Eilenberg 1974] Eilenberg, S. *Automata, Languages, and Machines*, vol. A. Academic Press, 1974.
[Eilenberg, Elgot, and Shepherdson 1969] Eilenberg, S., Elgot, C. C., and Shepherdson, J. C. Sets recognized by n-tape automata. *J. Algebra* **13**, (1969) 447–464.
[Einsiedler, Katok, and Lindenstrauss 2006] Einsiedler, M., Katok, A., and Lindenstrauss, E. Invariant measures and the set of exceptions to Littlewood's conjecture. *Ann. of Math. (2)* **164**, (2006) 513–560.
[Einsiedler and Schmidt 2002] Einsiedler, M. and Schmidt, K. Irreducibility, homoclinic points and adjoint actions of algebraic $\mathbb{Z}^d$-actions of rank one. In *Dynamics and Randomness (Santiago, 2000)*, vol. 7 of *Nonlinear Phenom. Complex Systems*, pp. 95–124. Kluwer, 2002.
[Elgot and Mezei 1965] Elgot, C. C. and Mezei, J. E. On relations defined by generalized finite automata. *IBM J. Res. Develop.* **9**, (1965) 47–68.
[Elliott 1979] Elliott, P. D. T. A. *Probabilistic Number Theory. I*, vol. 239 of Grundlehren der Mathematischen Wissenschaften [Fundamental Principles of Mathematical Science]. Springer-Verlag, 1979. Mean-value theorems.
[Elliott 1980] Elliott, P. D. T. A. *Probabilistic Number Theory. II*, vol. 240 of Grundlehren der Mathematischen Wissenschaften [Fundamental Principles of Mathematical Sciences]. Springer-Verlag, 1980. Central limit theorems.
[Elliott 1985] Elliott, P. D. T. A. *Arithmetic functions and integer products*, vol. 272 of Grundlehren der Mathematischen Wissenschaften [Fundamental Principles of Mathematical Sciences]. Springer-Verlag, New York, 1985.
[Elsner 1995] Elsner, L. The generalized spectral-radius theorem: An analyticgeometric proof. *Linear Algebra Appl.* **220**, (1995) 151–159.
[Epstein, Cannon, Holt, et al. 1992] Epstein, D. B. A., Cannon, J. W., Holt, D. F., Levy, S. V., Paterson, M. S., and Thurston, W. P. *Word processing in groups*. Jones and Bartlett Publishers, 1992.
[Erdős 1939] Erdős, P. On a family of symmetric Bernoulli convolutions. *Amer. J. Math.* **61**, (1939) 974–976.
[Erdős 1940] Erdős, P. On the smoothness properties of a family of Bernoulli convolutions. *Amer. J. Math.* **62**, (1940) 180–186.
[Euler 1922] Euler, L. *Introductio in analysin infinitorum*. Teubner, 1922. Reprint of the 1748 original.
[Fabre 1995] Fabre, S. Substitutions et β-systèmes de numération. *Theoret. Comput. Sci.* **137**, (1995) 219–236.
[Fekete 1923] Fekete, M. Über die Verteilung der Wurzeln bei gewissen algebraischen Gleichungen mit ganzzahligen Koeffizienten. *Math. Zeitschrift* **17**, (1923) 228–249.
[Feng, Furukado, Ito, et al. 2006] Feng, D.-J., Furukado, M., Ito, S., and Wu, J. Pisot substitutions and the Hausdorff dimension of boundaries of atomic surfaces. *Tsukuba J. Math.* **30**, (2006) 195–223.
[Ferenczi 1995] Ferenczi, S. Les transformations de Chacon: combinatoire, structure géométrique, lien avec les systèmes de complexité $2n + 1$. *Bull. Soc. Math. France* **123**, (1995) 271–292.
[Ferenczi 1996] Ferenczi, S. Rank and symbolic complexity. *Ergod. Th. & Dynam. Sys.* **16**, (1996) 663–682.

[Ferenczi 1997] Ferenczi, S. Systems of finite rank. *Colloq. Math.* **73**(1), (1997) 35–65.
[Ferenczi, Fisher, and Talet 2009] Ferenczi, S., Fisher, A., and Talet, M. Some (non)-uniquely ergodic adic transformations. *J. Analyse Math.*, to appear.
[Ferenczi, Holton, and Zamboni 2001] Ferenczi, S., Holton, C., and Zamboni, L. Q. Structure of three interval exchange transformations I. An arithmetic study. *Ann. Inst. Fourier (Grenoble)* **51**, (2001) 861–901.
[Ferenczi, Holton, and Zamboni 2003] Ferenczi, S., Holton, C., and Zamboni, L. Q. Structure of three-interval exchange transformations. II. A combinatorial description of the trajectories. *J. Anal. Math.* **89**, (2003) 239–276.
[Ferenczi, Holton, and Zamboni 2004] Ferenczi, S., Holton, C., and Zamboni, L. Q. Structure of three-interval exchange transformations III: ergodic and spectral properties. *J. Anal. Math.* **93**, (2004) 103–138.
[Ferenczi and Mauduit 1997] Ferenczi, S. and Mauduit, C. Transcendence of numbers with a low complexity expansion. *J. Number Theory* **67**, (1997) 146–161.
[Ferenczi, Mauduit, and Nogueira 1996] Ferenczi, S., Mauduit, C., and Nogueira, A. Substitution dynamical systems: algebraic characterization of eigenvalues. *Ann. Sci. École Norm. Sup.* **29**, (1996) 519–533.
[Ferenczi and Zamboni 2008] Ferenczi, S. and Zamboni, L. Q. Eigenvalues and simplicity for interval exchange transformations, 2008. Preprint.
[Ferenczi and Zamboni 2009] Ferenczi, S. and Zamboni, L. Q. Structure of k-interval exchange transformations: induction, trajectories, and distance theorems. *J. Analyse Math.*, to appear.
[Fernique 2006] Fernique, T. Multidimensional Sturmian sequences and generalized substitutions. *Int. J. Found. Comput. Sci.* **17**, (2006) 575–600.
[Fernique 2009] Fernique, T. Generation and recognition of digital planes using multi-dimensional continued fractions. *Pattern Recogn.* **42**, (2009) 2229–2238.
[Fischler 2006] Fischler, S. Palindromic prefixes and episturmian words. *J. Combin. Theory. Ser. A* **113**, (2006) 1281–1304.
[Fisher 2009] Fisher, A. M. Nonstationary mixing and the unique ergodicity of adic transformations. *Stoch. Dyn.*, to appear.
[Flajolet and Golin 1993] Flajolet, P. and Golin, M. Exact asymptotics of divide-and-conquer recurrences. In *Automata, Languages and Programming (Lund, 1993)*, vol. 700 of Lecture Notes in Computer Science, pp. 137–149. Springer, 1993.
[Flajolet, Grabner, Kirschenhofer, et al. 1994] Flajolet, P., Grabner, P., Kirschenhofer, P., Prodinger, H., and Tichy, R. F. Mellin transforms and asymptotics: digital sums. *Theoret. Comput. Sci.* **123**, (1994) 291–314.
[Flatto, Lagarias, and Pollington 1995] Flatto, L., Lagarias, J., and Pollington, A. On the range of fractional parts $\{\xi(p/q)^n\}$. *Acta Arith.* **70**, (1995) 125–147.
[Flatto, Lagarias, and Poonen 1994] Flatto, L., Lagarias, J., and Poonen, B. The zeta function of the beta transformation. *Ergod. Th. & Dynam. Sys.* **14**, (1994) 237–266.
[Forrest 1997] Forrest, A. H. K-groups associated with substitution minimal systems. *Israel J. Math.* **98**, (1997) 101–139.
[Foster 1987] Foster, D. M. E. Estimates for a remainder term associated with the sum of digits function. *Glasgow Math. J.* **29**, (1987) 109–129.
[Foster 1991] Foster, D. M. E. A lower bound for a remainder term associated with the sum of digits function. *Proc. Edinburgh Math. Soc.* **34**, (1991) 121–142.
[Foster 1992] Foster, D. M. E. Averaging the sum of digits function to an even base. *Proc. Edinburgh Math. Soc.* **35**, (1992) 449–455.

[Fouvry and Mauduit 1996a] Fouvry, E. and Mauduit, C. Méthodes de crible et fonctions sommes des chiffres. *Acta Arith.* **77**, (1996) 339–351.

[Fouvry and Mauduit 1996b] Fouvry, E. and Mauduit, C. Somme des chiffres et nombres presque premiers. *Math. Annalen* **305**, (1996) 571–599.

[Fraenkel 1985] Fraenkel, A. S. Systems of numeration. *Amer. Math. Monthly* **92**, (1985) 105–114.

[Fretlöh and Sing 2007] Fretlöh, D. and Sing, B. Computing modular coincidences for substitution tilings and point sets. *Discrete Comput. Geom.* **37**, (2007) 381–401.

[Frougny 1992] Frougny, Ch. Representations of numbers and finite automata. *Math. Systems Theory* **25**, (1992) 37–60.

[Frougny 1997] Frougny, Ch. On the sequentiality of the successor function. *Inform. Comput.* **139**, (1997) 17–38.

[Frougny 2002] Frougny, Ch. On multiplicatively dependent linear numeration systems, and periodic points. *Theoret. Informatics Appl.* **36**, (2002) 293–314.

[Frougny and Sakarovitch 1993] Frougny, Ch. and Sakarovitch, J. Synchronized relations of finite and infinite words. *Theoret. Comput. Sci.* **18**, (1993) 45–82.

[Frougny and Sakarovitch 1999] Frougny, Ch. and Sakarovitch, J. Automatic conversion from Fibonacci representation to representation in base φ, and a generalization. *Internat. J. Algebra Comput.* **9**, (1999) 351–384.

[Frougny and Solomyak 1992] Frougny, Ch. and Solomyak, B. Finite beta-expansions. *Ergod. Th. & Dynam. Sys.* **12**, (1992) 713–723.

[Frougny and Solomyak 1996] Frougny, Ch. and Solomyak, B. On representation of integers in linear numeration systems. In M. Pollicott and K. Schmidt, eds., *Ergodic Theory of $\mathbb{Z}^d$ Actions (Warwick, 1993–1994)*, vol. 228 of London Mathematical Society Lecture Note Series, pp. 345–368. Cambridge University Press, 1996.

[Frougny and Solomyak 1999] Frougny, Ch. and Solomyak, B. On the context-freeness of the theta-expansions of the integers. *Internat. J. Algebra Comput.* **9**, (1999) 347–350.

[Frougny and Steiner 2008] Frougny, Ch. and Steiner, W. Minimal weight expansions in Pisot bases. *J. Math. Crypt.* **2**, (2008) 365–392.

[Fuchs and Tijdeman 2006] Fuchs, Ch. and Tijdeman, R. Substitutions, abstract number systems and the space filling property. *Ann. Inst. Fourier (Grenoble)* **56**(7), (2006) 2345–2389. Numération, pavages, substitutions.

[Furstenberg 1961] Furstenberg, H. Strict ergodicity and transformation of the torus. *Amer. J. Math.* **83**, (1961) 573–601.

[Gantmacher 1960] Gantmacher, F. R. *The Theory of Matrices.* Chelsea, 1960.

[Gazeau, Nešetřil, and B. Rovan, eds 2007] Gazeau, J., Nešetřil, J., and B. Rovan, eds. *Physics and Theoretical Computer Science.* IOS Press, 2007.

[Gelfond 1968] Gelfond, A. O. Sur les nombres qui ont des propriétés additives et multiplicatives données. *Acta Arith.* **13**, (1968) 259–265.

[Gilbert 1981] Gilbert, W. J. Radix representations of quadratic fields. *J. Math. Anal. Appl.* **83**, (1981) 264–274.

[Gilbert 1991] Gilbert, W. J. Gaussian integers as bases for exotic number systems, 1991. Unpublished manuscript.

[Giordano, Putnam, and Skau 1995] Giordano, T., Putnam, I., and Skau, C. F. Topological orbit equivalence and C^*-crossed products. *Internat. J. Math.* **469**, (1995) 51–111.

[Gjerde and Johansen 2000] Gjerde, R. and Johansen, O. Bratteli-Vershik models for Cantor minimal systems: applications to Toeplitz flows. *Ergod. Th. & Dynam. Sys.* **20**, (2000) 1687–1710.

[Gjerde and Johansen 2002] Gjerde, R. and Johansen, O. Bratteli-Vershik models for Cantor minimal systems associated to interval exchange transformations. *Math. Scand.* **90**, (2002) 87–100.

[Glasner and Weiss 1995] Glasner, E. and Weiss, B. Weak orbit equivalence of Cantor minimal systems. *Internat. J. Math.* **6**, (1995) 559–579.

[Grabner 1993] Grabner, P. J. Completely q-multiplicative functions: the Mellin transform approach. *Acta Arith.* **65**, (1993) 85–96.

[Grabner 1997] Grabner, P. J. Functional iterations and stopping times for Brownian motion on the Sierpiński gasket. *Mathematika* **44**, (1997) 374–400.

[Grabner 2004] Grabner, P. J. Minima of digital functions related to large digits in q-adic expansions. *Quaest. Math.* **27**, (2004) 75–87.

[Grabner and Heuberger 2006] Grabner, P. J. and Heuberger, C. On the number of optimal base 2 representations of integers. *Des. Codes Cryptogr.* **40**, (2006) 25–39.

[Grabner, Heuberger, and Prodinger 2005] Grabner, P. J., Heuberger, C., and Prodinger, H. Counting optimal joint digit expansions. *Integers* **5**, (2005) A9, (electronic).

[Grabner and Hwang 2005] Grabner, P. J. and Hwang, H.-K. Digital sums and divide-and-conquer recurrences: Fourier expansions and absolute convergence. *Constr. Approx.* **21**, (2005) 149–179.

[Grabner, Kirschenhofer, and Prodinger 1998] Grabner, P. J., Kirschenhofer, P., and Prodinger, H. The sum-of-digits function for complex bases. *J. London Math. Soc.* **57**, (1998) 20–40.

[Grabner and Liardet 1999] Grabner, P. J. and Liardet, P. Harmonic properties of the sum-of-digits function for complex bases. *Acta Arith.* **91**, (1999) 329–349.

[Grabner, Liardet, and Tichy 1995] Grabner, P. J., Liardet, P., and Tichy, R. F. Odometers and systems of numeration. *Acta Arith.* **70**, (1995) 103–123.

[Grabner, Liardet, and Tichy 2005] Grabner, P. J., Liardet, P., and Tichy, R. F. Spectral disjointness of dynamical systems related to some arithmetic functions. *Publ. Math. Debrecen* **66**, (2005) 213–243.

[Grabner and Rigo 2003] Grabner, P. J. and Rigo, M. Additive functions with respect to numeration systems on regular languages. *Monatsh. Math.* **139**, (2003) 205–219.

[Grabner and Rigo 2007] Grabner, P. J. and Rigo, M. Distribution of additive functions with respect to numeration systems on regular languages. *Theory Comput. Syst.* **40**, (2007) 205–223.

[Grabner and Tichy 1990] Grabner, P. J. and Tichy, R. F. Contributions to digit expansions with respect to linear recurrences. *J. Number Theory* **36**, (1990) 160–169.

[Grabner and Tichy 1991] Grabner, P. J. and Tichy, R. F. α-expansions, linear recurrences, and the sum-of-digits function. *Manuscripta Math.* **70**, (1991) 311–324.

[Graham, Knuth, and Patashnik 1989] Graham, R. L., Knuth, D. E., and Patashnik, O. *Concrete Mathematics*. Addison-Wesley, 1989.

[Granville 1997] Granville, A. Arithmetic properties of binomial coefficients. I. Binomial coefficients modulo prime powers. In *Organic Mathematics (Burnaby, BC, 1995)*, vol. 20 of CMS Conf. Proc., pp. 253–276. Amer. Math. Soc., Providence, RI, 1997.

[Greene and Knuth 1990] Greene, D. H. and Knuth, D. E. *Mathematics for the Analysis of Algorithms*. Birkhäuser Boston, 1990.

[Grillenberger 1972] Grillenberger, C. Construction of strictly ergodic systems

I. Given entropy. *Z. Wahrscheinlichkeitstheorie und verw. Gebiete* **25**, (1972/73) 323–334.
[Gripenberg 1996] Gripenberg, G. Computing the Joint Spectral Radius. *Linear Algebra Appl.* **234**, (1996) 43–60.
[Gröchenig and Haas 1994] Gröchenig, K. and Haas, A. Self-similar Lattice Tilings. *J. Fourier Anal. Appl.* **1**, (1994) 131–170.
[Grünwald 1885] Grünwald, V. Intorno all'aritmetica dei sistemi numerici a base negativa con particolare riguardo al sistema numerico a base negativo-decimale per lo studio delle sue analogie coll'aritmetica (decimale). *Giorn. Mat. Battaglini* **23**, (1885) 203–221. Errata, p. 367.
[Guglielmi, Wirth, and Zennaro 2005] Guglielmi, N., Wirth, F., and Zennaro, M. Complex polytope extremality results for families of matrices. *SIAM J. Matrix Anal. Appl.* **27**(3), (2005) 721–743.
[Gurvits 1991] Gurvits, L. Stability and observability of discrete linear inclusion – Finite automata approach. In *Book of Abstracts of the International Symposium on the Mathematical Theory of Networks and Systems*, pp. 166–167. Kobe, 1991.
[Gurvits 1995] Gurvits, L. Stability of discrete Linear Inclusions. *Linear Algebra Appl.* **231**, (1995) 47–85.
[Gurvits 1996] Gurvits, L. Stability of Linear Inclusions - Part 2. NECI technical report TR pp. 96–173.
[Hahn and Katznelson 1967] Hahn, F. and Katznelson, Y. On the entropy of uniquely ergodic transformations. *Trans. Amer. Math. Soc.* **126**, (1967) 335–360.
[Hama and Imahashi 1997] Hama, M. and Imahashi, T. Periodic β-expansions for certain classes of Pisot numbers. *Comment. Math. Univ. St. Paul.* **46**(2), (1997) 103–116.
[Hansel 1982] Hansel, G. A propos d'un théorème de Cobham. In D. Perrin, ed., *Actes de la Fête des Mots*, pp. 55–59. Greco de Programmation, CNRS, Rouen, 1982.
[Hansel 1998] Hansel, G. Systèmes de numération indépendants et syndéticité. *Theoret. Comput. Sci.* **204**, (1998) 119–130.
[Hansel and Safer 2003] Hansel, G. and Safer, T. Vers un théorème de Cobham pour les entiers de Gauss. *Bull. Belg. Math. Soc. Simon Stevin* **10**, (2003) 723–735.
[Harborth 1977] Harborth, H. Number of odd binomial coefficients. *Proc. Amer. Math. Soc.* **62**, (1977) 19–22.
[Hardy and Riesz 1964] Hardy, G. H. and Riesz, M. *The General Theory of Dirichlet's Series*. Cambridge Tracts in Mathematics and Mathematical Physics, No. 18. Stechert-Hafner, Inc., 1964.
[Hardy and Wright 1985] Hardy, G. H. and Wright, E. M. *An Introduction to the Theory of Numbers*. Oxford University Press, 5th edn., 1985.
[Harju and Karhumäki 1997] Harju, T. and Karhumäki, J. Morphisms. In G. Rozenberg and A. Salomaa, eds., *Handbook of Formal Languages*, vol. 1, pp. 439–510. Springer-Verlag, 1997.
[Harju and Linna 1986] Harju, T. and Linna, M. On the periodicity of morphisms on free monoids. *RAIRO Inform. Théor. App.* **20**, (1986) 47–54.
[Hbaib and Mkaouar 2006] Hbaib, M. and Mkaouar, M. Sur le bêta-développement de 1 dans le corps des séries formelles. *Int. J. Number Theory* **2**, (2006) 365–378.
[Heinis 2002] Heinis, A. The $P(n)/n$ function for bi-infinite words. *Theoret. Comput. Sci.* **273**, (2002) 35–46.
[Herman, Putnam, and Skau 1992] Herman, R. H., Putnam, I., and Skau, C. F.

Ordered Bratteli diagrams, dimension groups and topological dynamics. *Internat. J. Math.* **3**, (1992) 827–864.

[Heuberger 2004] Heuberger, C. Minimal expansions in redundant number systems: Fibonacci bases and greedy algorithms. *Period. Math. Hungar.* **49**(2), (2004) 65–89.

[Heuberger and Prodinger 2006] Heuberger, C. and Prodinger, H. Analysis of alternative digit sets for nonadjacent representations. *Monatsh. Math.* **147**, (2006) 219–248.

[Hollander 1996] Hollander, M. Linear numeration systems, finite beta expansions, and discrete spectrum of substitution dynamical Systems. Ph.D. thesis, University of Washington, 1996.

[Hollander 1998] Hollander, M. Greedy numeration systems and regularity. *Theory Comput. Systems* **31**, (1998) 111–133.

[Hollander and Solomyak 2003] Hollander, M. and Solomyak, B. Two-symbol Pisot substitutions have pure discrete spectrum. *Ergod. Th. & Dynam. Sys.* **23**(2), (2003) 533–540.

[Holton and Zamboni 1999] Holton, C. and Zamboni, L. Q. Descendants of primitive substitutions. *Theory Comput. Systems* **32**, (1999) 133–157.

[Honkala 1986] Honkala, J. A decision method for the recognizability of sets defined by number systems. *RAIRO Inform. Théor. App.* **20**, (1986) 395–403.

[Honkala 1997] Honkala, J. A decision method for Parikh slenderness of context-free languages. *Discrete Appl. Math.* **73**, (1997) 1–4.

[Honkala 1998] Honkala, J. Decision problem concerning thinness and slenderness of formal languages. *Acta Inform.* **35**, (1998) 625–636.

[Honkala 2000] Honkala, J. A short solution for the HDT0L sequence equivalence problem. *Theoret. Comput. Sci.* **244**, (2000) 267–270.

[Honkala 2001a] Honkala, J. A note on uniform HDT0L systems. *Bull. European Assoc. Theor. Comput. Sci.* **75**, (2001) 220–223.

[Honkala 2001b] Honkala, J. On Parikh slender context-free languages. *Theoret. Comput. Sci.* **255**, (2001) 667–677.

[Honkala 2005] Honkala, J. The class of HDT0L sequences is closed with respect to rational functions. *Inform. Process. Lett.* **94**, (2005) 155–158.

[Honkala 2007a] Honkala, J. The D0L ω-equivalence problem. *Internat. J. Found. Comp. Sci.* **18**, (2007) 181–194.

[Honkala 2007b] Honkala, J. A new bound for the D0L sequence equivalence problem. *Acta Inform.* **43**, (2007) 419–429.

[Honkala 2008] Honkala, J. Cancellation and periodicity properties of iterated morphisms. *Theoret. Comput. Sci.* **391**, (2008) 61–64.

[Honkala 2009a] Honkala, J. The equality problem for infinite words generated by primitive morphisms. *Inform. Comput.* **207**, (2009) 900–907.

[Honkala 2009b] Honkala, J. On the simplification of infinite morphic words. *Theoret. Comput. Sci.* **410**, (2009) 997–1000.

[Honkala and Rigo 2004] Honkala, J. and Rigo, M. Decidability questions related to abstract numeration systems. *Discrete Math.* **285**, (2004) 329–333.

[Hopcroft, Motwani, and Ullman 2006] Hopcroft, J. E., Motwani, R., and Ullman, J. D. *Introduction to Automata Theory, Languages, and Computation.* Addison-Wesley, 2006. 3rd Edition.

[Hopcroft and Ullman 1979] Hopcroft, J. E. and Ullman, J. D. *Introduction to Automata Theory, Languages, and Computation.* Addison-Wesley, 1979.

[Host 1986] Host, B. Valeurs propres des systèms dynamiques définis par des substitutions de longueur variable. *Ergod. Th. & Dynam. Sys.* **6**, (1986) 529–540.

[Hubert and Messaoudi 2006] Hubert, P. and Messaoudi, A. Best simultaneous

Diophantine approximations of Pisot numbers and Rauzy fractals. *Acta Arith.* **124**(1), (2006) 1–15.

[Hwang 1998] Hwang, H.-K. Asymptotics of divide-and-conquer recurrences: Batcher's sorting algorithm and a minimum Euclidean matching heuristic. *Algorithmica* **22**(4), (1998) 529–546. Average-case analysis of algorithms.

[Ito, Fujii, Higashino, et al. 2003] Ito, S., Fujii, J., Higashino, H., and Yasutomi, S.-I. On simultaneous approximation to (α, α^2) with $\alpha^3 + k\alpha - 1 = 0$. *J. Number Theory* **99**(2), (2003) 255–283.

[Ito and Ohtsuki 1993] Ito, S. and Ohtsuki, M. Modified Jacobi-Perron algorithm and generating Markov partitions for special hyperbolic toral automorphisms. *Tokyo J. Math.* **16**(2), (1993) 441–472.

[Ito and Rao 2006] Ito, S. and Rao, H. Atomic surfaces, tilings and coincidences I. Irreducible case. *Israel J. Math.* **153**, (2006) 129–156.

[Ito and Takahashi 1974] Ito, S. and Takahashi, Y. Markov subshifts and realization of β-expansions. *J. Math. Soc. Japan* **26**, (1974) 33–55.

[Iwaniec and Kowalski 2004] Iwaniec, H. and Kowalski, E. *Analytic number theory*, vol. 53 of *American Mathematical Society Colloquium Publications*. American Mathematical Society, 2004.

[Iwanik 1996] Iwanik, A. Toeplitz flows with pure point spectrum. *Studia Math.* **118**, (1996) 27–35.

[Jacobs and Keane 1969] Jacobs, K. and Keane, M. 0 − 1-sequences of Toeplitz type. *Z. Wahrscheinlichkeitstheorie und verw. Gebiete* **13**, (1969) 123–131.

[Jewett 1969] Jewett, R. The prevalence of uniquely ergodic systems. *J. Math. Mech.* **19**, (1969/70) 717–729.

[Johnson 1999] Johnson, K. Beta-shift, dynamical systems and their associated languages. Ph.D. thesis, University of North Carolina at Chapel Hill, 1999.

[del Junco, Rahe, and Swanson 1980] del Junco, A., Rahe, M., and Swanson, L. Chacon's automorphism has minimal self-joinings. *J. Analyse Math.* **37**, (1980) 276–284.

[Jungers 2005] Jungers, R. On the growth of codes whose differences avoid forbidden patterns. Master's thesis, Université Catholique de Louvain, 2005.

[Jungers 2008] Jungers, R. M. *Infinite matrix products, from the joint spectral radius to combinatorics*. Ph.D. thesis, Université Catholique de Louvain, 2008.

[Jungers 2009] Jungers, R. M. *The Joint Spectral Radius, Theory and Applications*, vol. 385 of *LNCIS*. Springer-Verlag, 2009.

[Jungers and Blondel 2008] Jungers, R. M. and Blondel, V. D. On the finiteness property for rational matrices. *Linear Algebra Appl.* **428**(10), (2008) 2283–2295.

[Jungers, Protasov, and Blondel 2008a] Jungers, R. M., Protasov, V., and Blondel, V. D. Computing the growth of the number of overlap-free words with spectra of matrices. In *Lecture Notes in Computer Science, Proceedings of LATIN08*, vol. 4957, pp. 84–93. Buzios, Rio de Janeiro, Brazil, 2008.

[Jungers, Protasov, and Blondel 2008b] Jungers, R. M., Protasov, V., and Blondel, V. D. Efficient algorithms for deciding the type of growth of products of integer matrices. *Linear Algebra Appl.* **428**(10), (2008) 2296–2311.

[Jungers, Protasov, and Blondel 2009] Jungers, R. M., Protasov, V., and Blondel, V. D. Overlap-free words and spectra of matrices. *Theoret. Comput. Sci.* **410**, (2009) 3670–3684.

[Kalle and Steiner 2009] Kalle, C. and Steiner, W. Beta-expansions, natural extensions and multiple tilings, 2009. Preprint.

[Kamae 1977] Kamae, T. Mutual singularity of spectra of dynamical systems given by 'sums of digits' to different bases. In *Dynamical Systems I – Warsaw*, vol. 49 of *Astérisque*, pp. 109–114. Société Mathématique de France, 1977.

[Kamae 1978] Kamae, T. Sum of digits to different bases and mutual singularity of their spectral measures. *Osaka J. Math.* **15**, (1978) 569–574.

[Kamae 1987] Kamae, T. Cyclic extensions of odometer transformations and spectral disjointness. *Israel J. Math.* **59**, (1987) 41–63.

[Kari, Rozenberg, and Salomaa 1997] Kari, L., Rozenberg, G., and Salomaa, A. L systems. In G. Rozenberg and A. Salomaa, eds., *Handbook of Formal Languages*, vol. 1, pp. 253–328. Springer-Verlag, 1997.

[Kátai 1992] Kátai, I. Distribution of q-additive function. In *Probability Theory and Applications*, vol. 80 of *Math. Appl.*, pp. 309–318. Kluwer Acad. Publ., 1992.

[Kátai and Kovács 1981] Kátai, I. and Kovács, B. Canonical number systems in imaginary quadratic fields. *Acta Math. Acad. Sci. Hung.* **37**, (1981) 159–164.

[Kátai and Szabó 1975] Kátai, I. and Szabó, J. Canonical number systems for complex integers. *Acta Sci. Math. (Szeged)* **37**, (1975) 255–260.

[Katok 1973] Katok, A. B. Invariant measures of flows on orientable surfaces. *Dokl. Akad. Nauk SSSR* **211**, (1973) 775–778.

[Keane 1975] Keane, M. S. Interval exchange transformations. *Math. Zeitschrift* **141**, (1975) 25–31.

[Keane 1977] Keane, M. Non-ergodic interval exchange transformations. *Israel J. Math.* **26**(2), (1977) 188–196.

[Kenyon and Vershik 1998] Kenyon, R. and Vershik, A. Arithmetic construction of sofic partitions of hyperbolic toral automorphisms. *Ergod. Th. & Dynam. Sys.* **18**(2), (1998) 357–372.

[Keynes and Newton 1976] Keynes, H. B. and Newton, D. A "minimal", non-uniquely ergodic interval exchange transformation. *Math. Z.* **148**(2), (1976) 101–105.

[Kfoury 1988] Kfoury, A. J. A linear time algorithm to decide whether a binary word contains an overlap. *Theoret. Informatics Appl.* **22**, (1988) 135–145.

[Khintchine 1926] Khintchine, A. Y. Über eine Klasse linearer diophantischer Approximationen. *Rendiconti Circ. Mat. Palermo* **50**, (1926) 170–195.

[Khintchine 1963] Khintchine, A. Y. *Continued Fractions*. Translated by Peter Wynn. P. Noordhoff Ltd., Groningen, 1963.

[Kim 1999] Kim, D.-H. On the joint distribution of q-additive functions in residue classes. *J. Number Theory* **74**(2), (1999) 307–336.

[Kirschenhofer and Tichy 1984] Kirschenhofer, P. and Tichy, R. F. On the distribution of digits in Cantor representations of integers. *J. Number Theory* **18**, (1984) 121–134.

[Kitchens 1998] Kitchens, B. P. *Symbolic Dynamics*. Springer-Verlag, 1998.

[Kleene 1956] Kleene, S. C. Representation of events in nerve nets and finite automata. In *Automata Studies*, pp. 3–42. Princeton University Press, 1956.

[Knuth 1998] Knuth, D. E. *The Art of Computer Programming. Volume 2: Seminumerical Algorithms*. Addison-Wesley, 1998. 3rd edition.

[Knuth 2000] Knuth, D. E. *Selected Papers on the Analysis of Algorithms*. CSLI Lecture Notes. CSLI Publications, Stanford, CA, 2000.

[Kobayashi 1988] Kobayashi, Y. Enumeration of irreducible binary words. *Discrete Appl. Math.* **20**, (1988) 221–232.

[Kovács 1981] Kovács, B. Canonical number systems in algebraic number fields. *Acta Math. Hung.* **37**, (1981) 405–407.

[Kozyakin 1990] Kozyakin, V. S. Algebraic unsolvability of problem of absolute stability of desynchronized systems. *Automat. Rem. Contr.* **51**(6), (1990)

754–759.

[Kozyakin 2005] Kozyakin, V. A dynamical systems construction of a counterexample to the finiteness conjecture. In *Proceedings of the 44th IEEE Conference on Decision and Control and ECC 2005*, pp. 2338–2343. 2005.

[Krieger 1972] Krieger, W. On unique ergodicity. In *Proceedings of the Sixth Berkeley Symposium on Mathematical Statistics and Probability (Univ. California, Berkeley, Calif., 1970/1971), Vol. II: Probability theory*, pp. 327–346. University of California Press, Berkeley, Calif., 1972.

[Krieger, Miller, Rampersad, et al. 2009] Krieger, D., Miller, A., Rampersad, N., Ravikumar, B., and Shallit, J. Decimations of languages and state complexity. *Theoret. Comput. Sci.* **410**, (2009) 2401–2409.

[Kubilius 1949] Kubilius, J. On the application of I. M. Vinogradov's method to the solution of a problem of the metric theory of numbers. *Doklady Akad. Nauk SSSR (N. S.)* **67**, (1949) 783–786.

[Kubilius 1964] Kubilius, J. *Probabilistic methods in the theory of numbers*. Translations of Mathematical Monographs, Vol. 11. American Mathematical Society, 1964.

[Kuipers and Niederreiter 1974] Kuipers, L. and Niederreiter, H. *Uniform Distribution of Sequences*. Wiley, 1974.

[Kůrka 2003] Kůrka, P. *Topological and Symbolic dynamics*, vol. 11 of *Cours spécialisés [Specialised Courses]*. Société Mathématique de France, 2003.

[Lagarias and Pleasants 2002] Lagarias, J. C. and Pleasants, P. A. B. Local complexity of Delone sets and crystallinity. *Canad. Math. Bull.* **45**(4), (2002) 634–652.

[Lagarias and Pleasants 2003] Lagarias, J. C. and Pleasants, P. A. B. Repetitive Delone sets and quasicrystals. *Ergod. Th. & Dynam. Sys.* **23**(3), (2003) 831–867.

[Lagarias and Wang 1995] Lagarias, J. C. and Wang, Y. The finiteness conjecture for the generalized spectral radius of a set of matrices. *Linear Algebra Appl.* **214**, (1995) 17–42.

[Lang 1983] Lang, S. *Real analysis*. Addison-Wesley, second edn., 1983.

[Lang 1993] Lang, S. *Real and functional analysis. Third Edition*, vol. 142 of *Graduate Texts in Mathematics*. Springer-Verlag, 1993.

[Lang 1995] Lang, S. *Introduction to Diophantine Approximations*. Springer-Verlag, 1995.

[Larcher 1996] Larcher, G. On the number of odd binomial coefficients. *Acta Math. Hung.* **71**, (1996) 183–203.

[Lecomte and Rigo 2001] Lecomte, P. B. A. and Rigo, M. Numeration systems on a regular language. *Theory Comput. Systems* **34**, (2001) 27–44.

[Lecomte and Rigo 2002] Lecomte, P. and Rigo, M. On the representation of real numbers using regular languages. *Theory Comput. Systems* **35**, (2002) 13–38.

[Lecomte and Rigo 2004] Lecomte, P. and Rigo, M. Real numbers having ultimately periodic representations in abstract numeration systems. *Inform. Comput.* **192**, (2004) 57–83.

[Lee 2007] Lee, J.-Y. Substitution Delone sets with pure point spectrum are intermodel sets. *J. Geom. Phys.* **57**(11), (2007) 2263–2285.

[Lee, Moody, and Solomyak 2003] Lee, J.-Y., Moody, R. V., and Solomyak, B. Consequences of pure point diffraction spectra for multiset substitution systems. *Discrete Comput. Geom.* **29**, (2003) 525–560.

[Lepistö 1995] Lepistö, A. A characterization of 2+-free words over a binary alphabet. Master's thesis, University of Turku, Finland, 1995.

[Leroux 2005] Leroux, J. A polynomial time Presburger criterion and synthesis for number decision diagrams. In *IEEE Symposium on Logic in Computer Science (LICS 2005)*, IEEE Computer Society, pp. 147–156. 2005.

[Lew, Morales, and Sánchez-Flores 1996] Lew, J. S., Morales, L. B., and Sánchez-Flores, A. Diagonal polynomials for small dimensions. *Math. Systems Theory* **29**, (1996) 305–310.

[Liardet 1978] Liardet, P. Répartition et ergodicité. In *Séminaire Delange-Pisot-Poitou, 19e année: 1977/78, Théorie des nombres, Fasc. 1*, pp. Exp. No. 10, 12. Secrétariat Mathématique, Paris, 1978.

[Liberzon, Hespanha, and Morse 1999] Liberzon, D., Hespanha, J. P., and Morse, A. S. Stability of switched systems: a Lie-algebraic condition. *Syst. Control Lett.* **37**(3), (1999) 117–122.

[Lind 1984] Lind, D. The entropies of topological Markov shifts and a related class of algebraic integers. *Ergod. Th. & Dynam. Sys.* **4**, (1984) 283–300.

[Lind and Marcus 1995] Lind, D. and Marcus, B. *An Introduction to Symbolic Dynamics and Coding*. Cambridge University Press, 1995.

[Lindenstrauss and Schmidt 2005] Lindenstrauss, E. and Schmidt, K. Symbolic representations of nonexpansive group automorphisms. *Israel J. Math.* **149**, (2005) 227–266.

[Linna 1977] Linna, M. The decidability of the D0L prefix problem. *Internat. J. Comput. Math.* **6**, (1977) 127–142.

[Liouville 1844] Liouville, J. Sur des classes très étendues de quantités dont la valeur n'est ni algébrique, ni même reductible à des irrationelles algébriques. *C. R. Acad. Sci. Paris* **18**, (1844) 883–885, 910–911.

[Littlewood 1968] Littlewood, J. E. *Some Problems in Real and Complex Analysis*. Heath, 1968. See problem 19.

[Livshits 1987] Livshits, A. N. On the spectra of adic transformations of Markov compact sets. *Uspekhi. Mat. Nauk* **42**(3(255)), (1987) 189–190. In Russian. English translation in *Russian Math. Surveys* **42** (1987), 222–223.

[Loraud 1995] Loraud, N. β-shift, systèmes de numération et automates. *J. Théorie Nombres Bordeaux* **7**, (1995) 473–498.

[Lothaire 1983] Lothaire, M. *Combinatorics on Words*, vol. 17 of Encyclopedia of Mathematics and Its Applications. Addison-Wesley, 1983.

[Lothaire 2002] Lothaire, M. *Algebraic Combinatorics on Words*, vol. 90 of Encyclopedia of Mathematics and Its Applications. Cambridge University Press, 2002.

[Lothaire 2005] Lothaire, M. *Applied Combinatorics on Words*, vol. 105 of Encyclopedia of Mathematics and Its Applications. Cambridge University Press, 2005.

[Loxton and van der Poorten 1988] Loxton, J. H. and van der Poorten, A. J. Arithmetic properties of automata: regular sequences. *J. Reine Angew. Math.* **392**, (1988) 57–69.

[Maes 1998] Maes, A. Morphisms and almost-periodicity. *Discrete Appl. Math.* **86**, (1998) 233–248.

[Maes 1999] Maes, A. An automata-theoretic decidability proof for first-order theory of $\langle \mathbb{N}, <, P \rangle$ with morphic predicate P. *J. Automata, Languages, and Combinatorics* **4**, (1999) 229–245.

[Maes 2000] Maes, A. More on morphisms and almost-periodicity. *Theoret. Comput. Sci.* **231**, (2000) 205–215.

[Mahler 1927] Mahler, K. On the translation properties of a simple class of arithmetical functions. *J. Math. and Phys.* **6**, (1927) 158–163.

[Mahler 1929] Mahler, K. Arithmetische Eigenschaften der Lösungen einer Klasse von Funktionalgleichungen. *Math. Annalen* **101**, (1929) 342–366. Corrigendum, **103** (1930), 532.

[Mahler 1968] Mahler, K. An unsolved problem on the powers of 3/2. *J. Austral. Math. Soc.* **8**, (1968) 313–321.

[Mahler 1984] Mahler, K. Some suggestions for further research. *Bull. Austral. Math. Soc.* **29**, (1984) 101–108.

[Manstavičius 1997] Manstavičius, E. Probabilistic theory of additive functions related to systems of numeration. In *New trends in probability and statistics, Vol. 4 (Palanga, 1996)*, pp. 413–429. VSP, Utrecht, 1997.

[Marcus and Monteil 2006] Marcus, S. and Monteil, T. Quasiperiodic infinite words : multi-scale case and dynamical properties, 2006. Preprint.

[Marmi, Moussa, and Yoccoz 2005] Marmi, S., Moussa, P., and Yoccoz, J.-C. The cohomological equation for Roth-type interval exchange maps. *J. Amer. Math. Soc.* **18**(4), (2005) 823–872 (electronic).

[Martensen 2004] Martensen, B. F. Generalized balanced pair algorithm. *Topology Proc.* **28**(1), (2004) 163–178. Spring Topology and Dynamical Systems Conference.

[Masur 1982] Masur, H. Interval exchange transformations and measured foliations. *Ann. of Math. (2)* **115**(1), (1982) 169–200.

[Masur 1992] Masur, H. Hausdorff dimension of the set of nonergodic foliations of a quadratic differential. *Duke Math. J.* **66**(3), (1992) 387–442.

[Mauduit 1986] Mauduit, C. Morphismes unispectraux. *Theoret. Comput. Sci.* **46**, (1986) 1–11.

[Mauduit 1988] Mauduit, C. Sur l'ensemble normal des substitutions de longueur quelconque. *J. Number Theory* **29**, (1988) 235–250.

[Mauduit 1992] Mauduit, C. Propriétés arithmétiques des substitutions. In S. David, ed., *Séminaire de Théorie des Nombres, Paris, 1989–90*, pp. 177–190. Birkhäuser, 1992.

[Mauduit and Rivat 2009a] Mauduit, C. and Rivat, J. La somme des chiffres des carrés. *Acta Math.* **203**, (2009) 107–148.

[Mauduit and Rivat 2009b] Mauduit, C. and Rivat, J. Sur un problème de Gelfond: la somme des chiffres des nombres premiers. *Ann. of Math.*, to appear.

[Mauduit and Sárközy 1997] Mauduit, C. and Sárközy, A. On the arithmetic structure of the integers whose sum of digits is fixed. *Acta Arith.* **81**, (1997) 145–173.

[Mauldin and Williams 1988] Mauldin, R. D. and Williams, S. C. Hausdorff dimension in graph directed constructions. *Trans. Amer. Math. Soc.* **309**(2), (1988) 811–829.

[McCulloch and Pitts 1943] McCulloch, W. S. and Pitts, W. A logical calculus of the ideas immanent in nervous activity. *Bull. Math. Biophysics* **5**, (1943) 115–133.

[Medynets 2006] Medynets, K. Cantor aperiodic systems and Bratteli diagrams. *C. R. Acad. Sci. Paris* **342**, (2006) 43–46.

[Méla and Petersen 2005] Méla, X. and Petersen, K. Dynamical properties of the Pascal adic transformation. *Ergodic Th. & Dynam. Sys.* **25**(1), (2005) 227–256.

[Mendès France 1967] Mendès France, M. Nombres normaux, applications aux fonctions pseudoaléatoires. *J. Analyse Math.* **20**, (1967) 1–56.

[Mendès France 1970] Mendès France, M. Fonctions g-additives et les suites à spectre vide. In *Séminaire Delange-Pisot-Poitou*, pp. 10.01–10.06., Secrétariat Mathématique, 1970/71.

[Mendès France 1973a] Mendès France, M. Les suites additives et leur répartition (mod. 1). In *Séminaire de Théorie des Nombres de Bordeaux*, pp. 8.01–8.06. Lab. Théorie des Nombres, Centre Nat. Recherche Sci., Talence, 1973/74.

[Mendès France 1973b] Mendès France, M. Les suites à spectre vide et la répartition modulo 1. *J. Number Theory* **5**, (1973) 1–15.

[Messaoudi 1998] Messaoudi, A. Propriétés arithmétiques et dynamiques du fractal de Rauzy. *J. Théorie Nombres Bordeaux* **10**, (1998) 135–162.

[Messaoudi 2000] Messaoudi, A. Frontière du fractal de Rauzy et systèmes de numération complexes. *Acta Arith.* **95**, (2000) 195–224.

[Michel 1978] Michel, P. Coincidence values and spectra of substitutions. *Z. Wahrscheinlichkeitstheorie und verw. Gebiete* **42**(3), (1978) 205–227.

[Moision, Orlitsky, and Siegel 1999] Moision, B. E., Orlitsky, A., and Siegel, P. H. Bounds on the rate of codes which forbid specified difference sequences. In *Proceedings of the IEEE Global Telecommunication Conference (GLOBECOM'99)*. 1999.

[Moision, Orlitsky, and Siegel 2001] Moision, B. E., Orlitsky, A., and Siegel, P. H. On codes that avoid specified differences. *IEEE Trans. Inform. Theory* **47**, (2001) 433–442.

[Moision, Orlitsky, and Siegel 2007] Moision, B. E., Orlitsky, A., and Siegel, P. H. On codes with local joint constraints. *Linear Algebra Appl.* **422**(2-3), (2007) 442–454.

[Monteil 2005] Monteil, T. Illumination dans les billards polygonaux et dynamique symbolique. Ph.D. thesis, Institut de Mathématiques de Luminy, Université de la Méditerranée, 2005. Chapter 5.

[Monteil 2009] Monteil, T. An upper bound for the number of ergodic invariant measures of a minimal subshift with linear complexity, 2009. In preparation.

[Moody 1997] Moody, R. V. Meyer sets and their duals. In R. V. Moody, ed., *The Mathematics of Long-Range Aperiodic Order*, vol. 489 of *NATO ASI Ser., Ser. C., Math. Phys. Sci.*, pp. 403–441. Kluwer, 1997.

[Morse 1921] Morse, M. Recurrent geodesics on a surface of negative curvature. *Trans. Amer. Math. Soc.* **22**, (1921) 84–100.

[Morse and Hedlund 1938] Morse, M. and Hedlund, G. A. Symbolic dynamics. *Amer. J. Math.* **60**, (1938) 815–866.

[Morse and Hedlund 1940] Morse, M. and Hedlund, G. A. Symbolic Dynamics II. Sturmian trajectories. *Amer. J. Math.* **62**, (1940) 1–42.

[Mossé 1992] Mossé, B. Puissances de mots et reconnaissabilité des points fixes d'une substitution. *Theoret. Comput. Sci.* **99**, (1992) 327–334.

[Nicolay and Rigo 2007] Nicolay, S. and Rigo, M. About frequencies of letters in generalized automatic sequences. *Theoret. Comput. Sci.* **374**, (2007) 25–40.

[Niederreiter 1992] Niederreiter, H. *Random Number Generation and Quasi-Monte Carlo Methods*, vol. 63 of CBMS-NSF Regional Conference Series in Applied Mathematics. Society for Industrial and Applied Mathematics (SIAM), 1992.

[Nivat 1978] Nivat, M. Sur les ensembles de mots infinis engendrés par une grammaire algébrique. *RAIRO Inform. Théor. App.* **12**, (1978) 259–278.

[Nogueira and Rudolph 1997] Nogueira, A. and Rudolph, D. Topologically weak-mixing of interval exchange maps. *Ergod. Th. & Dynam. Sys.* **17**, (1997) 1183–1209.

[Okada, Sekiguchi, and Shiota 1995] Okada, T., Sekiguchi, T., and Shiota, Y. Applications of binomial measures to power sums of digital sums. *J. Number Theory* **52**, (1995) 256–266.

[Oseledec 1966] Oseledec, V. I. The spectrum of ergodic automorphisms. *Dokl. Akad. Nauk SSSR* **168**, (1966) 1009–1011.

[Oseledec 1968] Oseledec, V. I. A multiplicative ergodic theorem. Lyapunov characteristic numbers for dynamical systems. *Trans. Moscow Math. Soc.* **19**, (1968) 197–231.

[Oxtoby 1952] Oxtoby, J. C. Ergodic sets. *Bull. Amer. Math. Soc.* **58**, (1952) 116–136.
[Pansiot 1983] Pansiot, J.-J. Hiérarchie et fermeture de certaines classes de tag-systèmes. *Acta Inform.* **20**, (1983) 179–196.
[Pansiot 1984] Pansiot, J.-J. Complexité des facteurs des mots infinis engendrés par morphismes itérés. In J. Paredaens, ed., *Proc. 11th Int'l Conf. on Automata, Languages, and Programming (ICALP)*, vol. 172 of Lecture Notes in Computer Science, pp. 380–389. Springer-Verlag, 1984.
[Pansiot 1985] Pansiot, J.-J. Subword complexities and iteration. *Bull. European Assoc. Theor. Comput. Sci.* **26**, (1985) 55–62.
[Pansiot 1986] Pansiot, J.-J. Decidability of periodicity for infinite words. *RAIRO Inform. Théor. App.* **20**, (1986) 43–46.
[Papadimitriou 1994] Papadimitriou, C. M. *Computational Complexity*. Addison-Wesley, 1994.
[Parrilo 2000] Parrilo, P. A. Structured semidefinite programs and semialgebraic geometry methods in robustness and optimization. Ph.D. thesis, California Institute of Technology, 2000.
[Parrilo and Jadbabaie 2008] Parrilo, P. and Jadbabaie, A. Approximation of the joint spectral radius using sum of squares. *Linear Algebra Appl.* **428**(10), (2008) 2385–2402.
[Parry 1960] Parry, W. On the β-expansions of real numbers. *Acta Math. Acad. Sci. Hung.* **11**, (1960) 401–416.
[Pascal 1654] Pascal, B. *Œuvres complètes*. Seuil, 1963. The treatise *De numeris multiplicibus*, written with the other arithmetical treatises before 1654, was published by Guillaume Desprez in 1665.
[Păun and Salomaa 1995] Păun, G. and Salomaa, A. Thin and slender languages. *Discrete Appl. Math.* **61**, (1995) 257–270.
[Penney 1964] Penney, W. A numeral system with a negative base. *Math. Student Journal* **11**(4), (1964) 1–2.
[Peres, Schlag, and Solomyak 2000] Peres, Y., Schlag, W., and Solomyak, B. Sixty years of Bernoulli convolutions. In *Fractal Geometry and Stochastics, II (Greifswald/Koserow, 1998)*, vol. 46 of *Progr. Probab.*, pp. 39–65. Birkhäuser, 2000.
[Perrin 1990] Perrin, D. Finite automata. In J. van Leeuwen, ed., *Handbook of Theoretical Computer Science, Volume B: Formal Models and Semantics*, pp. 1–57. Elsevier/MIT Press, 1990.
[Perrin 1995a] Perrin, D. Les débuts de la théorie des automates. *Technique et Science Informatique* **14**, (1995) 409–443.
[Perrin 1995b] Perrin, D. Symbolic dynamics and finite automata. In J. Wiedermann and P. Hájek, eds., *Proc. 20th Symposium, Mathematical Foundations of Computer Science 1995*, vol. 969 of Lecture Notes in Computer Science, pp. 94–104. Springer-Verlag, 1995.
[Perrin and Pin 2003] Perrin, D. and Pin, J.-E. *Infinite Words: Automata, Semigroups, Logic and Games*, vol. 141 of Pure and Applied Mathematics. Academic Press, 2003.
[Perron 1929] Perron, O. *Die Lehre von den Ketterbruechen*. Teubner, 1929.
[Pollington 1981] Pollington, A. Progressions arithmétiques généralisées et le problème des $(3/2)^n$. *C.R. Acad. Sc. Paris série I* **292**, (1981) 383–384.
[Pollington and Velani 2000] Pollington, A. D. and Velani, S. L. On a problem in simultaneous Diophantine approximation: Littlewood's conjecture. *Acta Math.* **185**, (2000) 287–306.
[Praggastis 1999] Praggastis, B. Numeration systems and Markov partitions from self-similar tilings. *Trans. Amer. Math. Soc.* **351**, (1999) 3315–3349.

[Protasov 1996] Protasov, V. Y. The joint spectral radius and invariant sets of linear operators. *Fundam. Prikl. Mat.* **2**(1), (1996) 205–231.

[Protasov 1997] Protasov, V. Y. The generalized spectral radius. A geometric approach. *Izv. Math.* **61**(5), (1997) 995–1030.

[Protasov 2000] Protasov, V. Y. Asymptotic behaviour of the partition function. *Sb. Math.* **191**(3-4), (2000) 381–414.

[Protasov, Jungers, and Blondel 2009] Protasov, V. Y., Jungers, R. M., and Blondel, V. D. Joint spectral characteristics of matrices: a conic programming approach. Submitted.

[Putnam 1989] Putnam, I. F. The C^*-algebras associated with minimal homeomorphisms of the Cantor set. *Pacific J. Math.* **136**, (1989) 329–353.

[Pytheas Fogg 2002] Pytheas Fogg, N. *Substitutions in Dynamics, Arithmetics and Combinatorics*, vol. 1794 of Lecture Notes in Mathematics. Springer-Verlag, 2002. Ed. by V. Berthé, S. Ferenczi, C. Mauduit and A. Siegel.

[Queffélec 1987] Queffélec, M. *Substitution Dynamical Systems – Spectral Analysis*, vol. 1294 of Lecture Notes in Mathematics. Springer-Verlag, 1987.

[Queffélec 1998] Queffélec, M. Transcendance des fractions continue de Thue-Morse. *J. Number Theory* **73**, (1998) 201–211.

[Queffélec 2000] Queffélec, M. Irrational numbers with automaton-generated continued fraction expansion. In J.-M. Gambaudo, P. Hubert, P. Tisseur, and S. Vaienti, eds., *Dynamical Systems: From Crystal to Chaos*, pp. 190–198. World Scientific, 2000.

[Rabin and Scott 1959] Rabin, M. O. and Scott, D. Finite automata and their decision problems. *IBM J. Res. Develop.* **3**, (1959) 115–125.

[Raney 1973] Raney, G. N. On continued fractions and finite automata. *Math. Annalen* **206**, (1973) 265–283.

[Rao and Wen 2009] Rao, H. and Wen, Z. Invertible substitutions with a common periodic point, 2009. Preprint.

[Ratner 1983] Ratner, M. Horocycle flows, joinings and rigidity of products. *Ann. of Math. (2)* **118**(2), (1983) 277–313.

[Ratner 1991] Ratner, M. On Raghunathan's measure conjecture. *Ann. of Math. (2)* **134**(3), (1991) 545–607.

[Rauzy 1982] Rauzy, G. Nombres algébriques et substitutions. *Bull. Soc. Math. France* **110**, (1982) 147–178.

[Rauzy 1983] Rauzy, G. Suites à termes dans un alphabet fini. In *Séminaire de Théorie des Nombres de Bordeaux*, pp. 25.01–25.16. 1982/83.

[Rauzy 1990] Rauzy, G. Sequences defined by iterated morphisms. In R. M. Capocelli, ed., *Sequences: Combinatorics, Compression, Security, and Transmission*, pp. 275–287. Springer-Verlag, 1990.

[Reingold and Tarjan 1981] Reingold, E. M. and Tarjan, R. E. On a greedy heuristic for complete matching. *SIAM J. Comput.* **10**(4), (1981) 676–681.

[Reitwiesner 1960] Reitwiesner, G. W. Binary arithmetic. In F. L. Alt, ed., *Advances in Computers*, vol. 1, pp. 231–308. Academic Press, 1960.

[Rényi 1957] Rényi, A. Representations for real numbers and their ergodic properties. *Acta Math. Acad. Sci. Hung.* **8**, (1957) 477–493.

[Restivo and Salemi 1985] Restivo, A. and Salemi, S. Overlap free words on two symbols. In M. Nivat and D. Perrin, eds., *Automata on Infinite Words*, vol. 192 of Lecture Notes in Computer Science, pp. 198–206. Springer-Verlag, 1985.

[Reveillès 1991] Reveillès, J.-P. Géométrie discrète, calcul en nombres entiers et algorithmique. Thèse de Doctorat, Université Louis Pasteur, Strasbourg, 1991.

[Reznick 1990] Reznick, B. Some binary partition functions. In B. Berndt, H. Diamond, H. Halberstam, and A. Hildebrand, eds., *Analytic Number Theory, Proceedings of a Conference in Honor of Paul T. Bateman*, pp. 451–477. Birkhäuser, Boston, 1990.

[Ridout 1957] Ridout, D. Rational approximations to algebraic numbers. *Mathematika* **4**, (1957) 125–131.

[Rigo 2000] Rigo, M. Generalization of automatic sequences for numeration systems on a regular language. *Theoret. Comput. Sci.* **244**, (2000) 271–281.

[Rigo 2002] Rigo, M. Construction of regular languages and recognizability of polynomials. *Discrete Math.* **254**, (2002) 485–496.

[Rigo and Maes 2002] Rigo, M. and Maes, A. More on generalized automatic sequences. *J. Autom. Lang. Comb.* **7**, (2002) 351–376.

[Rigo and Steiner 2005] Rigo, M. and Steiner, W. Abstract β-expansions and ultimately periodic representations. *J. Théorie Nombres Bordeaux* **17**, (2005) 283–299.

[Rigo and Waxweiler 2006] Rigo, M. and Waxweiler, L. A note on syndeticity, recognizable sets and Cobham's theorem. *Bull. European Assoc. Theor. Comput. Sci.*, No. 88, (2006), 169–173.

[Robinson 2004] Robinson, E. A., Jr. Symbolic dynamics and tilings of $\mathbb{R}^d$. In *Symbolic dynamics and its applications*, vol. 60 of Proc. Sympos. Appl. Math., American Mathematical Society, pp. 81–119. 2004.

[Rockett and Szüsz 1992] Rockett, A. M. and Szüsz, P. *Continued Fractions*. World Scientific, 1992.

[Rosenberg 1967] Rosenberg, A. A machine realization of the linear context-free languages. *Inform. Control* **10**, (1967) 175–188.

[Rosenthal 1988] Rosenthal, A. Strictly ergodic models for noninvertible transformations. *Israel J. Math.* **64**(1), (1988) 57–72.

[Rota and Strang 1960] Rota, G. C. and Strang, G. A note on the joint spectral radius. *Proc. Neth. Acad.* **22**, (1960) 379–381.

[Rote 1994] Rote, G. Sequences with subword complexity $2n$. *J. Number Theory* **46**, (1994) 196–213.

[Roth 1955] Roth, K. F. Rational approximations to algebraic numbers. *Mathematika* **2**, (1955) 1–20. Corrigendum, p. 168.

[Roy 2003a] Roy, D. Approximation simultanée d'un nombre et de son carré. *C. R. Acad. Sci. Paris* **336**, (2003) 1–6.

[Roy 2003b] Roy, D. Approximation to real numbers by cubic algebraic integers. II. *Ann. of Math. (2)* **158**, (2003) 1081–1087.

[Roy 2004] Roy, D. Approximation to real numbers by cubic algebraic integers. I. *Proc. Lond. Math. Soc.* **88**, (2004) 42–62.

[Rozenberg and Salomaa 1980] Rozenberg, G. and Salomaa, A. *The Mathematical Theory of L Systems*, vol. 90 of *Pure and Applied Mathematics*. Academic Press, 1980.

[Rudin 1987] Rudin, W. *Real and Complex Analysis*. McGraw-Hill Book Co., third edn., 1987.

[Rudin 1991] Rudin, W. *Functional Analysis*. International Series in Pure and Applied Mathematics. McGraw-Hill, second edn., 1991.

[Rudolph 1979] Rudolph, D. J. An example of a measure preserving map with minimal self-joinings, and applications. *J. Analyse Math.* **35**, (1979) 97–122.

[Ruohonen 1986a] Ruohonen, K. Equivalence problems for regular sets of word morphisms. In G. Rozenberg and A. Salomaa, eds., *The Book of L*, pp. 393–401. Springer, 1986.

[Ruohonen 1986b] Ruohonen, K. Test sets for iterated morphisms. Tech. rep., Tampere University of Technology, Tampere, 1986.

[Ruzsa 1984] Ruzsa, I. Z. Generalized moments of additive functions. *J. Number Theory* **18**(1), (1984) 27–33.

[Sadahiro 2006] Sadahiro, T. Multiple points of tilings associated with Pisot numeration systems. *Theoret. Comput. Sci.* **359**(1-3), (2006) 133–147.

[Sadun 2008] Sadun, L. *Topology of tiling spaces*, vol. 46 of University Lecture Series. American Mathematical Society, 2008.

[Safer 1998] Safer, T. Radix representations of algebraic number fields and finite automata. In M. Morvan, C. Meinel, and D. Krob, eds., *STACS 98, Proc. 15th Symp. Theoretical Aspects of Comp. Sci.*, vol. 1373 of Lecture Notes in Computer Science, pp. 356–365. Springer-Verlag, 1998.

[Sakarovitch 2003] Sakarovitch, J. *Éléments de théorie des automates*. Vuibert, 2003. English corrected edition: *Elements of Automata Theory*, Cambridge University Press, 2009.

[Salomaa and Soittola 1978] Salomaa, A. and Soittola, M. *Automata-theoretic aspects of formal power series*. Texts and Monographs in Computer Science. Springer-Verlag, 1978.

[Salon 1987] Salon, O. Suites automatiques à multi-indices et algébricité. *C. R. Acad. Sci. Paris* **305**, (1987) 501–504.

[Sano, Arnoux, and Ito 2001] Sano, Y., Arnoux, P., and Ito, S. Higher dimensional extensions of substitutions and their dual maps. *J. Anal. Math.* **83**, (2001) 183–206.

[Sataev 1975] Sataev, E. A. The number of invariant measures for flows on orientable surfaces. *Izv. Akad. Nauk SSSR Ser. Mat.* **39**(4), (1975) 860–878. English translation in *Math. USSR-Izv.* **9**(4) (1975), 813–830.

[Scheicher 2007] Scheicher, K. Beta-expansions in algebraic function fields over finite fields. *Finite Fields Appl.* **13**, (2007) 394–410.

[Scheicher and Thuswaldner 2002] Scheicher, K. and Thuswaldner, J. M. Canonical number systems, counting automata and fractals. *Math. Proc. Camb. Philos. Soc.* **133**, (2002) 163–182.

[Scheicher and Thuswaldner 2003] Scheicher, K. and Thuswaldner, J. M. Digit systems in polynomial rings over finite fields. *Finite Fields Appl.* **9**, (2003) 322–333.

[Scheicher and Thuswaldner 2004] Scheicher, K. and Thuswaldner, J. M. On the characterization of canonical number systems. *Osaka J. Math.* **41**, (2004) 1–25.

[Schlickewei 1976] Schlickewei, H. P. On products of special linear forms with algebraic coefficients. *Acta Arith.* **31**, (1976) 389–398.

[Schmidt 1980a] Schmidt, K. On periodic expansions of Pisot numbers and Salem numbers. *Bull. London Math. Soc.* **12**(4), (1980) 269–278.

[Schmidt 1980b] Schmidt, W. M. *Diophantine Approximation*, vol. 785 of Lecture Notes in Mathematics. Springer-Verlag, 1980.

[Schmidt 2000] Schmidt, K. Algebraic coding of expansive group automorphisms and two-sided beta-shifts. *Monatsh. Math.* **129**(1), (2000) 37–61.

[Seneta 1981] Seneta, E. *Nonnegative Matrices and Markov Chains*. Springer-Verlag, 1981. Second ed.

[Shallit 1994] Shallit, J. O. Numeration systems, linear recurrences, and regular sets. *Inform. Comput.* **113**, (1994) 331–347.

[Shallit 2008] Shallit, J. *A Second Course in Formal Languages and Automata Theory*. Cambridge University Press, 2008.

[Sidorov 2003] Sidorov, N. Arithmetic dynamics. In S. Bezuglyi and S. Kolyada, ed., *Topics in dynamics and ergodic theory*, vol. 310 of London Math. Soc. Lect. Note Ser., pp. 145–189. Cambridge University Press, 2003.

[Siegel 2004] Siegel, A. Pure discrete spectrum dynamical system and periodic tiling associated with a substitution. *AIFG* **54**(2), (2004) 288–299.
[Siegel and Thuswaldner 2010] Siegel, A. and Thuswaldner, J. M. Topological properties of Rauzy fractals. *Mémoire de la Société Mathématique de France*, to appear.
[Silva 2008] Silva, C. E. *Invitation to ergodic theory*, vol. 42 of Student Mathematical Library. American Mathematical Society, 2008.
[Sing 2006] Sing, B. Pisot substitutions and beyond. Ph.D. thesis, Universität Bielefeld, 2006.
[Singmaster 1974a] Singmaster, D. Notes on binomial coefficients I – a generalization of Lucas' congruence. *J. London Math. Soc.* **8**, (1974) 545–548.
[Singmaster 1974b] Singmaster, D. Notes on binomial coefficients II – the least n such that p^e divides an r-nomial coefficient of rank n. *J. London Math. Soc.* **8**, (1974) 549–554.
[Singmaster 1974c] Singmaster, D. Notes on binomial coefficients III – any integer divides almost all binomial coefficients. *J. London Math. Soc.* **8**, (1974) 555–560.
[Sirvent and Solomyak 2002] Sirvent, V. F. and Solomyak, B. Pure discrete spectrum for one-dimensional substitution systems of Pisot type. *Canad. Math. Bull.* **45**(4), (2002) 697–710. Dedicated to Robert V. Moody.
[Sirvent and Wang 2002] Sirvent, V. F. and Wang, Y. Self-affine tiling via substitution dynamical systems and Rauzy fractals. *Pacific J. Math.* **206**(2), (2002) 465–485.
[Sloane] Sloane, N. J. A. On-Line Encyclopedia of Integer Sequences. URL www.research.att.com/~njas/sequences.
[Solomyak 1994] Solomyak, B. Conjugates of beta-numbers and the zero-free domain for a class of analytic functions. *Proc. Lond. Math. Soc.* **68**, (1994) 477–498.
[Solomyak 1997] Solomyak, B. Dynamics of self-similar tilings. *Ergod. Th. & Dynam. Sys.* **17**, (1997) 695–738.
[Stein 1989] Stein, A. H. Binomial coefficients not divisible by a prime. In D. V. Chudnovsky, G. V. Chudnovsky, H. Cohn, and M. B. Nathanson, eds., *Number Theory (New York, 1985/1988)*, vol. 1383 of Lecture Notes in Mathematics, pp. 170–177. Springer-Verlag, 1989.
[Steiner 2002] Steiner, W. Parry expansions of polynomial sequences. *Integers* **2** (2002), A14 (electronic), www.integers-ejcnt.org/vol2.html
[Strang 2001] Strang, G. The joint spectral radius, commentary by Gilbert Strang on paper number 5. In *Collected Works of Gian-Carlo Rota*. 2001.
[Sudkamp 2005] Sudkamp, T. A. *Languages and Machines: An Introduction to the Theory of Computer Science*. Pearson, 2005.
[Sugisaki 2003] Sugisaki, F. The relationship between entropy and strong orbit equivalence for the minimal homeomorphisms. I. *Internat. J. Math.* **14**, (2003) 735–772.
[Sugisaki 2007] Sugisaki, F. On the subshift within a strong orbit equivalence class for minimal homeomorphisms. *Ergod. Th. & Dynam. Sys.* **27**, (2007) 971–990.
[Szilard, Yu, Zhang, et al. 1994] Szilard, A., Yu, S., Zhang, K., and Shallit, J. Characterizing regular languages with polynomial densities. In V. K. I. M. Havel, ed., *Proc. 19th Symposium, Mathematical Foundations of Computer Science 1994*, vol. 629 of Lecture Notes in Computer Science, pp. 494–503. Springer-Verlag, 1994.
[Takahashi 1980] Takahashi, Y. A Formula for Topological Entropy of One-

dimensional Dynamics. *Sci. Papers College Gen. Ed. Univ. Tokyo* **30**, (1980) 11–22.
[Tapsoba 1994] Tapsoba, T. Automates calculant la complexité de suites automatiques. *J. Théorie Nombres Bordeaux* **6**, (1994) 127–134.
[Tenenbaum 1995] Tenenbaum, G. *Introduction to Analytic and Probabilistic Number Theory*. Cambridge University Press, 1995.
[Tenenbaum 1997] Tenenbaum, G. Sur la non-dérivabilité de fonctions périodiques associées à certaines formules sommatoires. In R. L. Graham and J. Nešetřil, eds., *The Mathematics of Paul Erdős*, pp. 117–128. Springer-Verlag, 1997.
[Theys 2005] Theys, J. Joint spectral radius : Theory and approximations. Ph.D. thesis, Université catholique de Louvain, 2005.
[Thomas 1990] Thomas, W. Finite automata. In J. van Leeuwen, ed., *Handbook of Theoretical Computer Science, Volume B: Formal Models and Semantics*, pp. 133–191. Elsevier/MIT Press, 1990.
[Thue 1906] Thue, A. Über unendliche Zeichenreihen. *Norske vid. Selsk. Skr. Mat. Nat. Kl.* **7**, (1906) 1–22. Reprinted in *Selected Mathematical Papers of Axel Thue*, T. Nagell, editor, Universitetsforlaget, Oslo, 1977, pp. 139–158.
[Thue 1912] Thue, A. Über die gegenseitige Lage gleicher Teile gewisser Zeichenreihen. *Norske vid. Selsk. Skr. Mat. Nat. Kl.* **1**, (1912) 1–67. Reprinted in *Selected Mathematical Papers of Axel Thue*, T. Nagell, editor, Universitetsforlaget, Oslo, 1977, pp. 413–478.
[Thurston 1989] Thurston, W. Groups, Tilings and Finite State Automata, 1989. AMS Colloquium Lecture Notes.
[Thuswaldner 1998] Thuswaldner, J. M. Elementary properties of canonical number systems in quadratic fields. In G. E. Bergsun, A. N. Philippou and A. F. Horadau, eds., *Applications of Fibonacci Numbers*, vol. 7, pp. 405–414. Kluwer Academic Publisher, 1998.
[Thuswaldner 2006] Thuswaldner, J. M. Unimodular Pisot substitutions and their associated tiles. *J. Théorie Nombres Bordeaux* **18**(2), (2006) 487–536.
[Thuswaldner and Tichy 2005] Thuswaldner, J. M. and Tichy, R. F. Waring's problem with digital restrictions. *Israel J. Math.* **149**, (2005) 317–344. Probability in mathematics.
[Titchmarsh 1986] Titchmarsh, E. C. *The theory of the Riemann zeta-function*. The Clarendon Press Oxford University Press, New York, second edn., 1986. Edited and with a preface by D. R. Heath-Brown.
[Tsitsiklis and Blondel 1997] Tsitsiklis, J. N. and Blondel, V. D. The Lyapunov exponent and joint spectral radius of pairs of matrices are hard – when not impossible – to compute and to approximate. *Math. Control Signal* **10**, (1997) 31–40.
[Varga 2000] Varga, R. S. *Matrix Iterative Analysis*, vol. 27 of Springer Series in Computational Mathematics. Springer-Verlag, second edn., 2000.
[Veech 1969] Veech, W. A. Strict ergodicity in zero dimensional dynamical systems and the Kronecker-Weyl theorem mod 2. *Trans. Amer. Math. Soc.* **140**, (1969) 1–33.
[Veech 1982] Veech, W. A. Gauss measures for transformations on the space of interval exchange maps. *Ann. of Math. (2)* **115**(1), (1982) 201–242.
[Veech 1984] Veech, W. A. The metric theory of interval exchange transformations. I. Generic spectral properties. *Amer. J. Math.* **106**(6), (1984) 1331–1359.
[Veech 1999] Veech, W. A. Measures supported on the set of uniquely ergodic directions of an arbitrary holomorphic 1-form. *Ergod. Th. & Dynam. Sys.* **19**(4), (1999) 1093–1109.
[Vershik 1985] Vershik, A. M. A theorem on the Markov periodical approximation

in ergodic theory. *J. Sov. Math.* **28**, (1985) 667–674.
[Vershik and Livshits 1992] Vershik, A. M. and Livshits, A. N. Adic models of ergodic transformations, spectral theory, substitutions, and related topics. In A. M. Vershik, ed., *Representation Theory and Dynamical Systems*, vol. 9 of *Advances in Soviet Mathematics*, pp. 185–204. American Mathematical Society, 1992.
[Waldschmidt 2006] Waldschmidt, M. Diophantine analysis and words. In *Diophantine Analysis and Related Fields 2006*, vol. 35 of Sem. Math. Sci., pp. 203–221. Keio University, 2006.
[Waldschmidt 2008] Waldschmidt, M. Words and transcendence. In *Analytic Number Theory - Essays in Honour of Klaus Roth*, pp. 449–470. Cambridge University Press, 2008.
[Walters 1982] Walters, P. *An Introduction to Ergodic Theory.* Springer-Verlag, 1982.
[Weiss 1973] Weiss, B. Subshifts of finite type and sofic systems. *Monatsh. Math.* **77**, (1973) 462–474.
[Wiener 1927] Wiener, N. The spectrum of an array and its applications to the study of the translation properties of a simple class of arithmetical functions. *J. Math. and Phys.* **6**, (1927) 145–157.
[Wirsing 1960] Wirsing, E. Approximation mit algebraischen Zahlen beschraenkten Grades. *J. Reine Angew. Math.* **206**, (1960) 67–77.
[Wirth 2000] Wirth, F. The generalized spectral radius and extremal norms. *Linear Algebra Appl.* **342**, (2000) 17–40.
[Wirth 2005] Wirth, F. The generalized spectral radius is strictly increasing. *Linear Algebra Appl.* **395**, (2005) 141–153.
[Wolfram 1984] Wolfram, S. Geometry of binomial coefficients. *Amer. Math. Monthly* **91**, (1984) 566–571.
[Yoccoz 2005] Yoccoz, J.-C. Échanges d'intervalles, 2005. Cours du Collège de France.
[Yoccoz 2006] Yoccoz, J.-C. Continued fraction algorithms for interval exchange maps: an introduction. In *Frontiers in number theory, physics, and geometry. I*, pp. 401–435. Springer, 2006.
[Yu 1997] Yu, S. Regular languages. In G. Rozenberg and A. Salomaa, eds., *Handbook of Formal Languages*, vol. 1, pp. 41–110. Springer-Verlag, 1997.

Notation index

(a, b) (greatest common divisor), 474

$\|\cdot\|$ (distance to the nearest integer), 365, 444

$\|\sigma\|$ (width), 11, 142, 507

$|\cdot|_\ell$ (ℓ-adic absolute value), 420

$f \ll g$, 2

$g \gg f$, 2

$\|\cdot\|_1$ (Manhattan norm), 2, 191

$\|\cdot\|_2$ (Euclidean norm), 2

$\|\cdot\|_\infty$ (maximum norm), 2

$\mathbb{1}_S$ (indicator function), 6, 455

$A^\mathbb{N}$ (set of infinite words), 47

$\mathcal{A}_{\sigma,a}$ (automaton associated with the morphism σ), 142

$\mathcal{A}_{\sigma,a,\tau}$ (automaton associated with the morphisms σ, τ), 142

Adh(L) (adherence of L), 152

A_p (canonical alphabet in base p), 37

alph(u) (alphabet of u), 6

alph(L) (alphabet of L), 14

$A^{\leq n}$ (words of length at most n), 4

A^n (words of length n), 4

A^+ (free semigroup), 4

A^* (free monoid), 4

A_U (canonical alphabet), 110

$b(n)$ (second difference of $p(n)$), 173

Bad (set of badly approximable real numbers), 444

BAL(σ, τ), 526

$B(\mathbf{x}, R)$ (open ball), 2

$\overline{k}$ (the signed digit $-k$), 41

B_d (symmetrical digit alphabet with largest digit d), 40

B_k (language of the numeration in base k), 109

$b_{q,i}(w)$ (coefficients in the decomposition of val$_\mathcal{S}(w)$), 120

$[w]_x$ (cylinder), 29

$[u]$ (cylinder), 375

$[u]_X$ (cylinder), 375

$BS_n(u)$ (bispecial factors), 171

$BS'_n(u)$ (bispecial factors and exceptional prefix), 172

$\chi_{[u]}$ (characteristic function), 375

$\mathcal{C}_p(C \times A)$ (the converter between C and A (in base p)), 42

$C(X, \mathbb{Z})$ (continuous maps), 352

CYCLIC(σ) (set of cyclic letters), 510

d (distance on words), 7

$d^-(w)$ (left valence), 171

$d^+(w)$ (right valence), 171

Notation index 595

δ_x (Dirac measure), 378, 474
∂X (boundary), 2
$DG(X,T)$ (dimension group), 352
$u \oplus v$ (digitwise addition), 42
$u \ominus v$ (digitwise subtraction), 42
D_p (set of all p-expansions of reals in $[0,1)$), 50

$e(t) = e^{2\pi i t}$, 476
$\mathbf{e}$ (row vector in which all coordinates equal 1), 512
$E^-(w)$ (left extensions), 171, 231
$E^+(w)$ (right extensions), 171
$E(w)$ (extension type), 172
$E_y(f_1, f_2)$ (sandwich set), 188
$e_n(\mu)$, 387
ε (empty word), 4
$f \sim g$, 3
$\mathcal{E}_u$ (self-similar tiling), 261
$\mathbb{E}X$ (mean value of random variable X), 483

$\mathbb{F}_q$ (finite field with q elements), 453
$(F_j)_{j \geq 0}$ (Fibonacci sequence), 417
$\lfloor x \rfloor$ (floor function), 2
$\{x\}$ (fractional part), 2
$f_w(x)$ (frequency), 376, 380

Γ_c (self-replicating translation set), 263
Γ_e (self-similar translation set), 261
$[\gamma, i]^*$ (tip), 263
$[\gamma, i]_g^*$ (projected face), 263
$G_{\mu,f}$, 379
G_n (Rauzy graph), 176
$G_n(X)$ (Rauzy graph), 384
$G_n(x)$ (Rauzy graph), 384
$\mathcal{G}_\mathcal{O}$ (graph of overlaps), 300

$\mathcal{G}_\mathcal{O}(\lambda)$ (graph of overlaps), 301
$\mathcal{G}_\sigma$ (prefix-suffix graph), 256

$\mathbb{H}_c$ (contracting space), 252
$\mathbb{H}_e$ (expanding line), 252
h_σ (contraction), 252

$\cap$ (intersection), 1
$[\![i,j]\!]$ (interval of integers), 2
$\mathcal{I}_\sigma$ (self-replicating multiple tiling), 271

$K_n^\sigma(x,y,z)$, 201
$\mathbb{K}_{\geq a}$, 1
$\mathbb{K}_{<a}$, 2
$\mathbb{K}_{>a}$, 1
$\mathbb{K}_{\leq a}$, 2
$K_m(a)$ (continuant), 429

$L_n(u)$ (factors of length n), 164
$L^\sigma(x,y,z)$ (centric factors), 197
$L(u)$ (factors), 164
Λ_r, 391
$\langle \sigma_1, \sigma_2 \rangle$ (monoid generated by σ_1, σ_2), 519
$\mathcal{L}_b$, 415
$\mathbf{L}$ (set of pairs of real numbers satisfying Littlewood's conjecture), 444
$u < v$ (lexicographic order), 9
$u \leq_p v$ (u is a prefix of v), 513
$u \preceq v$ (lexicographic order), 9
$u \prec v$ (radix order), 9
$u \preceq v$ (radix order), 9
$L^{\leq n}$ (concatenation of at most n words in L), 14
$L[n \mapsto n^k]$ (language s.t. $\mathcal{U}(n) = n^k$), 129
$L(\mathcal{A})$ (language recognised by $\mathcal{A}$), 16
$L(X)$ (language), 374
$L(x)$ (language), 7

$L_n(X)$ (words of L of length n), 374
$L_n(x)$ (factors of length n), 7
log (logarithm), 3
$\log_2$ (binary logarithm), 3
L_p (set of all p-expansions), 39
$L_{\frac{p}{q}}$ (set of all $\frac{p}{q}$-expansions), 86
L_q (language accepted from state q), 117
$LS_n(u)$ (left special factors), 171
$LS'_n(u)$ (left special factors and unioccurent prefix), 172
L^* (Kleene star), 14
L^n (power of a language), 14
L_u (broken line), 253
$L(x)$ (language), 375

$\mathbf{M}(\mathcal{A})$ (adjacency matrix), 22
$\mathbf{M}_\sigma$ (incidence matrix), 22, 191
$m(w)$ (bilateral multiplicity), 172
$\mathcal{E}(X,T)$ (ergodic invariant measures), 377
$\mathcal{M}f(s)$ (Mellin transform of f), 458
M_β (minimal polynomial), 62
m (Lebesgue measure), 392
$\mathcal{M}(X)$ (Borel measures), 30
μ_k (Lebesgue measure), 252
$\mathcal{M}(X,T)$ (invariant measures), 377

$\mathcal{N}_p(C)$ (the normaliser over the alphabet C (in base p)), 43
$\langle N \rangle_p$ (p-expansion of N), 39
$\langle N \rangle_{\frac{p}{q}}$ ($\frac{p}{q}$-expansion of N), 86
$\nu_{A,p}$ (normalisation), 26

$\mathcal{O}(f)$, 2
$o(f)$, 3
$\Omega(f)$, 2
u^ω (concatenation), 8

$\omega(G)$ (infinite word generated by G), 506
$\omega(H)$, 508
$\mathcal{O}(x)$ (orbit), 28, 374
$\overline{\mathcal{O}(x)}$ (orbit closure), 28

$\mathbb{P}$ (probability), 482
$\mathbf{P}$ (abelianisation map), 6
$p_u(n)$ (factor complexity), 164
$\mathbf{P}(x)$ (Parikh vector), 191
PER(w) (period of w), 516
$\Phi(y)$ (normal distribution function), 487
φ (Golden Ratio), 12
π (permutation), 391
π_c (projection on the contracting space), 252
π_e (projection on the expanding line), 252
π_p (evaluation map), 37
$\pi_{\frac{p}{q}}$ (evaluation map in the $\frac{p}{q}$ numeration system), 86
$P(n)$ (paths in Bratteli diagrams), 358
$u \wedge v$ (longest common prefix), 48
$x \wedge y$ (longest common prefix), 7
P_σ (prefix-suffix edges), 256
$p_X(n)$ (complexity function), 383
$p_x(n)$ (complexity function), 8

rep_k (representation in base k), 109
$\mathrm{rep}_\mathcal{S}$ ($\mathcal{S}$-representation), 114, 117
$\mathrm{rep}_q = \mathrm{rep}_{\mathcal{S}_q}$, 118
rep_U (U-representation), 109
$\rho(\mathbf{A})$ (spectral radius), 25, 512, 531
$\rho(\Sigma)$ (joint spectral radius), 533
$\check{\rho}(\Sigma)$ (joint spectral subradius), 534
$\check{\rho}_t$, 533

$\hat{\rho}_t$, 531
ρ_t, 531
$\underline{\rho}_t$, 533
$RS_n(u)$ (right special factors), 171

$s(n)$ (first difference of $p(n)$), 173
$\setminus$ (set difference), 1
S (shift map), 374
$\sqcup\!\sqcup$ (shuffle), 129
$\sigma^\omega(a)$, 11
$\sigma^\omega(b).\sigma^\omega(a)$, 11
$S_q = (L_q, A, <)$, 118
(X, S) (subshift), 374
(X_x, S) (subshift generated by an infinite word), 375
$\mathcal{S}(X, T)$ (states of the dimension group), 362

$\mathbb{T} = \mathbb{R}/\mathbb{Z}$ (circle group), 453
$\Theta(g)$, 2
$\tilde{u}$ (mirror image), 4, 446
$\tilde{L}$ (mirror image), 14
$T_{\lambda,\pi}$ (interval exchange map), 391
$\mathcal{T}_\sigma$ (central tile), 253
$\mathcal{T}_\sigma(i)$ (subtile), 253
(T, S, λ) (overlap), 297
$[T, S, \lambda]$ (overlap equivalence class), 298

U (lower unit cube), 274
$\mathbf{u}_\beta$ (right eigenvector), 251
$\mathcal{U}_L(n) = \mathcal{U}(n)$ (number of words of length n in L), 118
$U_{n,\varepsilon}$, 393
$\cup$ (union), 1
$\mathcal{U}_q(n)$ (number of words of length n accepted from q), 117
$\mathcal{U}_{q,r}(n)$ (number of directed paths of length n from q to r), 110

$v_p(n)$ (p-adic valuation), 474

$\mathrm{val}_\mathcal{S}$ ($\mathcal{S}$-numerical value), 114
$\mathrm{val}_q = \mathrm{val}_{\mathcal{S}_q}$, 118
$\mathbf{v}_\beta$ (left eigenvector), 251
$(V, E, \geq)$ (ordered Bratteli diagram), 327
$\mathcal{V}_L(n) = \mathcal{V}(n)$ (number of words of length at most n in L), 118
$\mathcal{V}_q(n)$ (number of words of length at most n accepted from q), 117
V_σ (seed patch), 276
$\mathbf{v}_\sigma$ (Perron–Frobenius eigenvector), 24, 513
$\mathbb{V}X$ (variance of random variable X), 490

$\|u\|$ (weight of u), 47
$\mathcal{W}(X, S)$ (weight functions), 379
W_σ (two-piece seed patch), 282

X_B (infinite path space associated with an ordered Bratteli diagram B), 329
$X_B^{\max}$, 329
$X_B^{\min}$, 329
$[\mathbf{x}, i]$ (basic formal strand), 291
ξ_a (real number whose b-ary expansion is given by the word a), 412
$[\mathbf{x}, i]_g$ (basic geometric strand), 253
X_σ (substitutive dynamical system), 32

$\mathcal{Z}_{\beta,d}$ (the zero-automaton in base β over the alphabet B_d), 62
$\mathcal{Z}_p$ (the evaluator in base p), 40
$\mathcal{Z}_{\frac{p}{q}}$ (the evaluator in base $\frac{p}{q}$), 89
$\mathcal{Z}_{p,d}$ (the zero-automaton in base p over the alphabet B_d), 41

$\zeta(s)$ (Riemann zeta function), 459
$\zeta_{f'}$ (Fibonacci continued fraction), 411
ζ_a (real number whose continued fraction expansion is given by the word a), 430
$\zeta_{t'}$ (Thue–Morse continued fraction), 411
$\zeta(s,\alpha)$ (Hurwitz zeta function), 459

General index

1-system
   definition, 507
   Z-balanced, 519

a.e., *see* almost everywhere
abelianisation map, 6
abstract numeration system, 114
accessible state, 16
Adamczewski, B., 423
additive function, 162, *see* q-additive
adherence, 152
adic, 399
   dynamical system, 399
   Pascal, 404
   transformation, 324
adjacency matrix, 22
Adjan, S. I., 33
Aho–Corasick
   algorithm, 550
   automaton, 549
Akiyama, S., 61, 83, 84, 249
d'Alembert ratio test, 196
algebraic
   coincidence, 319
   conjugate, 23, 24
   integer, 24
Allouche, J.-P., 20, 161, 508, 512
almost everywhere, 30
alphabet, 3
ancestor, 276

Angrand, P.-Y., 128
ANS, *see* abstract numeration system
approximation algorithm
   (k,l)-, 540
   non-existence of, 540
Arnoux, P., 8, 319
atoms, 330
automatic sequence, 19, 214, 450
   q-automatic, 19, 138, 214, 452
automatic word, *see* automatic sequence
automaton
   Aho–Corasick, 549
   Büchi, 54
   complete, 17
   deterministic, 17
   deterministic with output, 19
   finite, 414
   local, 162
   trim, 16
   underlying input, 20
Avila, A., 366

badly approximable number, 444
balanced pair
   coincidence, 310
   algorithm, 310
   combinatorial, 310
   irreducible, 310
   one-letter, 310

Barabanov, N., 560
Barat, G., 474
Barbolosi, D., 463
Barge, M., 258, 311, 319
base, 37, 330
   base-b expansion, 412
   odometer, 336
Bell, J., 161
Berend, D., 63
Berger, M. A., 538
Bernoulli shift, 395
Berry–Esseen inequality, 478
Berstel, J., 8
Berthé, V., 162, 527
Bertrand, A., 59, 158
Bertrand-Mathis, A., 70
Bès, A., 73
β-admissible, 57
β-transformation, 56, 162
Bezuglyi, S., 351
bilateral multiplicity, 172
binomial numeration system, 159
Birkhoff ergodic theorem, 30, 377, 486
Birkhoff, G., 377
bispecial factor, 171
   bound on their number, 216
BK-property, 489, 491
block growth, 163
block triangular matrices, 536
block-additive function, 472
block-multiplicative function, 472
Blondel, V. D., 539, 540
Boasson, L., 152
Borbély, T., 83, 84
Borel measure, 596
Borel, É., 413, 444, 449
Boshernitzan, M., 335, 383, 384, 387
bounded
   gap, 7

   letter, 206
   word, 198
Boyle, M., 354, 357, 358, 360
Bratteli compactum, 329
Bratteli diagram, 325
   equivalence relation, 327
   incidence matrix, 326
   infinite path, 329
   isomorphic, 326
   morphism, 328
   ordered, 327
   properly ordered, 329
   range map, 326
   simple, 327
   source map, 326
   stationary, 337
   substitution, 337
   telescoping, 326, 328
Bratteli, O., 324, 325
Bratteli–Vershik
   BV, 329
   dynamical system, 329, 399
Bressaud, X., 369, 370, 404
broken line, 253
Brunotte, H., 83, 84
Bruyère, V., 116, 153, 160
Büchi automaton, 54
Bugeaud, Y., 423, 449

canonical alphabet, 37
canonical numeration system, 78
Cantor
   dynamical system, 325, 329
      linearly recurrent, 347
   set, 7, 449
   space, 325, 329
   version of interval exchange, 350
capacity of codes, 546
Carroll, C. R., 335
Carton, O., 129, 148

Cassaigne, J., 403
central limit theorem, 487, 489, 494
central tile, 253
centric factor, 197
Chacon
    morphism, 24
    substitution, 24, 340, 342
    word, 342
Chacon, R. V., 342
Chaika, J., 402
Champernowne word, 148, 165, 395, 413
Champernowne, D. G., 413
characteristic word, 5, 221
Charlier, E., 137, 150, 161
Chebyshev norm, 188
Cheung, Y., 397
Choffrut, Ch., 72, 103, 159
Chomsky hierarchy, 109
Chomsky, N., 96
Christol, G., 450
circular shift, 170
clopen
    partition, 330
    set, 29, 325
CNS, *see* canonical numeration system
co-accessible state, 16
co-sequential, 102
Cobham's conjecture, 414
Cobham's theorem, 181
Cobham, A., 20, 27, 109, 138, 181, 215, 414, 508, 526
coboundary, 352
coboundary condition, 396
code, 5, 183, 338
    capacity, 546
    cirular, 338
    constrained, 546
    prefix, 5

coded subshift, 54
coding, 10
cofinal, 329
coincidence
    algebraic, 319
    combinatorial strong coincidence, 258
    geometric, 319
    geometric strong coincidence, 293
    half-coincidence overlap, 298
    modular, 319
    overlap, 298
    strong, 318
    strong overlap, 300
    super, 319
combinatorial strong coincidence condition, 258
common reducibility, 536
compactum, 329
comparable edges, 327
complete automaton, 17
complexity function, 118, 163, 164, 383
    action of a letter-to-letter morphism, 182
    action of a non-erasing morphism, 183
    action of an injective morphism, 183
    computation, 221
    exponential, 169
    linear, 219
    maximal, 165
    non-decreasing, 165
    of a language, 166
    of a morphic word, 185, 209
    of a periodic word, 164, 239
    of a purely morphic word, 185, 209
    of a sparse word, 179

of an automatic word, 214
of an interval translation map, 227
of the Fibonacci word, 222
of the Thue–Morse word, 224
subadditivity, 168
complexity of a real number, 413
concatenation, 4
cone
  positive, 351
conjecture
  finiteness property for the capacity, 562
  Pisot, 272
conjugacy map, 31
conjugate
  algebraic, 23
  words, 516
consecutive, 358
constant of expansivity, 345
constrained codes, 546
context-free language, 89
continuant, 429
continued fractions, 427
convergent, 428
converter, 41
convex combination method, 555
Coquet, J., 473
Cornfeld, I. P., 349
Cortez, M. I., 369, 370
counting
  function, 118, 129
Culik II, K., 505, 515
cycle, 278
cyclic
  letter, 510
  morphism, 510
cylinder, 29, 329, 375

D0L
  ω-equivalence problem, 506

language, 13
ω-equivalent systems, 506
prefix problem, 509
system, 13, 149, 505
  nearly primitive, 507
  prolongable, 506
Damanik, D., 343
Dartnell, P., 348
Daubechies, I., 532
Davenport, H., 438
De Bruijn graph, 547
decimation, 123, 161
deconnectable, 384
Dekking, F. M., 318, 508
Delange, H., 457
Delone set, 260
density, 470
  logarithmic, 471
de Bruijn, N. G., 384
DFA, 17
DFAO, 19
Diamond, B., 258
digit-conversion transducer, 41
digitwise
  addition, 42
  subtraction, 42
dimension group, 351, 352
Diophantine approximation
  uniform, 437
directive language, 144
Dirichlet series, 465
Dirichlet's theorem, 437
Dirichlet, P. G. L., 436
discrete hyperplane, 262
distance, 7
  ultrametric, 7
division algorithm, 38
dominant root condition, 67
Downarowicz, T., 347, 349, 366
Drmota, M., 493, 496, 500, 503
Duchêne, E., 161

Dumont–Thomas numeration, 254, 319, 502, 513
Durand, F., 149, 335, 337, 338, 344, 345, 348, 369, 370
dynamical system
  Bratteli–Vershik, 329
  Cantor, 325, 329
  conjugacy, 30
  induced, 325
  measure-theoretic, 30, 376
  stability, 537
  symbolic, 28, 53, 374
  topological, 377
    equicontinuous, 347
  topological isomorphism, 30

edge, 325
  maximal, 329
  minimal, 329
Ehrenfeucht, A., 163, 201, 209, 509, 511
Ehrenfeucht, Lee and Rozenberg's theorem, 209
Ei, H., 527
eigenfunction
  $L^2(\mu)$, 364
  continuous, 364
eigenvalue
  $L^2(\mu)$, 364
  continuous, 364
  Perron–Frobenius, 23
Eisiedler, M., 445
elementary morphism, 509
endomorphism, 10, *see* substitution
  everywhere-growing, 199
  exponentially diverging, 200
  polynomially diverging, 200
  quasi-uniform, 199
entropy, 358
  topological, 60, 169

episturmian, 407
equicontinuous, 347
equivalent norms, 187
equivalent orbit, 354
erasing morphism, 10
ergodic, 30, 377
  Birkhoff theorem, 30, 377, 486
  individual ergodic theorem, 30, 377, 486
  theorem, 30, 377, 486
Euler–Lagrange's theorem, 430
evaluation map, 26, 37
eventually periodic word, 8, 164
Evertse, J.-H., 449
everywhere-growing endomorphism, 199
exceptional prefix, 172
exduction, 397
expansion
  p-, 26, 39
  base-b, 412
  $\frac{p}{q}$-, 86
expansive dynamical system, 345, 399
expansive morphism, 199
exponential language, 136
exponentially diverging endomorphism, 200
exponentially growing word, 194
extension, 171
extension type, 172
extraction, 381
extremal norm theorem, 560
extremal number, 439

factor, 6, 163, 325
  centric, 197
  complexity, 163
  map, 325
  special, 171
factor graph, 176

factorial language, 15
Fekete's lemma, 169, 532
Fekete, M., 532
Ferenczi, S., 342, 365, 366, 420
fiber, 148
Fibonacci
  continued fraction, 430
  sequence, 417
  word, 12, 168, 222, 223, 388, 410
final state, 16
finite automaton, 17, 414
finite difference, 173
finiteness property
  for capacity, 562
  for joint spectral radius, 550
  geometric, 275
  property (F), 60, 321
  weak (W), 321
first entrance time map, 325
first finite difference, 173
Fischler, S., 443
fixed point, 10
folded β-expansion, 77
Fomin, S. V., 349
Forni, G., 366
Forrest, A. H., 336
fractional part, 49
Fraenkel, A. S., 27, 159, 161
frequency, 9
Fretlöh, D., 319
Frid, A., 404
Frougny, Ch., 27, 64, 71, 73, 75, 77, 111, 116, 128, 153, 160
full shift, 28, 395
function
  additive, 162
  block-additive, 472
  block-multiplicative, 472
  co-sequential, 102
  completely

  q-additive, 454
  q-multiplicative, 454
  complexity, 118
  counting, 118
  generating, 96
  q-additive, 453
  q-automatic, 452
  q-multiplicative, 454
  q-regular, 453
  rational, 96, 508
  sequential, 102
fundamental lemma, 482
Furstenberg, H., 396

Gambaudo, J.-M., 370
gap, 7
Gauss Lemma, 38
Gelfand, I. M., 187
Gelfond, A. O., 499
genealogical order, 9
generalised spectral radius, 532
generalised spectral subradius, 534
generating function, 96
generation of bispecial factors, 236
generic, 380
geometric strong coincidence, 293
GIFS, 255
  substitution, 265
Giordano, T., 351, 354
Gjerde, R., 349, 350
Glasner, E., 354
Golden Ratio, 12, 436
Goldwurm, M., 159
Grabner, P. J., 162, 463, 474
graph
  overlaps, 300
  prefix-suffix, 256
  two-piece ancestor, 282

graph-directed iterated function system, 255
greedy
  β-expansion, 56
  algorithm, 38, 49
Grillenberger, C., 169, 391
group ordered, 351
growing word, 198

Haar measure, 486
Hahn, F., 391
half-coincidence overlap, 298
Hamming weight, 47
Handelman, D., 357, 358, 360
Hansel, G., 27, 116, 148, 149, 153, 160
Harju, T., 161, 505, 515, 529
Hausdorff dimension, 445
HD0L
  language, 13
  sequence, 507
  system, 149, 507
    prolongable, 507
Hedlund, G. A., 163, 348
height, 253
Heinis, A., 220
Herman, R. H., 324, 330, 333, 335
Hollander, M., 68, 69, 111, 272, 321
Holton, C., 335, 366
homomorphism of monoids, 4
Honkala, J., 160, 505, 508, 526, 529
Horner scheme, 39
Host, B., 337, 338, 344, 345, 364, 365, 369
hyperplane
  discrete, 262
  stepped, 262

i.d.o.c. property, 392

immortal letter, 246
incidence matrix, 22, 191, 200
  of a Bratteli diagram, 326
independence condition, 396
indicator function, 6, 356
individual ergodic theorem, 30, 377, 486
induced
  dynamical system, 325
  map, 325, 398
infinite word
  automatic, 214
  q-automatic, 214
  Champernowne, 165
  D0L, 506
  Fibonacci, 168, 222, 223
  HD0L, 508
  left-infinite, 180
  lexicographically shift maximal, 54
  lsm-word, 54
  morphic, 181
  paperfolding, 182
  period-doubling, 244
  periodic, 164
  purely morphic, 181
  recurrent, 177, 186
  sparse, 179
  Sturmian, 167, 177, 223
  Thue–Morse, 224
inflation factor, 23
injective morphism, 183
integer
  multiplicatively
    dependent, 27
    independent, 27
  representation, 86
integral part, 49
internal alphabet, 214
interval exchange
  Cantor version, 350

map, 391
  transformation, 349
invariance
  under similarity, 535
invariant
  measure, 30
  subset, 377
irreducible
  matrix, 23
  morphism, 23
  permutation, 392
  set of matrices, 536
  substitution, 23
isomorphism, 325
  measure-theoretic, 31
  topological, 30, 325
iterated function system
  graph-directed, 255
Ito, S., 319, 527
Iwanik, A., 366

Jacobs, K., 349
Jewett, R., 381
Johansen, O., 349, 350
joint spectral radius, 533
  capacity, 548
  computation, 552
  finiteness property, 550
  introduction, 531
  partition function, 541
  repetition-free words, 543
  undecidability, 540
joint spectral subradius
  definition, 533, 534
Jordan normal form, 187
Jungers, R., 538

k-kernel, *see* q-kernel
Kaboré, I., 403
Kakutani equivalent, 335
Kakutani, S., 330

Kakutani–Rokhlin partition, 401
Kärki, T., 150
Kátai, I., 488
Katok, A., 392, 445
Katznelson, Y., 391
Keane, M., 349, 393
kernel
  k-kernel, 147
  q-kernel, 452
  S-kernel, 148
Khintchine, A. Ya., 430
Kleene star, 14
Kleene, S., 18
Kozyakin, V. S., 539, 560
Krieger, D., 161
Krieger, W., 381
Kronecker's theorem, 264
Kubilius model, 481
Kubilius, J., 438
Kwapisz, J., 311, 319
Kwiatkowski, J., 351
Kůrka, P., 345, 348

Lacroix, L., 366
Lagarias, J. C., 532, 551
Lang, S., 361
language, 13, 341, 374
  adherence, 152
  context-free, 89
  D0L, 13
  directive, 144
  exponential, 136
  extendable, 374
  factorial, 15, 374
  finite, 14
  HD0L, 13
  infinite, 14
  of an infinite word, 164
  polynomial, 136
  prefix-closed, 15
  ray, 96

slender, 15, 134, 160, 525
sparse, 136
suffix-closed, 15
with bounded growth, 15, 96
Larcher, G., 464
Lecomte, P., 137, 152
Lee, J.-Y., 319
Lee, K. P., 163, 201, 209
left
  extension, 171
  special factor, 171
  valence, 171
left-infinite word, 180
length, 4
Lenz, D., 343
Leroux, J., 161
Lesigne, E., 389
letter, 3
  cyclic, 510
letter-to-letter
  morphism, 10, 182
  transducer, 20
level, 330
Lévy metric, 488
lexicographic
  map, 329
  order, 9, 48, 186, 328
Liardet, P., 162
Lindenmayer systems, 149
Lindenmayer, A., 505
Lindenstauss, E., 445
linearly recurrent
  Cantor dynamical system, 347
Linna, M., 161, 509, 529
Liouville's inequality, 415
Liouville, J., 415
Littlewood's conjecture, 444
Livshits, A. N., 309, 340
local automaton, 162
locally finite, 259
logarithmic density, 471

logarithmically syndetic set, 227
looping morphism, 517
Loraud, N., 67, 111
Lothaire, M., 3, 349, 509
Luca, F., 423

Maass, A., 347, 348, 369, 370
Maes, A., 137, 148, 151
Mahler, K., 423, 449
Manhattan norm, 2, 191
map
  abelianisation, 6
  factor, 325
  first entrance time, 325
  induced, 325
  lexicographic, 329
  range, 326
  source, 326
  Vershik, 329
Markov compacta, 324
Masur, H., 393
matrix
  adjacency, 22
  incidence, 22
  irreducible, 23
  primitive, 23
Mauduit, Ch., 148, 161, 365, 420
measure invariant, 30
measure-theoretic
  dynamical system, 30
  factor, 31
  isomorphism, 31
Medynets, K., 350, 351
Mellin–Perron summation formula, 466
Mendès France, M., 502
Michaux, C., 160
Miller, A., 161
Miller, G. A., 96
minimal, 375
  dynamical system, 29

word, 7
mirror, 4, 446
    formula, 429
Modified Division algorithm, 85
monoid, 4
    free, 4
    morphism, 4
    of matrices, 531
Monteil, T., 383, 384
Moody, R.V., 319
morphic word, 11, 181
morphism, 10, *see also* endomorphism
    cyclic, 510
    elementary, 509
    erasing, 10
    expansive, 199
    fixed point, 10
    growing, 517
    injective, 183
    invertible, 527
    irreducible, 23
    letter-to-letter, 10, 182
    loop-free, 517
    looping, 517
    non-erasing, 10
    non-trivial, 511
    of ordered groups with order unit, 351
    Pisot, 24
    primitive, 23
    prolongable, 10, 11
    proper, 338
    read on a Bratteli diagram, 328
    simplifiable, 509
    uniform, 10
    unit, 25
Morse and Hedlund theorem, 166, 175, 199
Morse substitution, 340
Morse, M., 163, 224, 340, 348

mortal letter, 246
Mossé, B., 365
multi-graph, 110
multi-scale quasiperiodic, 408
multiple tiling, 259
multiplicative function, *see* q-multiplicative
multiplicatively
    dependent integers, 27
    independent integers, 27
multiplicity
    of a bispecial factor, 172
Méla, X., 405

natural coding, 392
neutral bispecial factor, 173
Nicolay, S., 161
Nivat, M., 152
Nogueira, A., 365, 366
non-algebraicity, 539
non-defective, 560
non-erasing morphism, 10
non-periodic word, 8
non-transient letter, 246
non-trivial morphism, 511
norm, 186
    submultiplicative, 25, 531
normal
    matrices, 554
    number, 413
normalisation, 26, 71, 112, 160
normaliser, 43
Novikov, P. S., 33
Nowakowski, R. J., 161
number
    Parry, 58
    Perron, 104
    Pisot, 24, 35, 115
    Salem, 104
    triangular, 117
numeration

scale, 65
numeration system
   p-ary, 26
   abstract, 114
   adic, 115
   canonical (CNS), 78
   Dumont–Thomas, 319, 502
   linear, 67
   positional, 110
   scale, 65, 110
numerical value, 26
   S-numerical value, 114

occurrence, 341
odometer, 162, 336, 401, 486, 501
   base, 336
ω-equivalent
   D0L systems, 506
   HDOL systems, 149
one-sided shift, 28
orbit, 374
   equivalent, 354
      strongly, 354
   of a word, 28
order
   genealogical, 9
   lexicographic, 9, 328
   radix, 9
   unit, 351, 352
ordered, 327
   Bratteli diagram, 327
   group, 351
   properly, 329
ordering
   consecutive, 358
ordinary bispecial factor, 173, 234
Oseledec, V. I., 391
overlap, 297, 419, 543
   graph, 300
   coincidence, 298
   equivalent, 298
   half-coincidence, 298
Oxtoby, J., 381

palindrome, 4, 434
palindromic density, 443
Pansiot's theorem, 185
Pansiot, J.-J., 161, 185, 508, 509, 529
paperfolding word, 182
Parikh mapping, 6
Parikh vector, 191
Parry number, 58
Parry, W., 58
partial quotients, 428
partition
   clopen, 330
   Kakutani–Rohlin, 330
   KR, 330
patch, 260
   ancestor, 281
   equivalent, 260
path, 16
   Bratteli diagram, 329
   cofinal, 329
   label, 16
   space, 329
   successful, 16
Păun, G., 134
Pell equation, 137
period, 8
period cycle, 164
period-doubling word, 244
periodic word, 8, 164
Perron number, 104
Perron–Frobenius
   eigenvalue, 23
   normalised eigenvector, 24
   theorem, 23, 250
Petersen, K., 405
Pethö, A., 83, 84
p-expansion, 26, 39

$\frac{p}{q}$-expansion, 86
Pisot
　conjecture, 272
　morphism, 24
　number, 24, 35, 115
　　unit, 24
　substitution, 24
　Vijayaraghavan number, 24
polynomial language, 136
polynomially bounded word, 194
polynomially diverging endomorphism, 200
positional numeration system, 110
positive cone, 351
positive uniform frequencies, 390
power of a set, 535
powers of two, 221
prefix, 6
　-closed language, 15
　exceptional, 172
　proper, 6
　unioccurrent, 171
prefix-suffix graph, 256
preperiod, 8, 164
primitive
　matrix, 23
　morphism, 23
　substitution, 23
　word, 164, 178
problem
　D0L ω-equivalence, 506
　D0L prefix, 509
　HD0L ω-equivalence, 149
prolongable
　D0L system, 506
　morphism, 10
proper
　substitution, 338
properly
　ordered, 329

property (F), *see* finiteness property
pumping lemma, 18, 89
purely
　morphic word, 11, 181
　substitutive word, 11, 505
Putnam, I. F., 324, 330, 333, 335, 351, 354
Pytheas Fogg, N., 509

q-additive function, 453
　completely, 454
q-automatic function, 452
q-automatic sequence, 452
q-kernel, 452
q-multiplicative function, 454
　completely, 454
q-regular function/sequence, 453
quasi-greedy expansion, 58
quasi-recurrent word, 506
quasi-uniform endomorphism, 199
Queffélec, M., 309, 434

Rabin, M. O., 17
radix
　order, 9, 39
　point, 48
Rampersad, N., 161
range map, 326
rank
　topological, 347
　　infinite, 347
ranking, 159
Rao, H., 319, 527
rational function, 96, 508
Ratner, M., 406
Rauzy
　fractal, 253
　graph, 176, 384
Rauzy, G., 176, 248, 338, 384

Ravikumar, B., 161
recurrence
  linear, 408
  uniform, 375
recurrent word, 7, 171, 177, 186
reducible
  set of matrices, 536
redundancy transducer, 44
regular function/sequence, see q-regular
regular language, 88
relatively dense, 260
repetition, 419, 543
repetitive, 260
representation
  U-, 65, 109
  S-, 114, 117
  p-ary, 26, see expansion
  BV, 331
  greedy, 26
  integer, 86
return word, 178, 341
reversal, 4
Ridout's Theorem, 420
Ridout, D., 420
right
  extension, 171
  special factor, 171
  valence, 171
right transducer, 21
right context, 88
Rigo, M., 137, 148, 150, 158, 317, 526
Rohlin, V. A., 330
Rosenthal, A., 381
Rota, G. C., 531, 536
Roth's Theorem, 416
Roth, K. F., 416
Roy, D., 438
Rozenberg, G., 163, 201, 209, 509, 511

Rudin–Shapiro word, 388
Rudolph, D. J., 366, 406
Ruohonen, K., 508

s-adic construction, 226
S-recognisable set, 117
Séébold, P., 8
Sadun, L., 321
Sakarovitch, J., 77, 128
Salem number, 104
Salomaa and Soittola's theorem, 191
Salomaa, A., 134, 191, 509
Salon, O., 150
sandwich set, 180
sandwich set theorem, 188
Sataev, E., 397
scale, 65, 110
scaling property, 535
Schützenberger, M.-P., 33
Schlickewei, H. P., 450
Schmidt's subspace theorem, 422
Schmidt, K., 59, 158
Schmidt, W. M., 434, 438
Scott, D., 17
second finite difference, 173
seed patch, 276, 282
self-replicating multiple tiling, 271
semigroup, 4
  of matrices, 531
sequence
  2-regular, 545
  q-automatic, 19, 138, 452
  q-regular, 453
  S-automatic, 138
sequential, 102
  transducer, 20
set
  Cantor, 7
  eventually periodic, 8

recognisable, 27, 109
  S-recognisable, 117
  U-recognisable, 73, 110
syndetic, 7
Shallit, J., 20, 67, 111, 161, 508, 512
shape of a Rauzy graph, 178
shift, 28, 139, 374
  circular, 170
  full, 28
  one-sided, 28
  two-sided, 28
shift radix system, 83
shuffle, 129
Siegel, A., 249
simple
  Bratteli diagram, 327
  hat, 327
simplifiable morphism, 509
Sinai, Y. V., 349
Sing, B., 319
sink, 18
Sirvent, V., 296
Skau, C. F., 324, 330, 333, 335, 337, 338, 344, 345, 351, 354
S-kernel, 148
skew-product, 498
slender language, 15, 134, 160, 525
SOE, *see* strongly orbit equivalent
Soittola, M., 191
Solomyak, B., 71, 77, 153, 272, 296, 319, 370
source map, 326
sparse language, 136
sparse word, 179
special factor, 171
  bound on their number, 219
spectral radius, 25, 187, 512
square, 4

SRS, *see* shift radix system
stammering, 419
state, 362
  accessible, 16
  co-accessible, 16
  final, 16
  terminal, 16
stationary, 337
  Bratteli diagram, 337
Steiner, W., 137, 158
stepped hyperplane, 262
strand
  basic formal, 291
  basic geometric, 253
  formal, 291
  geometric, 291
Strang, G., 531, 536
strictly ergodic, 390
strong bispecial factor, 172
strong mixing, 408
strongly orbit equivalent, 354
Sturmian
  expansion, 421
  subshift, 348
  word, 8, 167, 177, 223, 388
subadditive function, 168
subadditivity, 532
submultiplicative norm, 25, 531
subshift, 28, 53, 374
  aperiodic, 28, 337, 385
  coded, 54
  conjugate, 30
  entropy, 60
  finite type, 28, 53, 395
  periodic, 28, 337
  sofic, 28, 54, 395
  substitution, 343
  Toeplitz, 349
substitution, 10, *see* endomorphism, *see* morphism
  Chacon, 342

GIFS, 265
 invertible, 527
 irreducible, 23
 Morse, 340
 Pisot, 24
  irreducible, 25
  reducible, 25
 primitive, 23
 proper, 338
 read on a stationary Bratteli diagram, 337
 subshift, 343
 unit, 25
substitutive
 word, 11
 word sequence, 507
subword
 complexity, 163
 scattered, 6
successor, 128
suffix, 6
 -closed language, 15
 proper, 6
Sugisaki, F., 360
sum-of-digits, 162, 452, 457, 467, 499, 502
switched linear system, 538
symbol, 3
symbolic dynamical system, 28
synchronisation lemma, 221, 229
syndetic, 227
 set, 7
system
 D0L, 13, 149, 505
  nearly primitive, 507
  prolongable, 506
 HD0L, 149, 507

tag sequence, 138
telescoping, 326, 328
Tenenbaum, G., 457

terminal state, 16
theorem
 Cobham, 181
 Ehrenfeucht, Lee and Rozenberg, 209
 Grillenberger, 169
 Kronecker, 264
 Morse and Hedlund, 166, 175, 199
 Pansiot, 185
 Perron–Frobenius, 23, 250
 Salomaa and Soittola, 191
 sandwich set, 188
Theys, J., 534, 539
Thomas, W., 129, 148
Thue, A., 33, 224
Thue–Morse
 Bratteli diagram, 340
 continued fraction, 434
 word, 12, 224, 388, 410, 560
Thurston, W., 248
Thuswaldner, J., 83, 84
Tichy, R. F., 162
tiles, 259
tiling, 162, 259
 synchronisation, 298
 lattice multiple, 320
 multiple, 259
 patch, 260
 property, 272
 self-replicating multiple, 271
 self-similar, 261
 tip, 263
Toeplitz
 subshift, 349
 word, 170, 349
topological
 conjugacy, 30
 dynamical system, 29
 entropy, 60, 169, 390
 isomorphism, 30

total irrationality, 392
tower, 330
  level, 330
Transcendence criterion, 423
transducer, 20
  co-sequential, 102
  digit-conversion, 41
  letter-to-letter, 20, 101
  right, 21
  sequential, 20, 102
transformation
  β-, 56, 162
  adic, 324
transient part, 164
transition
  function, 17
  relation, 16
  structure, 141
triangular
  matrices, 536
    joint spectral radius of, 554
  number, 117
trie, 116
trim, 16
Tsitsiklis, J. N., 539, 540
Turán–Kubilius inequality, 484
two-piece ancestor graph, 282

U-recognisable, 73, 110
U-representation, 65
underlying input automaton, 20
uniform Diophantine approximation, 437
uniform frequencies, 376
uniformly recurrent word, 7
uniformly discrete, 260
unioccurrent prefix, 171
unique ergodicity, 30, 32, 380
unit, 24
  morphism, 25
  order, 351, 352

substitution, 25
universal counter-example, 179

valence, 171
Veech, W., 393, 396
Vershik map, 329
Vershik, A. M., 324, 329, 330, 340, 399
vertex, 325
Villemaire, R., 160

Walters, P., 357, 358
Wang, Y., 538, 551
weak (W), see finiteness property
weak bispecial factor, 172
weight, 47, 475
weight function, 379
  on a graph, 393
Weiss, B., 354
Wen, Z., 527
width of a morphism, 11
Wirsing, E., 451
word, 3, see sequence
  automatic, 19
    S-automatic, 138
  β-admissible, 57
  bi-infinite, 5
  bounded, 198
  Chacon, 24, 342
  Champernowne, 148, 395, 413
  characteristic, 5
  comparable, 522
  concatenation, 4
  D0L infinite word, 506
  distance, 7
  empty, 4
  eventually periodic, 8
  exponentially growing, 194
  factor, 6
  Fibonacci, 12, 410
  growing, 198

infinite
  lexicographically shift maximal, 54
  lsm-word, 54
  one-sided, 5
  two-sided, 5
  length, 4
  minimal, 7
  mirror, 4
  morphic, 11
    purely, 11
  nearly periodic, 515
  non-periodic, 8
  ω-equivalent, 149
  period, 8
  periodic, 8
  polynomially bounded, 194
  prefix, 6
  preperiod, 8
  primitive, 164, 178
  purely substitutive, 11, 505
  quasi-recurrent, 506
  recurrent, 7
  reversal, 4
  Sturmian, 8
  substitutive, 11
    purely, 11, 505
  subword (scattered), 6
  suffix, 6
  Thue–Morse, 12, 410, 560
  Toeplitz, 349
  uniformly recurrent, 7
Wythoff's game, 161

Yu, S., 13

Zamboni, L., 335, 366
Z-balanced 1-systems, 519
zero automaton, 41, 62, 288
zero spectral radius, 553